W9-BBX-582

Practical Antenna Handbook

2nd Edition

Other Books by Joseph J. Carr

Mastering IC Electronics
Mastering Oscillator Circuits through Projects and Experiments
Mastering Solid-State Amplifiers
Old Time Radios! Restoration and Repair
Secrets of RF Circuit Design
Mastering Radio Frequency Circuits through Projects & Experiments

Practical Antenna Handbook

2nd Edition

Joseph J. Carr

TAB Books
Division of McGraw-Hill, Inc.
New York San Francisco Washington, D.C. Auckland Bogotá
Caracas Lisbon London Madrid Mexico City Milan
Montreal New Delhi San Juan Singapore
Sydney Tokyo Toronto

SECOND EDITION
FIRST PRINTING

Library of Congress Cataloging-in-Publication Data

Carr, Joseph J.
Practical antenna handbook / by Joseph J. Carr—2nd ed.
p. cm.
Includes index.
ISBN 0-07-011104-9 ISBN 0-07-011105-7 (pbk.)
1. Radio—Antenna—Design and construction. I. Title.
TK6565.A6C37 1994
621.384'135—dc20 93-44428
CIP

Aquisitions editor: Roland S. Phelps
Managing editor: Andrew Yoder
Technical editor: William Taylor, N3LRA
Production Team: Katherine G. Brown, Director
Jan Fisher, Desktop Operator
Cindi Bell, Proofreading
Design team: Jaclyn J Boone, Designer
Brian Allison, Associate Designer
Cover: Cindy Staub

EL1
0111057

Dedicated in memoriam to Johnnie Harper Thorne, K4NFU:
a friend and colleague for nearly 30 years who is sorely missed.
Johnnie was a genius who knew some real smoke about antennas.
Killed by a drunk driver. . . .

Contents

Introduction to the second edition

WHEN I WAS PLANNING THE ORIGINAL EDITION OF THE *PRACTICAL ANTENNA HANDBOOK,* the intent was to provide the reader with a practical (yet also theoretical) book that could be used with only a minimal effort to actually design and install radio antennas. It was assumed that the readership would possess a wide range in levels of sophistication from the novice to the professional, and that assumption has proven to be correct.

This second edition maintains the focus of the original edition, but adds the material that readers of this book (or my columns in *Popular Electronics*, *73*, *Radio-Fun* and *Popular Communications*) have requested. The level of approach is the same, but the number of examples is expanded. I've also increased the coverage of radio propagation phenomena (chapter 2) because that knowledge is critical to selecting and building proper antennas for specific uses.

The popularity of the original edition has astounded both me and the publisher. We have written and expanded this book because we are all radio people who like antennas. While the "labor of love" routine might sound trite, the amount of money normally made by technical books tends to support this idea. Really! But the *Practical Antenna Handbook* proved to be wildly successful; and although no one, least of all myself, got rich (I've not yet quit my "day job"), it was a very satisfying success. I've had the experience of walking into a radio equipment dealer and being pointed out as the author of "3270" (the original catalog number). The sales people had sold so many copies of the book that they had the catalog number memorized!

While the sales of the original book are an honor, there was one comment that made it well worthwhile. The salesmen at the radio store introduced me to an instructor from a U.S. Government communications facility who typically bought the book twenty copies at a time for use by his students in a training class. He told me that the reason why he selected my book over the others was ". . . it's the *only* book on the market that people can give to a secretary, or clerk-typist, and expect them to be able to put up a working half wave dipole two hours later." And, he averred, in his business that could literally happen.

Joseph J. Carr, MSEE

Introduction to the first edition

WHY ANOTHER ANTENNA BOOK? THE ANSWER TO THIS QUESTION IS ROOTED IN MY OWN experience, both as a technical writer and in electronics. My interest in radio dates from about 1958, when a neighbor first loaned me a *Knightkit* regenerative shortwave receiver, and then later a surplus World War II BC-342 military shortwave receiver. Licensed as an amateur radio operator since 1959, and working as an electronics technician for 16 years, practicing as an electronics engineer since 1978, I have a wide variety of experience in antenna use . . . as well as a wide familiarity with the available literature. Why another book on antennas? The reason is simple: it is necessary to blend together, in one source, the theoretical concepts that the engineers and others need to design practical antennas, and the hard-learned practical lessons derived from actually building and using antennas—real antennas made of real metal—not merely theoretical constructs on a blackboard.

This book is, therefore, a blend of the *practical* nuts-and-bolts stuff that you need to make antennas work for you, the *theoretical* material needed to understand what you did, and the means to extend that work into new projects that are not presented in a book elsewhere.

Thus, the purpose of this book is to give you some projects, to be sure, but further, and more importantly, to empower you to make your own antennas for the cases that the author thoughtlessly failed to cover. Empowerment, that's the game, and the underlying reason why another antenna book is needed. Several types of readers were kept in mind when this book was compiled:

- Radio technicians.
- Electronics technicians.
- Amateur radio operators.
- Citizen's banders.
- Shortwave listeners (SWL).
- Monitoring hobbyists.
- Radio enthusiasts and professionals of all types.

In my view, a book is neither a tribute to its author (although I gladly accept the royalties), nor a decorative ornament for your den library; it is, rather, a tool first and foremost. Never forget that the tools are designed to be used. As the author of scores of technical books, and more than 450 magazine articles, I feel that the greatest honor that an author of such books can receive is to find a well-worn copy—beaten up, tattered, annotated with the owner's personal notes on the subject matter, and obviously used to the fullest extent possible—lying in the bottom of the toolbox and splattered with coffee and solder. Enjoy, and use this tool in good health.

Joseph J. Carr, MSEE

1
CHAPTER

Introduction to radio broadcasting and communications

RADIO BROADCASTING AND COMMUNICATIONS SEEMS TO HOLD A STRANGE KIND OF magical allure that attracts a wide variety of people and holds them for years. There is something fascinating about the ability to project yourself over vast intercontinental distances.

Radio communications have been with us now for the entire twentieth century. Experiments are on record as early as 1867, and by the turn of the century "wireless telegraphy" (as radio was called then) sparked the imaginations of countless people across the world. Radio communications began in earnest, however, when Guglielmo Marconi successfully demonstrated wireless telegraphy as a commercially viable entity. The "wireless" aspect to radio so radically changed communications that the word is still used to denote radio communications in many countries of the world. Marconi made a big leap to international fame on a cold December day in 1903, when he and a team of colleagues successfully demonstrated transatlantic wireless telegraphy. Until that time, wireless was a neighborhood—or crosstown at best—endeavor that was of limited usefulness. Of course, ships close to shore, or each other, could summon aid in times of emergency, but the ability to communicate over truly long distances was absent. All that changed on that fateful day in Newfoundland when Marconi heard the Morse letter "S" tickle his ears.

Wireless telegraphy was pressed into service by shipping companies because it immediately provided an element of safety that was missing in the pre-wireless days. Indeed, a ship that sank, leaving its crew and passengers afloat on a forbidding sea, was alone. Any survivors often succumbed to the elements before a chance encounter with a rescue vessel. Some early shipping companies advertised that their ships were safer because of wireless aboard. It was not until 1909, however, that wireless telegraphy proved its usefulness on the high seas. Two ships collided in the foggy Atlantic Ocean and were sinking. All passengers and crew members of both ships were in imminent danger of death in the icy waters. But radio operator Jack

Binns became the first man in history to send out a maritime distress call. There is some debate over which distress call Binns transmitted, but one thing is certain: It was not "SOS" (today's distress call), because SOS was not adopted until later. Binns probably transmitted either "CQD" or "CQE," both of which were recognized in those days before standardization. Regardless of which call was sent, however, it was received and relayed from ship to ship, allowing another vessel to come to the aid of the stricken pair of ships.

All radio prior to about 1916 was carried on via telegraphy (i.e., the on-off keying of a radio signal in the Morse code). But in 1916 some more magic happened. On a little hill in Arlington, Virginia, on a site that now overlooks the Pentagon and the U.S. Marine Corps base called *Henderson Hall*, there were (and still are) a pair of two-story brick buildings housing the Naval Research Laboratory radio station NAA (callsign since reassigned to the VLF station at Cutler, ME). On a night in 1916, radio operators and monitors up and down the Atlantic seaboard—from the midwest to the coast and out to sea for hundreds of miles—heard something that must have startled them out of their wits, for crackling out of the "ether," amidst the whining of Alexanderson alternators and "ZZZCHHT" of spark-gap transmitters, came a new sound—a human voice. Engineers and scientists at NRL had transmitted the first practical amplitude-modulated (AM) radio signal. Earlier attempts, prior to 1910, had been successful at scientific experiments, but they did not use commercially viable equipment.

Although radio activity in the early years was unregulated and chaotic, today it is quite heavily regulated. Order was brought to the bands (don't laugh, ye who tune the shortwaves) that was lacking before. Internationally radio is regulated by the International Telecommunications Union (ITU) in Geneva, Switzerland through the treaties arising from World Administrative Radio Conferences (WARC) held every 10 to 15 years. In the United States, radio communications are regulated by the Federal Communications Commission (FCC), headquartered in Washington, D.C.

Amateur radio has grown from a few thousand "hams" prior to World War I to more than 900,000 today; about one-third of them in the United States. Amateur operators were ordered off the air during World War I, and almost did not make a comeback after the war. There were, by that time, many powerful commercial interests that greedily coveted the frequencies used by amateurs, and they almost succeeded in keeping post-war amateurs off the air. But the amateurs won the dispute, at least partially. In those days, it was the low frequencies with wavelengths longer than 200 meters (i.e., 20 kHz to 1500 kHz) that were valuable for communications.

The cynical attitude attributed to the commercial interests regarding amateurs was, "put 'em on 200 meters and below . . . they'll never get out of their backyards there!" But there was a surprise in store for those commercial operators, because the wavelengths shorter than 200 meters are in the high-frequency region that we now call "shortwaves." Today, the shortwaves are well-known for their ability to communicate over transcontinental distances, but in 1919 that ability was not suspected.

I once heard an anecdote from an amateur operator "who was there." In the summer of 1921 this man owned a large, beautiful wire "flattop" antenna array for frequencies close to 200 meters on his family's farm in southwestern Virginia. Using those frequencies he was used to communicating several hundred miles into eastern Ohio and down to the Carolinas. But, in September 1921 he went to college to study electrical engineering at the University of Virginia in Charlottesville. When he returned home for Thanksgiving he noticed that his younger brother had replaced the long flattop array with a short dipole antenna. He was furious, but managed through great effort to contain the anger until after dinner. Confronting his brother over the incredible sacrilege, he was told that they no longer used 150 to 200 meters, but rather were using 40 meters instead. Everyone "knew" that 40 meters was useless for communications more than a few blocks, so (undoubtedly fuming) the guy took a turn at the key. He sent out a "CQ" (general call) and was answered by a station with a callsign like "8XX." Thinking that the other station was in the 8th U.S. call district (WV, OH, MI) he asked him to relay a message to a college buddy in Cincinnati, OH. The other station replied in the affirmative, but suggested that ". . . you are in a better position to reach Cincinnati than me, I am *FRENCH* 8XX." (Callsigns in 1921 did not have national prefixes that are used today.) The age of international Amateur communications had arrived!

During the 1930s, radio communications and broadcasting spread like wildfire as technology and techniques improved. World War II became the first war to be fought with extensive electronics. Immediately prior to the war, the British developed a new weapon called *RADAR* (radio detection and ranging). This tool allowed them to see and be forewarned of German aircraft streaming across the English Channel to strike targets in the United Kingdom. The German planes were guided by (then sophisticated) wireless highways in the sky, while British fighters defended the home island by radio vectoring from ground controllers. With night fighters equipped with the first "centimetric" (i.e., microwave) RADAR, the Royal Air Force was able to strike the invaders accurately—even at night. The first kill occurred one dark, foggy, moonless night when a *Beaufighter* closed on a spot in the sky where the RADAR in the belly of the plane said an enemy plane was flying. Briefly thinking he saw a form, the pilot cut loose a burst from his quad mount of 20-mm guns slung in the former bomb bay. Nothing. Thinking that the new toy had failed, the pilot returned to base—only to be told that ground observers had reported that a German *Heinkle* bomber fell from the overcast sky at the exact spot where the pilot had his ghostly encounter.

Radio, television, RADAR, and a wide variety of services, are available today under the general rubric "radio communications and broadcasting." Homeowners, and other non-professionals in radio, can own a receiver system in their backyard that picks up television and radio signals from satellites in geosynchronous orbit 23,000 miles out in space. Amateur operators are able to communicate worldwide on low power, and have even launched their own "OSCAR" satellites.

Some people had written off the HF radio spectrum in recent years, citing satellite technology as the reason. But the no-code license for amateur radio operators, which does not carry HF privileges, has proven to be a stepping stone to higher-class licenses, which do. Also, the shortwave broadcasting market received a tremendous boost during the Gulf War. When the troops of *Operation Desert Shield* and *Desert Storm* were assembling to take back Kuwait from the Iraqis, the sales of shortwave receivers jumped dramatically. And, following January 16th, when the forces started pouring across the border into the actual fight, the sales skyrocketed out of sight. One dealer told me that he couldn't keep receivers in stock, and that he had sold out most models. That interest seems to have matured into long-term interest on the part of a significant number of listeners and new ham operators.

The antenna is arguably one of the most important parts of the receiving and/or transmitting station. That is what this book is all about.

1-1 This AM/FM broadcast antenna tower bristles with two-way antennas.

2

CHAPTER

Radio wave propagation

THE PROPAGATION OF RADIO SIGNALS IS NOT THE SIMPLE MATTER THAT IT SEEMS AT first glance. Intuitively, radio signal propagation seems similar to light propagation; after all, light and radio signals are both electromagnetic waves. But simple inverse square law predictions, based on the optics of visible light, fall down radically at radio frequencies. In the microwave region of the spectrum, the differences are more profound because atmospheric pressure and water vapor content become more important than for light. For similar reasons, the properties of microwave propagation differ from lower VHF and HF propagation. In the HF region, solar ionization of the upper reaches of the atmosphere causes the kind of effects that lead to long-distance "skip" communications and intercontinental broadcasting. This chapter examines radio propagation phenomena so that you have a better understanding of what an antenna is used for and what parameters are important to ensure the propagation results that you desire.

Radio waves

Although today it is well recognized that radio signals travel in a wave-like manner, that fact was not always so clear. It was well known in the first half of the 19th century that wires carrying electrical currents produced an *induction field* surrounding the wire, which is capable of causing action over short distances. It was also known that this induction field is a magnetic field, and that knowledge formed the basis for electrical motors. In 1887, physicist Heinrich Hertz demonstrated that radio signals were *electromagnetic waves*, like light. Like the induction field, the electromagnetic wave is created by an electrical current moving in a conductor (e.g., such a wire). Unlike the induction field, however, the *radiated field* leaves the conductor and propagates through space as an electromagnetic wave.

The *propagation* of waves is easily seen in the "water analogy." Although not a perfect match to radio waves, it serves to illustrate the point. Figure 2-1 shows a body of water into which a ball is dropped (Fig. 2-1A). When the ball hits the water (Fig. 2-1B), it displaces water at its point of impact, and pushes a leading wall of water away from itself. The ball continues to sink and the wave propagates away from it until the energy is dissipated. Although Fig. 2-1 shows the action in only one dimension (a side view), the actual waves propagate outward in all directions, forming concentric circles when viewed from above.

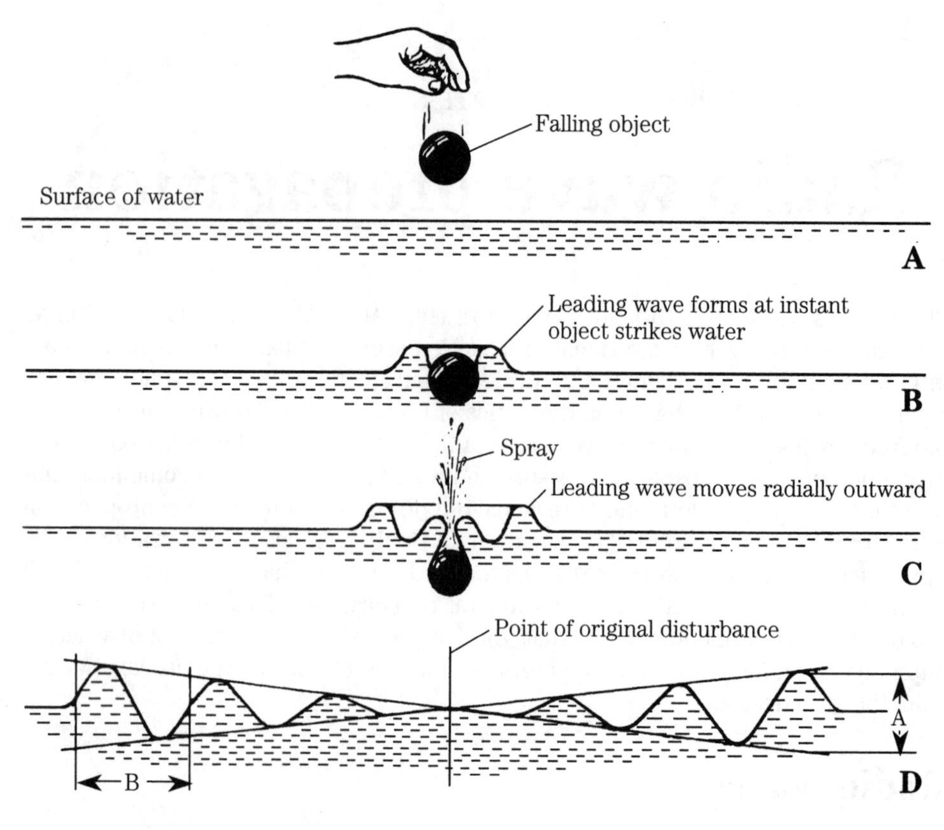

2-1 A ball dropped into water generates a wavefront that spreads out from the point of original disturbances.

The wave produced by a dropped ball is not continuous, but rather is *damped* (i.e., it will reduce in amplitude on successive crests until the energy is dissipated and the wave ceases to exist). But to make the analogy to radio waves more realistic, the wave must exist in a continuous fashion. Figure 2-1B shows how this is done: a ball is dipped up and down in a rhythmic, or cyclic manner, successively reinforcing new wave crests on each dip. The waves continue to radiate outward as long as the ball continues to oscillate up and down. The result is a *continuous wave train*.

There are two related properties of all waves that are important to radio waves as well: *frequency (f)* and *wavelength (λ)*. The frequency is the number of oscillations (or cycles), per unit of time. In radio waves, the unit of time is the second, so frequency is an expression of the number of *cycles per second (cps)*. If the period of time required for the leading wave to travel from point "A" to "B" is one second (1 s), and there are two complete wave cycles in that space, then the frequency of the wave created by the oscillating ball is 2 cps.

At one time, radio frequencies (along with the frequencies of other electrical and acoustical waves) were expressed in cps, but in honor of Heinrich Hertz, the unit was renamed the *Hertz (Hz)* many years ago. Because the units are equal (1 Hz = 1 cps), the wave in Fig. 2-2 has a frequency of 2 Hz.

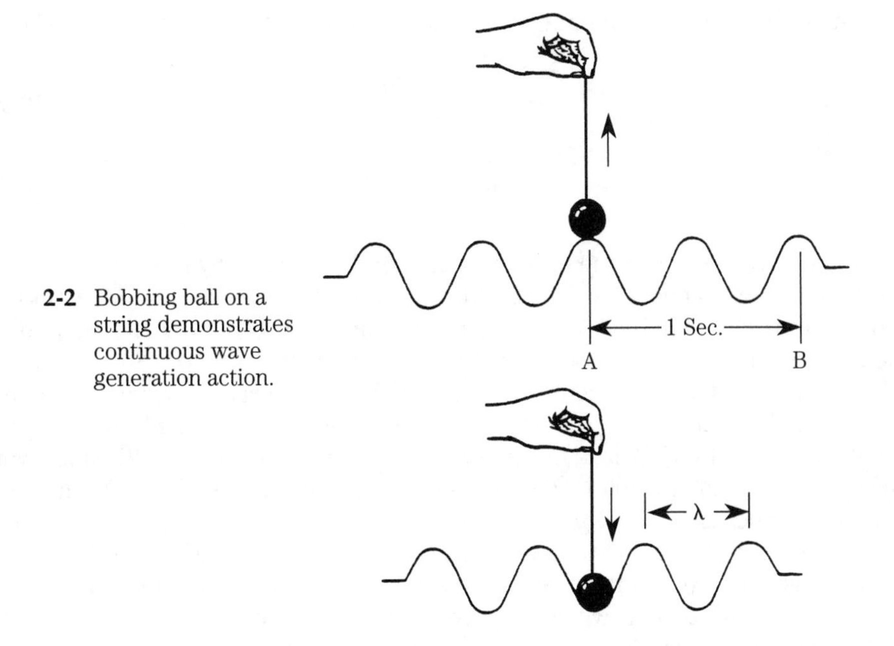

2-2 Bobbing ball on a string demonstrates continuous wave generation action.

Because radio frequencies are so high, the frequency is usually expressed in *kilohertz* (kHz—1000s of Hz) and *megahertz* (MHz—1,000,000s of Hz). Thus, the frequency of a station operating in the middle of the AM broadcasting band can be properly expressed as 1,000,000 Hz, or 1,000 kHz, or 1 MHz, all of which are equivalent to each other. Radio dials in North America are usually calibrated in kHz or MHz. In Europe, on the other hand, it is not uncommon to find radio dials calibrated in *meters*, the unit of wavelength, as well as in frequency. In most equations used in radio antenna design, the proper units are Hertz.

The *wavelength* of any wave is the distance between like features on the waveform. In the case of Fig. 2-2, the wavelength (λ) is the distance between successive positive peaks. We could also measure the same distance between successive negative peaks, or between any two similar features on successive waves. In radio work, the wavelength of the signal is expressed in *meters*.

The wavelength is proportional to the reciprocal of the frequency. The wavelength of any wave is related to the frequency so that $f\lambda = v$, where f is the frequency in hertz (Hz), λ is the wavelength in meters (m), and v is the *velocity of propagation* in meters per second (m/s). Because radio waves propagate at the speed of light (which is also an electromagnetic wave), approximately 300,000,000 m/s in both free space and the earth's atmosphere, the lowercase letter c is used to represent velocity, so you can rewrite this expression in the form:

$$f_{Hz} = \frac{C}{\lambda_{meters}} = \frac{300{,}000{,}000}{\lambda_{meters}} \qquad \textbf{[2.1]}$$

These equations are sometimes abbreviated for use with the units kHz and MHz:

$$F_{kHz} = \frac{300{,}000}{\lambda_{meters}} \qquad \textbf{[2.2]}$$

$$f = \frac{300}{\lambda_{meters}} \qquad \textbf{[2.3]}$$

You can get an idea of the order of length of these waves by solving Eq. 2.3 for several different frequencies: 1 MHz (in the AM broadcast band), 10 MHz (in the shortwave bands) and 1000 MHz (microwave bands). If you work the equations, then you will find that these wavelengths are 300 meters (1 MHz), 30 meters (10 MHz), and 0.3 meters, or 30 centimeters (1,000 MHz). You can see from these numbers why 1 MHz is in what is called the *medium wave band*, 10 MHz is in the *shortwave band*, and 1,000 MHz is in the *microwave* ("very small" wave) band. If you make the calculation for 100 kHz, which is 0.1 MHz, the wavelength is 3,000 meters, so this frequency is in the *longwave band*.

The place where the water analogy falls down most profoundly is in the nature of the medium of propagation. Water waves move by moving water molecules; water is said to be the *medium* in which the wave propagates. At one time, scientists could not conceive of the "action at a distance" provided by radio waves, so they invented a hypothetical medium called *ether* (or *aether*) for propagating electromagnetic waves (such as radio waves and light). It was not until the late 19th century, that physicists proved that the ether does not exist. Nonetheless, radio enthusiasts still refer to the "stuff" out of which radio waves arrive as the "ether." This terminology is merely an archaic, linguistic echo of the past.

The electromagnetic field: a brief review

A great deal of heavily mathematical material can be presented about electromagnetic waves. Indeed, developing Maxwell's equations is a complete field of study for specialists. In this section, you will not use this rigorous treatment because you can refer to engineering textbooks for that depth of information. The purpose here is to present a descriptive approach that is designed to present you with a basic understanding of the phenomena. The approach here is similar to the learning of a "conversational" foreign language, rather than undertaking a deep study of its grammar, syntax, and context. For those whose professional work routinely involves electromagnetic waves, this treatment is hopelessly simplistic. For that I make no apology, because it serves a greater audience. The goal here is to make you more comfortable when thinking about the propagation of electromagnetic fields in the radio portion of the electromagnetic spectrum.

Radio signals are electromagnetic (EM) waves exactly like light, infrared (IR) and ultraviolet (UV), except for frequency. Radio waves have much lower frequencies than light, IR or UV, hence they have much longer wavelengths. The EM wave consists of two mutually perpendicular oscillating fields (see Fig. 2-3) traveling together. One of the fields is an electric field and the other is a magnetic field.

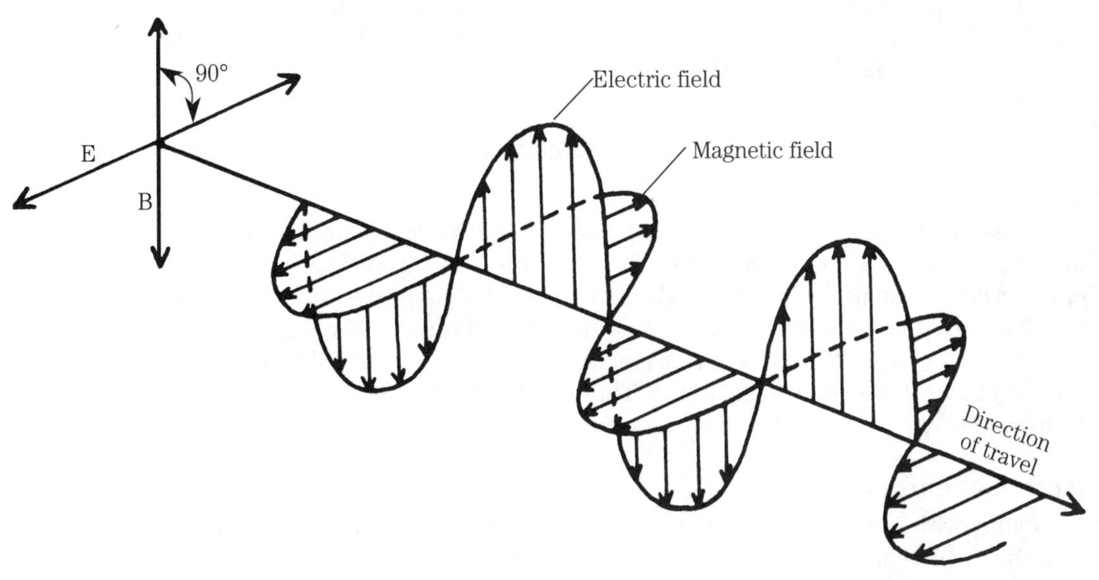

2-3 Electromagnetic wave consists of right angle electric and magnetic fields.

Radio wave intensity

The radio wave is *attenuated* (i.e. reduced in apparent power) as it propagates from the transmitter to the receiver. Although at some very high microwave frequencies, there is additional path loss as a result of the oxygen and water vapor content of the air. Radio waves at all frequencies suffer losses due to the *inverse square law*. Let's take a look at that phenomenon.

The electric field vector falls off in direct proportion to the distance traveled. The E-field is measured in terms of *volts per meter (V/m)*, or in the subunits *millivolts per meter (mV/m)* or *microvolts per meter (μV/m)*. That is, if an E-field of 10

V/m crosses your body from head to toe, and you are about 2 meters tall, then an electrical voltage is generated of (2 m)×(10 V/m), or 20 volts. The reduction of the E-field is linearly related to distance (i.e., if the distance doubles the E-field voltage vector halves). Thus, a 100 mV/m E-field that is measured one mile from the transmitter will be 50 mV/m at two miles.

The power in any electrical system is related to the voltage by the relationship:

$$P = \frac{E^2}{R} \quad [2.4]$$

Where:

P is the power in watts (W)
R is the resistance in ohms (Ω)
E is the electrical potential in volts (V)

In the case of a radio wave, the R term is replaced with the impedance (Z) of free space, which is on the order of 377 ohms. If the E-field intensity is, for example, 10 V/m, then the *power density* of the signal is:

$$P = \frac{(10\ \text{V/m})^2}{377\ \Omega} = 0.265\ \text{watts/cm}^2 = 26.5\ \text{mw/m}^2 \quad [2.5]$$

The power density, measured in *watts per square meter (w/m²)*, or the subunits (e.g., mw/cm²), falls off according to the square of the distance. This phenomenon is shown graphically in Fig. 2-4. Here, you can see a lamp emitting a light beam that falls on a surface (A), at distance L, with a given intensity. At another surface (B), that is $2L$ from the source, the same amount of energy is distributed over an area (B) that is twice as large as area A. Thus, the power density falls off according to $1/D^2$, where D is the difference in distance.

Isotropic sources

In dealing with both antenna theory and radio wave propagation, a theoretical construct called an *isotropic source* is sometimes used for the sake of comparison, and for simpler arithmetic. You will see the isotropic model several places in this book. An isotropic source assumes that the radiator (i.e., an "antenna") is a very tiny spherical source that radiates energy equally well in all directions. The radiation pattern is thus a sphere with the isotropic antenna at the center. Because a spherical source is uniform in all directions, and its geometry is easily determined mathematically, the signal intensities at all points can be calculated from basic geometric principles.

For the isotropic case, you can calculate the average power in the extended sphere from:

$$P_{av} = \frac{P_t}{4\pi d^2} \quad [2.6]$$

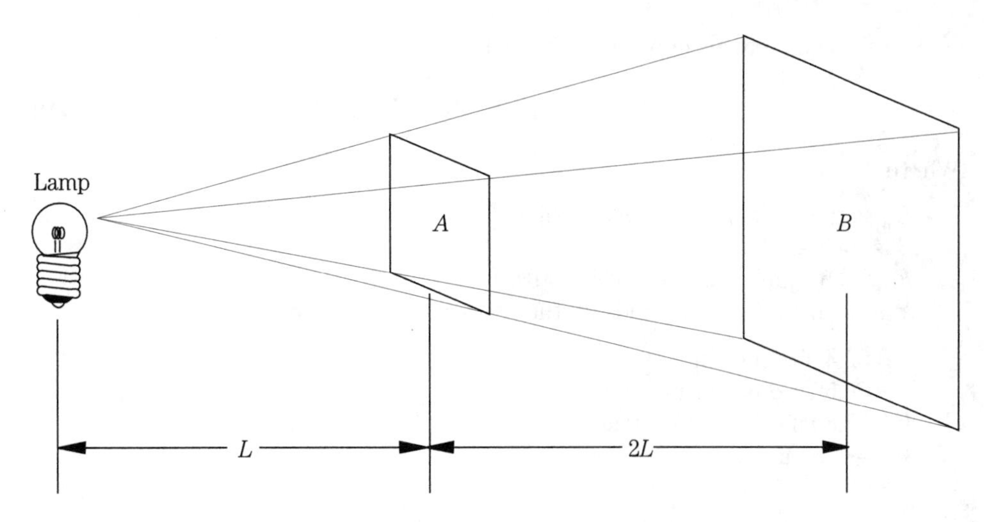

2-4 As a wave propagates, it spreads out according to the inverse square law, i.e., the area of "B" is four times that of "A" because it is twice as far from the source.

Where:

P_{av} is the average power per unit area
P_t is the total power
d is the radius of the sphere in meters (i.e., the distance from the radiator to the point in question)

The *effective aperture* (A_e) of the receiving antenna relates to its ability to collect power from the EM wave and deliver it to the load. Although typically smaller than the surface area in real antennas, for the theoretical isotropic case $A_e = \lambda^2/4\pi$. The power delivered to the load is:

$$P_L = P_{av} A_e \quad \textbf{[2.7]}$$

By combining the two previous equations, the power delivered to a load at distance d is given by:

$$P_L = \frac{P_t \lambda^2}{(4\pi)^2 d^2} \quad \textbf{[2.8]}$$

Where:

P_L is the power to the load
λ is the wavelength (c/F) of the signal

From these expressions, there can then be derived an expression for ordinary path losses between an isotropic transmitter antenna and a receiver antenna:

$$L_{dB} = 10 \text{ LOG} \left(\frac{P_t}{P_L} \right) \quad \textbf{[2.9]}$$

or, by rearranging to account for individual terms:

$$L_{dB} = [20 \text{ LOG}d] + [20 \text{ LOG}F_{MHz}] + k \qquad \textbf{[2.10]}$$

Where:

L_{dB} is the path loss in *decibels* (dB)
d is the path length
F_{MHz} is frequency in megahertz (MHz)
k is a constant that depends on the units of d as follows:

k = 32.4 if d in kilometers
k = 36.58 if d in statute miles
k = 37.80 if d in nautical miles
k = –37.87 if d in feet
k = –27.55 if d in meters

The radiated sphere of energy gets ever larger as the wave propagates away from the isotropic source. If, at a great distance from the center, you take a look at a small slice of the advancing wavefront you can assume that it is essentially a flat plane, as in Fig. 2-5. This situation is analogous to the apparent flatness of the prairie, even though the surface of the earth is nearly spherical. You would be able to "see" the electric and magnetic field vectors at right angles to each other in the flat plane wavefront.

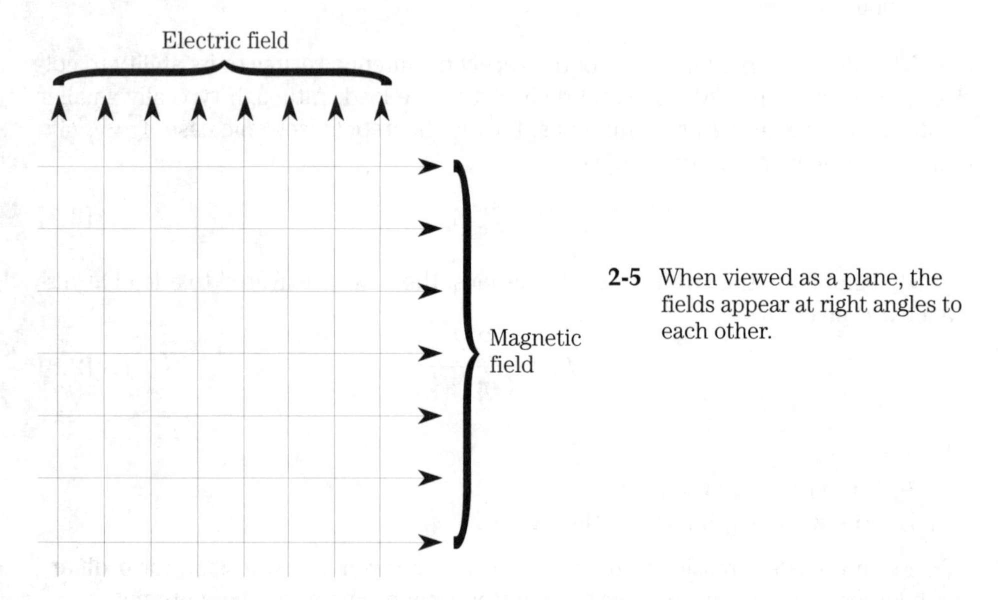

2-5 When viewed as a plane, the fields appear at right angles to each other.

The *polarization* of an EM wave is, by definition, the direction of the electric field. Figure 2-6A shows vertical polarization, because the electric field is vertical with respect to the earth. If the fields were exchanged (as in Fig. 2-6B), then the EM wave would be horizontally polarized.

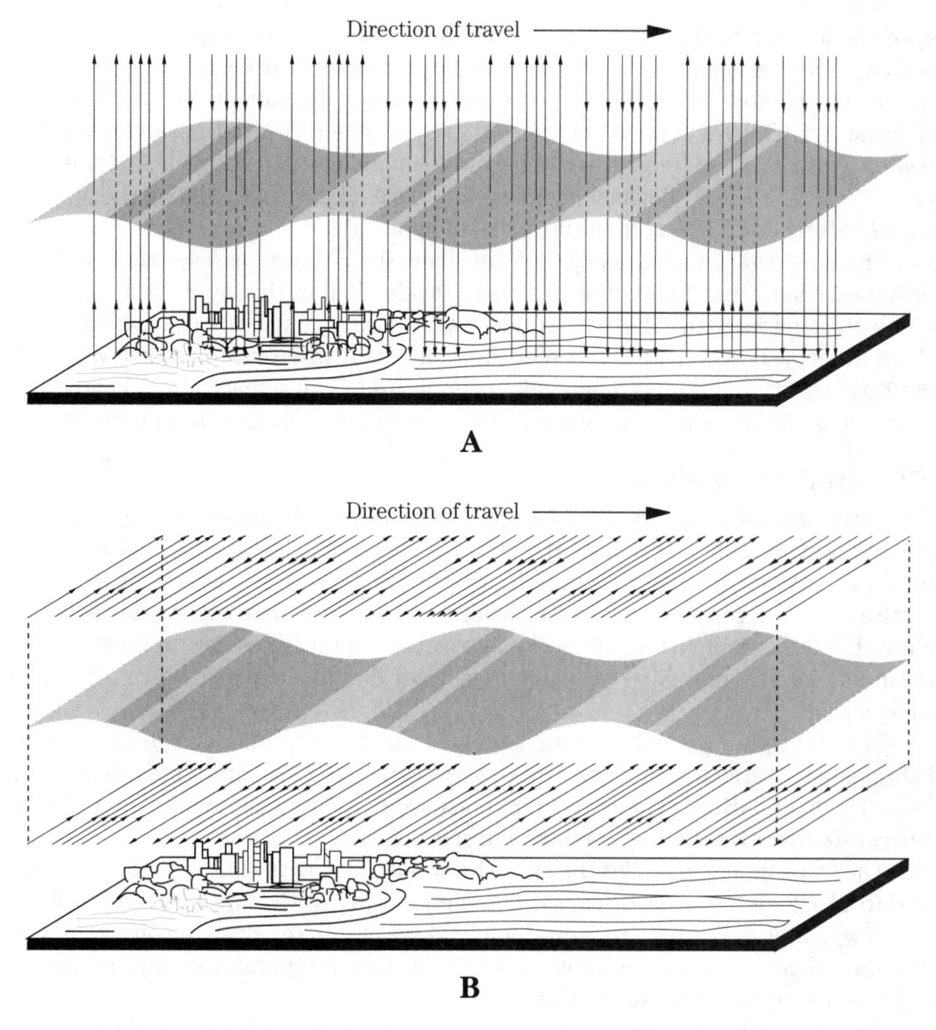

2-6 Wave polarization is determined by the direction of the electric field lines of force: A) Vertical polarized electromagnetic wave; B) horizontally polarized wave.

These designations are especially convenient because they also show the type of antenna used: vertical antennas produce vertically polarized signals, while horizontal antennas produce horizontally polarized signals. At least one text erroneously states that antennas will not pick up signals of the opposite polarity. Such is not the case, especially in the HF and lower VHF regions. At microwave frequencies a loss of approximately 20 dB can be observed due to cross-polarization.

An EM wave travels at the speed of light, designated by the letter *c*, which is about 300,000,000 meters per second (or 186,000 miles per second). To put this velocity in perspective, a radio signal originating on the sun's surface would reach earth in about eight minutes. A terrestrial radio signal can travel around the earth seven times in one second. The velocity of the wave slows in dense media, but in air the speed is so close to the "free-space" value of *c*, that the same figures are used for both air and the near vacuum of outer space in practical problems. In pure water, which is much denser than air, the speed of radio signals is about ⅑ that of the free-space speed. This same phenomenon shows up in practical work in the form of the *velocity factor (V)* of transmission lines. In foam dielectric coaxial cable, for example, the value of *V* is 0.80, which means that the signal propagates along the line at a speed of 0.80*c*, or 80 percent of the speed of light.

This coverage of radio propagation considers the EM wave as a very narrow "ray" or "pencil beam" that does not diverge as it travels. That is, the ray remains the same width all along its path. This convention makes it easy to use ray tracing diagrams. Keep in mind, however, that the real situation, even when narrow beamwidth microwave signals are used, is much more complicated. Real signals, after all, are sloppier than textbook examples: they are neither infinitesimally thin, nor nondivergent.

The earth's atmosphere

The electromagnetic waves do not need an atmosphere in order to propragate, as you will undoubtedly realize from the fact that space vehicles can transmit radio signals back to earth in a near vacuum. But when a radio wave does propagate in the earth's atmosphere, it interacts with the atmosphere, and its path of propagation is altered. A number of factors affect the interaction, but it is possible to break the atmosphere into several different categories according to their respective effects on radio signals.

The atmosphere, which consists largely of oxygen (O_2) and nitrogen (N) gases, is broken into three major zones: *troposphere*, *stratosphere*, and *ionosphere* (Fig. 2-7). The boundaries between these regions are not very well defined, and change *diurnally* (over the course of a day) and seasonally.

The troposphere occupies the space between the earth's surface and an altitude of 6 to 11 kilometers (4 to 7 miles). The temperature of the air in the troposphere varies with altitude, becoming considerably lower at greater altitude compared with ground temperature. For example, a +10°C surface temperature could reduce to –55°C at the upper edges of the troposphere.

The stratosphere begins at the upper boundary of the troposphere (6 to 11 km), and extends up to the ionosphere (≈50 km). The stratosphere is called an *isothermal region* because the temperature in this region is somewhat constant, despite altitude changes.

The ionosphere begins at an altitude of about 50 km (31 miles) and extends up to approximately 300 km (186 miles). The ionosphere is a region of very thin atmosphere. Cosmic rays, electromagnetic radiation of various types (including ultraviolet light from the sun), and atomic particle radiation from space (most of these from the sun also), have sufficient energy to strip electrons away from the gas molecules of the atmosphere. These freed electrons are called *negative ions*, while the O_2 and N molecules that lost electrons are called *positive ions*. Because the density of the air is so low at those altitudes, the ions can travel long distances before neutralizing each other by recombining. Radio propagation on some bands varies markedly between daytime and nighttime because the sun keeps the level of ionization high dur-

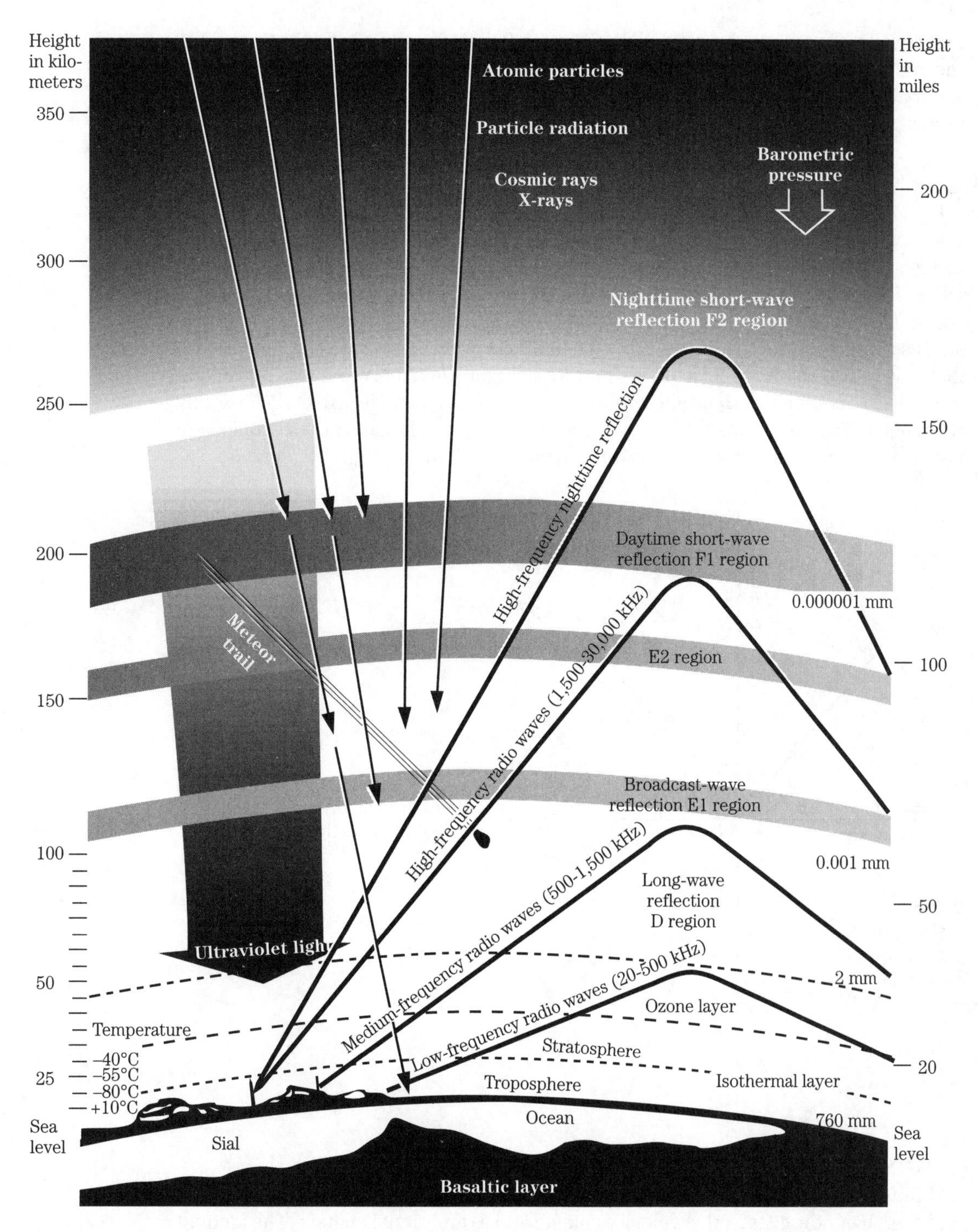

2-7 Radio propagation in the ionosphere is affected by a number of different physical factors: cosmic rays, atomic particles, solar radiation.

ing daylight hours, but the ionization begins to fall off rapidly after sunset, altering the radio propagation characteristics after dark. The ionization does not occur at lower altitudes because the air density is such that the positive and negative ions are numerous and close together, so recombination occurs rapidly.

EM wave propagation phenomena

Because EM waves are waves, they behave in a wave-like manner. Figure 2-8 illustrates some of the wave behavior phenomena associated with light and radio waves: *reflection, refraction*, and *diffraction*. All three play roles in radio propagation.

Reflection and refraction are shown in Fig. 2-8A. Reflection occurs when a wave strikes a denser reflective medium, as when a light wave strikes a glass mirror. The incident wave (shown as a single ray) strikes the interface between less dense and more dense mediums at a certain *angle of incidence* (a_i), and is reflected at exactly the same angle (now called the *angle of reflection* (a_r). Because these angles are equal, a reflected radio signal can often be traced back to its origin.

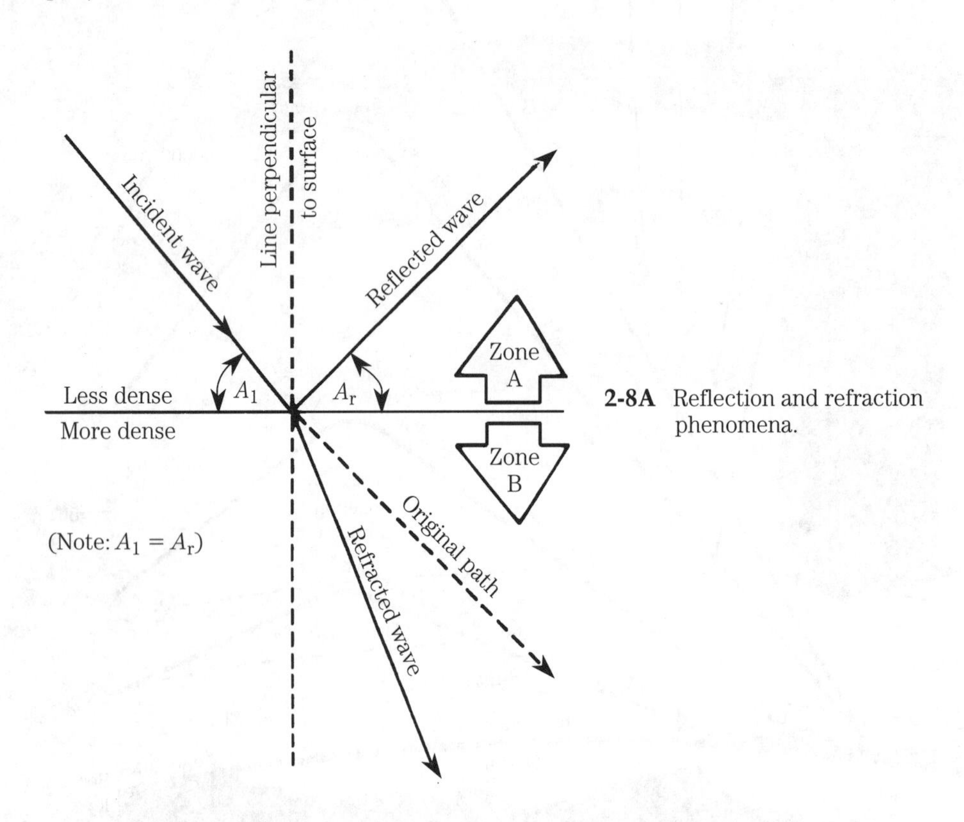

2-8A Reflection and refraction phenomena.

Refraction occurs when the incident wave enters a region of different density, and thereby undergoes both a velocity change and a directional change. The amount and direction of the change are determined by the ratio of the densities between the two media. If Zone B is much different from Zone A, then bending is great. In radio systems, the two media might be different layers of air with different densities. It is possible for both reflection and refraction to occur in the same system. Indeed, more than one form of refraction might be present. These topics will be covered in greater depth shortly.

Diffraction is shown in Fig. 2-8B. In this case, an advancing wavefront encounters an opaque object (e.g., a steel building). The shadow zone behind the building is not simply perpendicular to the wave, but takes on a cone shape as waves bend around the object. The "umbra region" (or diffraction zone) between the shadow zone ("cone of silence") and the direct propagation zone is a region of weak (but not zero) signal strength. In practical situations, signal strength in the cone of silence never reaches zero. A certain amount of reflected signals scattered from other sources will fill in the shadow a little bit. The degree of diffraction effect seen in any given case is a function of the wavelength of the signal, the size of the object, and its electromagnetic properties.

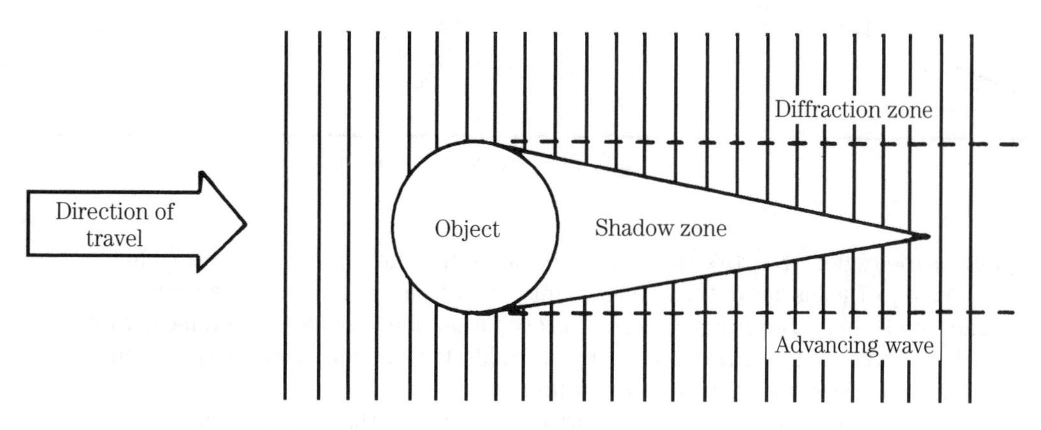

2-8B Diffraction phenomena.

Propagation paths

There are four major propagation paths: *surface wave*, *space wave*, *tropospheric* and *ionospheric*. The ionospheric path is important to MW and HF propagation, but is not important to VHF, UHF, or microwave propagation. The space wave and surface wave are both *ground waves*, but they behave differently enough to warrant separate consideration. The surface wave travels in direct contact with the earth's surface and it suffers a severe frequency-dependent attenuation caused by absorption into the ground.

The space wave is also a ground wave phenomenon, but it is radiated from an antenna many wavelengths above the surface. No part of the space wave normally travels in contact with the surface; VHF, UHF, and microwave signals are usually space waves. There are, however, two components of the space wave in many cases: *direct* and *reflected* (see Fig. 2-9).

The tropospheric wave is lumped with the direct space wave in some texts, but it has properties that actually make it different in practical situations. The tropos-

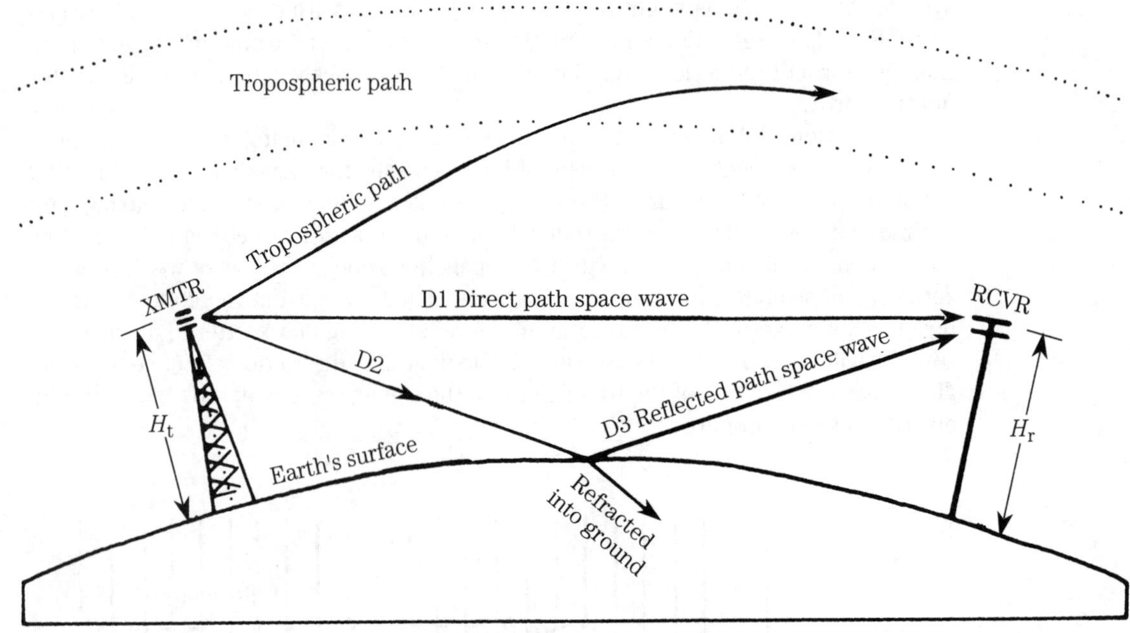

2-9 Space wave propagation.

phere is the region of earth's atmosphere between the surface and the stratosphere, or about 4 to 7 miles above the surface. Thus, most forms of ground wave propagate in the troposphere. But because certain propagation phenomena (caused mostly by weather conditions) only occur at higher altitudes, tropospheric propagation should be differentiated from other forms of ground wave.

The ionosphere is the region of earth's atmosphere that is above the stratosphere. The peculiar feature of the ionosphere is that molecules of air gas (O_2 and N) can be ionized by stripping away electrons under the influence of solar radiation and certain other sources of energy. In the ionosphere, the air density is so low that ions can travel relatively long distances before recombining with oppositely charged ions to form electrically neutral atoms. As a result, the ionosphere remains ionized for long periods of the day—even after sunset. At lower altitudes, however, air density is greater and recombination thus occurs rapidly. At those altitudes, solar ionization diminishes to nearly zero immediately after sunset, or never achieves any significant levels even at local noon.

Ionization and recombination phenomena in the ionosphere add to the noise level experienced at VHF, UHF, and microwave frequencies. The properties of the ionosphere are important to microwave technology because of the noise contribution. In satellite communications, there are some additional transionospheric effects.

Groundwave propagation

The *groundwave*, naturally enough, travels along the ground, or at least in close proximity to it. There are three basic forms of ground wave: *space wave*, *surface wave*, and *tropospheric wave*. The space wave does not actually touch the ground. As a result, space wave attenuation with distance in clear weather is about the same as in free space (except above about 10 GHz, where H_2O and O_2 absorption increases dramatically). Of course, above the VHF region, weather conditions add attenuation not found in outer space.

The surface wave is subject to the same attenuation factors as the space wave, but in addition it also suffers ground losses. These losses are caused by ohmic resistive losses in the conductive earth. In other words, the signal heats up the ground! Surface wave attenuation is a function of frequency, and it increases rapidly as frequency increases. Both of these forms of ground wave communications are affected by the following factors: wavelength, height of both the receiving and transmitting antennas, distance between the antennas,and the terrain and weather along the transmission path. Figure 2-10 is a nomograph that can be used to calculate the line of sight distances in miles from a knowledge of the receiver and transmitter antenna heights. Similarly, Figs. 2-11A and 2-11B show power attenuation with frequency and distance (Fig. 2-11A) and power attenuation in terms of field intensity (Fig. 2-11B).

Groundwave communications also suffer another difficulty, especially at VHF, UHF, and microwave frequencies. The space wave is like a surface wave, but it is radiated many wavelengths above the surface. It is made up of two components (see Fig. 2-9 again): the *direct* and *reflected* waves. If both of these components arrive at the receiving antenna, they will add algebraically to either increase or decrease signal strength. There is always a phase shift between the two components because the two signal paths have different lengths (i.e., D_1 is less than $D_2 + D_3$). In addition, there may possibly be a 180 degree (π radians) phase reversal at the point of reflection (especially if the incident signal is horizontally polarized), as in Fig. 2-12. The following general rules apply in these situations:

- A phase-shift of an odd number of half wavelengths causes the components to add, increasing signal strength (*constructive interference*);
- A phase-shift of an even number of half wavelengths causes the components to subtract (Fig. 2-12), thus reducing signal strength (*destructive interference*); and
- Phase-shifts other than half wavelength add or subtract according to relative polarity and amplitude.

You can characterize the loss of signal over path D_1 with a parametric term, n, that is defined as follows:

$$n = \frac{S_r}{S_f} \qquad \textbf{[2.11]}$$

Where:

n is the signal loss coefficient
S_r is the signal level at the receiver in the presence of a ground reflection component
S_f is the free-space signal strength over path D_1 if no reflection took place.

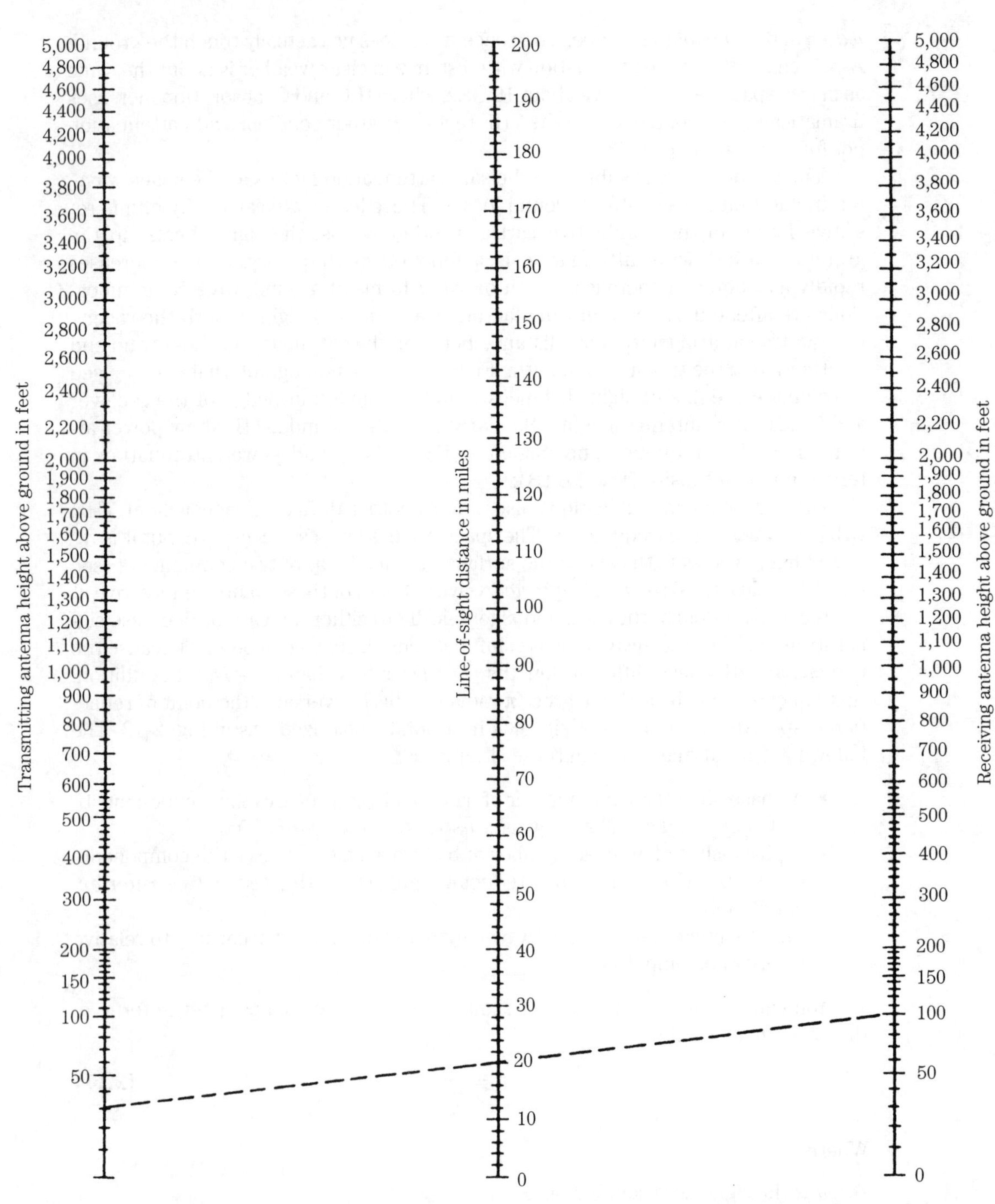

2-10 Nomograph showing the line-of-sight transmission distance as a function of receiving and transmitting antenna heights.

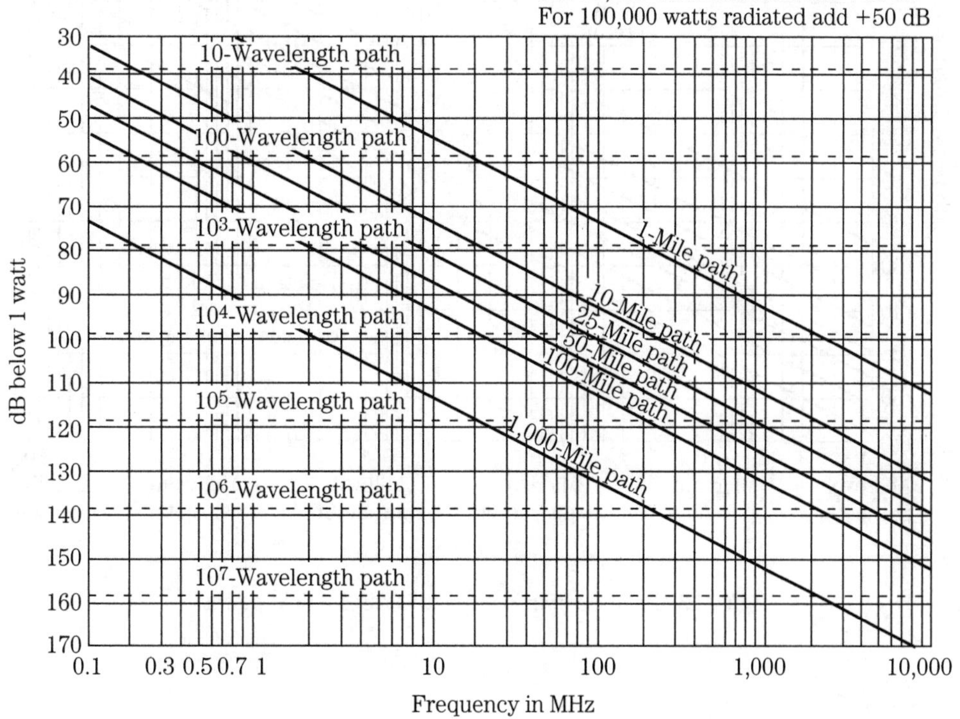

2-11A Power in a free-space field (normalized to 1 watt).

You can calculate as follows:

$$n^2 = 4 \sin^2\left(\frac{2\pi h_t h_r}{\lambda D_1}\right) \qquad \textbf{[2.12]}$$

or,

$$n = 2 \sin\left(\frac{2\pi h_t h_r}{\lambda D_1}\right) \qquad \textbf{[2.13]}$$

The reflected signal contains both amplitude change and phase change. The phase change is typically π radians (180 degrees). The amplitude change is a function of frequency and the nature of the reflecting surface. The *reflection coefficient* can be characterized as:

$$\gamma = pe^{j\phi} \qquad \textbf{[2.14]}$$

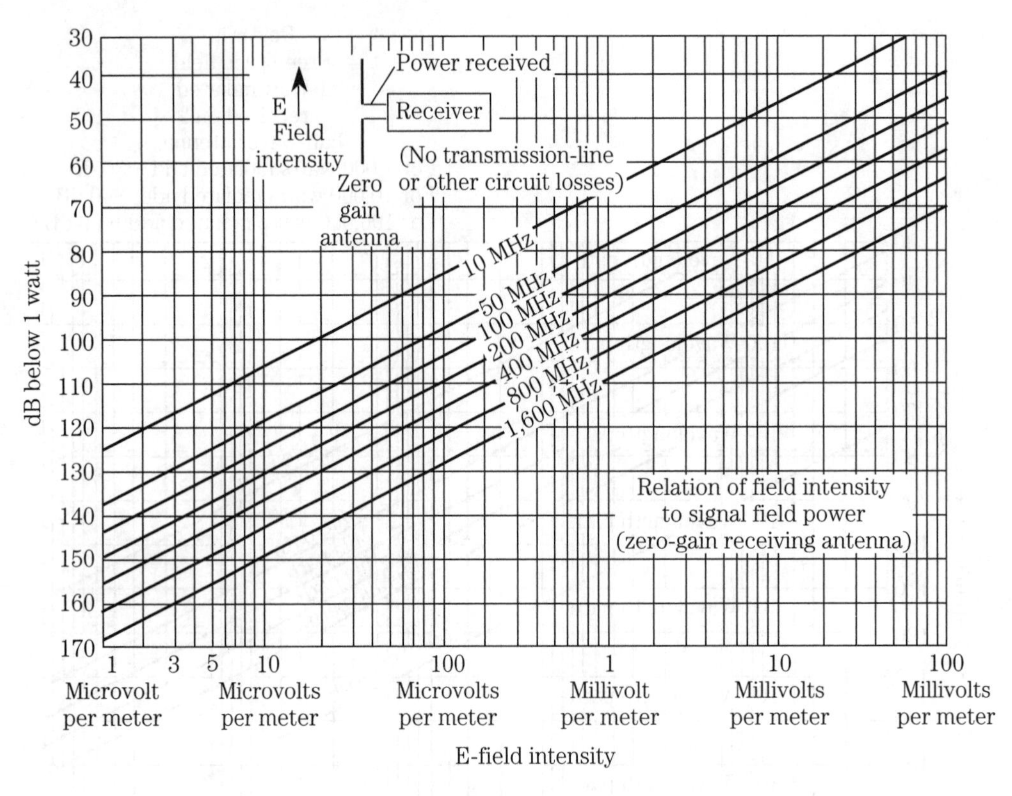

2-11B Relation of field strength to signal field power.

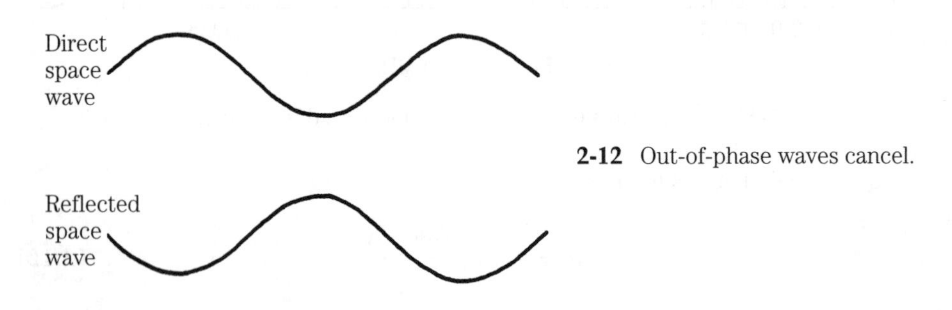

2-12 Out-of-phase waves cancel.

Where:

γ is the reflection coefficient
p is the amplitude change
ϕ is the phase change
j is the imaginary operator

For smooth, high-reflectivity surfaces, a horizontally polarized microwave signal that has a shallow angle of incidence, the value of the reflection coefficient is close to –1.

The phase change of the reflected signal at the receiving antenna is at least π radians because of the reflection. Added to this change is an additional phase shift that is a function of the difference in path lengths. This phase shift can be expressed in terms of the two antenna heights and path length.

$$s = \pi + \left(\frac{4\pi h_t h_r}{\lambda D_1}\right) \qquad [2.15]$$

A category of reception problems called *multipath phenomena* exists because of interference between the direct and reflected components of the space wave. The form of multipath phenomena that is, perhaps, most familiar to many readers is ghosting in television reception. Some multipath events are transitory in nature (as when an aircraft flies through the transmission path), while others are permanent (as when a large building, or hill, reflects the signal). In mobile communications, multipath phenomena are responsible for reception dead zones and "picket fencing." A *dead zone* exists when destructive interference between direct and reflected (or multiply reflected) waves drastically reduces signal strengths. This problem is most often noticed at VHF and above when the vehicle is stopped; and the solution is to move the antenna one half wavelength. *Picket fencing* occurs as a mobile unit moves through successive dead zones and signal enhancement (or normal) zones, and it sounds like a series of short noise bursts.

At VHF, UHF, and microwave frequencies, the space wave is limited to so-called "line-of-sight" distances. The horizon is theoretically the limit of communications distance, but the radio horizon is actually about 15 percent farther than the optical horizon (Fig. 2-13). This phenomenon is caused by refractive bending in the atmosphere around the curvature of the earth, and it makes the geometry of the situation look as if the earth's radius is ⁴⁄₃ the actual radius.

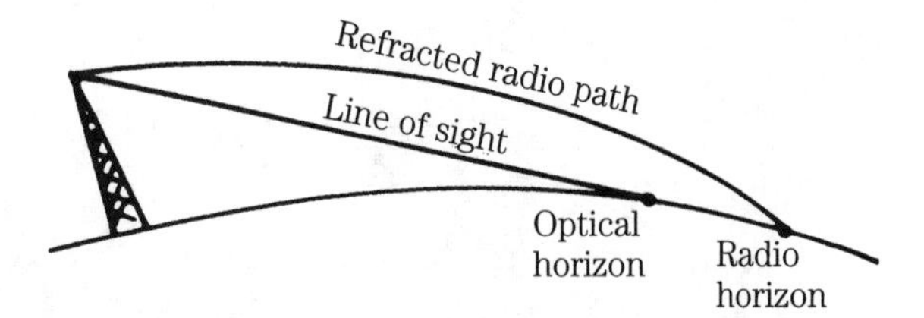

2-13 Phenomenon by which "greater than line of sight" communications occurs.

The refraction phenomenon occurs at VHF through microwave frequencies, but not in the visible light spectrum, because water and atmospheric pressure (which relates to the effects of atmospheric gases on microwave signals) become important contributors to the phenomenon. The *K-factor* expresses the degree of curvature along any given path, while the *index of refraction (n)* measures the differential properties between adjacent zones of air.

The K-factor, also called the *effective earth's radius factor*, is defined according to the relationship of two hypothetical spheres, both centered at the exact center of the earth. The first sphere is the earth's surface, which has a radius r_o (3,440 nmi or 6,370 km). The second sphere is larger than the first by the curvature of the signal "ray path," and has a radius r. The value of K is approximately:

$$K = \frac{r}{r_o} \qquad [2.16]$$

A value of $K = 1$ indicates a straight path (Fig. 2-14); a value of $K > 1$ indicates a positively curved path; and a value of $K < 1$ indicates a negatively curved path. The actual value of K varies with local weather conditions, so one can expect variation not only between locations, but also seasonally. In the arctic regions, K varies approximately over the range 1.2 to 1.34. In the "lower-48" states of the USA, K varies from 1.25 to 1.9 during the summer months (especially in the south and southeast), and from 1.25 to 1.45 in the winter months. The *index of refraction (n)* can be defined in two ways, depending on the situation. When a signal passes across boundaries between adjacent regions of distinctly different properties (as occurs in temperature inversions, etc.), the index of refraction is the ratio of the signal velocities in the two regions. In a homogeneous region n can be expressed as the ratio of the free-space velocity (c) to the actual velocity in the atmosphere (V):

$$n = \frac{c}{V} \qquad [2.17]$$

At the surface, near sea level, under standard temperature and pressure conditions, the value of n is approximately 1.0003, and in homogeneous atmospheres it will decrease by 4×10^{-8} per mile of altitude. The units of n are a bit cumbersome in equations, so the UHF/microwave communities tend to use a derivative parameter, N, called the *refractivity of the atmosphere*:

$$N = (n - 1) \times 10^6 \qquad [2.18]$$

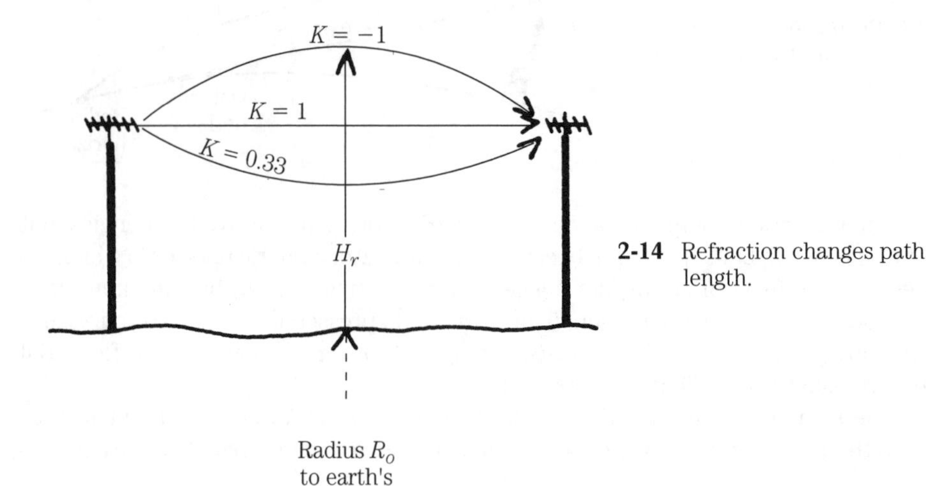

2-14 Refraction changes path length.

The value of N tends to vary from about 280 to 320, and both n and N vary with altitude. In nonhomogeneous atmospheres (the usual case), these parameters will vary approximately linearly for several tenths of a kilometer. All but a few microwave relay systems can assume an approximately linear reduction of n and N with increasing altitude, although airborne radios and radars cannot. There are two methods for calculating N:

$$N = \left(\frac{77.6}{T}\right)\left(P + \frac{4810 e_s H_{rel}}{T}\right) \quad \textbf{[2.19]}$$

and,

$$N = \frac{77.6}{T}\left(\frac{3.73 \times 10^5 e_s}{T^2}\right) \quad \textbf{[2.20]}$$

Where:

P is the atmospheric pressure in millibars (1 Torr = 1.3332 mbar)
T is temperature in degrees Kelvin
e_s is saturation vapor pressure of atmospheric water, in millibars
H_{rel} is the relative humidity expressed as a decimal fraction (rather than a percentage)

Ray path curvature (K) can be expressed as a function of either n or N, provided that the assumption of a linear gradient d_n/d_h holds true:

$$K = \frac{1}{\left(1 + \frac{r_o d_n}{d_h}\right)} \quad \textbf{[2.21]}$$

or,

$$K = \frac{1}{\left(1 + \frac{d_N/d_h}{157}\right)} \quad \textbf{[2.22]}$$

For the near-surface region, where d_n/d_h varies at about 3.9×10^{-8} per meter, the value of K is 1.33. For most terrestial microwave paths, this value ($K = 4/3 = 1.33$) is called a *standard refraction*, and is used in calculations *in the absence of additional data*. For regions above the linear zone close to the surface, you can use another expression of refractivity:

$$N_a = N_s e^{[-C_e(h_r - h_t)]} \quad \textbf{[2.23]}$$

Where:

N_1 is the refractivity at 1-km altitude
h_r is the height of the receiver antenna
h_t is the height of the transmit antenna

$C_e = L_n(N_s/N_1)$
N_a is the refractivity at altitude
N_s is the refractivity at the earth's surface

For simple models close to the surface, you can use the geometry shown in Fig. 2-15. Distance d is a curved path along the surface of the earth. But because the earth's radius r_o is about 4000 statute miles, and it is thus very much larger than a practical antenna height h, you can simplify the model to that shown in Figs. 2-15A and 2-15B. The underlying assumption is that the earth has a radio radius equal to about ⅘ ($K = 1.33$) of the actual physical radius of its surface.

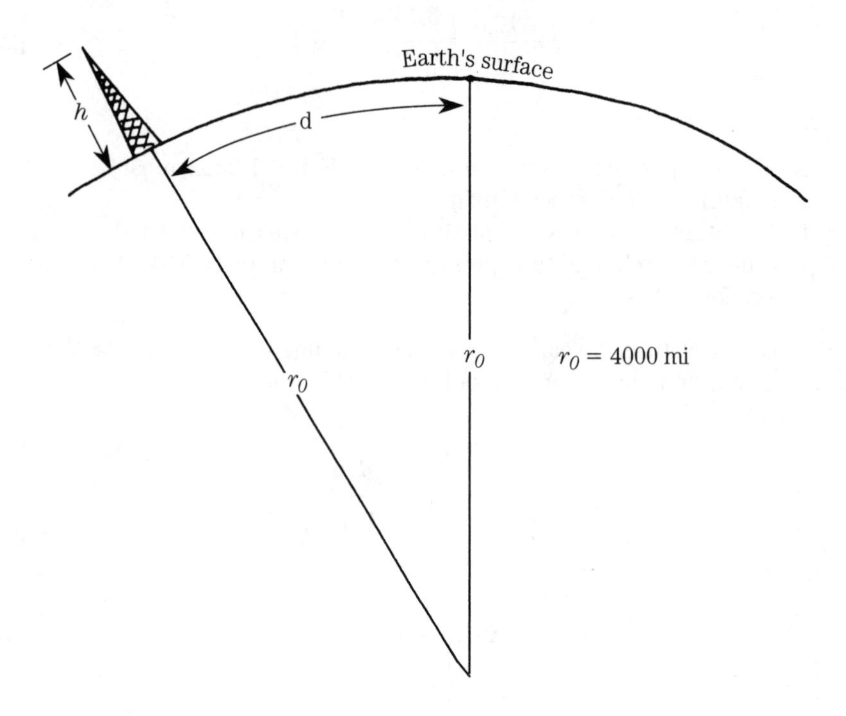

2-15A Geometry for calculating radio distances.

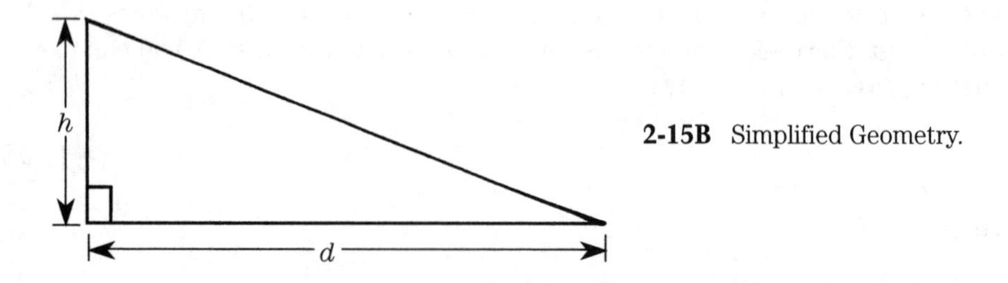

2-15B Simplified Geometry.

The value of distance d is found from the expression:

$$d = \sqrt{2R_o h_{\text{ft}}} \qquad \textbf{[2.24]}$$

Where:

d is the distance to the radio horizon in statute miles
r_o is the normalized radius of the earth
h_t is the antenna height in feet

Accounting for all constant factors, the expression reduces to:

$$d = 1.414 \sqrt{\text{h}_{\text{ft}}} \qquad \textbf{[2.25]}$$

(All factors being the same as defined above.)

Example 2-1 A radio tower has a UHF radio antenna that is mounted 150 feet above the surface of the earth. Calculate the radio horizon (in statute miles) for this system.

Solution:

$$d = 1.414\ (h_t)t$$
$$d = (1.414)(150\ \text{ft})^{1/2}$$
$$d = (1.414)(12.25)$$
$$d = 17.32\ \text{miles}$$

For other units of measurement:

$$d_{nmi} = 1.23 \sqrt{h_{\text{ft}}} \qquad \textbf{[2.26]}$$

and,

$$d_{km} = 130 \sqrt{h_t} \qquad \textbf{[2.27]}$$

Surface wave communications The surface wave travels in direct contact with the earth's surface, and it suffers a severe frequency-dependent attenuation because of absorption by the ground.

The surface wave extends to considerable heights above the ground level, although its intensity drops off rapidly at the upper end. The surface wave is subject to the same attenuation factors as the space wave, but in addition, it also suffers ground losses. These losses are caused by ohmic resistive losses in the conductive earth, and to the dielectric properties of the earth. In other words, the signal heats up the ground. Horizontally polarized waves are not often used for surface wave communications because the earth tends to short-circuit the E-field component. On vertically polarized waves, however, the earth offers electrical resistance to the E-field and returns currents to following waves (Fig. 2-16). The conductivity of the soil determines how much energy is returned. Table 2-1 shows the typical conductivity values for several different forms of soil.

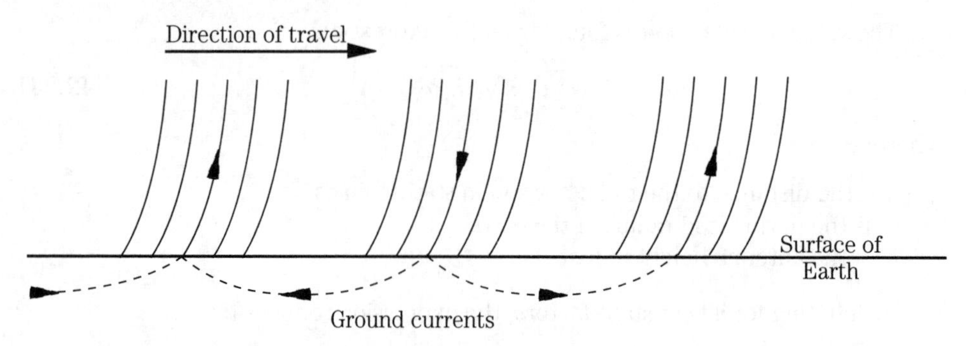

2-16 Distortion of vertically polarized electric field by lossy ground resistance.

Table 2-1. Sample soil conductivity values

Type of soil	Dielectric constant	Conductivity (siemans/meter)	Relative quality
Salt water	81	5	Best
Fresh water	80	0.001	Very poor
Pastorial hills	14–20	0.03–0.01	Very good
Marshy, wooded	12	0.0075	Average/poor
Rocky hills	12–14	10	Poor
Sandy	10	0.002	Poor
Cities	3–5	0.001	Very poor

The wavefront of a surface wave is tilted because of the losses in the ground that tend to retard the wavefront's bottom (also in Fig. 2-16). The tilt angle is a function of the frequency, as shown in Table 2-2.

Table 2-2. Tilt angle as a function of frequency

Frequency	Tilt angle ratio	Earth/sea water (degrees)
20 kHz	2.07	4.3/2.08
200 kHz	104	13.4/0.13
2,000 kHz	64	32.3/0.5
20,000 kHz	25	35/1.38

Surface wave attenuation is a function of frequency, and increases rapidly as frequency increases. For both forms of the ground wave, reception is affected by these factors: *wavelength, height* of both the receiving and transmitting antennas, *distance* between antennas, and both *terrain* and *weather* along the transmission path. In addition, the surface wave is affected by the ground losses in Table 2-1. Because of the ground loss effects, the surface wave is attenuated at a much faster rate than the inverse square law.

Ground wave propagation frequency effects The frequency of a radio signal in large measure determines its surface wave behavior. In the very low frequency (VLF) band (<300 kHz), ground losses are small for vertically polarized signals, so medium distance communications (up to several hundred miles) are possible. In the Medium Wave band (300 to 3,000 kHz, including the AM broadcast band), distances of 1,000 miles are possible with regularity—especially at night. In the High Frequency (HF) band, ground losses are more considerable, so the surface wave distance reduces drastically. It is possible, in the upper end of the HF band (3,000 to 30,000 kHz) for surface wave signals to die out within a few dozen miles. This phenomenon is often seen in the 15-meter and 10-meter amateur radio bands, as well as the 11-meter (27 MHz) Citizen's Band. Stations only 20 miles apart cannot communicate, but both can talk to a third station across the continent via ionospheric skip. Thus, the two stations close together must have a station more than 2,000 miles away relay messages between them.

Tropospheric propagation The troposphere is the portion of the atmosphere between the surface of the earth and the stratosphere (or about 4 to 7 miles above the surface). Some older texts group tropospheric propagation with ground wave propagation, but modern practice requires separate treatment. The older grouping overlooks certain common propagation phenomena that simply don't happen with space or surface waves.

Refraction is the mechanism for most tropospheric propagation phenomena. The dielectric properties of the air, which are set mostly by the moisture content (Fig. 2-17), is a primary factor in tropospheric refraction. Recall from earlier that refraction occurs in both light or radio wave systems when the wave passes between mediums of differing density. Under that situation, the wave path will bend an amount proportional to the difference in density.

Two general situations are typically found—especially at UHF and microwave frequencies. First, because air density normally decreases with altitude, the top of a beam of radio waves typically travels slightly faster than the lower portion of the beam. As a result, those signals refract a small amount. Such propagation provides slightly longer surface distances than are normally expected from calculating the distance to the radio horizon. This phenomenon is called *simple refraction*, and was discussed in a preceding section.

A special case of refraction called *super refraction* occurs in areas of the world where warmed land air goes out over a cooler sea (Fig. 2-18). Examples of such areas have deserts that are adjacent to a large body of water: the Gulf of Aden, the southern Mediterranean, and the Pacific Ocean off the coast of Baja California are examples. VHF/UHF/microwave communications up to 200 miles are reported in such areas.

The second form of refraction is weather related. Called *ducting*, this form of propagation (Fig. 2-19) is actually a special case of super refraction. Evaporation of sea water causes temperature inversion regions to form in the atmosphere. That is, layered air masses in which the air temperature is greater than in the layers below it (note: air temperature normally decreases with altitude, but at the boundary with an

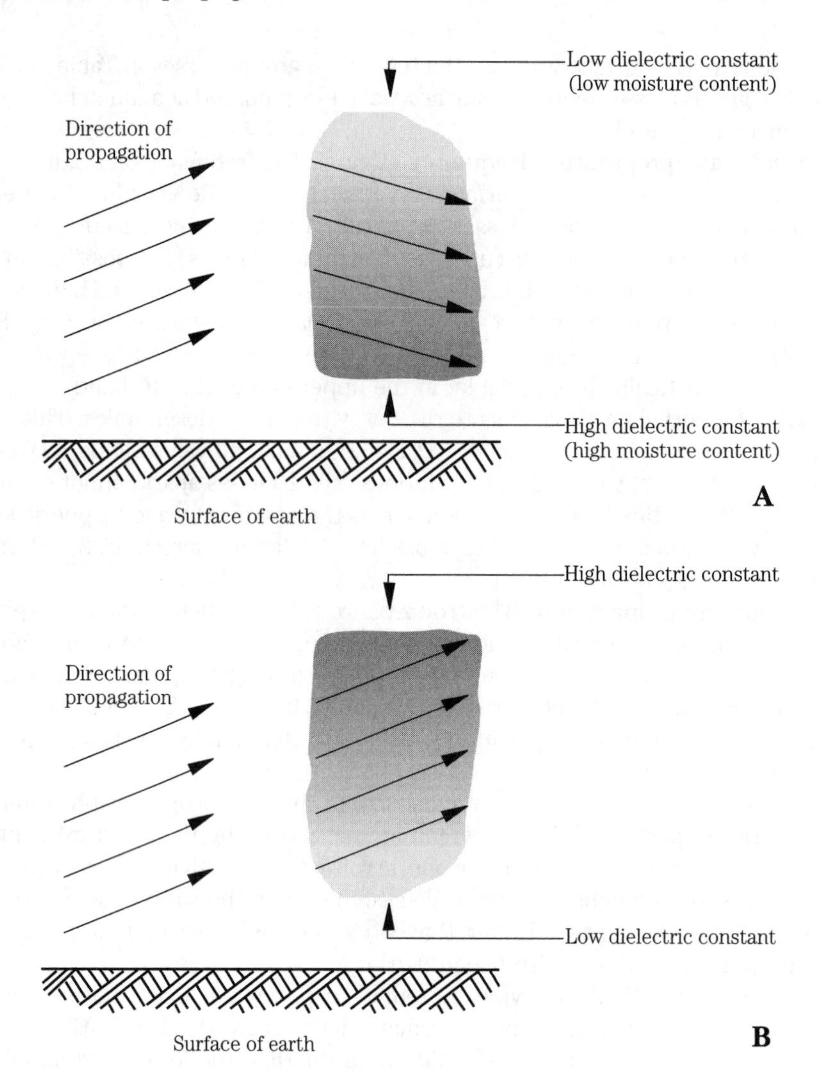

2-17 Refraction in the troposphere.

inversion region, it increases). The inversion layer forms a "duct" that acts similarly to a waveguide. In Fig. 2-19, the distance D_1 is the normal "radio horizon" distance, and D_2 is the distance over which duct communications can occur.

Ducting allows long distance communications from lower VHF through microwave frequencies; with 50 MHz being a lower practical limit, and 10 GHz being an ill-defined upper limit. Airborne operators of radio, radar, and other electronic equipment can sometimes note ducting at even higher microwave frequencies.

Antenna placement is critical for ducting propagation. Both the receiving and transmitting antennas must be either: (a) inside the duct physically (as in airborne cases), or (b) able to propagate at an angle such that the signal gets trapped inside the duct. The latter is a function of antenna radiation angle. Distances up to 2,500

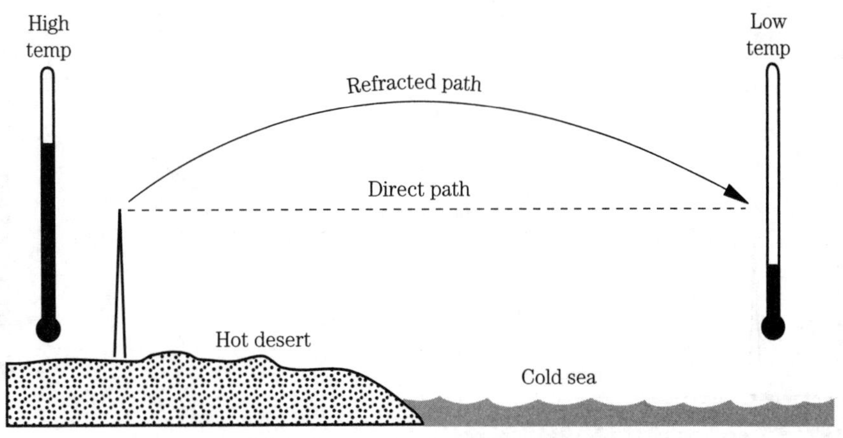

2-18 Superrefraction phenomena.

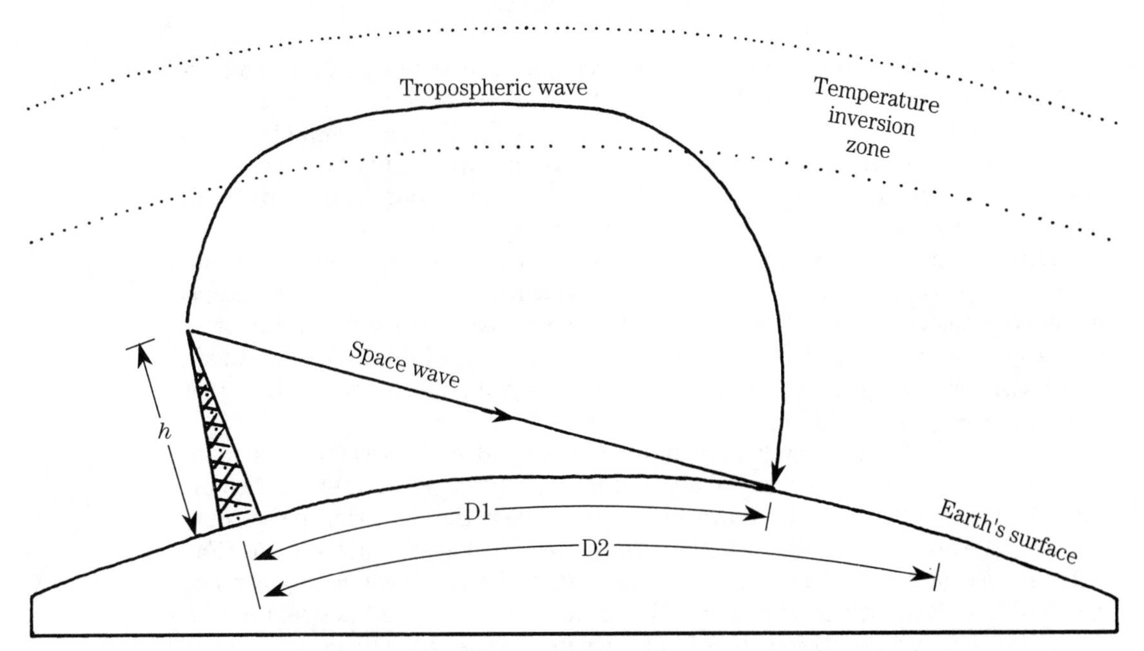

2-19 Ducting phenomena.

miles or so are possible through ducting. Certain paths, where frequent ducting occurs, have been identified: the Great Lakes to the Atlantic seaboard; Newfoundland to the Canary Islands; across the Gulf of Mexico from Florida to Texas; Newfoundland to the Carolinas; California to Hawaii; and Ascension Island to Brazil.

Another condition is noted in the polar regions, where colder air from the land mass flows out over warmer seas (Fig. 2-20). Called *subrefraction*, this phenomena bends EM waves away from the earth's surface—thereby *reducing* the radio horizon by about 30 to 40 percent.

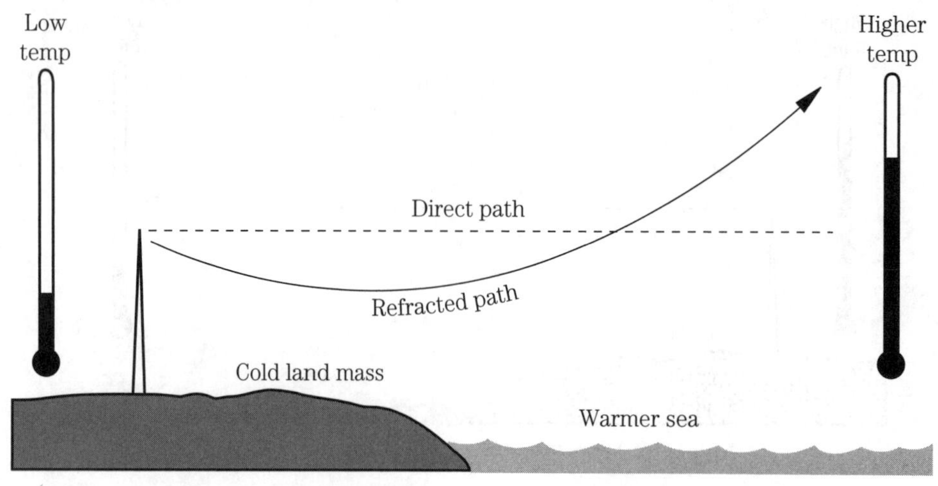

2-20 Subrefraction phenomena.

All tropospheric propagation that depends upon air-mass temperatures and humidity, shows diurnal (i.e., over the course of the day) variation caused by the local rising and setting of the sun. Distant signals may vary 20dB in strength over a 24-hour period. These tropospheric phenomena explain how TV, FM broadcast, and other VHF signals can propagate great distances, especially along seacoast paths, at some times while being weak or nonexistent at others.

Diffraction phenomena Electromagnetic *waves* diffract when they encounter a radio-opaque object. The degree of diffraction, and the harm it causes, is frequency related. Above 3 GHz, wavelengths are so small compared to object sizes that large attenuation of the signal occurs. In addition, beamwidths (a function of antenna size compared with wavelength) tend to be small enough above 3 GHz that blockage of propagation by obstacles is much greater.

Earlier in this chapter, large-scale diffraction around structures (such as buildings) was discussed. The view presented was from above, so it represented the horizontal plane. But there is also a diffraction phenomena in the vertical plane. Terrain, or man-made objects, intervening in the path between UHF microwave stations (Fig. 2-21A) cause diffraction, and some signal attenuation. There is a minimum clearance required to prevent severe attenuation (up to 20–30 dB) from diffraction. Calculation of the required clearance comes from Huygens-Fresnel wave theory.

Consider Fig. 2-21B. A wave source (A), which might be a transmitter antenna, transmits a wavefront to a destination [(C) receiver antenna]. At any point along path AC, you can look at the wavefront as a partial spherical surface (Bl-B2) on which all wave rays have the same phase. This plane can be called an *isophase plane*. You can assume that the d_n/d_h refraction gradient over the height extent of the wavefront is small enough to be considered negligible.

Using ray tracing we see rays r_a incoming to plane [Bl-B2], and rays r_b out-going from plane [Bl-B2]. The signal seen at C is the algebraic sum of all rays r_b. The signal pattern will have the form of an optical interference pattern with wave cancellation occurring between r_b waves that are a half-wavelength apart on [Bl-B2]. The ray im-

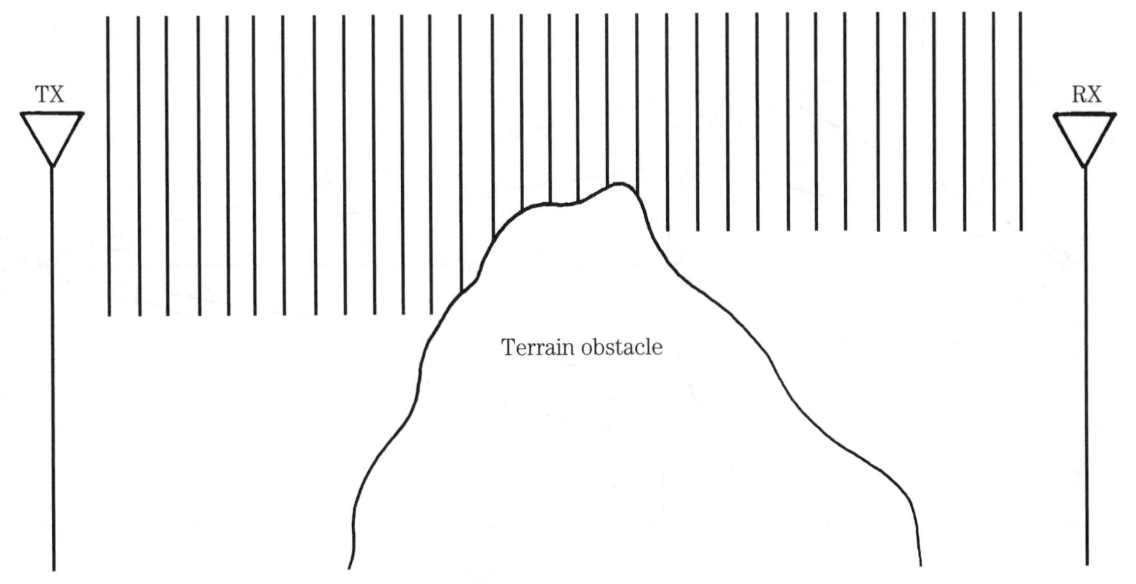

2-21A Terrain masking of VHF and up signals.

pact points on plane [Bl-B2] form radii R_n called Fresnel zones. The lengths of the radii are a function of the frequency, and the ratio of the distances D_1 and D_2 (see Fig. 2-21B). The general expression is:

$$R_n = M\sqrt{\frac{N}{F_{\text{GHz}}}\left(\frac{D_1 D_2}{D_1 + D_2}\right)} \quad \textbf{[2.28]}$$

Where:

R_m is the radius of the nth Fresnel zone
F_{GHz} is the frequency in GHz
D_1 is the distance from the source to plane $[B_1\text{-}B_2]$
D_2 is the distance from destination to plane $[B_1\text{-}B_2]$
N is an integer (1, 2, 3, . . .)[1st, 2nd, 3rd, . . .]
M a constant of proportionality as follows:

$m = 17.3$ if R_m is in meters, and D_1, D_2 are in kilometers

$m = 72.1$ if R_m is in feet, and D_1,D_2 are in statute miles.

If we first calculate the radius of the critical first Fresnel zone (R_1), then you can calculate the nth Fresnel zone from:

$$R_n = R_1\sqrt{n} \quad \textbf{[2.29]}$$

Example 2-2 Calculate the radius of the first Fresnel zone for a 2.5 GHz signal at a point that is 12 km from the source and 18 km from the destination.

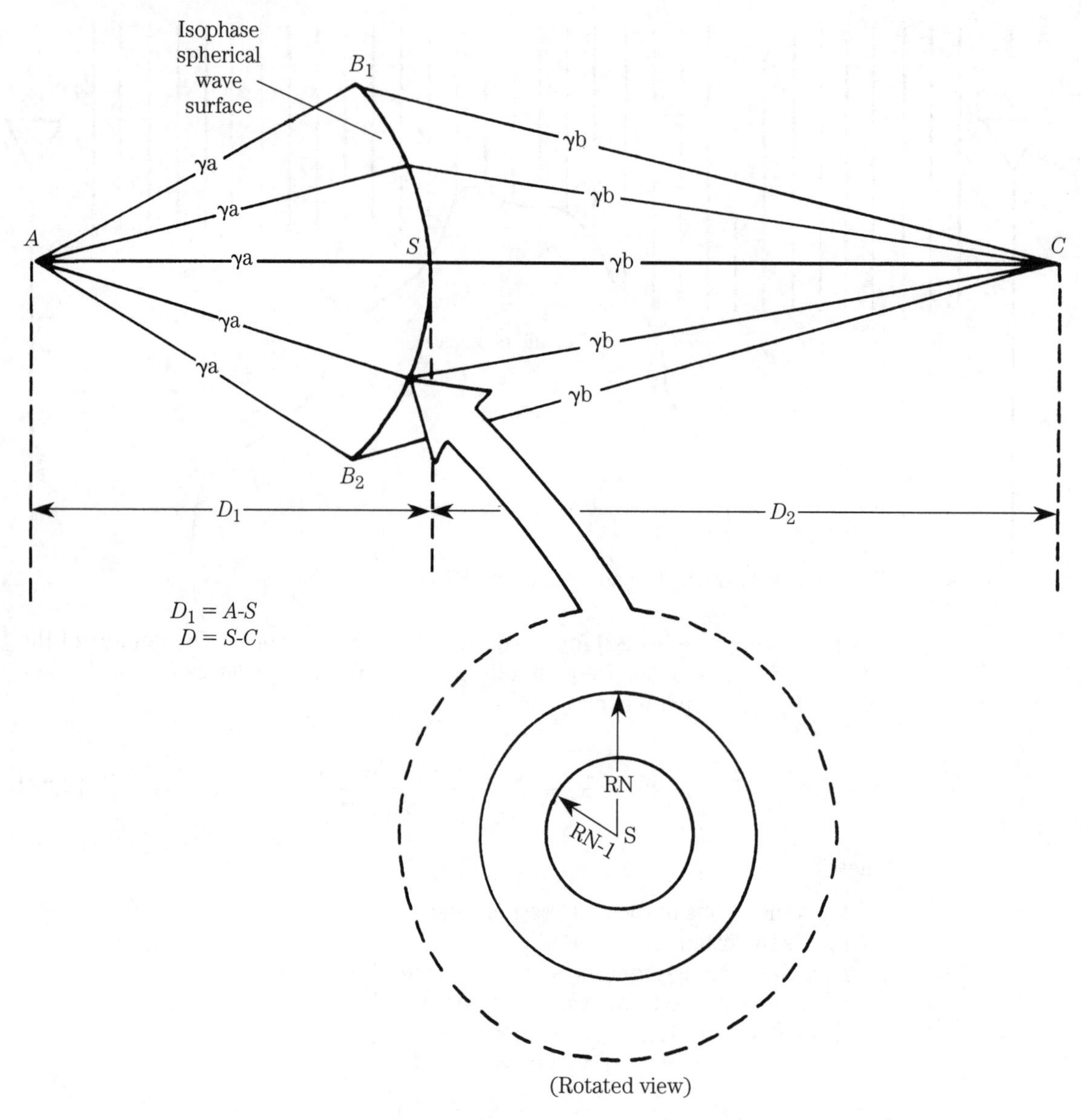

2-21B Fresnel zone geometry.

Solution:

$$R_1 = M\sqrt{\frac{N}{F_{\text{GHz}}}\left(\frac{D_1 D_2}{D_1 + D_2}\right)}$$

$$R_1 = (17.3)\sqrt{\frac{1}{2.5}\left(\frac{12 \times 18}{12 + 18}\right)}$$

$$R_t = (17.3)\sqrt{(0.4)\,(7.2)}$$

$$R_1 = (17.3)\,(2.88) = 29.4 \text{ meters}$$

For most terrestial microwave systems an obstacle clearance of $0.6R_1$ is required to prevent diffraction attenuation under most normal conditions. However, there are conditions in which the clearance zone should be more than one Fresnel zone.

Fading mechanisms Fading is defined as a reduction in amplitude caused by reduced received signal power, changes in phase or polarization, wave cancellation, or other related problems that are not caused by a change in the output power level or other parameters associated with either the transmitter or its antenna system. You would not ordinarily think that line-of-sight radio relay links would experience fading, but that is not true. Fading does, in fact, occur, and it can reach levels of 30dB in some cases (20dB is relatively common). In addition, fading phenomena in the VHF-and-up range can last several hours, with some periods being reported of several days in duration (although very rare). There are several mechanisms of fading, and these will be dealt with in this section. HF fading caused by ionospheric mechanisms will be covered later.

Any or all of the mechanisms shown in Fig. 2-22 can occur in a given system. In all cases, two or more signals arrive at the receiver antenna (R_x). Ray A represents the direct path signal that is, ideally, the only signal to reach the destination. But it is also possible that a signal, in an elevated layer or other atmospheric anomaly, will cause refraction or subrefraction of the wave creating a second component, *B*. If this second signal arrives out of phase with *A*, then fading will occur (signal reinforce-

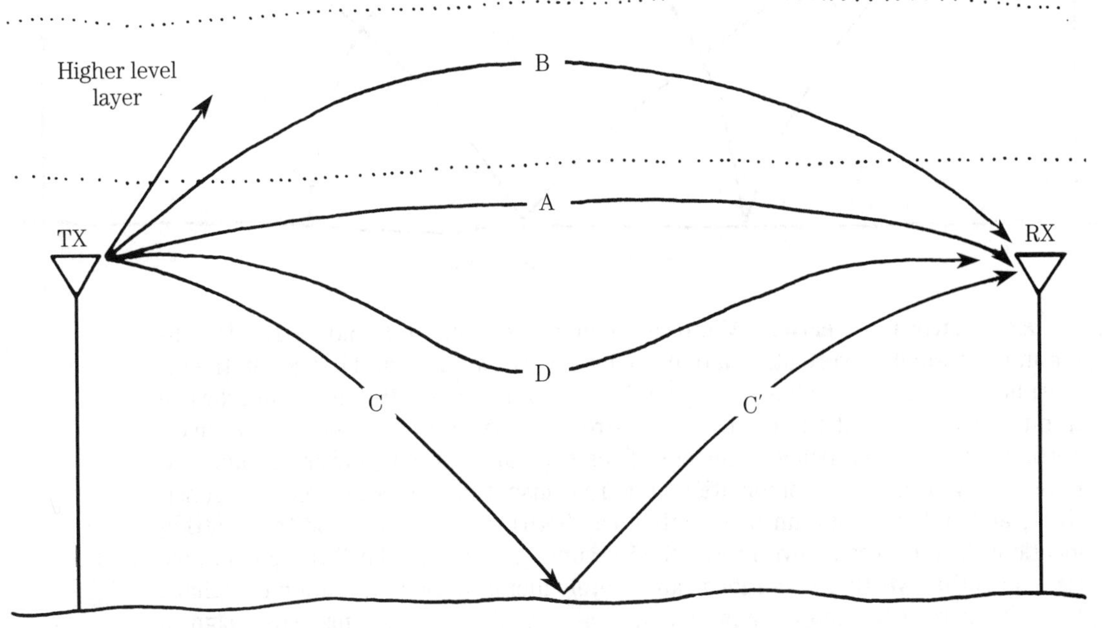

2-22 Multiple paths for signal to take between transmitter and receiver.

ment—in phase—can also occur). It is also possible to see subrefraction fading, as in D. The classical multipath situation represented by ray C and its reflected component C' is also a source of fading.

These mechanisms are frequency sensitive, so a possible countermeasure is to use *frequency diversity*. Hopping over a 5 percent frequency change will help eliminate fading. In cases where either system constraints, or local spectrum, usage prevents a 5 percent delta (change), then try for at least 2 or 3 percent.

Over ocean areas or other large bodies of water there is a possibility of encountering fair weather surface ducting as a cause of fading. These ducts form in the midlatitudes, starting about 2 to 3 km from shore, up to heights of 10 to 20 miles; wind velocities are found in the 10 to 60 km/hr range. The cause of the problem is a combination of power fading due to the presence of the duct, and surface reflections (see Fig. 2-23). Power fading alone can occur when there is a superrefractive duct elevated above the surface. The duct has a tendency to act as a waveguide, and focus the signal (Fig. 2-24). Although the duct shown is superrefractive, it is also possible to have a subrefractive duct.

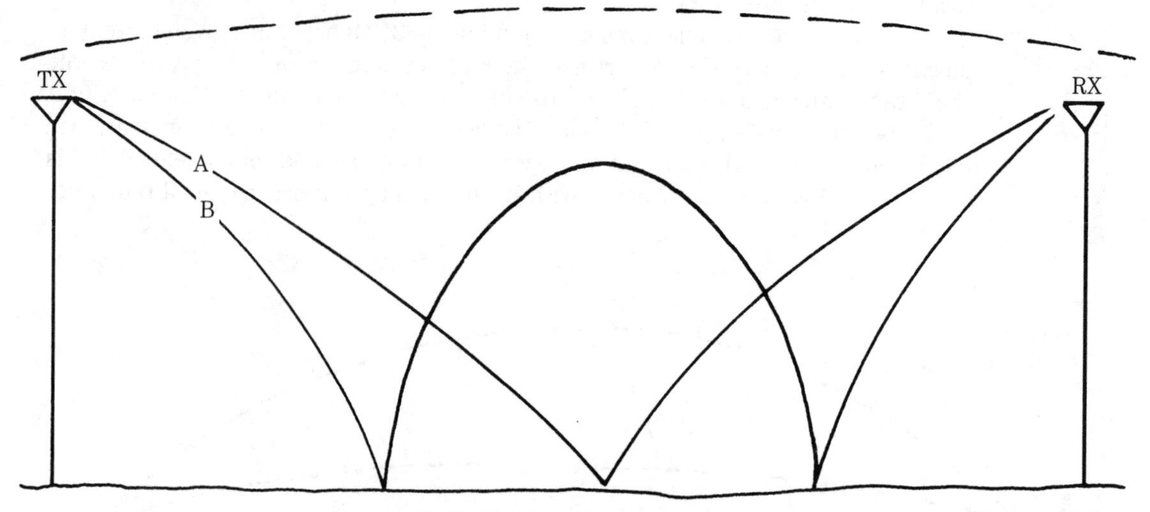

2-23 Multi-hop interference.

Attenuation in weather Microwave communications above about 10 GHz suffer an increasingly severe attenuation because of water vapor and oxygen in the atmosphere. Figure 2-25 shows the standard attenuation in dB/km for microwave frequencies. Note that there are several strong peaks in an ever-increasing curve. Setting a system frequency in these regions can cause either poorer communications; or it will require a combination of more transmit power, better receiver sensitivity, and/or better antennas on either (or both) receiving and or transmitting locations. The curves shown in Fig. 2-25 assume certain standardizing conditions. Rain and other weather conditions can severely increase the attenuation of signals. In addition to attenuation, radar exhibits severe clutter problems when signals backscatter from rain cells.

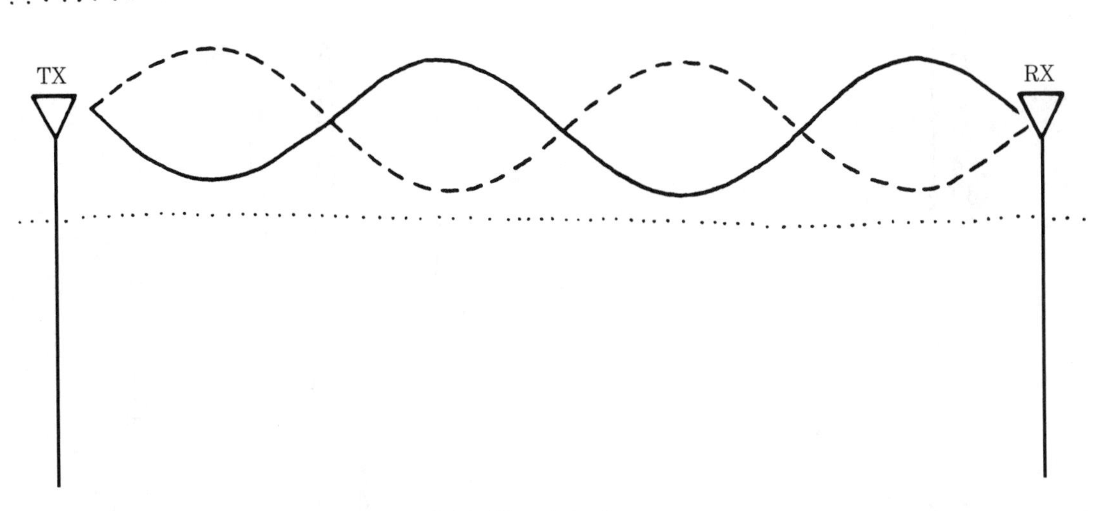

2-24 Wave interference.

Ionospheric propagation

Now let's turn your attention to the phenomenon of skip communications. Ionospheric propagation is responsible for the ability to do intercontinental broadcasting and communications. Long distance radio transmission is carried out on the high frequency (HF) bands (3 to 30 MHz), also called the "*shortwave*" bands. These frequencies are used because of the phenomenon called *skip*. Under this type of propagation, the earth's ionosphere acts as though it is a "radio mirror." Although the actual phenomenon is based on refraction (not reflection, as is frequently believed), the appearance to the casual observer is that shortwave and low-VHF radio signals are reflected from the ionosphere. The actual situation is a little different.

The key lies in the fact that this radio mirror is produced by ionization of the upper atmosphere. The upper portion of the atmosphere is called the "ionosphere" because it tends to be easily ionized by solar and cosmic radiation phenomena. The reason for the ease with which that region (30 to 300 miles above the surface) ionizes, is that the air density is very low. Energy from the sun strips away electrons from the outer shells of oxygen and nitrogen molecules. The electrons become negative ions, while the remaining portion of the atom forms positive ions. Because the air is so rarified at those altitudes, those ions can travel great distances before recombining to form electrically neutral atoms again. As a result, the average ionization level remains high in that region.

Several sources of energy will cause ionization of the upper atomosphere. Cosmic radiation from outer space causes some degree of ionization, but the vast majority of ionization is caused by solar energy.

The role of cosmic radiation was first noticed during World War II, when British radar operators discovered that the distance at which their equipment could detect German aircraft was dependent upon whether or not the Milky Way was above the horizon. Intergalactic radiation raised the background microwave noise level, thereby adversely affecting the signal-to-noise ratio.

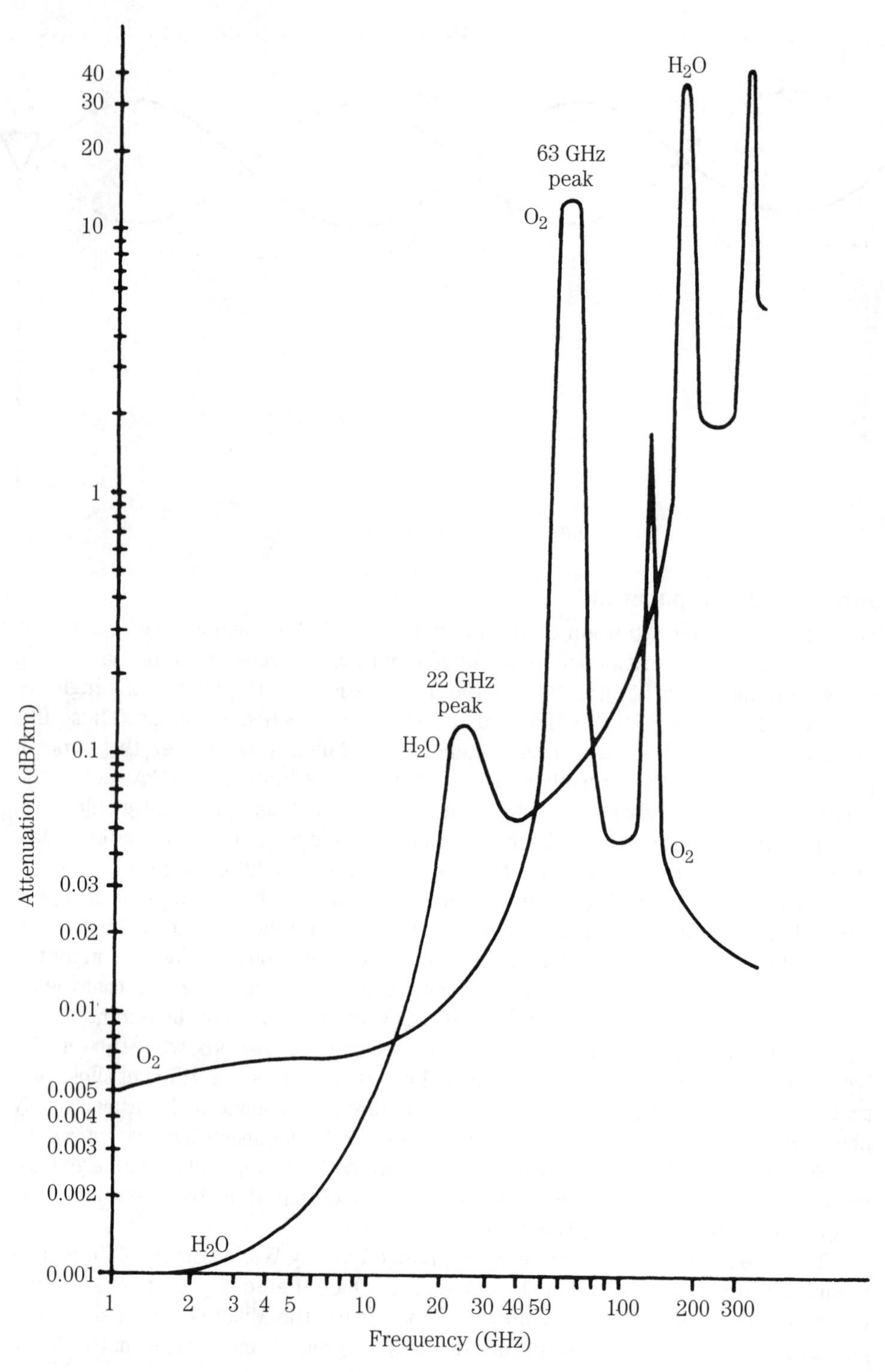

2-25 Atmospheric absorption of radio signals at microwave frequencies.

Events on the surface of the sun sometimes cause the radio mirror to seem to be almost perfect, and this situation makes spectacular propagation possible. At other times, however, solar disturbances (Fig. 2-26A) disrupt radio communications for days at a time.

2-26A Solar event that can affect radio propagation on earth.

There are two principal forms of solar energy that affect shortwave communications: *electromagnetic radiation* and *charged solar particles*. Most of the radiation is above the visible spectrum, in the ultraviolet and X-ray/gamma-ray region of the spectrum. Because electromagnetic radiation travels at the speed of light, solar events that release radiation cause changes to the ionosphere about eight minutes later. Charged particles, on the other hand, having a finite mass must travel at a considerably slower velocity. They require two or three days to reach earth.

Various sources of both radiation and particles exist on the sun. Solar flares can release huge amounts of both radiation and particles. These events are unpredictable and sporadic. Solar radiation also varies over an approximately 27 day period, which is the rotational period of the sun. The same source of radiation will face the earth once every 27 days, and so events tend to be somewhat repetitive.

Solar and galactic noise affects the reception of weak signals. Solar noise can also affect radio propagation and act as a harbinger of changes in propagation patterns. Solar noise can be demonstrated by using an ordinary radio receiver and a directional antenna, preferably operating in the VHF/UHF regions of the spectrum. Aim the antenna at the sun on the horizon at either sunset or sunrise. A dramatic change in background noise will be noted as the sun slides across the horizon.

Sunspots A principal source of solar radiation, especially the periodic forms, is sunspots (Fig. 2-26B). Sunspots can be as large as 70,000 to 80,000 miles in diameter, and generally occur in clusters. The number of sunspots varies over a period of approximately 11 years, although the actual periods since 1750 (when records were first kept) have varied from 9 to 14 years. The sunspot number is reported daily as the statistically massaged *Zurich Smoothed Sunspot Number*, or *Wolf Number*. The number of sunspots greatly affects radio propagation via the ionosphere. The low was in the range of 60 (in 1907), and the high was about 200 (1958).

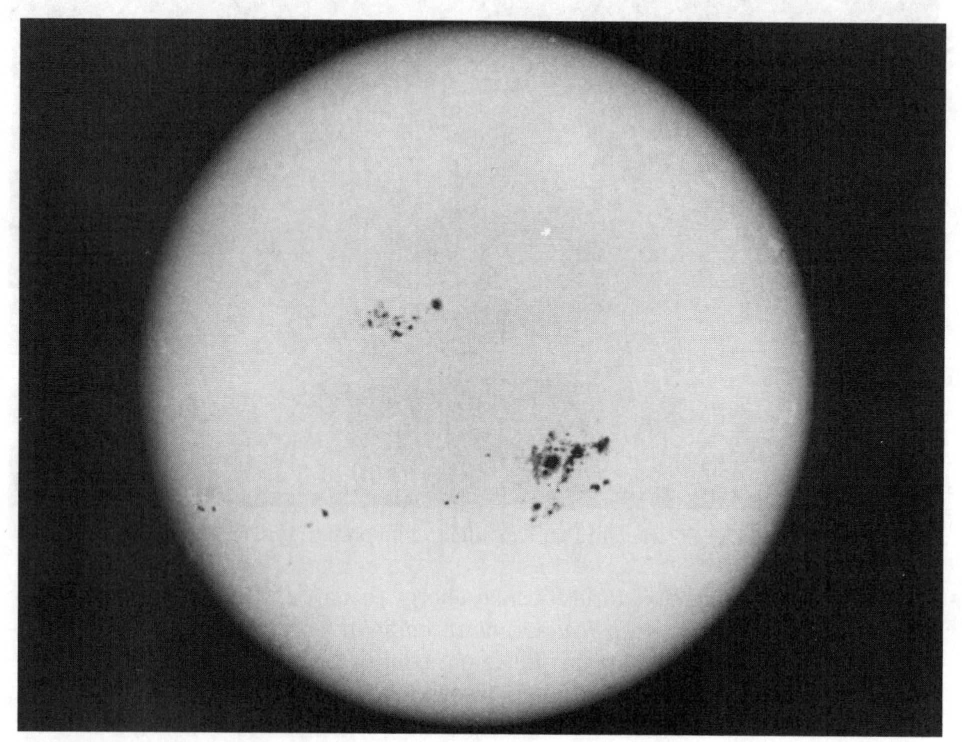

2-26B Sunspots.

Another indicator of ionospheric propagation potential is the *Solar Flux Index* (SFI). This measure is taken in the microwave region (wavelength of 10.2 cm, or 2.8 GHz), at 1700 UTC at Ottawa, Canada. The SFI is reported by the National Bureau of Standards (NBS) radio stations WWV (Fort Collins, CO) and WWVH (Maui, Hawaii).

The ionosphere offers different properties that affect radio propagation at different times. Variations occur not only over the 11 year sunspot cycle, but also diurnally and seasonally. Obviously, if the sun affects propagation in a significant way, then differences between nighttime and daytime, and between summer and winter, must cause variations in the propagation phenomena observed.

The ionosphere is divided, for purposes of radio propagation studies, into various layers that have somewhat different properties. These layers are only well-defined in textbooks. However, even there you find a variation in the location above the earth's surface, where these layers are located. In addition, the real physical situation is such that layers don't have sharply defined boundaries, but rather fade one into another instead. Thus, the division into layers is somewhat arbitrary. These layers (Fig. 2-27) are designated D, E, and F (with F being further sub-divided into F1 and F2 sub-layers).

D-layer The D-layer is the lowest layer in the ionosphere, and exists from approximately 30 to 50 miles above the surface. This layer is not ionized as much as the higher layers, because all forms of solar energy that cause ionization are severely attenuated by the higher layers above the D-layer. The reason for this is that the D-layer is much more dense than the E and F layers, and that density of air molecules allows ions to recombine to form electroneutral atoms very quickly.

The extent of D-layer ionization is proportional to the elevation of the sun, so it will achieve maximum intensity at midday. The D-layer exists mostly during the warmer months of the year because of both the greater height of the sun above the horizon and the longer hours of daylight. The D-layer almost completely disappears after local sunset. Some observers have reported sporadic incidents of D-layer activity for a considerable time past sunset. The D-layer exhibits a large amount of absorption of medium wave and shortwave signals (to such an extent that signals below 4 to 6 MHz are completely absorbed by the D-layer).

E-layer The E-layer exists from approximately 50 to 70 miles above the earth's surface, and it is considered the lowest region of the ionosphere that is important to radio communications. Like the D-layer, this region is ionized only during the daylight hours with ionization levels peaking at midday. The ionization level drops off sharply in the late afternoon, and almost completely disappears after local sunset.

During most of the year, the E-layer is absorptive and it will not reflect radio signals. During the summer months, however, E-layer propagation does occur. A phenomenon called "short skip" (i.e., less than 100 miles for medium wave and 1000 miles for shortwave signals) occurs in the E-layer during the summer months, and in equatorial regions at other times.

A propagation phenomenon associated with the E-layer is called *sporadic-E* propagation. This phenomenon is caused by scattered zones of intense ionization in the E-layer region of the ionosphere. Sporadic-E phenomenon varies seasonally, and it is believed to be caused by the bombardment of solar particles. Sporadic-E propagation affects the upper HF and lower VHF region. It is observed most frequently in the lower VHF spectrum (50 to 100 MHz), but it is also sometimes observed at higher frequencies. The VHF bands occasionally experience sporadic-E propagation. Skip distances on VHF can reach 500 to 1500 miles on one hop—especially in the lower VHF region (including the 6-meter band).

F-layer The F-layer of the ionosphere is the region that is the principal cause of long-distance shortwave communications. This layer is located from about 100 to 300 miles above the earth's surface. Unlike the lower layers, the air density in the F-layer is low enough that ionization levels remain high all day, and decay slowly after local sunset. Minimum levels are reached just prior to local sunrise. Propagation in the F-layer is capable of skip distances up to 2,500 miles on a single hop. During the day there are actually two identifiable and distinct sub-layers in the F-layer region, and these are designated the "Fl" and "F2" layers. The F1 layer is found ap-

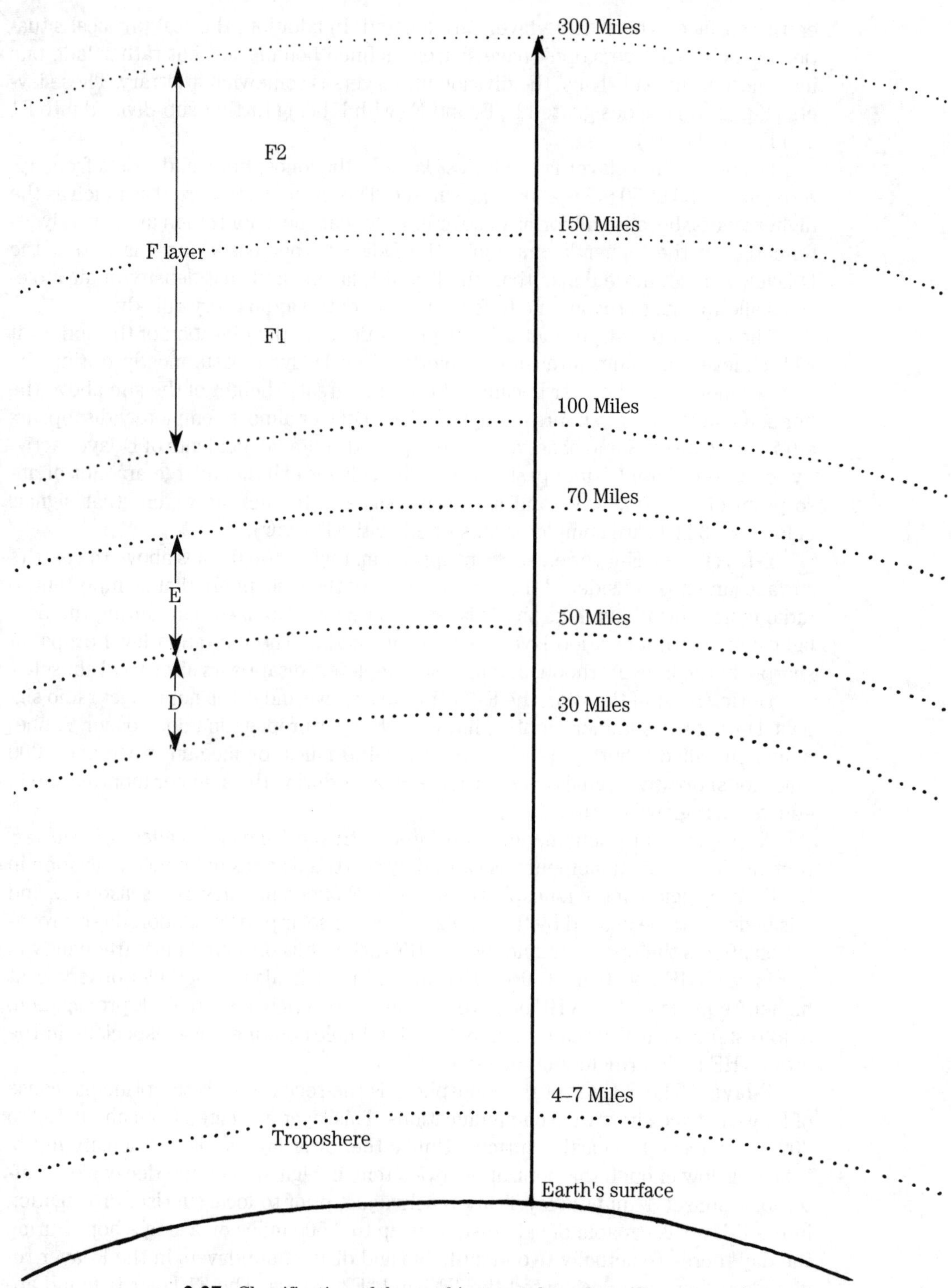

2-27 Classification of the earth's atmosphere for radio propagation.

proximately 100 to 150 miles above the earth's surface, and the F2 layer is above the F1 extending up to the 270–300 mile limit. Beginning at local sundown, however, the lower regions of the F1 layer begin to de-ionized because of recombination of positive and negative ions. At some time after local sunset, the F1 and F2 layers have effectively merged to become a single reduced layer beginning at about 175 miles.

The height and degree of ionization of the F2 layer varies over the course of the day, with the season of the year, and with the 27 day sunspot cycle. The F2 layer begins to form shortly after local sunrise and reaches maximum shortly before noon. During the afternoon, the F2 layer ionization begins to decay in an exponential manner until, for purposes of radio propagation, it disappears sometime after local sunset. There is some evidence that ionization in the F-layer does not completely disappear, but its importance to HF radio communication does disappear.

Measures of ionospheric propagation

There are several different measures by which the ionosphere is characterized at any given time. These measures are used in making predictions of radio activity and long-distance propagation.

The *Critical Frequency* and *Maximum Usable Frequency* (MUF) are indices that tell us something of the state of ionization and communications ability. These frequencies increase rapidly after sunrise and international communications usually begin within 30 minutes.

Critical frequency, F_c The *critical frequency*, designated by F_c, is the highest frequency that can be reflected when a signal strikes the ionosphere as a vertical (90 degrees with respect to the surface) incident wave. The critical frequency is determined from an *ionogram*, which is a cathode ray tube (CRT) oscilloscope display of the height of the ionosphere as a function of frequency. The ionogram is made by firing a pulse vertically (Fig. 2-28) at the ionosphere from the transmitting station. The critical frequency is that frequency that is just sufffficient to be reflected back to the transmitter site. Values of F_c can be as low as 3 MHz during the nightime hours, and as high as 10 to 15 MHz during the day.

Virtual height Radio waves are refracted in the ionosphere, and those above a certain critical frequency are refracted so much that they return to earth. Such waves appear to have been reflected from an invisible radio "mirror." An observer on the earth's surface could easily assume the existence of such a mirror by noting the return of the "reflected" signal. The height of this apparent "mirror" is called the *virtual height* of the ionosphere. Figure 2-29 shows the refraction phenomenon by which a radio wave is bent sufficiently to return to earth. Virtual height is determined by measuring the time interval required for an ionosonde pulse (similar to that used to measure critical frequency) to travel between the transmitting station and a receiving station (Fig. 2-30). A radio signal travels at a velocity of 300,000,000 meters per second (the speed of light). By observing the time between transmitting the pulse and receiving it, you can calculate the virtual height of the ionosphere.

Maximum usable frequency (MUF) The *Maximum Usable Frequency* (MUF), is the highest frequency at which communications can take place via the ionosphere over a given path. The MUF between a fixed transmitter site and two different, widely separated, receivers need not be the same. Generally, however, the MUF is approxi-

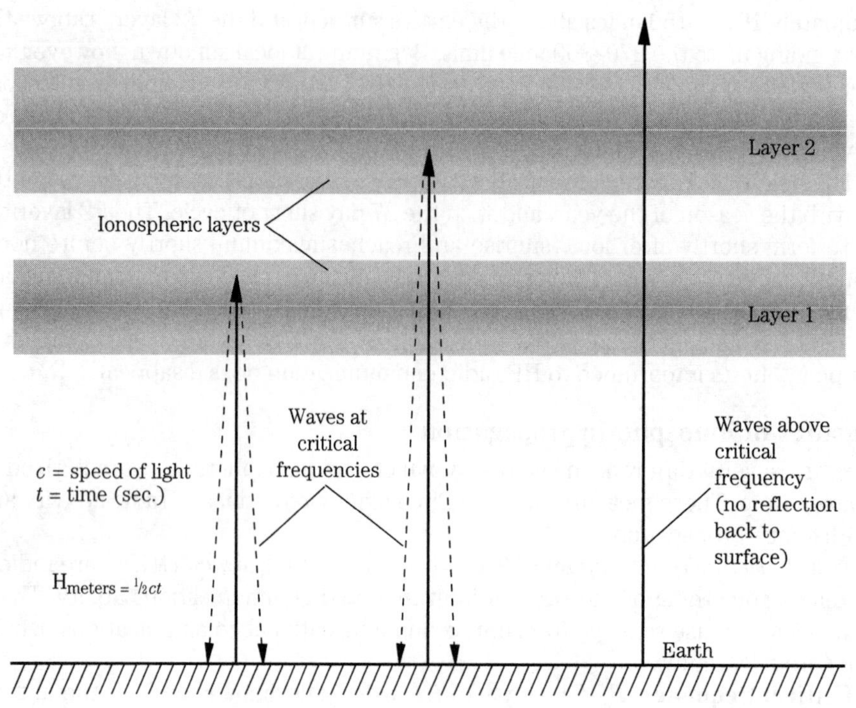

2-28 Finding critical height of the ionosphere.

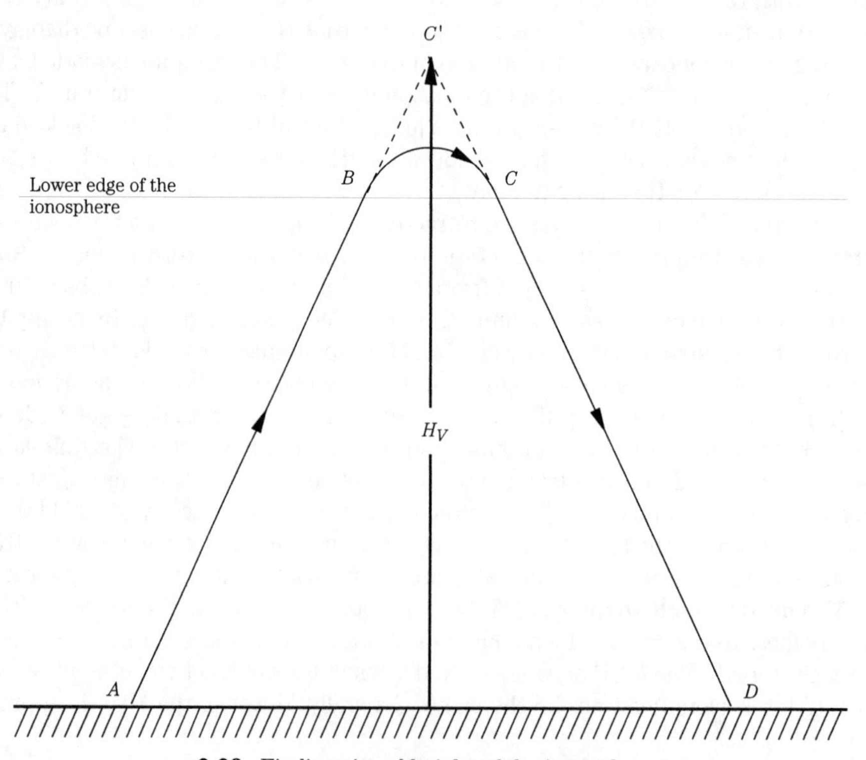

2-29 Finding virtual height of the ionosphere.

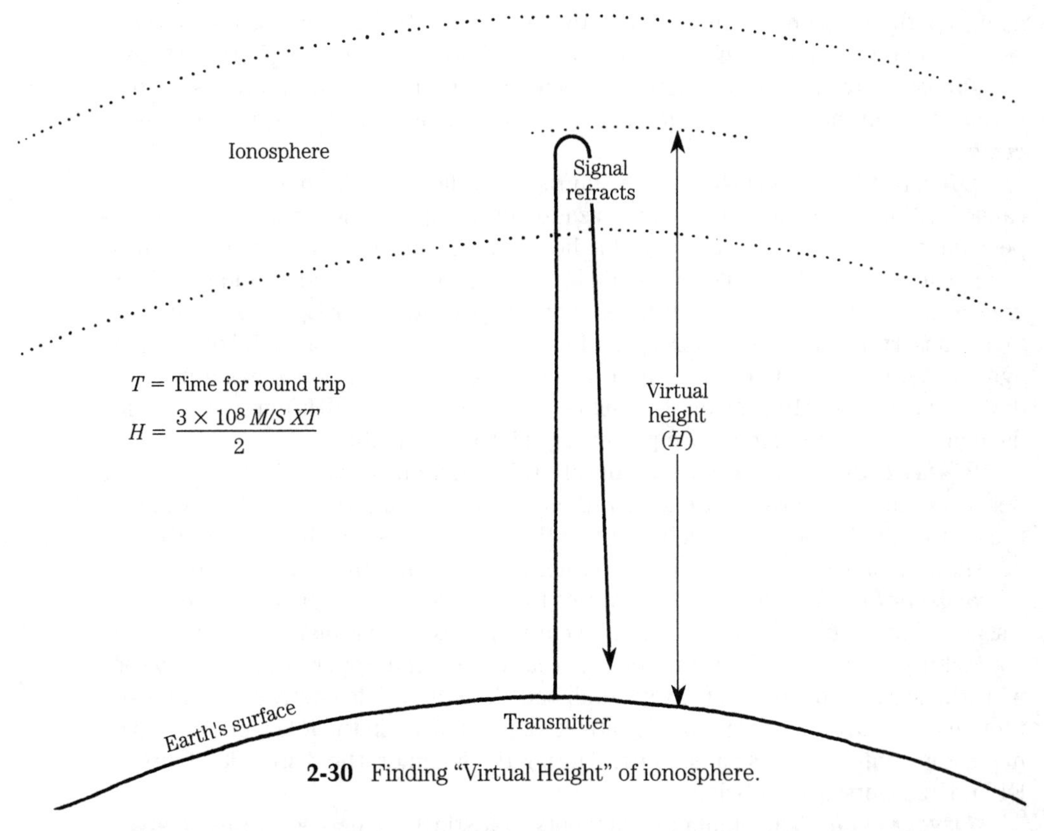

2-30 Finding "Virtual Height" of ionosphere.

mately three times higher than the critical frequency. Both the MUF and the critical frequency, vary geographically, and they become higher at latitudes close to the equator.

It is a general rule that the best propagation occurs at frequencies just below the MUF. In fact, there is a so-called *frequency of optimum traffic* (FOT) that is approximately 85 percent of the MUF. Both noise levels and signal strengths are improved at frequencies near the FOT.

Lowest usable frequency (LUF) At certain low frequencies, the combination of ionospheric absorption, atmospheric noise, miscellaneous static, and/or receiver signal-to-noise ratio requirements conspire to reduce radio communications. The lowest frequency that can be used for communications, despite these factors, is the *lowest usable frequency*, or LUF.

Unlike the MUF, the LUF is not totally dependent on atmospheric physics. The LUF of a system can be varied by controlling the signal-to-noise ratio. Although certain factors that contribute to SNR are beyond our control, the effective radiated power (ERP) of the transmitter can be changed; a 2 MHz decrease in LUF is available for every 10 dB increase in the ERP of the transmitter.

Ionospheric variation and disturbances

The ionosphere is an extremely dynamic region of the atmosphere, especially from a radio operator's point of view, because it significantly alters radio propagation. The dynamics of the ionosphere are conveniently divided into two general classes: *regular variation* and *disturbances*. This section covers both types of ionospheric change.

Ionospheric variation There are several different forms of variation seen on a regular basis in the ionosphere: *diurnal*, *27-day* (monthly), *seasonal*, and *11-year cycle*.

Diurnal (daily) variation The sun rises and falls on a 24-hour cycle, and because it is the principal source of iononization of the upper atmosphere, you can expect diurnal variation. During daylight hours the E and D levels exist, but these disappear at night. The height of the F2 layer increases until midday, and then it decreases until evening, when it disappears or merges with other layers. As a result of higher absorption in the E and D layers, lower frequencies are not useful during daylight hours. On the other hand, the F layers reflect higher frequencies during the day. In the 1 to 30 MHz region, the higher frequencies (>11 MHz) are used during daylight hours, and the lower frequencies (<11 MHz) at night.

27-day cycle Approximately monthly, this variation is caused by the rotational period of the sun. Sunspots are localized on the surface of the sun, so they will face the earth only during a portion of the month. As new sunspots are formed, they do not show up on the earthside face until their region of the sun rotates earthside.

Seasonal cycle The earth's tilt varies the exposure of the planet to the sun on a seasonal basis. In addition, the earth's yearly orbit is not circular; it is elliptical. As a result, the intensity of the sun's energy that ionizes the upper atmosphere varies with the seasons of the year. In general, the E, D, and F layers are affected—although the F2 layer is only minimally affected. Ion density in the F2 layer tends to be highest in winter, and less in summer. During the summer, the distinction between F1 and F2 layers is less obvious.

11-year cycle The number of sunspots, statistically averaged, varies on an approximately 11-year cycle. As a result, the ionospheric effects that affect radio propagation also vary on an 11-year cycle. Radio propagation, in the shortwave bands, is best when the average number of sunspots is at its highest.

Disturbances Disturbances in the ionosphere can have a profound effect on radio communications . . . and most of them (but not all) are bad. This section will briefly examine some of the more common forms.

Sporadic E-layer A reflective cloud of ionization sometimes appears in the E-layer of the ionosphere; this layer is sometimes called the *E_s layer*. It is believed that the E_s layer forms from the effects of wind shear between masses of air moving in opposite directions. This action appears to redistribute ions into a thin layer that is radio reflective.

Sporadic-E propagation is normally thought of as a VHF phenomenon, with most activity between 30 and 100 MHz, and decreasing activity up to about 200 MHz. However, about 25 to 50 percent of the time, sporadic-E propagation is possible on frequencies down to 10 or 15 MHz. Reception over paths of 1400 to 2600 miles are possible in the 50 MHz region when sporadic-E is present. In the northern hemisphere, the months of June and July are the most prevalent sporadic-E months. On most days when sporadic-E is present, it lasts only a few hours.

Sudden ionospheric disturbances (SIDs) The SID, or *Dellinger fade*, mechanism occurs suddenly, and rarely gives any warning. The SID can last from a few minutes to many hours. It is believed that SIDs occur in correlation with solar flares, or "bright solar eruptions," that produce an immense amount of ultraviolet radiation that impinges the upper atmosphere. The SID causes a tremendous increase in D-layer ionization, which accounts for the radio propagation effects. The ionization is so intense that all receiver operators on the sunny side of the earth experience profound loss of signal strength above about 3 MHz. It is not uncommon for receiver owners to think their receivers are malfunctioning when this occurs. The sudden loss of signals on sunny side receivers is called *Dellinger fade*. The SID is often accompanied by variations in terrestial electrical currents and magnetism levels.

Ionospheric storms The *ionospheric storm* appears to be produced by an abnormally large rain of atomic particles in the upper atmosphere, and is often preceded by SIDs 18 to 24 hours earlier. These storms tend to last from several hours, to a week or more, and are often preceded by two days or so by an abnormally large collection of sunspots crossing the solar disk. They occur, most frequently, and with greatest severity, in the higher latitudes, decreasing toward the equator. When the ionospheric storm commences, shortwave radio signals may begin to flutter rapidly and then drop out altogether. The upper ionosphere becomes chaotic; turbulence increases and the normal stratification into "layers," or zones, diminishes.

Radio propagation may come and go over the course of the storm, but it is mostly dead. The ionospheric storm, unlike the SID, which affects the sunny side of the earth, is worldwide. It is noted that the MUF and critical frequency tend to reduce rapidly as the storm commences.

An ionospheric disturbance observed by me in November 1960 was preceded by about 30 minutes of extremely good, but abnormal propagation. At 1500 hours EST, European stations were noted with S9+ signal strengths in the 7000 to 7300 kHz region of the spectrum, which is an extremely rare occurance. After about 30 minutes, the bottom dropped out and even AM broadcast band skip (later that evening) was nonexistent. At the time, the National Bureau of Standards radio station, WWV, was broadcasting a "W2" propagation prediction at 19 and 49 minutes after each hour. It was difficult to hear even the 5 MHz WWV frequency in the early hours of the disturbance, and it disappeared altogether for the next 48 hours.

Ionospheric sky wave propagation

Sky wave propagation occurs because signals in the ionosphere are refracted so much that they are bent back toward the earth's surface. To observers on the surface, it looks like the signal was reflected from a radio mirror at the virtual height of the ionosphere. The *skip distance* is the surface distance between the transmitter point (A in Fig. 2-31) and the point where it returns to earth (point C in Fig. 2-31). The *ground wave zone* is the distance from the transmitter site (A in Fig. 2-31) to where the ground wave fades to a low level below usefulness (point B in Fig. 2-31). The *skip zone* is the distance from the outer edge of the ground wave zone and the skip distance, or the distance from B to C in Fig. 2-31.

It is possible for the sky wave and the ground wave to interfere with each other at some frequencies, under some circumstances. When this happens, the sky wave has a relative phase that depends on its path length (among other things), so it will arrive at some seemingly random phase relative to the ground

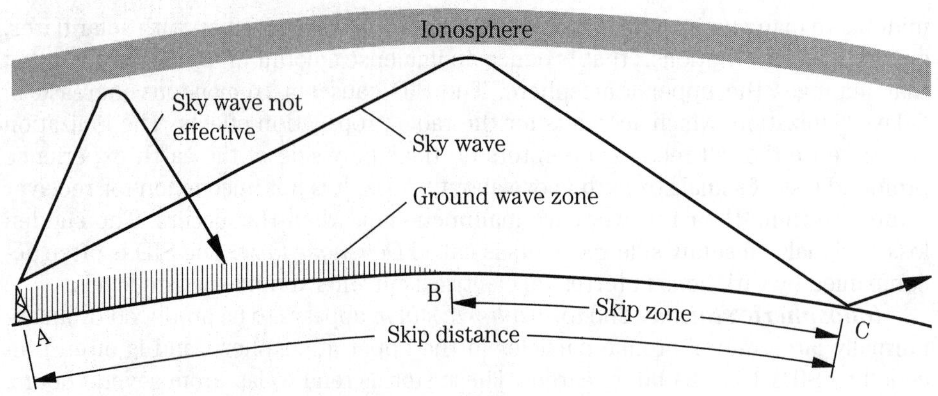

2-31 Relationship between ground wave and different cases of sky wave.

wave. Thus, the sky wave can selectively strengthen or cancel the ground wave, giving rise to a type of fading.

Incident angle One of the factors that affects the length of the skip distance is the incident angle of the radio wave. This angle is partially a function of the frequency, and partially a function of the natural radiation angle of the antenna (α_r). The frequency effects are seen as a function of the iononization level, and how much a given frequency is refracted.

The antenna radiation angle is the angle of the vertical lobe with respect to the earth's surface; and it is partially a function of its design, and partially its installation configuration. For example, a ⅝ wavelength vertical antenna tends to have a lower angle of radiation than a ¼ wave vertical antenna. Similarly, the dipole's angle of radiation is a function of its height above ground.

Figure 2-32 shows how the angle of radiation affects skip distance. Low angle of radiation signals tend to travel farther with respect to the earth's surface before refracting, so they produce the longest skip distances. Higher angles of radiation have shorter skip distances because they tend to return to earth more rapidly. In summertime, some high frequency bands (e.g,. the 11-meter—27 MHz—Citizen's Band) offer high angle "short skip" during the summer, and longer skip during the other months. At higher angles of radiation, there will be a *critical angle* wave and *escape angle* waves that are not returned to earth. These waves are not used for terrestial communications or broadcasting. Figure 2-33 shows the difference between single-hop, skip and multi-hop skip, as a function of incident angle. It is generally true that a multi-hop transmission is more subject to fading, and it is weaker, than a single-hop transmission.

Using the ionosphere

The refraction of high frequency and some medium wave radio signals back to earth via the ionosphere gives rise to intercontinental HF radio communications. This phenomena becomes possible during daylight hours and for awhile after sunset when the ionosphere is ionized. Figure 2-34 reiterates the mechanism of long-distance skip communications. The transmitter is located at point T, while receiving stations are located at sites R1 and R2. Signals 1 and 2 are not refracted sufficiently to be re-

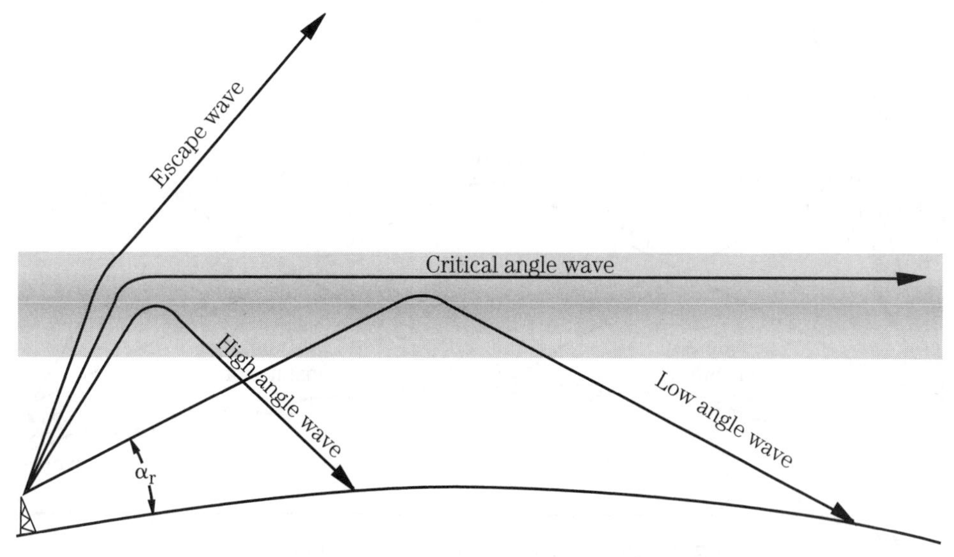

2-32 Sky wave propagation as a function of antenna radiation angle.

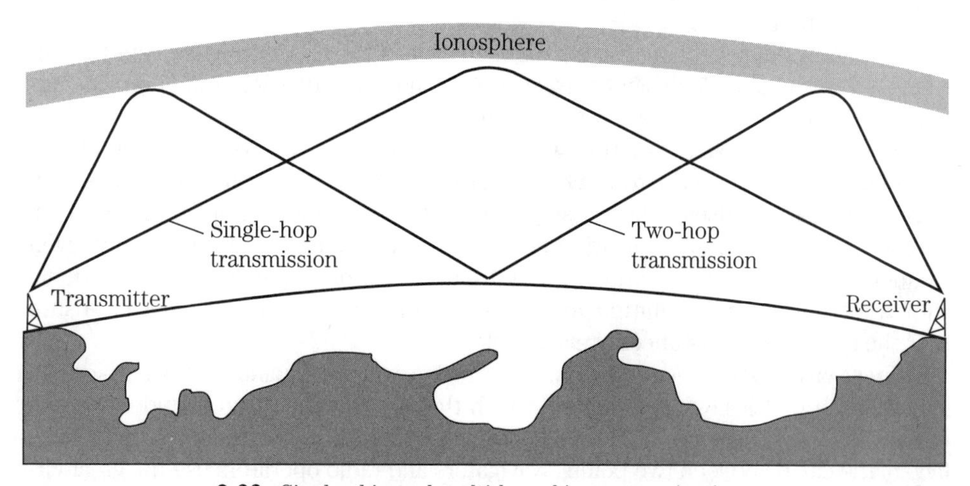

2-33 Single-skip and multi-hop skip communications.

fracted back to earth, so they are lost in space. Signal 3, however, is refracted enough to return to earth, so it is heard at station Rl. The skip distance for signal 3 is the distance from T to Rl. At points between T and R1, signal 3 is inaudible, except within groundwave distance of the transmitter site (T). This is the reason why two stations 40 miles apart hear each other only weakly, or not at all, while both stations can communicate with a third station 2000 miles away. In amateur radio circles, it is common for South American stations to relay between two U.S. stations only a few miles apart. For an example of this problem, listen to the Inter-American and Halo Mis-

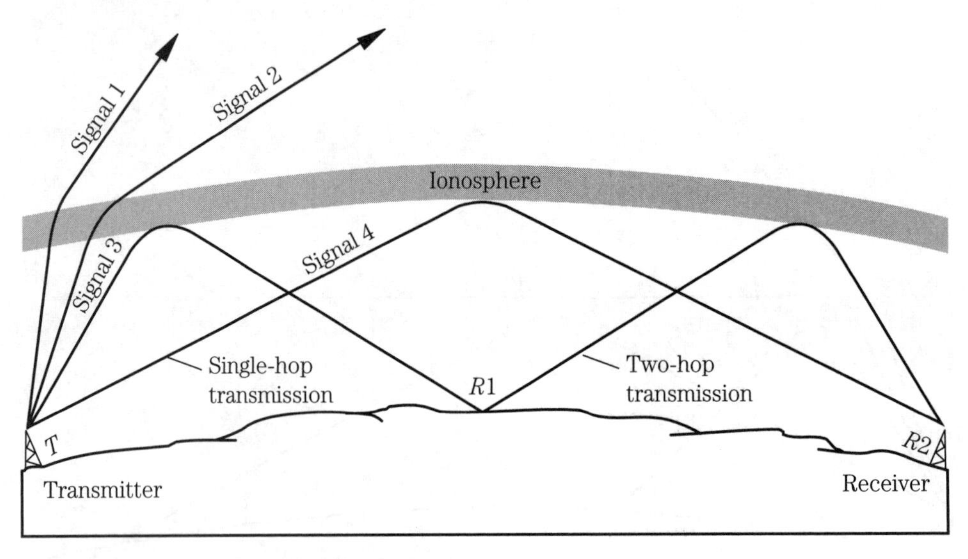

2-34 Effects of radiation angle on distance.

sionary Nets on 21.290 MHz (16 M) daily from about 1700Z to 2100Z (ending time dependent upon traffic).

Multi-hop skip is responsible for the reception of the signal from transmitter T at site R3. The signal *reflects* (not refracts) from the surface at R1, and is retransmitted into the ionosphere, where it is again refracted back to earth.

Figure 2-35 shows a situation where skip signals are received at different distances depending upon the *angle of radiation* of the transmitting antenna. A high angle of radiation causes a shorter skip zone, but a lower angle of radiation results in a longer skip zone. Communication between any particular locations on any given frequency requires adjustment of the antenna radiation angle. Some international shortwave stations have multiple antennas with different radiation angles to ensure that the correct skip distances are available.

Great circle paths A *great circle* is a line between two points on the surface of a sphere, such that it lays on a plane through the earth's center and includes the two points. When translated to "radiospeak," a great circle is the shortest path on the surface of the earth between two points. Navigators and radio operators use the great circle for similar, but different, reasons. The navigator's reason is in order to get from here to there, and the radio operator's is to get a transmission path from here to there.

The heading of a directional antenna is normally aimed at the receiving station along its great circle path. Unfortunately, many people do not understand the concept well enough, for they typically aim the antenna in the wrong direction. For example, I live near Washington, DC, which is on approximately the same latitude as Lisbon, Portugal. If I catch a lift on Superman's back, and he flies due East, we'll have dinner in Lisbon, right? Wrong. If you head due East from Washington, DC, across the Atlantic, the first landfall would be west Africa, somewhere near Zaire or Angola. Why? Because the great circle bearing 90 degrees takes us far south. The *geometry of spheres*, not flat planes, governs the case.

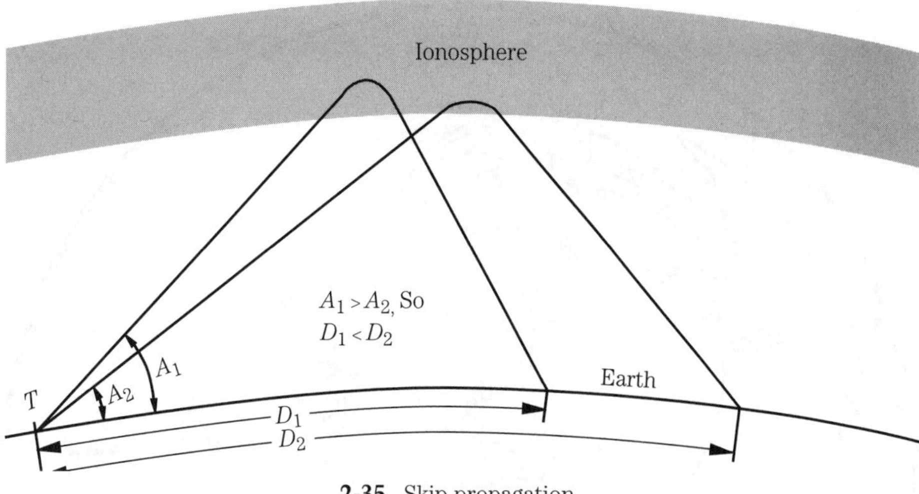

2-35 Skip propagation.

Figure 2-36 shows a great circle map centered on the Washington, DC area. These maps, or computer tabulations of the same data, can often be purchased for your own location by supplying your latitude and longitude to the service company that does the job. By drawing a line from your location at the center of the chart, to the area you want to hear, and then extend it to the edge of the chart, you will obtain the beam heading required.

Long path vs. short path The earth is a sphere (or more precisely, an "oblique spheroid"), so from any given point to any other point there are two great circle paths: the *long path* (major arc) and the *short path* (minor arc). In general, the best reception occurs along the short path. In addition, short path propagation is more nearly "textbook," compared with long path reception. However, there are times when long path is better, or is the only path that will deliver a signal to a specific location from the geographic location in question.

Gray line propagation The *gray line* is the twilight zone between the night and daytime halves of the earth. This zone is also called the *planetary terminator* (Fig. 2-37). It varies up to +23 degrees either side of the north-south longitudinal lines, depending on the season of the year (it runs directly north-south only at the vernal and autumnal equinoxes). The D-layer of the ionosphere absorbs signals in the HF region. This layer disappears almost completely at night, but it builds up during the day. Along the gray line, the D-layer is rapidly decaying west of the line, and has not quite built up east of the line.

Brief periods of abnormal propagation occur along the gray line. Stations on either side of the line can be heard from regions, and at distances, that would otherwise be impossible on any given frequency. For this reason, radio operators often prefer to listen at dawn and dusk for this effect.

Scatter propagation modes

Ionospheric scatter propagation occurs when clouds of ions exist in the atmosphere. These clouds can exist in both the ionosphere and the troposphere, although

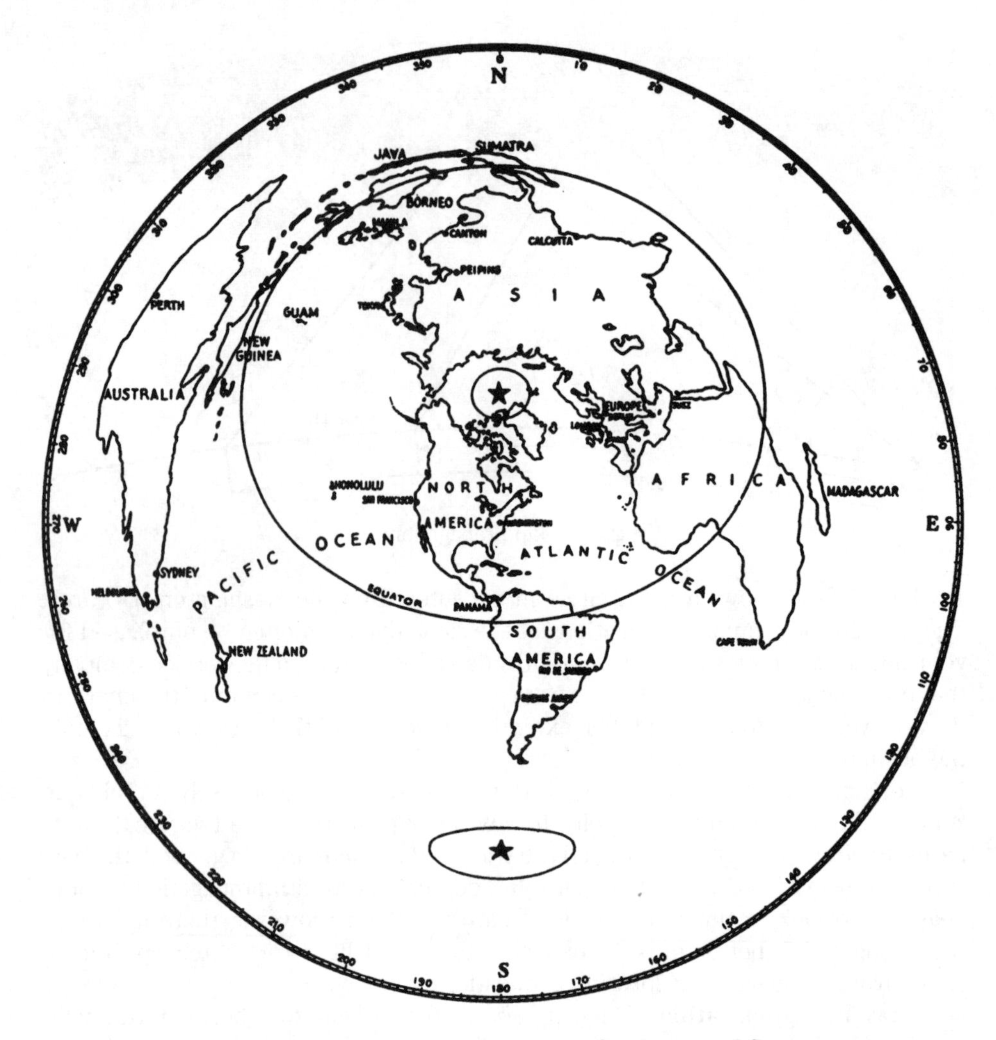

2-36 Azimuthal map centered on Washington, DC.

the tropospheric model is more reliable for communications. Figure 2-38 shows the mechanism for scatter propagation. Radio signals from the transmitter are reflected from the cloud of ions, to a receiver location that otherwise might not receive it. Scatter propagation occurs mostly in the VHF region, and it allows communications over extended paths that are not normally available.

There are at least three different modes of scatter from ionized clouds: *backscatter*, *side scatter*, and *forward scatter*. The backscatter mode is a bit like radar, in that signal is returned back to the transmitter site, or in regions close to the transmitter. Forward scatter occurs when the reflected signal continues in the same azimuthal direction (with respect to the transmitter), but is redirected toward the earth's surface. Side scatter is similar to forward scatter, but the azimuthal direction might change.

Unfortunately, there are often multiple reflections from the ionized cloud, and these are shown as "multiple scatter" in Fig. 2-38. When these reflections are able to

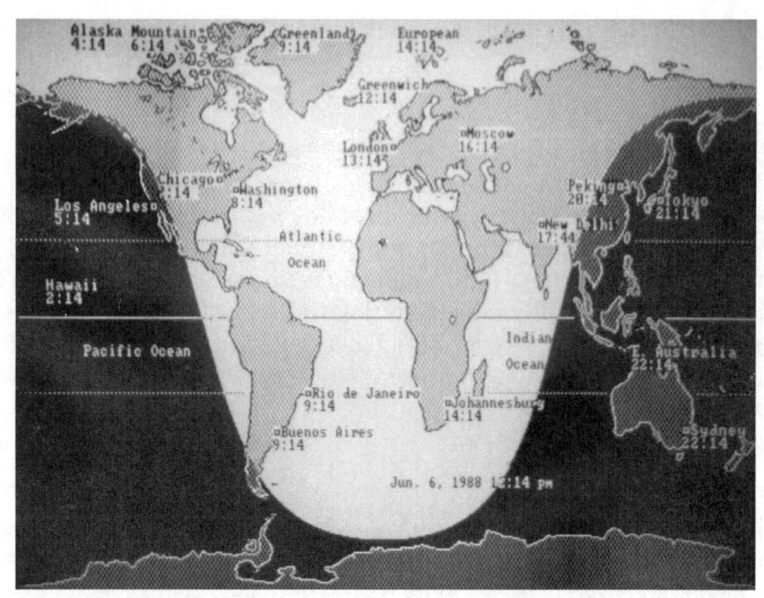

2-37 The planetary terminator ("grey line") provides some unusual propagation effects.

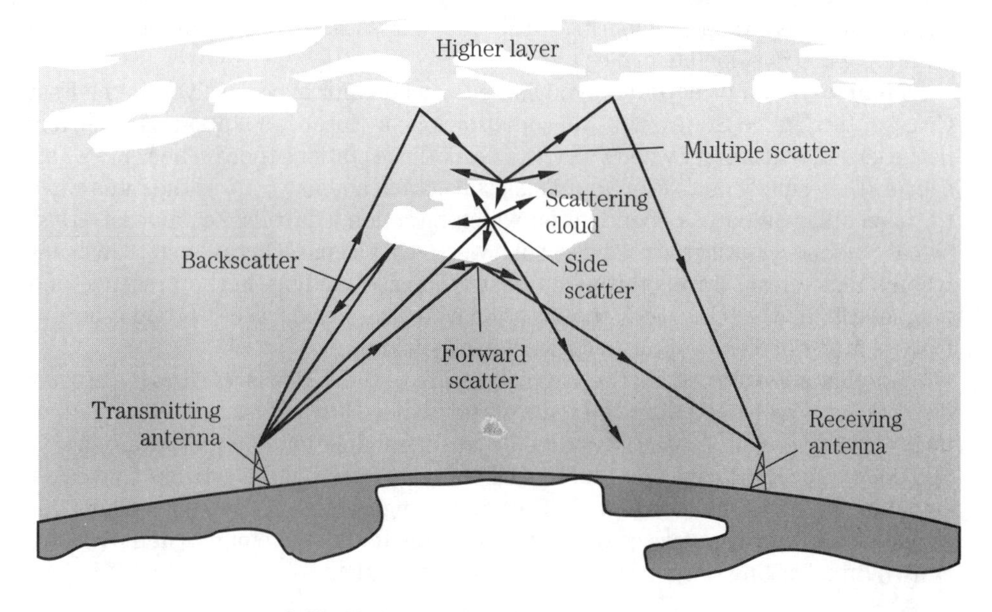

2-38 Various modes of scatter propagation.

reach the receiving site, the result is a rapid, fluttery fading that can be of quite profound depths.

Auroral propagation

The *auroral effect* produces a luminescence in the upper atmosphere resulting from bursts of particles released from the sun 18 to 48 hours earlier. The light emitted is called the *northern lights* and the *southern lights*. The ionized regions of the atmosphere that create the lights form a radio reflection shield, especially at VHF and above, although 15 to 20 MHz effects are known. Auroral propagation effects are normally seen in the higher latitudes, although listeners in the southern tier of states in the USA are often treated to the reception of signals from the north being reflected from auroral clouds.

Meteor scatter propagation

When meteors enter the earth's atmosphere, they do more than simply heat up to the point of burning. The burning meteor leaves a wide, but very short duration, transient cloud of ionized particles in its path. These ions act as a radio mirror that permits short bursts of reception between sites correctly situated. Meteor scatter reception is not terribly reliable, although at least two companies offer meter scatter communcations services for users in the higher latitudes.

Other propagation anomalies

The ionosphere is a physically complex place, and even the extensive coverage in this chapter is not sufficient to do it justice. Indeed, entire books are available on the subject, and it is a valid engineering subspecialty. It is therefore not surprising that a number of propagation anomalies are known.

Nonreciprocal direction If you listen to an amateur band receiver on the East Coast of the United States, you will sometimes hear European stations—especially in the late afternoon. But when you try to work those stations there is no reply whatsoever. They simply don't hear you! This propagation anomaly causes the radio wave to travel different paths dependent on which direction it travels; i.e., an east→west signal is not necessarily the reciprocal of a west→east signal. This anomaly can occur when a radio signal travels through a heavily ionized medium in the presence of a magnetic field, which is exactly the situation when the signal travels through the ionisphere in the presence of the earth's magnetic field.

Another anomaly seen in the radio literature of the 1930s is the *Radio Luxemburg Effect*. It is named after the radio station where it was first noticed. In a nonlinear ionosphere, it is sometimes noted that the modulation of superpower (i.e., > 500,000 watts) shortwave broadcasters will be transferred to the carrier of a weaker signal in the same or nearby band. The interchange noted in the 1930s when this phenomenon was first discovered was between Radio Luxemburg and Britain's British Broadcasting Corporation (BBC) overseas outlets.

Propagation predictions

Propagation predictions for the VLF through low VHF bands are published each month in several magazines. Ham radio operators often use those in *QST* magazine (Fig. 2-39), while SWLs and others tend to prefer those published in magazines such as *Monitoring Times*. These charts relate the time of day in Universal Coordinated Time (UTC, formerly, GMT) and the frequency for transmission to different parts of the world.

Fading

Skip communications are not without problems. One phenomena is fading (i.e., a variation in signal strength as perceived at the receiver site). This problem can sometimes be overcome by using one of several diversity reception systems. Three forms of diversity technique are used: *frequency diversity*, *spacial diversity*, and *polarity diversity*.

In the frequency diversity system (Fig. 2-40), the transmitter will send out two or more frequencies simultaneously with the same modulating information. Because the two frequencies will fade differentially, one will always be strong.

The spacial diversity system (Fig. 2-41) assumes that a single transmitter frequency is used. At the receiving site, two or more receiving antennas are used, spaced one-half wavelength apart. The theory is that the signal will fade at one antenna while it increases at the other. A three-antenna system is often used. Three separate, but identical, receivers, often tuned by the same Master Local Oscillator, are connected to the three antennas. Audio mixing, based on the strongest signal, keeps the audio output constant while the RF signal fades.

Polarity diversity reception (Fig. 2-42) uses both vertical and horizontal polarization antennas to receive the signal. Like the space diversity system, the outputs of the vertical and horizontal receivers are combined to produce a constant level output.

Another form of fading, *selective fading*, derives from the fact that fading is a function of frequency. The carrier and upper and lower sidebands of an AM signal have slightly different frequencies, so they arrive out of phase with each other. Although this type of fading is lessened by using single-sideband transmission, that does not help AM users. In those systems, some people use a filtering system that eliminates the carrier and one sideband, it then reconstitutes the AM signal with a product detector.

SSB receivers with stable local and product detector oscillators, and a sharp IF bandpass filter, can be used to reduce the effects of differential fading of AM signals because of the phasing of the LSB, USB, and carrier components. Carefully tune the receiver to only one sideband of the signal, and note when the heterodyne beatnote disappears. The correct point is characterized by the fact that you can then switch among USB, LSB, and CW modes without changing the received signal output.

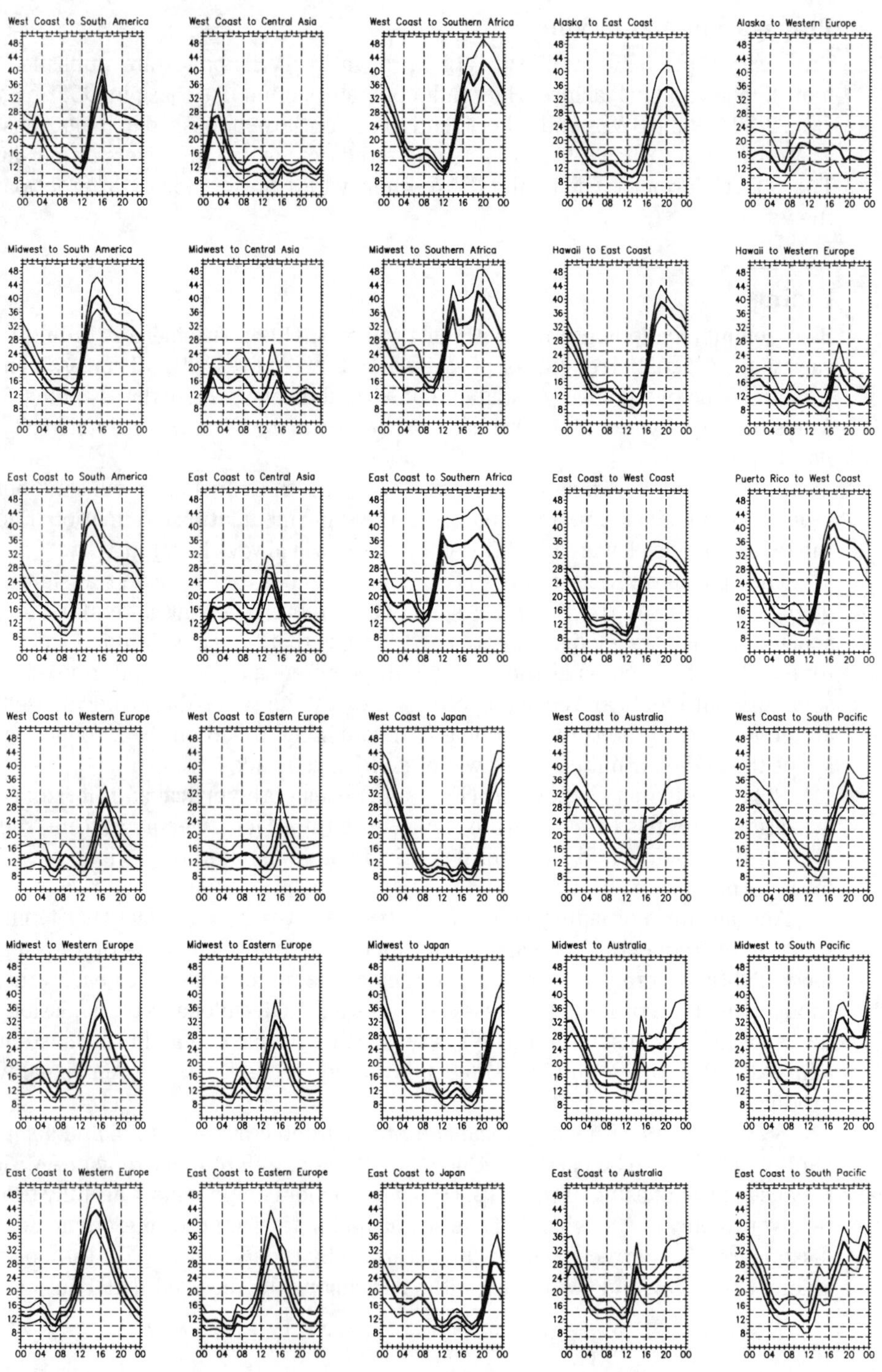

2-39 Propagation prediction charts appear in magazines, such as *QST* every month.
Courtesy American Radio Relay League, 225 Main Street, Newington, CT 06111

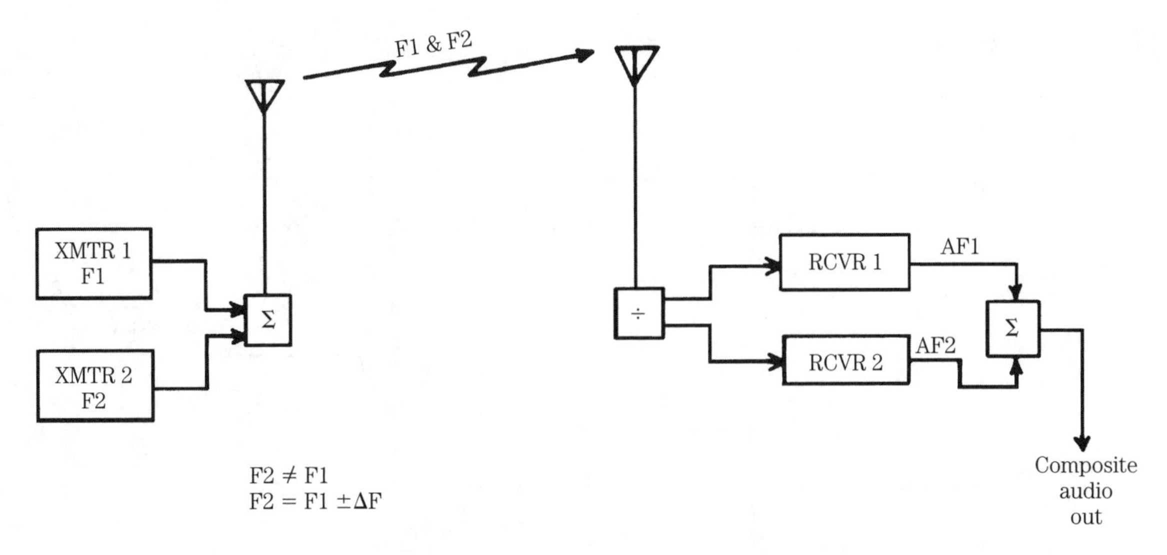

2-40 Frequency diversity reduces fading.

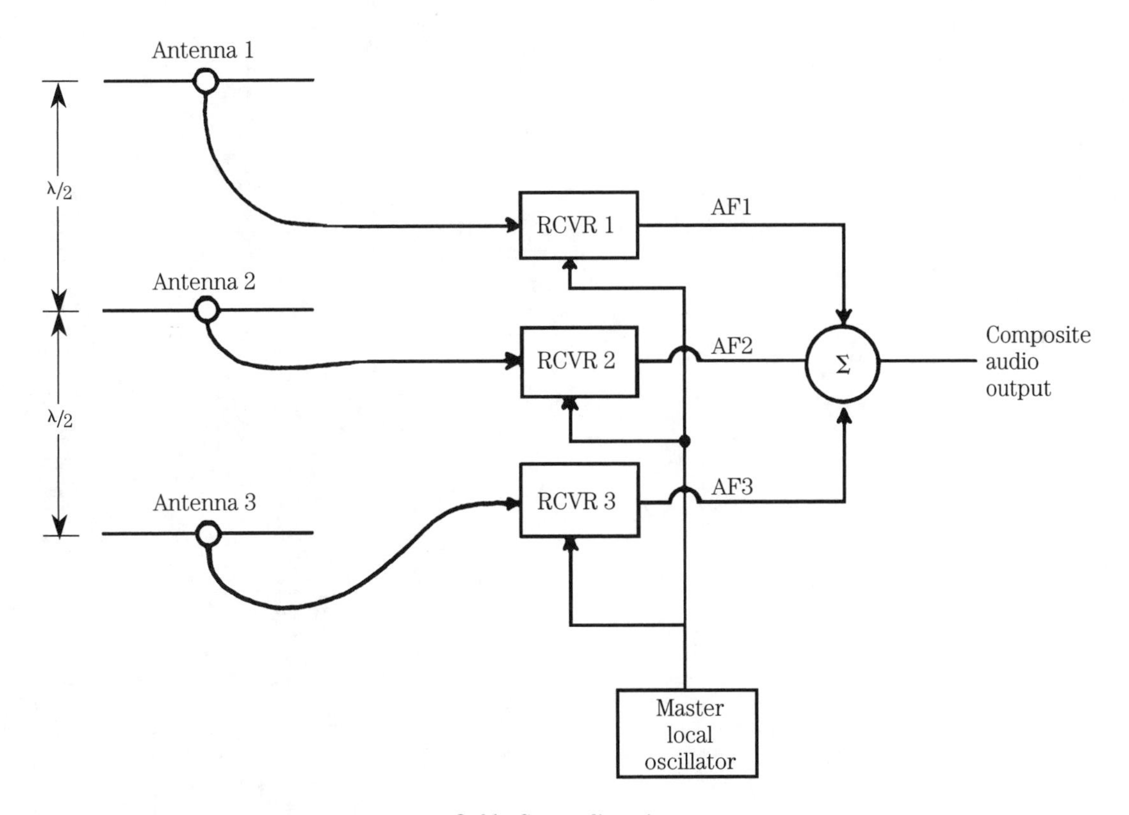

2-41 Space diversity.

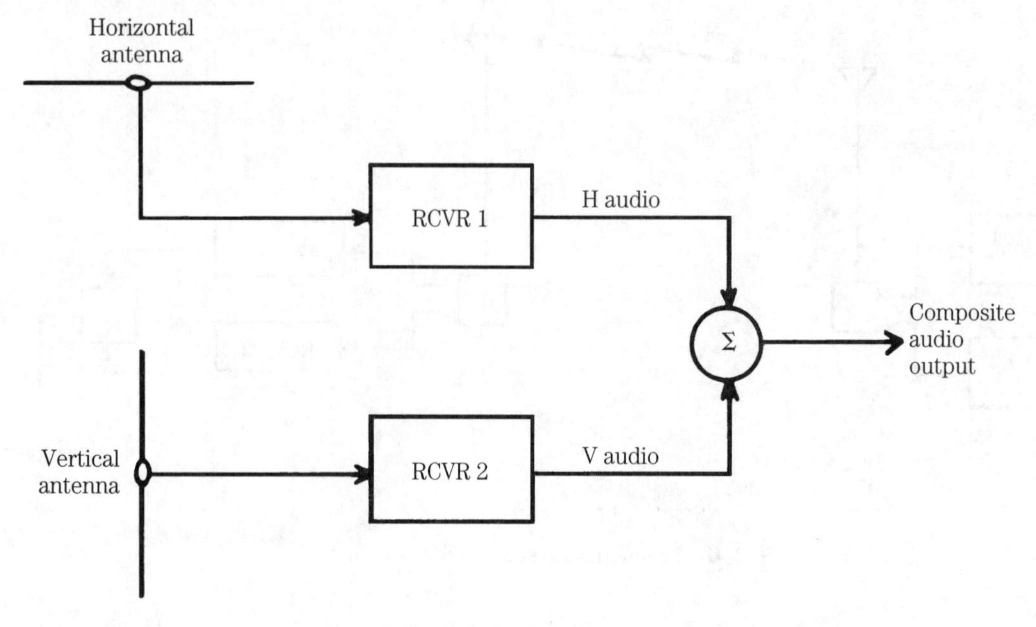

2-42 Polarization diversity.

3
CHAPTER

Transmission lines

TRANSMISSION LINES AND *WAVEGUIDES* ARE CONDUITS FOR TRANSPORTING RF SIGNALS between elements of a system. For example, transmission lines are used between an exciter output and transmitter input, and between the transmitter input, and between the transmitter output and the antenna. Although often erroneously characterized as a "length of shielded wire," transmission lines are actually complex networks containing the equivalent of all the three basic electrical components: resistance, capacitance, and inductance. Because of this fact, transmission lines must be analyzed in terms of an RLC network.

Parallel and coaxial lines

This chapter will consider several types of transmission lines. Both step-function and sine-wave ac responses will be studied. Because the subject is both conceptual and analytical, both analogy and mathematical approaches to the theory of transmission lines will be used.

Figure 3-1 shows several basic types of transmission line. Perhaps the oldest and simplest form is the *parallel line* shown in Figs. 3-1A through 3-1D. Figure 3-1A shows an end view of the parallel conductor transmission line. The two conductors, of diameter d, are separated by a dielectric (which might be air) by a spacing S. These designations will be used in calculations later. Figure 3-1B shows a type of parallel line called *twin-lead*. This is the old-fashioned television antenna transmission line. It consists of a pair of parallel conductors separated by a plastic dielectric. TV-type twin-lead has a characteristic impedance of 300 Ω, while certain radio transmitting antenna twin-lead has an impedance of 450 Ω. Another form of twin-lead is *open-line*, shown in Fig. 3-1C. In this case, the wire conductors are separated by an air dielectric, with support provided by stiff (usually ceramic) insulators. A tie wire (only one shown) is used to fasten each insulator end to the main conductor. Some users of open-line prefer the form of insulator or supporter shown in Fig. 3-1D. This

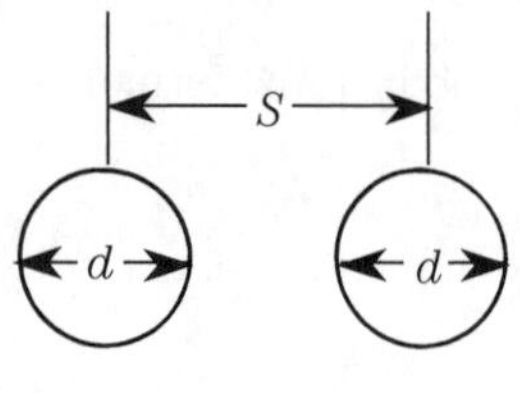

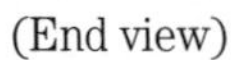

3-1A Parallel line transmission line (end view).

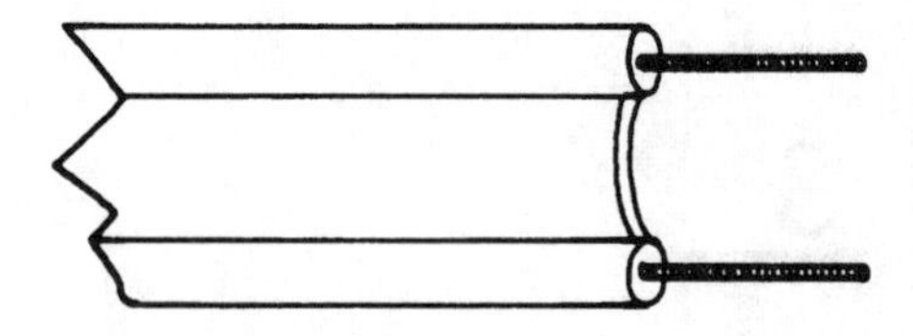

3-1B Twin-lead transmission line.

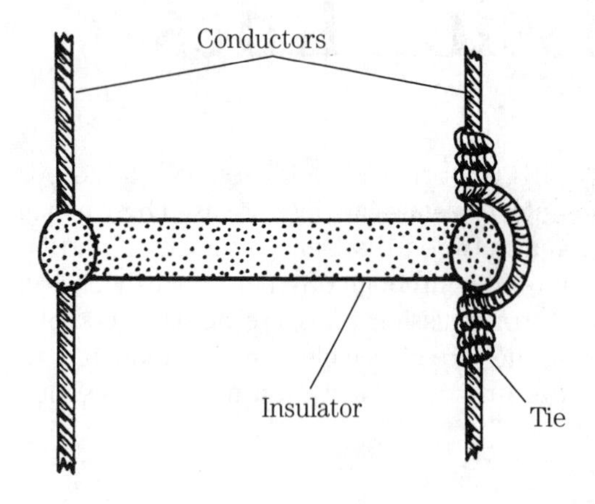

3-1C Parallel line construction details.

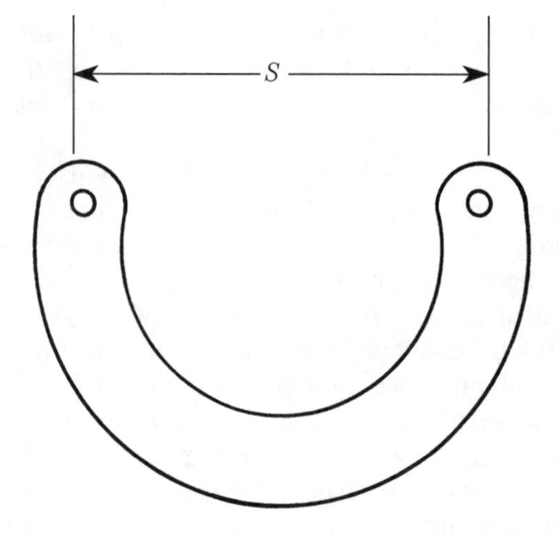

3-1D Horseshoe parallel line spreader.

form of insulator is made of either plastic or ceramic, and is in the form of a "U." The purpose of this shape is to reduce losses, especially in rainy weather, by increasing the leakage currents path relative to spacing S.

Parallel lines have been used at VLF, MW, and HF frequencies for decades. Even antennas into the low VHF are often found using parallel lines. The higher impedance of these lines (relative to coaxial cable) yields lower loss in high-power applications. For years, the VHF, UHF, and microwave application of parallel lines was limited to educational laboratories, where they are well suited to performing experiments (to about 2 GHz) with simple, low-cost instruments. Today, however, printed circuit and hybrid semiconductor packaging has given parallel lines a new lease on life, if not an overwhelming market presence.

Figure 3-1E shows a form of parallel line called *shielded twin-lead*. This type of line uses the same form of construction as TV-type twin-lead, but it also has a braided shielding surrounding it. This feature makes it less susceptible to noise and other problems.

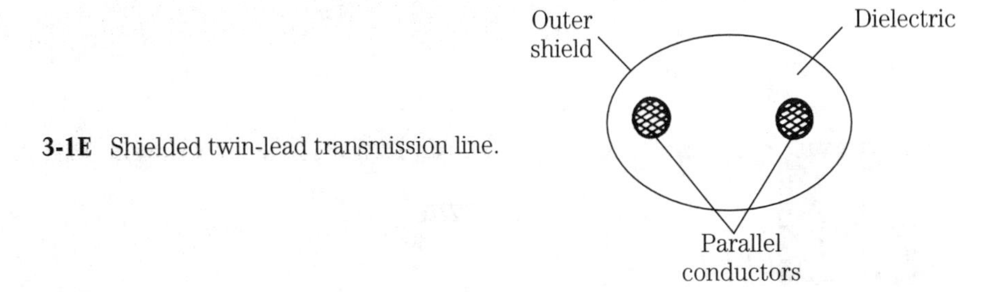

3-1E Shielded twin-lead transmission line.

The second form of transmission line, which finds considerable application at microwave frequencies, is *coaxial cable* (Figs. 3-1F through 3-1L). This form of line consists of two cylindrical conductors sharing the same axis (hence "co-axial"), and separated by a dielectric (Fig. 3-1F). For low frequencies (in flexible cables) the dielectric may be polyethylene or polyethylene foam, but at higher frequencies *Teflon* and other materials are used. Also used, in some applications, are dry air and dry nitrogen.

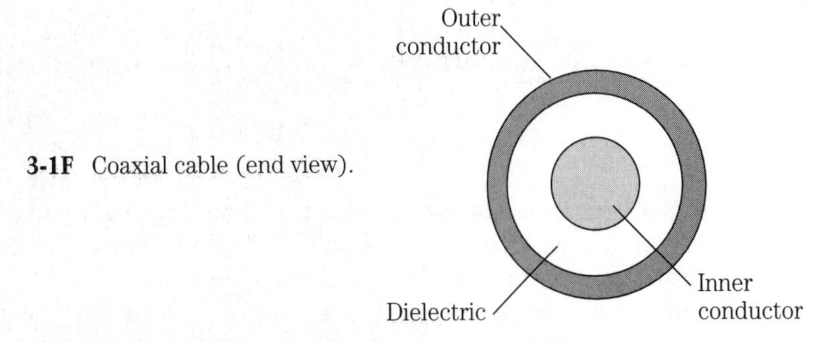

3-1F Coaxial cable (end view).

Several forms of coaxial line are available. Flexible coaxial cable is perhaps the most common form. The outer conductor in such cable is made of either braid or foil (Fig. 3-1G). Television broadcast receiver antennas provide an example of such cable from common experience. Another form of flexible or semi-flexible coaxial line is *helical line* (Fig. 3-1H) in which the outer conductor is spiral wound. *Hardline* (Fig. 3-1I) is coaxial cable that uses a thinwall pipe as the outer conductor. Some hardline coax used at microwave frequencies uses a rigid outer conductor and a solid dielectric.

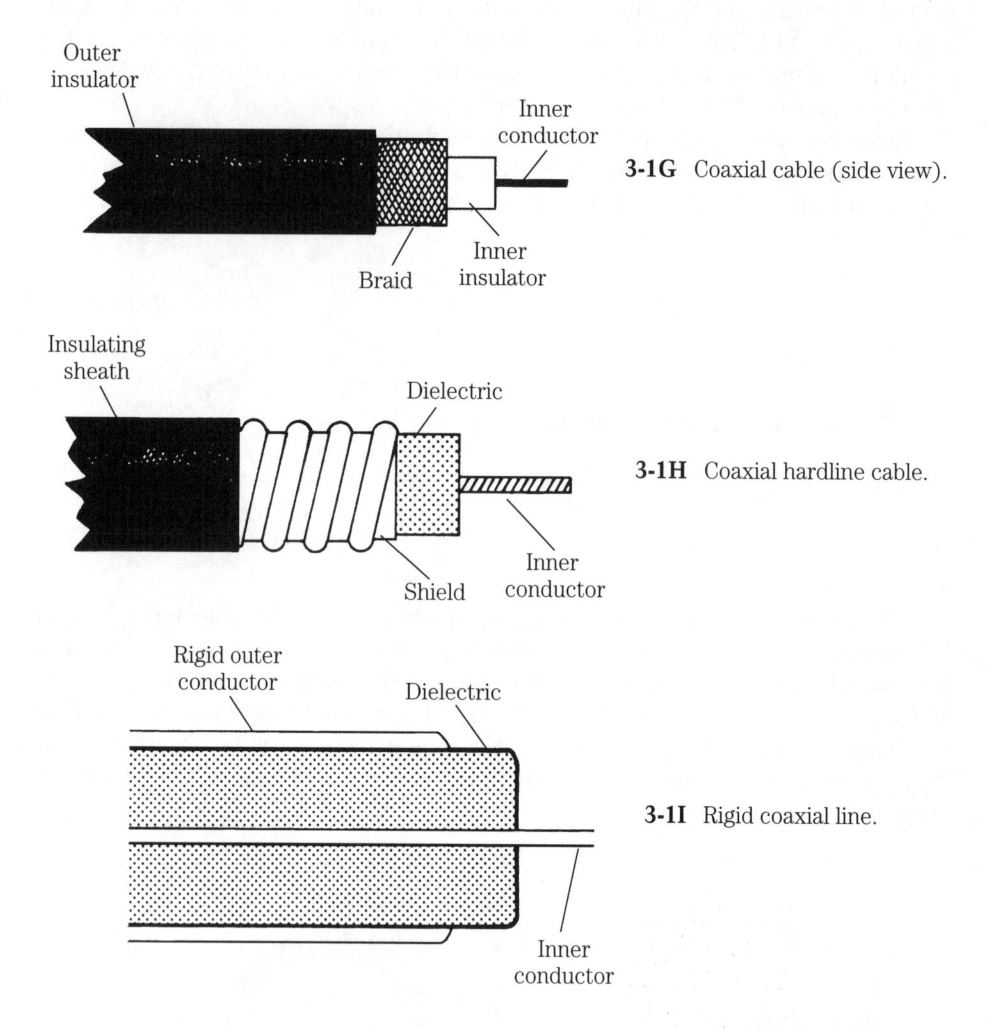

3-1G Coaxial cable (side view).

3-1H Coaxial hardline cable.

3-1I Rigid coaxial line.

Gas-filled line is a special case of hardline that is hollow (Fig. 3-1J), the center conductor is supported by a series of thin ceramic or *Teflon* insulators. The dielectric is either anhydrous (i.e., dry) nitrogen or some other inert gas.

Some flexible microwave coaxial cable uses a solid "air-articulated" dielectric (Fig. 3-1K), in which the inner insulator is not continuous around the center conductor, but rather is ridged. Reduced dielectric losses increase the usefulness of the

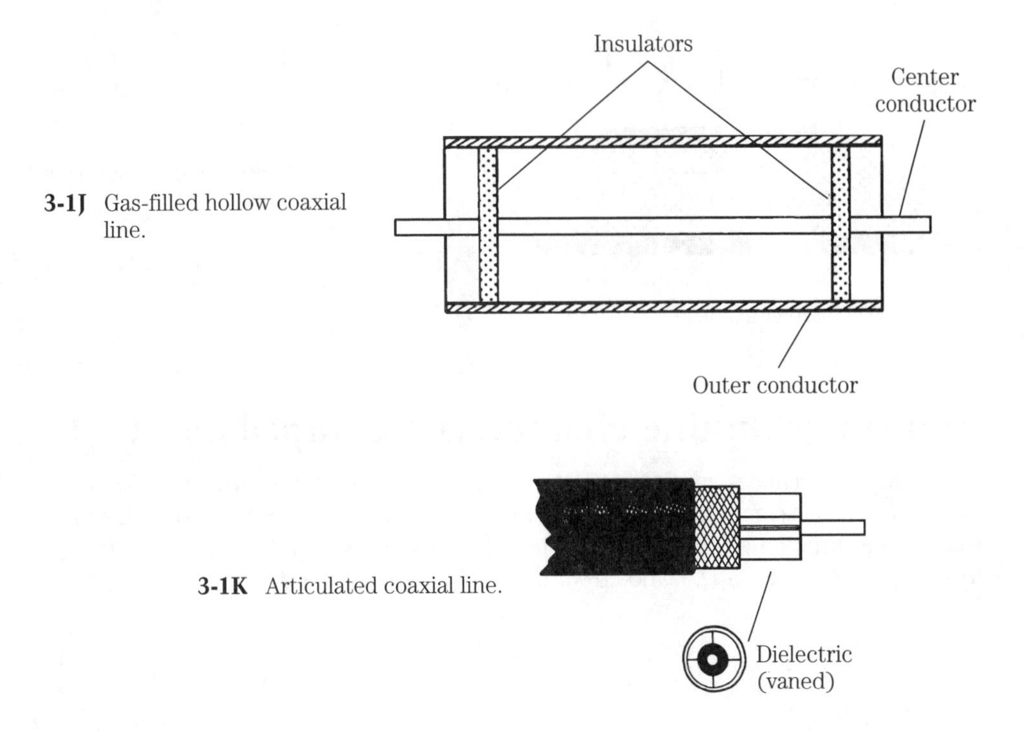

3-1J Gas-filled hollow coaxial line.

3-1K Articulated coaxial line.

cable at higher frequencies. Double-shielded coaxial cable (Fig. 3-1L) provides an extra measure of protection against radiation from the line, and EMI from outside sources, from getting into the system.

Stripline, also called *microstripline*, (Fig. 3-1M) is a form of transmission line used at high UHF and microwave frequencies. The stripline consists of a critically sized conductor over a ground-plane conductor, and separated from it by a dielectric. Some striplines are sandwiched between two groundplanes and are separated from each by a dielectric.

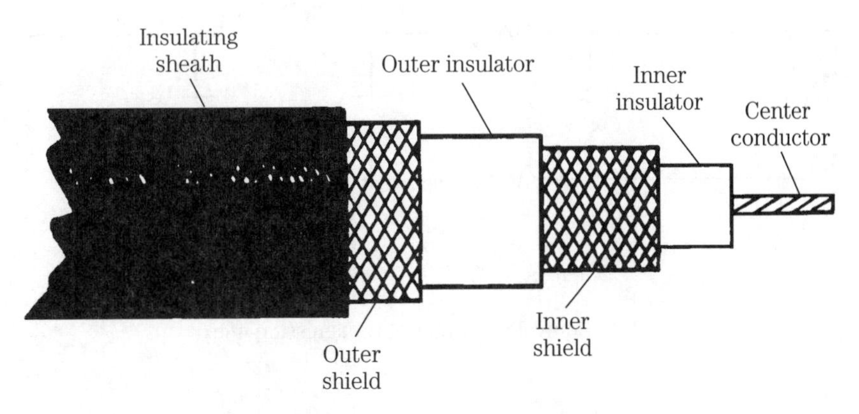

3-1L Double shielded coaxial line.

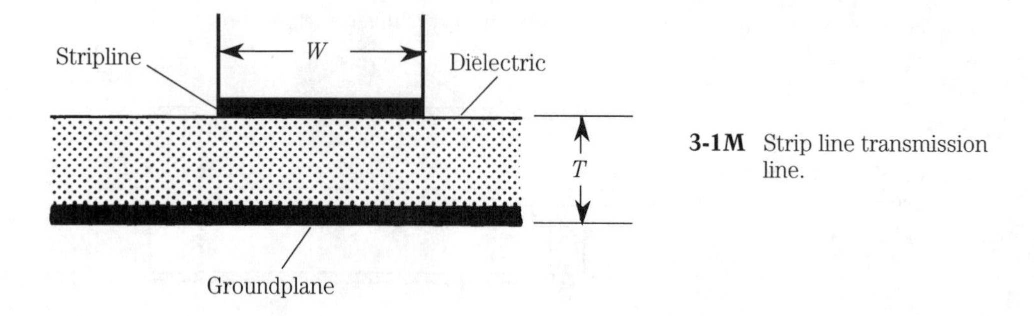

3-1M Strip line transmission line.

Transmission line characteristic impedance (Z_o)

The transmission line is an RLC network (see Fig. 3-2), so it has a *characteristic impedance*, Z_o, also sometimes called a *surge impedance*. Network analysis will show that Z_o is a function of the per unit of length parameters *resistance* (R), *conductance* (G), *inductance* (L), and *capacitance* (C), and is found from:

$$Z_o = \sqrt{\frac{R + j\omega L}{G + j\omega C}} \qquad [3.1]$$

Where:

Z_o is the characteristic impedance, in ohms
R is the resistance per unit length, in ohms
G is the conductance per unit length, in mhos
L is the inductance per unit length, in Henrys
C is the capacitance per unit length, in farads
ω is the angular frequency in radians per second ($2\pi F$)

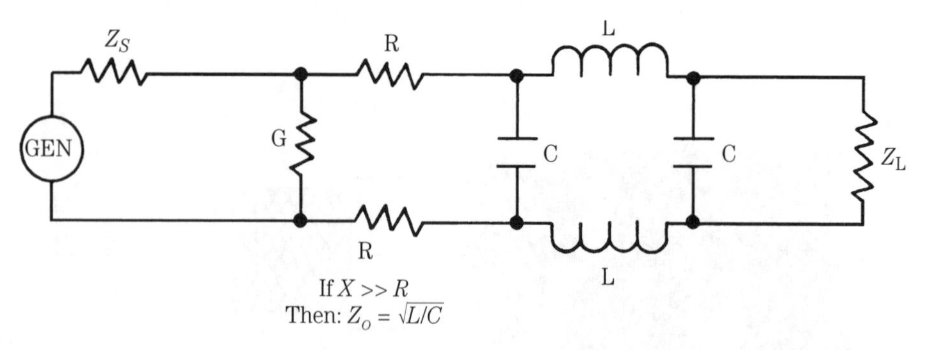

3-2 Equivalent circuit of transmission line.

In microwave systems the resistances are typically very low compared with the reactances, so Eq. 3.1 can be reduced to the simplified form:

$$Z_o = \sqrt{\frac{L}{C}} \quad \textbf{[3.2]}$$

Example 3-1 A nearly lossless transmission line (R is very small) has a unit length inductance of 3.75 nanohenrys, and a unit length capacitance of 1.5 pF. Find the characteristic impedance, Z_o.

Solution:

$$Z_o = \sqrt{\frac{L}{C}}$$

$$Z_o = \sqrt{\frac{\left[3.75 \text{ nH} \times \dfrac{1 \text{ H}}{10^9 \text{ nH}}\right]}{\left[1.5 \text{ pF} \times \dfrac{1 \text{ F}}{10^{12} \text{ pF}}\right]}}$$

$$Z_o = \sqrt{\frac{3.75 \times 10^{-9} \text{ H}}{1.5 \times 10^{-12} \text{ F}}}$$

$$Z_o = \sqrt{2.5 \times 10^3} = 50\ \Omega$$

The characteristic impedance for a specific type of line is a function of the conductor size, the conductor spacing, the conductor geometry (see again Fig. 3-1), and the dielectric constant of the insulating material used between the conductors. The dielectric constant (e) is equal to the reciprocal of the velocity (squared) of the wave when a specific medium is used:

$$e = \frac{1}{v^2} \quad \textbf{[3.3]}$$

Where:

e is the dielectric constant
v is the velocity of the wave in the medium
(note: for a perfect vacuum $e = 1.000$)

a) *Parallel Line*

$$Z_o = \frac{276}{\sqrt{e}} \text{ LOG}\left[\frac{2S}{d}\right] \quad \textbf{[3.4]}$$

Where:

Z_o is the characteristic impedance, in ohms
e is the dielectric constant
S is the center-to-center spacing of the conductors
d is the diameter of the conductors

b) *Coaxial Line*

$$Z_o = \frac{138}{\sqrt{e}} \text{ LOG} \left[\frac{D}{d} \right] \quad \textbf{[3.5]}$$

Where:

D is the diameter of the outer conductor
d is the diameter of the inner conductor

c) *Shielded Parallel Line*

$$Z_o = \frac{276}{\sqrt{e}} \text{ LOG} \left(2A \frac{(1-B^2)}{(1+B^2)} \right) \quad \textbf{[3.6]}$$

Where:

$A = s/d$
$B = s/D$

d) *Stripline*

$$Z_o = \frac{377}{\sqrt{e_t}} \left(\frac{T}{W} \right) \quad \textbf{[3.7A]}$$

Where:

e_t is the relative dielectric constant of the printed wiring board (PWB)
T is the thickness of the printed wiring board
W is the width of the stripline conductor

The relative dielectric constant (e_t) used above differs from the normal dielectric constant of the material used in the PWB. The relative and normal dielectric constants move closer together for larger values of the ratio W/T.

Example 3-2 A stripline transmission line is built on a 4-mm thick printed wiring board that has a relative dielectric constant of 5.5. Calculate the characteristic impedance if the width of the strip is 2 mm.

Solution:

$$Z_o = \frac{377}{\sqrt{e_t}} \left(\frac{T}{W} \right)$$

$$Z_o = \frac{377}{\sqrt{5.5}} \left(\frac{4 \text{ mm}}{2 \text{ mm}} \right)$$

$$Z_o = \frac{377}{2.35} (2) = 321\ \Omega$$

In practical situations, we usually don't need to calculate the characteristic impedance of a stripline, but rather design the line to fit a specific system impedance (e.g., 50 Ω). We can make some selection choices of printed circuit material (hence dielectric constant) and thickness, but even these are usually limited in practice by the availability of standardized boards. Thus, stripline *width* is the variable parameter. Equation 3.2 can be arranged to the form:

$$W = \frac{377\ T}{Z_o \sqrt{e}} \qquad \textbf{[3.7B]}$$

The impedance of 50 Ω is accepted as standard for RF systems, except in the cable-TV industry. The reason for this diversity is that power handling ability and low-loss operation don't occur at the same characteristic impedance. For example, the maximum power handling ability for coaxial cables occurs at 30 Ω, while the lowest loss occurs at 77 Ω; 50 Ω is therefore a reasonable trade-off between the two points. In the cable-TV industry, however, the RF power levels are minuscule, but lines are long. The trade-off for TV is to use 75 Ω as the standard system impedance in order to take advantage of the reduced attenuation factor.

Transmission line characteristics

Velocity factor

In the section preceding this section, we discovered that the velocity of the wave (or signal) in the transmission line is less than the free-space velocity (i.e., less than the speed of light). Further, we discovered in Eq. 3.3 the velocity is related to the dielectric constant of the insulating material that separates the conductors in the transmission line. Velocity factor (v) is usually specified as a decimal fraction of c, the speed of light (3×10^8 m/s). For example, if the velocity factor of a transmission line is rated at "0.66," then the velocity of the wave is 0.66c, or (0.66) (3×10^8 m/s) = 1.98×10^8 m/s.

Velocity factor becomes important when designing things like transmission line transformers, or any other device in which the length of the line is important. In most cases, the transmission line length is specified in terms of *electrical length*, which can be either an angular measurement (e.g., 180 degrees or π radians), or a relative measure keyed to wavelength (e.g., one-half wavelength, which is the same as 180 degrees). The *physical length* of the line is longer than the equivalent electrical length. For example, let's consider a 1-GHz half-wavelength transmission line.

A rule of thumb tells us that the length of a wave (in meters) in free-space is $0.30/F$, where frequency (F) is expressed in gigahertz; therefore, a half-wavelength line is 0.15/*F*. At 1 GHz, the line must be 0.15 meters/1 GHz = 0.15 meters. If the velocity factor is 0.80, then the *physical length* of the transmission line that will achieve the desired *electrical length* is [(0.15 meters) (v)]/F = [(0.15 meters) (0.80)]/1 GHz = 0.12 meters. The derivation of the "rule of thumb" is left as an exercise for the student (hint: it comes from the relationship between wavelength, frequency, and velocity of propagation for any form of wave).

There are certain practical considerations regarding velocity factor that result from the fact that the physical and electrical lengths are not equal. For example, in a certain type of phased array antenna design, radiating elements are spaced a half-wavelength apart, and must be fed 180 degrees (half-wave) out of phase with each other. The simplest interconnect is to use a half-wave transmission line between the 0-degree element and the 180-degree element. According to the standard wisdom, the transmission line will create the 180-degree phase delay required for the correct operation of the antenna. Unfortunately, because of the velocity factor, the physical length for a one-half electrical wavelength cable is *shorter* than the free-space half-wave distance between elements. In other words, the cable will be too short to reach between the radiating elements by the amount of the velocity factor!

Clearly, velocity factor is a topic that must be understood before transmission lines can be used in practical situations. Table 3-1 shows the velocity factors for several types of popular transmission line.

Table 3-1. Transmission line characteristics

Type of line	Z_o (ohms)	Vel. factor (v)
½-in. TV parallel line (air dielectric)	300	0.95
1-in. TV parallel line (air dielectric)	450	0.95
TV twin-lead	300	0.82
UHF TV twin-lead	300	0.80
Polyethylene coaxial cable	*	0.66
Polyethylene foam coaxial cable	*	0.79
Air-space polyethylene foam coaxial cable	*	0.86
Teflon	*	0.70

* Various impedances depending upon cable type.

Transmission line noise

Transmission lines are capable of generating noise and spurious voltages that are seen by the system as valid signals. Several such sources exist. One source is the coupling between noise currents flowing in the outer conductor, and the inner conductor. Such currents are induced by nearby electromagnetic interference and other sources (e.g., connection to a noisy groundplane). Although coaxial design reduces noise pick-up, compared with parallel line, the potential for EMI exists. Selection of high-grade line, with a high degree of shielding, reduces the problem.

Another source of noise is thermal noises in the resistances and conductances. This type of noise is proportional to resistance and temperature.

There is also noise created by mechanical movement of the cable. One species results from the movement of the dielectric against the two conductors. This form of noise is caused by electrostatic discharges in much the same manner as the spark created by rubbing a piece of plastic against woolen cloth.

A second species of mechanically generated noise is *piezoelectricity* in the dielectric. Although more common in cheap cables, one should be aware of it. Mechanical deformation of the dielectric causes electrical potentials to be generated.

Both species of mechanically generated noise can be reduced or eliminated by proper mounting of the cable. Although rarely a problem at lower frequencies, such noise can be significant at microwave frequencies when signals are low.

Coaxial cable capacitance

A coaxial transmission line possesses a certain capacitance per unit of length. This capacitance is defined by:

$$C = \frac{24\, e}{\mathrm{LOG}\ (D/d)} \ \frac{\mathrm{pF}}{\mathrm{meter}} \qquad \textbf{[3.8A]}$$

A long run of coaxial cable can build up a large capacitance. For example, a common type of coax is rated at 65 pF/meter. A 150-meter roll thus has a capacitance of 65 pF/m × (150 m), or 9,750 pF. When charged with a high voltage, as is done in performing breakdown voltage tests at the factory, the cable *acts like a charged high voltage capacitor*. Although rarely (if ever) lethal to humans, the stored voltage in new cable can deliver a nasty electrical shock and can irreparably damage electronic components.

Coaxial cable cut-off frequency (F_o)

The normal mode in which a coaxial cable propagates a signal is as a transverse electromagnetic (TEM) wave, but others are possible—and usually undesirable. There is a maximum frequency above which TEM propagation becomes a problem, and higher modes dominate. Coaxial cable should *not* be used above a frequency of:

$$F = \frac{6.75}{(D + d)\sqrt{e}}\ \mathrm{GHz} \qquad \textbf{[3.8B]}$$

Where:

F is the TEM-mode cut-off frequency
D is the diameter of the outer conductor in inches
d is the diameter of the inner conductor in inches
e is the dielectric constant

When maximum operating frequencies for cable are listed, it is the TEM mode that is cited. Beware of attenuation, however, when making selections for microwave frequencies. A particular cable may have a sufficiently high TEM-mode frequency, but still exhibit a high attenuation per unit length at X or Ku-bands.

Transmission line responses

In order to understand the operation of transmission lines, we need to consider two cases: *step-function response* and the *steady-state ac response*. The step function case involves a single event when a voltage at the input of the line snaps from zero (or a steady value) to a new (or non-zero) value, and remains there until all action dies out. This response tells us something of the behavior of pulses in the line, and in fact is used to describe the response to a single-pulse stimulus. The steady-state ac response tells us something of the behavior of the line under stimulation by a sinusoidal RF signal.

Step-function response of a transmission line

Figure 3-3 shows a parallel transmission line with characteristic impedance (Z_o) connected to a load impedance (Z_L). The "generator" at the input of the line consists of a voltage source (V) in series with a "source impedance" (Z_s) and a switch (S1). Assume for the present that all impedances are pure resistances (i.e., $R + j0$). Also, assume that $Z_s = Z_o$.

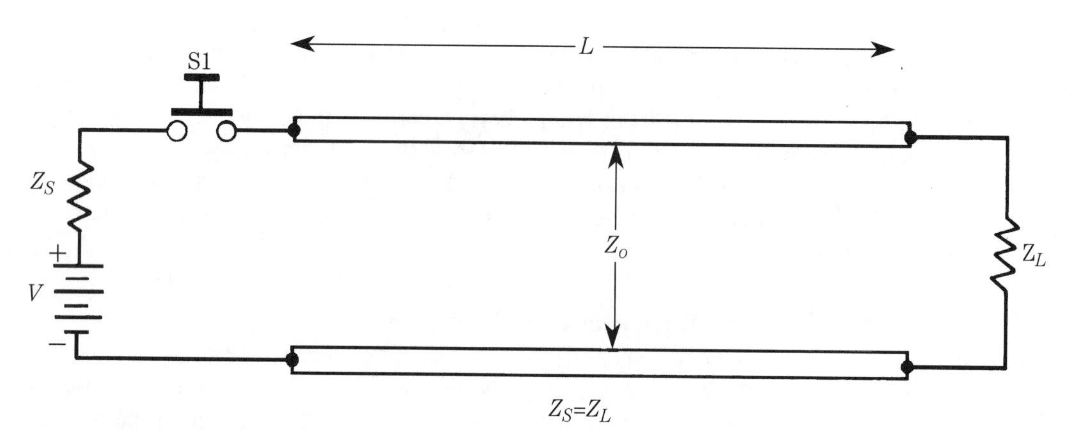

3-3 Schematic example of transmission line.

When the switch is closed at time T_o (Fig. 3-4A), the voltage at the input of the line (V_{in}) jumps to $V/2$. In Fig. 3-2, you may have noticed that the LC circuit resembles a delay line circuit. As might be expected, therefore, the voltage wavefront propagates along the line at a velocity (v) of:

$$v = \frac{1}{\sqrt{LC}} \tag{3.9}$$

Where:

v is the velocity in meters per second
L is the inductance in Henrys
C is the capacitance in farads

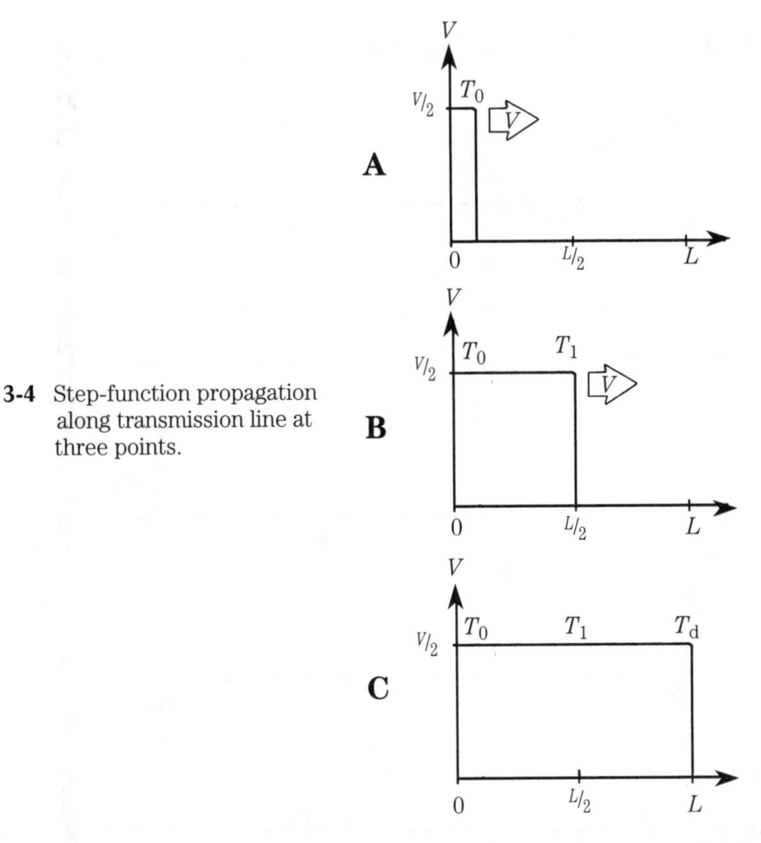

3-4 Step-function propagation along transmission line at three points.

At time T_1 (Fig. 3-4B), the wavefront has propagated one-half the distance L, and by T_d it has propagated the entire length of the cable (Fig. 3-4C).

If the load is perfectly matched (i.e., $Z_L = Z_o$), then the load absorbs the wave and no component is reflected. But in a mismatched system (Z_L is not equal to Z_o), a portion of the wave is reflected back down the line toward the generator.

Figure 3-5 shows the "rope analogy" for reflected pulses in a transmission line. A taut rope (Fig. 3-5A) is tied to a rigid wall that does not absorb any of the energy in the pulse propagated down the rope. When the free end of the rope is given a vertical displacement (Fig. 3-5B), a wave is propagated down the rope at velocity v (Fig. 3-5C). When the pulse hits the wall (Fig. 3-5D), it is reflected (Fig. 3-5E) and propagates back down the rope toward the free end (Fig. 3-5F).

If a second pulse is propagated down the line before the first pulse dies out, then there will be two pulses on the line at the same time (Fig. 3-6A). When the two pulses interfere, the resultant will be the algebraic sum of the two. In the event that a pulse train is applied to the line, the interference pattern will set up *standing waves*, an example of which is shown in Fig. 3-6B.

Reflection coefficient (Γ)

The *reflection coefficient* (Γ) of a circuit containing a transmission line and load impedance is a measure of how well the system is matched. The absolute value of the re-

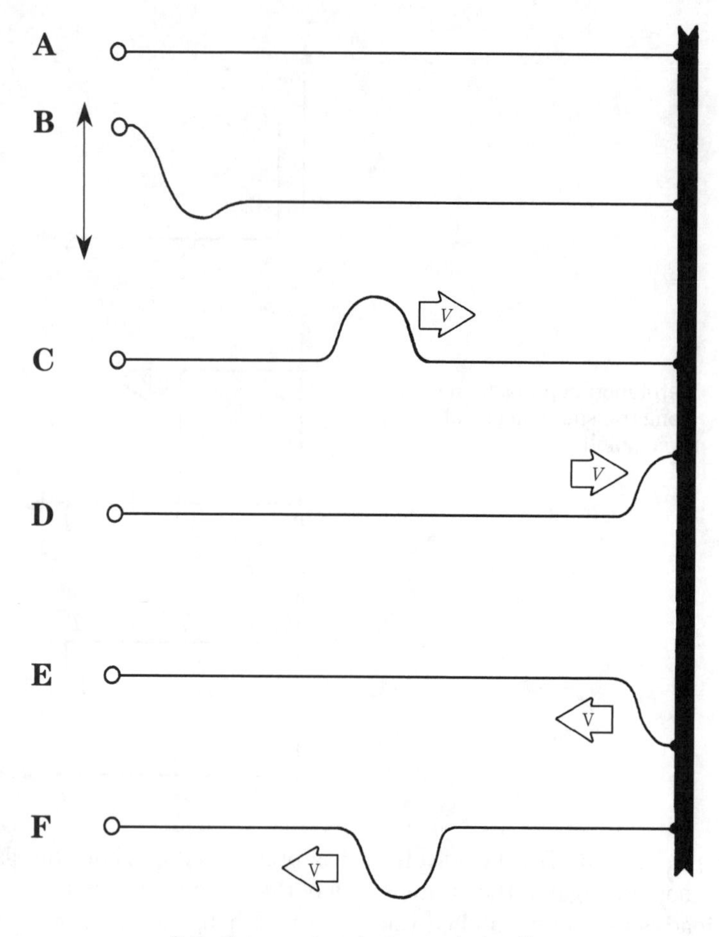

3-5 Rope analogy to transmission line.

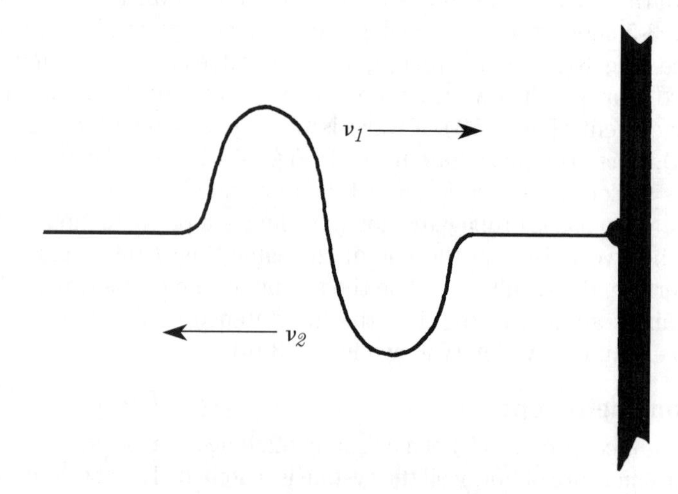

3-6A Interfering opposite waves.

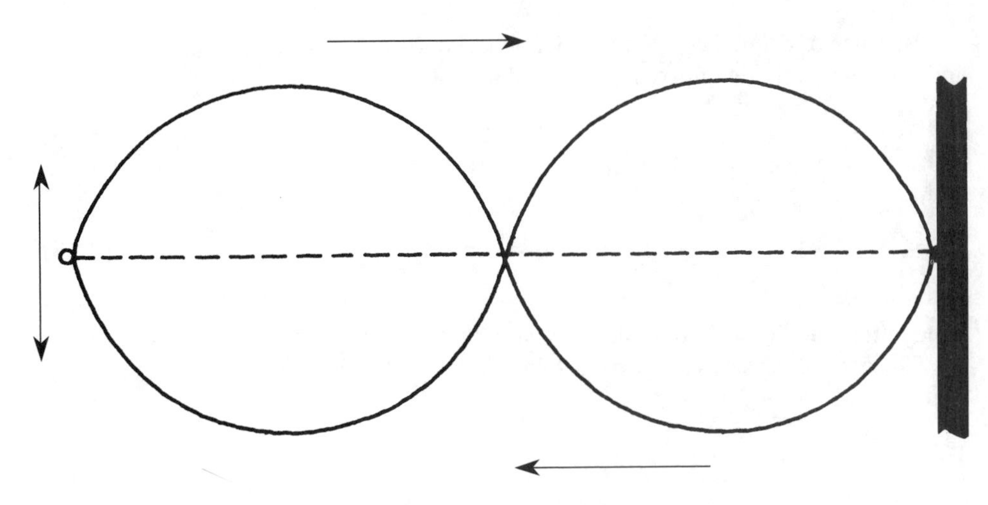

3-6B Standing waves.

flection coefficient varies from –1 to +1, depending upon the magnitude of reflection; $\Gamma = 0$ indicates a perfect match with no reflection, while –1 indicates a short circuited load, and +1 indicates an open circuit. To understand the reflection coefficient, let's start with a basic definition of the resistive load impedance $Z = R + j0$:

$$Z_L = \frac{V}{I} \qquad [3.10]$$

Where:

Z_L is the load impedance $R + j0$
V is the voltage across the load
I is the current flowing in the load

Because there are both reflected and incident waves, we find that V and I are actually the sum of incident and reflected voltages and currents, respectively. Therefore:

$$Z_L = \frac{V}{I} \qquad [3.11A]$$

$$Z_L = \frac{V_{inc} + V_{ref}}{I_{inc} + I_{ref}} \qquad [3.11B]$$

Where:

V_{inc} is the incident (i.e., forward) voltage
V_{ref} is the reflected voltage
I_{inc} is the incident current
I_{ref} is the reflected current

Because of Ohm's law, you can define the currents in terms of voltage, current, and the characteristic impedance of the line:

$$I_{inc} = \frac{V_{inc}}{Z_o} \qquad [3.12]$$

and,

$$I_{inc} = \frac{-V_{ref}}{Z_o} \qquad [3.13]$$

(The minus sign in Eq. 3.13 indicates that a direction reversal has taken place.)

The two expressions for current (Eqs. 3.12 and 3.13 may be substituted into Eq. 3.11 to yield:

$$Z_L = \frac{(V_{inc} + V_{ref})}{\left[\frac{V_{inc}}{Z_o} - \frac{V_{ref}}{Z_o}\right]} \qquad [3.14]$$

The reflection coefficient (Γ) is defined as the ratio of reflected voltage to incident voltage:

$$\frac{V_{ref}}{V_{inc}}$$

so, by solving Eq. 3.14 for this ratio:

$$\Gamma = \frac{V_{ref}}{V_{inc}} \qquad [3.15]$$

and,

$$\Gamma = \frac{Z_L - Z_o}{Z_L + Z_o} \qquad [3.16]$$

Example 3-3 A 50-Ω transmission line is connected to a 30-Ω resistive load. Calculate the reflection coefficient, Γ.

Solution:

$$\Gamma = \frac{Z_L - Z_o}{Z_L + Z_o}$$

$$\Gamma = \frac{(50\ \Omega) - (30\ \Omega)}{(50\ \Omega) + (30\ \Omega)}$$

$$\Gamma = \frac{20}{80} = 0.25$$

Example 3-4 In Example 3-3, the incident voltage is 3 Volts rms. Calculate the reflected voltage.

Solution:

If,

$$\Gamma = \frac{V_{ref}}{V_{inc}}$$

Then,

$$V_{ref} = \Gamma \, V_{inc}$$

$$V_{ref} = (0.25)\ (3\ \text{Volts}) = 0.75\ \text{Volts}$$

The phase of the reflected signal is determined by the relationship of load impedance and transmission line characteristic impedance. For resistive loads (Z = R + j0): if the ratio ZL/Zo is 1.0, then there is no reflection; if ZL/Zo is less than 1.0, then the reflected signal is 180 degrees out of phase with the incident signal; if the ratio ZL/Zo is greater than 1.0 then the reflected signal is in-phase with the incident signal. In summary:

Ratio	**Angle of reflection**
$Z_L/Z_o = 1$	No reflection
$Z_L/Z_o < 1$	180 degrees
$Z_L/Z_o > 1$	0 degrees

The step-function (or pulse) response of the transmission line leads to a powerful means of analyzing the line, and its load, on an oscilloscope. Figure 3-7A shows (in schematic form) the test set-up for *time domain reflectrometry* (TDR) measurements. An oscilloscope, and a pulse (or square-wave) generator, are connected in parallel across the input end of the transmission line. Figure 3-7B shows a pulse test jig built by the author for testing lines at HF. The small shielded box contains a TTL square-wave oscillator circuit. Although a crystal oscillator can be used, an RC-timed circuit running close to 1000 kHz is sufficient. In Fig. 3-7B, you can see the test pulse generator box is connected in parallel with the cable under test and the input of the oscilloscope. A closer look is seen in Fig. 3-7C. A BNC "Tee" connector, and a double male BNC adapter, are used to interconnect the box with the 'scope.

If a periodic waveform is supplied by the generator, then the display on the oscilloscope will represent the sum of reflected and incident pulses. The duration of the pulse (i.e., pulse width), or one-half the period of the square wave, is adjusted so that the returning reflected pulse arrives approximately in the center of the incident pulse.

Figure 3-8 shows a TDR display under several circumstances. Approximately 30 meters of coaxial cable, with a velocity factor of 0.66, was used in a test set-up similar to Fig. 3-7. The pulse width was approximately 0.9 microseconds (μs). The hori-

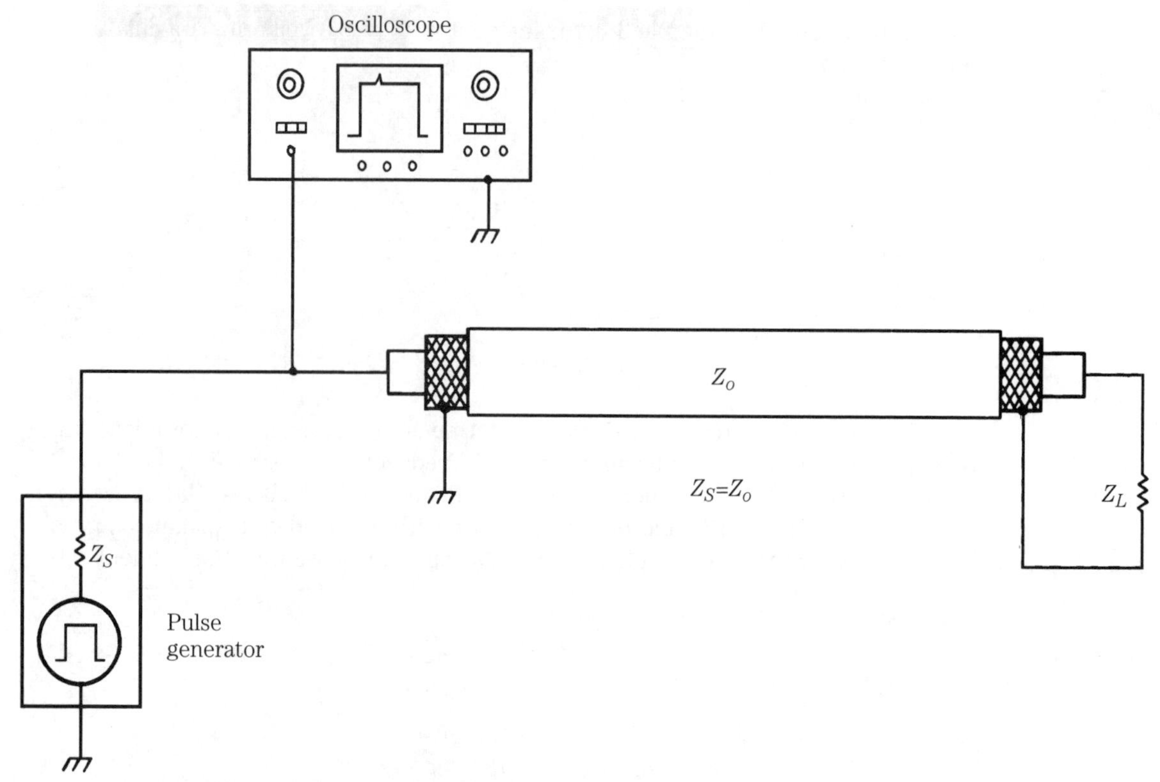

3-7A Time domain reflectometry set-up. Ham Radio Magazine

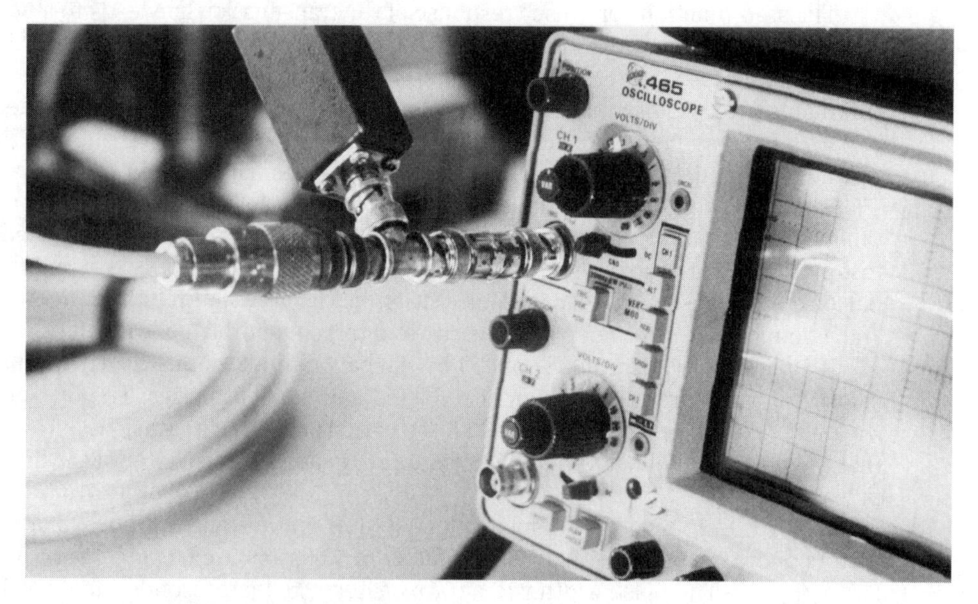

3-7B Test set-up for impromptu time domain reflectometry.

3-7C Close-up of RF connections.

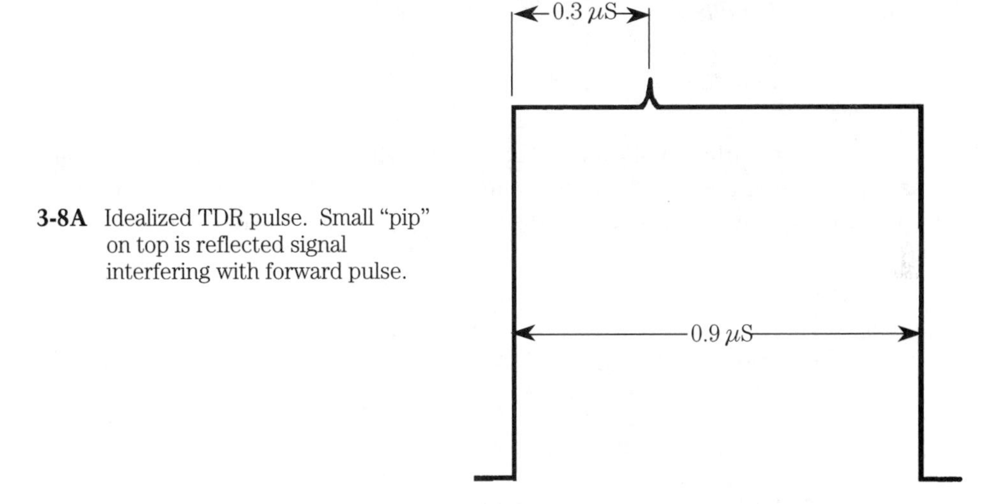

3-8A Idealized TDR pulse. Small "pip" on top is reflected signal interfering with forward pulse.

zontal sweep time on the 'scope was adjusted to show only one pulse—which, in this case, represented one-half of a 550-kHz square wave (Fig. 3-8B).

The displayed trace in Fig. 3-8B shows the pattern when the load is matched to the line ($Z_L = Z_o$). A slight discontinuity exists on the high side of the pulse, and this represents a small reflected wave. Even though the load and line were supposedly matched, the connectors at the end of the line presented a slight impedance discontinuity that shows up on the 'scope as a reflected wave. In general, any discontinuity in the line, any damage to the line, any too-sharp bend, or other anomaly, causes a slight impedance variation, and hence a reflection.

3-8B TDR pulse with no significant reflection.

Notice that the anomaly occurs approximately one-third of the 0.9 μs duration (or 0.3 μs) after the onset of the pulse. This fact tells us that the reflected wave arrives back at the source 0.3 μs after the incident wave leaves. Because this time period represents a round-trip, you can conclude that the wave required 0.3 μs/2, or 0.15 μs to propagate the length of the line. Knowing that the velocity factor is 0.66 for that type of line, you can calculate its approximate length:

$$Length = c\ v\ T \qquad [3.17]$$

$$Length = \frac{(3 \times 10^8 \text{ m/s})}{\text{sec}} \times (0.66) \times (1.5 \times 10^{-7} \text{ sec})$$

$$Length = 29.7 \text{ meters}$$

. . . which agrees "within experimental accuracy" with the 30 meters actual length prepared for the experiment ahead of time. Thus, the TDR set-up (or a TDR instrument) can be used to measure the length of a transmission line. A general equation is:

$$L_{meters} = \frac{c\ V\ T_d}{2} \qquad [3.18]$$

Where:

L is the length in meters
c is the velocity of light (3×10^8 m/s)
v is the velocity factor of the transmission line
T_d is the round-trip time between the onset of the pulse, and the first reflection

Figures 3-8C through 3-8H shows the behavior of the transmission line to the stepfunction when the load impedance is mismatched to the transmission line (Z_L not equal to Z_o). Figure 3-8C shows what happens when the load impedance is less than the line impedance (in this case, 0.5 Z_o). The reflected wave is inverted, and

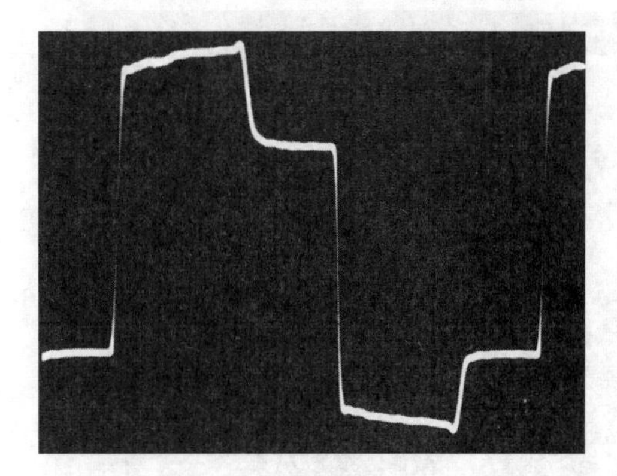

3-8C $Z_L < Z_o$

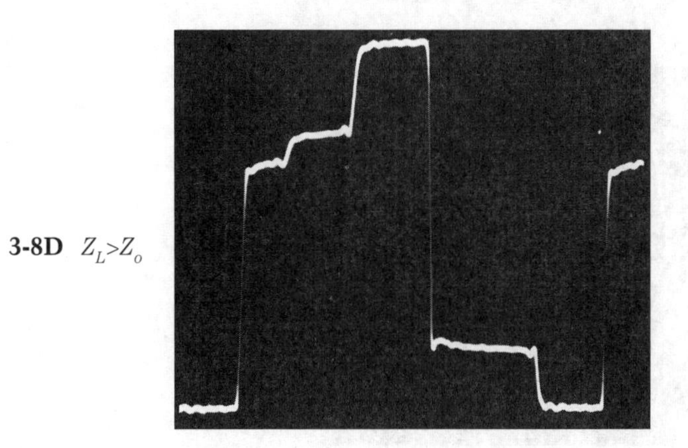

3-8D $Z_L > Z_o$

sums with the incident wave along the top of the pulse. The reflection coefficient can be determined by examining the relative amplitudes of the two waves.

The opposite situation, in which Z_L is $2Z_o$, is shown in Fig. 3-8D. In this case, the reflected wave is in-phase with the incident wave, so it adds to the incident wave as shown. The cases for a short circuited load, and an open circuited load, are shown in Figs. 3-8E and 3-8F, respectively. The cases of reactive loads are shown in Figs. 3-8G and 3-8H. The waveform in Fig. 3-8G resulted from a capacitance in series with a 50-Ω (matched) resistance; the waveform in Fig. 3-8H resulted from a 50-Ω resistance in series with an inductance.

The ac response of the transmission line

When a CW RF signal is applied to a transmission line, the excitation is sinusoidal (Fig. 3-9), so it becomes useful for us to investigate the steady-state ac response of the line. The term *steady-state* implies a sinewave of constant amplitude, phase, and

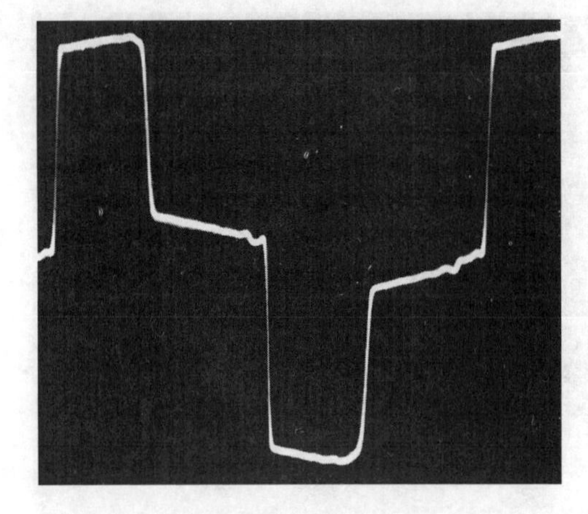

3-8E $Z_L = 0$

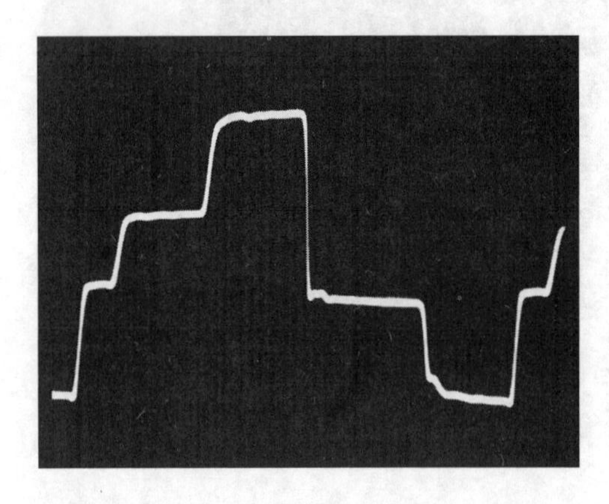

3-8F $Z_L = \infty$

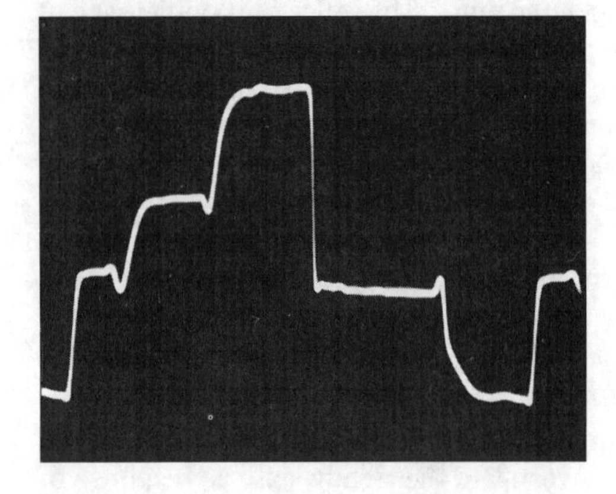

3-8G $Z_L = 50 - jX_C$

3-8H $Z_L = 50 + jX_L$

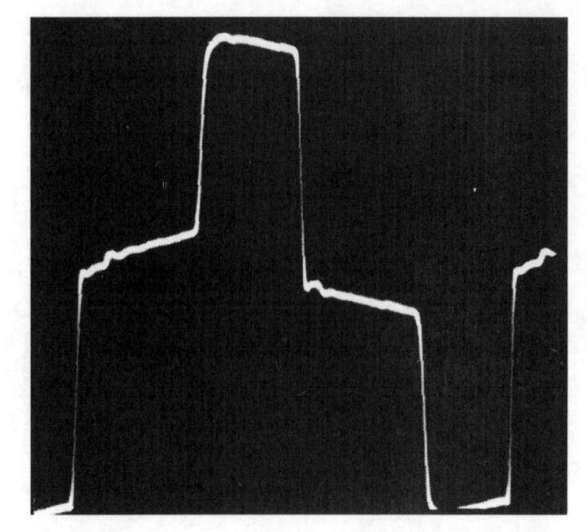

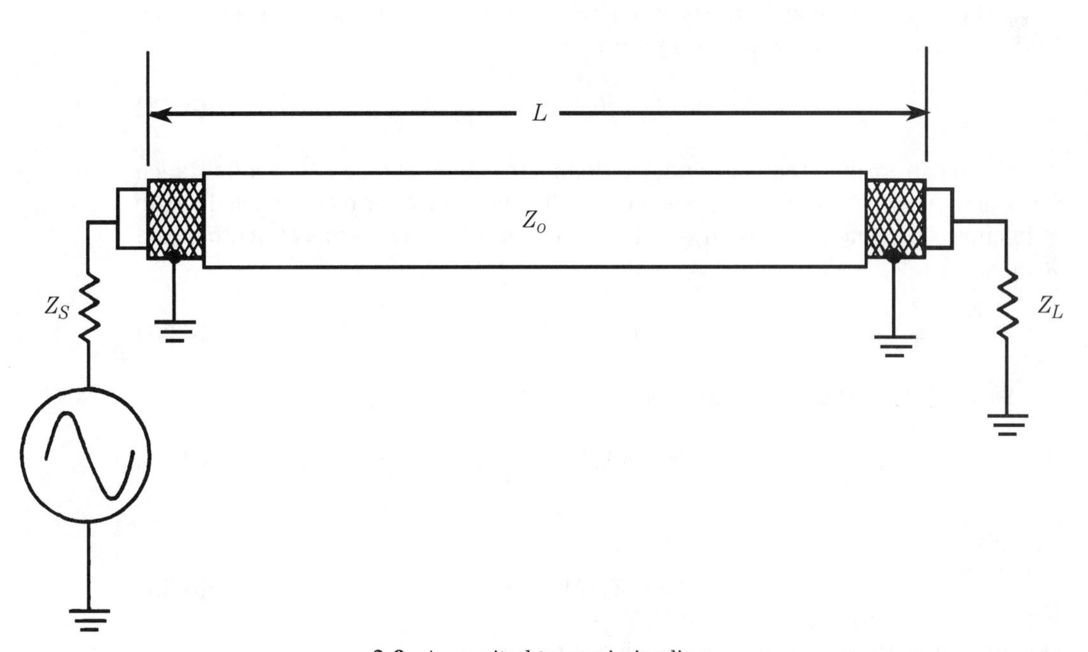

3-9 Ac-excited transmission line.

frequency. When ac is applied to the input of the line, it propagates along the line at a given velocity. The ac signal amplitude and phase will decay exponentially in the manner shown by Eq. 3-19:

$$V_R = Ve^{-\gamma l} \qquad \textbf{[3.19]}$$

Where:

V_R is the voltage received at the far end of the line
V is the applied voltage
l is the length of the line
y is the *propagation constant* of the line

The propagation constant (y) is defined in various equivalent ways, each of which serves to illustrate its nature. For example, the propagation constant is proportional to the product of impedance and admittance characteristics of the line:

$$y = \sqrt{[ZY]} \quad \textbf{[3.20]}$$

or, since $Z = R + j\,\omega\,L$ and $Y = G + j\,\omega\,C$, we may write:

$$y = \sqrt{(R + j\,\omega\,l)\;(G + j\,\omega\,C)} \quad \textbf{[3.21]}$$

You can also write an expression for the propagation constant in terms of the *line attenuation constant* (a) and *phase constant* (B):

$$y = a + jB \quad \textbf{[3.22]}$$

If you can assume that susceptance dominates conductance in the admittance term, and reactance dominates resistance in the impedance term (both usually true at microwave frequencies), then we may neglect the R and G terms altogether and write:

$$y = j\,\omega\,\sqrt{LC} \quad \textbf{[3.23]}$$

We may also reduce the phase constant (B) to:

$$B = \omega\,\sqrt{LC} \quad \textbf{[3.24]}$$

or,

$$B = \omega\,Z_o\,C \text{ rad/m} \quad \textbf{[3.25]}$$

and, of course, the characteristic impedance remains:

$$Z_o = \sqrt{LC} \quad \textbf{[3.26]}$$

Special cases

The impedance "looking-into" a transmission line (Z) is the impedance presented to the source by the combination of load impedance and transmission line characteristic impedance. Below are presented equations that define the "looking-in" impedance seen by a generator (or source) driving a transmission line.

The case where the load impedance, and line characteristic impedance, are matched is defined by:

$$Z_L = R_L + j0 = Z_o$$

In other words, the load impedance is resistive and equal to the characteristic impedance of the transmission line. In this case, the line and load are matched, and the impedance looking-in will be a simple $Z = Z_L = Z_o$. In other cases, however, we find different situations where Z_L is not equal to Z_o.

1. Z_L is not equal to Z_o in a random length lossy line:

$$Z = (Z_o)\left(\frac{Z_L + Z_o \operatorname{Tanh}(y\, l)}{Z_o + Z_L \operatorname{Tanh}(y\, l)}\right) \qquad \textbf{[3.27]}$$

Where:

Z is the impedance looking-in, in ohms
Z_L is the load impedance, in ohms
Z_o is the line characteristic impedance, in ohms
l is the length of the line in meters
y is the propagation constant

2. Z_L not equal to Z_o in a lossless, or very low loss, random length line:

$$Z = (Z_o)\left(\frac{(Z_L + jZ_o \operatorname{Tan}(B\, l)}{Z_o + jZ_L \operatorname{Tan}(B\, l)}\right) \qquad \textbf{[3.28]}$$

Equations 3.27 and 3.28 serve for lines of any random length. For lines that are either integer multiples of a half wavelength, or odd integer (i.e., 1, 3, 5, 7 . . . etc.) multiples of a quarter wavelength, special solutions for these equations are found—and some of these solutions are very useful in practical situations. For example, consider . . .

3. *Half wavelength lossy lines*:

$$Z = (Z_o)\left(\frac{(Z_L + Z_o \operatorname{Tanh}(a\, l)}{Z_o + Z_L \operatorname{Tanh}(a\, l)}\right) \qquad \textbf{[3.29]}$$

Example 3-5 A lossless 50-Ω (Z_o) transmission line is exactly one-half wavelength long, and is terminated in a load impedance of $Z = 30 + j0$. Calculate the input impedance "looking into" the line (note: in a lossless line $a = 0$).

Solution:

$$Z = (Z_o)\left(\frac{Z_L + Z_o \operatorname{Tanh}(a\, l)}{Z_o + Z_L \operatorname{Tanh}(a\, l)}\right)$$

$$Z = (50\ \Omega)\left(\frac{(30) + [(50)\ (\tanh((0)\ (\pi))]}{(50) + [(30)\ (\tanh((0)\ (\pi))]}\right)$$

$$Z = (50\ \Omega)\left(\frac{(30) + ((50)\ (\tanh(0)))}{(50) + [(30)\ (\tanh(0))]}\right)$$

$$Z = (50\ \Omega)\left(\frac{30 + 0}{50 + 0}\right)$$

$$Z = (50\ \Omega)\ (30/50) = 30\ \Omega$$

In Example 3-5 we discovered that the impedance looking into a lossless (or very low-loss) half-wavelength transmission line is the load impedance:

$$Z = Z_L \qquad \textbf{[3.30]}$$

The fact that line input impedance, equals load impedance, is very useful in certain practical situations. For example, a resistive impedance is not changed by the line length. Therefore, when an impedance is inaccessible for measurement purposes, the impedance can be measured through a transmission line that is an integer multiple of a half wavelength.

Our next special case involves a quarter-wavelength transmission line, and those that are *odd* integer multiples of quarter wavelengths (of course, *even* integer multiples of a quarter wavelength obey the half-wavelength criteria).

4. Quarter wavelength lossy lines:

$$Z = (Z_o)\left[\frac{Z_L + Z_o \operatorname{Coth}(a\ l)}{Z_o + Z_L \operatorname{Coth}(a\ l)}\right] \qquad \textbf{[3.31]}$$

and,

5. Quarter wavelength lossless or very low loss lines:

$$Z = \frac{[Z_o]^2}{Z_L} \qquad \textbf{[3.32]}$$

From Eq. 3.32, you can discover an interesting property of the quarter-wavelength transmission line. First, divide each side of the equation by Z_o:

$$\frac{Z}{Z_o} = \frac{[Z_o]^2}{Z_L Z_o} \qquad \textbf{[3.33]}$$

$$\frac{Z}{Z_o} = \frac{Z_o}{Z_L} \qquad \textbf{[3.34]}$$

The ratio Z/Z_o shows an inversion of the load impedance ratio Z_L/Z_o, or, stated another way:

$$\frac{Z}{Z_o} = \frac{1}{Z_L/Z_o} \quad \textbf{[3.35]}$$

Again, from Eq. 3.32, you can deduce another truth about quarter-wavelength transmission lines:
If,

$$Z = \frac{\sqrt{Z_o}}{Z_L} \quad \textbf{[3.36]}$$

Then,

$$Z\,Z_L = \sqrt{Z_o} \quad \textbf{[3.37]}$$

Which means,

$$Z_o = \sqrt{Z\,Z_L} \quad \textbf{[3.38]}$$

Equation 3.38 shows that a quarter-wavelength transmission line can be used as an *impedance matching network*. Called a *Q-section*, the quarter-wavelength transmission line used for impedance matching requires a characteristic impedance Z_o—if Z is the source impedance, and Z_L is the load impedance.

Example 3-6 A 50-Ω source must be matched to a load impedance of 36 Ω. Find the characteristic impedance required of a Q-section matching network.

Solution:

$$Z = \sqrt{Z\,Z_L}$$

$$Z = \sqrt{(50\ \Omega)\ (36\ \Omega)}$$

$$Z = \sqrt{1800\ \Omega^2} = 42\ \Omega$$

6. Transmission line as a reactance: Reconsider Eq. 3.28, which related impedance looking-in, to load impedance and line length:

$$Z = (Z_o)\left(\frac{Z_L + jZ_o \operatorname{Tan}(B\,l)}{Z_o + jZ_L \operatorname{Tan}(B\,l)}\right) \quad \textbf{[3.39]}$$

Now, for the case of a shorted line (i.e., $Z_L = 0$), the solution is:

$$Z = (Z_o)\left(\frac{(0) + jZ_o \operatorname{Tan}(B\,l)}{Z_o + j(0) \operatorname{Tan}(B\,l)}\right) \quad \textbf{[3.40]}$$

$$Z = (Z_o)\left(\frac{jZ_o \text{ Tan } (B\,l)}{Z_o}\right) \quad [3.41]$$

$$Z = j\,Z_o \text{ Tan } (B\,l) \quad [3.42]$$

Recall from Eq. 3.25 that:

$$B = W\,Z_o\,C \quad [3.43]$$

Substituting Eq. 3.43 into Eq. 3.42 produces:

$$Z = j\,Z_o \text{ Tan } (\omega\,Z_o\,C\,l) \quad [3.44]$$

or,

$$Z = j\,Z_o \text{ Tan } (2\,\pi\,F\,Z_o\,C\,l) \quad [3.45]$$

Because the solutions to Eqs. 3.44 and 3.45 are multiplied by the j-operator, the impedance is actually a reactance ($Z = 0 + jX$). It is possible to achieve almost any possible reactance (within certain practical limitations) by adjusting the length of the transmission line and shorting the "load" end. This fact leads us to a practical method for impedance matching.

Figure 3-10A shows a circuit in which an unmatched load is connected to a transmission line with characteristic impedance Z_o. The load impedance Z_L is of the form $Z = R \pm jX$, and, in this case, it is equal to $50 - j20$.

A complex impedance load can be matched to its source by interposing the complex conjugate of the impedance. For example, in the case where $Z = 50 - j20$, the matching impedance network will require an impedance of $50 + j20\ \Omega$. The two impedances combine to produce a result of 50 Ω. The situation of Fig. 3-10A shows a matching stub with a reactance equal in magnitude, but opposite in sign, with respect to the reactive component of the load impedance. In this case, the stub has a reactance of $+j20\ \Omega$ to cancel a reactance of $-j20\ \Omega$ in the load.

A *quarter-wavelength shorted stub* is a special case of the stub concept that finds particular application in microwave circuits. *Waveguides* (chapter 17) are based on the properties of the quarter-wavelength shorted stub. Figure 3-10B shows a quarter-wave stub and its current distribution. The current is maximum across the short, but wave cancellation forces it to zero at the terminals. Because $Z = V/I$, when I goes to zero, the impedance becomes infinite. Thus, a quarter-wavelength stub has an infinite impedance at its resonant frequency, and redundant acts as an insulator. This concept may be hard to swallow, but the stub is a "metal insulator." You will see this concept developed further in chapter 17.

Standing wave ratio

The reflection phenomena that was noted earlier during the coverage of the step-function and single pulse response of a transmission line; the same phenomena also applies when the transmission line is excited with an ac signal. When a transmission

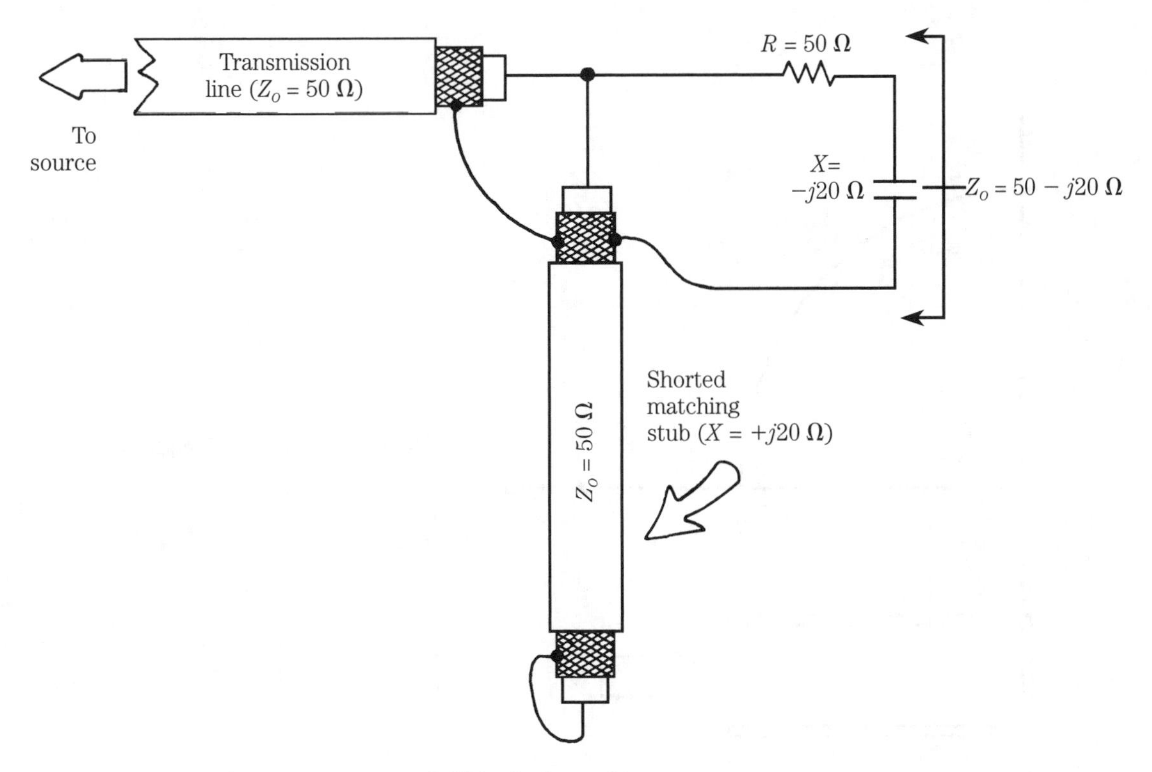

3-10A Stub matching system.

line is not matched to its load, some of the energy is absorbed by the load and some is reflected back down the line toward the source. The interference of incident (or "forward") and reflected (or "reverse") waves creates *standing waves* on the transmission line.

If the voltage or current is measured along the line, it will vary, depending on the load, according to Fig. 3-11. Figure 3-11A shows the voltage-vs-length curve for a matched line (i.e., where $Z_L = Z_o$). The line is said to be "flat" because the voltage (and current) is constant all along the line. But now consider Figs. 3-11B and 3-11C.

Figure 3-11B shows the voltage distribution over the length of the line when the load end of the line is *shorted* (i.e., $Z_L = 0$). Of course, at the load end the voltage is zero, which results from zero impedance. The same impedance and voltage situation is repeated every half wavelength down the line from the load end toward the generator. Voltage *minima* are called *nodes*, and voltage *maxima* are called *antinodes*.

The pattern in Fig. 3-11C results when the line is unterminated (open) (i.e., $Z_L = \infty$). Note that the pattern is the same shape as Fig. 3-11B (shorted line), but the phase is shifted 90 degrees. In both cases, the reflection is 100 percent, but the phase of the reflected wave is opposite.

Figure 3-11D shows the situation in which Z_L is not equal to Z_o, but is neither zero nor infinite. In this case, the nodes represent some finite voltage, V_{min}, rather

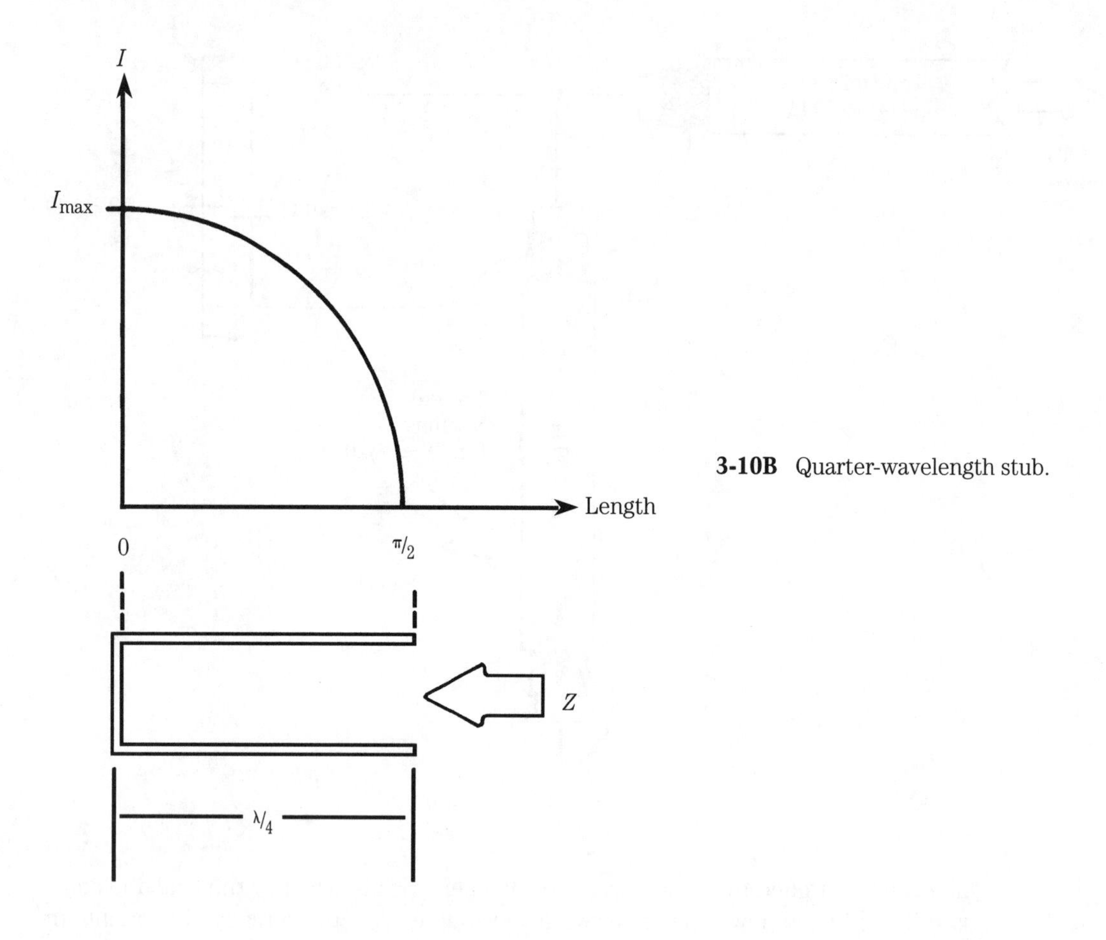

3-10B Quarter-wavelength stub.

than zero. The *standing wave ratio* (SWR) reveals the relationship between load and line.

If the current along the line is measured, the pattern will resemble the patterns of Fig. 3-11. The SWR is then called *ISWR*, to indicate the fact that it came from a current measurement. Similarly, if the SWR is derived from voltage measurements it is called *VSWR*. Perhaps because voltage is easier to measure, VSWR is the term most commonly used in microwave work.

VSWR can be specified in any of several equivalent ways:

1. *From incident voltage (V_i) and reflected voltage (V_i):*

$$VSWR = \frac{V_i + V_r}{V_i - V_r} \quad \textbf{[3.46]}$$

2. *From transmission line voltage measurements (Fig. 3-11D):*

$$VSWR = \frac{V_{max}}{V_{min}} \quad \textbf{[3.47]}$$

3. *From load and line characteristic impedances:*

$$(Z_L > Z_o)\ VSWR = Z_L/Z_o \quad \textbf{[3.48]}$$

$$(Z_L < Z_o)\ VSWR = Z_o/Z_L \quad \textbf{[3.49]}$$

4. *From incident (P_i) and reflected (P_r) power:*

$$VSWR = \frac{1 + \sqrt{P_r/P_i}}{1 - \sqrt{P_r/P_i}} \quad \textbf{[3.50]}$$

5. *From reflection coefficient (P):*

$$VSWR = \frac{1 + \Gamma}{1 - \Gamma} \quad \textbf{[3.51]}$$

It is also possible to determine the reflection coefficient (Γ) from a knowledge of VSWR:

$$P = \frac{VSWR - 1}{VSWR + 1} \quad \textbf{[3.52]}$$

The relationship between reflection coefficient (Γ) and VSWR is shown in Fig. 3-11D.

VSWR is usually expressed as a *ratio*. For example, when Z_L is 100 Ω and Z_o is 50 Ω, the VSWR is Z_L/Z_o = 100 Ω/50 Ω = 2, which is usually expressed as "*VSWR* = 2:1." VSWR can also be expressed in decibel form:

$$VSWR = 20\ \text{LOG (VSWR)} \quad \textbf{[3.53]}$$

Example 3-7 A transmission line is connected to a mismatched load. Calculate both the VSWR and VSWR decibel equivalent if the reflection coefficient (Γ) is 0.25

Solution:

a) $$VSWR = \frac{1 + \Gamma}{1 - \Gamma}$$

$$VSWR = \frac{1 + 0.25}{1 - 0.25}$$

$$VSWR = \frac{1.25}{0.75} = 1.67{:}1$$

b) $$VSWR_{\text{dB}} = 20\ \text{LOG VSWR}$$

$$VSWR_{\text{dB}} = (20)\ (\text{LOG } 1.67)$$

$$VSWR_{\text{dB}} = (20)\ (0.22) = 4.3\ \text{dB}$$

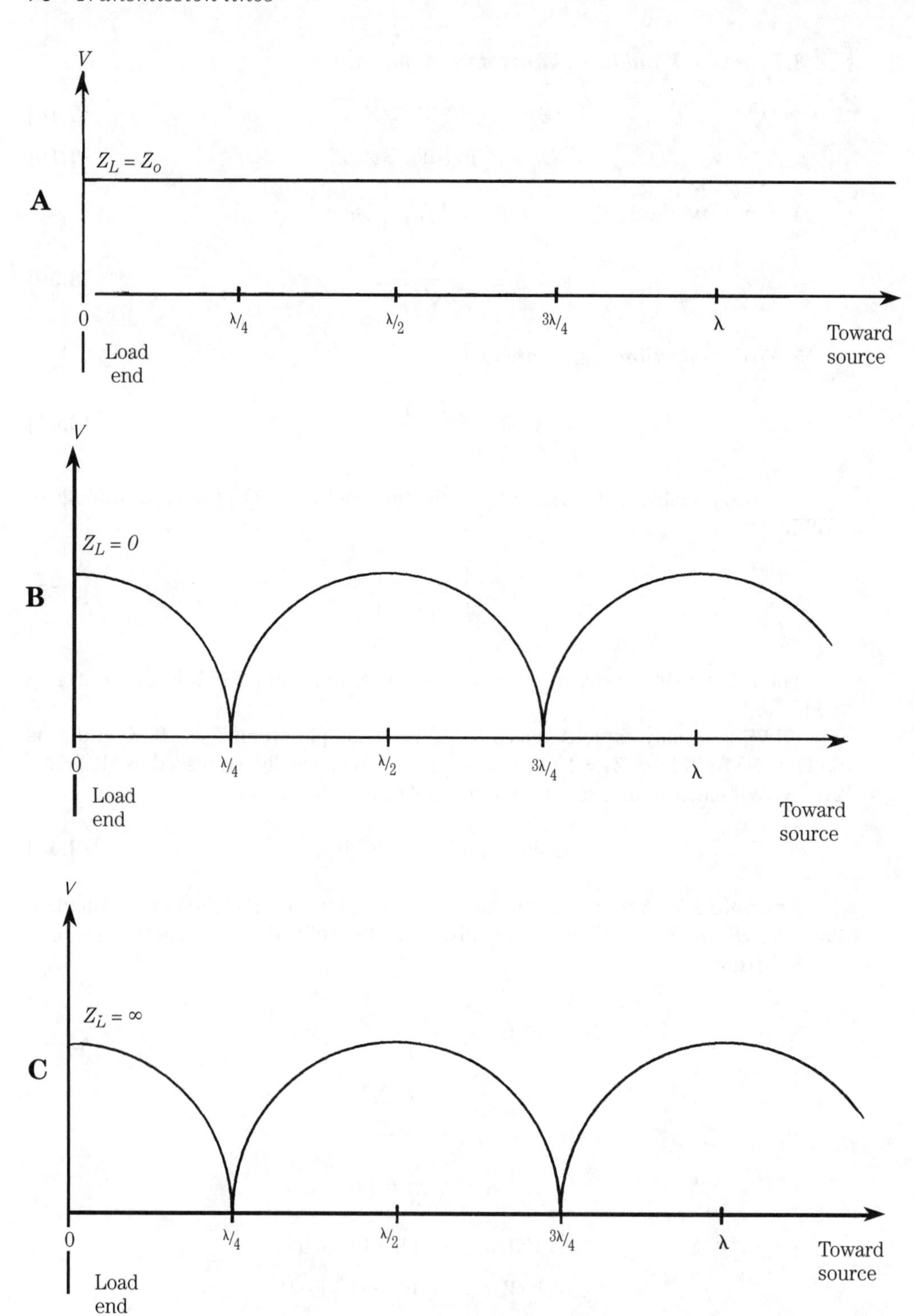

3-11 Voltage vs. electrical length A) Matched impedances, B) $Z_L = o$, C) Z_L = infinite.

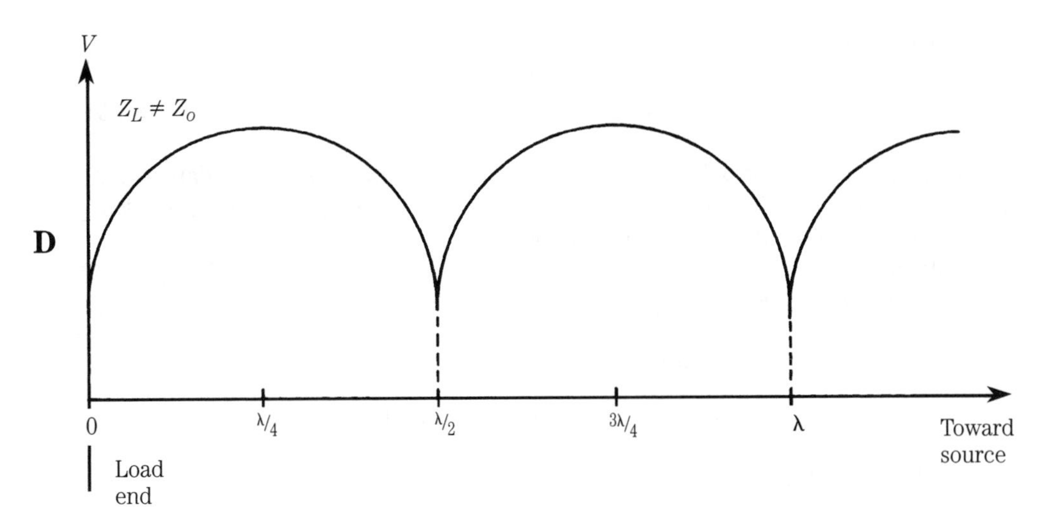

3-11 Continued. D) Z_L not equal to Z_o

The SWR is regarded as important in systems for several reasons. The base of these reasons is the fact that the reflected wave represents energy lost to the load. For example, in an antenna system, less power is radiated if some of its input power is reflected back down the transmission line; because the antenna feedpoint impedance does not match the transmission line characteristic impedance. The next section covers the problem of mismatch losses.

Mismatch (VSWR) losses

The power reflected from a mismatched load represents a loss, and will have implications that range from negligible to profound, depending on the situation. For example, one result might be a slight loss of signal strength at a distant point from an antenna. A more serious problem can result in the destruction of the output device in a transmitter. The latter problem so plagued early solid-state transmitters, that designers opted to include shut-down circuitry to sense high VSWR, and turn down output power proportionally.

In microwave measurements, VSWR on the transmission lines (that interconnect devices under test, instruments and signal sources) can cause erroneous readings—and invalid measurements.

Determination of VSWR losses must take into account *two* VSWR situations. Figure 3-9 shows a transmission line of impedance Z_o interconnecting a load impedance (Z_L), and a source with an output impedance, Z_s. There is a potential for impedance mismatch at *both ends* of the line.

In the case where one end of the line is matched (either Z_s or Z_L), the *mismatch loss* caused by SWR at the mismatched end is:

$$ML = -10 \text{ LOG} \left[1 - \left[\frac{SWR - 1}{SWR + 1} \right]^2 \right] \qquad \textbf{[3.54]}$$

Which from Eq. 3.52 is:

$$ML = -10 \text{ LOG } (1 - \Gamma^2) \qquad \textbf{[3.55]}$$

Example 3-8 A coaxial transmission line with a characteristic impedance of 50 Ω is connected to the 50-Ω output (Z_o) of a signal generator, and also to a 20-W load impedance (Z_L). Calculate the mismatch loss.

Solution:

a) First find the VSWR:

$$VSWR = Z_o/Z_L$$

$$VSWR = (50\ \Omega)/(20\ \Omega) = 2.5{:}1$$

b) Mismatch loss:

$$ML = -10 \text{ LOG}\left[1 - \left[\frac{SWR - 1}{SWR + 1}\right]^2\right]$$

$$ML = -10 \text{ LOG}\left[1 - \left[\frac{2.5 - 1}{2.5 + 1}\right]^2\right]$$

$$ML = -10 \text{ LOG}\left[1 - \left[\frac{1.5}{3.5}\right]^2\right]$$

$$ML = -10 \text{ LOG } [1 - (0.43)^2]$$

$$ML = -10 \text{ LOG } [1 - 0.185]$$

$$ML = -10 \text{ LOG } [0.815]$$

$$ML = (-10)\ (-0.089) = 0.89$$

When both ends of the line are mismatched a different equation is required:

$$ML = 20 \text{ LOG } [1 \pm (\Gamma_1 \times \Gamma_2)] \qquad \textbf{[3.56]}$$

Where:

Γ_1 is the reflection coefficient at the source end of the line $(VSWR_1 - 1)/(VSWR_1 + 1)$

Γ_2 is the reflection coefficient at the load end of the line, $(VSWR_2 - 1)/(VSWR_2 + 1)$

Note that the solution to Eq. 3.56 has two values: $[1+(\Gamma_1\Gamma_2)]$ and $[1-(\Gamma_1\Gamma_2)]$

The equations reflect the mismatch loss solution for low-loss or "lossless" transmission lines. This is a close approximation, in some situations, however it is insufficient when the line is lossy. Although not very important at low frequencies, loss becomes higher at microwave frequencies. Interference between incident and reflected waves produces increased current at certain antinodes—which increases ohmic losses—and increased voltage at certain antinodes—which increases dielectric losses. It is the latter that increases with frequency. Equation 3.57 relates reflection coefficient and line losses to determine total loss on a given line.

$$Loss = 10 \text{ LOG}\left[\frac{n^2 - \Gamma^2}{n - n\Gamma^2}\right] \quad \textbf{[3.57]}$$

Where:

Loss is the total line loss in decibels
Γ is the reflection coefficient
n is the quantity $10^{(A/10)}$
A is the total attenuation presented by the line, in dB, when the line is properly matched ($Z_L = Z_o$)

Example 3-9 A 50-Ω transmission line is terminated in a 30-Ω resistive impedance. The line is rated at a loss of 3 dB/100 ft at 1 GHz. Calculate a) loss in 5 ft of line, b) reflection coefficient, and c) total loss in a 5 foot line mismatched per above.

Solution:

$$\text{a) } A = \frac{3 \text{ dB}}{100 \text{ ft}} \times 5 \text{ ft} = 0.15 \text{ dB}$$

$$\text{b) } \Gamma = \frac{Z_L - Z_o}{Z_L + Z_o}$$

$$\Gamma = \frac{50 - 30}{50 + 30}$$

$$\Gamma = 20/80 = 0.25$$

$$\text{c) } n = 10^{(A/10)}$$

$$n = 10^{(0.15/10)}$$

$$n = 10^{(0.015)} = 1.04$$

$$Loss = 10 \text{ LOG}\left(\frac{n^2 - \Gamma^2}{n - n\Gamma^2}\right)$$

$$Loss = 10 \text{ LOG}\left(\frac{(1.04)^2 - (0.25)^2}{(1.04) - ((1.04)\ (0.25)^2)}\right)$$

$$Loss = 10 \text{ LOG}\left(\frac{1.082 - 0.063}{1.04 - ((1.04)\ (0.063))}\right)$$

$$Loss = 10 \text{ LOG}\left(\frac{1.019}{1.04 - 0.066}\right)$$

$$Loss = 10 \text{ LOG} \left(\frac{1.019}{0.974} \right)$$

$$Loss = 10 \text{ LOG } (1.046)$$

$$Loss = (10)\ (0.02) = 0.2 \text{ dB}$$

Compare the matched line loss ($A = 0.15$ dB) with the total loss (Loss = 0.2 dB), which includes mismatch loss and line loss. The difference (i.e., Loss – A) is only 0.05 dB. If the VSWR was considerably larger, however, the loss would rise.

4
CHAPTER
The Smith chart

THE MATHEMATICS OF TRANSMISSION LINES, AND CERTAIN OTHER DEVICES, BECOMES cumbersome at times, especially when dealing with complex impedances and "nonstandard" situations. In 1939, Phillip H. Smith published a graphical device for solving these problems; followed in 1945 by an improved version of the chart. That graphic aid, somewhat modified over time, is still in constant use in microwave electronics, and other fields where complex impedances and transmission line problems are found. The *Smith chart* is indeed a powerful tool for the RF designer.

Smith chart components

The modern Smith chart is shown in Fig. 4-1, and consists of a series of overlapping orthogonal circles (i.e., circles that intersect each other at right angles). This chapter will dissect the Smith chart, so that the origin and use of these circles is apparent. The set of orthogonal circles makes up the basic structure of the Smith chart.

The normalized impedance line

A baseline is highlighted in Fig. 4-2, and bisects the Smith chart outer circle. This line is called the *pure resistance line*, and forms the reference for measurements made on the chart. Recall that a complex impedance contains both resistance and reactance, and is expressed in the mathematical form:

$$Z = R \pm jX \qquad \textbf{[4.1]}$$

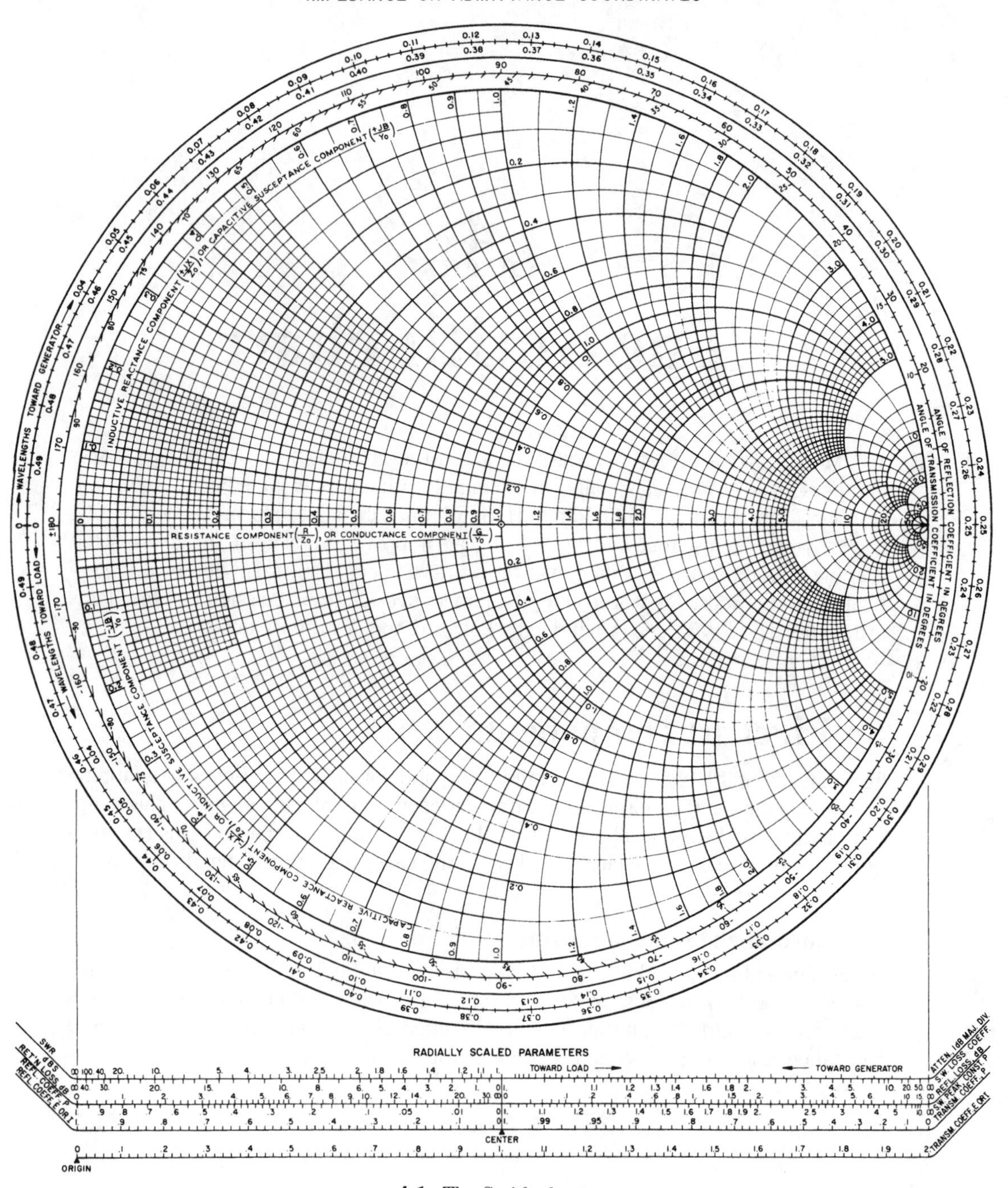

4-1 The Smith chart.

Where:

Z is the complex impedance
R is the resistive component of the impedance
X is the reactive component of the impedance*

The pure resistance line represents the situation where $X = 0$, and the impedance is therefore equal to the resistive component only. In order to make the Smith chart universal, the impedances along the pure resistance line are *normalized* with reference to system impedance (e.g., Z_o in transmission lines); for most microwave RF systems the system impedance is standardized at 50 Ω. In order to normalize the actual impedance, divide it by the system impedance. For example, if the load impedance of a transmission line is Z_L, and the characteristic impedance of the line is Z_o, then $Z = Z_L/Z_o$. In other words:

$$Z = \frac{R \pm jX}{Z_o} \quad \textbf{[4.2]}$$

The pure resistance line is structured such that the system standard impedance is in the center of the chart, and has a normalized value of 1.0 (see point "A" in Fig. 4-2). This value derives from the fact that $Z_o/Z_o = 1.0$.

To the left of the 1.0 point are decimal fraction values used to denote impedances less than the system impedance. For example, in a 50-Ω transmission line system with a 25-Ω load impedance, the normalized value of impedance is 25 Ω/50 Ω or 0.50 ("B" in Fig. 4-2). Similarly, points to the right of 1.0 are greater than 1 and denote impedances that are higher than the system impedance. For example, in a 50-Ω system connected to a 100-Ω resistive load, the normalized impedance is 100 Ω/50 Ω, or 2.0; this value is shown as point "C" in Fig. 4-2. By using normalized impedances, you can use the Smith chart for almost any practical combination of system, and load and/or source, impedances, whether resistive, reactive, or complex.

Reconversion of the normalized impedance to actual impedance values is done by multiplying the normalized impedance by the system impedance. For example, if the resistive component of a normalized impedance is 0.45, then the actual impedance is:

$$Z = (Z_{\text{normal}})\,(Z_o) \quad \textbf{[4.3]}$$

$$Z = (0.45)\,(50\ \Omega) \quad \textbf{[4.4]}$$

$$Z = 22.5\ \Omega \quad \textbf{[4.5]}$$

The constant resistance circles

The *isoresistance circles*, also called the *constant resistance circles*, represent points of equal resistance. Several of these circles are shown highlighted in Fig. 4-3. These circles are all tangent to the point at the righthand extreme of the pure resis-

*According to the standard sign convention the inductive reactance (X_L) is positive (+) and the capacitive reactance (X_c) is negative (–). The term X, in Eq. 4.1 above, is the difference between the two reactances ($X = X_L - X_c$).

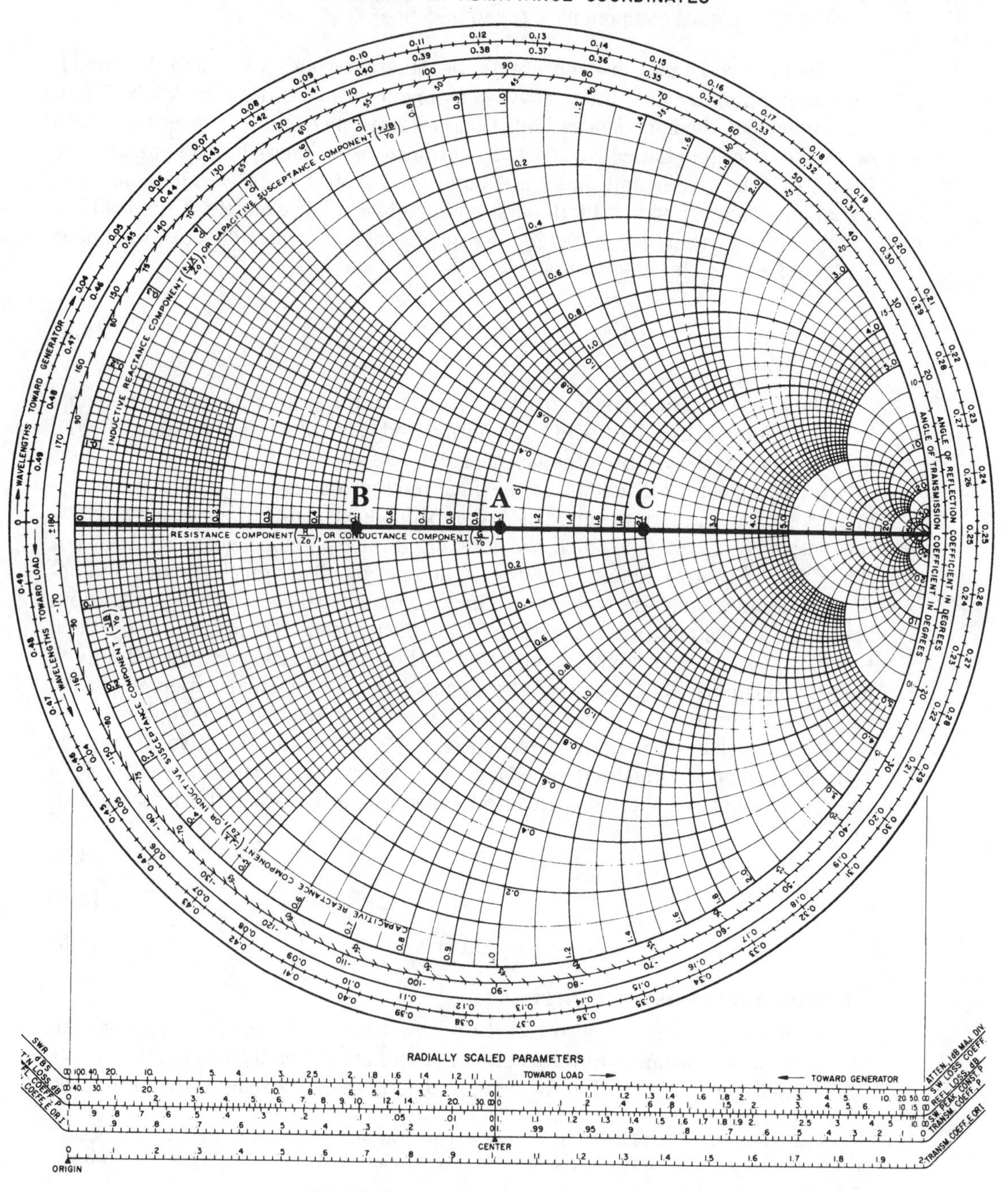

4-2 Normalized impedance line. Kay Elementrics

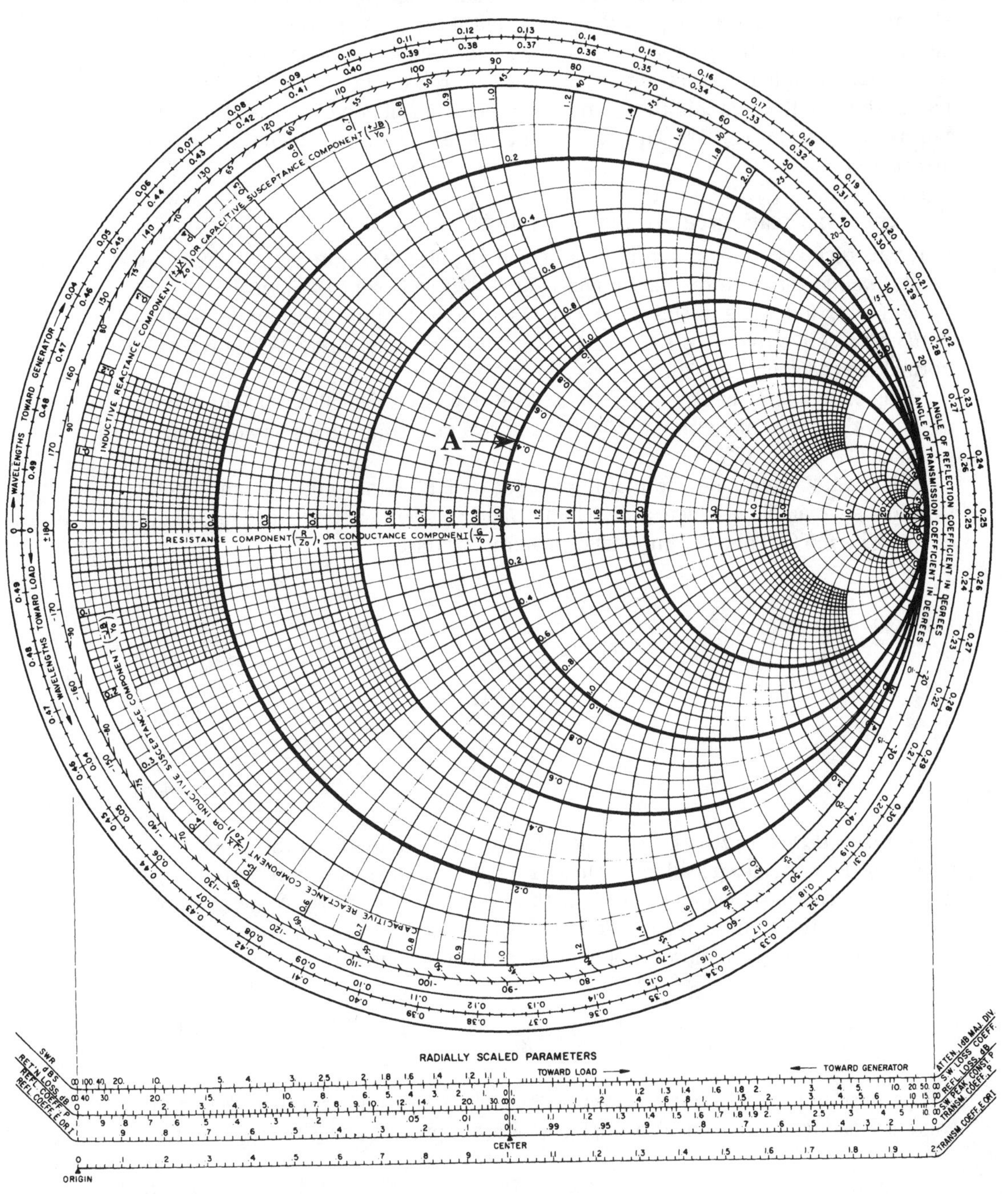

4-3 Constant resistance circles. Kay Elementrics

tance line, and are bisected by that line. When you construct complex impedances (for which X = nonzero) on the Smith chart, the points on these circles will all have the same resistive component. Circle "A," for example, passes through the center of the chart, so it has a normalized constant resistance of 1.0. Note that impedances that are pure resistances (i.e., $Z = R + j0$) will fall at the intersection of a constant resistance circle and the pure resistance line, and complex impedances (i.e., X not equal to zero) will appear at any other points on the circle. In Fig. 4-2, circle "A" passes through the center of the chart, so it represents all points on the chart with a normalized resistance of 1.0. This particular circle is sometimes called the *unity resistance circle.*

The constant reactance circles

Constant reactance circles are highlighted in Fig. 4-4. The circles (or circle segments) *above* the pure resistance line (Fig. 4-4A) represent the *inductive reactance* ($+X$), and those circles (or segments) *below* the pure resistance line (Fig. 4-4B) represent *capacitive reactance* ($-X$). In both cases, circle "A" represents a normalized reactance of 0.80.

One of the outer circles (i.e., circle "A" in Fig. 4-4C) is called the *pure reactance circle*. Points along circle "A" represent reactance only; in other words, an impedance of $Z = 0 \pm jX$ ($R = 0$).

Figure 4-4D shows how to plot impedance and admittance on the Smith chart. Consider an example in which system impedance Z_o is 50 Ω, and the load impedance is $Z_L = 95 + j55$ Ω. This load impedance is normalized to:

$$Z = \frac{Z_L}{Z_o} \quad [4.6]$$

$$Z = \frac{95 + j55\ \Omega}{50\ \Omega} \quad [4.7]$$

$$Z = 1.9 + j1.1 \quad [4.8]$$

An *impedance radius* is constructed by drawing a line from the point represented by the normalized load impedance, $1.9 + j1.1$, to the point represented by the normalized system impedance (1.0) in the center of the chart. A circle is constructed from this radius, and is called the *VSWR circle*

Admittance is the reciprocal of impedance, so it is found from:

$$Y = \frac{1}{Z} \quad [4.9]$$

Because impedances in transmission lines are rarely pure resistive, but rather contain a reactive component also, impedances are expressed using complex notation:

$$Z = R \pm jX \quad [4.10]$$

NAME	TITLE	DWG. NO.
SMITH CHART FORM 82-BSPR (9-66)	KAY ELECTRIC COMPANY, PINE BROOK, N.J., ©1966 PRINTED IN U.S.A.	DATE

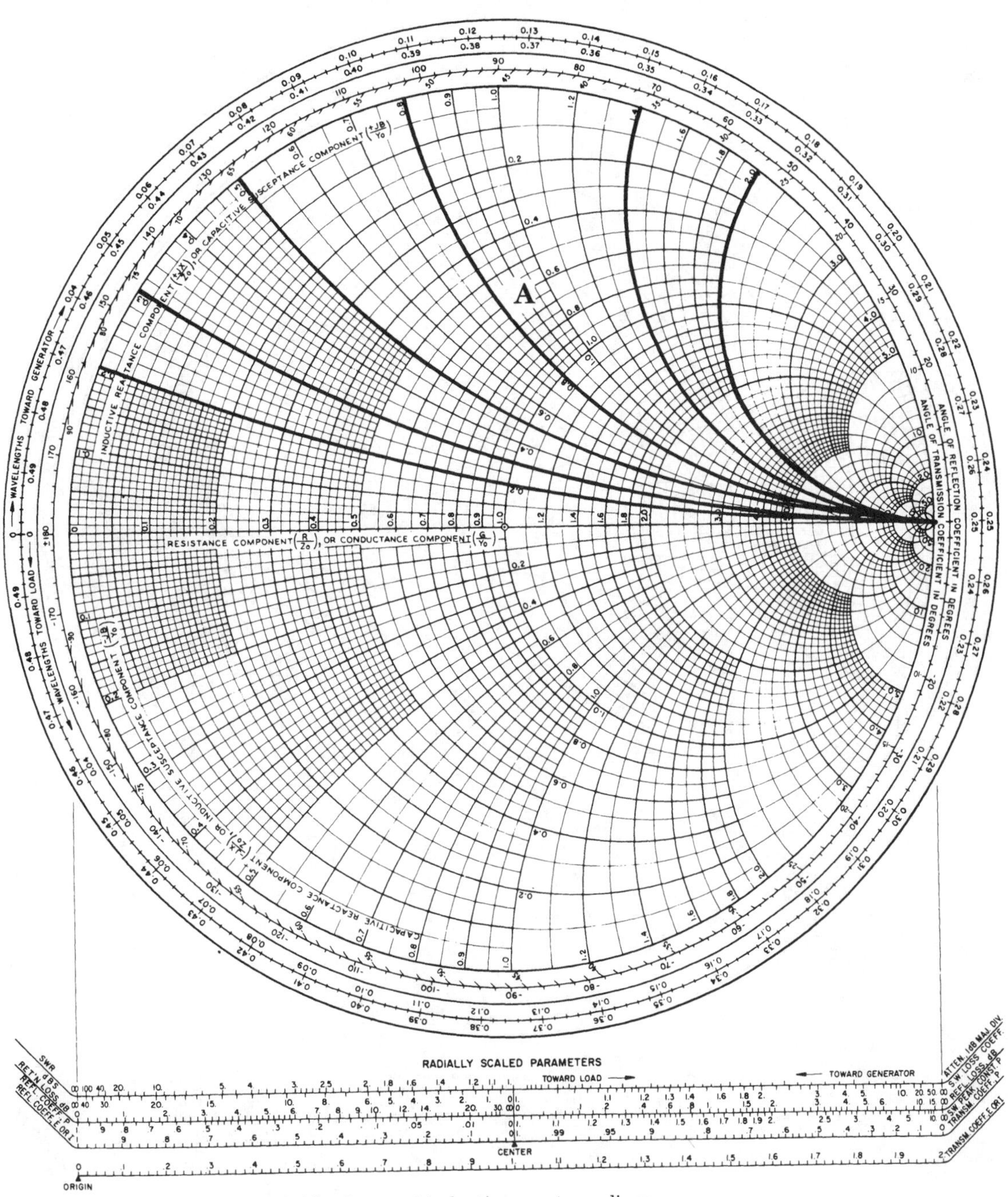

4-4A Constant inductive reactance lines. Kay Elementrics

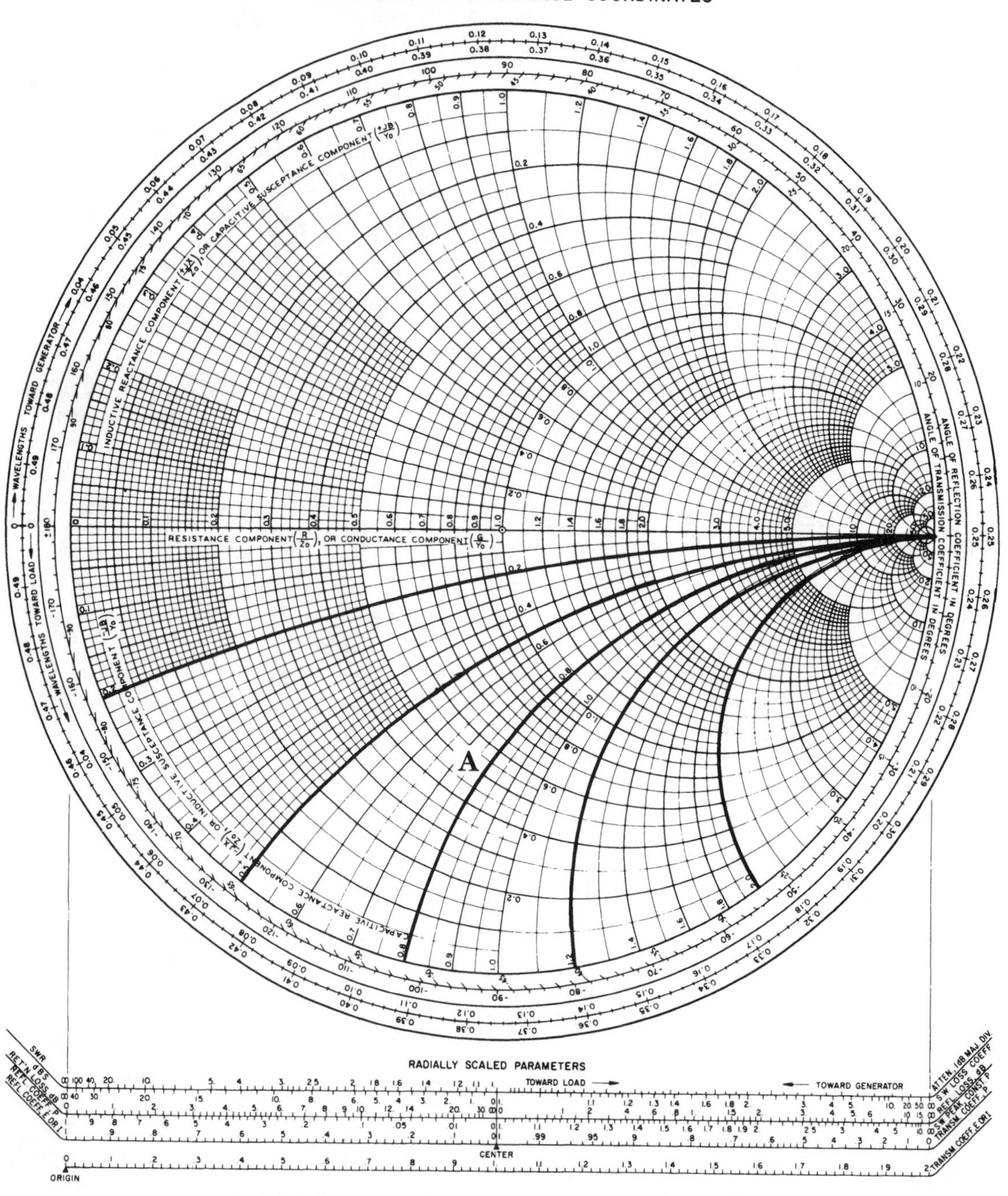

4-4B Constant capacitive reactance lines. Kay Elementrics

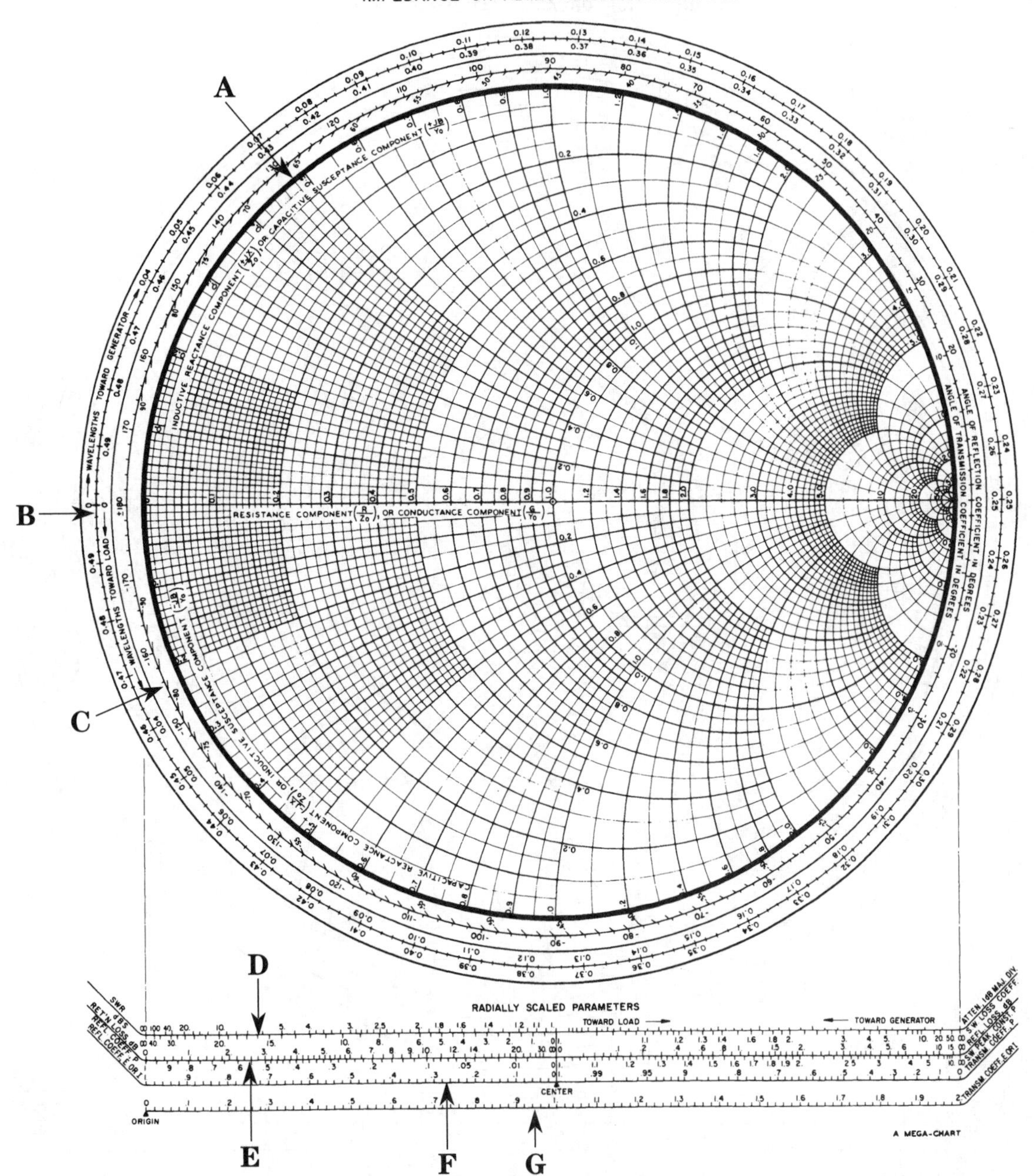

4-4C Angle of transmission coefficient circle. Kay Elementrics

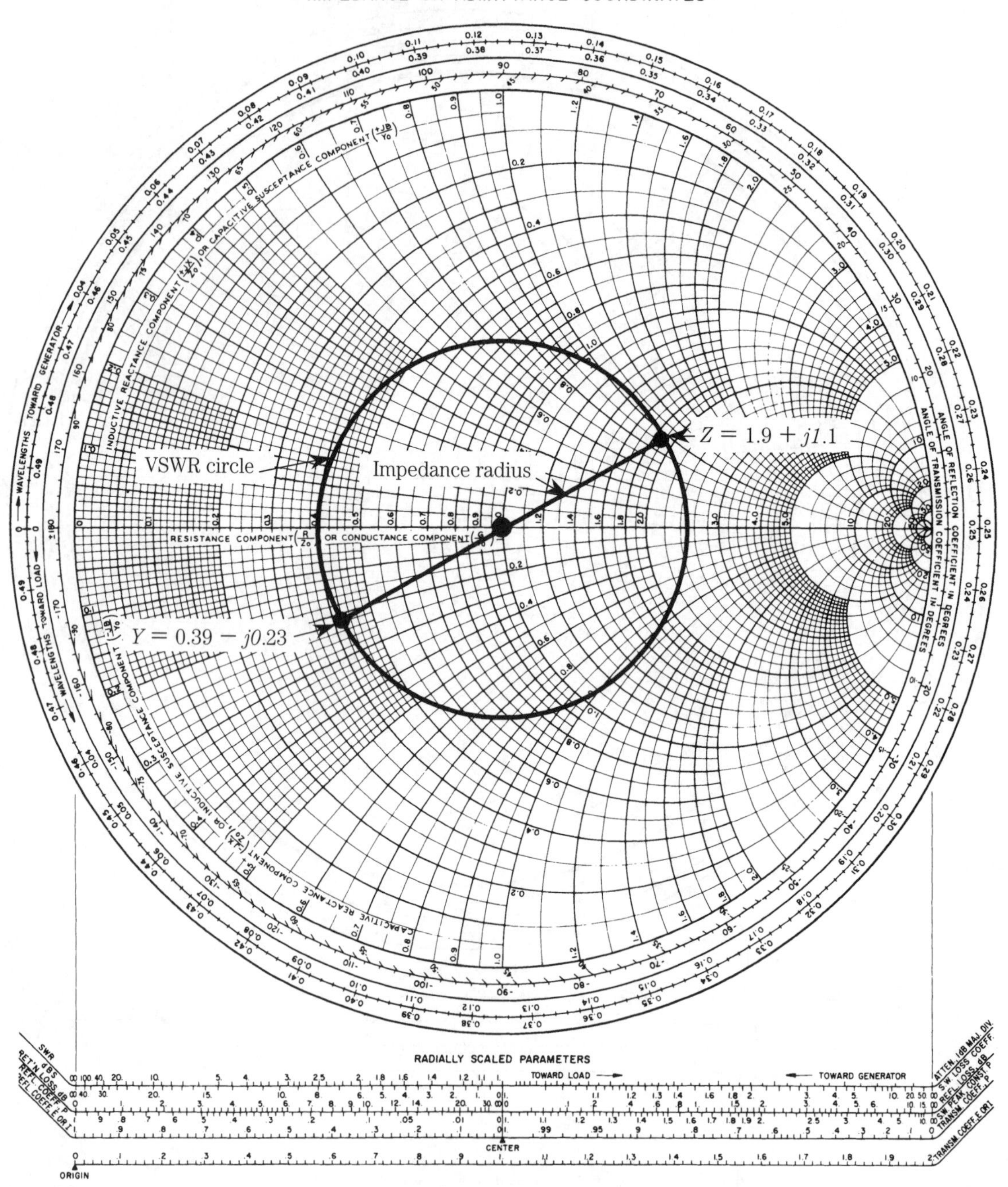

4-4D VSWR circles. Kay Elementrics

Where:

Z is the complex impedance
R is the resistive component
X is the reactive component

In order to find the *complex admittance*, take the reciprocal of the complex impedance by multiplying the simple reciprocal by the complex conjugate of the impedance. For example, when the normalized impedance is $1.9 + j1.1$, the normalized admittance will be:

$$Y = \frac{1}{Z} \quad [4.11]$$

$$Y = \frac{1}{1.9 + j1.1} \times \frac{1.9 - j1.1}{1.9 - j1.1} \quad [4.12]$$

$$Y = \frac{1.9 - j1.1}{3.6 + 1.2} \quad [4.13]$$

$$Y = \frac{1.9 - j1.1}{4.8} = 0.39 - j0.23 \quad [4.14]$$

One of the delights of the Smith chart is that this calculation is reduced to a quick graphical interpretation! Simple extend the impedance radius through the 1.0 center point until it intersects the VSWR circle again. This point of intersection represents the normalized admittance of the load.

Outer circle parameters

The standard Smith chart shown in Fig. 4-4C contains three concentric calibrated circles on the outer perimeter of the chart. Circle "A" has already been covered and it is the pure reactance circle. The other two circles define the wavelength distance ("B") relative to either the load or generator end of the transmission line; and either the transmission, or reflection, coefficient angle in degrees ("C").

There are two scales on the *wavelengths* circle ("B" in Fig 4-4C), and both have their zero origin on the left-hand extreme of the pure resistance line. Both scales represent *one-half wavelength for one entire revolution*, and are calibrated from 0 through 0.50 such that these two points are identical with each other on the circle. In other words, starting at the zero point and travelling 360 degrees around the circle brings one back to zero, which represents one-half wavelength, or 0.5 λ.

Although both wavelength scales are of the same magnitude (0 – 0.50), they are opposite in direction. The outer scale is calibrated clockwise and it represents *wavelengths toward the generator*; the inner scale is calibrated counter-clockwise and represents *wavelengths toward the load*. These two scales are complementary at all points. Thus, 0.12 on the outer scale corresponds to (0.50 – 0.12) or 0.38 on the inner scale.

The *angle of transmission coefficient* and *angle of reflection coefficient* scales are shown in circle "C" in Fig. 4-4C. These scales are the relative phase angle between reflected and incident waves. Recall from transmission line theory (see chapter 4), that a short circuit at the load end of the line reflects the signal back toward the generator 180 degrees out of phase with the incident signal; an open line (i.e., infinite impedance) reflects the signal back to the generator in-phase (i.e., 0 degrees) with the incident signal. These facts are shown on the Smith chart by the fact that both scales start at 0 degrees on the right-hand end of the pure resistance line, which corresponds to an infinite resistance, and it goes half-way around the circle to 180 degrees at the 0 end of the pure resistance line. Note that the upper half-circle is calibrated 0 to +180 degrees, and the bottom half-circle is calibrated 0 to −180 degrees, reflecting indictive or capacitive reactance situations, respectively.

Radially scaled parameters

There are six scales laid out on five lines ("D" through "G" in Fig. 4-4C and in expanded form in Fig. 4-5) at the bottom of the Smith chart. These scales are called the *radially scaled parameters*—and they are both very important, and often overlooked. With these scales, we can determine such factors as VSWR (both as a ratio and in decibels), return loss in decibels, voltage or current reflection coefficient, and the power reflection coefficient.

The *reflection coefficient* (Γ) is defined as the *ratio of the reflected signal to the incident signal*. For voltage or current:

$$\Gamma = \frac{E_{ref}}{E_{inc}} \qquad \textbf{[4.15]}$$

and

$$\Gamma = \frac{I_{ref}}{I_{inc}} \qquad \textbf{[4.16]}$$

Power is proportional to the square of voltage or current, so:

$$P_{pwr} = \Gamma^2 \qquad \textbf{[4.17]}$$

or,

$$\Gamma_{pwr} = \frac{P_{ref}}{P_{inc}} \qquad \textbf{[4.18]}$$

Example 10 watts of microwave RF power is applied to a lossless transmission line, of which 2.8 watts is reflected from the mismatched load. Calculate the reflection coefficient.

$$\Gamma_{pwr} = \frac{P_{ref}}{P_{inc}} \qquad \textbf{[4.19]}$$

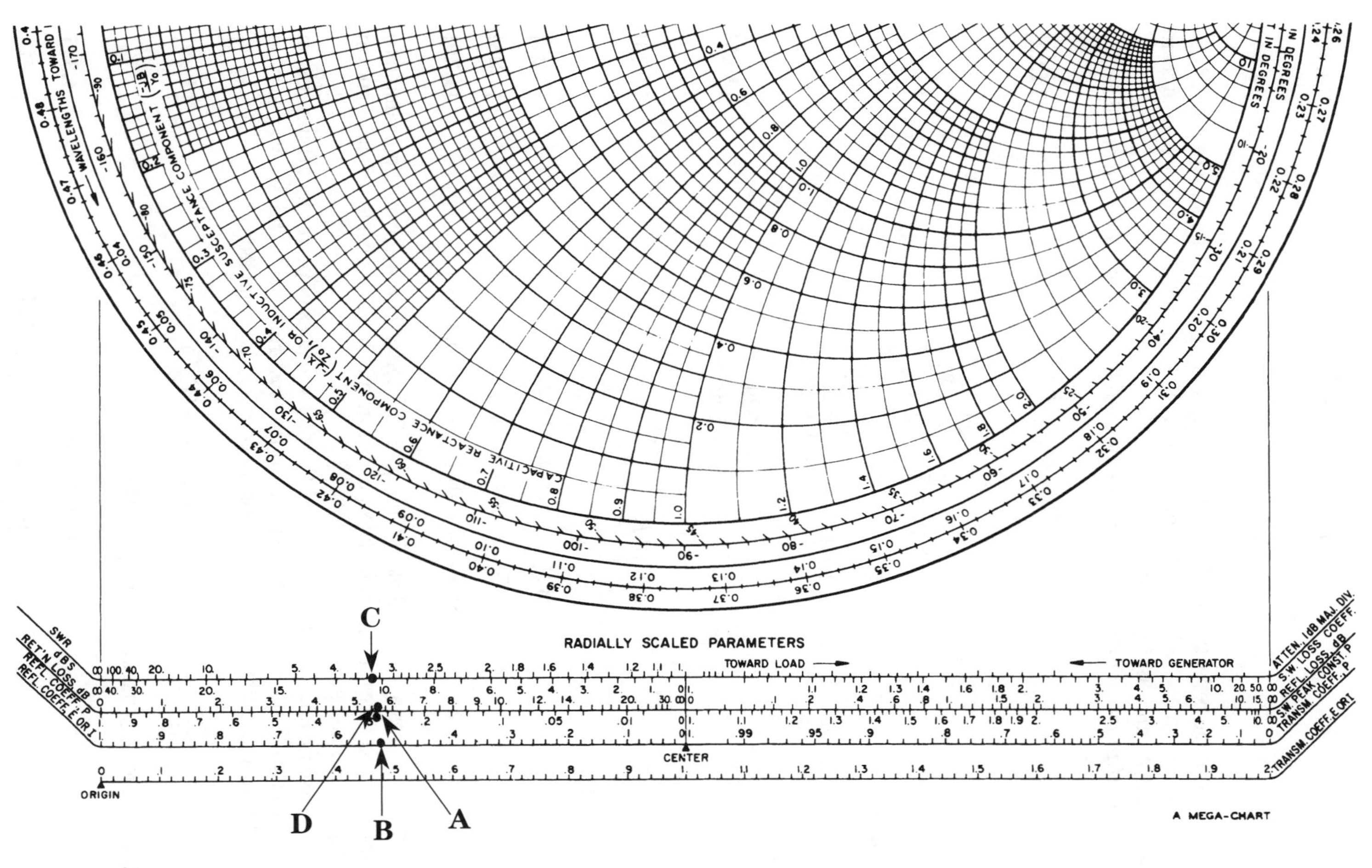

4-5 Radially scaled parameters.

$$\Gamma_{pwr} = \frac{2.8 \text{ watts}}{10 \text{ watts}} \quad [4.20]$$

$$\Gamma_{pwr} = 0.28 \quad [4.21]$$

The voltage reflection coefficient (Γ) is found by taking the square root of the power reflection coefficient, so in this example it is equal to 0.529. These points are plotted at "A" and "B" in Fig. 4-5.

Standing wave ratio (SWR) can be defined in terms of reflection coefficient:

$$VSWR = \frac{1 + \Gamma}{1 - \Gamma} \quad [4.22]$$

or,

$$VSWR = \frac{1 + \sqrt{\Gamma_{pwr}}}{1 - \sqrt{\Gamma_{pwr}}} \quad [4.23]$$

or, in our example:

$$VSWR = \frac{1 + \sqrt{0.28}}{1 - \sqrt{0.28}} \quad [4.24]$$

$$VSWR = \frac{1 + 0.529}{1 - 0.529} \quad [4.25]$$

$$VSWR = \frac{1.529}{0.471} = 3.25{:}1 \quad [4.26]$$

or, in decibel form:

$$VSWR_{dB} = 20 \text{ LOG (VSWR)} \quad [4.27]$$

$$VSWR_{dB} = 20 \text{ LOG } (20) \quad [4.28]$$

$$VSWR_{dB} = (20)\ (0.510) = 10.2 \text{ dB} \quad [4.29]$$

These points are plotted at "C" in Fig. 4-5. Shortly, you will work an example to show how these factors are calculated in a transmission line problem from a known complex load impedance.

Transmission loss is a measure of the one-way loss of power in a transmission line because of reflection from the load. *Return loss* represents the two-way loss, so it is exactly twice the transmission loss. Return loss is found from:

$$Loss_{ret} = 10 \text{ LOG } (\Gamma_{pwr}) \quad [4.30]$$

and, for our example in which $\Gamma_{pwr} = 0.28$:

$$Loss_{ret} = 10 \text{ LOG } (0.28) \quad [4.31]$$

$$Loss_{ret} = (10)\ (-0.553) = -5.53 \text{ dB} \quad [4.32]$$

This point is shown as "D" in Fig. 4-5.

The *transmission loss coefficient* can be calculated from:

$$TLC = \frac{1 + \Gamma_{pwr}}{1 - \Gamma_{pwr}} \quad \textbf{[4.33]}$$

or, for our example:

$$TLC = \frac{1 + (0.28)}{1 - (0.28)} \quad \textbf{[4.34]}$$

$$TLC = \frac{1.28}{0.72} = 1.78 \quad \textbf{[4.35]}$$

The TLC is a correction factor that is used to calculate the attenuation caused by mismatched impedance in a lossy, as opposed to the ideal "lossless," line. The TLC is found from laying out the impedance radius on the *Loss Coefficient* scale on the radially scaled parameters at the bottom of the chart.

Smith chart applications

One of the best ways to demonstrate the usefulness of the Smith chart is by practical example. The following sections look at two general cases: *transmission line problems* and *stub matching systems*.

Transmission line problems

Figure 4-6 shows a 50-Ω transmission line connected to a complex load impedance Z_L of $36 + j40\ \Omega$. The transmission line has a velocity factor (v) of 0.80, which means that the wave propagates along the line at $\frac{8}{10}$ the speed of light (c = 300,000,000 m/s). The length of the transmission line is 28 centimeters. The generator (V_{in}) is operated at a frequency of 4.5 GHz and produces a power output of 1.5 watts. See what you can glean from the Smith chart (Fig. 4-7).

First, normalize the load impedance. This is done by dividing the load impedance by the systems impedance (in this case, $Z_o = 50\ \Omega$):

$$Z = \frac{36 + j40\ \Omega}{50\ \Omega} \quad \textbf{[4.36]}$$

$$Z = 0.72 + j0.8 \quad \textbf{[4.37]}$$

The resistive component of impedance Z is located along the "0.72" pure resistance circle (see Fig. 4-7). Similarly, the reactive component of impedance Z is located by traversing the 0.72 constant resistance circle until the $+j0.8$ constant reactance circle is intersected. This point graphically represents the normalized load impedance $Z = 0.72 + j0.80$. A VSWR circle is constructed with an impedance radius equal to the line between "1.0" (in the center of the chart) and the "$0.72 + j0.8$" point.

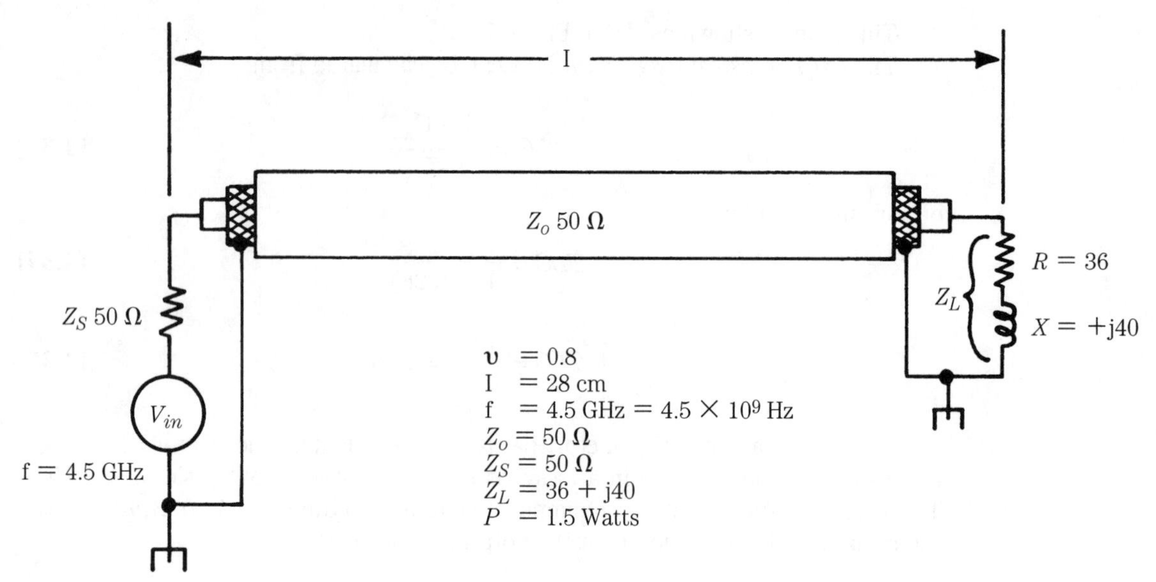

4-6 Transmission line and load circuit.

At a frequency of 4.5 GHz, the length of a wave propagating in the transmission line, assuming a velocity factor of 0.80, is:

$$\lambda_{\text{line}} = \frac{c\,v}{F_{\text{Hz}}} \qquad \textbf{[4.38]}$$

$$\lambda_{\text{line}} = \frac{(3 \times 10^8 \text{ m/s})\ (0.80)}{4.5 \times 10^9 \text{ Hz}} \qquad \textbf{[4.39]}$$

$$\lambda_{\text{line}} = \frac{2.4 \times 10^8 \text{ m/s}}{4.5 \times 10^9 \text{ Hz}} \qquad \textbf{[4.40]}$$

$$\lambda_{\text{line}} = 0.053 \text{ m} \times \frac{100 \text{ cm}}{m} = 5.3 \text{ cm} \qquad \textbf{[4.41]}$$

One wavelength is 5.3 cm, so a half wavelength is 5.3 cm/2, or 2.65 cm. The 28 cm line is 28 cm/5.3 cm, or 5.28 wavelengths long. A line drawn from the center (1.0) to the load impedance is extended to the outer circle, and it intersects the circle at 0.1325. Because one complete revolution around this circle represents one-half wavelength, 5.28 wavelengths from this point represents ten revolutions plus 0.28 more. The residual 0.28 wavelengths is added to 0.1325 to form a value of (0.1325 + 0.28), or 0.413. The point "0.413" is located on the circle, and is marked. A line is then drawn from 0.413 to the center of the circle, and it intersects the VSWR circle at $0.49 - j0.49$, which represents the input impedance (Z_{in}) looking into the line.

To find the actual impedance represented by the normalized input impedance, you have to "de-normalize" the Smith chart impedance by multiplying the result by Z_o:

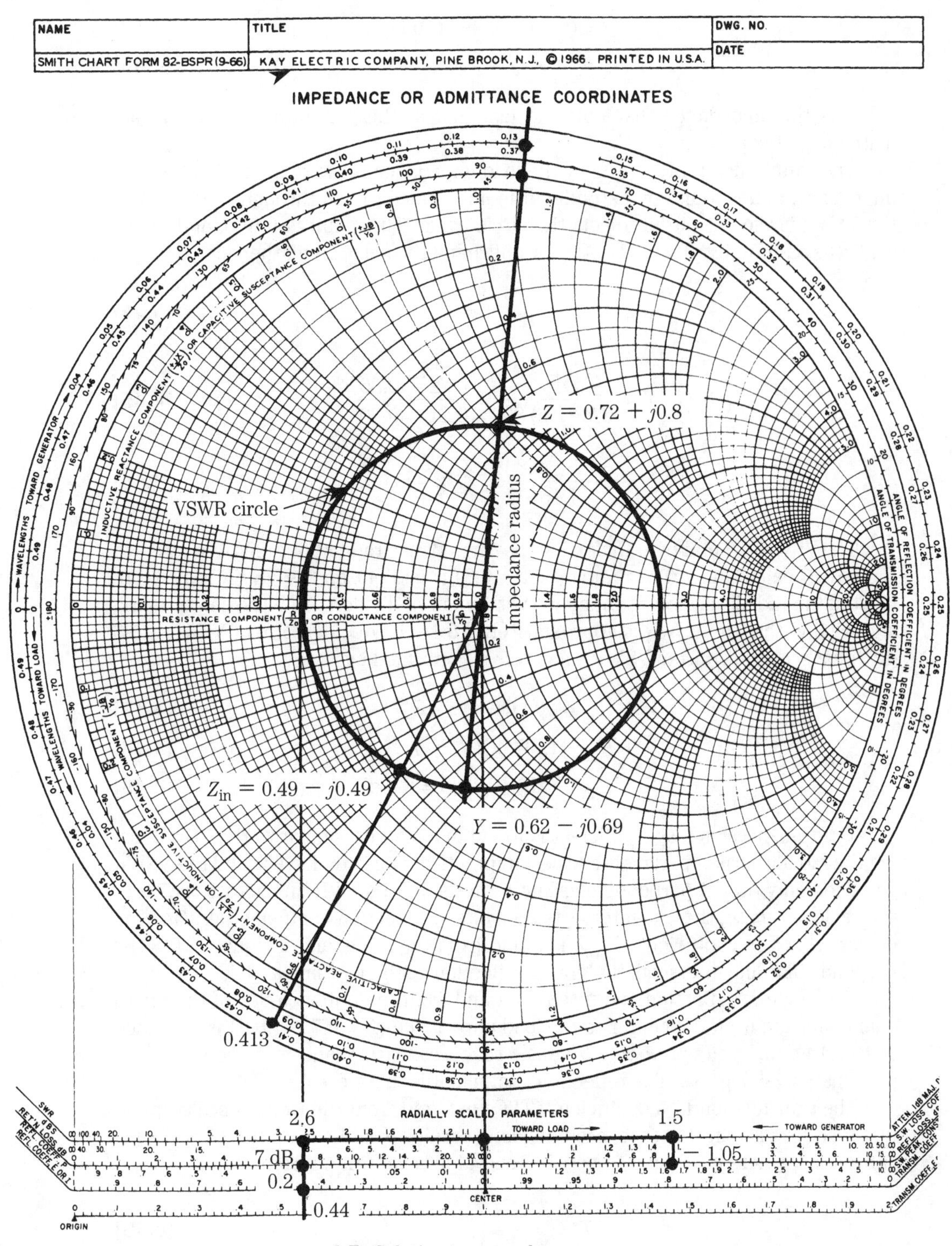

4-7 Solution to example. Kay Elementrics

$$Z_{in} = (0.49 - j0.49)\ (50\ \Omega) \quad \textbf{[4.42]}$$

$$Z_{in} = 24.5 - j24.5\ \Omega \quad \textbf{[4.43]}$$

It is this impedance that must be matched at the generator by a conjugate matching network.

The admittance represented by the load impedance is the reciprocal of the load impedance, and is found by extending the impedance radius through the center of the VSWR circle until it intersects the circle again. This point is found, and represents the admittance $Y = 0.62 - j0.69$. Confirming the solution mathematically:

$$Y = \frac{1}{Z} \quad \textbf{[4.44]}$$

$$Y = \frac{1}{0.72 + j0.80} \times \frac{0.72 - j0.80}{0.72 - j0.80} \quad \textbf{[4.45]}$$

$$Y = \frac{0.72 - j0.80}{1.16} = 0.62 - j0.69 \quad \textbf{[4.46]}$$

The VSWR if found by transferring the "impedance radius" of the VSWR circle to the radial scales below. The radius (0.72 – 0.8) is laid out on the VSWR scale (topmost of the radially scaled parameters) with a pair of dividers from the center mark, and we find that the VSWR is approximately 2.6:1. The decibel form of VSWR is 8.3 dB (next scale down from VSWR), and this is confirmed by:

$$VSWR_{dB} = 20\ \text{LOG}\ (\text{VSWR}) \quad \textbf{[4.47]}$$

$$VSWR_{dB} = (20)\ \text{LOG}\ (2.7) \quad \textbf{[4.48]}$$

$$VSWR_{dB} = (20)\ (0.431) = 8.3\ \text{dB} \quad \textbf{[4.49]}$$

The transmission loss coefficient is found in a manner similar to the VSWR, using the radially scaled parameter scales. In practice, once you have found the VSWR you need only drop a perpendicular line from the 2.6:1 VSWR line across the other scales. In this case, the line intersects the voltage reflection coefficient at 0.44. The power reflection coefficient (Γ_{pwr}) is found from the scale, and is equal to Γ^2. The perpendicular line intersects the power reflection coefficient line at 0.20.

The angle of reflection coefficient is found from the outer circles of the Smith chart. The line from the center to the load impedance ($Z = 0.72 + j0.8$) is extended to the *Angle of Reflection Coefficient in Degrees* circle, and intersects it at approximately 84 degrees. The reflection coefficient is therefore 0.44/84°.

The transmission loss coefficient (TLC) is found from the radially scaled parameter scales also. In this case, the impedance radius is laid out on the *Loss Coefficient* scale, where it is found to be 1.5. This value is confirmed from:

$$TLC = \frac{1 + \Gamma_{pwr}}{1 - \Gamma_{pwr}} \quad \textbf{[4.50]}$$

$$TLC = \frac{1 + (0.20)}{1 - (0.21)} \quad \textbf{[4.51]}$$

$$TLC = \frac{1.20}{0.79} = 1.5 \quad \textbf{[4.52]}$$

The *Return Loss* is also found by dropping the perpendicular from the VSWR point to the *RET'N LOSS, dB* line, and the value is found to be approximately 7 dB, which is confirmed by:

$$Loss_{ret} = 10 \text{ LOG } (\Gamma_{pwr}) \text{ dB} \quad \textbf{[4.53]}$$

$$Loss_{ret} = 10 \text{ LOG } (0.21) \text{ dB} \quad \textbf{[4.54]}$$

$$Loss_{ret} = (10)\ (-0.677) \text{ dB} \quad \textbf{[4.55]}$$

$$Loss_{ret} = 6.77 \text{ dB} = -6.9897 \text{ dB} \quad \textbf{[4.56]}$$

The reflection loss is the amount of RF power reflected back down the transmission line from the load. The difference between incident power supplied by the generator (1.5 watts in this example), $P_{inc} - P_{ref} = P_{abs}$ and the reflected power, is the *absorbed power* (P_a); or in the case of an antenna, the radiated power. The reflection loss is found graphically by dropping a perpendicular from the TLC point (or by laying out the impedance radius on the REFL. Loss, dB scale), and in this example (Fig. 4-7) is –1.05 dB. You can check the calculations:

The return loss was –7 dB, so:

$$-7 \text{ dB} = 10 \text{ LOG} \left(\frac{P_{ref}}{P_{inc}} \right) \quad \textbf{[4.57]}$$

$$-7 = 10 \text{ LOG} \left(\frac{P_{ref}}{1.5 \text{ watts}} \right) \quad \textbf{[4.58]}$$

$$\frac{-7}{10} = \text{LOG} \left(\frac{P_{ref}}{1.5 \text{ watts}} \right) \quad \textbf{[4.59]}$$

$$10^{\left(\frac{-7}{10}\right)} = \frac{P_{ref}}{1.5 \text{ watts}} \quad \textbf{[4.60]}$$

$$0.2 = \frac{P_{ref}}{1.5 \text{ watts}} \quad \textbf{[4.61]}$$

$$(0.2)(1.5 \text{ watts}) = P_{ref} \quad \textbf{[4.62]}$$

$$0.3 \text{ watts} = P_{ref} \quad \textbf{[4.63]}$$

The power absorbed by the load (P_a) is the difference between incident power (P_{inc}) and reflected power (P_{ref}). If 0.3 watts is reflected, then that means the absorbed power is (1.5 – 0.3), or 1.2 watts.

The reflection loss is –1.05 dB, and can be checked from:

$$-1.05 \text{ dB} = 10 \text{ LOG}\left(\frac{P_a}{P_{inc}}\right) \quad \textbf{[4.64]}$$

$$\frac{-1.05}{10} = \text{LOG}\left(\frac{P_a}{1.5 \text{ watts}}\right) \quad \textbf{[4.65]}$$

$$10^{\left(\frac{-1.05}{10}\right)} = \frac{P_a}{1.5 \text{ watts}} \quad \textbf{[4.66]}$$

$$0.785 = \frac{P_a}{1.5 \text{ watts}} \quad \textbf{[4.67]}$$

$$(1.5 \text{ watts}) \times (0.785) = P_a \quad \textbf{[4.68]}$$

$$1.2 \text{ watts} = P_a \quad \textbf{[4.69]}$$

Now check what you have learned from the Smith chart. Recall that 1.5 watts of 4.5-GHz microwave RF signal were input to a 50-Ω transmission line that was 28 cm long. The load connected to the transmission line has an impedance of 36 + *j*40. From the Smith chart:

Admittance (load):	0.62 – j0.69
VSWR:	2.6:1
VSWR(dB):	8.3 dB
Refl. coef. (*E*):	0.44
Refl. coef. (*P*):	0.2
Refl. coef. angle:	84 degrees
Return loss:	–7 dB
Refl. loss:	–1.05 dB
Trans. loss. coef.:	1.5

Note that in all cases the mathematical interpretation corresponds to the graphical interpretation of the problem, within the limits of accuracy of the graphical method.

Stub matching systems

A properly designed matching system will provide a conjugate match to a complex impedance. Some sort of matching system or network is needed any time the load impedance (Z_L) is not equal to the characteristic impedance (Z_o) of the transmission line. In a transmission line system, it is possible to use a *shorted stub* connected in parallel with the line, at a critical distance back from the mismatched load, in order to affect a match. The stub is merely a section of transmission line that is shorted at the end not connected to the main transmission line. The reactance (hence also susceptance) of a shorted line can vary from $-\lambda$ to $+\lambda$, depending upon length, so you can use a line of critical length L2 to cancel the reactive component of the load impedance. Because the stub is connected in parallel with the line it is a bit easier to work with admittance parameters, rather than impedance.

Consider the example of Fig. 4-8, in which the load impedance is $Z = 100 + j60$, which is normalized to $2.0 + j1.2$. This impedance is plotted on the Smith chart in Fig. 4-9, and a VSWR circle is constructed. The admittance is found on the chart at point $Y = 0.37 - j0.22$.

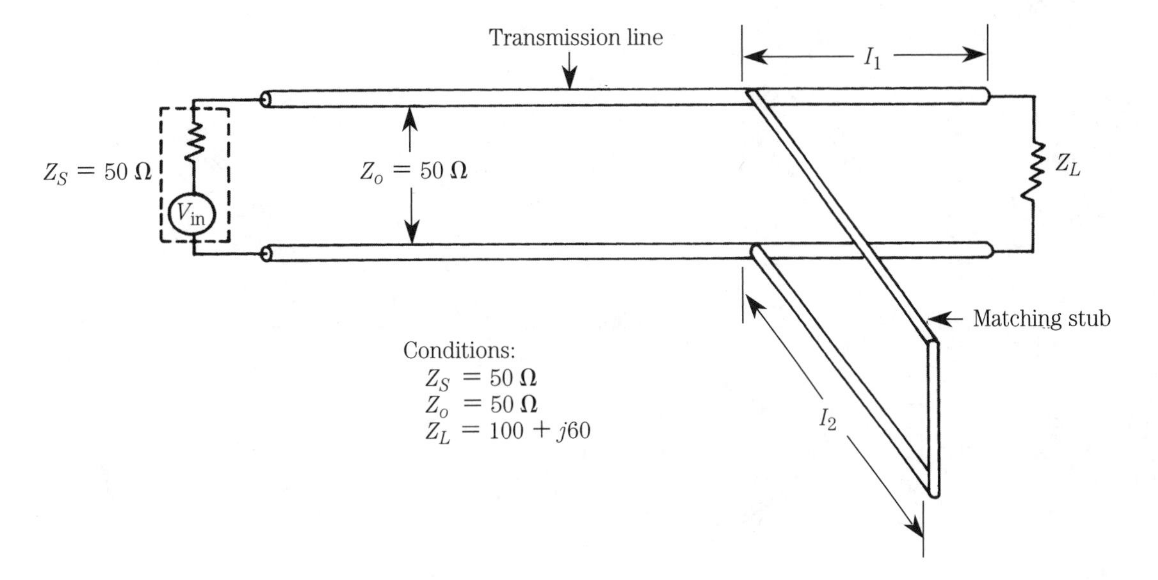

4-8 Matching stub length and position.

In order to provide a properly designed matching stub, you need to find two lengths. L1 is the length (relative to wavelength) from the load toward the generator (see L1 in Fig. 4-8); L2 is the length of the stub itself.

The first step in finding a solution to the problem is to find the points where the unit conductance line (1.0 at the chart center) intersects the VSWR circle; there are two such points shown in Fig. 4-9: *1.0 + j1.1* and *1.0 – j1.1*. Select one of these (choose $1.0 + j1.1$) and extend a line from the center 1.0 point through the $1.0 + j1.1$ point to the outer circle (WAVELENGTHS TOWARD GENERATOR). Similarly, a line is drawn from the center through the admittance point 0.37 – 0.22 to the outer circle. These two lines intersect the outer circle at the points 0.165 and 0.461. The distance of the stub back toward the generator is found from:

$$L_1 = 0.165 + (0.500 - 0.461)\ \lambda \qquad \textbf{[4.70]}$$

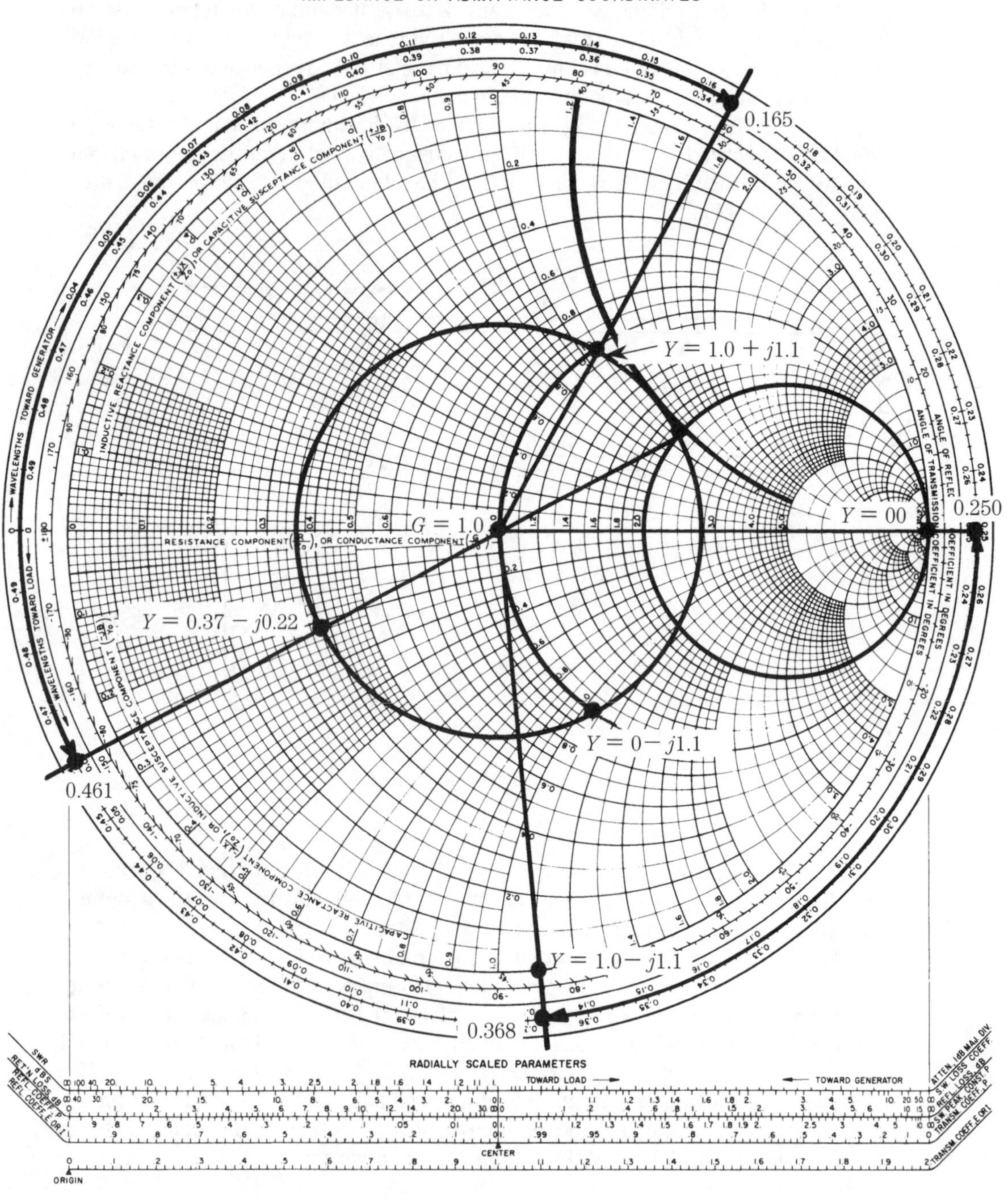

4-9 Solution to problem. Kay Elementrics

$$L_1 = 0.165 + 0.039\ \lambda \qquad \textbf{[4.71]}$$

$$L_1 = 0.204\ \lambda \qquad \textbf{[4.72]}$$

The next step is to find the length of the stub required. This is done by finding two points on the Smith chart. First, locate the point where admittance is infinite (far right side of the pure conductance line); second, locate the point where the admittance is $0 - j1.1$ (note that the susceptance portion is the same as that found where the unit conductance circle crossed the VSWR circle). Because the conductance component of this new point is 0, the point will lay on the $-j1.1$ circle at the intersection with the outer circle. Now draw lines from the center of the chart through each of these points to the outer circle. These lines intersect the outer circle at 0.368 and 0.250. The length of the stub is found from:

$$L_2 = (0.368 - 0.250)\ \lambda \qquad \textbf{[4.73]}$$

$$L_2 = 0.118\ \lambda \qquad \textbf{[4.74]}$$

From this analysis you can see that the impedance, $Z = 100 + j60$, can be matched by placing a stub of a length $0.118\ \lambda$ at a distance $0.204\ \lambda$ back from the load.

The Smith chart in lossy circuits

Thus far, you have dealt with situations in which loss is either zero (i.e., ideal transmission lines), or so small as to be negligible. In situations where there is appreciable loss in the circuit or line, however, you see a slightly modified situation. The VSWR circle, in that case, is actually a spiral, rather than a circle.

Figure 4-10 shows a typical situation. Assume that the transmission line is 0.60λ long, and is connected to a normalized load impedance of $Z = 1.2 + j1.2$. an "ideal" VSWR circle is constructed on the impedance radius represented by $1.2 + j1.2$. A line ("A") is drawn, from the point where this circle intersects the pure resistance baseline ("B"), perpendicularly to the ATTEN 1 dB/MAJ. DIV. line on the radially scaled parameters. A distance representing the loss (3 dB) is stepped off on this scale. A second perpendicular line is drawn, from the –3 dB point, back to the pure resistance line ("C"). The point where line "C" intersects the pure resistance line becomes the radius for a new circle that contains the actual input impedance of the line. The length of the line is 0.60λ, so you must step back $(0.60 - 0.50)\lambda$ or 0.1λ. This point is located on the WAVELENGTHS TOWARD GENERATOR outer circle. A line is drawn from this point to the 1.0 center point. The point where this new line intersects the new circle is the actual input impedance (Z_{in}).The intersection occurs at $0.76 + j0.4$, which (when de-normalized) represents an input impedance of $38 + j20$ ohms.

Frequency on the Smith chart

A complex network may contain resistive, inductive reactance, and capacitive reactance components. Because the reactance component of such impedances is a function of frequency, the network or component tends to also be frequency sensitive.

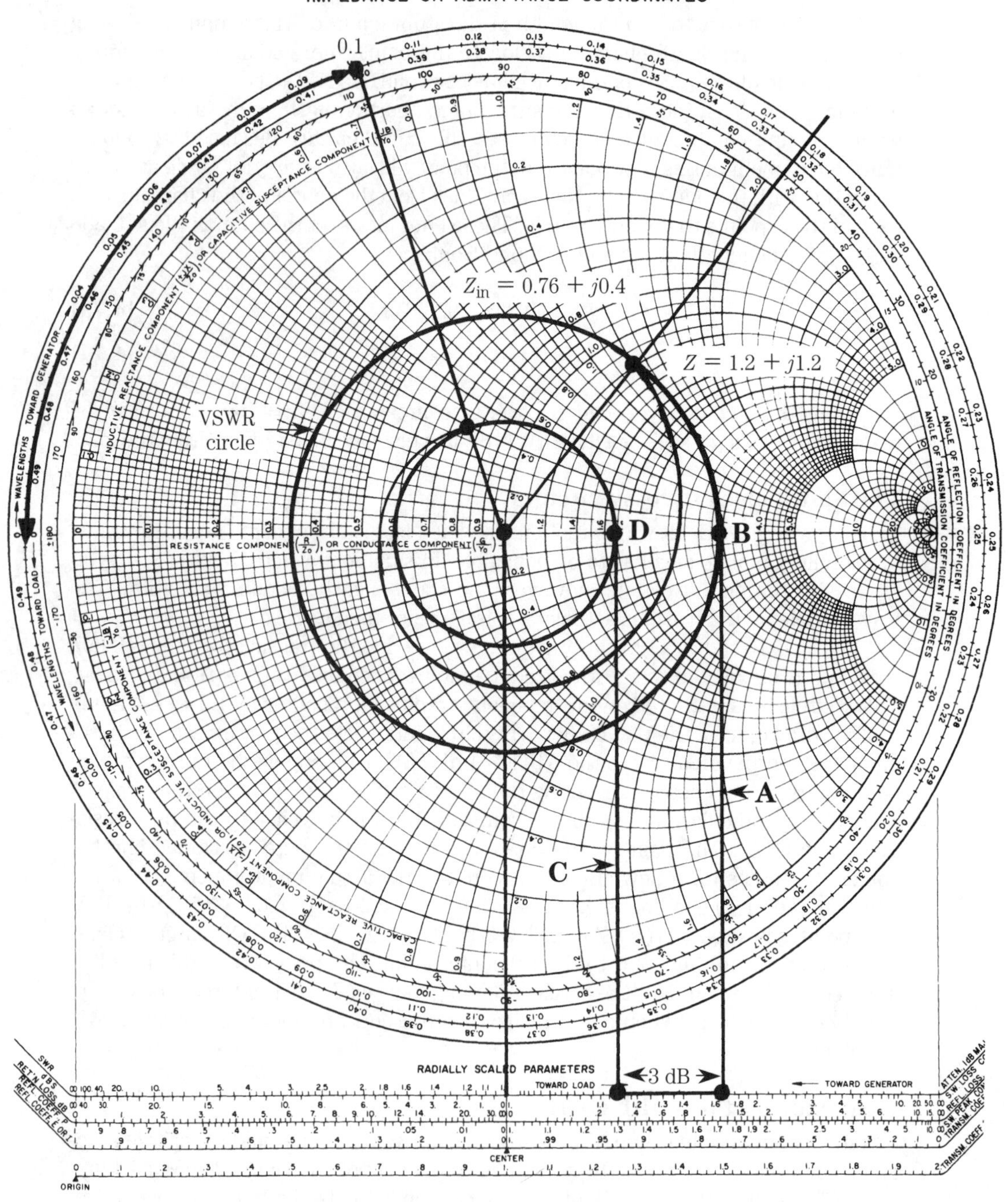

4-10 Solution. Kay Elementrics

You can use the Smith chart to plot the performance of such a network with respect to various frequencies. Consider the load impedance connected to a 50-Ω transmission line in Fig. 4-11. In this case, the resistance is in series with a 2.2-pF capacitor, which will exhibit a different reactance at each frequency. The impedance of this network is:

$$Z = R - j\left(\frac{1}{\omega C}\right) \quad \textbf{[4.75]}$$

or,

$$Z = 50 - j\left(\frac{1}{(2\pi F C)}\right) \quad \textbf{[4.76]}$$

And, in normalized form:

$$Z' = 1.0 - \frac{j}{((2\pi F C)\times 50)} \quad \textbf{[4.77]}$$

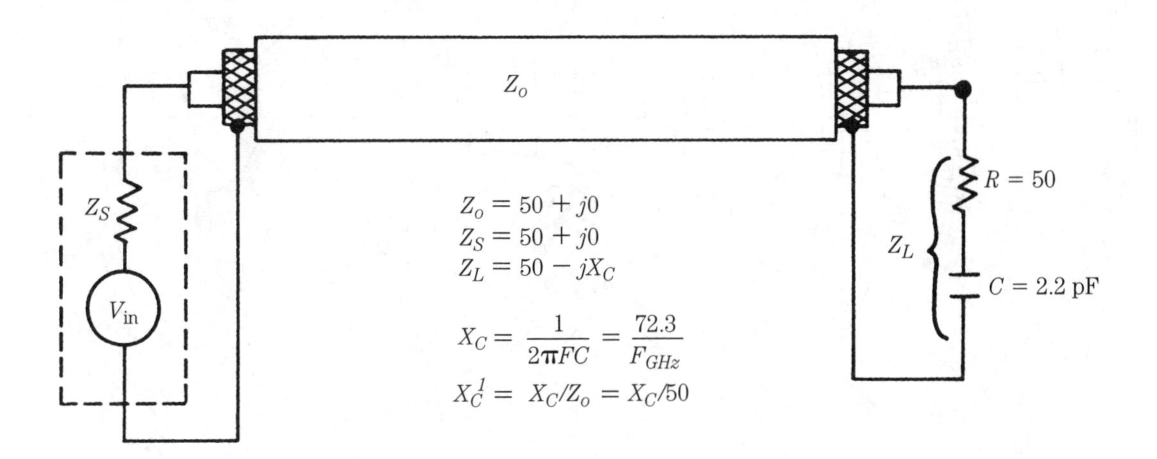

Freq. (GHz)	X_C	
1	$-j72.3$	$-j1.45$
2	$-j36.2$	$-j0.72$
3	$-j24.1$	$-j0.48$
4	$-j18$	$-j0.36$
5	$-j14.5$	$-j0.29$
6	$-j12$	$-j0.24$

4-11 Load and source impedance transmission line circuit.

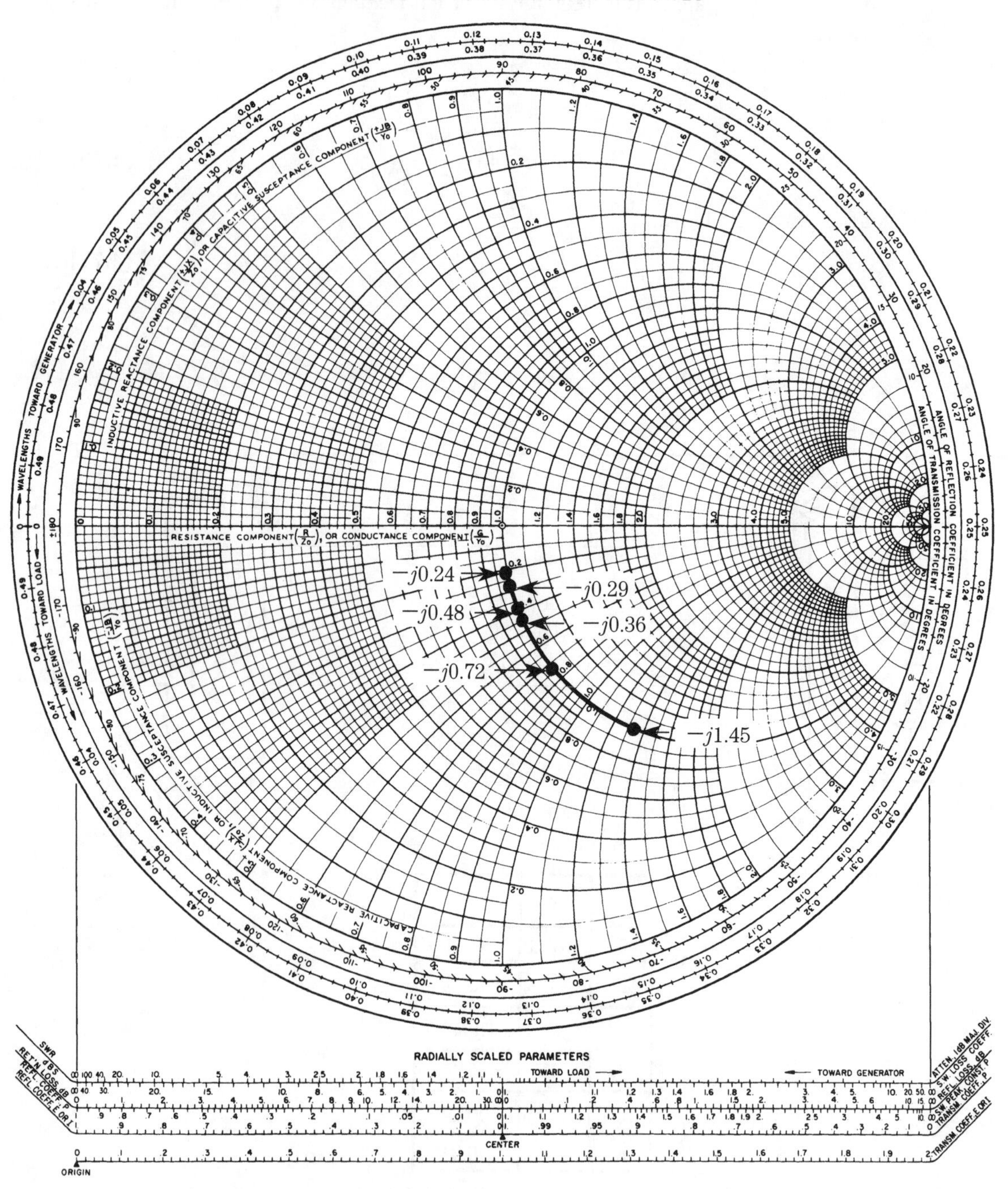

4-12 Solution. Kay Elementrics

$$Z' = 1.0 - \frac{j}{(6.9 \times 10^{-10} F)} \quad \textbf{[4.78]}$$

$$Z' = 1.0 - \frac{(j \times 7.23 \times 10^{10})}{F} \quad \textbf{[4.79]}$$

Or, converted to GHz:

$$Z' = 1.0 - \left(\frac{j72.3}{F_{\text{GHz}}}\right) \quad \textbf{[4.80]}$$

The normalized impedances for the sweep of frequencies from 1 to 6 GHz is therefore:

$$Z = 1.0 - j1.45 \quad \textbf{[4.81]}$$

$$Z = 1.0 - j0.72 \quad \textbf{[4.82]}$$

$$Z = 1.0 - j0.48 \quad \textbf{[4.83]}$$

$$Z = 1.0 - j0.36 \quad \textbf{[4.84]}$$

$$Z = 1.0 - j0.29 \quad \textbf{[4.85]}$$

$$Z = 1.0 - j0.24 \quad \textbf{[4.86]}$$

These points are plotted on the Smith chart in Fig. 4-12. For complex networks, in which both inductive and capacitive reactance exist, take the difference between the two reactances (i.e., $X = X_L - X_C$).

5
CHAPTER

Fundamentals of radio antennas

AN UNFORTUNATE OVERSIGHT IN MANY BOOKS ON RADIO ANTENNAS IS A LACK OF coverage on the most basic fundamentals of antenna theory. Most books, including the first draft of this one, start with a discussion of dipoles, but overlook that certain physical mechanisms are at work. An antenna is basically a *transducer* that converts electrical alternating current oscillations at a radio frequency, to an electromagnetic wave of the same frequency. This chapter looks at the physics of how that job is accomplished.

The material in this chapter was adapted from a U.S. Army training manual on antennas and radio propagation. Although unfortunately no longer in print, the manual contained the best coverage of basics the author of this book could find. Given that U.S. government publications are not protected by copyright, this information can be brought to you in full.

Antenna fundamentals

The electric and magnetic fields radiated from an antenna form the electromagnetic field, and this field is responsible for the transmission and reception of electromagnetic energy through free space. An antenna, however, is also part of the electrical circuit of a transmitter (or a receiver); and, because of its distributed constants, it acts as a circuit containing inductance, capacitance, and resistance. Therefore, it can be expected to display definite voltage and current relationships in respect to a given input. A current through it produces a magnetic field, and a charge on it produces an electrostatic field. These two fields taken together form the induction field. To gain a better understanding of antenna theory, a review of the basic electrical concepts of voltage and electric field, and current and magnetic field is necessary.

Voltage and electric field

When a capacitor is connected across a source of voltage, such as a battery (Fig. 5-1), it is charged some amount, depending on the voltage and the value of capacitance. Because of the emf (electromotive force) of the battery, negative charges flow to the lower plate, leaving the upper plate positively charged. Accompanying the accumulation of charge is the building up of the electric field. The flux lines are directed from the positive to the negative charges and at right angles to the plates.

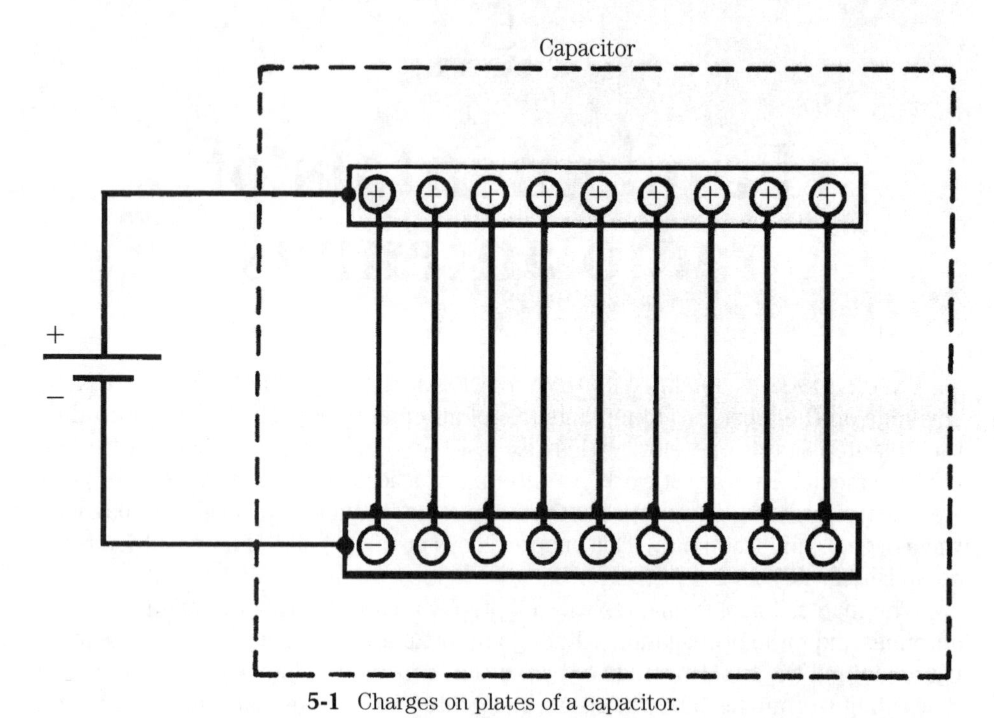

5-1 Charges on plates of a capacitor.

If the two plates of the capacitor are spread farther apart, the electric field must curve to meet the plates at right angles (Fig. 5-2). The straight lines in A become arcs in B, and approximately semicircles in C, where the plates are in a straight line. Instead of flat metal plates, as in the capacitor, the two elements can take the form of metal rods or wires. The three-dimensional view in Fig. 5-3 depicts the electric field more accurately. In A of Fig. 5-3, the wires are approximately 30° apart, and the flux lines are projected radially from the positively charged wire to the negatively charged wire. In B of Fig. 5-3, the two wires lie in a straight line, and the flux lines from a pattern similar to the lines of longitude around the earth. To bring out the picture more clearly, only the lines in one plane are given.

Assume that the sphere marked *E* in Fig. 5-3B, is a transmitter supplying RF energy. The two wires then can serve as the antenna for the transmitter. RF energy is radiated from the antenna and charges move back and forth along the wires, alternately compressing and expanding the flux lines of the electric field. The reversals in polarity of the transmitter signal also reverse the direction of the electric field.

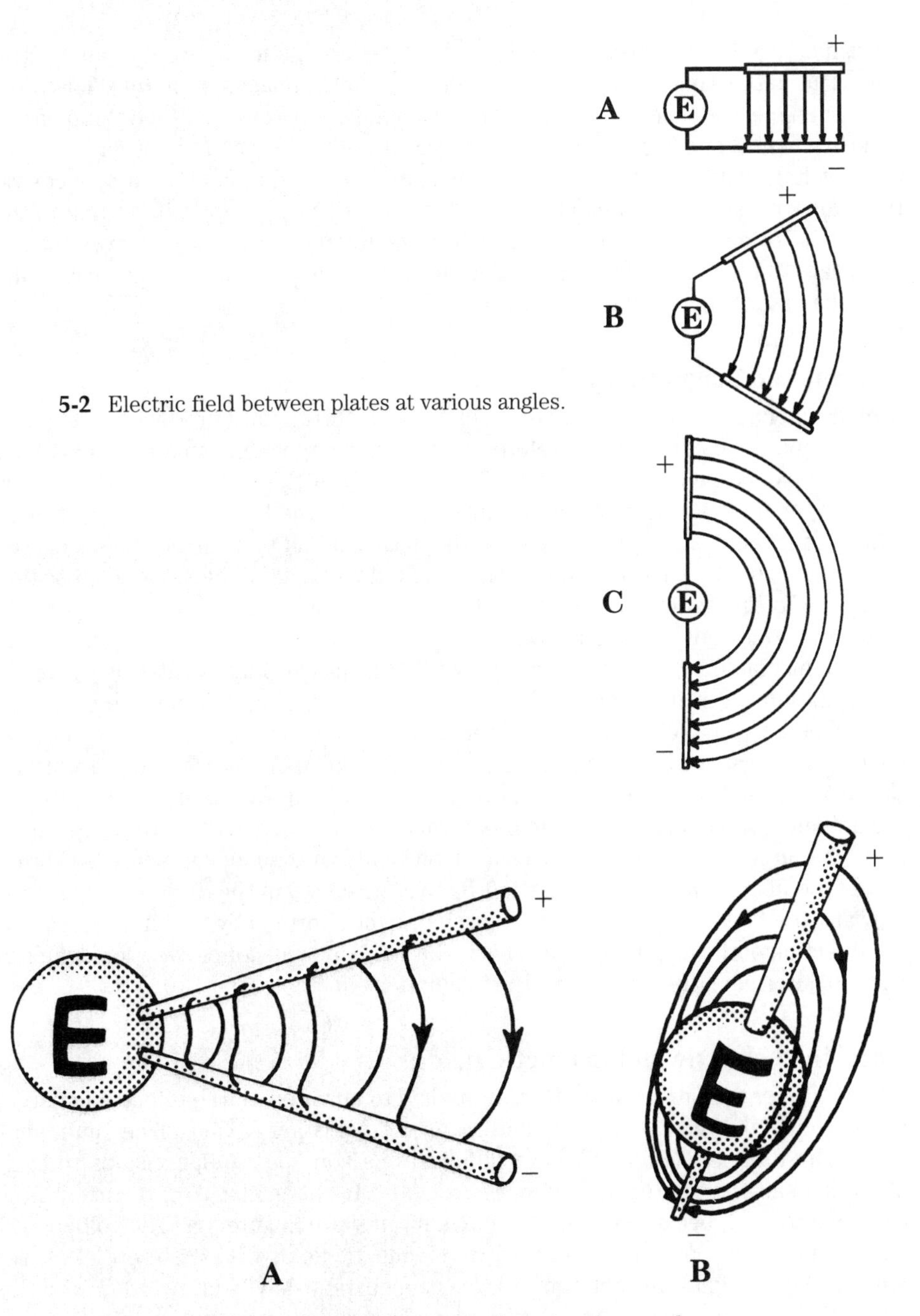

5-2 Electric field between plates at various angles.

5-3 Electric field between wires at various angles.

When a charge is put on the plates of a capacitor by means of a battery, an electric field is set up between its plates. The flow of charge from source to capacitor ceases when the capacitor is fully charged, and the capacitor is said to be charged to a *voltage* equal, and of opposite polarity, to that of the source. The charged capacitor can be used as a source of emf since it stores energy in the form of an electric field. This is the same as saying that *an electric field indicates voltage*. The presence of an electric field about an antenna also indicates voltage. Since the polarity, and the amount of charge, depend on the nature of the transmitter output, the antenna voltage also depends on the energy source. For example, if a battery constitutes the source, the antenna charges to a voltage equal and opposite to that of the battery. If RF energy is supplied to a half-wave antenna, the voltage across the antenna lags the current by 90°. The half-wave antenna acts as if it was a capacitor, and it can be described as being *capacitive* .

Current and magnetic field

A *moving* charge along a conductor constitutes a current and produces a magnetic field around the conductor. Therefore, the flow of charge along an antenna also will be accompanied by a magnetic field. The intensity of this field is directly proportional to the flow of charge. When the antenna is uncharged, the current flow is maximum, since there is no opposing electric field. Because of this current flow, a charge accumulates on the antenna, and an electric field builds up in increasing opposition to the emf of the source. The current flow decreases and when the antenna is fully charged, the current no longer flows.

The magnetic field in the space around a current-carrying device has a specific configuration, with the magnetic flux lines drawn according to a definite rule (Fig. 5-4). Whereas, in the electric field, the electric lines are drawn from a positive charge to a negative charge, in the magnetic field, the flux lines are drawn according to the *left-hand rule*. The direction of current flow is upward along both halves of the antenna. The lines of magnetic flux form concentric loops that are perpendicular to the direction of current flow. The arrowheads on the loops indicate the direction of the field. If the thumb of the left hand is extended in the direction of current flow and the fingers clenched, then the rough circles formed by the fingers indicate the direction of the magnetic field. This is the left-hand rule, or convention, which is used to determine the direction of the magnetic field.

Combined electric and magnetic fields

When RF energy from a transmitter is supplied to an antenna, the effects of charge, voltage, current, and the electric and magnetic fields are taking place simultaneously. These effects (Fig. 5-5) have definite time and space relationships to each other. If a half-wave antenna is used, the relations between charge and current flow can be predicted, because of the capacitive nature of this antenna. The voltage will lag the current by 90°, and the electric and magnetic fields will be 90° out of phase. With no electric field present (no charge), the current flow is unimpeded, and the magnetic field is maximum. As charge accumulates on the antenna, the electric field builds up in opposition to current flow and the magnetic field decreases in intensity. When the electric field reaches its maximum strength, the magnetic field has decayed to zero.

A reversal in polarity of the source, reverses the direction of current flow as well as the polarity of the magnetic field, and the electric field aids the flow of current by

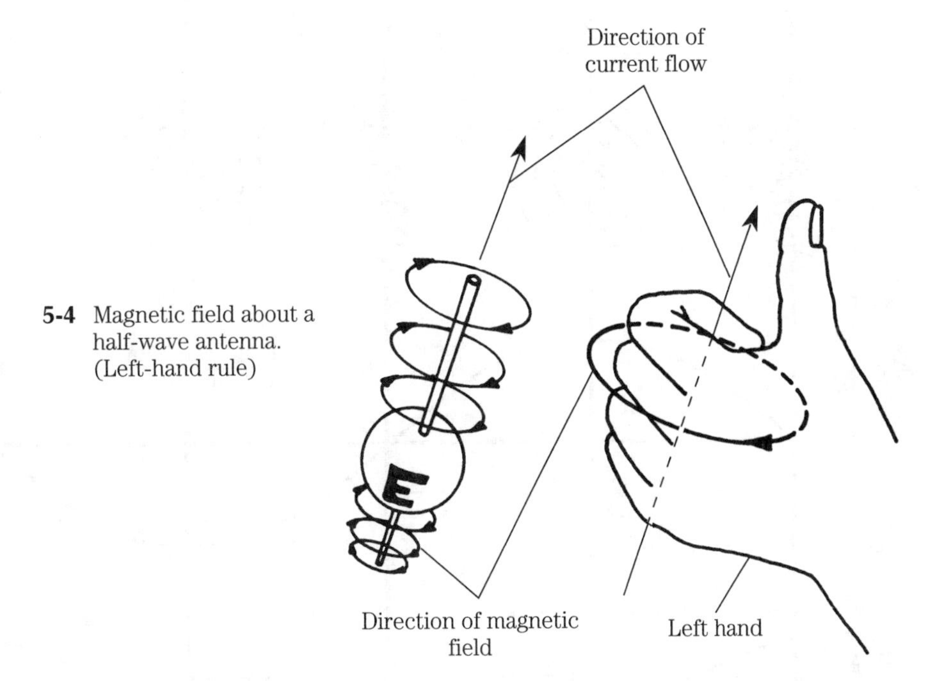

5-4 Magnetic field about a half-wave antenna. (Left-hand rule)

discharging. The magnetic field builds up to a maximum, and the electric field disappears as the charge is dissipated. The following half-cycle is a repetition of the first half-cycle, but in the reverse direction. This process continues as long as energy is supplied to the antenna. The fluctuating electric and magnetic fields combine to form the induction field, in which the electric and magnetic flux maximum intensities occur 90° apart in time, or in *time quadrature*. Physically, they occur at right angles to each other, or in *space quadrature*. To sum up, the electric and magnetic fields about the antenna are in space and time quadrature.

Standing waves

Assume that it is possible to have a wire conductor with one end extending infinitely, with an RF transmitter connected to this wire. When the transmitter is turned on, an RF current in the form of sine waves of RF energy moves down the wire. These waves of energy are called *traveling waves*. The resistance of the conductor gradually diminishes the amplitude of the waves, but they continue to travel so long as the line does not come to an end.

The antenna, however, has some finite length. Therefore, the traveling waves are halted when they reach the end of the conductor. Assume that the RF transmitter is turned on just long enough to allow one sine wave of energy to get on the line (Fig. 5-6A). This traveling wave is moving down the antenna toward the end. When the wave reaches the end of the conductor, the current path is broken abruptly. With the stoppage of current flow, the magnetic field collapses. A voltage is induced at the end of the conductor that causes current to flow *back toward the source*, as in Fig. 5-6B. The wave is reflected back to the source; and, if a continual succession of

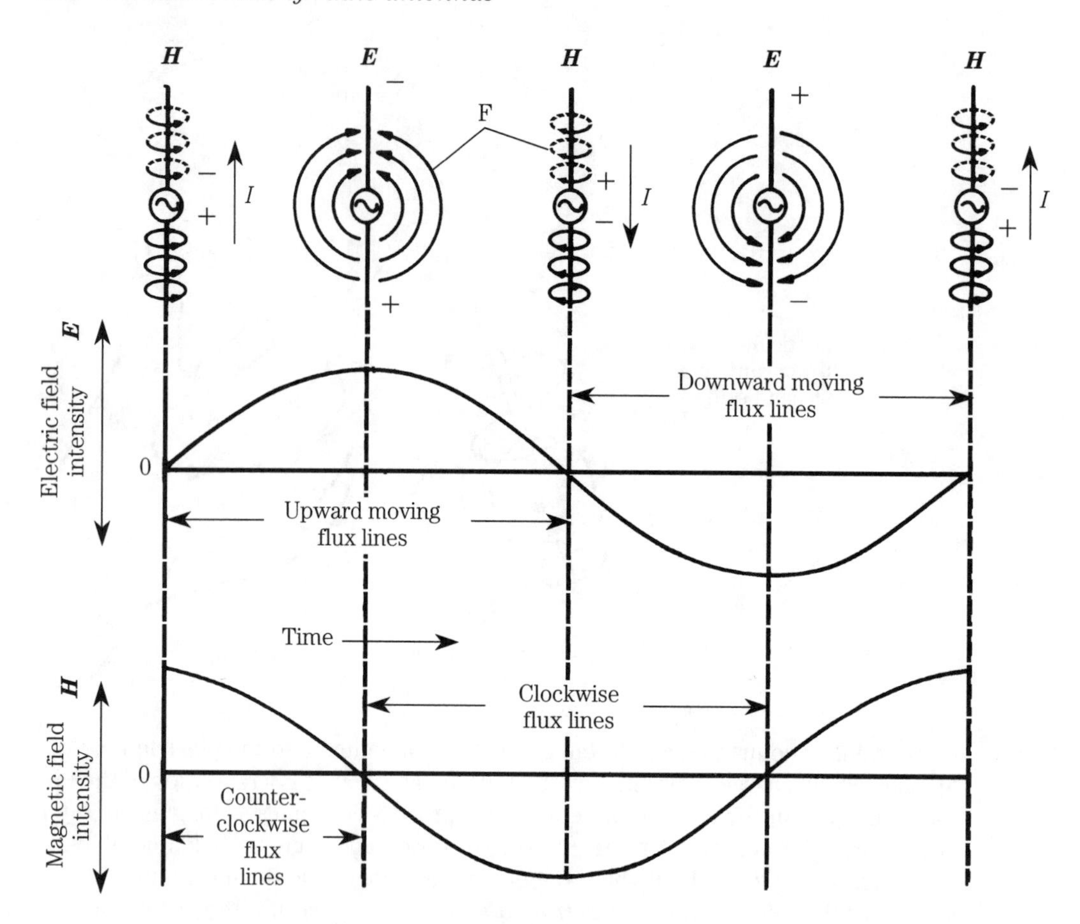

5-5 Electric and magnetic fields 90° out of phase.

waves is sent down the line, they will be reflected in the same continual pattern. The wave moving from the transmitter toward the end is called the *incident* wave, and its reflection is called the *reflected* wave.

A continuous flow of incident waves results in a continuous flow of reflected waves. Because there is only one conductor, the two waves must pass each other. Electrically, the only current that actually flows is the *resultant* of both of these waves. The waves can reinforce or cancel each other as they move.

When they reinforce, the resultant wave is maximum; when they cancel, the resultant wave is minimum. In a conductor that has a finite length, such as an antenna, the points at which the maxima and minima of the resultant wave occur (Fig. 5-6C) are *stationary*. In other words, the maximum and minimum points stand still, although both the incident and reflected waves are moving. The resultant wave stands still on the line, only its amplitude being subject to change. Because of this effect, the resultant is referred to as a *standing wave*.

The development of the standing wave on an antenna by actual addition of the traveling waves is illustrated in Fig. 5-7. At the instant pictured in A, the incident and reflected waves just coincide. The result is a standing wave having twice the amplitude of either traveling wave. In B, the waves move apart in opposite directions, and

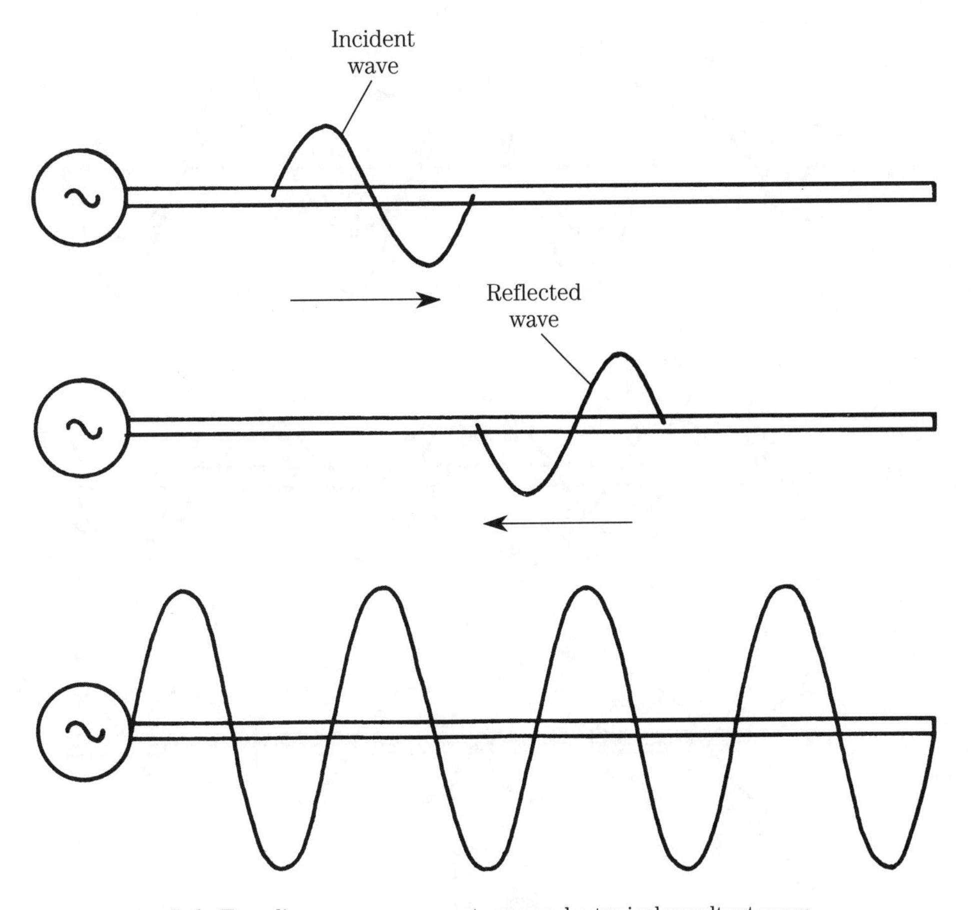

5-6 Traveling waves on an antenna and a typical resultant wave.

the amplitude of the resultant decreases, but the points of maximum and minimum do not move.

When the traveling waves have moved to a position of 180° phase difference, the resultant is zero along the entire length of the antenna, as shown in C. At this instant, there can be no current flow in the antenna. The continuing movement of the traveling waves, shown in D, builds up a resultant in a direction opposite to that in A. The in-phase condition of the traveling waves results in a standing wave, in E, equal in amplitude, but 180° out of phase with the standing wave in A.

If the progressive pictures of the standing wave are assembled on one set of axes, the result is as in Fig. 5-8. The net effect of the incident and reflected waves is apparent. The curves are lettered with reference to Fig. 5-7. As the traveling waves move past each other, the standing wave changes only its amplitude. The fixed minimum points are called *nodes*, and the curves representing the amplitude are called *loops*.

The concept of the standing wave can be applied to the half-wave antenna with reference to either current or voltage distribution at any instant. This application is possible because there are traveling waves of both voltage and current. Because voltage and current are out of phase on the half-wave antenna, the standing waves also are found to be out of phase.

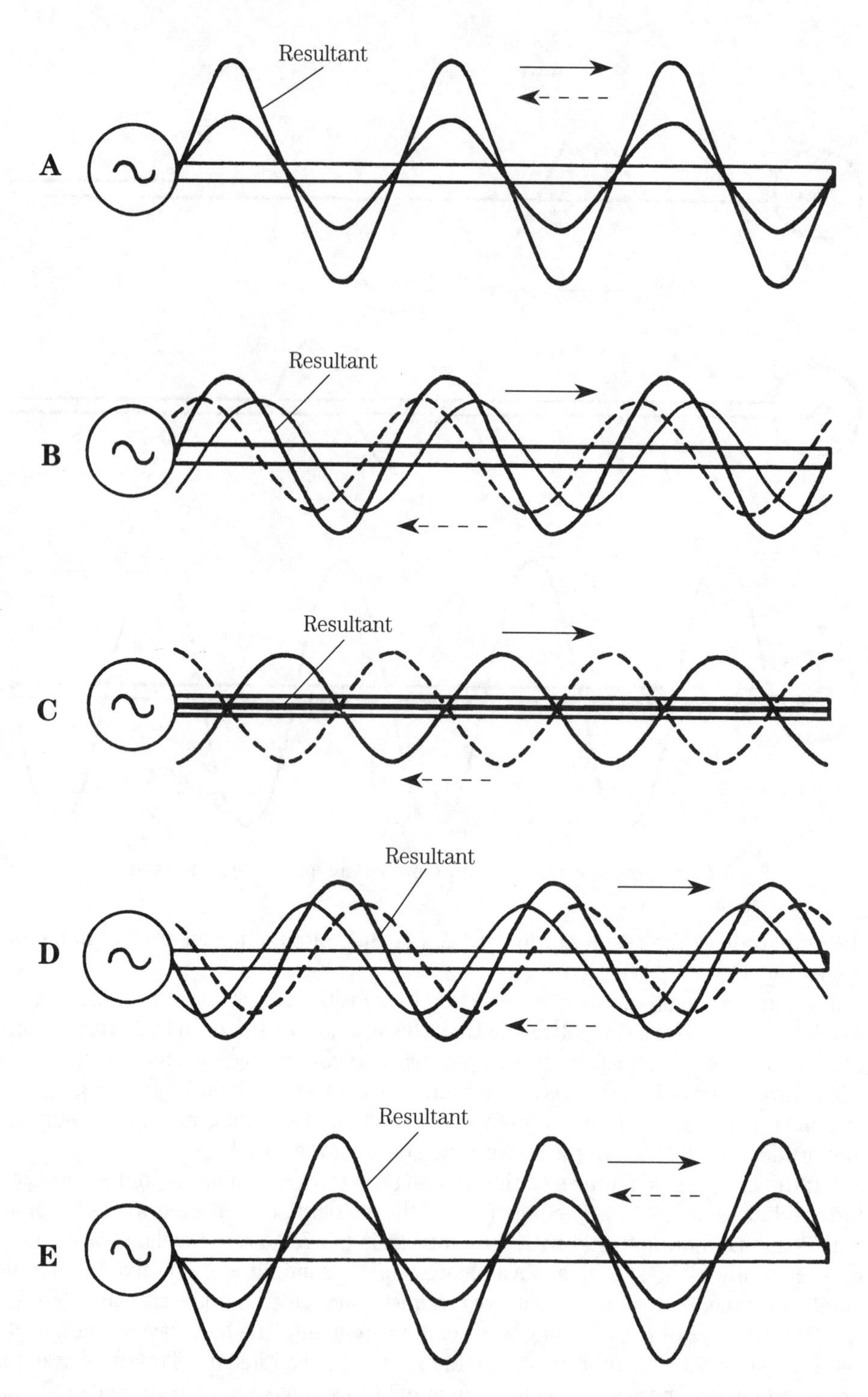

5-7 Development of standing wave from traveling wave.

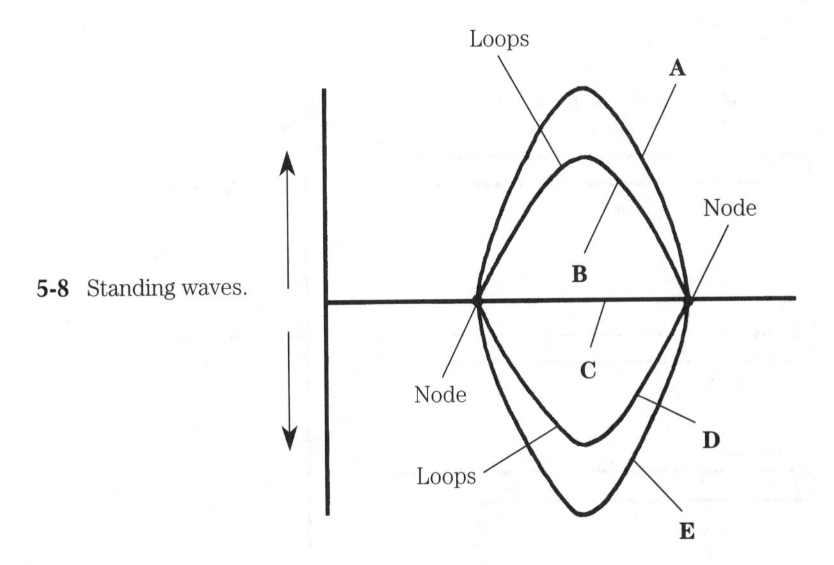

5-8 Standing waves.

Voltage and current distribution on half-wave antenna

When an RF transmitter is feeding a half-wave antenna, positive and negative charges move back and forth along the antenna (Figs. 5-9 and 5-10). The first picture shows the position of the charges at some arbitrary time, T_0. The RF charges being observed are at the ends of the antenna, and there is a maximum difference in potential between the ends, A and B. The remaining illustrations show the instantaneous positions of the charges at regular intervals of 22.5° throughout a complete cycle.

To the right of each instantaneous position of the charges are curves representing the current and voltage *at that particular time* for any point on the antenna. For example, at time T_0, the positive and negative charges are at points A and B on the antenna. The voltage between these points represents a maximum difference of potential. The current, being 90° out of phase in respect to the voltage, is everywhere zero. These distribution curves are *standing waves* derived in the same manner as those covered in the previous paragraph.

The next illustration shows the position of the charges at time T_1. The standing wave of current is a relative maximum at the center of the antenna. This current loop has nodes that remain at the ends of the antenna, and it is, therefore, 90° out of phase with the standing wave of voltage.

At T_2 and T_3, the charges move closer together, and the standing wave of voltage slowly decreases in amplitude. Conversely, the current loop increases in magnitude. When the charges meet after 90° of the RF cycle (T_4), the effect is that of having the positive and negative charges cancel. The voltage loop accordingly is zero everywhere on the antenna, and the current loop rises to its maximum value, unimpeded by any charge on the antenna.

At time T_5, the charges have passed each other, each charge having moved past the center point of the antenna. The polarity of the voltage loops is reversed, and they build up in the opposite direction, keeping the node always at the center point of the antenna. The reversal of polarity is shown in the charge positions at T_3, T_4, and

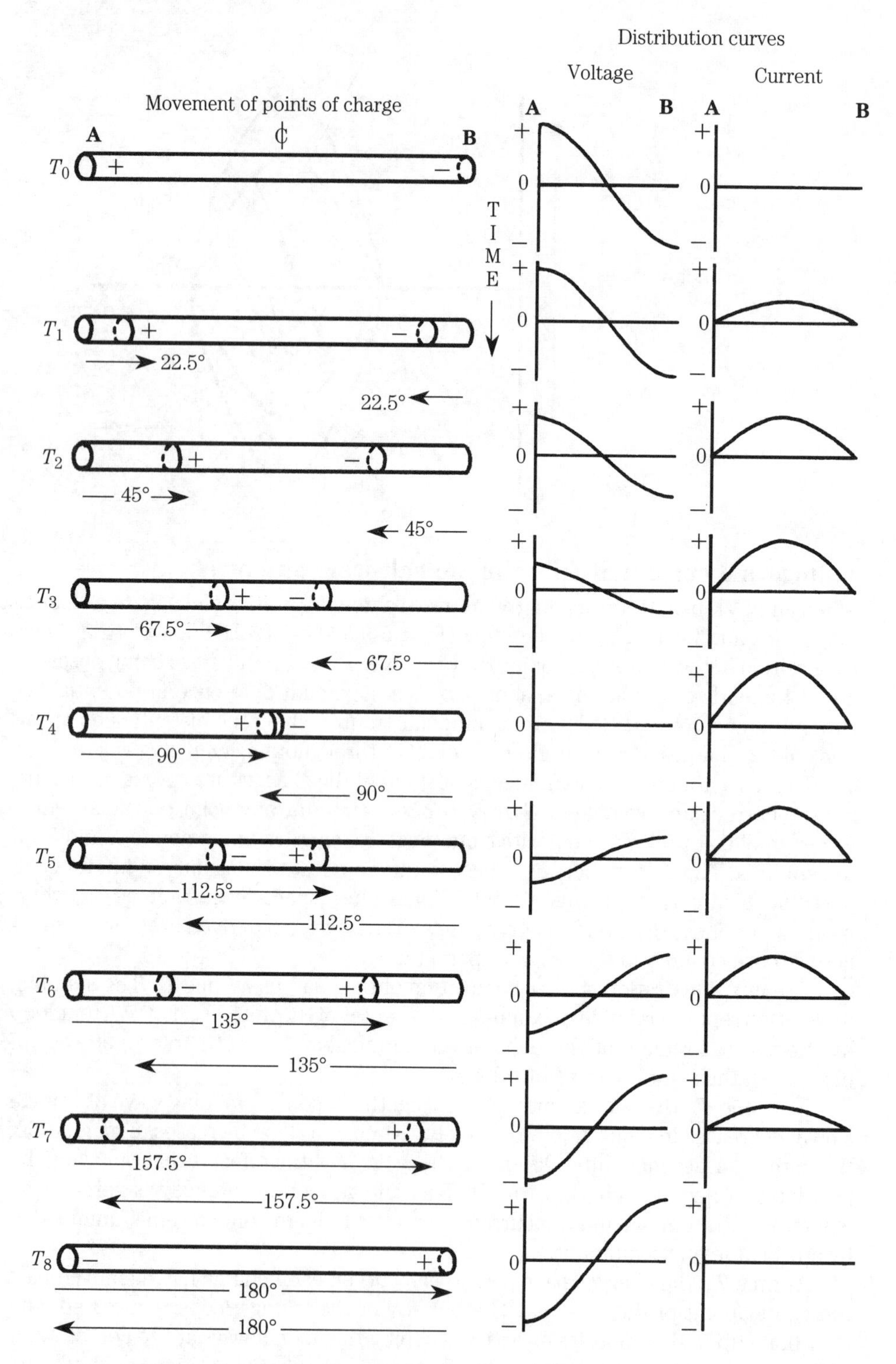

5-9 Voltage and current distribution in terms of positive and negative charges.

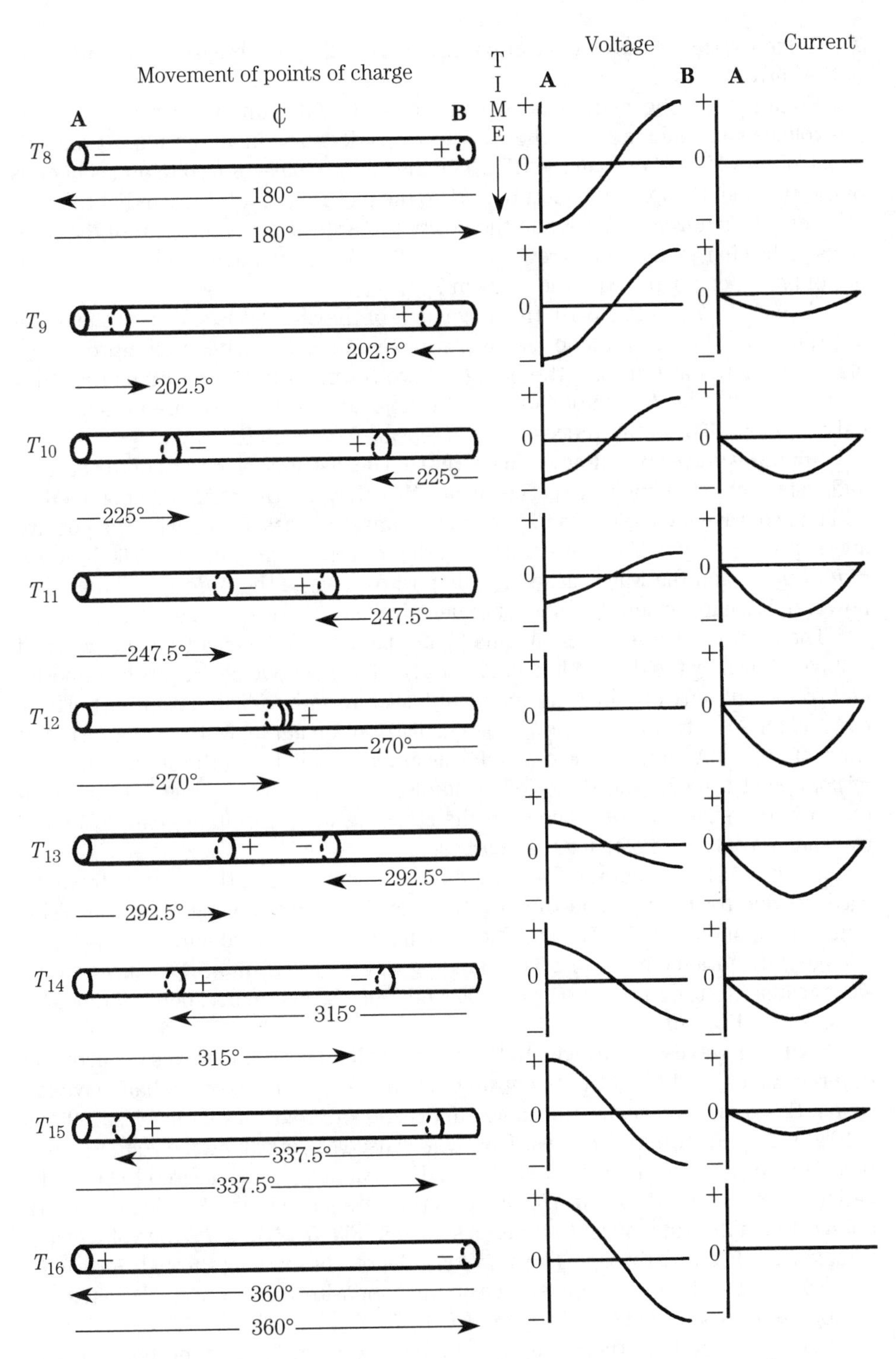

5-10 Voltage and current distribution in terms of positive and negative charges.

T_5. The separation of the charges also is accompanied by a decrease in the amplitude of the current loop.

From T_5 to T_8, the charges move out to the ends of the antenna. During this time, the voltage loops increase and the current loops decrease in amplitude. At time T_8, which occurs 180° after T_0 in the RF cycle, the charges have moved to opposite ends of the antenna. Compare the picture in T_0 to the picture in T_8. It is seen that the negative charge is now at point A and the positive charge at point B. Because the positions of the charges have been reversed from T_0 to T_8, the voltage loops in T_8 are 180° out of phase, compared with the loops in T_0.

From T_8 to T_{16} in Fig. 5-10, the movement of the charges is shown in the opposite direction, the current loop reaching a maximum at T_{12}. When the entire RF cycle is completed at time T_{16}, the charges have returned to the positions that they occupied at T_0. The distribution curves of voltage and current also are in their original conditions. The entire process then is repeated for each RF cycle.

Standing waves of voltage and current The distribution curves of the current and voltage are standing waves. This means that they are the resultants obtained by adding two traveling waves. The two traveling waves are associated with the positive and negative charges. The wave caused by the negative charge can be called the *incident wave* and the wave caused by the positive charge the *reflected wave*. This, however, is clearer when the concept of negative and positive charges is used.

The positive charge, taken at time T_0 in Fig. 5-9, produces a traveling wave of voltage, shown by the dashed line in Fig. 5-11A. The negative charge at the opposite end of the antenna produces an identical traveling wave (dash-dot curve). These two add together to produce the T_0 voltage distribution curve, which is the resultant wave of Fig. 5-9A. Both of these waveforms are identical, being the *standing wave of voltage* at time T_0. All of the following distribution curves of Fig. 5-11 are produced in the same manner. They are the standing wave resultants caused by the traveling waves accompanying the charges.

In Fig. 5-11B, each of the traveling waves has moved 45°, the positive traveling wave moving to the right and the negative traveling wave moving to the left. This time corresponds to T_2 in Fig. 5-9. The standing wave produced corresponds to the voltage distribution curve at T_2. The standing waves of current are produced in the same manner. The current curves at D, E, and F of Fig. 5-11 correspond to times T_0, T_2, and T_4 of Fig. 5-9.

Standing waves of voltage In Fig. 5-12A, voltage standing waves occurring at different times are brought together on one axis, AB, representing a half-wave antenna. Essentially, these are the same curves shown progressively in Figs. 5-9 and 5-10 as voltage distribution curves. They can be used to determine the voltage at any point on the antenna, at any instant of time. For example, if it is desired to know the variations of voltage occurring at point Y on the antenna over the RF cycle, the variations are graphed in respect to time, as shown in Fig. 5-12B. At T_0, the voltage at Y is maximum. From T_0 through T_3, the voltage decreases, passing through zero at T_4. The voltage builds up to a maximum in the opposite direction at T_8, returning through zero to its original position from T_8 to T_{16}.

Between T_0 and T_{16}, therefore, an entire sine-wave cycle, Y, is reproduced. This is also true of any other point on the antenna, with the exception of the node at X.

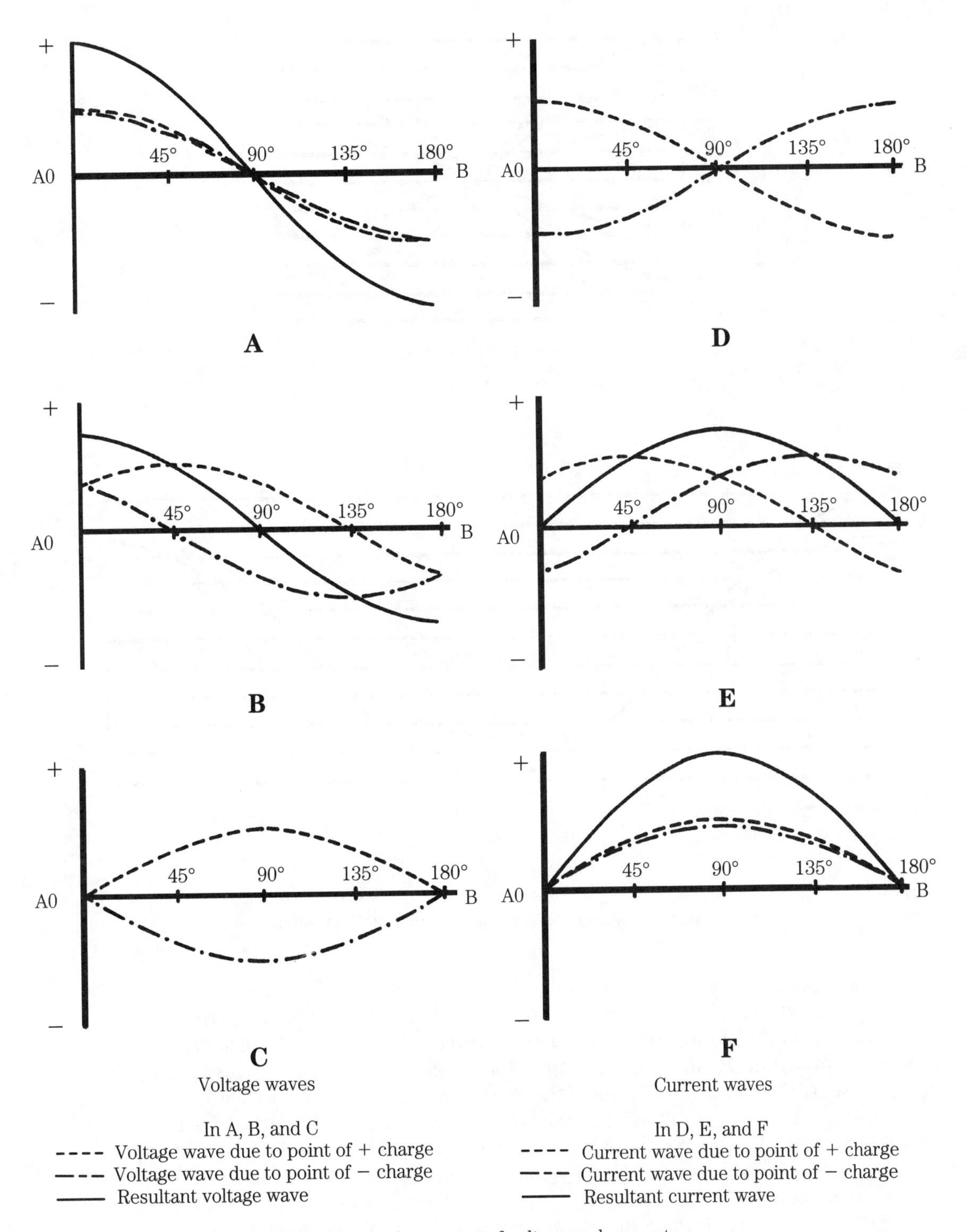

5-11 Standing waves of voltage and current.

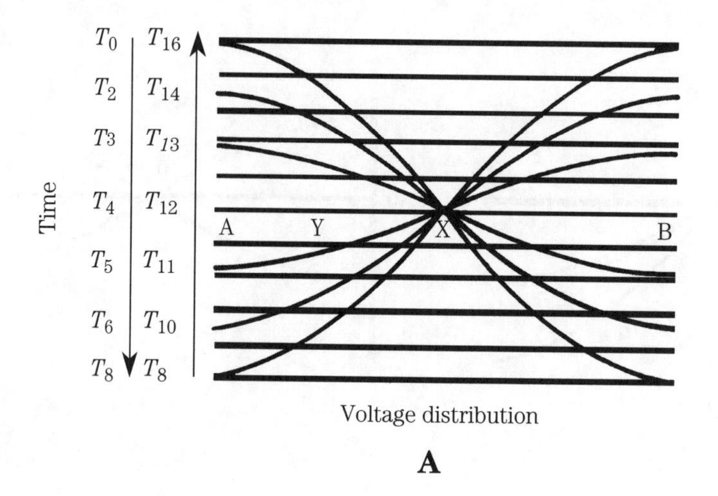

A

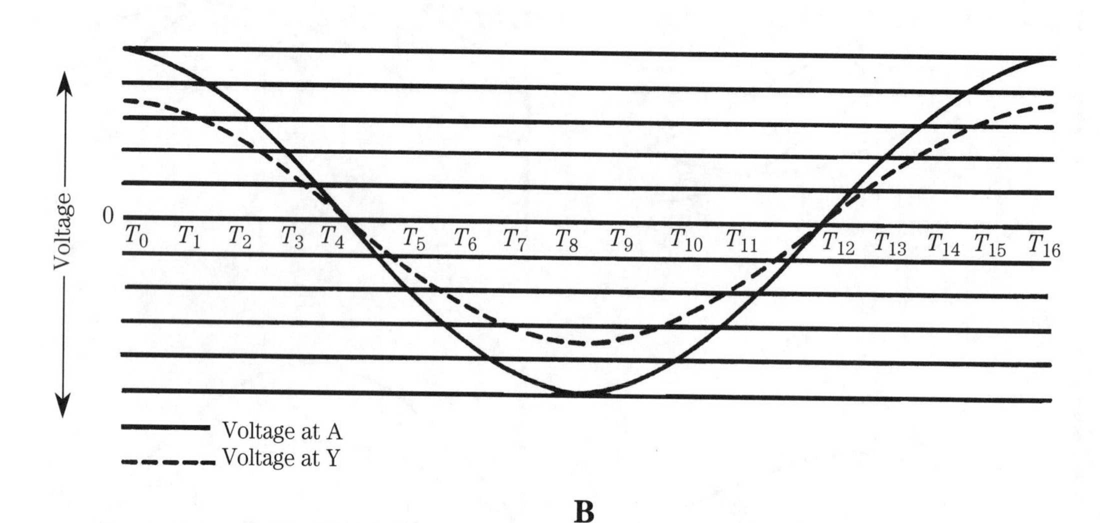

B

5-12 Standing waves of voltage at a point on the antenna.

The peak amplitude of the sine wave produced at any point depends on its position on the antenna. The nearer the point is to either end, the greater its peak amplitude.

Standing waves of current The standing waves of current occurring at various times through the RF cycle are assembled on a single axis in Fig. 5-13. This axis, AB, represents the half-wave antenna. If the current variations at point Y from T_0 to T_{16} are graphed in respect to time, the result is the sine wave in Fig. 5-13B. This is true for any point along the antenna with the exception of the nodes at the ends. The current has its greatest swing at X, the center of the antenna. Comparison of the voltage variation curve (Fig. 5-12A) with the current variation curve (Fig. 5-13A) shows the voltage curve leading the current curve by 90° at Y. This relation can be expected on any half-wave device.

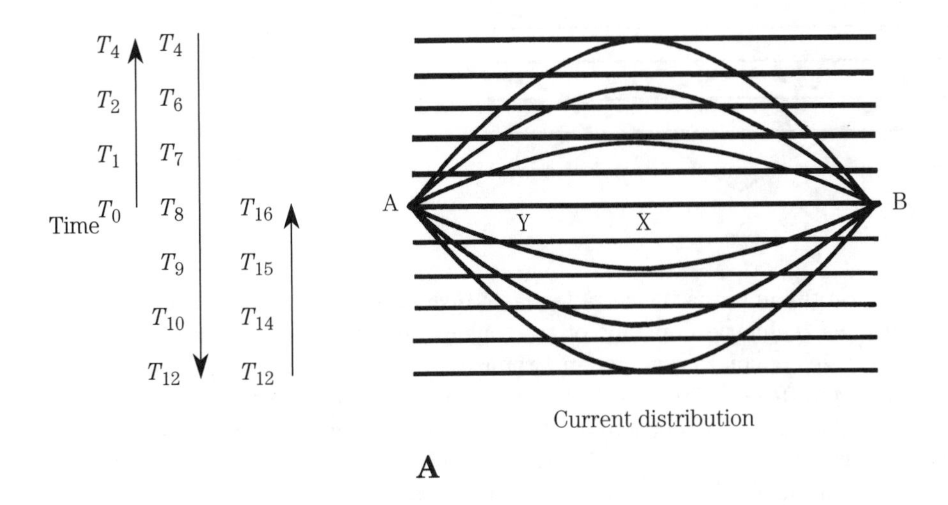

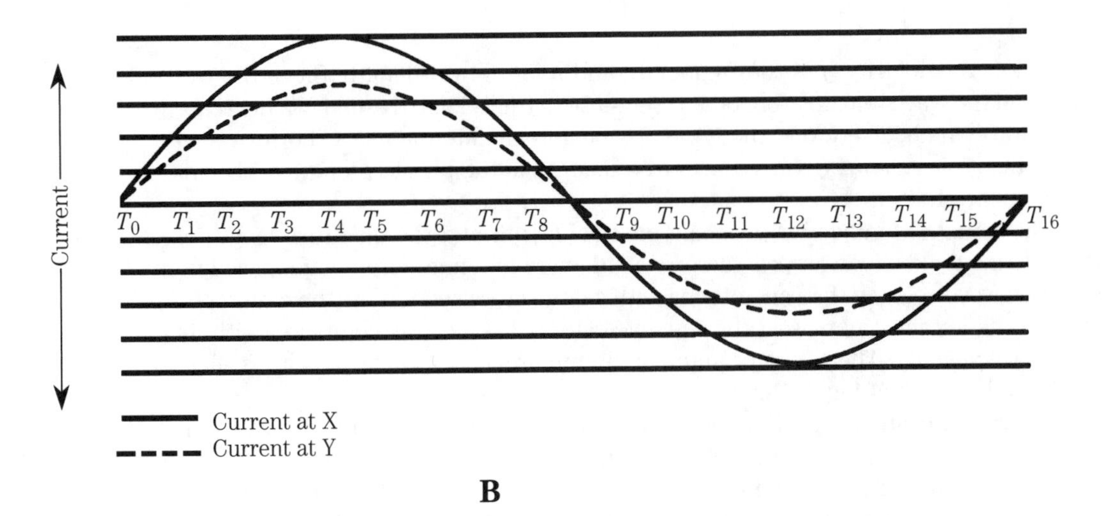

5-13 Standing waves of current at a point on the antenna.

Measurement of standing waves In Fig. 5-14, the standing waves of voltage, E, and current, I, are indicated along the antenna. There are current nodes at A and B and a voltage node at X. These standing waves are found on any half-wave antenna. A meter that indicates the effective value (0.707 of peak) of the ac signal can be used to measure the standing waves present on the half-wave antenna.

Velocity of propagation and antenna length

In free space, electromagnetic waves travel at a constant velocity of 300,000 kilometers (or approximately 186,000 miles) per second. The RF energy on an antenna, however, moves at a velocity considerably less than that of the radiated energy in

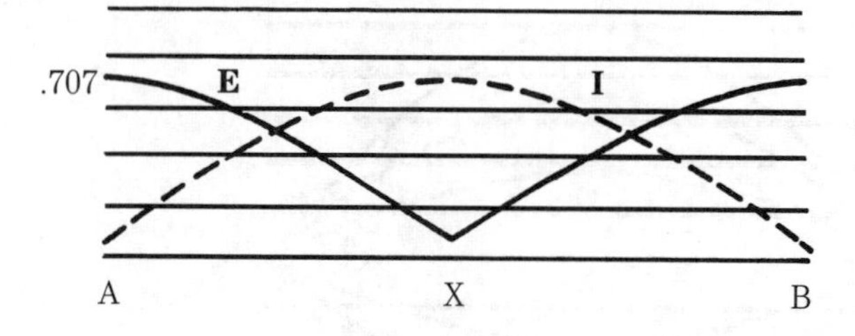

5-14 Standing waves measured with a meter.

free space because the antenna has a *dielectric constant* greater than that of free space. Because the dielectric constant of free space (air or vacuum) is approximately 1, a dielectric constant greater than 1 retards electromagnetic-wave travel.

Because of the difference in velocity between the wave in free space, and the wave on the antenna, the *physical* length of the antenna no longer corresponds to its *electrical* length. The antenna is a half-wavelength electrically, but somewhat shorter than this physically. This is shown in the formula for the velocity of electromagnetic waves,

$$V = f\lambda \tag{5.1}$$

where V is the velocity, f is the frequency, and λ is the wavelength. Since the frequency of the wave remains constant, a decrease in the velocity results in a decrease in the wavelength. Therefore, the wave traveling in an antenna has a shorter wavelength than the same wave traveling in free space, and the physical length of the antenna can be shorter.

The actual difference between the physical length and the electrical length of the antenna depends on several factors. A thin wire antenna, for example, has less effect on wave velocity than an antenna with a large cross section. As the circumference of the antenna increases, the wave velocity is lowered, as compared with its free-space velocity. The effect of antenna circumference on wave velocity is illustrated in the graph of Fig. 5-15.

Other factors are involved that lower wave velocity on the antenna. Stray capacitance, for example, increases the dielectric constant and lowers wave velocity. This capacitance can be caused by the line connecting the antenna to the transmitter, the insulators used to give physical support to the antenna, or nearby objects made of metallic or dielectric materials. The change in velocity resulting from stray capacitance is called *end effect* because the ends of the antenna are made farther apart electrically, than they are physically. End effect is counteracted by making the physical length about 5 percent shorter than the electrical length, as expressed in the formula.

$$\begin{aligned} L &= 0.95\left(\frac{492}{f}\right) \\ &= \frac{468}{f} \end{aligned} \tag{5.2}$$

where L is the physical length in feet and f is the frequency in megacycles. This formula is accurate for all practical purposes in determining the physical length of a half-wavelength antenna at the operating frequency.

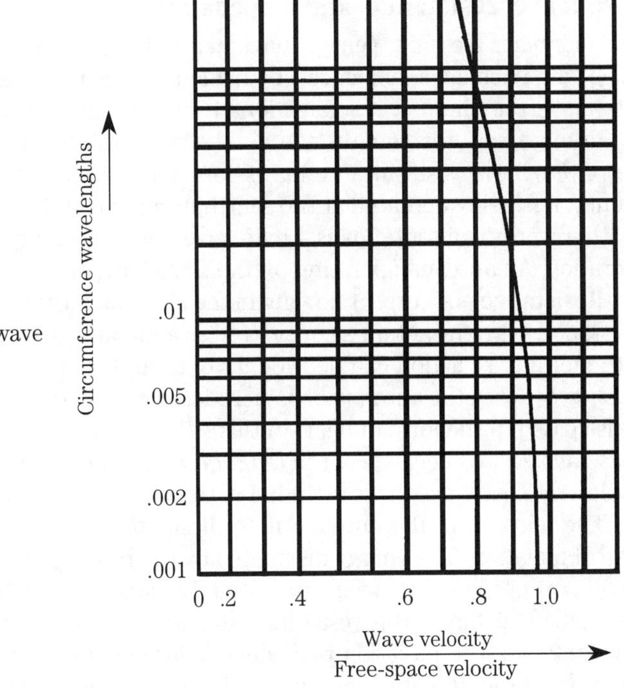

5-15 Effect of antenna circumference on wave velocity.

The capacitive end effect also slightly changes the standing waves of voltage and current. When the standing waves are measured, it is found that the nodes have some value and do not reach zero, because some current is necessary to charge the stray capacitance. The standing waves measured in Fig. 5-16 show the results of end effect.

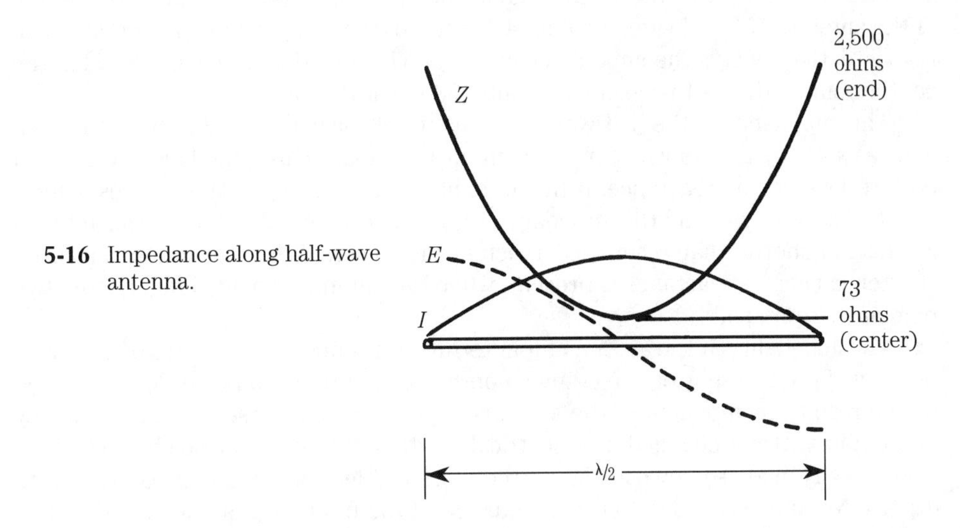

5-16 Impedance along half-wave antenna.

Resonance, resistance, and impedance

The antenna is a circuit element having distributed constants of inductance, capacitance, and resistance, which can be made to form a resonant circuit. The half-wave antenna is the shortest resonant length of antenna. However, antennas which are two or more half-wavelengths can also be resonant. Such antennas are said to operate on *harmonics*. If an antenna is four half-wavelengths at the transmitter frequency, it is being operated at the fourth harmonic of its lowest resonant frequency. In other words, this antenna is a half-wavelength at one-quarter of the frequency of operation. An antenna operating on the third harmonic is shown in Fig. 5-7.

Resistance A current flowing in the antenna must contend with three kinds of resistance. With the antenna considered as a radiator of energy, the power expended in the form of radiation can be thought of as an I^2R_t, loss. R_t is called the *radiation resistance*. With the antenna considered as a conductor, a certain amount of energy is dissipated in the form of heat. In this I^2R_0 loss, R_0 is the *ohmic resistance*. There is also an I^2R loss because of the *leakage resistance* of dielectric elements, such as insulators. This R usually is included in the ohmic resistance.

The purpose of the antenna is to dissipate as much energy as possible in the form of radiation. The energy dissipated by the radiation resistance, therefore, is the useful part of the total power dissipated. Because the actual power loss depends on the ohmic resistance, this resistance should be kept as low as possible. In the half-wave antenna, the radiation resistance is large compared to the ohmic resistance, and most of the available energy is radiated. The half-wave antenna is, therefore, a very efficient radiator for most purposes.

For a half-wave antenna fed at the center point, the radiation resistance is equal to 73 Ω. The reference point is the center of the antenna at the time of peak current flow. Ohmic resistance is referred to this point. The total resistance is of importance in matching the antenna to a transmission line.

Impedance Because the half-wave antenna has different conditions of voltage and current at different points, and because impedance is equal to the voltage across a circuit divided by the current through it, the impedance will vary along the length of the antenna. If E is divided by I at each point of the voltage and current curves in Fig. 5-16, the result is the impedance curve, Z. The impedance is about 73 Ω at the center point and rises to a value of about 2,500 Ω at the ends.

The impedance of the half-wave antenna usually is considered to be the impedance as *seen* by the transmitter at the input terminals. This impedance consists of both resistance and reactance. If the antenna is cut to a length of exact resonance, the reactance is zero and the impedance is purely resistive. However, if the antenna is longer or shorter than resonance, reactance is present. When the antenna is made shorter, capacitive reactance is present; when the antenna is made longer, inductive reactance is present.

The impedance at the antenna input terminals is important in terms of power efficiency. If the transmitter is feeding a nonresonant antenna, a power loss is caused by the reactive component of the antenna impedance. Conversely, if the frequency of the transmitter is changed, the electrical length of the antenna also changes. If the frequency is made somewhat higher, the electrical length is made greater, and inductive reactance is added to the impedance. If the frequency is lowered, the electrical length is shortened, and capacitive reactance is added to the impedance.

6

CHAPTER

High-frequency dipole antennas

A MYTH AROSE IN RADIO COMMUNICATIONS CIRCLES SOME TIME AGO. PEOPLE CAME TO believe (especially in the ham and CB communities) that large antenna arrays are absolutely necessary for effective communications, especially over long distances. Overlooked, almost to the point of disdain, were effective (but simple) antennas that can be erected by inexperienced people and made to work well. The simple *dipole*, or *doublet*, is a case in point. This antenna is also sometimes called the *Hertz*, or *Hertzian*, antenna because radio pioneer Heinrich Hertz reportedly used this form in his experiments.

The dipole is a balanced antenna consisting of two radiators (Fig. 6-1) that are each a quarter wavelength, making a total of a half wavelength. The antenna is usually installed horizontally with respect to the earth's surface, so it produces a horizontally polarized signal.

In its most common configuration (Fig. 6-1), the dipole is supported at each end by rope and end insulators. The rope supports are tied to trees, buildings, masts, or some combination of such structures.

The length of the antenna is a half wavelength. Keep in mind that the physical length of the antenna, and the theoretical electrical length, are often different by about five percent. A free space, half wavelength is found from:

$$L = \frac{492}{F_{\text{MHz}}} \text{ feet} \qquad \textbf{[6.1]}$$

In a perfect antenna, that is self-supported many wavelengths away from any object, Eq. 6.1 will yield the physical length. But in real antennas, the length calculated

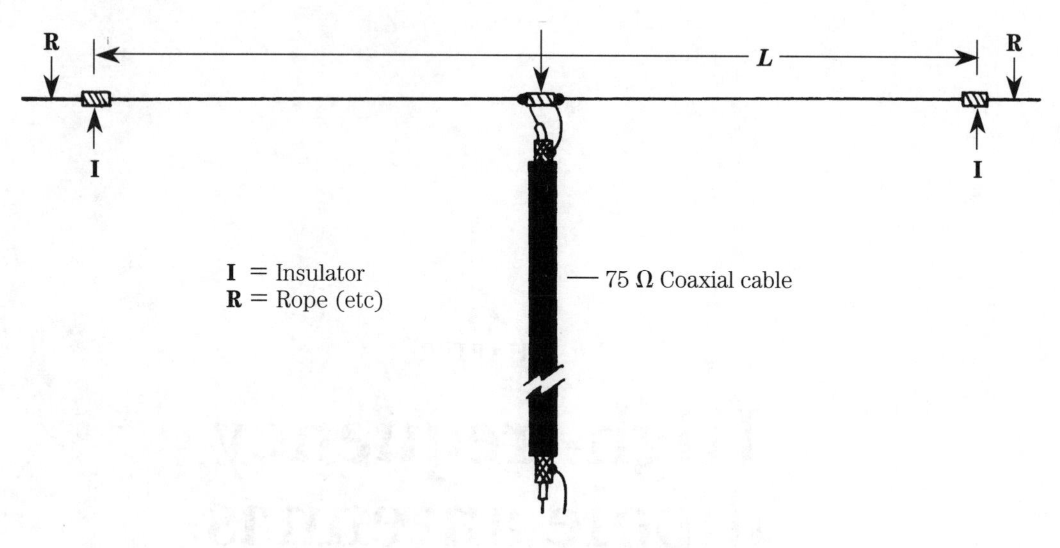

6-1 Simple halfwave dipole antenna.

above is too long. The physical length is shortened about five percent because of the capacitive effects of the end insulators. A more nearly correct approximation (remember that word, it's important) of a half-wavelength antenna is:

$$L = \frac{468}{F_{MHz}} \text{ feet} \qquad \textbf{[6.2]}$$

Where:

L is the length of a half-wavelength radiator, in feet
F_{MHz} is the operating frequency in megahertz

Example Calculate the approximate physical length for a half wavelength dipole operating on a frequency of 7.25 MHz.
Solution:

$$L = \frac{468}{F_{MHz}} \text{ feet}$$

$$L = \frac{468}{7.25} \text{ feet} = 64.55 \text{ feet}$$

or, restated another way:

$$L = 64'\ 6.6''$$

It is unfortunate that a lot of people accept Eq. 6.2 as a universal truth. Perhaps abetted by books and articles on antennas that fail to reveal the full story, too many people install dipoles without regard for reality. The issue is *resonance*. An antenna is a complex RLC network. At some frequency, it will appear like an inductive reactance ($X_L = +jX$), and at others it will appear like a capacitive reactance ($X_c = -jX$). At a specific frequency, the reactances are equal in magnitude, but opposite in sense, so they cancel each other out: $X_L - X_c = 0$. At this frequency, the impedance is purely resistive, and the antenna is said to be resonant.

The goal in erecting a dipole is to make the antenna resonant at a frequency that is inside the band of interest, and preferably in the portion of the band most often used by the particular station. Some of the implications of this goal are covered later on, but for the present, assume that the builder will have to custom tailor the length of the antenna. Depending on several local factors (among them, nearby objects, the shape of the antenna conductor, and the length/diameter ratio of the conductor) it might prove necessary to add, or trim, the length a small amount to reach resonance.

The dipole feedpoint

The dipole is a half wavelength antenna fed in the center. Figure 6-2 shows the voltage (V) and current (I) distributions along the length of the half-wavelength radiator element. The feedpoint is at a voltage minima and a current maxima, so you can assume that the feedpoint is a current antinode.

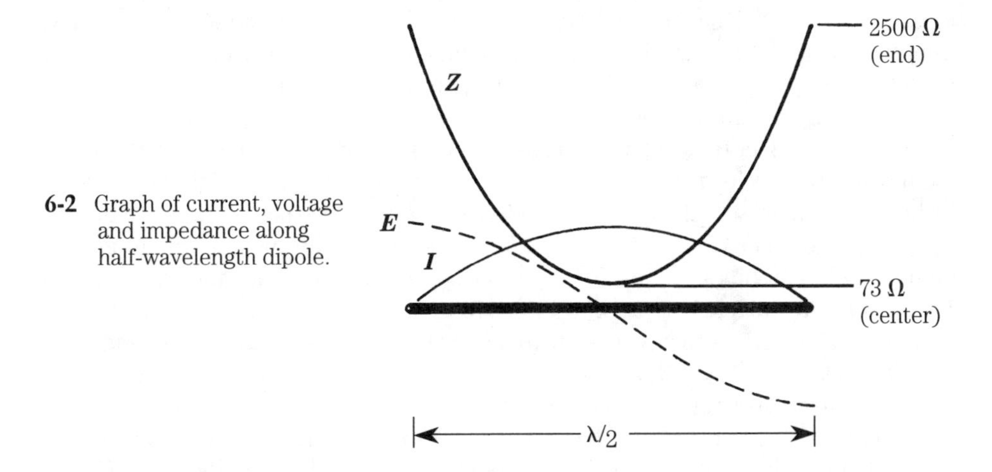

6-2 Graph of current, voltage and impedance along half-wavelength dipole.

At resonance, the impedance of the feedpoint is $R_o = V/I$. There are two resistances that make up R_o. The first, is the ohmic losses that generate nothing but heat when the transmitter is turned on. These ohmic losses come from the fact that conductors have electrical resistance and electrical connections are not perfect (even when properly soldered). Fortunately, in a well made dipole these losses are almost negligible. The second contributor is the *radiation resistance*, R_r, of the antenna. This resistance is a hypothetical concept that accounts for the fact that RF power is radiated by the antenna. The radiation resistance is the fictional resistance that would dissipate the amount of power that is radiated away from the antenna.

For example, suppose we have a large diameter conductor used as an antenna, and it has negligible ohmic losses. If 1,000 watts of RF power is applied to the feedpoint, and a current of 3.7 amperes is measured, what is the radiation resistance?

$$R_r = \frac{P}{I^2} \quad \textbf{[6.3]}$$

$$R_r = \frac{(1{,}000 \text{ watts})}{(3.7)^2} \quad \textbf{[6.4]}$$

$$R_r = 73 \text{ ohms} \quad \textbf{[6.5]}$$

It is always important to match the feedpoint impedance of an antenna to the transmission-line impedance. Maximum power transfer always occurs (in any system) when the source and load impedances are matched. In addition, if some applied power is not absorbed by the antenna (as happens in a mismatched system), then the unabsorbed portion is reflected back down the transmission line towards the transmitter. This fact gives rise to standing waves, and the so-called standing wave ratio (SWR or VSWR) discussed in chapter 3. This is a problem to overcome.

Matching antenna feedpoint impedance seems to be simplicity itself because the free space feedpoint impedance of a simple dipole is about 73 Ω, seemingly a good match to 75-Ω coaxial cable. Unfortunately, the 73-Ω feedpoint impedance is almost a myth. Figure 6-3 shows a plot of approximate radiation resistance (R_r) versus height above ground (as measured in wavelengths). As before, we deal in approximations in Fig. 6-3; in this case, the ambiguity is introduced by ground losses.

Despite the fact that Fig. 6-3 is based on approximations, you can see that radiation resistance varies from less than 10 Ω, to almost 100 Ω, as a function of height. At heights of many wavelengths, this oscillation of the curve settles down to the free space impedance (72 Ω). At the higher frequencies, it might be possible to install a dipole at a height of many wavelengths. In the 2-meter amateur radio band (144 to 148 MHz), one wavelength is around 6.5 feet (i.e., 2 meters × 3.28 ft/meter), so "many wavelengths" is relatively easy to achieve at reasonably attainable heights. In the 80-meter band (3.5 to 4.0 MHz), however, one wavelength is on the order of 262 feet, so "many wavelengths" is a practical impossibility.

There are three tactics that can be followed. First, ignore the problem altogether. In many installations, the height above ground will be such that the radiation resistance will be close enough to present only a slight impedance mismatch to a standard coaxial cable. The VSWR is calculated (among other ways) as the ratio:

1. $Z_o > R_r$:

$$VSWR = \frac{Z_o}{R_r} \quad \textbf{[6.6]}$$

2. $Z_o < R_r$:

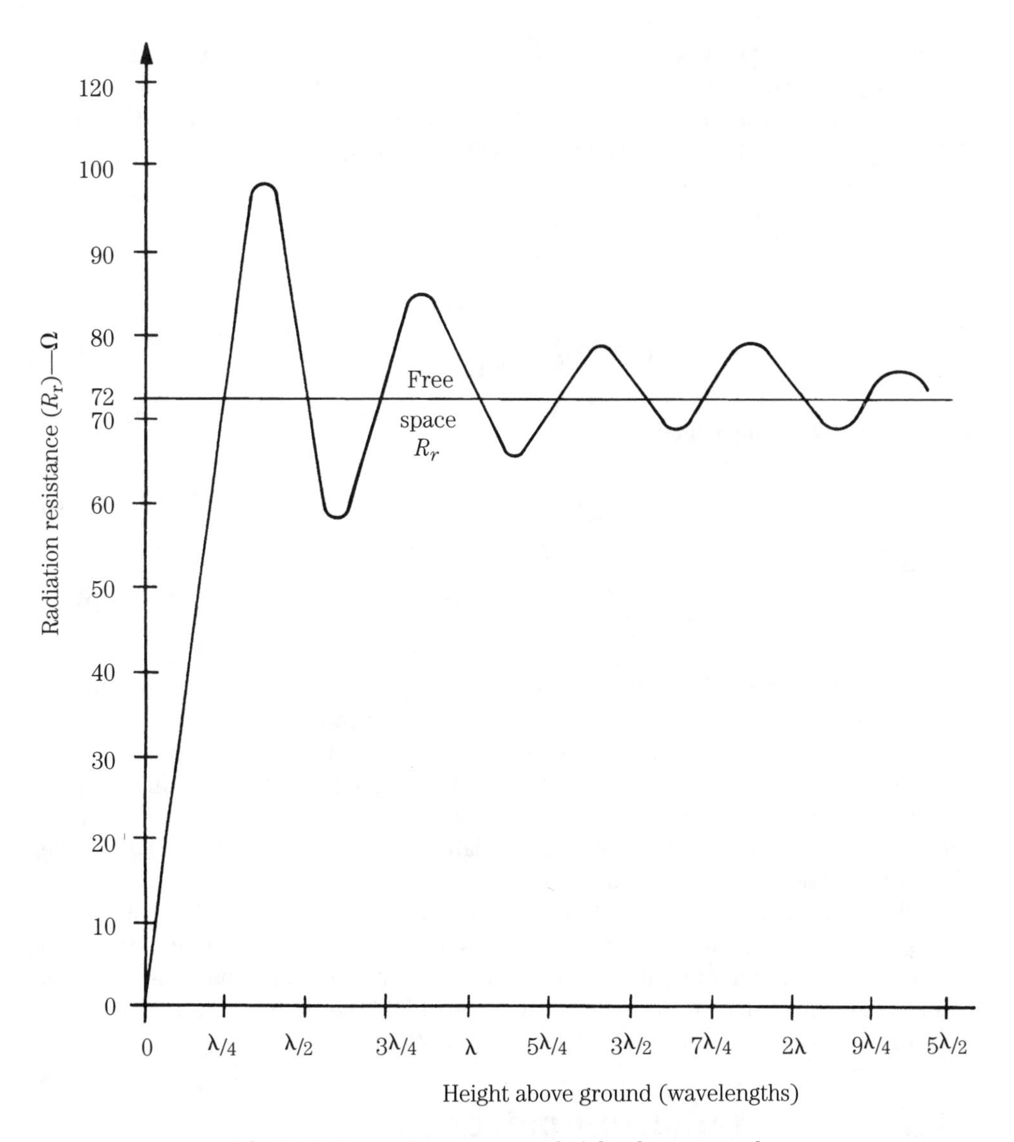

6-3 Radiation resistance versus height above ground.

$$VSWR = \frac{R_r}{Z_o} \quad \textbf{[6.7]}$$

Where:

Z_o is the coaxial-cable characteristic impedance
R_r is the radiation resistance of the antenna

Consider an antenna mounted at a height somewhat less than a quarter wavelength, such that the radiation resistance is 60 **Ω**. Although not recommended as good engineering practice (there are sometimes practical reasons) it is nonetheless necessary to install a dipole at less than optimum height. So, if that becomes necessary, what are the implications of feeding a 60-Ω antenna with either 52-Ω or 75-Ω standard coaxial cable? Some calculations are revealing:

For 75-Ω coaxial cable:

$$VSWR = \frac{Z_o}{R_r} \quad \textbf{[6.8]}$$

$$VSWR = \frac{75\ \Omega}{60\ \Omega} = 1.25{:}1 \quad \textbf{[6.9]}$$

For 52-Ω coaxial cable:

$$VSWR = \frac{R_r}{Z_o} \quad \textbf{[6.10]}$$

$$VSWR = \frac{60\ \Omega}{52\ \Omega} = 1.15{:}1 \quad \textbf{[6.11]}$$

In neither case is the VSWR created by the mismatch too terribly upsetting.

The second approach is to mount the antenna at a convenient height, and use an impedance matching scheme to reduce the VSWR. In chapter 19, you will find information on various suitable (relatively) broadbanded impedance matching methods including Q-sections, coaxial impedance transformers, and broadband RF transformers. Homebrew and commercially available transformers are available to cover most impedance transformation tasks.

The third approach is to mount the antenna at a height (Fig. 6-3) at which the expected radiation resistance crosses a standard coaxial cable characteristic impedance. The best candidate seems to be a height of half wavelength because the radiation resistance is close to the free space value of 72 Ω . . . and is thus a good match for 75-Ω coaxial cable (such as RG-11/U or RG-59/U).

The dipole radiation pattern

The radiation patterns of various antennas are covered in this book. Some basic theory of patterns is repeated at each section, not to fill space, but rather to drive home a point and refresh reader memories. We keep harking back to the concepts *directivity* and *gain* (which are actually different expressions of the same thing).

Antenna theory recognizes a point of reference called the *isotropic radiator*. This device is a theoretical construct consisting of a spherical point source of RF radiation in all directions (see chapter 2). It is truly *omnidirectional* ("all directions") because it creates an ever expanding sphere as the RF wavefront propagates outward. Antenna gain is a measure of how the antenna focuses available power away from a spherical wavefront to a limited number of directions (two, in the case of the dipole). Thus are related the concepts *directivity* and *gain*.

Always keep in mind that *directivity and gain are specified in three dimensions*. Too many times, authors (including me) simplify the topic too much by publishing only part of the radiation pattern (azimuth aspect). In other words, the reader is given a pattern viewed from above that shows the directivity in the horizontal plane. But signal does not propagate away from an antenna in an infinitely thin sheet, as such presentations seem to imply; but, rather, has an elevation extent in addition to the azimuth extent. Thus, proper evaluation of an antenna takes into consideration both horizontal and vertical plane patterns.

Figure 6-4 shows the radiation pattern of a dipole antenna in free space "in the round." In the horizontal plane (6-4A), when viewed from above, the pattern is a "fig-

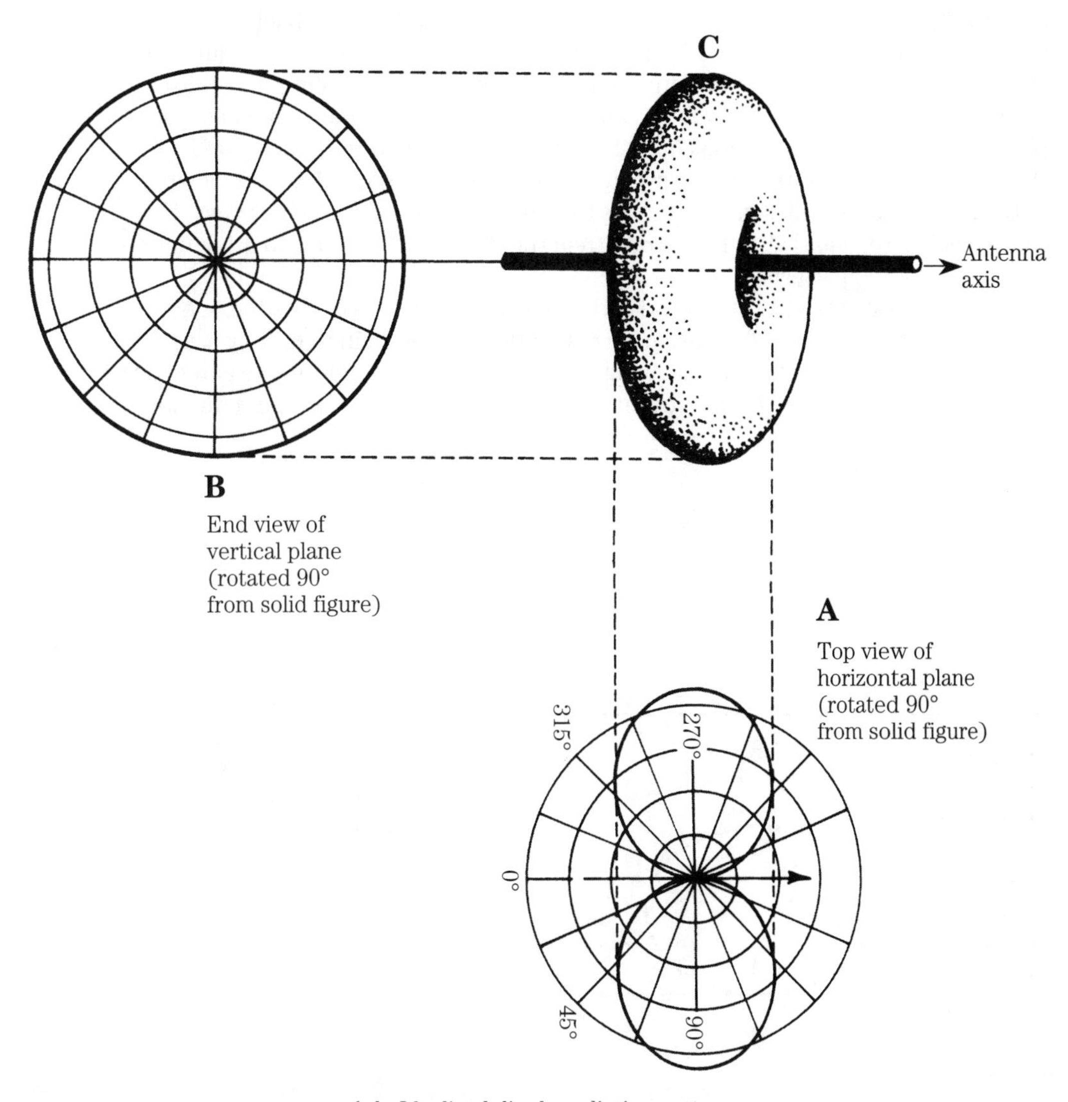

6-4 Idealized dipole radiation pattern.

ure-8" that exhibits bidirectional radiation. Two main "lobes" contain the RF power from the transmitter, with sharp nulls of little or no power off the ends of the antenna axis. This pattern is the classical dipole pattern that is published in most antenna books.

Also shown, however, is the vertical plane pattern for a dipole antenna in free space. Note that when sliced in this aspect the radiation pattern is circular (Fig. 6-4B). When the two patterns are combined in-the-round, you can see the three-dimensional "doughnut" shaped pattern (Fig. 6-4C) that most nearly approximates the true pattern of an unobstructed dipole in free space.

When a dipole antenna is installed close to the earth's surface, not in free space, as is the case at most stations, the pattern is distorted from that of Fig. 6-4. Two effects must be taken into consideration. First, and most important, is the fact that the signal from the antenna is reflected from the surface and bounces back into space. This signal will be phase shifted both by the reflection and by the time required for the transit to occur. At points where the reflected wave combines in-phase with the radiated signal, the signal is reinforced; and in places where it combines out of phase, the signal is attenuated. Thus, the reflection of signal from the ground alters the pattern from the antenna. The second factor is that the ground is lossy, so not all of the signal is reflected; some of it heats the ground underneath the antenna. Thus, the signal is attenuated at a greater rate than the inverse square law, so it further alters the expected pattern.

Figure 6-5 shows patterns typical of dipole antennas installed close to the earth's surface. The views in this illustration correspond to Fig. 6-4B, in that, they are looking at the vertical plane from a line along the antenna axis. Therefore, the antenna is represented by "R" in each case shown. Figure 6-5A shows the case for a dipole in-

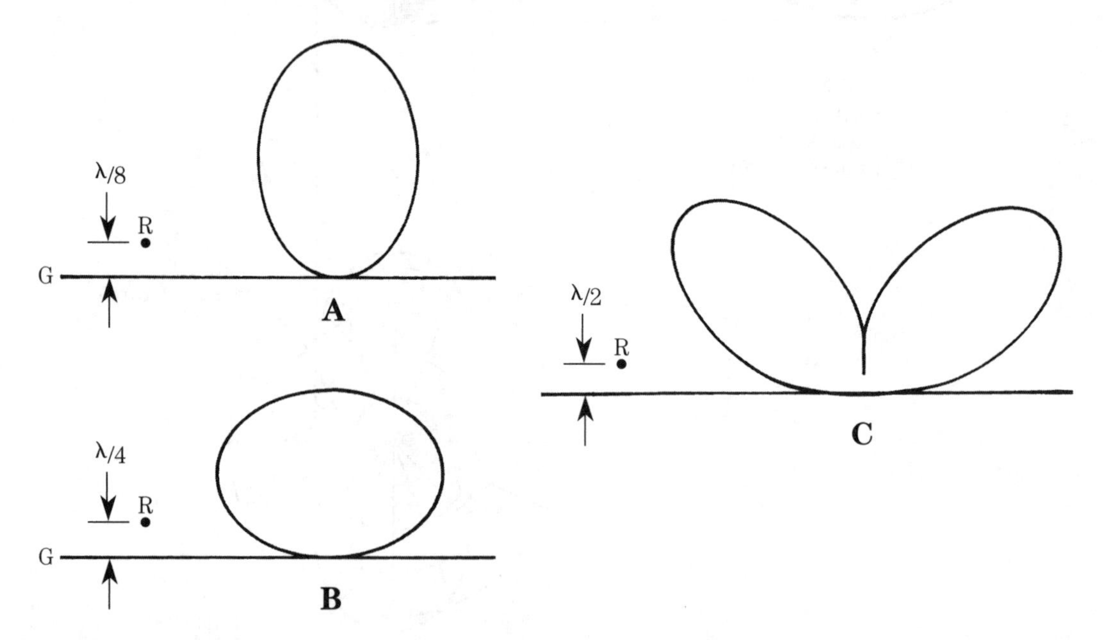

6-5 Vertical extent of dipole antenna at A) ⅛ wavelength, B) ¼ wavelength, and C) ½ wavelength above ground.

stalled at one-eighth wavelength above the surface. For this antenna, most of the RF energy is radiated almost straight up (not very useful). This type of antenna is basically limited to ground wave and very short skip (when available). The second case (Fig. 6-5B) shows the pattern when the antenna is a quarter wavelength above the earth's surface. Here the pattern is flattened, but it still shows considerable energy in the vertical direction (where it is useless). Finally, you can see the pattern obtained when the antenna is installed a half wavelength above the surface. In this case, the pattern is best for long-distance work because energy is redirected away from straight up into lobes at relatively shallow angles.

Dipole construction and installation techniques

The ideal dipole antenna should be installed at a very high altitude where its performance resembles the free-space model . . . or so says the "standard wisdom." Unfortunately, complying with the standard wisdom it is impossible even for antennas in the higher end of the HF spectrum. Given that the dipole feedpoint impedance is a good match for 75-Ω coaxial cable, and that the pattern is ideal for long-distance work, when the antenna is installed at a height of a half wavelength above the surface, it is a good idea to try installing the antenna at that height.

Building and installing simple dipoles is not a terribly difficult task. Figure 6-6 shows the method for building the antenna. First, cut the wire radiator elements to the approximate length demanded by Eq. 6.2 plus 12 to 24 inches extra; each element will finally be quarter wavelength long. The wire can be either hard-drawn copper wire or *Copperweld*®. The latter is a special tough-service antenna made with a copper coating over a steel core. The RF resistance of this wire at frequencies above 1 MHz is the same as for solid copper wire because of the *skin effect* (alternating currents such as RF flow on the outer surface of the conductor only). At 160 meters, the "skin effect" depth is only 50 microns (2 mils), while at 10 meters it is only 12 microns (0.5 mils). Thus, we are able to have the advantage of copper conductivity at the same time as we gain the strength of steel wire.

Two end insulators are required, but assembly of both are the same. Pass the wire through the hole in the insulator (refer to Fig. 6-6) to a length of about 6 to 12 inches. Wrap the wire back on itself and wind it around the portion of the wire that is left on the other side of the insulator. Make this a permanent connection by soldering it and clipping off excess wire. The solder will not provide mechanical strength, so don't even try to use it for such. The purpose of the solder is to make a good electrical connection in the presence of corrosion.

Similarly, fix the antenna wires to the center insulator unless you plan to use one of the special center insulators that are now on the market. Make these connections temporary until after the antenna is tuned and tested. Tuning of a dipole might require either lengthening or shortening of the radiators.

The transmission line (usually coaxial cable) is connected to the antenna wire at the center insulator in the manner shown in Fig. 6-6. The center conductor is connected to one radiator element, and the shield of the coax is connected to the other radiator element. Strain relief is necessary for the coaxial cable. If relief is not provided, the cable will break after only a short period of service. The simplest form of

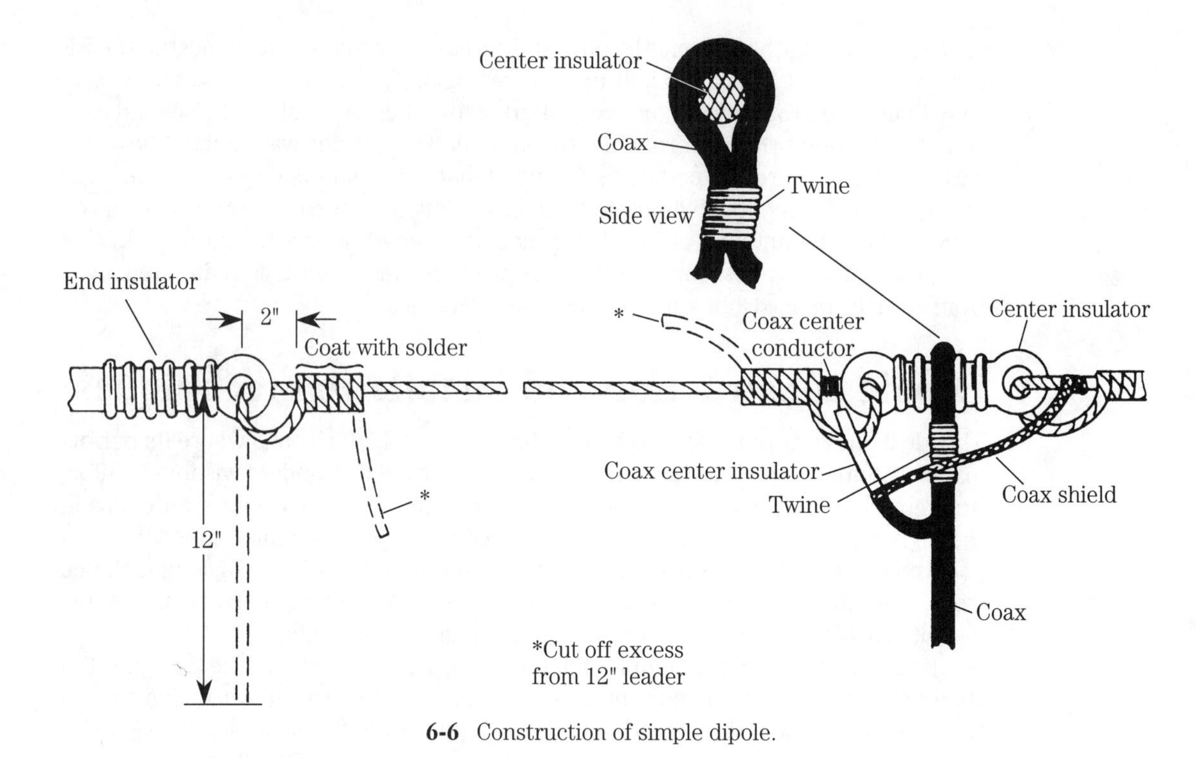

6-6 Construction of simple dipole.

strain relief is shown in Fig. 6-6: wrap the cable once around the insulator and tie it off with twine. Some commercial center insulators offer a strain relief hole or other mechanism.

Many people prefer to use a 1:1 BALUN transformer at the feedpoint of the dipole antenna (see Fig. 6-7). This transformer has a 1:1 impedance ratio, so it does not provide any matching. But rather, it is said to balance the currents flowing in the two radiators, and prevents radiation from the feedline. Although this claim has been controversial for some time, and the issue is still not firmly resolved, the best current evidence suggests that the pattern of a dipole close to the earth's surface is most nearly like the ideal pattern if a 1:1 BALUN transformer is used at the feedpoint. in Fig. 6-7, the BALUN transformer also acts as the center insulator, so no other arrangement is needed.

Tuning the dipole antenna

There are two issues to address when tuning an antenna (any antenna, not just the dipole): *resonance* and *impedance matching*. Although frequently treated in the literature as the same issue, they are not. This section deals mostly with the process of tuning the antenna to resonance. Although not all forms of antenna are resonant, the dipole *is* an example of a resonant antenna.

There is a lot of misinformation abroad concerning the tuning of antennas. Perhaps much of what is believed comes from the fact that VSWR is used as the indicator of both impedance matching and resonance. Quite a few people honestly, but

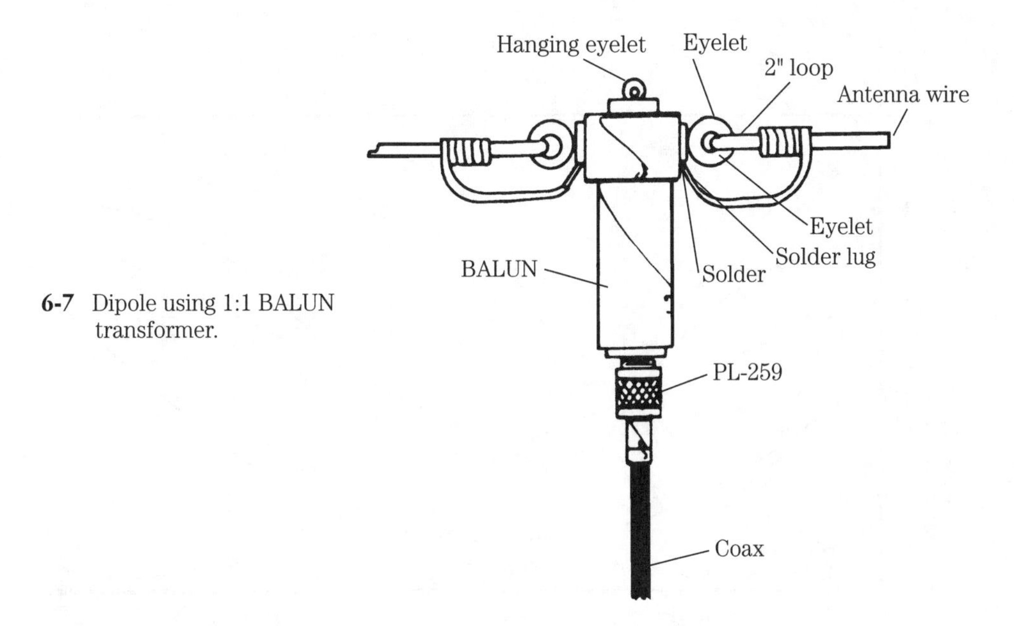

6-7 Dipole using 1:1 BALUN transformer.

erroneously, believe that the VSWR can be "tuned out" by adjusting the length of the feedline. That myth probably derives from the fact that voltage or current sensing instruments are used for VSWR measurement, and these are affected by transmission line length. But that fact is caused by a weakness in the instruments, not a fact of radio physics.

There is only one proper way to tune a dipole antenna: *adjust the length of the antenna elements*, not the transmission line. It was in order to make these adjustments, that we purposely did not initially tell you to solder the electrical connections at the center insulator.

Resonance The indicator of resonance is the minimum point in the VSWR curve. Figure 6-8 shows a graph of VSWR vs. frequency for several different cases. Curve "A" represents a disaster: a high VSWR all across the band. The actual value of VSWR can be anything from about 3.5:1 to 10:1, or thereabouts, but the cause is nonetheless the same: the antenna is either open or shorted; or it is so far off resonance as to appear to be open or shorted to the VSWR meter.

Curves "B" and "C" represent antennas that are resonant within the band of interest. Curve "B" represents a broadbanded antenna that is relatively flat all across the band, and does not exhibit excessive VSWR until the frequency is outside of the band. Curve "C" is also resonant within the band, but this antenna has a much higher Q than curve "B." In the naive sense, the broadbanded antenna is best, but that statement is true only if the broadness is not purchased at the expense of efficiency. Losses tend to broaden the antenna, but also reduce its effectiveness. So, if broadbandedness is purchased at the risk of increased loss, then it is less than desirable.

Curves "D" and "E" are resonant outside the band of interest. The curve marked "D" is resonant at a frequency on the low side of the band so that dipole is too long. In this case, you need to shorten the antenna a bit to raise the resonant point inside the band. Curve "E" represents an antenna that is resonant outside the upper limit

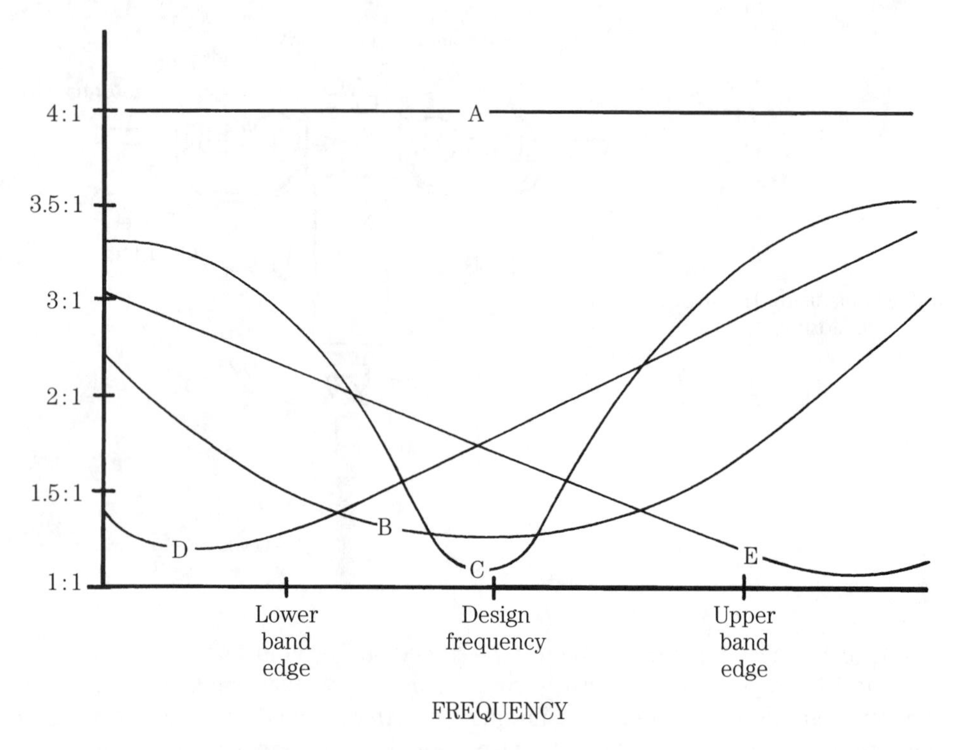

6-8 VSWR-vs-frequency for several cases.

of the band, so this antenna is too short, and must be lengthened. Because this situation is possible, the antenna elements are typically made longer than needed when they are first cut.

How much to cut? That depends on two factors: how far from the desired frequency the resonant point is found, and which band is being used. The latter requirement comes from the fact that the "frequency per unit length" varies from one band to another. Look at an example of how to calculate this figure. The procedure is simple:

1. Calculate the length required for the upper end of the band.
2. Calculate the length required for the bottom end of the band.
3. Calculate the difference in lengths for the upper and lower ends of the band.
4. Calculate the width of the band in kilohertz by taking the difference between the upper frequency limit and the lower frequency limit.
5. Divide the length difference by the frequency difference; the result is in kilohertz per unit length.

Example Calculate the frequency change per unit of length for 80 meters and for 15 meters.

Solution For 80 meters (3.5 to 4.0 MHz):

1. $L_{\text{ft}} = \frac{468}{4 \text{ MHz}} = 117$ feet

2. $L_{\text{ft}} = \frac{468}{3.5 \text{ MHz}} = 133.7$ feet

3. Difference in length: 133.7 ft – 117 ft = 16.7 feet
4. Frequency difference: 4000 kHz – 3500 kHz = 500 kHz
5. Calculate $\frac{\text{frequency}}{\text{unit length}} : \frac{500 \text{ kHz}}{16.7 \text{ feet}} = 30$ kHz/ft

For 15 meters (21.0 – 21.45 MHz):

1. $L_{\text{ft}} = \frac{468}{21.45} = 21.82$ ft

2. $L_{\text{ft}} = \frac{468}{21} = 22.29$ ft

3. Difference in length: 22.29 ft – 21.82 ft = 0.47 ft
4. Convert to inches: 0.47 ft × 12 in/ft = 5.64 inches
5. Frequency difference: 21,450 kHz – 21,000 kHz = 450 kHz
6. Calculate $\frac{\text{frequency}}{\text{unit length}} : \frac{450 \text{ kHz}}{5.64 \text{ in}} = 80$ kHz/in

At 80 meters, the frequency change per foot is small, but at 15 meters small changes can result in very large frequency shifts. You can calculate approximately how much to add (or subtract) from an antenna under construction from this kind of calculation. If, for example, you design an antenna for the so-called "net frequency" on 15 meters (21,390 kHz), and find the actual resonant point is 21,150 kHz, the frequency shift required is 21,390 – 21,150, or 240 kHz. To determine how much to add or subtract (as a first guess):

1. The factor for 15 meters is 80 kHz/inch, which is the same as saying 1 inch/80 kHz.
2. The required frequency shift is 240 kHz.
3. Therefore:

$$\textit{Length change} = 240 \text{ kHz} \times \frac{1 \text{ in.}}{80 \text{ kHz}} \qquad \textbf{[6.12]}$$

$$\textit{Length change} = 3 \text{ in.} \qquad \textbf{[6.13]}$$

Each side of the antenna must be changed by half of the length calculated above, or 1.5 inches. Because the first resonant frequency is less than the desired frequency, the length should be shortened 1.5 inches. Once the length is correct, as proven by the VSWR curve, the connections at the center insulator are soldered and made permanent, and the antenna rehoisted to the operating level.

Impedance matching The difference between resonance and impedance matching is seen in the value of the VSWR minima. While the minima indicates the resonant point, the value of that minima is a measure of the relationship between the feedpoint impedance of the antenna and the characteristic impedance of the transmission line. Earlier in this chapter you learned that:

1. $Z_o > R_r$:

$$VSWR = \frac{Z_o}{R_r} \quad [6.14]$$

2. $Z_o < R_r$:

$$VSWR = \frac{R_r}{Z_o} \quad [6.15]$$

Where:

Z_o is the coaxial-cable characteristic impedance
R_r is the radiation resistance of the antenna

Although knowledge of the VSWR will not show which situation is true, you can know that there is a high probability that *one* of them is true, and you can experiment to find which is the case. Of course, if the VSWR is less than about 1.5:1 or 2:1, then forget about it . . . the improvement is not generally worth the expense and cost. When coupled to a transmitter that is equipped with the tunable output network (most tube-type transmitters or final amplifiers), then it can accommodate a relatively wide range of reflected antenna impedances. But modern solid-state final amplifiers tend to be a little more picky about the load impedance. For these transmitters, a coax-to-coax *antenna tuning unit (ATU)* is needed.

Other dipoles

Thus far, the dipoles covered in this chapter have been the classic form, in which a half-wavelength single-conductor radiator element is connected to a coaxial transmission line. This antenna is typically installed horizontally at a half wavelength above the earth's surface (or wherever convenient if that is impossible). This section looks at other forms of the dipole. Some of these dipoles are in every way the equal of the horizontal dipole, and others are basically compensation antennas in that they are used when a proper dipole is not practical.

Inverted-vee dipole

The *inverted-vee dipole* is a half-wavelength antenna fed in the center like a dipole. By the rigorous definition, the inverted-vee is merely a variation on the dipole theme. But in this form of antenna (Fig. 6-9), the center is elevated as high as possible from the earth's surface, but the ends droop to very close to the surface. Angle a can be almost anything convenient, provided that $a > 90$ degrees; typically, most inverted-vee antennas use an angle of about 120 degrees. Although essentially a compensation antenna for use when the dipole is not practical, many operators believe that it is essentially a better performer on 40 and 80 meters in cases where the dipole cannot be mounted at a half wavelength (64 feet or so).

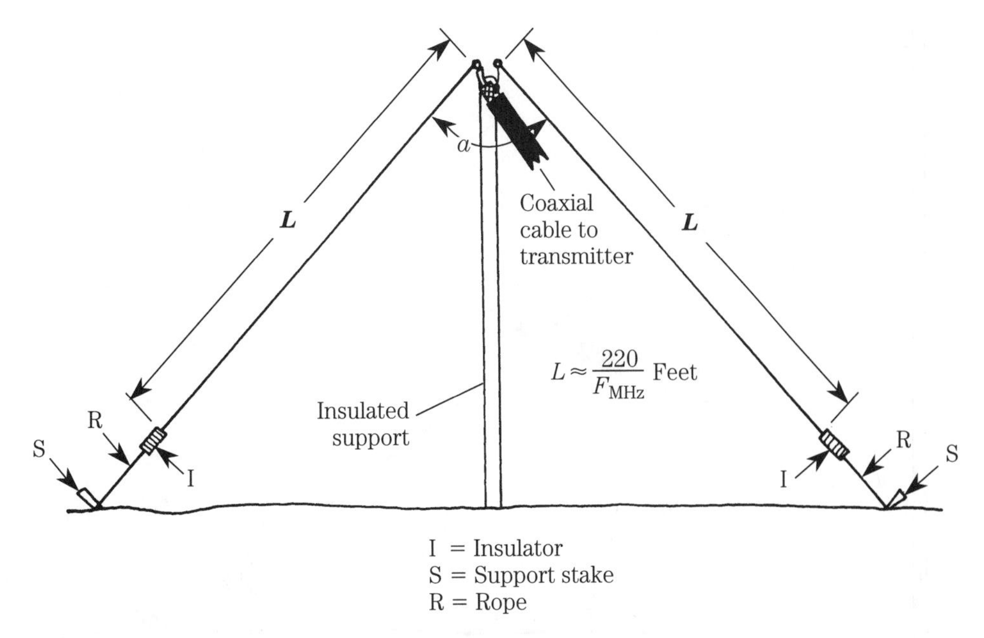

6-9 Inverted-vee dipole.

By sloping the antenna elements down from the horizontal to an angle (as shown in Fig. 6-9), the resonant frequency is effectively lowered. Thus, the antenna will need to be shorter for any given frequency than a dipole. There is no absolutely rigorous equation for calculation of the overall length of the antenna elements. Although the concept of "absolute" length does not hold for regular dipoles, it is even less viable for the inverted-vee. There is, however, a "rule of thumb" that can be followed for a starting point: make the antenna about 6 percent shorter than a dipole for the same frequency. The initial cut of the antenna element lengths (each quarter wavelength) is:

$$L = \frac{220}{F_{MHz}} \text{ feet} \qquad [6.16]$$

After this length is determined, the actual length is found from the same cut-and-try method used to tune the dipole in the previous section.

Bending the elements downwards also changes the feedpoint impedance of the antenna, and narrows its bandwidth. Thus, some adjustment in these departments is in order. You might want to use an impedance matching scheme at the feedpoint, or an antenna tuner at the transmitter.

Sloping dipole ("sloper" or "slipole")

The *sloping dipole* (Fig. 6-10) is popular with those operators who need a low angle of radiation, and are not overburdened with a large amount of land to install the antenna. This antenna is also called the *sloper* and the *slipole* in various texts. The author prefers the term "slipole," in order to distinguish this antenna from a sloping

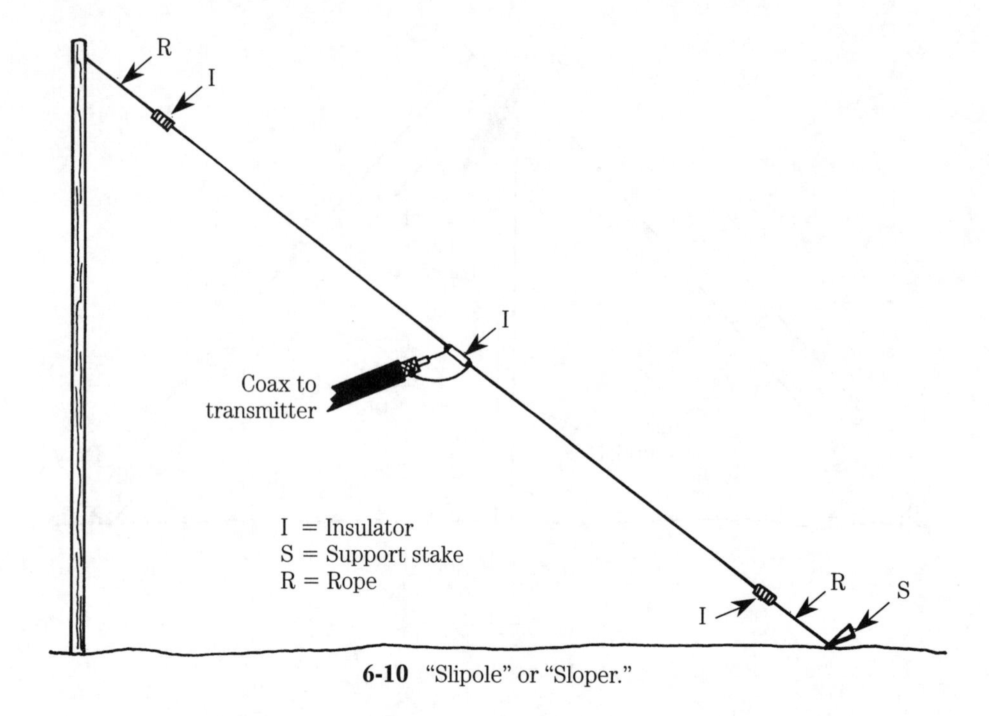

6-10 "Slipole" or "Sloper."

vertical of the same name. Whatever it is called, however, it is a half wavelength dipole that is built with one end at the top of a support, and the other end close to the ground, and being fed in the center by coaxial cable. Some of the same comments as obtained for the inverted-vee antenna also apply to the sloping dipole, so please see that section also.

Some operators like to arrange four sloping dipoles from the same mast such that they point in different directions around the compass (Fig. 6-11). A single four-position coaxial cable switch will allow switching a directional beam around the compass to favor various places in the world.

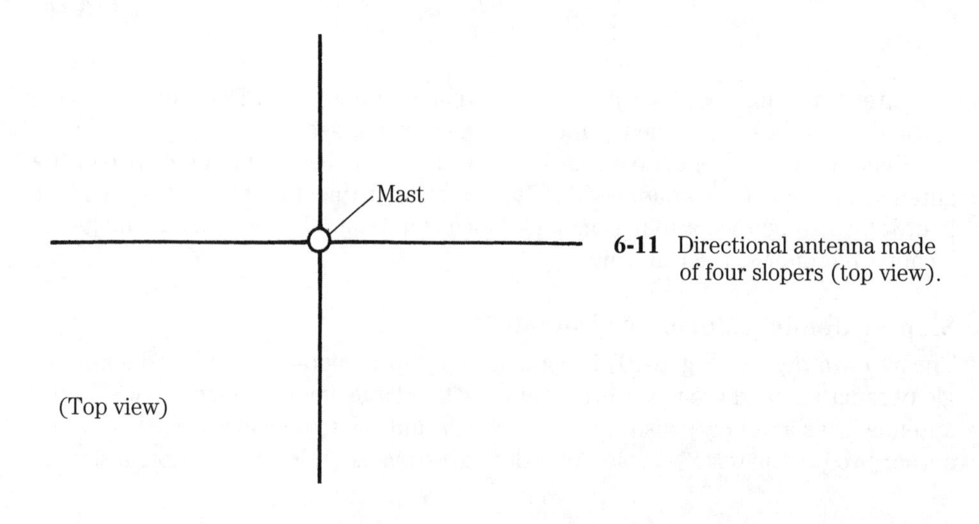

6-11 Directional antenna made of four slopers (top view).

Broadbanded dipoles

One of the rarely discussed aspects of antenna construction is that the length/diameter ratio of the conductor used for the antenna element is a factor in determining the bandwidth of the antenna. In general, the rule-of-thumb states that large crossectional area makes the antenna more broadbanded. In some cases, this rule suggests the use of aluminum tubing instead of copper wire for the antenna radiator. On the higher frequency bands that is a viable solution. Aluminum tubing can be purchased for relatively small amounts of money, and is both lightweight and easily worked with ordinary tools. But, as the frequency decreases, the weight becomes greater because the tubing is both longer and (for structural strength) must be of greater diameter. On 80 meters, aluminum tubing is impractical, and at 40 meters it is nearly so. Yet, 80 meters is a significant problem, especially for older transmitters, because the band is 500 kHz wide, and the transmitters often lack the tuning range for the entire band.

Some other solution is needed. Here are three basic solutions to the problem of wide bandwidth dipole antennas: *folded dipole*, *bowtie dipole*, and *cage dipole*.

Figure 6-12A shows the *folded dipole* antenna. This antenna basically consists of two half-wavelength conductors shorted together at the ends, and fed in the middle of one of them. The folded dipole is most often built from 300-Ω television antenna twinlead transmission line. Because the feedpoint impedance is nearly 300 Ω, the same type of twinlead can also be used for the transmission line. The folded dipole will exhibit excellent wide bandwidth properties, especially on the lower bands.

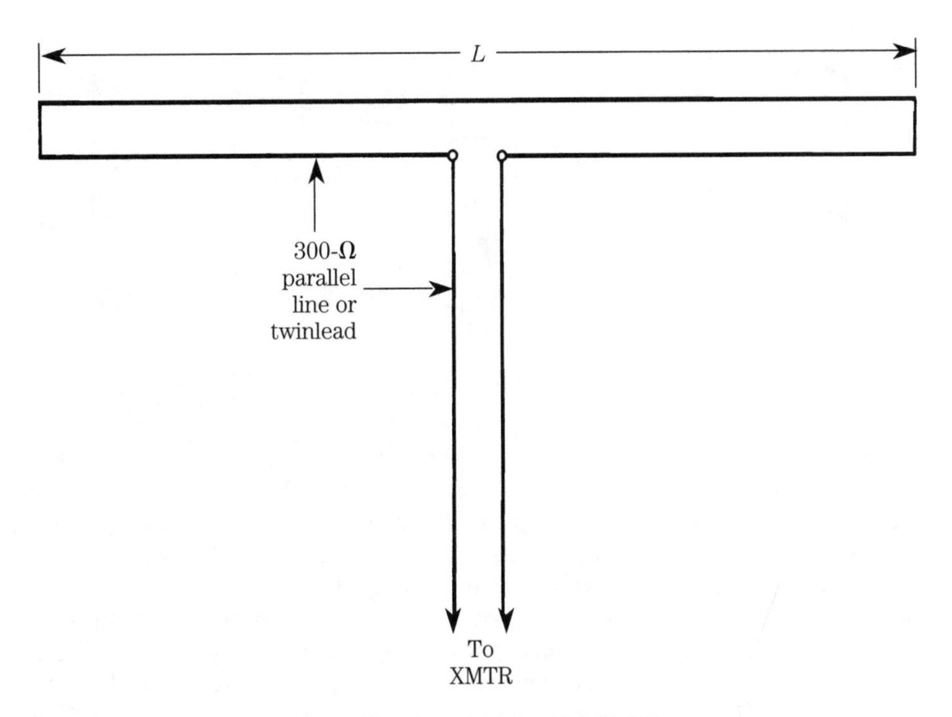

6-12A Folded dipole fed with 300-Ω line.

A disadvantage of this form of antenna is that the transmitter has to match the 300-Ω balanced transmission line. Unfortunately, most modern radio transmitters are designed to feed coaxial-cable transmission line. Although an antenna tuner can be placed at the transmitter end of the feedline, it is also possible to use a 4:1 BALUN transformer at the feedpoint (Fig. 6-12B). This arrangement makes the folded dipole a reasonable match to 52-Ω or 75-Ω coaxial-cable transmission line.

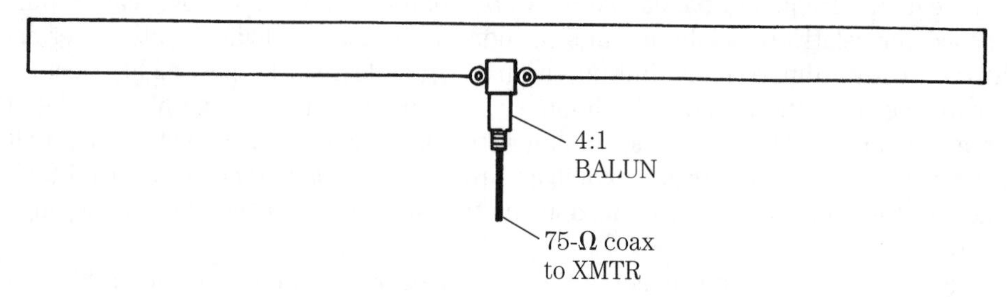

6-12B Folded dipole fed with coaxial cable.

Another method for broadbanding the dipole is to use two identical dipoles fed from the same transmission line, and arranged to form a "bowtie" as shown in Fig. 6-13. The use of two identical dipole elements on each side of the transmission line has the effect of increasing the conductor crossectional area so that the antenna has a slightly improved length/diameter ratio.

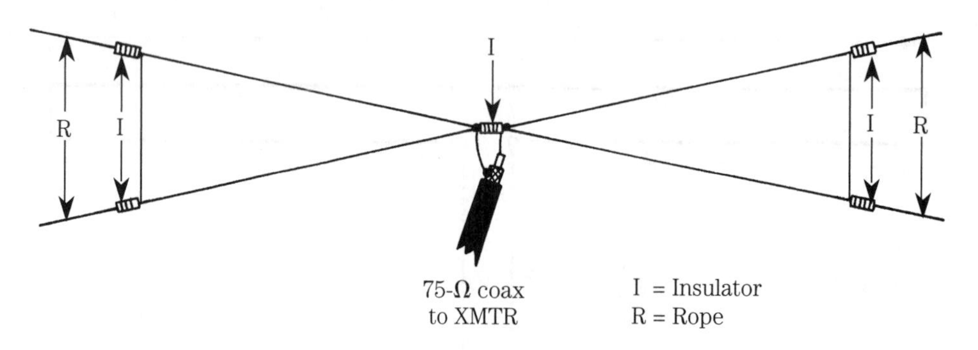

6-13 Bowtie dipole.

The *bowtie dipole* was popular in the 1930s and 1940s, and became the basis for the earliest television receiver antennas (TV signals are 3 to 5 MHz wide, so they require a broadbanded antenna). It was also popular during the 1950s as the so-called "Wonder Bar" antenna for 10 meters. It still finds use, but it has faded somewhat in popularity.

The *cage dipole* (Fig. 6-14) is similar in concept, if not construction, to the bowtie. Again, the idea is to connect several parallel dipoles together from the same transmission line in an effort to increase the apparent crossectional area. In the case of the cage diople, however, spreader disk insulators are constructed to keep the wires separated. The insulators can be built from plexiglass, lucite, or ceramic. They

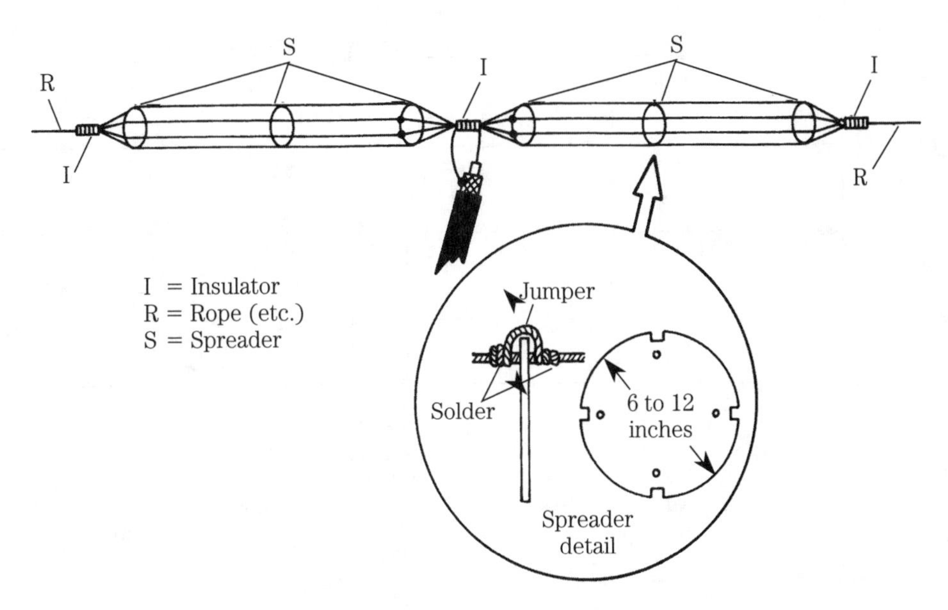

6-14 Cage dipole.

can also be constructed of materials such as wood, if the wood is properly treated with varnish, polyurethene, or some other material that prevents it from becoming waterlogged. The spreader disks are held in place with wire jumpers (see inset to Fig. 6-14) that are soldered to the main element wires.

A tactic used by some builders of both bowtie and cage dipoles is to make the elements *slightly* different lengths. This "stagger tuning" method forces one dipole to favor the upper end of the band, and the other to favor the lower end of the band. The overall result is a slightly flatter frequency response characteristic across the entire band. On the cage dipole, with four half-wavelength elements, it should be possible to overlap even narrower sections of the band in order to create an even flatter characteristic.

Shortened coil-loaded dipoles

The half-wavelength dipole is too long for some applications where real estate is at a premium. The solution for many operators is to use a coil-loaded shortened dipole such as shown in Fig. 6-15. A shortened dipole (i.e., one which is less than a half wavelength) is capacitive reactance. There is no reason why the loading coil cannot be any point along the radiator, but in Figs. 6-15A and 6-15B they are placed at 0 percent and 50 percent of the element length, respectively. The reason for this procedure is that it makes the calculation of coil inductances easier, and it also represents the most common practice.

Figure 6-15C shows a table of inductive reactances as a function of the percentage of a half wavelength, represented by the shortened radiator. It is likely that the percentage figure will be imposed on you by the situation, but the general rule is to pick the largest figure consistent with the available space. For example, suppose you

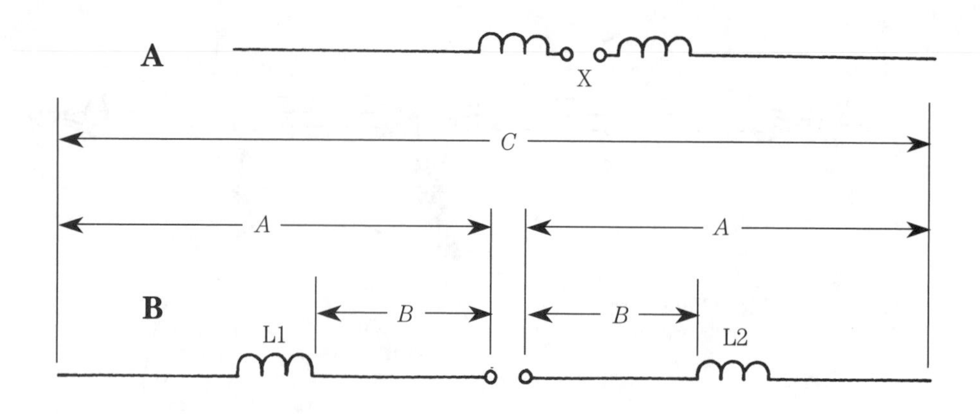

C

$L_1 = L_2 = L$

$C = 2A$

$A = \frac{1}{2}C$

$$C = \frac{468\,(M)}{F_{MHZ}}$$

$O \leq M \leq 1$

Percent of half wavelength	Coils at feedpoint	Coils at middle of radiators
20%	1800 Ω	2800 Ω
30%	950 Ω	1800 Ω
40%	700 Ω	1800 Ω
50%	500 Ω	1300 Ω
60%	360 Ω	950 Ω
70%	260 Ω	700 Ω
80%	160 Ω	500 Ω
90%	75 Ω	160 Ω
95%	38 Ω	80 Ω
98%	15 Ω	30 Ω

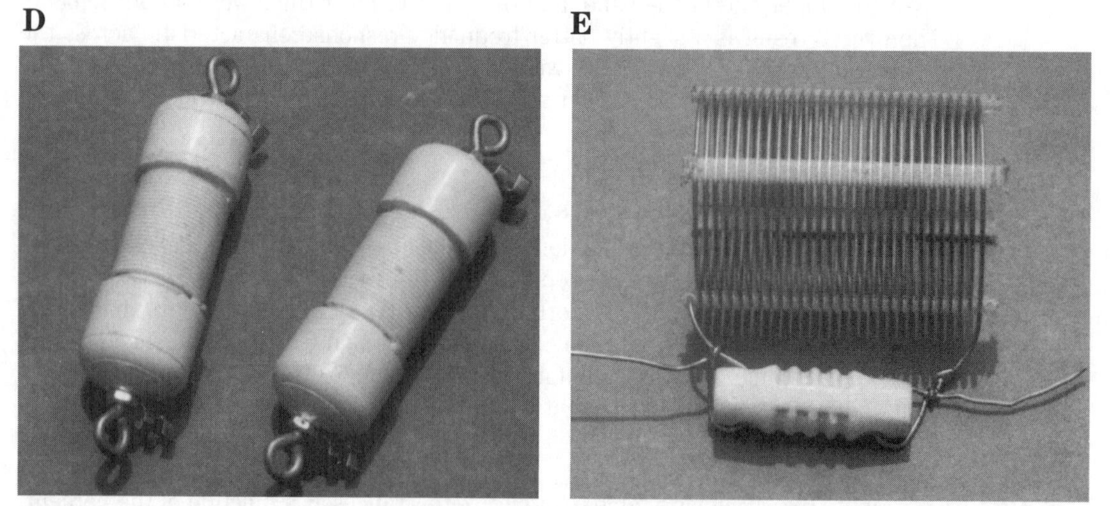

6-15 Shortened or loaded dipole A) inductors at feedpoint, B) inductors midway along elements, C) chart of reactances for coils, D) homemade coil based on B&W Miniductor, E) commercially available coils.

have about 40 feet available for a 40-meter antenna that normally needs about 65 feet for a half wavelength. Because 39 feet is 60 percent of 65 feet, you could use this value as the design point for this antenna. Looking on the chart, a 60-percent antenna with the loading coils at the midpoint of each radiator element wants to see an inductive reactance of 700 Ω. You can rearrange the standard inductive reactance equation ($X_L = 6.28\,FL$) to the form:

$$L_{\mu H} = \frac{X_L \times 10^6}{6.28\,F} \qquad \textbf{[6.17]}$$

Where:

$L_{\mu H}$ is the required inductance in microhenrys
F is the frequency in hertz (Hz)
X_L is the inductive reactance calculated from the table in Fig. 6-15C.

Example Calculate the inductance required for a 60 percent antenna operating on 7.25 MHz. The table requires a reactance of 700 Ω for a loaded dipole with the coils in the center of each element (Fig. 6-15B).
Solution:

$$L_{\mu H} = \frac{X_L \times 10^6}{6.28\,F} \qquad \textbf{[6.18]}$$

$$L_{\mu H} = \frac{(700)\ (10^6)}{(6.28)\ (7{,}250{,}000)} \qquad \textbf{[6.19]}$$

$$L_{\mu H} = \frac{7 \times 10^8}{4.6 \times 10^7} \qquad \textbf{[6.20]}$$

$$L_{\mu H} = 15.4\ \mu H \qquad \textbf{[6.21]}$$

The inductance calculated above is approximate, and it might have to be altered by cut-and-try methods.

The loaded dipole antenna is a very sharply tuned antenna. Because of this fact, you must either confine operation to one segment of the band, or provide an antenna tuner to compensate for the sharpness of the bandwidth characteristic. However, efficiency drops, markedly, far from resonance even with a transmission line tuner. The function of the tuner is to overcome the bad effects on the transmitter, but it does not alter the basic problem. Only a variable inductor in the antenna will do that trick (at least one commercial loaded dipole once used a motor-driven inductor at the center feedpoint).

Figures 6-15D and E show two methods for making a coil-loaded dipole antenna. Figure 6-15D shows a pair of commercially available loading coils especially designed for this purpose. The ones shown here are for 40 meters, but other models are also available. The inductor shown in Fig. 6-15E is a section of *B&W Miniductor* connected to a standard end or center insulator. No structural stress is assumed by the coil—all forces are applied to the insulator, which is designed to take it.

Inductance values for other length antennas can be approximated from the graph in Fig. 6-16. This graph contains three curves for coil-loaded, shortened dipoles that are 10, 50, and 90 percent of the normal half-wavelength size. Find the proposed location of the coil, as a percentage of the wire element length, along the horizontal axis. Where the vertical line from that point intersects with one of the three curves, that intersection yields the inductive reactance required (see along vertical axis). Inductances for other overall lengths can be "rough guessed" by interpolating between the three available curves, and then validated by "cut-and-try."

Tunable dipoles

Dipoles are resonant antennas, so they naturally tend to prefer one frequency over the others. The VSWR will be quite low at the resonant point (assuming no feedline mismatch problems), and will rise at frequencies above and below resonance. If the antenna is a *high-Q* model, then the effect is quite profound, and it might render the antenna nearly useless at frequencies on the other end of the same band (especially where modern transmitters equipped with VSWR shutdown circuitry are used). Figure 6-17 shows a method for overcoming this problem. Here is how it works.

An antenna that is too long for the desired resonant frequency will act inductive (i.e., it will show a feedpoint impedance of the form $Z = R + jX_L$). To counteract the inductive reactance component, $+jX_L$, it is necessary to add a bit of capacitive reactance ($-jX_c$). This approach is taken in tuning certain antenna forms and it can be used with dipoles to make the antenna tunable over a range of about 15 percent of the frequency.

The tunable dipole of Fig. 5-17 is longer than other dipoles and it has an overall length of:

$$L' = \frac{505}{F_{\text{MHz}}} \qquad \textbf{[6.22]}$$

Note that the velocity constant is 505, rather than the 468 that is used with ordinary dipoles.

The required capacitive reactance, which is used to electrically shorten the antenna, is provided by C1. In one effective design that I've tested, the capacitor was a 500-pF transmitting variable. When the Heath Company was making kits, they offered an antenna tuner based on this idea. It was a motor-driven 500-pF variable capacitor inside of a weather-resistant metal-shielded case. The Heath tuner can be used on inverted-vees and dipoles (if a center support is provided). A low-voltage (which is necessary for safety reasons) dc motor drives the shaft of the capacitor to tune the antenna to resonance.

Stacked dipoles

Figure 6-18 shows a *double dipole* (i.e., two half-wavelength dipole antennas spaced a half wavelength apart). The transmission lines are connected in parallel at the receiver. This antenna provides about 3 dB gain over a single dipole, plus it adds

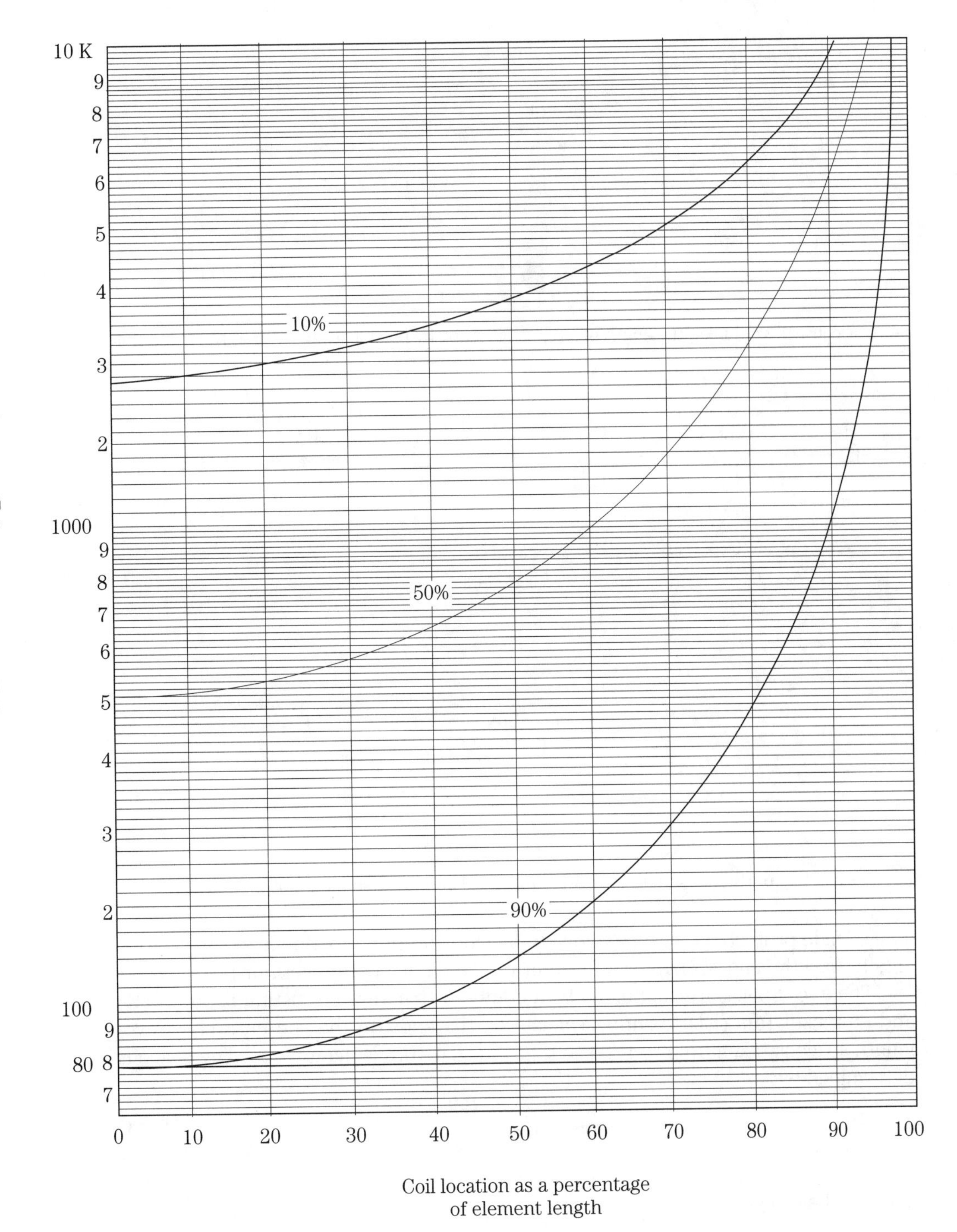

6-16 Inductive reactance vs. coil location for shortened inductance-loaded dipole.

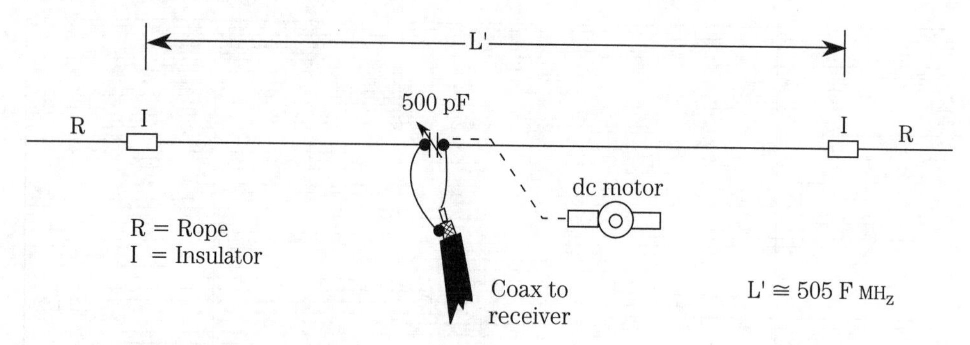

6-17 Electrically reducing thelength of a dipole antenna by placing a capacitor at the feedpoint.

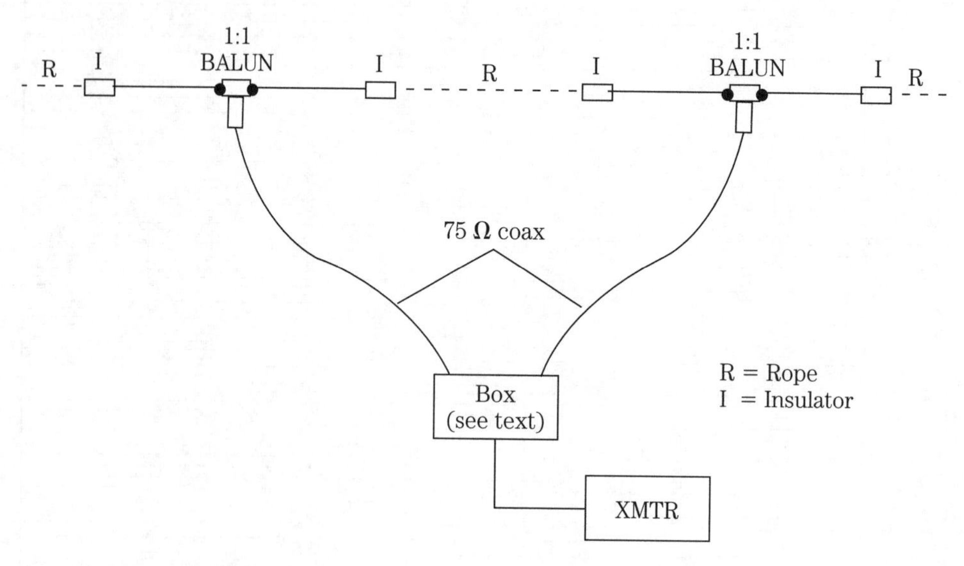

6-18 Phased dipoles provide a 3 dB gain over single dipole.

a bit of fade protection because two side-by-side antennas provide a bit of space diversity (see the end of chapter 2 for information on diversity reception).

The two dipoles are supported by a common structure consisting of ropes (R) and end insulators (EI) to support masts on the ends. In the center, a half-wavelength space is taken up by a rope so that the structure is maintained. The space is determined by:

$$Space_{feet} = \frac{492}{F_{MHz}} \qquad \textbf{[6.23]}$$

and, for the dipoles:

$$Length_{feet} = \frac{468}{F_{MHz}} \qquad \textbf{[6.24]}$$

The stacked dipole gets a bit lengthy on low frequency bands, but is easily achievable by most people on the upper HF bands.

The feedlines for the two dipoles can be connected directly in parallel and fed from the transmitter, provided that the antenna tuning unit will support one half the normal expected impedance. Alternatively, a box can be provided that includes a matching transformer for 1:2 ratio. These can be built like a BALUN transformer on a toroidal core. A trifilar winding is used (see Fig. 6-19). Alternatively, a phasing box can be built that will allow altering the directionality of the antenna by 90 degrees. This is done by using the switching circuit (shown in the inset of Fig. 6-19) to reverse the sense of L3.

Conclusion

The dipole antenna is easy to design, easy to build, and is well-behaved enough to permit even novice builders to successfully make the antenna work—and work well.

7 CHAPTER

Vertically polarized HF antennas

IN PREVIOUS CHAPTERS, YOU HAVE FOUND THAT THE POLARITY OF AN ANTENNA IS THE direction of the electrical (E) field. Because the transmitted signal is an orthogonal electromagnetic wave, the magnetic field radiated from the antenna is at right angles to the electric field. The direction of the electric field, which sets the polarity of the antenna, is a function of the geometry of the radiator element. If the element is vertical, then the antenna polarity is also vertical. The signal propagates out from the radiator in all directions of azimuth, making this antenna an "omnidirectional" radiator.

Figure 7-1A shows the basic geometry of the vertical antenna: an RF generator (transmitter or transmission line from a transmitter) at the base of a radiator of length L. Although most commonly encountered verticals are quarter wavelength ($L = \lambda/4$), that length is not the only permissible length. In fact, it may not even be the most desirable length. This chapter covers the standard quarter-wavelength vertical antenna (because it is so popular), and other-length verticals (both greater and less than quarter wavelength).

The quarter-wavelength vertical antenna is basically half of a dipole placed vertically, with the "other half" of the dipole being the ground. Because of this fact, some texts show the vertical with a double-line ghost radiator, or image antenna, in the earth beneath the main antenna element. Figure 7-1B shows the current and voltage distribution for the quarter-wavelength vertical. Like the dipole, the quarter-wavelength vertical is fed at a current node, so the feedpoint impedance is at a minimum (typically 35–55 ohms, depending upon nearby objects). As a result, the current is maximum and the voltage is minimum at the feedpoint. As you will see, however, not all vertical antennas are fed directly at the current node. As a result, some designs require antenna tuning units to make them match the antenna impedance to the transmitter output impedance.

Figure 7-1C and 7-1D show the two basic configurations for the HF vertical antenna. Figure 7-1C shows the ground-mounted vertical antenna. The radiator ele-

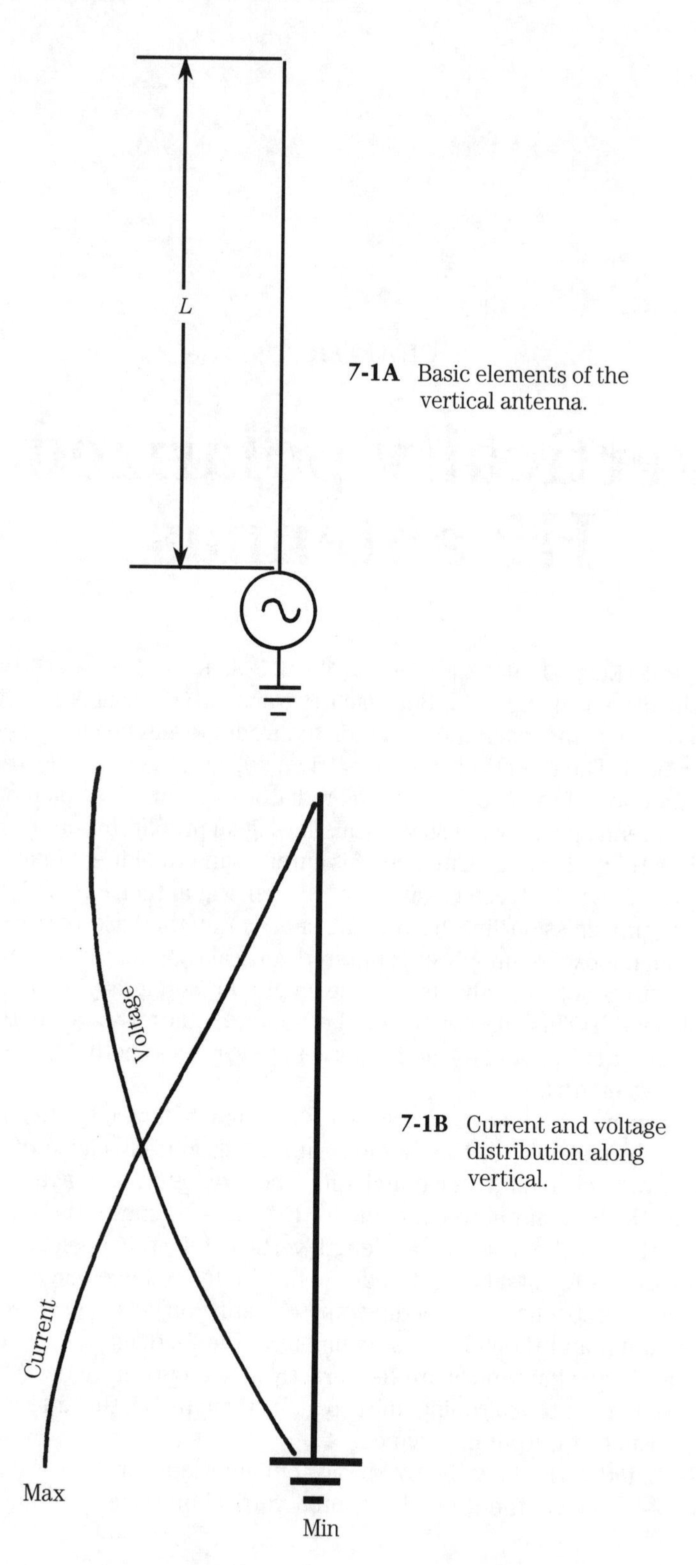

7-1A Basic elements of the vertical antenna.

7-1B Current and voltage distribution along vertical.

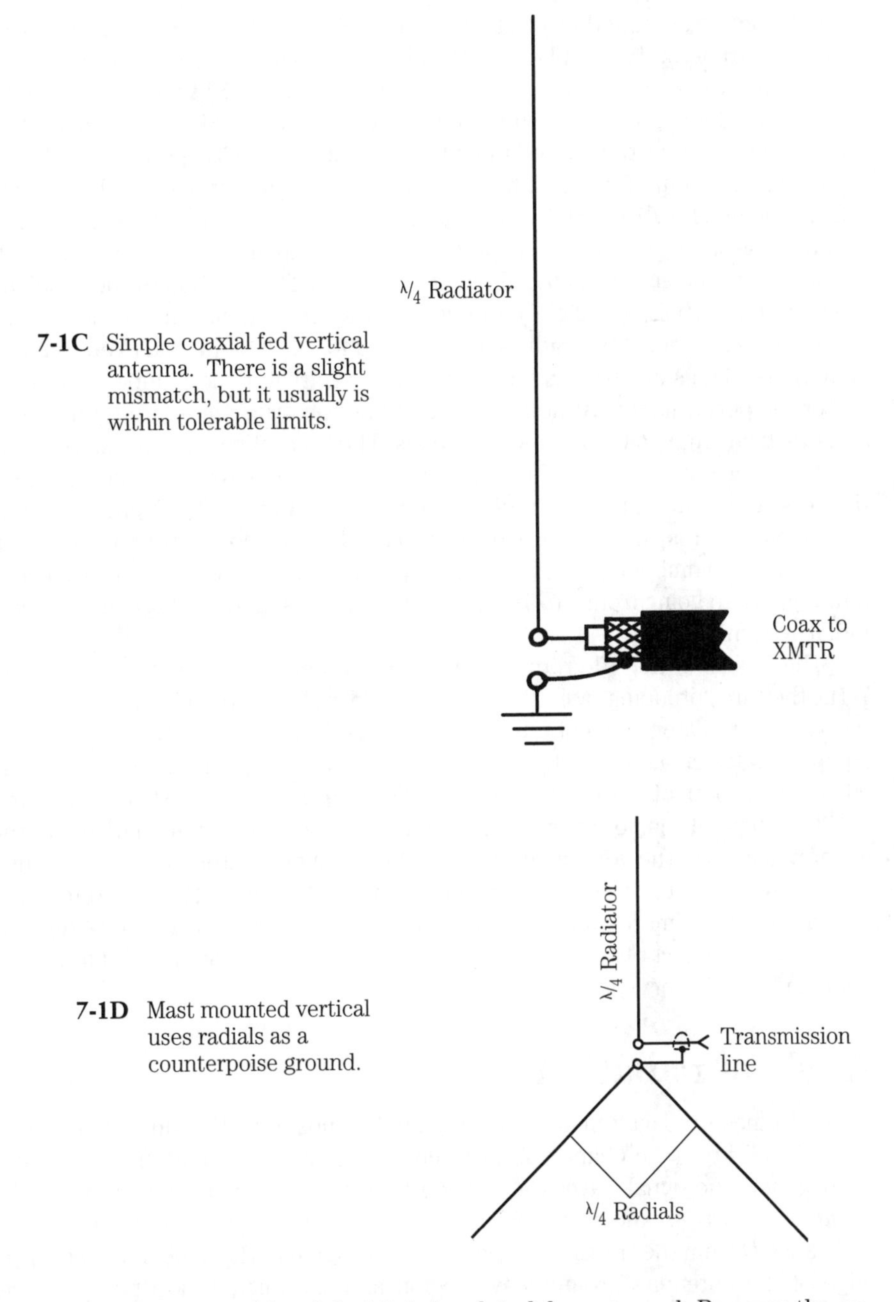

7-1C Simple coaxial fed vertical antenna. There is a slight mismatch, but it usually is within tolerable limits.

7-1D Mast mounted vertical uses radials as a counterpoise ground.

ment is mounted at ground level, but it is insulated from ground. Because the antenna shown is a quarter wavelength, it is fed at a current node with 52-Ω coaxial cable. The inner conductor of the coaxial cable is connected to the radiator element, and the coaxial cable shield is connected to the ground. As you will see shortly, the ground system for the vertical antenna is critical to its performance. Normally, the feedpoint impedance is not exactly 52 Ω, but rather is somewhat lower. As a result, without some matching there will be a slight VSWR, but in most cases, the VSWR is a tolerable trade-off for simplicity. If the antenna has a feedpoint impedance of 37 Ω, which is the value usually quoted, then the VSWR will be 52 Ω/37 Ω, or 1.41:1.

A vertical mounted above the ground level is shown in Fig. 7-1D. This antenna is equally as popular as the ground mounted. Amateurs and CB operators find it easy to construct this form of antenna because the lightweight vertical can be mounted at reasonable heights (15 to 60 feet) using television antenna slip-up telescoping masts that are reasonably low in cost. A problem with the nonground level vertical antenna is that there is no easy way to connect it to ground. The solution to the problem is to create a counterpoise (artificial) ground with a system of quarter-wavelength radials.

In general, at least two radials are required for each band, and even that number is marginal. The standard wisdom holds that the greater the number of radials, the better the performance. Although that statement is true, there are both theoretical and practical limits to the number of radials. The theoretical limit is derived from the fact that more than 120 radials returns practically no increase in operational effectiveness, and at more than 16 radials, the returned added effectiveness per new radial is less than is the case for fewer radials. That is, going from 16 to 32 radials (doubling the number) creates less of an increase in received field strength at a distant point than going from 8 to 16 radials (both represent doubling the density of the radial system).

The radials of the off-ground level vertical antenna can be at any angle. In Fig. 7-1D, they are "drooping radials," (i.e., the angle is greater than 90 degrees relative to the vertical radiator element). Similarly, Fig. 7-1E shows a vertical antenna that is equipped with radials at exactly 90 degrees (no common antenna has radials less than 90 degrees). Both of these antennas are called *ground plane vertical antennas*.

The angle of the verticals is said to affect the feedpoint impedance and the angle of radiation of the vertical antenna. Although those statements are undoubtedly true in some sense, there are other factors that also affect those parameters, and they are probably more important in most practical installations. Before digging further into the subject of vertical antennas, take a look at the subjects of angle of radiation and gain in vertical antennas.

Angle of radiation

Long-distance propagation in the HF region depends upon the ionospheric phenomena called "skip" (see chapter 2 for a more extensive explanation). In this type of propagation, the signal leaves the transmitting antenna at some angle (a), called the *angle of radiation*, and enters the ionosphere where it is refracted back to earth at a distance D from the transmitting station. The signal in the zone between the outer edge of the antenna's ground-wave region and the distant skip point is weak or nonexistent.

The distance covered by the signal on each skip is a function of the angle of radiation. Figure 7-2 shows a plot of the angle of radiation of the antenna, and the distance to the first skip zone. The angle referred to along the vertical axis is the angle of radiation away from the antenna relative to the horizon. For example, an angle of 10 degrees is elevated 10 degrees above the horizon. Shorter distances are found

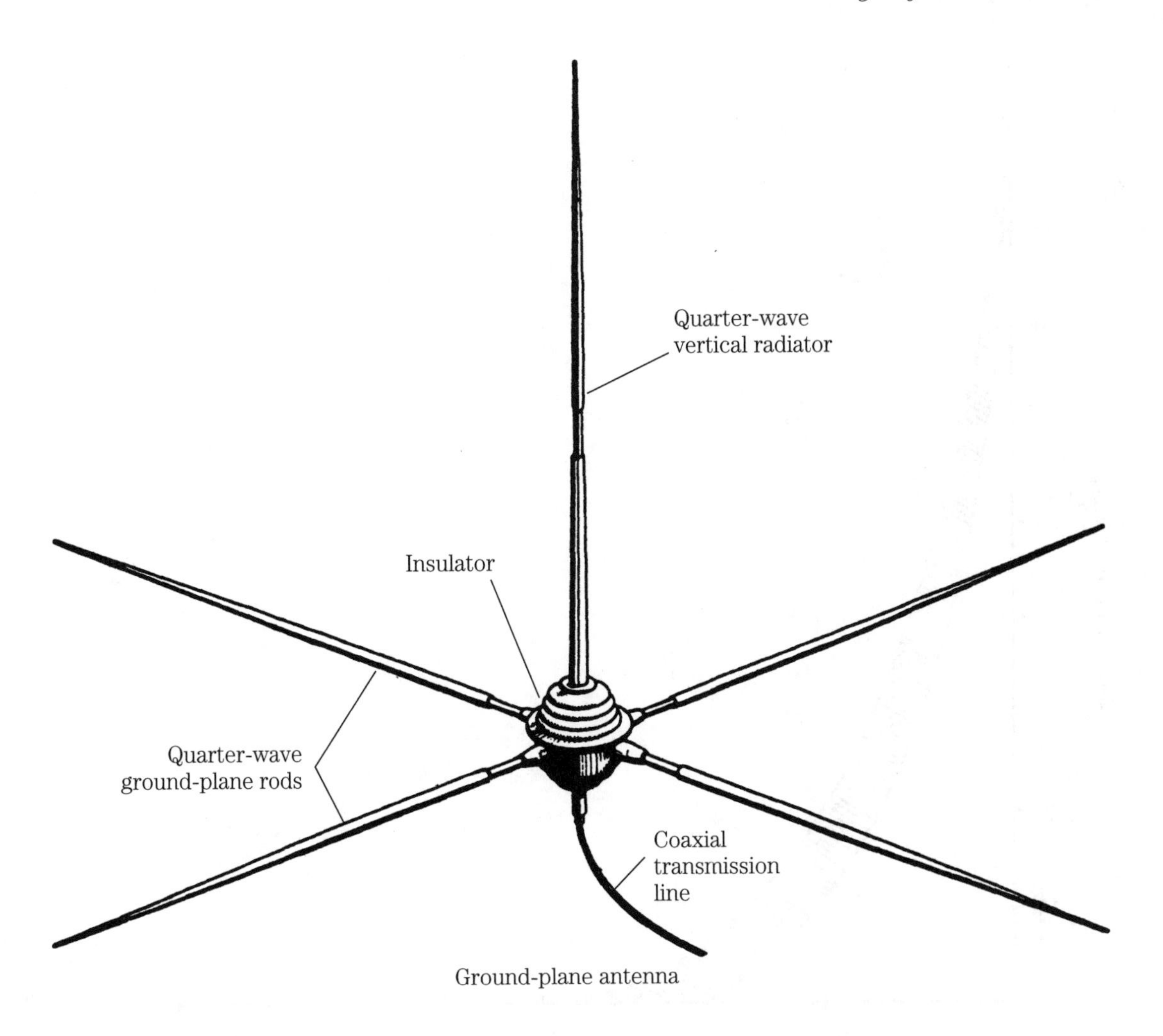

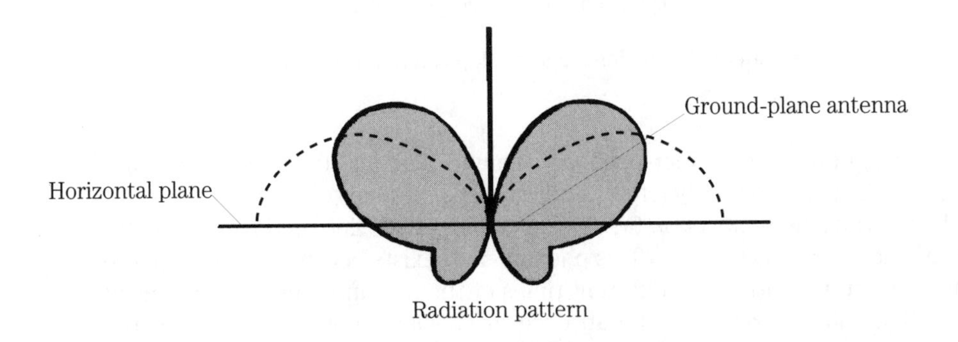

7-1E Ground plane vertical antenna.

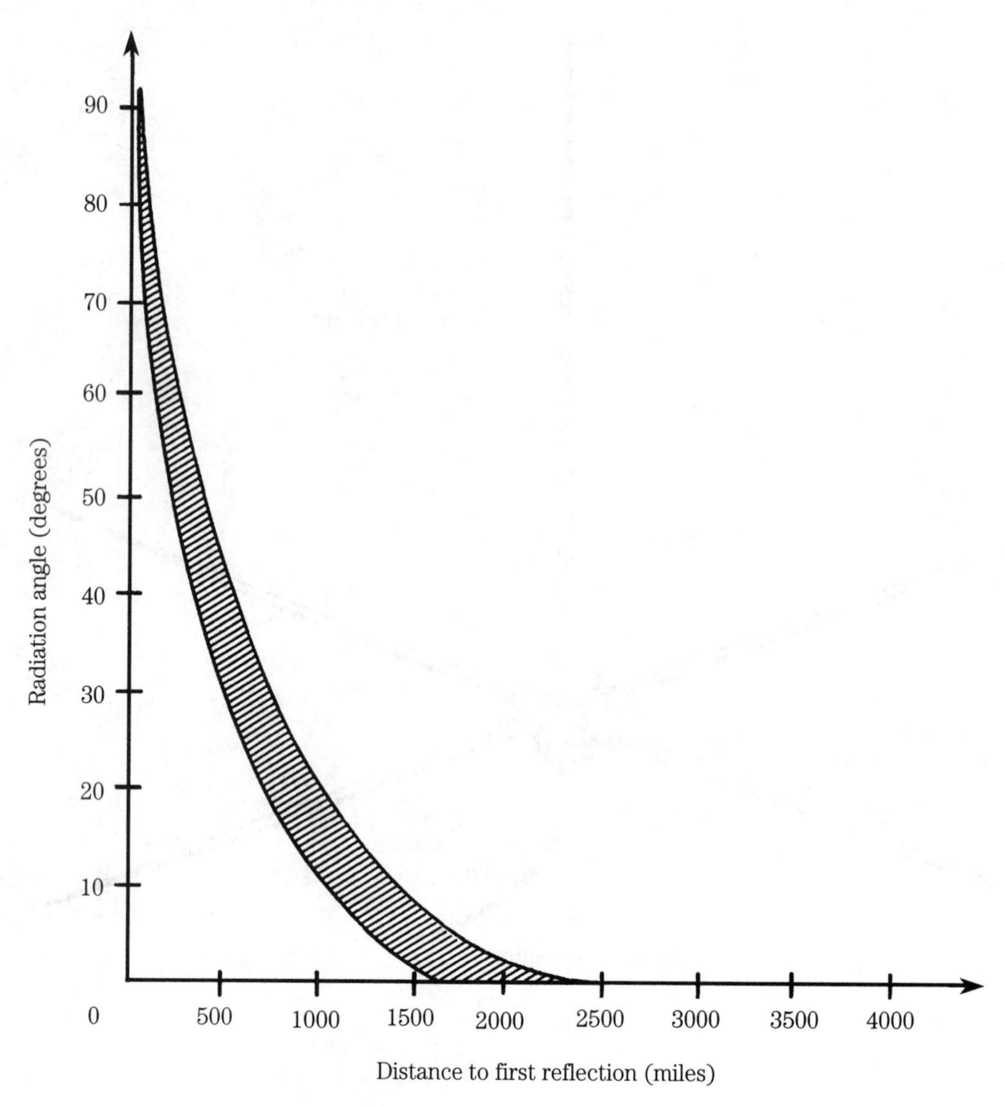

7-2 Effect of radiation angle on skip communications.

when the angle of radiation is increased. At an angle of about 30 degrees, for example, the distance per skip is only a few hundred miles.

Although you might expect on first blush to see a single line on the graph, there is actually a zone shown (shaded). This phenomenon exists because the ionosphere is found at different altitudes at different times of the day and different seasons of the year. Generally, however, in the absence of special event phenomena in the ionosphere, you can expect from 1,500 to 2,500 miles per bounce in the HF bands for low angles of radiation. Note, for example, that for a signal that is only a degree or two above the horizon the skip distance is maximum.

At distances greater than those shown in Fig. 7-2, the signal will make multiple hops. Given a situation where the skip distance is 2,500 miles, covering a distance of 7,500 miles requires three hops. Unfortunately, there is a signal strength loss on each hop of 3 to 6 dB, so you can expect the distant signal to be attenuated from making multiple hops between the earth's surface and the ionosphere. For maximizing distance, therefore, the angle of radiation needs to be minimized.

So what is the ideal angle of radiation? It is standard—but actually erroneous—wisdom among amateur radio operators (and even commercial operators, it turns out) that the lower the angle of radiation, the better the antenna. That statement is only true if long distance is wanted, so it reflects a strong bias toward the DX community. The correct answer to the question is: "It depends on where you want the signal to go." For example, I live in Virginia. If I want to communicate with stations in the Carolinas or New England, then it would behoove me to select a higher angle of radiation for radio conditions represented in Fig. 7-2 so that the signal will land in those regions. But if I wanted to work stations in Europe or Africa or South America, then a low angle of radiation is required. Because of the difference between performance of high and low angles of radiation, some stations have two antennas for each band: one each for high and low angles of radiation.

Figure 7-3 shows a signal from a hypothetical antenna located at point "0," in order to show what angle is meant by "angle of radiation." The beam from the antenna is elevated above the horizon (represented by the horizontal "tangent to horizon" line). The angle of radiation, (a) is the angle between the tangent line and the center of the beam. This angle is not to be confused with the beamwidth, which is also an angle. In the case of beamwidth, we are talking about the thickness of the main lobe of the signal between points where the field strength is –3 dB down from the maximum signal (which occurs at point P); these points are represented by points x and y in Fig. 7-3. Thus, angle b is the beamwidth, and angle a is the angle of radiation.

Gain in vertical antennas

Vertical antennas are known as being *omnidirectional* because they radiate equally well in all directions. "Gain" in an antenna is not the creation of power, but rather a simple refocusing of energy from all directions to a specific direction: gain therefore infers directivity. According to the convention, then, the vertical antenna cannot have any gain because it radiates in all directions equally . . . gain infers directivity. Right? No, not really. Let's develop the theme more carefully.

Again consider the idea of an isotropic radiator (the word *isotropic* means equal power in all directions). Consider a spherical point source radiator located at point "O" in Fig. 7-4. Whatever the level of power available from the transmitter, it will be spread equally well over the entire surface of the sphere as it radiates out into space away from point O. If you measure the power distributed over some area, A, at a distance, R, from the source, then the power available will be a fraction of the total power:

$$P_{avail} = \frac{Total\ Available\ Power \times A}{Total\ Surface\ Area\ of\ Sphere}$$

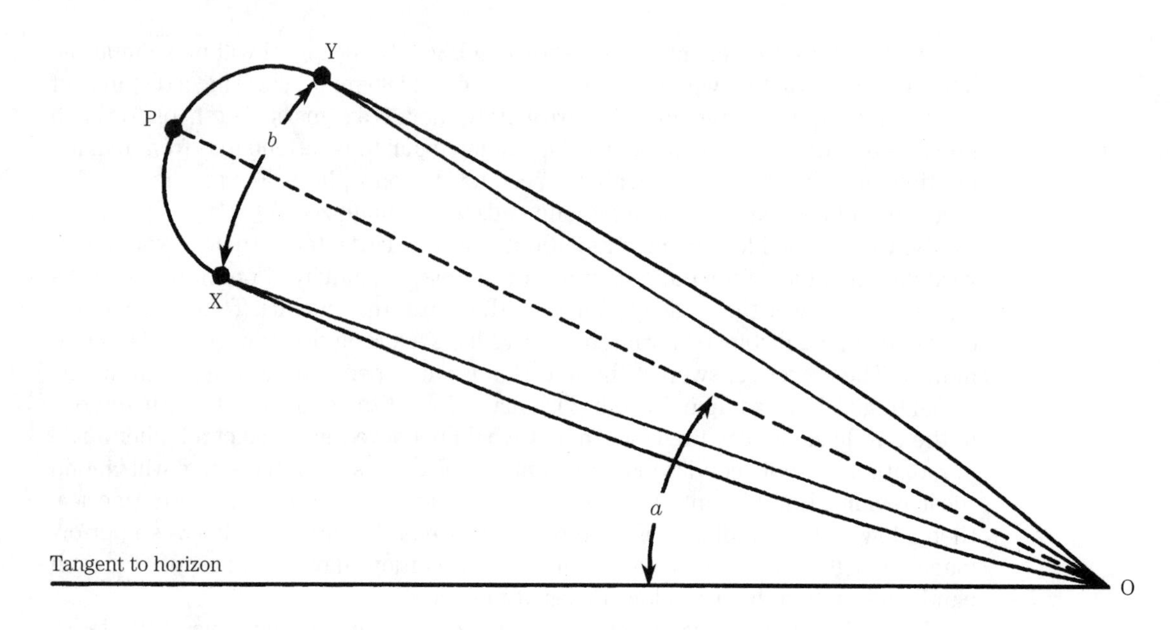

7-3 Side view of vertical extent radiation pattern.

or, in math symbols:

$$P_a = \frac{P_s}{4\pi r^2} \qquad \textbf{[7.1]}$$

Where:

P_a is the power available per solid degree
P_s is the total radiated power in watts
R is the radius of the sphere (i.e., the distance from O to P).

A practical rule of thumb for this problem is to calculate from the surface area of the sphere. If you perform the right calculations, you will find that there are approximately 41,253 square degrees on the surface of a sphere. By calculating the surface area of the beam front (also in square degrees), you can find the power within that region.

Now for the matter of gain in a vertical antenna. The vertical is not gainless because it does not, in fact, radiate equally well in all directions. In fact, the vertical is quite directional, except in the horizontal (azimuth) plane. Figure 7-5 shows the radiation pattern of the typical free-space vertical radiator. The pattern looks like a giant doughnut in free space (see solid pattern in Fig. 7-5). When sliced like a bagel, the pattern is the familiar, circular "omnidirectional" pattern. When examined in the vertical plane, however, the plane looks like a sliced figure-8. The "gain" comes from the fact that energy is not spread over an entire sphere, but rather it is concentrated

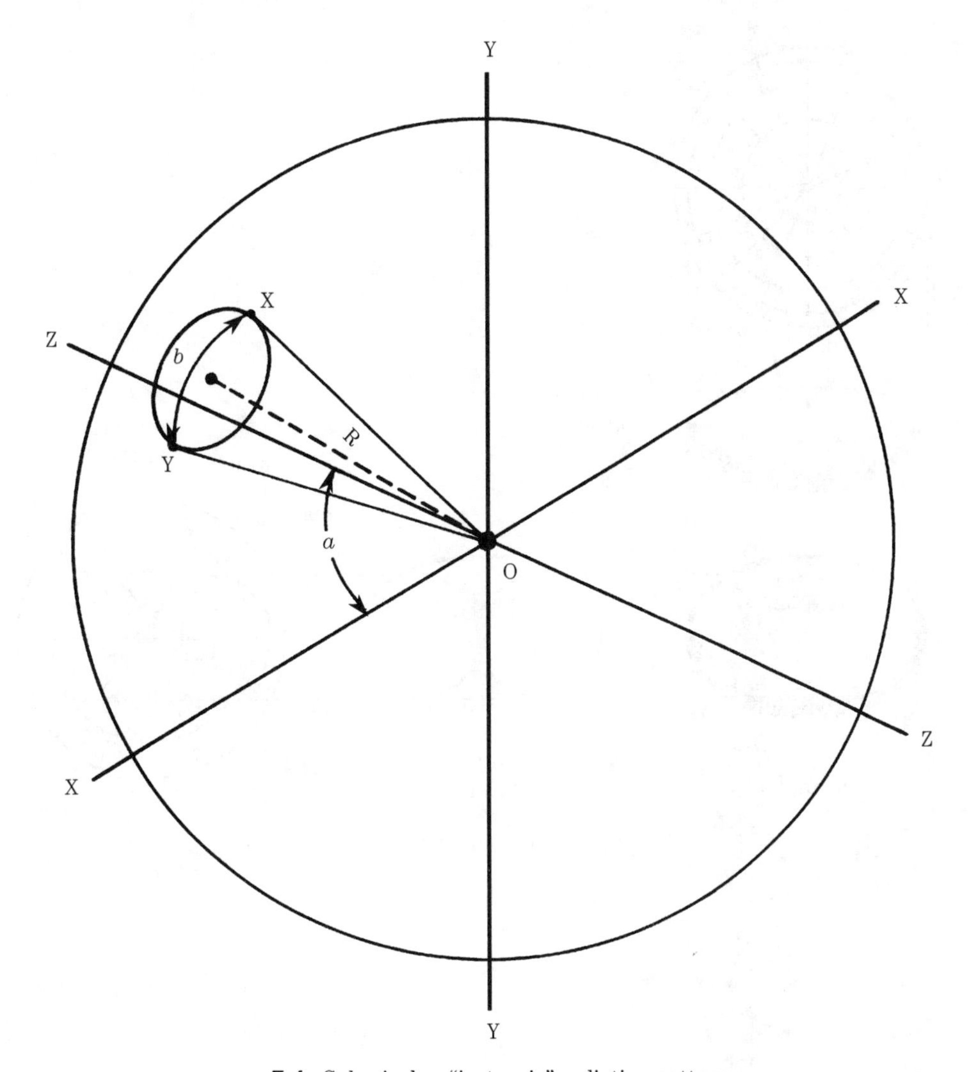

7-4 Spherical or "isotropic" radiation pattern.

to the toroidal "doughnut" shaped region shown. Therefore, the power per unit area is greater than for the isotropic (truly "omnidirectional") case.

Nonquarter-wavelength verticals

The angle of radiation for a vertical antenna, hence the shape of the hypothetical "doughnut" radiation pattern, is a function of the length of the antenna (note: "length" in terms of vertical antennas is the same as "height," and is sometimes expressed in degrees of wavelength, as well as feet and/or meters). Figure 7-6A shows the approximate patterns for three different length vertical antennas: quarter wavelength, half wavelength, and ⅝ wavelengths. Note that the quarter-wavelength antenna has the highest angle of radiation, as well as the lowest gain of the three cases.

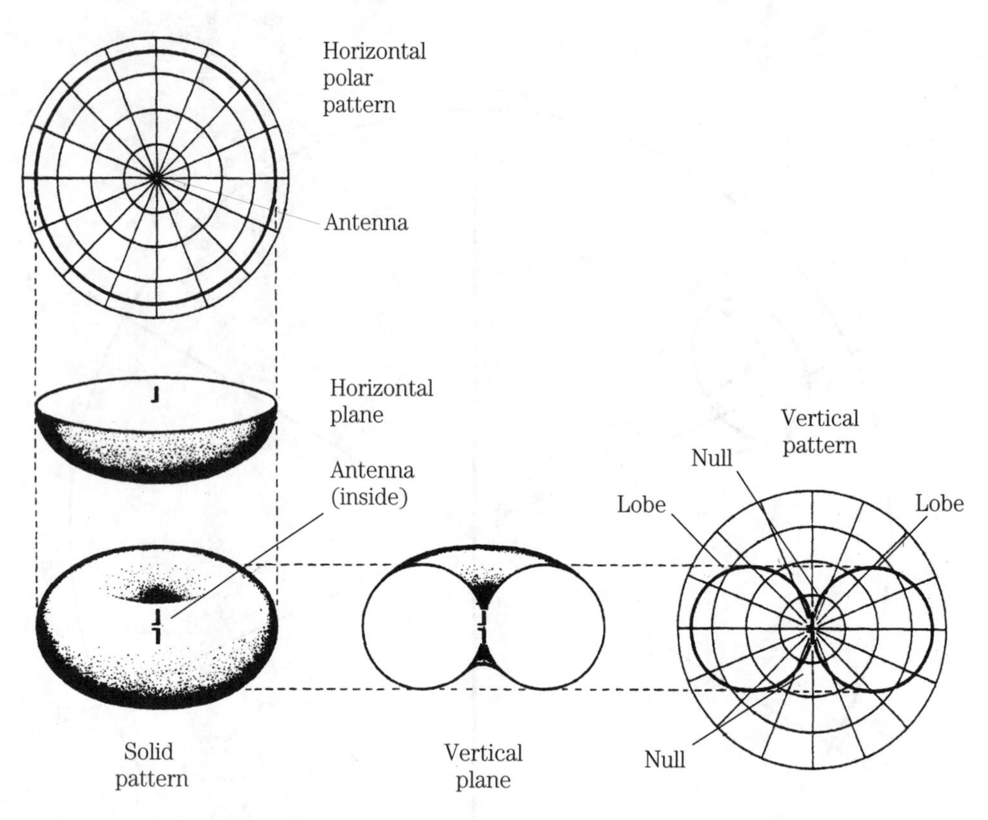

7-5 Vertical antenna radiation pattern.

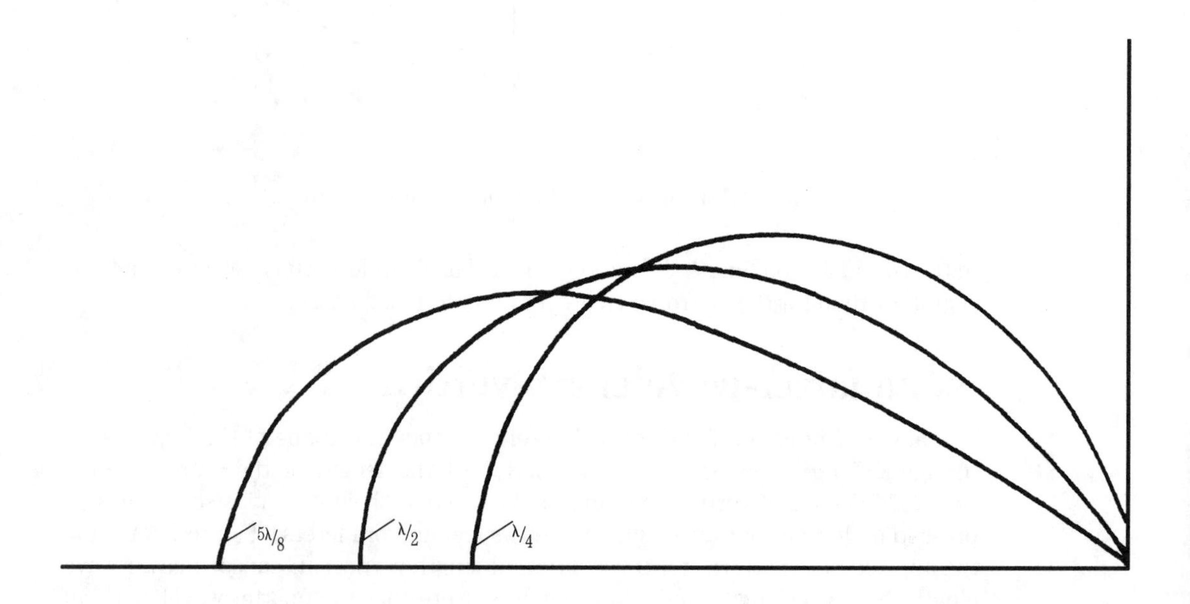

7-6A Vertical radiation pattern as a function of element length: Ideal.

The ⅝-wavelength antenna is both the lowest angle of radiation and the highest gain (compared with isotropic).

The patterns shown in Fig. 7-6A assume a perfectly conducting ground underneath the antenna. However, that is not a possible situation for practical antennas: all real grounds are lossy. The effect of ground losses is to pull in the pattern close to the ground (Fig. 7-6B). Although all of the patterns are elevated from those of Fig. 7-6A, the relationships still remain: the ⅝-wavelength radiator has the lowest angle of radiation and highest gain.

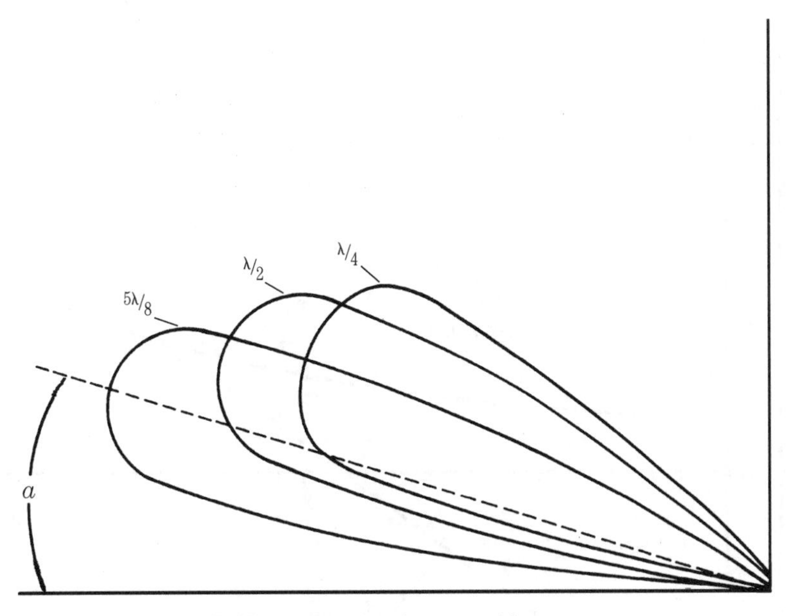

7-6B Accounting for ground losses.

The feedpoint impedance of a vertical antenna is a function of the length of the radiator. For the standard quarter-wavelength antenna, the feedpoint radiation resistance is a maximum of 37 Ω, with only a very small reactance component. Figure 7-7 shows the approximate feedpoint impedances for antennas from nearly zero effective length to 120 degrees of length.

Antenna length expressed in degrees derives from the fact that one wavelength is 360 degrees. Thus, a quarter-wavelength antenna has a length of 360 degrees/4 = 90 degrees. To convert any specific length from degrees to wavelength, divide the length in degrees by 360. Thus, for a 90-degree antenna: 90 degrees/360 degrees = ¼ wavelength. The graph in Fig. 7-7A shows the antenna feedpoint impedance, both reactance and radiation resistance, for antennas from 60 to 120 degrees; Fig. 7-7B shows the radiation resistance for antennas from near zero to 60 degrees. Note that the radiation resistance for such short antennas is extremely small. For example, an antenna that is 30 degrees long ($^{30}/_{360}$ = 0.083 wavelengths) has a resistance of approximately 3 Ω. It is generally the practice on such antennas to use a broadband impedance-matching transformer to raise the impedance of such antennas to a higher value (Fig. 7-8).

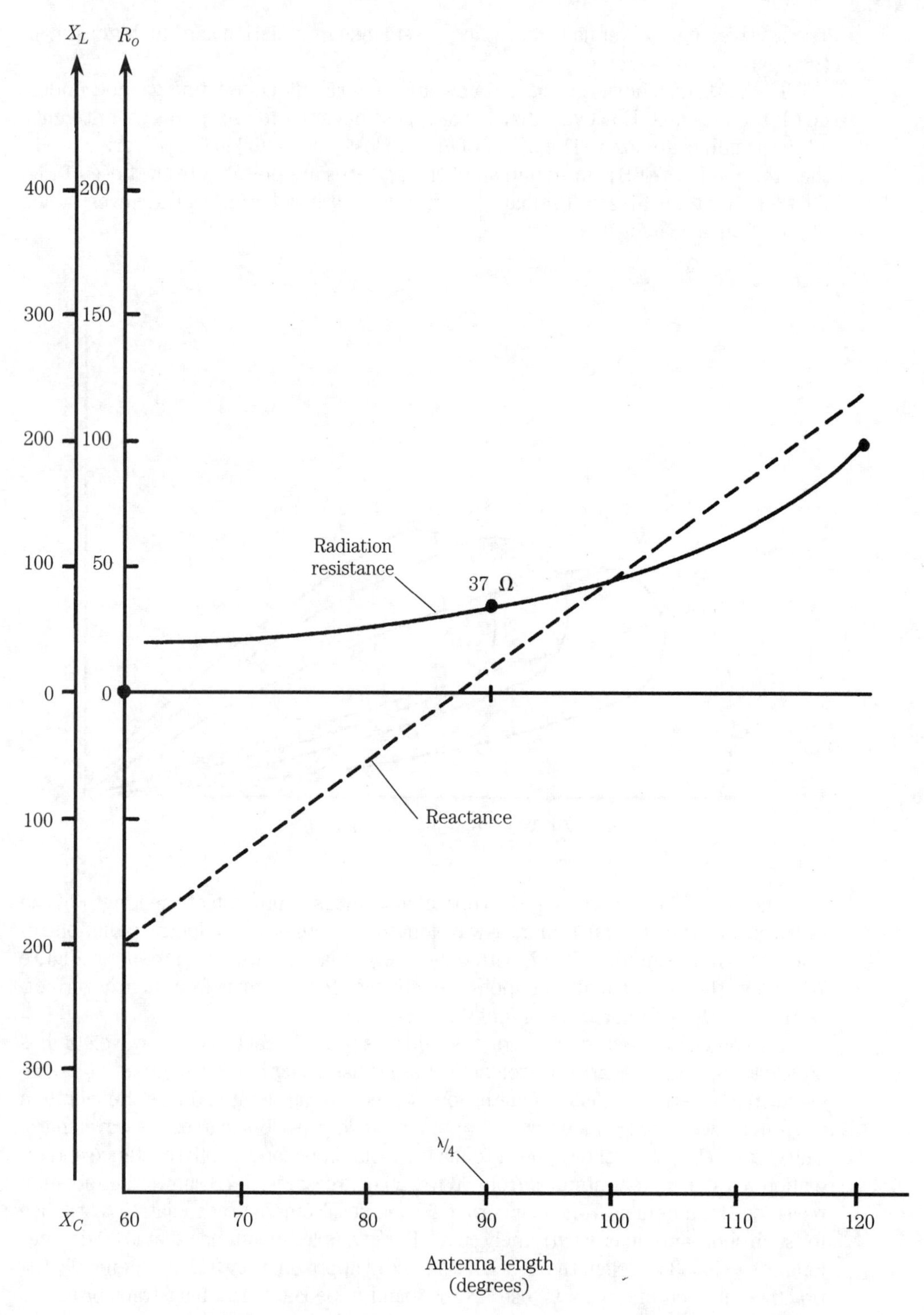

7-7A Antenna impedance as a function of antenna length.

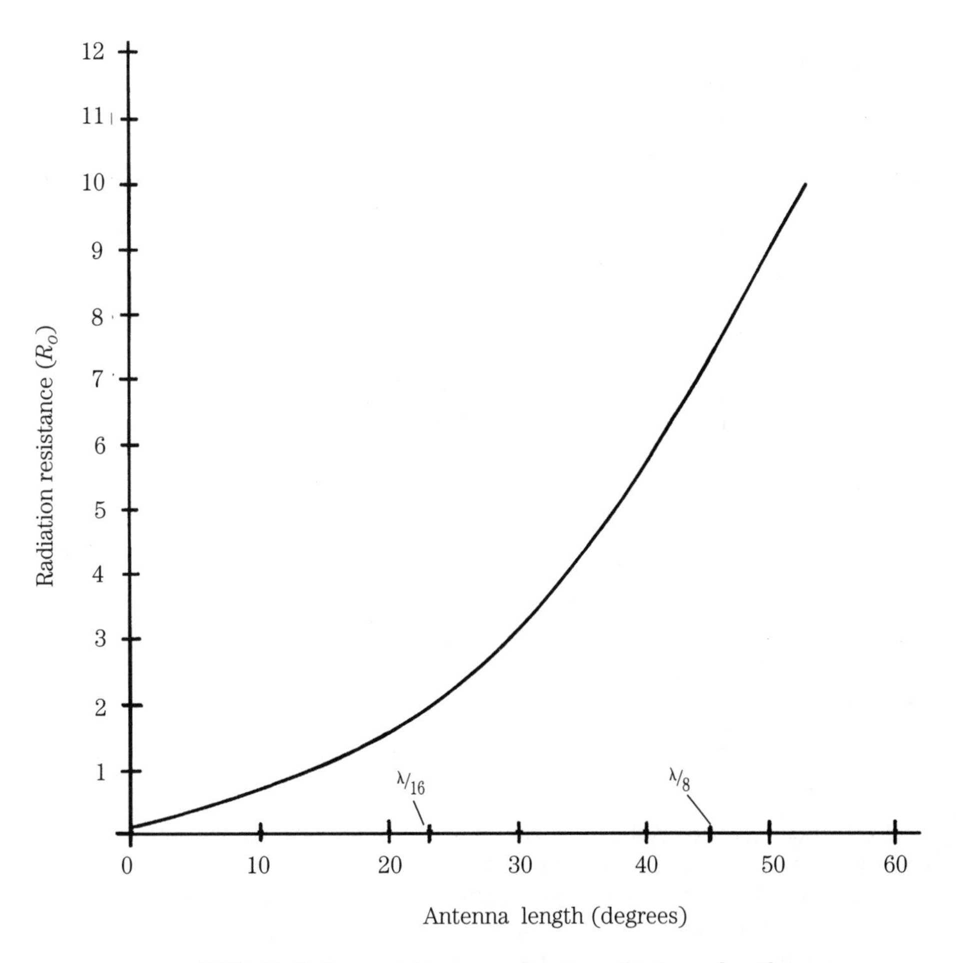

7-7B Radiation resistance as a function of antenna length.

The ground system for the vertical antenna

The vertical antenna works well only when placed over a good ground system. Chapter 24 gives details of proper ground systems for all radio antennas, verticals included, so only certain specifics are included here. The reader should also examine chapter 24, however, in order to get a firmer grasp of the problem.

The usual way to provide a good ground for a vertical is to use a system of radials such as Fig. 7-9A. This case shows a view (from above) of 16 quarter-wavelength radials arranged to cover the full circle around the antenna. Each radial is a quarter wavelength, so it will have a length (in fact) of $246/F_{MHz}$. All of the radials are connected together at the base of the antenna, and the ground side of the transmission line is connected to this system. The radials can be placed either on the surface, or underground. One friend of the author's built an extensive radial system on the bare dirt when his house was built, so when the sod was installed he had a very high-quality underground radial system.

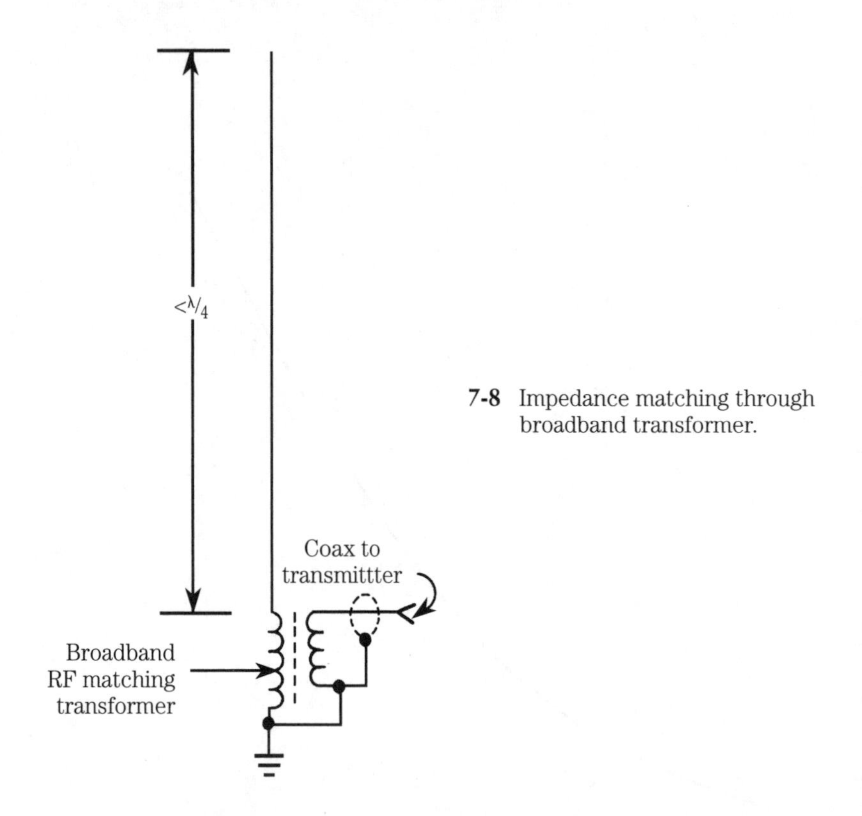

7-8 Impedance matching through broadband transformer.

If you decide to use an above-ground radial system, however, be sure to prevent people from tripping over it. There might be liability implications for people who trip and injure themselves, even when the person is an intruder or trespasser.

Some experts prefer to place a copper wire screen at the center of the radial system. The minimum size of this screen is about two-meters square (6 × 6 feet). Connect it to the radials at the points shown using solder. Other experts will drive ground stakes into the ground at these points. Still another method is shown in Fig. 7-9B. This case shows a "spider web" of conductors shorting the radials at points a meter or two from the antenna. Again, some authorities recommend that ground rods be driven into the earth at the indicated points.

The exact number of radials to use depends in part on practical matters (how many can you physically install?). Use at least two radials per band, with four per band preferred for simple, low-cost systems. However, even four is considered a compromise case. The general rule is: *the more the better*. But it's also true that there is a law of diminishing returns as the number of radials is increased. Figure 7-9C shows the approximate field intensity (mV/meter) as a function of the number of radials. Notice that the field intensity does not increase as rapidly with the number of radials above 20 or so. Note that the Federal Communications Commission requires AM band (550–1620 kHz) stations to use 120 radials, but that number is not necessary for amateur stations. A practical upper limit of 16 radials is usually accepted for amateur radio work.

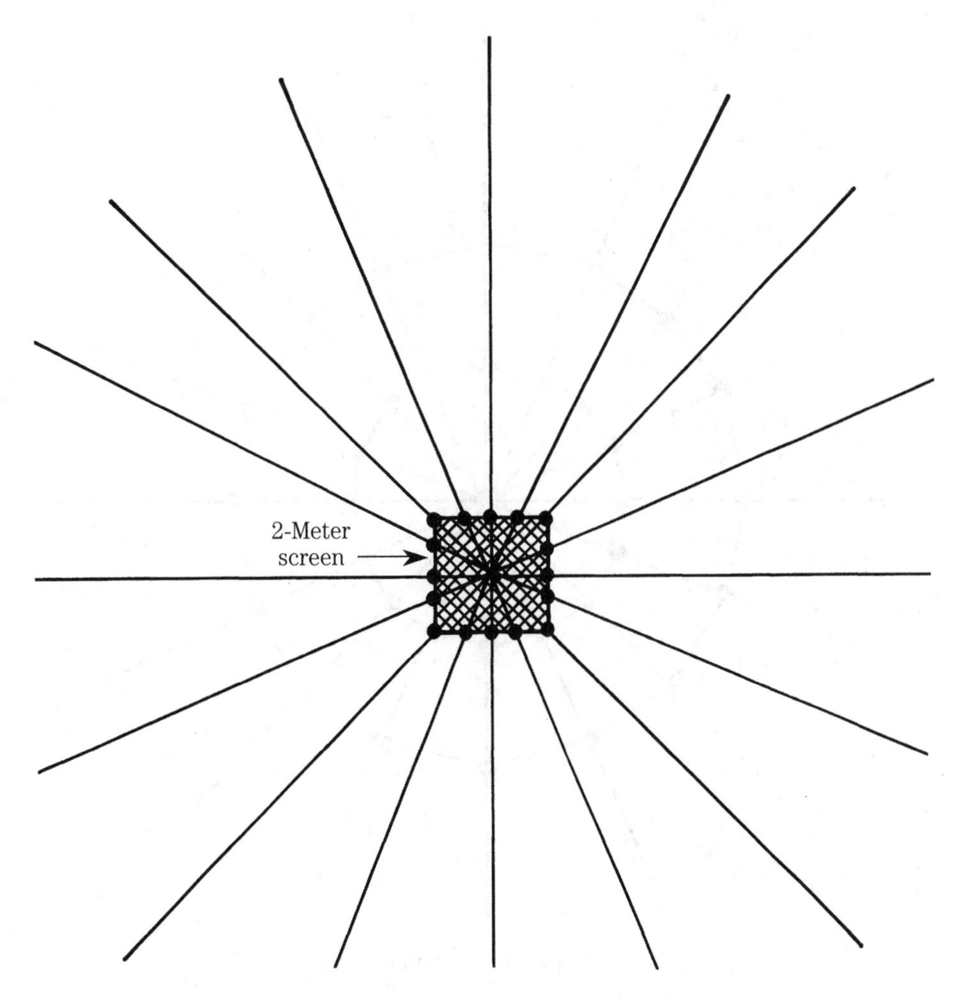

7-9A Comprehensive ground system for vertical antenna.

For vertical antennas mounted above ground, there is an optimum height above ground for the base of the antenna. This height is a quarter wavelength above the actual ground plane. Unfortunately, that distance might not be the height above the surface. Depending upon ground conductivity, and ground water content, the height can be exactly a quarter wavelength above the surface or slightly lower. The point is found from experimentation, and will, unfortunately, vary over the year if climatic changes are usual.

Variations on the vertical antenna theme

Thus far, the vertical antennas have been standard quarter- or ⅝-wavelength models. This section looks at several variations on the theme. Consider Fig. 7-10. This antenna is the vertical half-wavelength dipole. The vertical dipole is constructed in ex-

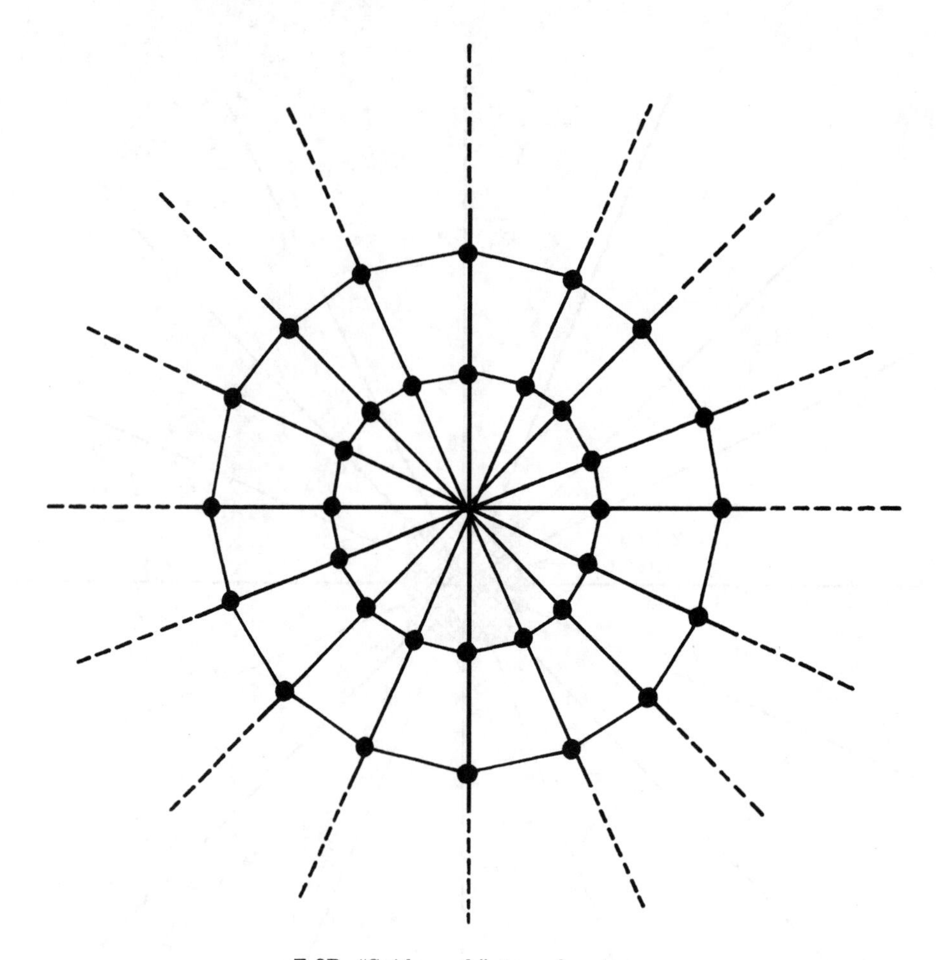

7-9B "Spider web" ground system.

actly the same manner as the horizontal dipole, but is mounted in the vertical plane. In general, the section of the radiator that is closest to the ground should be connected to the shield end of the coaxial cable transmission line.

Like the horizontal dipole, the approximate length of the vertical dipole is calculated from:

$$L_{ft} = \frac{468}{F_{MHz}} \tag{7.2}$$

Where:

L_{ft} is the length in feet
F_{MHz} is the operating frequency in megahertz

Example Calculate the length of half a wavelength vertical dipole for operation on a frequency of 14.250 MHz in the 20-meter amateur radio band.

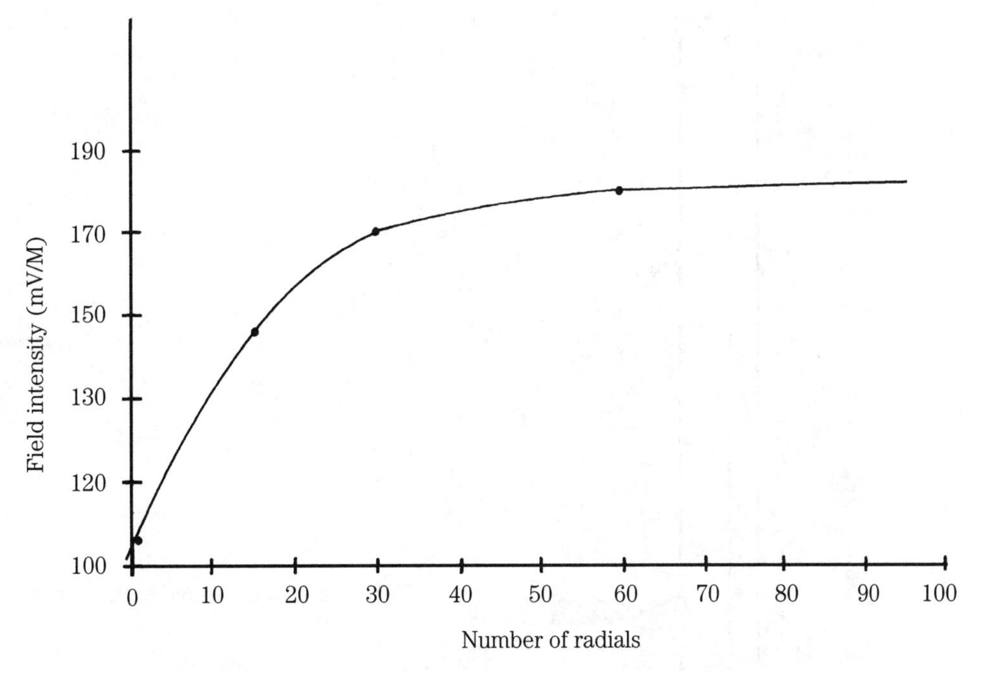

7-9C Effect of increasing the number of radials on field strength.

Solution:

$$L_{ft} = \frac{468}{F_{MHz}}$$

$$L_{ft} = \frac{468}{14.250} = 32.8 \text{ feet}$$

Note: the "0.8" feet part of this calculated length can be converted to inches by multiplying by 12: $0.8 \times 12 = 9.6$ inches.

Each leg of the vertical dipole is one half of the calculated length, or

$$\frac{32.8 \text{ feet}}{2} = 16.4 \text{ feet}$$

The vertical dipole antenna is used in many locations where it is impossible to properly mount a horizontal dipole, or where a roof or mast mounted antenna is impossible to install because of either logistics, or a hostile landlord and/or homeowners' association. Some row house and town house dwellers, for example, have been successful with the vertical dipole. In the 1950s and 1960s, the vertical dipole was popular amongst European amateurs because of space restrictions found in many of those locations.

The construction of the vertical dipole is relatively straight forward. You must find or build a vertical support structure. In the case shown in Fig. 7-10, the support

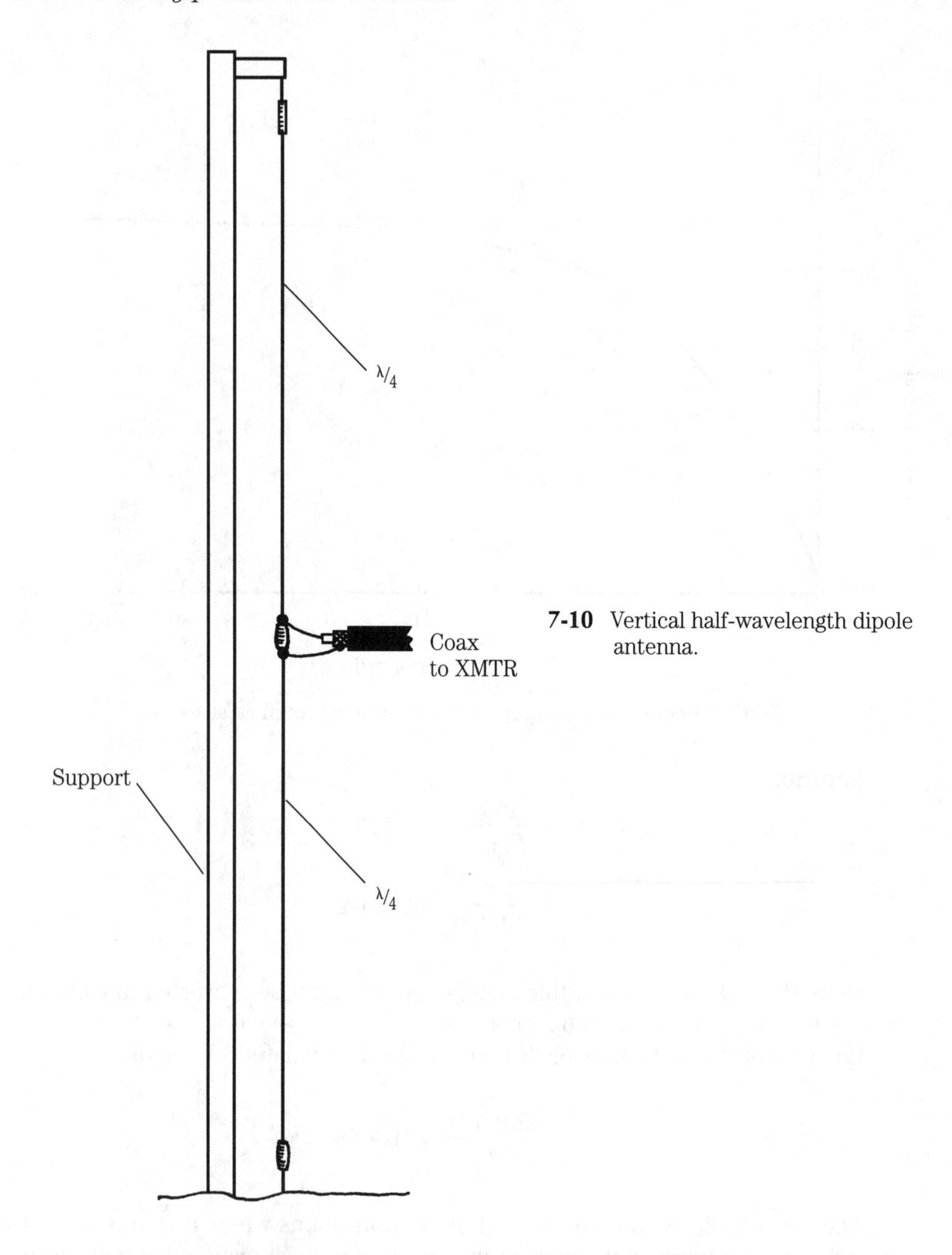

7-10 Vertical half-wavelength dipole antenna.

is a wooden or PVC mast erected for that purpose. Ropes and insulators at either end support the wire elements from the ends and keep the antenna taut. If the neighbors are a problem, then try to find some white PVC pipe that will make a fine flagpole and be patriotic—with a vertical dipole hidden inside of the pipe. In other cases, if your home is not metal sided, and if it is high enough, then a support from the roof structure (or soffits) will make a proper support.

One problem with the vertical dipole, and one that liability conscious people need to consider, is that a high impedance voltage node is found at the ends of a half wavelength dipole. Anyone touching the antenna will likely receive a nasty RF burn (or shock) from this antenna.

A *coaxial vertical* is made similar to the vertical dipole (and, in fact, it can be argued that it is a form of vertical dipole) in that it uses a pair of vertical radiator elements. In the case of the coaxial vertical antenna, however, the radiator that is closest to the ground is coaxial with the transmission line and the main radiator element. An example is shown in Fig. 7-11A. An insulator at the feedpoint separates the two halves of the radiator; in most cases, it is of smaller diameter than the coaxial sleeve (also called the *shield pipe* in some publications). The reasons for this arrangement are not entirely electrical, for the most part, but mechanical. The coaxial cable transmission line passes through the sleeve, and is itself coaxial to the sleeve.

The overall length of the coaxial vertical antenna is a half wavelength, which consists of two quarter-wavelength sections. Both the radiator and the sleeve are quarter wavelength. The length of each are found (approximately) from:

$$L_{\text{ft}} = \frac{234}{F_{\text{MHz}}} \quad \textbf{[7.3]}$$

or,

$$L_{\text{meters}} = \frac{72}{F_{\text{MHz}}} \quad \textbf{[7.4]}$$

You should recognize these as similar to the equation used previously to calculate half-wavelength antennas, but reduced by a factor of two.

The coaxial vertical antenna was once popular with CB operators, and as such was called the *colinear antenna*. In some cases, you can find hardware from these antennas on the hamfest or surplus markets, and the pieces can be modified for amateur radio use. In the situation where a 10-meter band antenna is being built, it is a simple matter to cut the 11-meter CB antenna for operation on a slightly higher frequency. In the case of the lower frequency bands, however, it is a little more difficult and it is likely that only the insulator and mounting assembly are salvageable. Keep in mind, however, the fact that adjacent sizes of aluminum tubing are designed such that the inside diameter (i.d.) of the larger piece is a slip-fit for the outside diameter (o.d.) of the smaller piece. You can, therefore, connect adjacent sizes of aluminum tubing together without the need for special couplers, etc. With that in mind, salvaged insulator assemblies can be cut off with just 6 to 10 inches of the former radiator and sleeve, and new radiators from "adjacent size" tubing installed.

The configuration shown in Fig. 7-11A is the manner of construction used by commercial antenna manufacturers for VHF and CB colinear vertical dipoles, but is a little difficult for amateurs (unless they happen to own a machine shop) to make the center insulator. For those people, some other method is indicated before this antenna is practical.

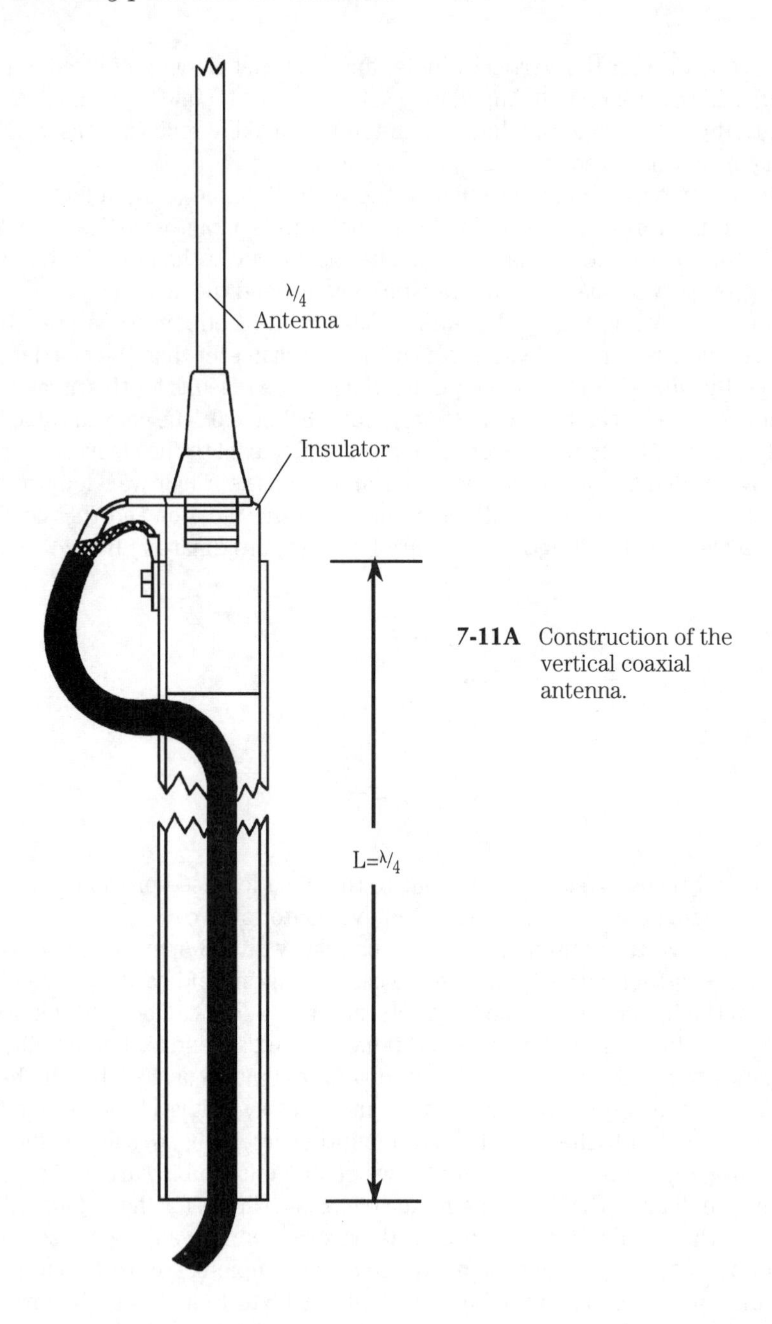

7-11A Construction of the vertical coaxial antenna.

Figure 7-11B shows a construction method that has been used by amateurs with good results. The radiator and shield pipe (sleeve) are joined together in an insulating piece of thick-wall PVC plumbing pipe, Lucite, or plexiglass tubing; 6 to 10 inches of tubing are needed.

A gap of about two inches is left between the bottom end of the radiator pipe and the top end of the shield pipe in order to keep them electrically insulated from each

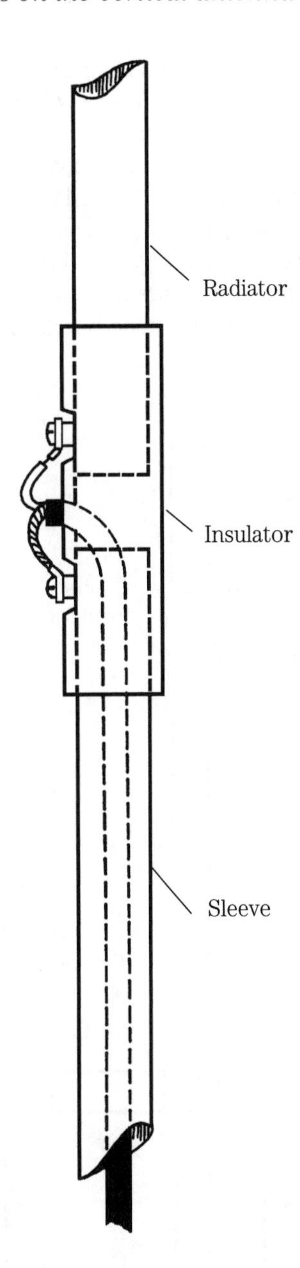

7-11B Vertical dipole made from tubing.

other, and to allow the coaxial cable to be passed through to the outside world. A hole in the insulator pipe is drilled for this purpose.

The aluminum tubing pieces for the radiator and the sleeve are fastened to the insulator with at least two heavy machine screws each. One of the machine screws on each can be used as the electrical connection between the coaxial cable and the pipes, provided that a larger hole is cut in the insulator at that point to admit the washer that provides the electrical pathway between the screw head and the aluminum pipe. If you depend upon the machine screw touching the pipe at the edges of the hole cut for it, then there will probably be intermittent connection and all of the aggravation that ensues.

Mounting of the homebrew coaxial vertical antenna can be a "pain in the neck." Normally, this antenna is mounted high in the air above ground, so some form of support is needed. Fortunately, you can use small-area metal supports connected to the sleeve for this purpose. Figure 7-11C shows one method for mounting that is popular. A pair of television antenna standoff mounting brackets are used to support the sleeve. Those brackets can be bought in sizes from 6 inches to 24 inches. Note that

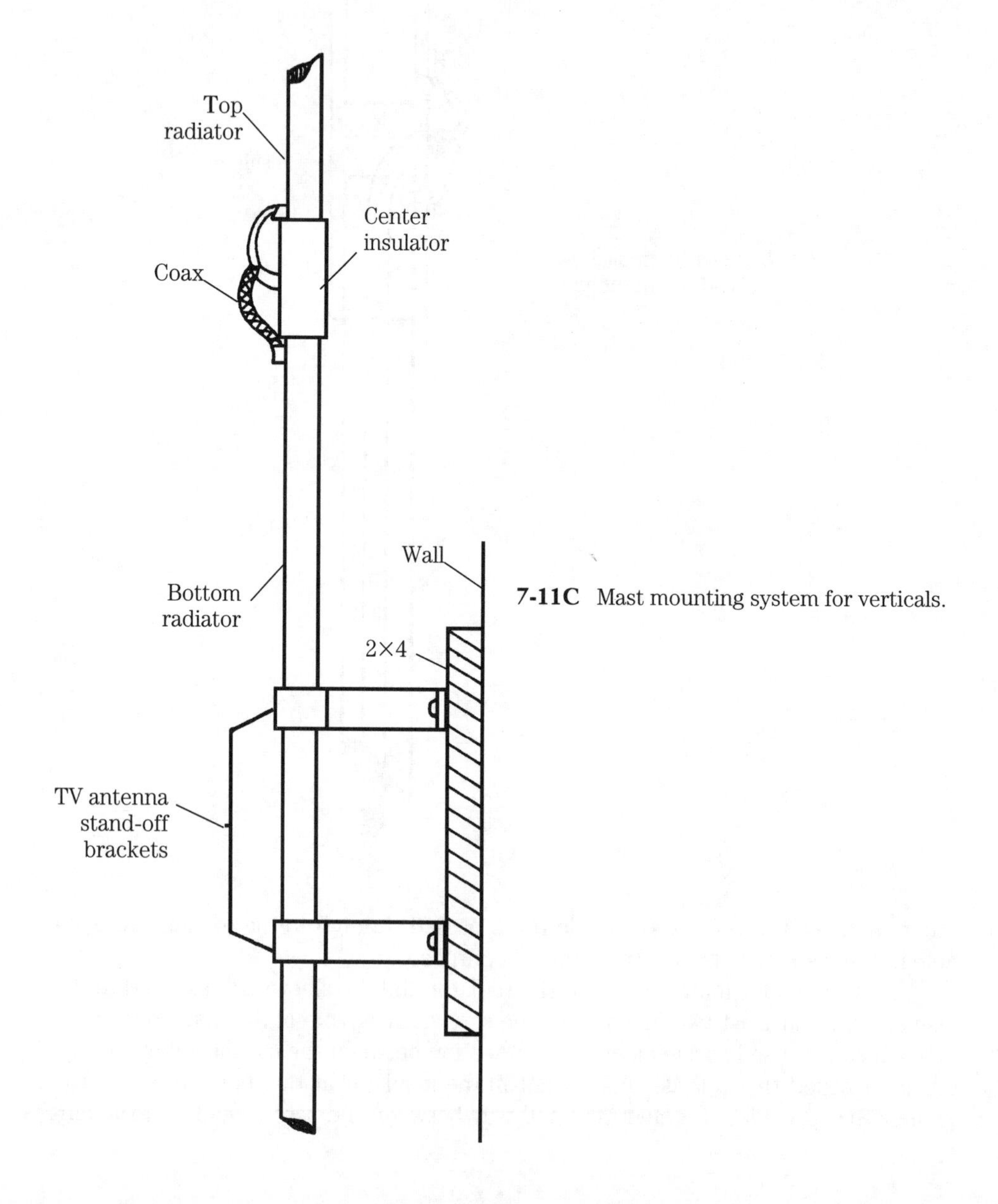

7-11C Mast mounting system for verticals.

a 2-×-4 piece of lumber is used between the building wall and the brackets. This wood serves as an insulator, so it should be varnished or painted. It is held to the wall with lag bolts, wing bolts, or some other competent method of anchoring. Keep in mind that the forces on the brackets increase tremendously during wind storms!

It is possible for the two vertical antennas shown previously to become a shock hazard to anyone who touches them. Both of these antennas are half-wavelength radiators and are of the dipole form of construction. The center point is used for feeding the antenna, so forms the low impedance point in the antenna. As a result, the ends of the antenna, one of which is close to the ground, are the high impedance points—hence the voltages at those points within reach of prying hands playing in the yard can be high. It is wise to either mount the antennas so far above the ground that they cannot be reached, or build a small nonconductive fence around the bottom end of the antenna.

Vertical antenna construction

There are two general cases for installing vertical antennas: ground-level mounted and nonground mounted. This section takes a brief look at both forms of mounting. We will concentrate on installation of homebrew verticals rather than commercial, because it is assumed that the vendors of such antennas will provide their own instructions.

The ground-level mounted vertical is shown in Fig. 7-12. The typical vertical antenna is 8 to 40 feet high. Thus, although the actual weight of the antenna is small, the forces applied to the mounting structure (especially during windstorms) can be quite high. Don't be fooled by the apparent light weight of the antenna in this respect.

The mounting structure for the vertical antenna can be a metal or wooden fence post buried in the ground. At least two feet of the fence post should be above ground. In the case of Fig. 7-12, a 4-×-4 wooden fence post is used as the mounting, but the principles are similar for all forms of post. A fence post hole is provided that is at least 2 feet deep. In some cases, it might be possible to use a 1-foot-deep gravel fill topped with back-filled dirt. In other cases, especially where a steel fence post is used, a concrete plug is placed at the bottom of the hole over a 4-inch layer of gravel.

The antenna radiator element is installed onto the fence post using stand-off insulators. Unfortunately, these insulators are difficult to find, so they might have to be deleted. Given that varnished or painted wood is not a terribly good conductor, it is not unreasonable to bolt the radiator directly to the 4-×-4 fence post. Use $\frac{5}{16}$-inch (or larger) bolts, and make them long enough to fit completely through both the antenna element and the 4-×-4 post. Thus, 6-inch, 7-inch, or 8-inch long $\frac{5}{16}$-inch bolts are the candidates for this job. Use at least two bolts, one at the bottom of the antenna radiator element, and one near the top of the fence post. A third bolt, halfway between the other two, would not be out of order.

If the antenna is quarter wavelength, then no matching is generally necessary. Although the feedpoint impedance is not exactly 52 Ω, it is close enough (37 Ω) to form a reasonable match for 52-Ω coaxial cable (with VSWR = 1.4:1). The center conductor of the coaxial cable is connected to the radiator element, while the shield is connected to the ground system. In the example shown in Fig. 7-12, two ground methods are used. First, is an 8-foot ground rod driven into the earth at the base of

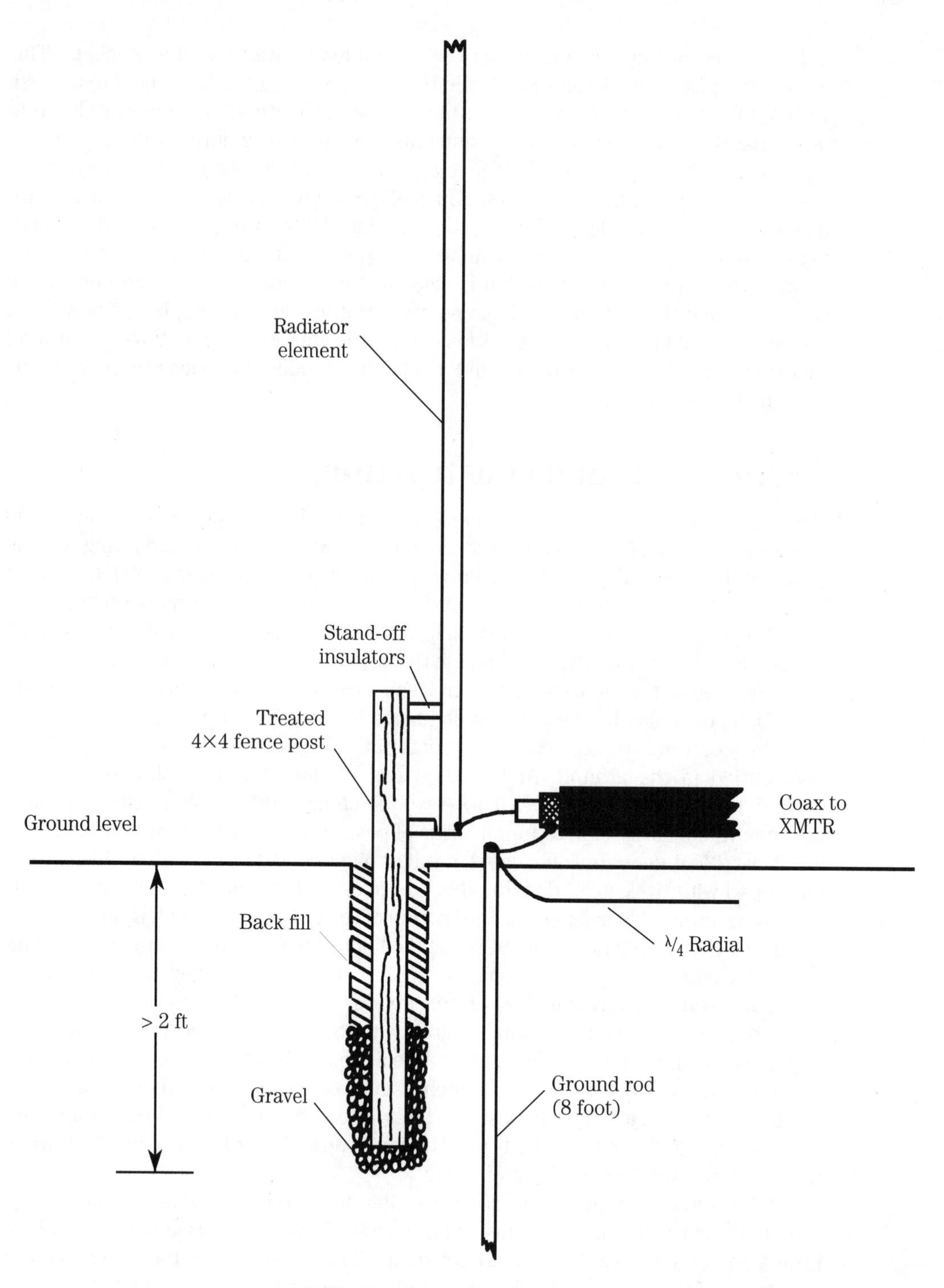

7-12 Ground mounted vertical.

the antenna; second, is a system of quarter-wavelength radials. Remember that the ground system is absolutely essential.

A method for installing a vertical antenna above ground is shown in Fig. 7-13. In this case, a wooden support (2-×-4 or 4-×-4) is installed in a manner similar to Fig. 7-12, but with a deeper hole to counter the longer length. Alternatively, the wooden support is affixed to the side of a building wall, shed, or other pre-existing structure. Once the support is arranged, however, the method of attachment of the radiator element is the same for the previous case, so that will not be repeated here.

Electrical connections to the antenna are also shown in Fig. 7-13. Because the antenna is above ground level, an electrical counterpoise ground consisting of a system of radials is absolutely essential; at least two radials per band must be provided. A small L-bracket is used to support the radials and to provide an SO-239 coaxial connector for the coax. This connector is a chassis-mounted type with its center conductor connected to the radiator element. The shield of the connector is connected to the bracket, so it is also connected to the radial system.

In some installations, the antenna support structure will require guy wires to keep the structure stable. Do not use the radials as guy wires. The type of wire that normally works well for radials is too soft, and too easily stretched, for guy wire service. Use regular steel guy line, available where TV antenna supplies are sold, for this antenna. Make the lengths nonresonant, and break the guy lines up with egg insulators, if necessary, to achieve the nonresonance.

⅝-Wavelength verticals

Figure 7-14 shows the configuration for the ⅝-wavelength vertical antenna. Such an antenna generally gives a lower angle of radiation than the more common ¼-wavelength radiator, so presumedly it works better for long distance.

The radiator of this antenna is made from 0.5-inch to 1.5-inch aluminum tubing. Again, remember that adjacent sizes fit together snugly to form longer sections. The physical length of the ⅝-wavelength radiator is found from:

$$L_{\text{ft}} = \frac{585}{F_{\text{MHz}}} \qquad \textbf{[7.5]}$$

or, in meters:

$$L_{\text{meters}} = \frac{180}{F_{\text{MHz}}} \qquad \textbf{[7.6]}$$

The radials are the usual quarter wavelength, and are made of #12 or #14 copper wire. These lengths are found from:

$$L_{\text{ft}} = \frac{246}{F_{\text{MHz}}} \qquad \textbf{[7.7]}$$

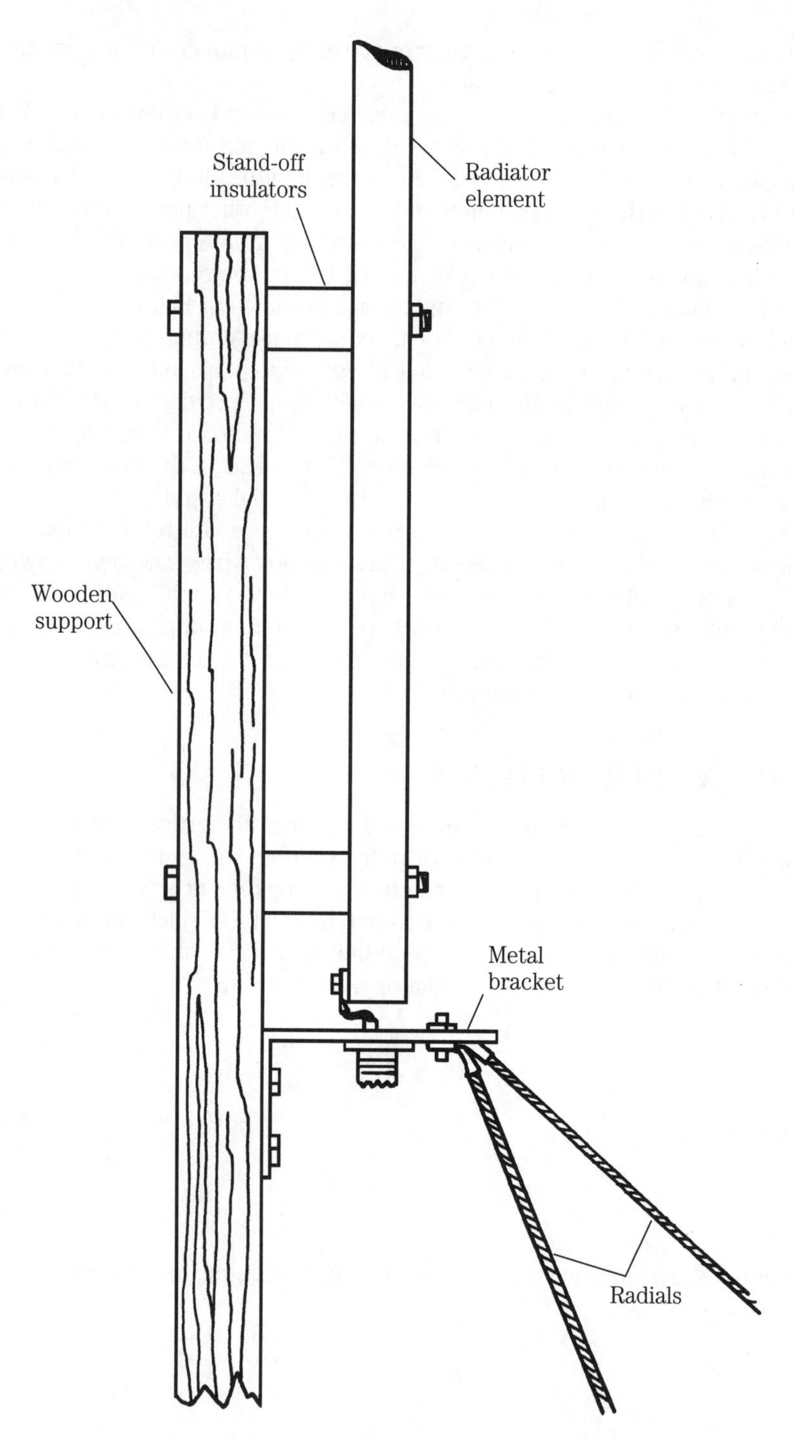

7-13 Details of feedpoint circuit on vertical.

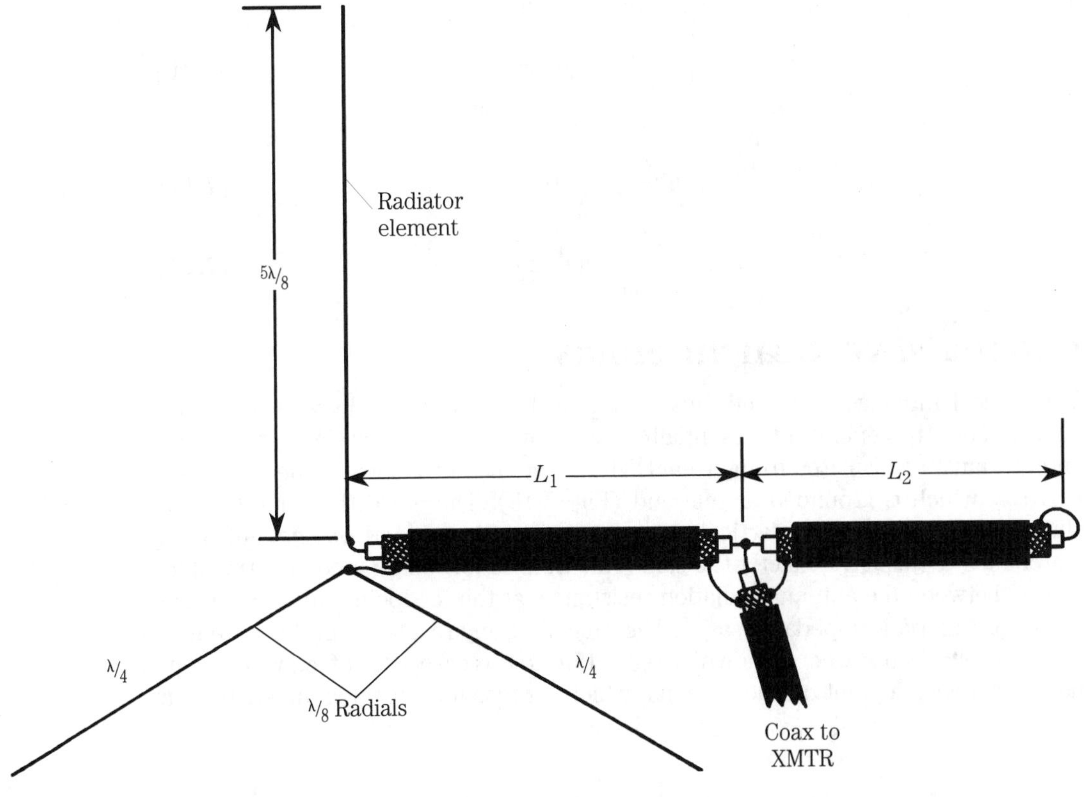

7-14 Using a Q-Section to match the feedpoint impedance of a vertical antenna.

or, in meters:

$$L_{\text{meters}} = \frac{75}{F_{\text{MHz}}} \qquad \textbf{[7.8]}$$

The feedpoint impedance of the ⅝-wavelength antenna is not a good match for the ordinary coaxial cables that are routinely available on the amateur market. Some form of impedance matching is needed.

One option is to use a broadbanded RF transformer, such as the *Palomar Engineers, Inc.* models shown in chapter 19. These transformers will work throughout the HF spectrum, and match a wide variety of impedances to the 50-**Ω** standard system impedance.

Another option, especially for a single-band antenna, is to use a coaxial cable impedance transformer, such as shown in Fig. 7-14. The transformer consists of two sections of coaxial cable joined together, shown as $L1$ and $L2$ in Fig. 7-14. The lengths are found from:

$$L_1 = \frac{122}{F_{\text{MHz}}} \text{ feet} \qquad \textbf{[7.9]}$$

or,

$$L_1 = \frac{38}{F_{\text{MHz}}} \text{ meters} \quad [7.10]$$

and,

$$L_2 = \frac{30}{F_{\text{MHz}}} \text{ feet} \quad [7.11]$$

$$L_2 = \frac{9}{F_{\text{MHz}}} \text{ meters} \quad [7.12]$$

Grounded vertical antennas

The vertical antennas presented thus far in this chapter are called *series-fed verticals*, because the generator is essentially in series with the radiator element. Such an antenna must be insulated from ground. The other class of vertical is the *shunt-fed vertical*, which is grounded at one end (Fig. 7-15). There are three methods of shunt-feeding a grounded vertical antenna: *delta, gamma* and *omega*. All three matching systems have exactly the same function: to form an impedance transformation between the antenna radiation resistance, at the feedpoint, and the coaxial cable characteristic impedance, as well as cancelling any reactance in the system.

The delta feed system is shown in Fig. 7-15A. In this case, a taut feed wire is connected between a point on the antenna, which represents a specific impedance on

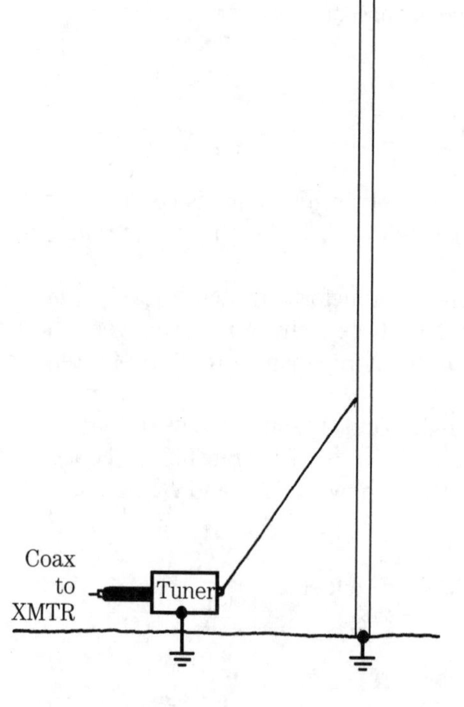

7-15A Delta fed grounded vertical.

the antenna, and an antenna tuner. This method of feed is common on AM broadcast antennas (which are usually—perhaps always—verticals). Although you would think that the sloping feed wire would distort the pattern, that is not the case. The distortion of the pattern, if any, is very minimal, hence it can be neglected.

The gamma feed system is shown in Fig. 7-15B. This method is commonly used by amateurs to feed Yagi beam antennas, so it is well familiar in the amateur radio world. The feed system consists of a variable capacitor to tune the system, and a matching rod that parallels the antenna radiator element. It is important that the rod not be anywhere near a quarter wavelength, or it would become a vertical antenna in its own right, and in fact would resemble the so-called *J-pole antenna*. A review of the gamma match is given in the chapter on beam antennas. The omega feed (shown in Fig. 7-15C) is similar to the gamma match except that a shunt capacitor is used.

Conclusion

The vertical antenna is a viable alternative for many situations, especially where real estate is at a premium. Contrary to popular opinion, the vertical antenna works well when installed properly and when due consideration has been given to matters such as the grounding and angle of radiation desired.

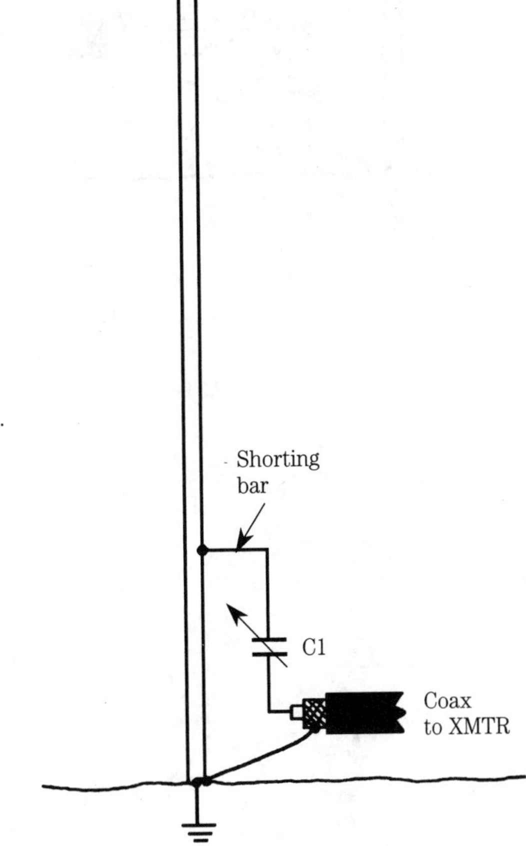

7-15B Gamma fed grounded vertical.

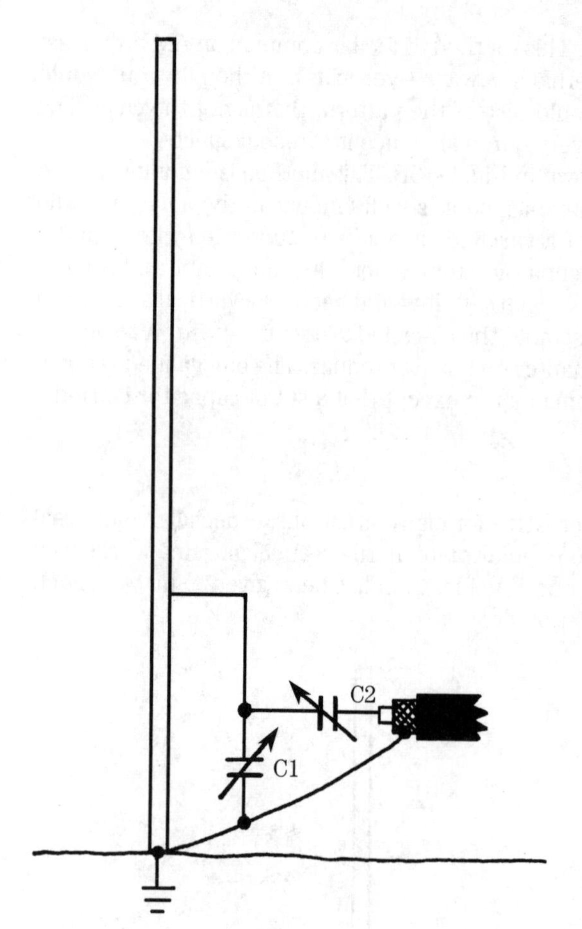

7-15C Omega fed grounded vertical.

8
CHAPTER

Multi-band and tunable-wire antennas

MOST COMMUNICATIONS OPERATORS REQUIRE MORE THAN ONE BAND, AND THAT MAKES the antenna problem exactly that—a problem to be solved. Amateur radio, commercial, and military operators are especially likely to need either multiple antennas for different bands, or a multi-band antenna that operates on any number of different bands. This situation is especially likely on the high-frequency (HF) bands from 3.5 to 29.7 MHz.

Another problem regards the tunability of an antenna. Some amateur bands are very wide (several hundred kilohertz), and that causes any antenna to be highly variable from one end of the band to another. It is typical for amateurs to design an antenna for the portion of the band that they use most often, and then tolerate a high VSWR at the other frequencies. Unfortunately, when you see an antenna that seems to offer a low VSWR over such a wide range, it is almost certain that some problem exists that reduces the Q, and the antenna efficiency, to broaden the response. However, it is possible to tune an antenna for a wide band. It is also possible (now that amateurs have new HF bands) to use a single antenna between them, and then tune the difference out. For example, designing a single antenna for 21/24 MHz, 14/18 MHz, or 7/10 MHz should prove possible.

In this chapter we will take a look at both problems: the multi-band and the tunable antenna.

Multi-band antennas

Although a tri-band Yagi or Quad beam antenna will undoubtedly work better than a wire antenna (when installed correctly!), the low-budget amateur operator need not lament any supposed inability to "get out" on wire antennas. To quote an old saying: "'Better' is the enemy of 'good enough.'" Or, to put it in terms of *Carr's Law:* "If it's good enough" then don't waste a lot of energy fretting over making it 'better' unless you really want to make it *a lot better*."

Trap dipoles

Perhaps the most common form of multi-band wire antenna is the *trap dipole* shown in Fig. 8-1A. In this type of antenna, one (or more) pairs of parallel resonant traps are placed in series with the quarter-wavelength elements of the dipole. The purpose of the traps is to block their own resonant frequency, while passing all other frequencies.

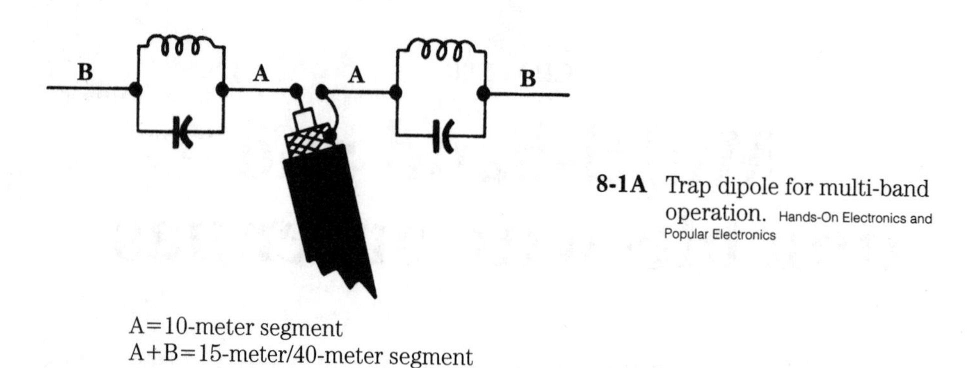

8-1A Trap dipole for multi-band operation. Hands-On Electronics and Popular Electronics

In the example of Fig. 8-1A, a 10-meter trap isolates the first eight feet or so (quarter wavelength on 10 meters) so that the antenna resonates on that band. A 40-meter or 15-meter signal, on the other hand, passes through the traps and uses the whole length of the antenna (note: a half wavelength 40-meter dipole works as a 3/2 wavelength antenna on 15 meters).

The overall length of the trap dipole will be a little less than the natural "non-trap" length for the lowest frequency of operation. At the low frequencies, the traps add a little inductance to the circuit so that the resonant point is lower than the natural resonant frequency. In general, most trap dipoles are just a few percent shorter than nontrap dipoles at the same band. The actual amount of shortening depends upon the values of the components in the traps, so consult the data for each trap purchased. Where more than one pair of traps are used in the antenna make sure they are of the same brand and are intended to work together.

Another solution to this problem is shown in Fig. 8-1B. This type of antenna actually has two or more half-wavelength dipoles fed from the same transmission line. In this illustration, a total of three dipoles are fed from the same 75 Ω transmission line. There is no theoretical limit to how many dipoles can be accommodated, although there is certainly a practical limit. For one thing, there is a mechanical limit to how many wires are supportable (or desirable) hanging from any given support. There is also an electrical limit, although it is less defined. Having a lot of dipoles increases the possibility of radiating harmonics and other spurious emissions from your transmitter.

The 75 Ω coaxial cable is connected to the center feedpoint of the multi-dipole either directly, or through a 1:1 BALUN transformer, as shown in Fig. 8-1B. Each an-

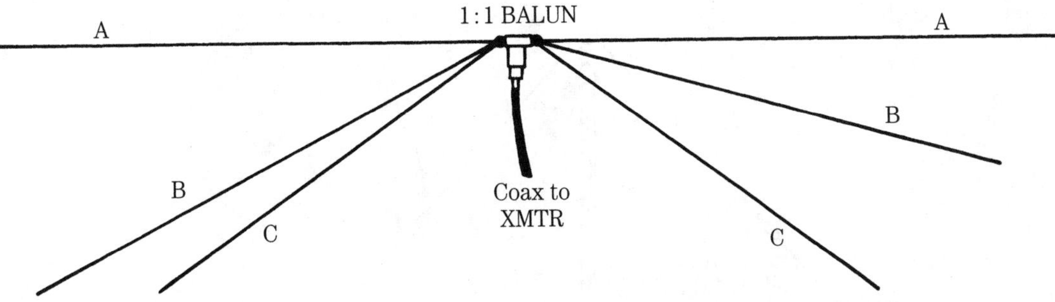

8-1B Multi-band dipole consists of several dipoles fed from a common feedline.

tenna (A-A, B-B, or C-C) is a half wavelength. Therefore, the overall length is found approximately from the standard dipole expressions:

Overall length (A+A, B+B or C+C):

$$L_{ft} = \frac{468}{F_{MHz}} \quad [8.1]$$

or, for each element alone (A, B, or C):

$$L_{ft} = \frac{234}{F_{MHz}} \quad [8.2]$$

As always, close to the earth's surface, these equations are approximations, and are not to be taken too literally. Some experimentation will probably be necessary to optimize resonance on each band. Also, be aware that the drooping dipoles (B and C in this case) may act more like an inverted-vee antenna (see chapter 7) than a straight dipole, so the equation length will be just a few percent too short. In any event, a little spritzing with this antenna will yield results.

Some amateurs build the multiple dipole from four- or five-wire TV rotator lead. That type of wire is used to control antenna rotators, and has either four or five parallel wires in a flat arrangement similar to lamp cord. Cut each wire to the length required for a band, and strip off any unused portions.

Another possibility is the link-tuned dipole shown in Fig. 8-1C. In this situation, a single conductor is used for each half of the dipole, or actually inverted vee. The conductors are broken into segments A, B, and C (or more, if desired). Each segment is separated from the two adjacent sections by inline insulators (standard end insulators are suitable). Segment A is a quarter wavelength on the highest frequency band of operation, A+B is a quarter wavelength on the next highest band of operation, and A+B+C is a quarter wavelength on the lowest frequency band of operation.

The antenna is "tuned" to a specific band by either connecting, or disconnecting, a similar wire (see inset) jumper across the insulator that breaks the connection between the segments. Either a switch or an alligator clip jumper will short out the insulator to effectively lengthen the antenna for a lower band. Some amateurs use single-pole 110-Vac power line switches to jumper the insulator. Although I have not tried this method, it should work.

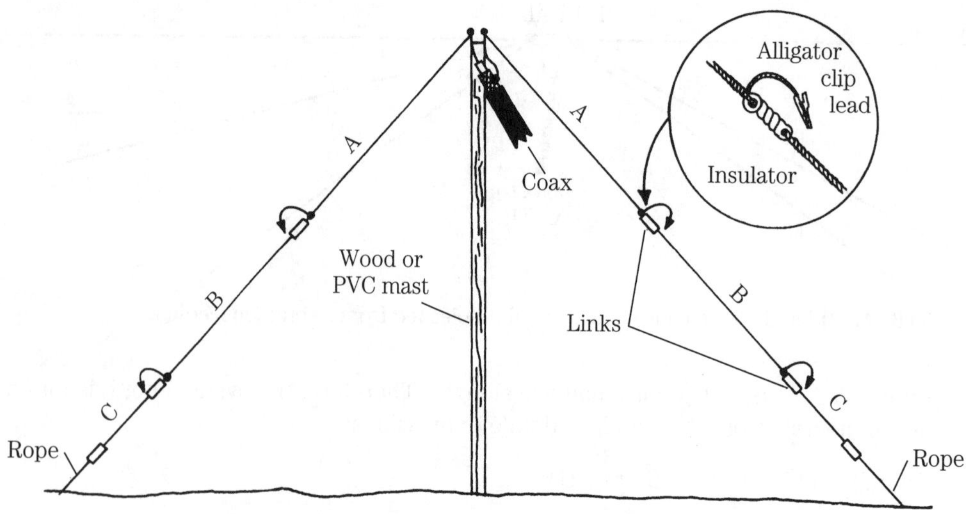

8-1C Multi-band inverted-vee uses shorting links to change bands.

A big disadvantage to this type of multi-band antenna is that you must go out into the yard and manually switch the links to change bands . . . which probably explains why other antennas are a lot more popular, especially in northern latitudes.

Tuned feeder antennas

Figure 8-2A shows the tuned feeder type of antenna. This antenna can be used from 80 through 10 meters, but it requires a special tuner and a length of parallel transmission line. There are two ways to get parallel transmission line: make it or buy it. Using #14 or #12 wire, and specially made insulators (also called "spreaders"), you can make 300 Ω, 450 Ω, or 600 Ω parallel transmission line. But that's a pain in the *ptusch* because you can also buy parallel line rather cheap. I paid $16 for 100-feet of 450 Ω line recently.

One form of parallel line is ordinary TV-type twin-lead, which has an impedance of 300 Ω. This line will take up to about 250 watts, although some people use it at higher powers (not recommended!). The antenna of Fig. 8-2A uses 450 Ω parallel line. You can buy insulated 450 Ω twin-lead (see Fig. 8-2B) that can be handled as easily as TV twin-lead . . . and a lot easier than open (uninsulated) parallel line.

The G5RV multi-band dipole

Figure 8-3 shows the popular G5RV antenna. Although not without some problems, this antenna is very popular at the moment. It can be used either as a horizontal dipole, a sloper or as an inverted-vee antenna (which is how I used it). The dipole elements are each 51 feet long. The feedline can be either 300 Ω or 450 Ω twin-lead. For 300 Ω cases, use 29 feet of line; and for 450 Ω line, use 34 feet. One end of the parallel transmission line is connected to the antenna, and the other end is connected

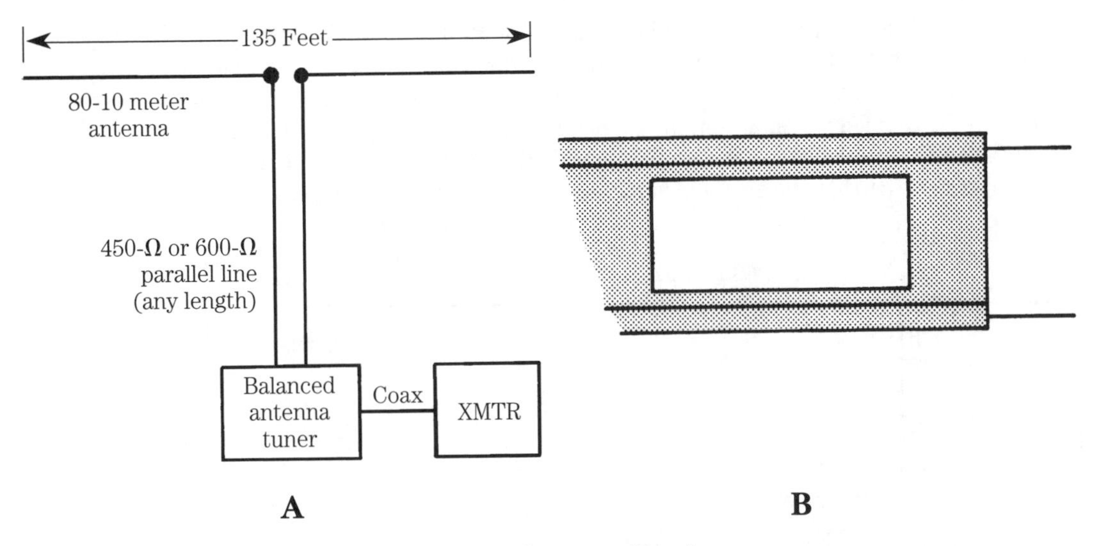

8-2 Tuned feeder antenna can be used on several bands. Hands-On Electronics and Popular Electronics

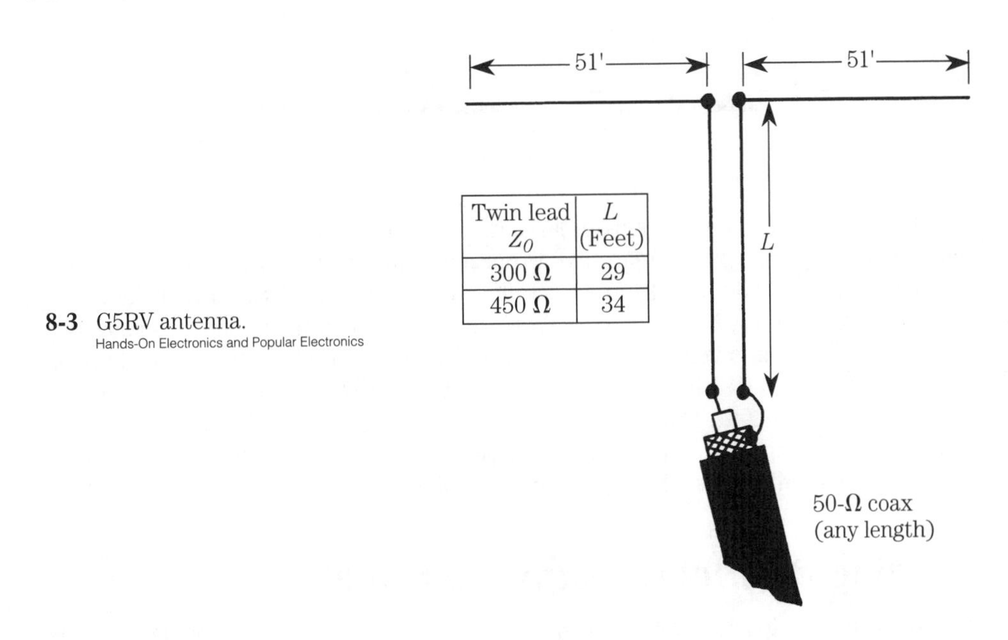

Twin lead Z_0	L (Feet)
300 Ω	29
450 Ω	34

8-3 G5RV antenna.
Hands-On Electronics and Popular Electronics

to a length of 50 Ω coaxial cable. Although most articles on the G5RV claim that any length of 50 Ω line will work, J.M. Haerle (*HF Antenna Systems: The Easy Way*) recommends that the 50 Ω segment should be at least 65 feet long.

Haerle is a little caustic in his comments on the G5RV, but his criticism is well taken. If you don't have a parallel transmatch, then the G5RV will work (especially if your rig can tolerate a 3:1 VSWR on some selected frequencies). Otherwise, use the antenna of Fig. 8-2A (or a transmatch with the G5RV).

Figure 8-4 shows the once-popular end-fed Zepp antenna. This antenna is a "monopole" in that it uses a half-wavelength radiator, but it is fed at a voltage node rather than a current node (i.e., the end of the antenna rather than the center). A 450 or 600 Ω parallel transmission line is used to feed the Zepp antenna. Although the line can theoretically be any length, practicality dictates a quarter wavelength because of the fact that the antenna is fed at a high RF voltage point.

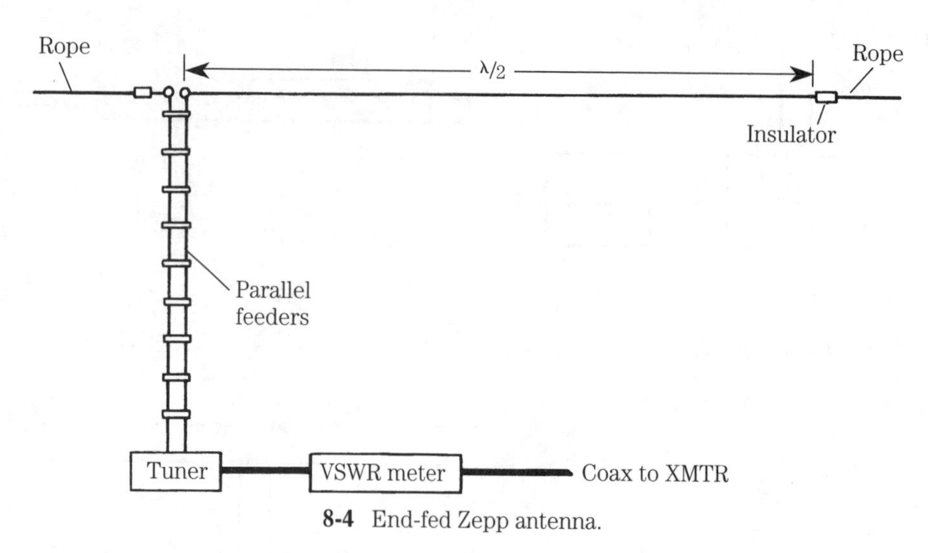

8-4 End-fed Zepp antenna.

The Zepp (as shown) is a single-band antenna unless the transmission line is fed with a good, widerange transmatch, or other antenna tuner unit. The antenna tuner required is a balanced type, although a standard transmatch with a 4:1 BALUN transformer at the output will also work well. Using the Zepp on many bands is easy, but keep in mind that it is voltage fed, and (at frequencies for which the parallel line is not an odd integer multiple of a quarter wavelength) there will be a high-voltage node at the transmatch. That raises the possibility of "RF in the shack"—"hot" snippets of RF on the grounded chassis of station equipment (including microphones, where it is uncomfortable when touched to the lips).

Feeding parallel transmission line

Parallel transmission line is *balanced* with respect to ground, but coaxial cable is *unbalanced* to ground (i.e., one side of the coax is usually grounded). As a result, the standard amateur transmitter output will not drive parallel (balanced) feedline properly. You need to do one of two things: 1) buy (or build) an antenna tuner that is balanced on the output and unbalanced on the input, or 2) convert a standard "coax-to-coax" transmatch or other form of antenna coupler to "coax-to-balanced" configuration. In some cases you can use a 4:1 BALUN transformer at the output of the coaxial cable tuner. Keep in mind, however, that some tuner manufacturers recommend against this practice. If you use the standard "Tee-network" (or SPC) transmatch then it is possible to make a BALUN for this purpose from some #12 formvar or enamel covered wire and either a ferrite toroid or a short piece of PVC plumbing pipe. Figure 8-5 shows a BALUN coil construction project. Use 12 turns of #12 formvar or enameled wire over a 1-inch outside diameter piece of PVC pipe or tubing.

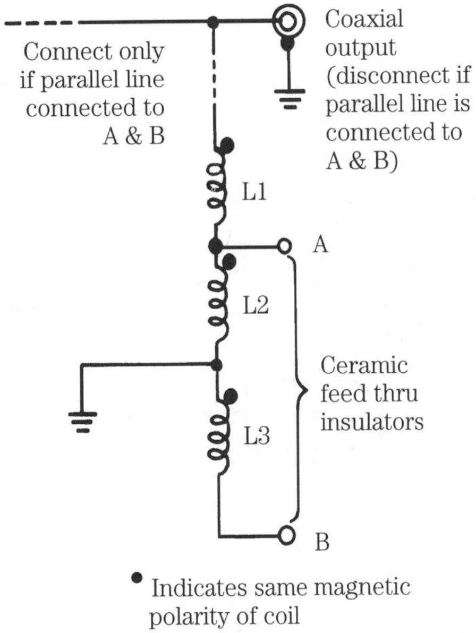

8-5 BALUN transformer converts coaxial line to parallel line feed system.

Next, we have the so-called "long-wire" antenna. I use the term "so-called" because not all long wire antennas are truly longwires. A true longwire is longer than two wavelengths at the lowest frequency of operation. In Fig. 8-6 we see a longwire, or "random length" antenna fed from a tuning unit. If the antenna length (L) is greater than a quarter wavelength, then the tuner consists of a single series capacitor (see below); and if it is shorter than a quarter wavelength, the tuner is a series inductor. The standard tuner for this type of antenna, regardless of length, is a simple L-section coupler (also shown), which can be selected for L-section, series-L or series-C operation.

The Windom antenna (Fig. 8-7) has been popular since the 1920s. Although Loren Windom is credited with the design, there were actually a number of contributors. Coworkers with Windom at the University of Illinois were John Byrne, E.F. Brooke, and W.L. Everett, and they are properly co-credited. The designation of Windom as the inventor was probably due to the publication of the idea (credited to Windom) in the July 1926 issue of *QST* magazine. Additional (later) contributions were rendered by G2BI and GM1IAA (Jim MacIntosh). We will continue the tradition of crediting Loren Windom, with the understanding that others also contributed to this antenna design.

The Windom is a roughly half-wavelength antenna that will also work on even harmonics of the fundamental frequency. The basic premise is that the antenna ra-

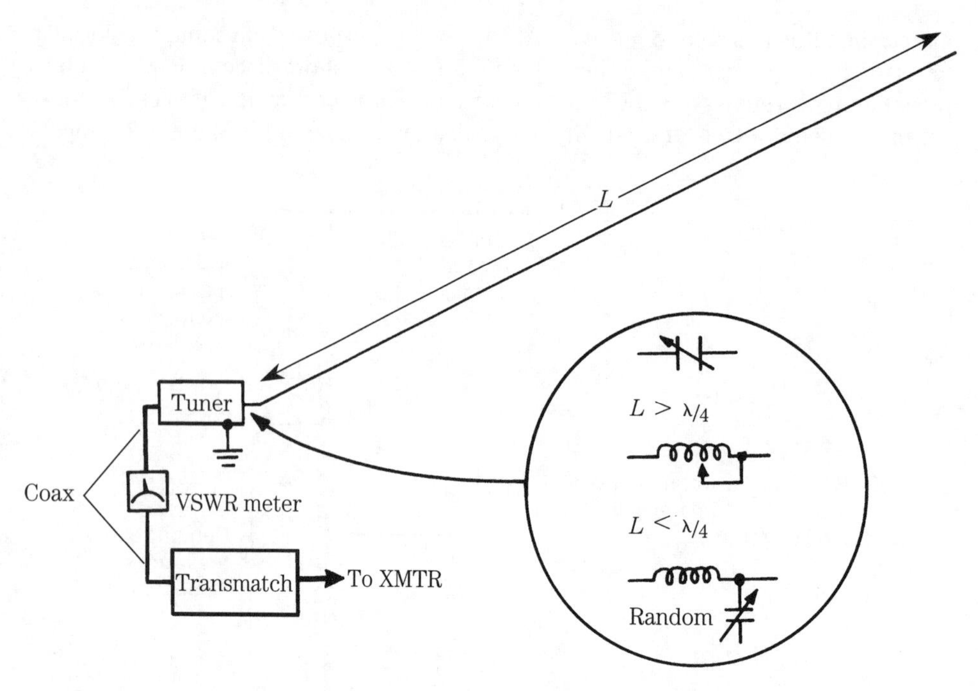

8-6 Random length (a.k.a. "longwire") antenna.

diation resistance varies from about 50 Ω, to about 5,000 Ω, depending upon the selected feedpoint. When fed in the exact center, a current node, the feedpoint impedance will be 50 Ω; similarly, end-feeding the antenna finds a feedpoint impedance of about 5,000 Ω. In Fig. 8-7A the feedpoint is tapped away from the center at a point that is about one-third the way from one end, at a point where the impedance is about 600 Ω.

The Windom antenna works well . . . but with some caveats. For example, the antenna has a tendency to put "RF in the shack" because of the fact that it is voltage fed. Second, there is some radiation loss from the feedline. Finally, the antenna works poorly on add harmonics of the fundamental frequency.

The antenna tuning unit can be either a parallel resonant, link-coupled, LC tank circuit (see inset to Fig. 8-7A); or a reversed pi-network. In the case of the Windom, the pi-network is turned around backwards from the usual configuration: C1 is at the low impedance end of the network, so it is larger than C2. Design a pi-network (see programs at end of book) to match 50 Ω on the transmitter end, and 600 Ω on the antenna end.

A reasonable compromise Windom, that reduces feedline radiation losses, is shown in Fig. 8-7B. In this antenna a 4:1 BALUN transformer is placed at the feedpoint, and this in turn is connected to 75 Ω coaxial transmission line to the transmitter. A transmatch, or similar antenna tuner, is then connected between the transmitter and the transmission line.

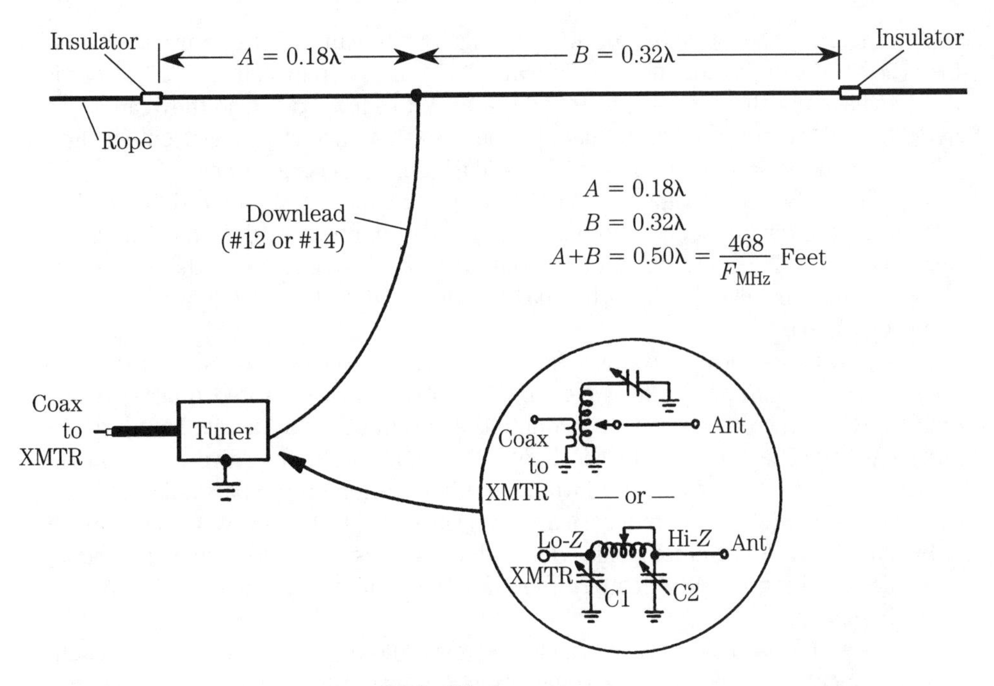

8-7A Windom antenna

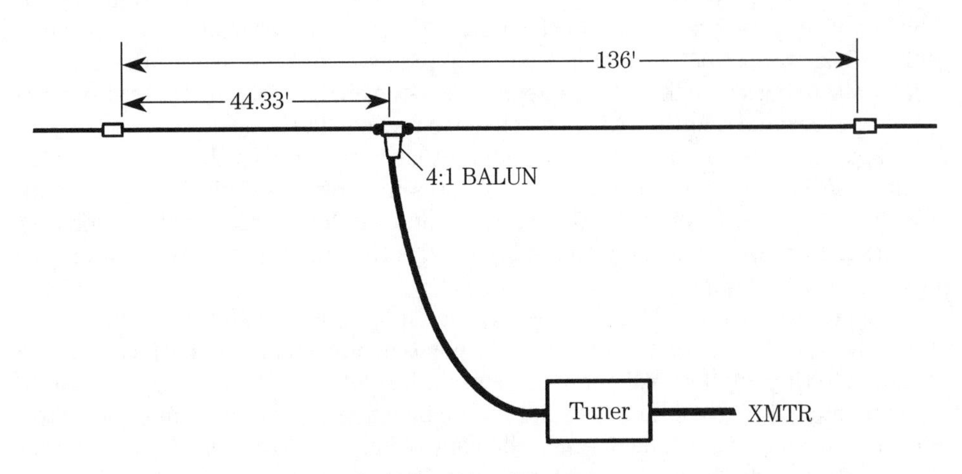

8-7B Coaxial fed Windom.

A "new" form of antenna tuner

There are a number of myths that are widely held among radio communications hobbyists . . . and amateur radio is no less infested with some of these myths than others (CB, for example). Twenty-five years ago I worked in a CB shop in Virginia, and we kept hearing one old saw over and over again: you can ". . . cut your coax to reduce the VSWR to 1" (actually, they meant "1:1" but routinely called it "1"). Hoards of CBers have cut the coax and watched the VSWR reduce to 1:1, so they cannot be talked out of the error. What actually happens in that case is a measurement difficulty that makes it appear to be true. We will address this issue later.

Of course, Hams are superior to CBers so don't believe that error, right? I'd like to think so; but having been in both the CB and the amateur worlds, and "Elmered" more than a few CBers studying for amateur licenses, I have to admit that at least as many amateurs believe the "cut-the-coax" error as CBers (sorry, fellows, but that's my observation).

The only proper way to reduce the VSWR to 1:1 is to *tune the antenna to resonance*. For a center-fed half wavelength dipole, or a bottom-fed quarter wavelength vertical, the proper way to resonate the antenna is to adjust its length to the correct point. The formulas in the books and magazines only give approximate lengths . . . the real length is found from experimentation on the particular antenna after it is installed. Even commercial antennas are adjusted this way. On certain CB mobile antennas, for example, this trick is done by raising (or lowering) the radiator while watching the VSWR meter. On amateur antennas similar tuning procedures are used.

Another ploy used by amateurs (including myself) is to connect an antenna matching unit (tuner) at the output of the transmitter. For my Kenwood TS-430, I use a Heath SA-2060A to "tune-out" the VSWR presented by my *Hustler* 4BTV and 75 feet of coax. But I don't even pretend to be tuning the antenna. The TS-430 is a solid-state rig, and the finals are, therefore, not terribly tolerant of VSWR—and will shut down with a high VSWR. The purpose of the antenna tuner is to reduce the VSWR seen by the transmitter . . . and to heck with the actual antenna mismatch on the roof. The tuner also serves to reduce harmonics further, thereby helping to prevent TVI.

The best form of antenna tuner is one that both reduces the VSWR (for the benefit of the transmitter), and also resonates the antenna. This form of antenna tuner is installed between the coaxial cable and the feedpoint of the antenna. To the best of my knowledge, the Heath *SA-2550* is the only tuner on the market that does this trick.

Figure 8-8A shows the *SA-2550* mast-mounted antenna tuner and its control box. The large black box is mounted on the antenna mast, and its output terminals connect to the antenna radiators. The control box sits inside the shack, for use by the operator. If your antenna does not use dc blocking capacitors, tuning capacitors or other means for blocking dc, then the control voltage to the tuner can be routed through the coaxial cable. Most antennas are designed in this manner. But if, for any reason, your antenna or transmission line will not pass dc, then a separate control cable between the two boxes can be used. Because either scenario is possible, the *SA-2550* is thus made more flexible.

Figure 8-8B reveals the internal circuitry of the *SA-2550*. This circuit has a large, high-voltage 500 pF variable capacitor driven by a small (but high torque) low-voltage dc motor. The two large eyelet terminals are used to connect the antenna radiators to the box. The use of this capacitor in horizontal and vertical antennas is shown in Fig. 8-9.

A

8-8 Heath antenna tuner for broadbanding an antenna.

B

L1 L2

$$L = L1 + L2 \approx \frac{538}{F_{MHz}} \text{ Feet}$$

8-9A Use of Heath tuner, or other variable capacitor, in a dipole.

At resonance the feedpoint impedance of an antenna is all resistive, and it has no reactance (inductive or capacitive). As the operating frequency departs from resonance, however, the reactive components become non-zero. For example, an antenna that is too long, exhibits an inductive reactance. Similarly, an antenna that is

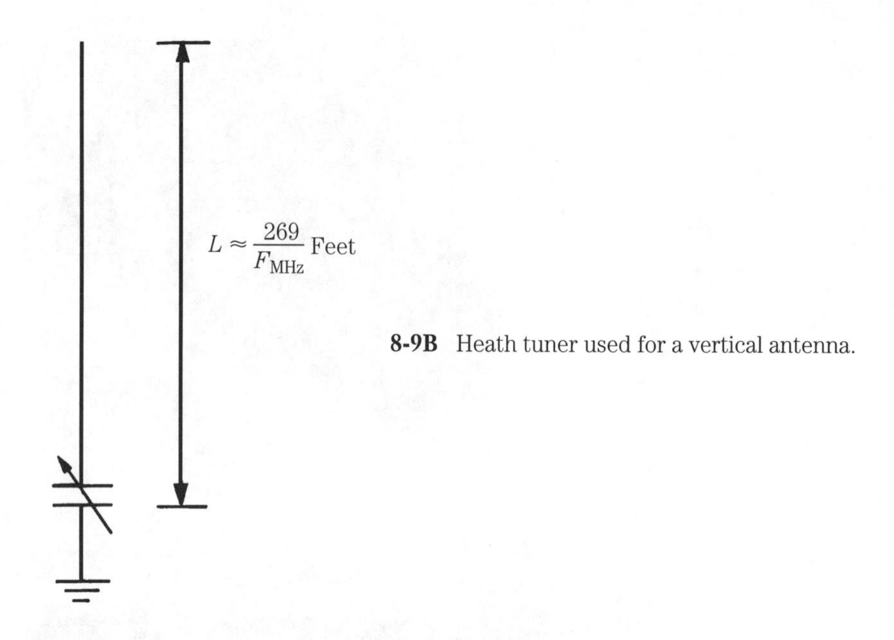

8-9B Heath tuner used for a vertical antenna.

too short, exhibits a capacitive reactance. For this type of antenna we insert a loading coil to provide inductance to cancel the antenna capacitance. This is the usual case for amateur and CB mobile antennas, and some vertical antennas operated on 80-meters. In the case of the *SA-2550* tuner, we make the antenna radiator (L1 and L2) in Fig. 8-9 slightly longer than is normally necessary . . . and use the capacitor to tune out the resulting inductive reactance. The result is an antenna that can be resonated over the entire amateur band . . . not just at a single point.

Two different antennas, using the *SA-2550*, are shown in Fig. 8-10. Figure 8-10A shows an inverted-vee with radiators that are about 15-percent longer than normal inverted-vees. The tuner is mounted at the top end of a mast, and the radiators are connected to the tuner eyelet connections. The case of Fig. 8-10B shows a vertical antenna. Once again, the length is 10 to 15 percent longer than most quarter-wavelength verticals. One eyelet terminal is connected to the radiator, while the other is connected to the ground, or system of radials (or both, if that is how you prefer to build the antenna).

Figure 8-1C shows the proper method for connecting the antenna wire to the eyelet terminals. Pass the conductor through the eyelet with sufficient length to wrap back around the wire 5 to 10 times, and then to the solder lug. Although some people don't bother to solder the multi-turn section of wire, I prefer to (and recommend you do also). The purpose is not strength, but rather keeping the electrical connection competent after weather sets in.

You can also build your own version using this same idea. A slow-speed motor (or stepper, as used to advance the platen on a computer printer), and a 500 to 1000 pF transmitting variable capacitor, are all that is needed. Alternatively, motor-driven 300- to 1500-pF vacuum variable capacitors are sometimes found on the surplus market.

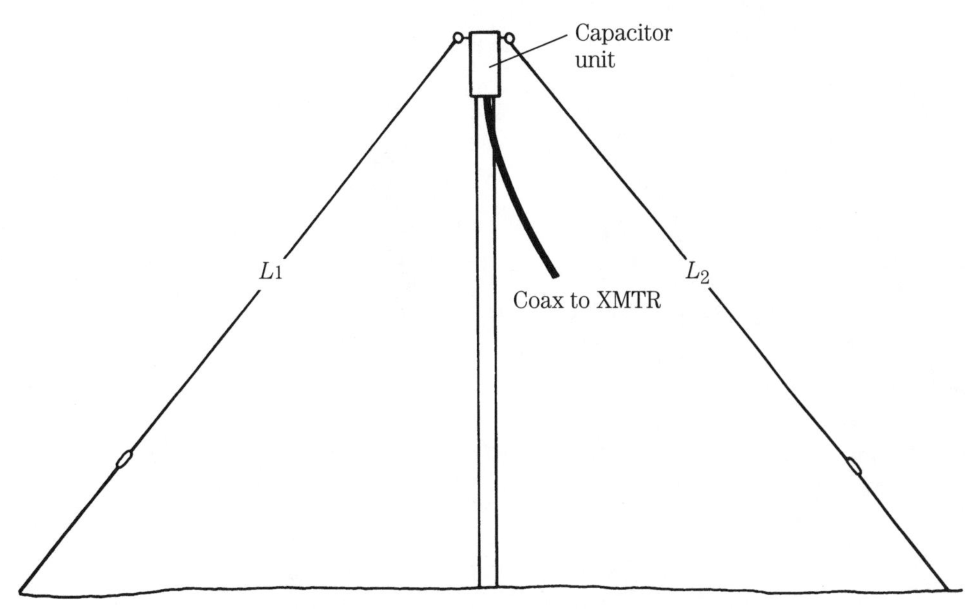

8-10A Tuner utilization for an inverted-vee antenna.

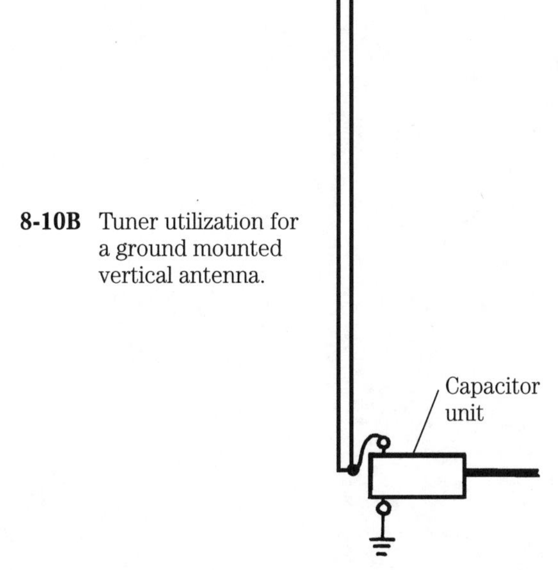

8-10B Tuner utilization for a ground mounted vertical antenna.

9 CHAPTER

Longwire directional antennas

YOU'VE HEARD IT DOZENS OF TIMES: THE RANDOM LENGTH LONGWIRE ANTENNA IS THE "perfect" solution to awkward antenna problems. Whether it's a lack of real estate, cranky landlords, or a profound lack of bucks, the longwire will do the job for you. Right? Well, now, that depends on who you ask, and what they did to make it work . . . or not work. One person says the longwire is not worth a plugged nickel; another is very ho-hum about it because his "kinda works;" still another is enthusiastic because hers is installed correctly and it works better than anything "since sliced pickle and liverwurst sandwiches." Over the years my various living arrangements have forced me to use longwires at many QTH's, as well as on Field Day. Why does the random length longwire have such a varied reputation? Before we answer that question let's find out (for those who came in late) just what is a longwire antenna.

As an aside, let me point out that random lengthwire antennas (less than 2 λ) are not true longwire antennas. However, common (if erroneous) usage compels the inclusion of both types.

Longwire antennas

Longwire antennas are any of several types of resonant and nonresonant antennas. Any given longwire antenna may be both resonant and nonresonant, depending upon the operating frequencies used. In the "old days," when I was first starting into amateur radio, most resonant longwires were resonant over all HF bands because those bands were harmonically related to each other. But with the addition of the 10-, 18-, and 24-MHz band segments, that relationship no longer holds true for all cases.

Figure 9-1 shows the classic random length, nonresonant longwire antenna. It consists of a wire radiator that is at least a quarter wavelength long, but is most often longer than a quarter wavelength. The specific length is not critical, but it must be *greater than a quarter wavelength at the lowest frequency of operation anticipated.* If you have a 90-foot wire, it will work on all HF bands above 3.5 MHz. In

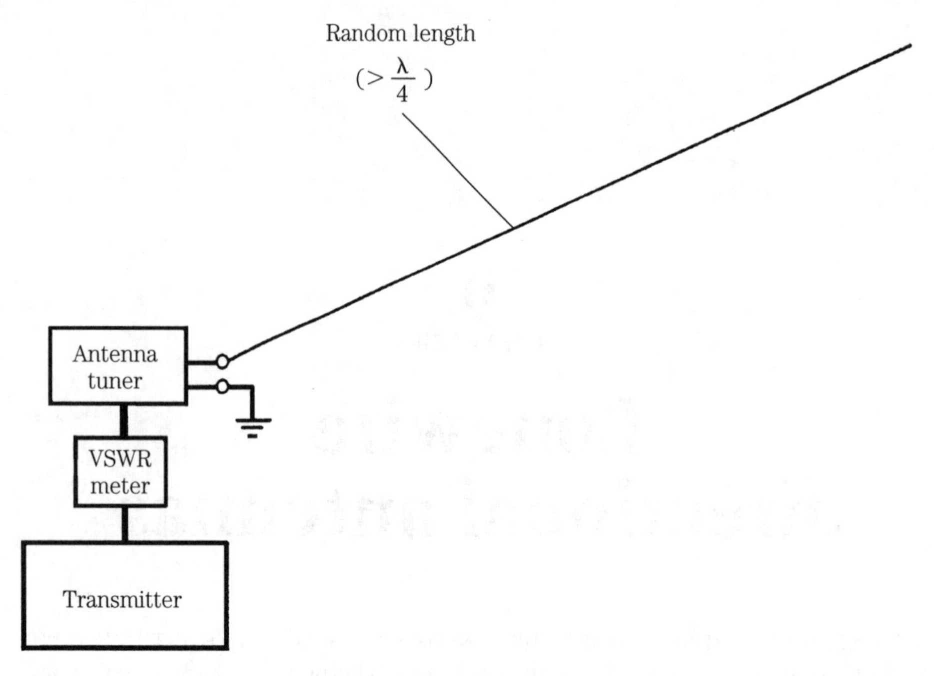

9-1 Random length antenna.

most installations, the wire is #12 and #14 *Copperweld*, or hard-drawn copper wire. I have successfully used both #12 and #14 house wire, but because it is solid (not stranded) it is not the best material. Stranded wire lasts longer in the wind, because solid wire fatigues and breaks quicker.

The longwire antenna is capable of providing gain over a dipole and a low angle of radiation (which is great for DX operators!). But these advantages are only found when the antenna is several wavelengths long, so it only occurs on typical HF antennas in the 21- through 29-MHz bands.

The longwire is end-fed, and therefore has a high impedance, except in those cases where the radiator happens to be quarter wavelength. Because of this fact, it is necessary to use an antenna tuner between the low-impedance transmitter output (usually limited to 50- to 75-Ω impedances) and the antenna. You can buy any of several commercial antenna tuners, or make one yourself. Figure 9-2 shows a typical antenna tuner for longwires. It is an L-section coupler consisting of a series inductance and a shunt capacitance, both variable. The inductor can be homemade, or it can be made from a *B&W* miniductor (3029 is suitable). Alternatively, you can buy a used roller inductor from a hamfest or *Fair Radio Sales* (Box 1105, Lima, OH, 45802), or a new roller inductor from *Radiokit* (Box 973, Pelham, NH, 03076). If you opt for the homemade or miniductor alternatives, then an alligator clip lead (short!) connected to one end of the coil can be used to short-out unneeded turns when adjusting the inductance. I personally prefer the roller inductor method, because it allows the whole "shootin' match" to be installed inside of a shielded cabinet, which helps the TVI/BCI situation.

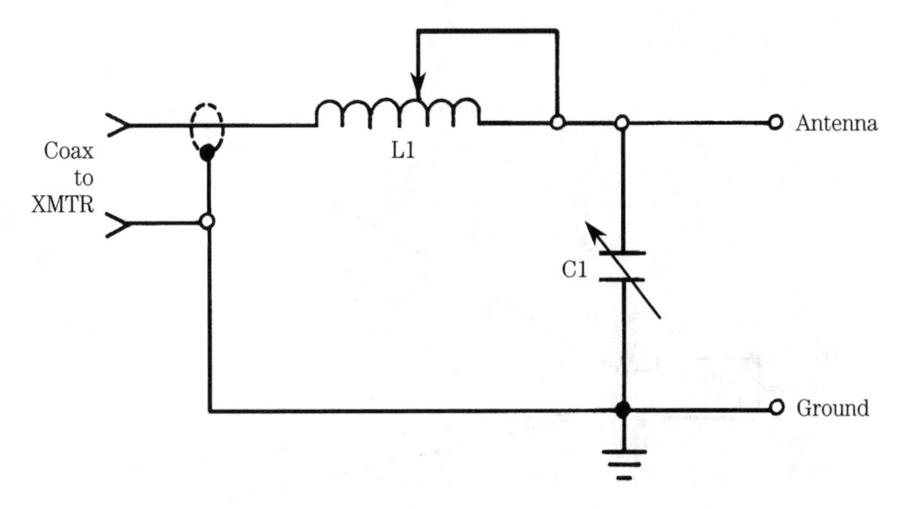

9-2 L-section coupler.

The capacitor (also available from the same sources) should be 150-pF to 250-pF maximum capacitance high voltage variable. So-called "transmitting variables" are usually ok if the plate spacing is at least ⅛ inch.

Tuning of the longwire is simplicity itself. If the tuner does not have a built-in VSWR meter, then install one in the line between the transmitter output and the tuner input. Adjust both L1 and C1 (they are a bit interactive so do it several times) for the lowest VSWR. If you use the alligator clip method, *turn off the darn transmitter before adjusting the clip position—RF burns are nasty!*

So what's the problem?

Ok, so we have decided to install a longwire. How do we make it work? First, make sure that it is long enough. The bare minimum lengths for HF bands are 70 feet for 3.5 MHz and up, 34 feet for 7 MHz and up, and so on. In general, the longer the better. The second problem is a *good ground.* The importance of a good ground cannot be underestimated, and it accounts for about 99 percent of the difference in reported performance of longwires.

A good ground consists of a short wire to either one very long ground rod, or multiple ground rods spaced a couple feet apart. An important factor is the length of the ground wire. It must be considerably less than a quarter wavelength. Use #10 stranded wire (several parallel lengths) or braid.

When I was in college I lived on the second floor of a whacky student boarding house in Norfolk, VA. The ground was 24 feet away, and that made "short ground wire" a joke . . . and my longwire nearly inoperative. But I figured out a way around the problem. A quarter-wave radial was dropped out the window (see Fig. 9-3). Use more than one radial if possible. In one case, when the landlord was particularly cranky, I tacked the radial to the baseboard of my room (insulated on standoffs at the far end to protect against the high RF voltages present). It worked well!

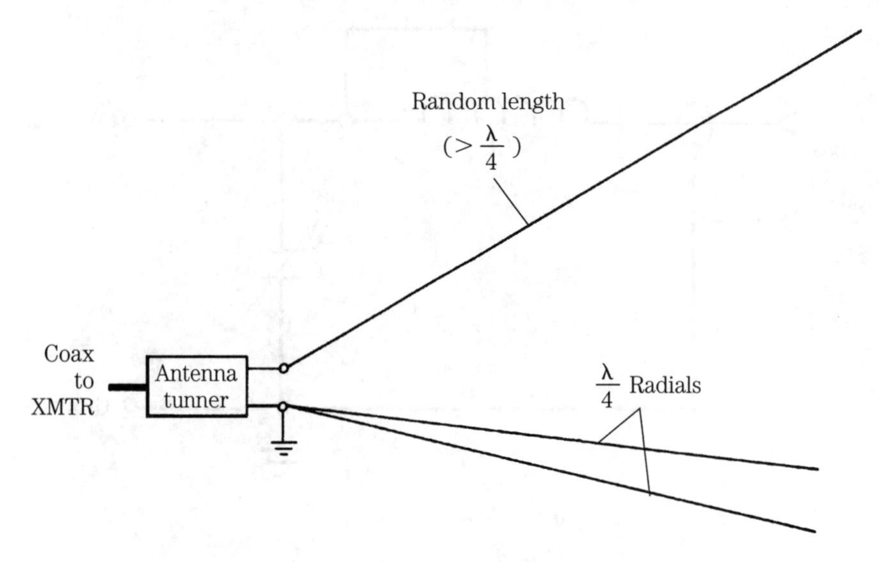

9-3 Radials improves the "ground" of random length antenna.

A new product also comes to the rescue of those reader-hams who cannot put up a good longwire. The *MFJ Enterprises* (Box 494, Miss. State, MS, 39762, 1-800-647-1800), Model MFJ-931 *artificial RF ground* (price: $79.95) is installed in the ground line (see Fig. 9-4) and is used to tune the ground wire. Adjust the capacitance and inductance controls for maximum ground current as shown on the built-in meter. Be sure to follow the instructions, however, because high RF voltages can appear on a nonresonant ground wire. I wish I'd had one of these when I was in that whacky boarding house . . . discussing existentialist poetry can't hold a candle to DXing.

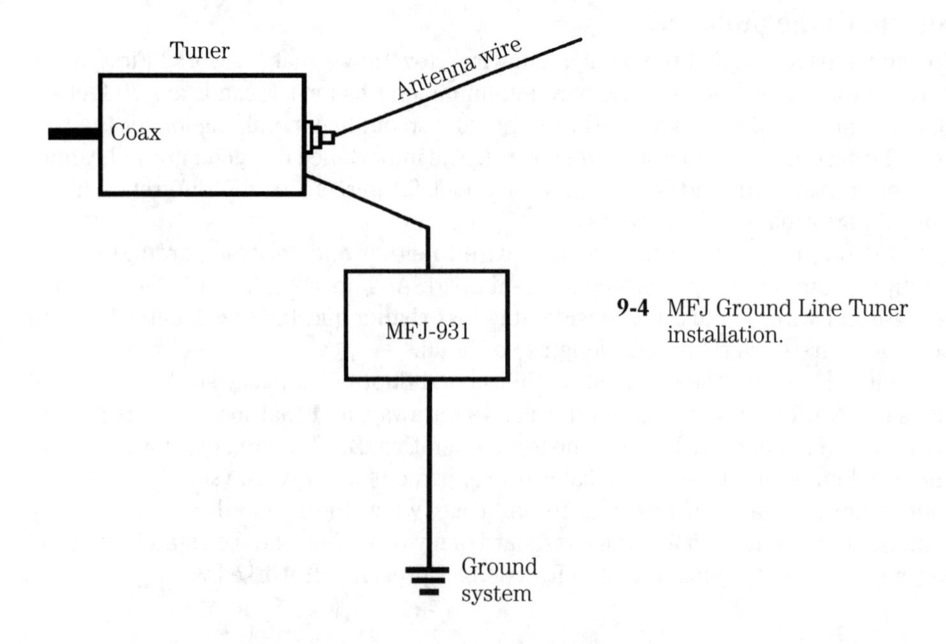

9-4 MFJ Ground Line Tuner installation.

When is a long wire a longwire?

For many years, the longwire has been a popular form of antenna. It is cheap, it is easy to construct, and—although reports vary—it has the potential to perform well. Properly constructed, it has a lot of utility. But what is a *real* longwire antenna? In past columns, I have used the term to mean an antenna such as Fig. 9-1. This form of antenna is popularly called a "longwire" if it is more than a quarter wavelength. I and other authors have used the term "longwire" to mean this antenna, but that's not rigorously correct. A *true* longwire is an antenna that is many wavelengths long, or to be a little more rigorous, an antenna that is *more than two wavelengths long*. Although I will still use the term "longwire" for both forms of longwire antenna, proper rigor requires Fig. 9-1 to be called a "random length antenna."

True longwire antennas

Figure 9-5 shows the true resonant longwire antenna. It is a horizontal antenna, and if properly installed, it is not simply attached to a convenient support (as is true with the random length antenna). Rather, the longwire is installed horizontally like a dipole. The ends are supported (dipole-like) from standard end insulators and rope.

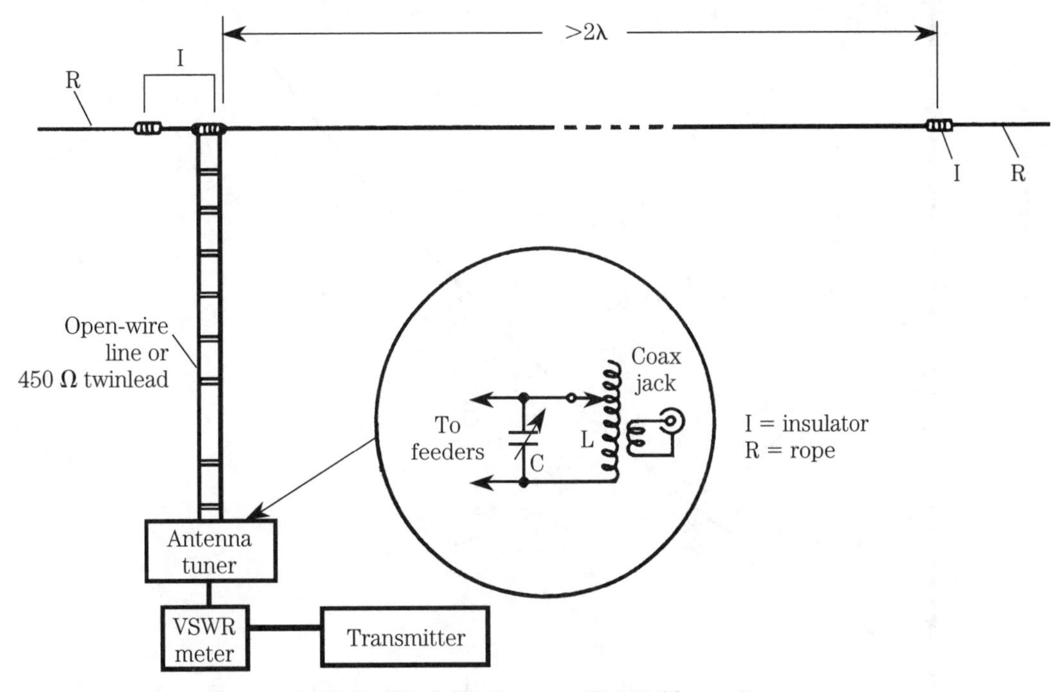

9-5 End-fed long-wire with suitable tuner.

The feedpoint of the longwire is one end, so we expect to see a voltage antinode where the feeder is attached. For this reason we do not use coaxial cable, but rather either parallel transmission line (also sometimes called "open-air" line or some such name), or 450-Ω twin lead. The transmission line is excited from any of several types of balanced antenna tuning unit (see Fig. 9-5). Alternatively, a standard antenna tuning unit (designed for coaxial cable) can be used if a 4:1 BALUN transformer is used between the output of the tuner and the input of the feedline.

What does "many wavelengths" mean? That depends upon just what you want the antenna to do. Figure 9-6 shows a fact about the longwire that excites many users of longwires: it has gain relative to a dipole! Although a two wavelength antenna only has a slight gain over a dipole; the longer the antenna, the greater the gain. In fact, it is possible to obtain gain figures greater than a three-element beam using a longwire, but only at nine or ten wavelengths.

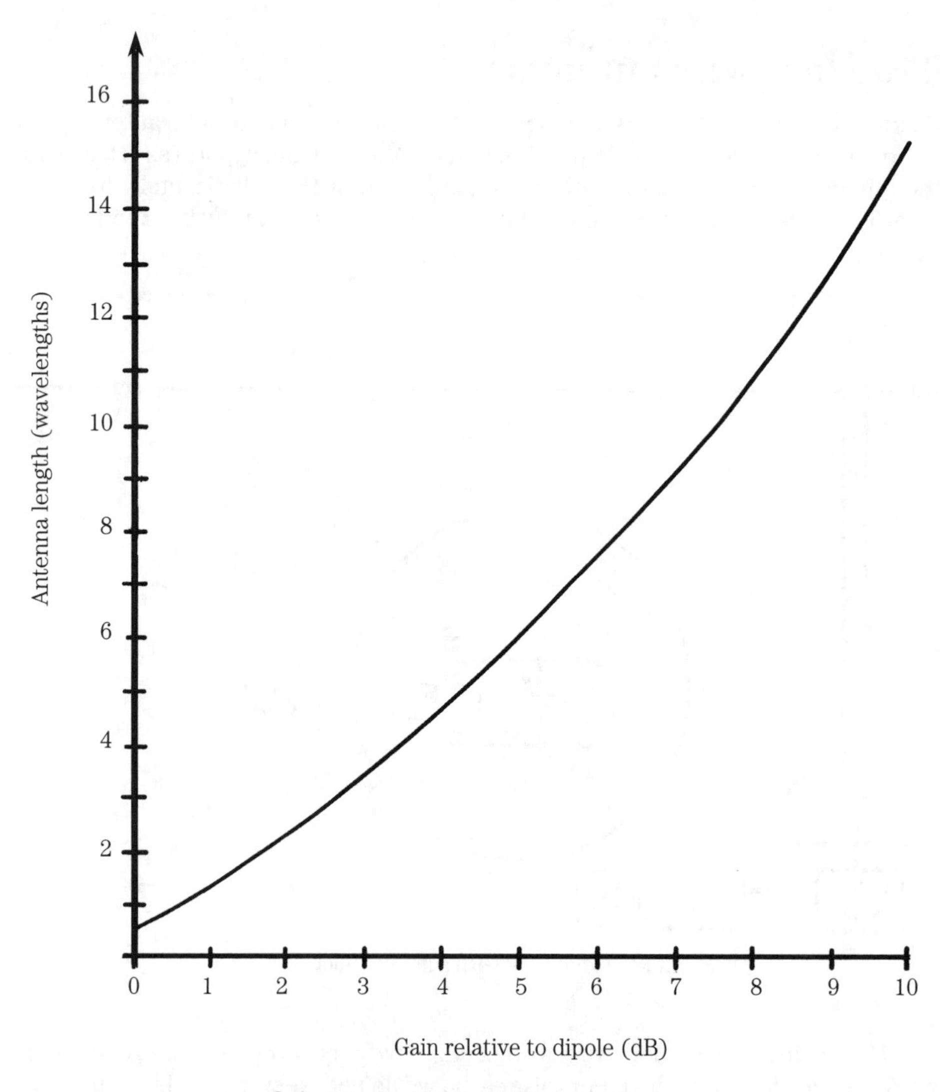

9-6 Antenna length vs. gain over dipole.

What does this mean? One wavelength is $984/F_{MHz}$ feet, so at 10 meters (29 MHz) one wavelength is about 34 feet; at 75 meters (3.8 MHz) one wavelength is 259 feet long. In order to meet the "two wavelengths" criterion a 10-meter antenna need only be 68 feet long, while a 75-meter antenna would be 518 feet long! For a ten wavelength antenna, therefore, we would need 340 feet for 10 meters; and for 75 meters, it is nearly 2,600 feet long. Ah me, now you see why the longwire is not more popular. The physical length of a nonterminated resonant longwire is on the order of:

$$L_{feet} = \frac{492(N - 0.025)}{F_{MHz}} \quad \textbf{[9.1]}$$

(N = Number of $\lambda/2$ in the radiator element.)

Of course, there are always people like my buddy (now deceased) John Thorne, K4NFU. He lived near Austin, TX on a multi-acre farmette that has a 1,400-foot property line along one side. John installed a 1,300 foot longwire and found it worked excitingly well. He fed the thing with home-brew 450-Ω parallel ("open-air") line and a Matchbox antenna tuner. John's longwire had an extremely low angle of radiation, so he regularly (much to my chagrin on my small suburban lot) worked ZL, VK, and other Southeast Asia and Pacific basin DX . . . with only 100 watts from a Kenwood transceiver.

Oddly enough, John also found a little bitty problem with the longwire that textbooks and articles rarely mention: electrostatic fields build up a high voltage dc charge on longwire antennas! Thunder storms as far as 20 miles away produces serious levels of electrostatic fields, and those fields can cause a build-up of electrical charge on the antenna conductor. The electric charge can cause damage to the receiver. John solved the problem by using a resistor at one end to ground. The "resistor" is composed of ten to twenty 10-MΩ resistors at 2 watts each. This resistor bleeds off the charge, preventing damage to the receiver.

A common misconception about longwire antennas regard the normal radiation pattern of these antennas. I have heard amateurs, on the air, claim that the maximum radiation for the longwire is:

a) broadside (i.e., 90 degrees) with respect to the wire run, or
b) in-line with the wire run

Neither is correct, although ordinary intuition would seem to indicate one or the other. Figure 9-7 shows the approximate radiation pattern of a longwire when viewed from above. There are four main lobes of radiation from the longwire (A, B, C, and D). There are also two or more (in some cases many) minor lobes (E and F) in the antenna pattern. The radiation angle with respect to the wire run (G–H) is a function of the number of wavelengths found along the wire. Also, the number, and extent, of the minor lobes is also a function of the length of the wire.

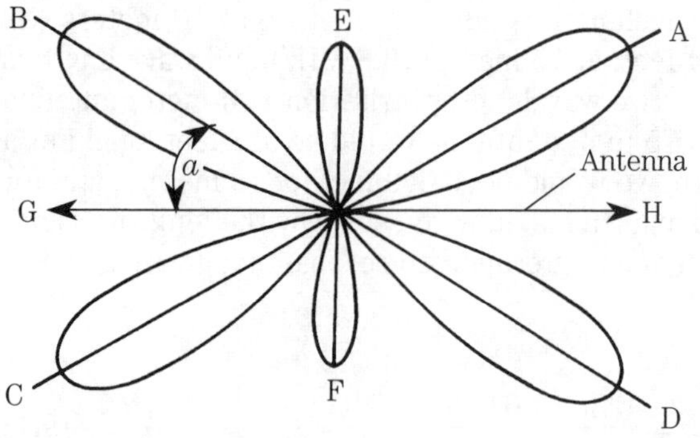

9-7 Radiation pattern of longwire antenna.

Nonresonant single wire longwire antennas

The resonant longwire antenna is a *standing wave antenna*, because it is unterminated at the far end. A signal propagating from the feedpoint, toward the open end, will be reflected back towards the source when it hits the open end. The interference between the forward and reflected waves sets up stationary standing current and voltage waves along the wire.

A *nonresonant longwire* is terminated at the far end in a resistance equal to its characteristic impedance. Thus, the incident waves are absorbed by the resistor, rather than being reflected. Such an antenna is called a *traveling wave antenna*. Figure 9-8 shows a terminated longwire antenna. The transmitter end is like the feed system for other longwire antennas, but the far end is grounded through a terminating resistor, R1, that has a resistance (R) equal to the characteristic impedance (Z_o) of the antenna (i.e., $R = Z_o$). When the wire is 20 to 30 feet above the ground, Z_o is about 500 to 600 Ω.

The radiation pattern for the terminated longwire is a unidirectional version of the multi-lobed pattern found on the unterminated longwires. The angles of the lobes vary with frequency, even though the pattern remains unidirectional. The directivity of the antenna is partially specified by the angles of the main lobes. It is interesting to note that gain rises almost linearly with $n\lambda$, while the directivity function changes rapidly at shorter lengths (above three or four wavelengths the rate of change diminishes considerably). Thus, when an antenna is cut for a certain low frequency, it will work at higher frequencies, but the directivity characteristic will be different at each end of the spectrum of interest.

A two-wavelength (2λ) pattern is shown in Fig. 9-7. There are four major lobes positioned at angles of ±36 degrees from the longwire. There are also four minor lobes—the strongest of which is –5 dB down from the major lobes—at angles of ±75 degrees from the longwire. Between all of the lobes, there are sharp nulls in which little reception is possible. As the wire length is made longer, the angle of the main lobes pulls in tighter (i.e., towards the wire). As the lobes pull in closer to the wire, the number of minor lobes increases. At 5λ, there are still four main lobes, but they are at angles of ±22 degrees from the wire. Also, the number of minor lobes increases to sixteen. The minor lobes are located at ±47, ±62, ±72, and ±83 degrees with respect to the wire. The minor lobes tend to be –5 to –10 dB below the major lobes.

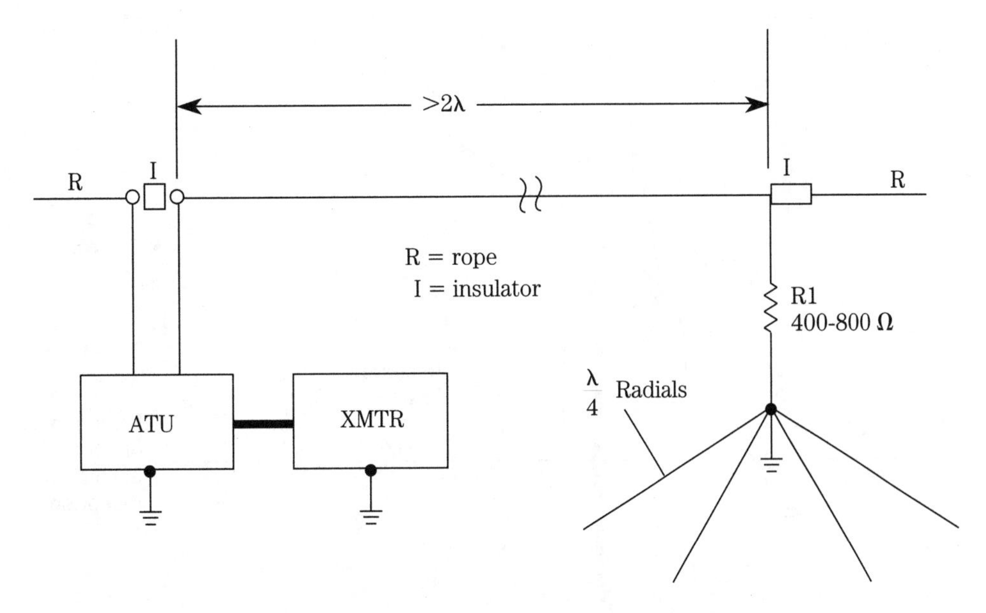

9-8 Longwire (>2) antenna is terminated in a resistance (typically 400 to 800 Ω). Radials under the resistor improve the grounding of the antenna.

When the longwire gets very much longer than 5λ, the four main lobes begin to converge along the length of the wire, and the antenna becomes bidirectional. This effect occurs at physical lengths greater than about 20λ.

In general, the following rules apply to longwire antennas:

- On each side of the antenna, there is at least one lobe, minor or major, for each half wavelength of the wire element. For the overall element, there is one lobe for every quarter wavelength.
- If there are an even number of lobes on either side of the antenna wire, then half of the total number of lobes are tilted backwards, and half are tilted forwards; symmetry is maintained.
- If there are an odd number of lobes on either side of the wire, then one lobe on either side will be perpendicular to the wire, with the other lobes distributed either side of the perpendicular lobe.

Vee beams and rhombic beams

Longwire antennas can be combined in several ways to increase gain, and sharpen directivity. Two of the most popular of these are the *vee beam* and the *rhombic* antennas. Both forms can be made in either resonant (unterminated), or nonresonant (terminated) versions.

Vee beams The *vee beam* (Fig. 9-9A) consists of two equal length longwire elements (wire 1 and wire 2), fed 180 degrees out of phase with each other, and spaced to produce an acute angle between them. The 180° phase difference is inherent in connecting the two wires of the vee to opposite conductors of the same parallel conductor feedline.

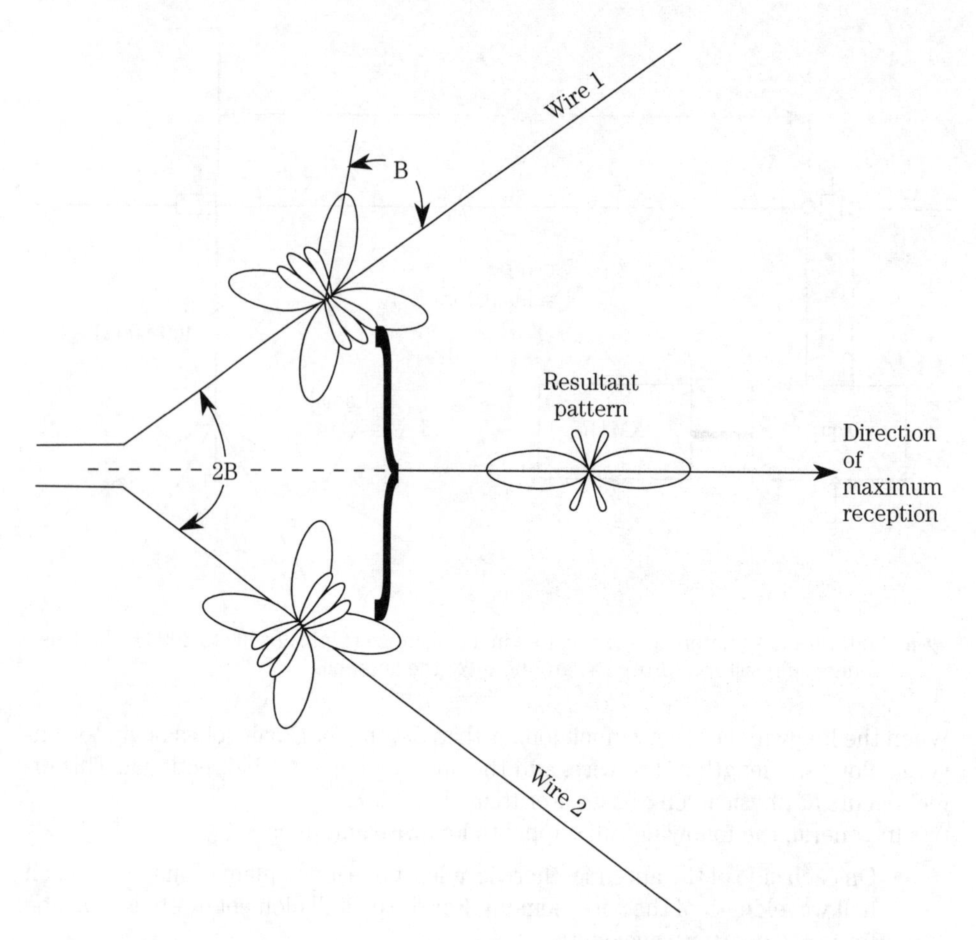

9-9A Radiation pattern of vee-beam antenna consists of the algebraic sum of the two long wire patterns that make up the antenna.

The unterminated vee beam (Fig. 9-9A) has a bidirectional pattern that is created by summing together the patterns of the two individual wires. Proper alignment of the main lobes of the two wires requires an included angle, between the wires, of twice the radiation angle of each wire. If the radiation angle of the wire is β, then the appropriate included angle is 2β. To raise the pattern a few degrees, the 2β angle should be slightly less than these values. It is common practice to design a vee beam for a low frequency (e.g., 75-/80- or 40-meter bands), and then to use it also on higher frequencies that are harmonics of the minimum design frequency. A typical vee beam works well over a very wide frequency range *only* if the included angle is adjusted to a reasonable compromise. It is common practice to use an included angle that is between 35 and 90 degrees, depending on how many harmonic bands are required.

Vee beam patterns are based on an antenna height that is greater than a half wavelength from the ground. At low frequencies, such heights will not be practical, and you must expect a certain distortion of the pattern because of ground reflection effects.

Gain on a vee beam antenna is about 3 dB higher than the gain of the single wire longwire antenna of the same size, and it is considerably higher than the gain of a dipole (see Fig. 9-9B). At three wavelengths, for example, the gain is 7 dB over a dipole. In addition, there may be some extra gain because of mutual impedance effects—which can be about 1 dB at 5λ, and 2 dB at 8λ.

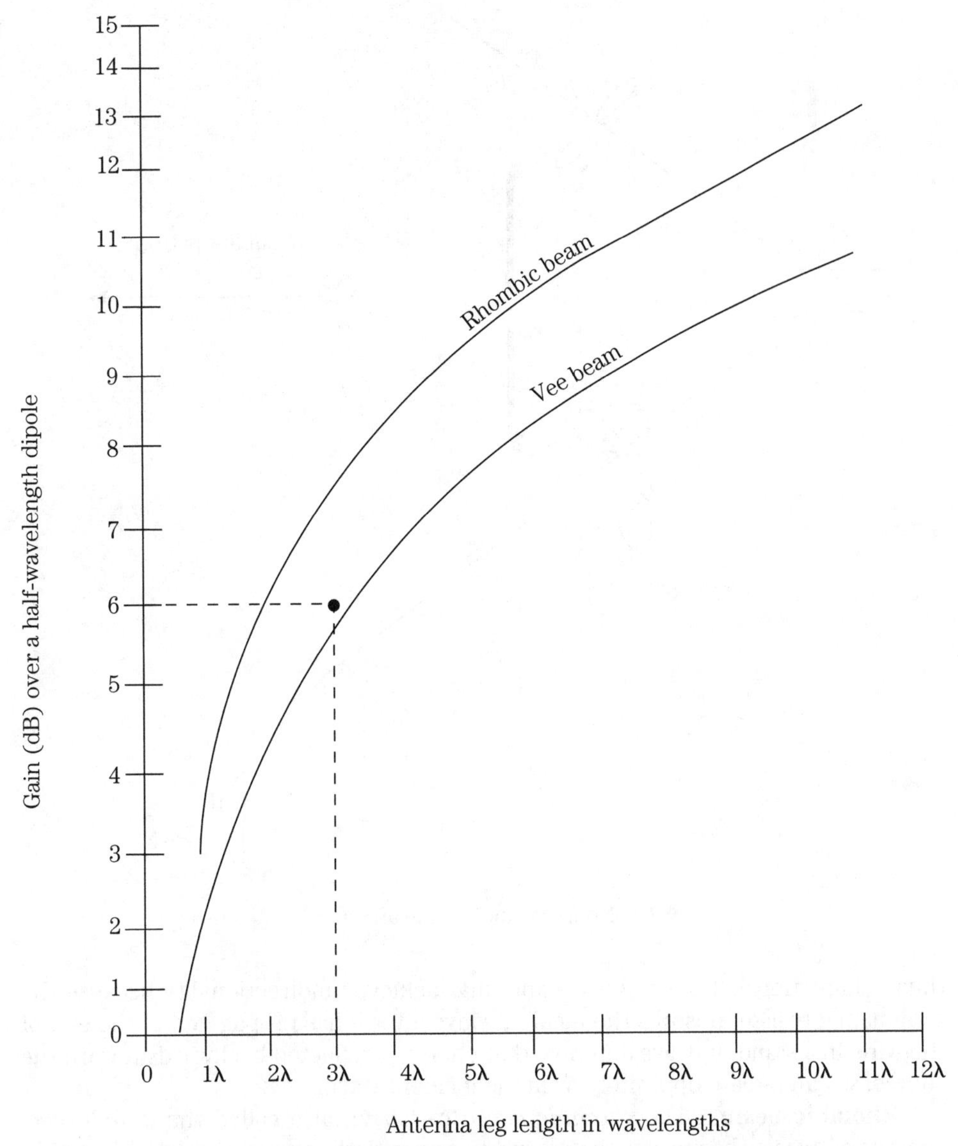

9-9B Gain-vs-length of vee-beam and rhombic beam antennas.

Nonresonant vee beams Like the single-wire longwire antennas, the vee beam can be made nonresonant by terminating each wire in a resistance that is equal to the antenna's characteristic impedance (Fig. 9-10). Although the regular vee is a standing wave antenna, the terminated version is a traveling wave antenna and is

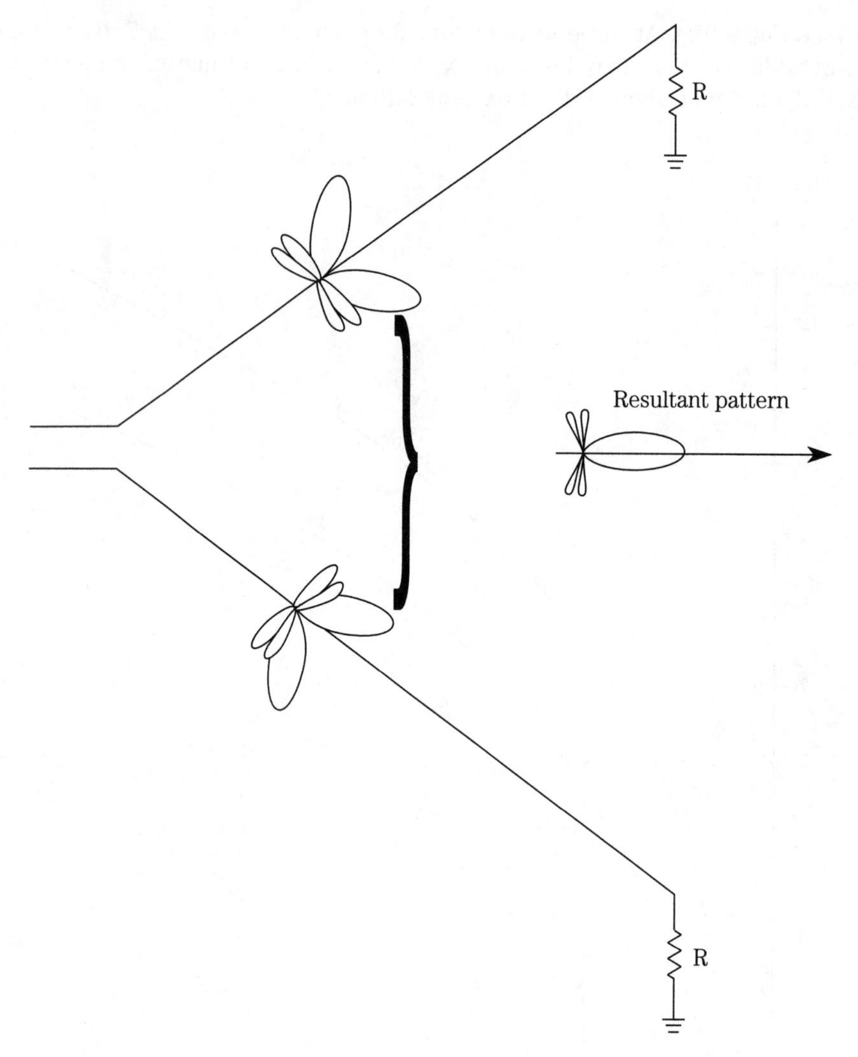

9-10 Nonresonant V-beam antenna.

thus unidirectional. Traveling wave antennas achieve unidirectionality because the terminating resistor absorbs the incident wave after it has propagated to the end of the wire. In a standing-wave antenna, that energy is reflected backwards toward the source, so can radiate oppositely from the incident wave.

Rhombic beams The *rhombic beam antenna*, also called the *double vee*, consists of two vee beams positioned end-to-end with the tips connected. The unidirectional, nonresonant (terminated) rhombic is shown in Fig. 9-11. The unterminated resonant form gives approximately the same gain and directivity as a vee beam of the same size. The nonresonant rhombic has a gain of about 3 dB over a vee beam of the same size (see Fig. 9-9 again).

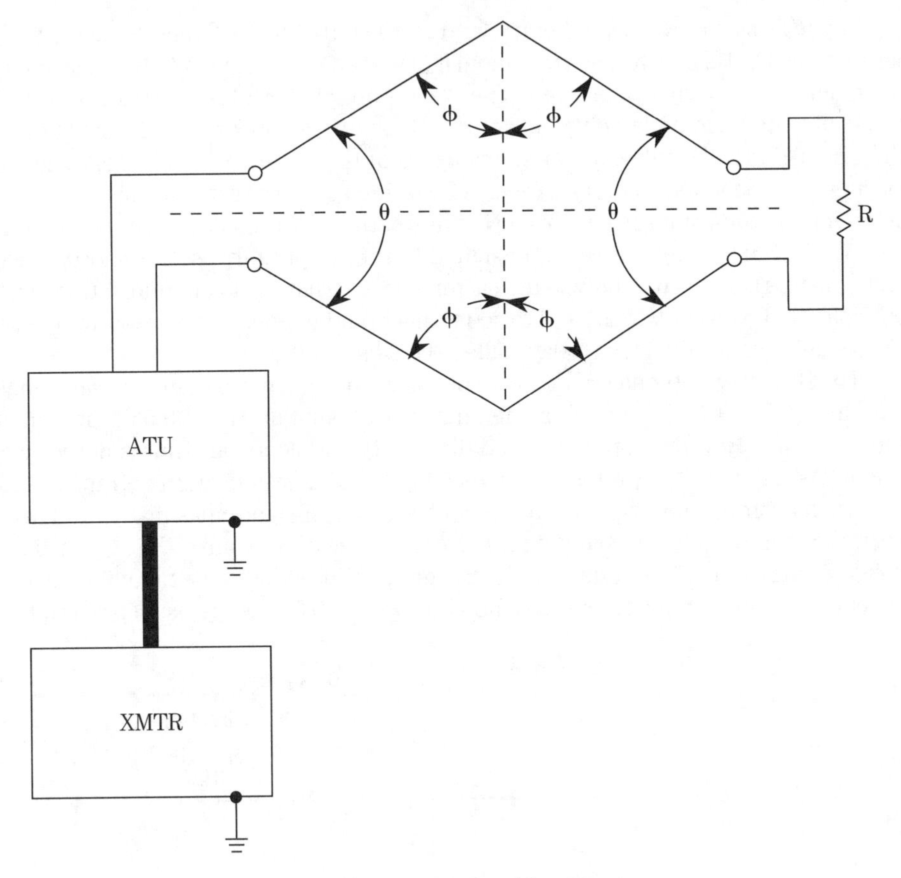

9-11 Terminated rhombic antenna.

Two angles are present on the rhombic antenna. One-half of the included angle of the two legs of one wire is the *tilt angle* (φ), while the angle between the two wires is the *apex angle* (θ). A common rhombic design uses a tilt angle of 70 degrees, a length of 6λ for each leg (two legs per side), and a height above the ground of 1.1λ.

The termination resistance for the nonresonant rhombic is 600 to 800 Ω, and must be noninductive. For transmitting rhombics, the resistor should be capable of dissipating at least one third of the average power of the transmitter. For receive-only rhombics, the termination resistor can be a 2-watt carbon composition or metal film type. Such an antenna works nicely over an octave (2:1) frequency range.

Beverage or "wave" antennas

The *Beverage* or *wave antenna* is considered by many people to be the best receive antenna available for Very Low Frequency (VLF), AM broadcast band (BCB), medium wave (MW), or Tropical Band (low HF region) DXing. The Beverage was used by RCA at its Riverhead, Long Island (NY) station in 1922, and a technical description by Dr. H.H. Beverage (for whom it is named) appeared in *QST* magazine for November 1922, in an article titled "The Wave Antenna for 200-Meter Reception." In 1984, an edited and updated version of the 1922 article appeared in the same magazine. In 1921, Paul Godley, under sponsorship of the American Radio Relay League journeyed to Scotland to erect a receiving station at Androssan. His mission was to listen for amateur radio signals from North America. As a result of politicking in the post World War I era, hams were consigned to the supposedly useless shortwave ($\lambda < 200$ meters), and it was not clear that reliable international communications were possible. Godley went to Scotland to see if that could happen . . . he reportedly used a wave antenna for the task (today, called the *Beverage*).

The Beverage antenna is a longwire of special design, more than one wavelength (1λ) long (Fig. 9-12), although some authorities maintain that $>0.5\lambda$ is minimally sufficient. The Beverage provides good directivity and good gain, but is not very efficient. As a result, it is preferred for receiving, and is less useful for transmitting. This is an example of how different attributes of various antennas make the Law of Reciprocity an unreliable sole guide to antenna selection. Unlike the regular longwire, which is of a different design, the Beverage is intended to be mounted close to the earth's surface (typically $< 0.1\lambda$); heights of 8 to 10 feet is the usual prescription.

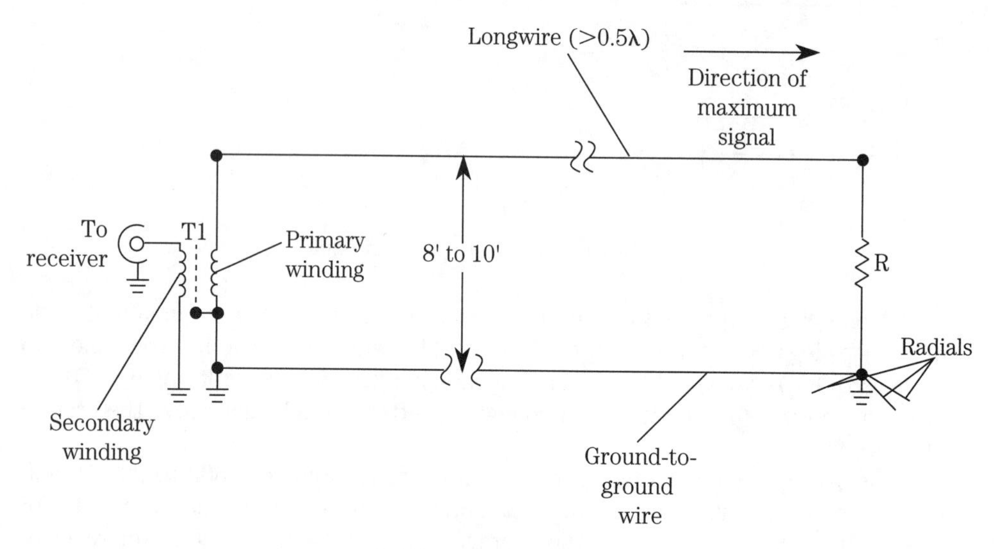

9-12 Beverage antenna.

Figure 9-12 shows the basic single-wire Beverage antenna. It consists of a single conductor (#16 to #8 wire, with #14 being most common) erected about 8 to 10 feet above ground. Some Beverages are unterminated (and bidirectional), but most of them are terminated at the far end in a resistance (*R*) equal to the antenna's characteristic impedance (Z_o). The receiver end is also terminated in its characteristic impedance, but generally requires an impedance matching transformer to reduce the antenna impedance to the 50-Ω standard impedance used by most modern transmitters.

The Beverage works best in the low-frequency bands (VLF through MW), although at least some results are reportedly relatively easy to obtain up to 25 meters (11.5 MHz). Some questionably successful attempts have been made at making Beverage antennas work as high as the 11-meter Citizen's Band or the 10-meter ham band (29.7 MHz).

The Beverage antenna works on vertically polarized waves arriving at low angles of incidence. These conditions are normal in the AM BCB, where nearly all transmitting antennas are vertically polarized. In addition, the ground and sky wave propagations found in these bands (VLF, BCB, and low HF) are relatively consistent. As the frequency increases, however, two factors become increasingly dominant. First, the likelihood of horizontal polarization increases because of the size of a wavelength at those frequencies. The polarization of the received signal not only changes in those bands, but does so constantly when conditions are unsettled. It is the strong dependence of the Beverage on relatively constant vertical polarization that makes me suspect the claims of Beverage-like performance above the 25—or even 31—meter bands.

The Beverage depends on being erected over poorly conductive soil, even though the terminating resistor needs a good ground. Thus, one source claimed that sand beaches adjacent to salty marshes make the best Beverage sites (a bit of an overstatement). Figure 9-13 shows why poorly conductive soil is needed. The E-field vectors are launched from the transmitting antenna perpendicular to the earth's surface. Over perfectly conducting soil, the vertical waves would remain vertical. But over imperfectly conducting soil, the field lines tend to bend close to the point of contact with the ground. As shown in the inset of Fig. 9-13, the bending of the wave provides a horizontal component of the E-field vector, and this provides the means of generating an RF current in the conductor wire.

A debate among Beverage antenna fans regards the best length for the antenna. Some sources state that the length can be anything $\geq 0.5\lambda$, yet others say $\geq 1\lambda$ is the minimum size. One camp says that the length should as long as possible, while others say it should be close to a factor called the *Maximum Effective Length* (MEL), which is:

$$MEL = \frac{\lambda}{4\left(\frac{100}{K} - 1\right)} \qquad \textbf{[9.2]}$$

Where:

MEL is the maximum effective length in meters (m)
λ is the wavelength in meters (m)
K is the velocity factor expressed as a percent

Misek, who may well be the leading exponent of the Beverage antenna, uses numbers like 1.6λ to 1.7λ over the 1.8- to 7.3-MHz region; and 0.53λ to 0.56λ on frequencies lower than 1.8 MHz. Doctor Beverage was once quoted as saying that the optimum length is 1λ.

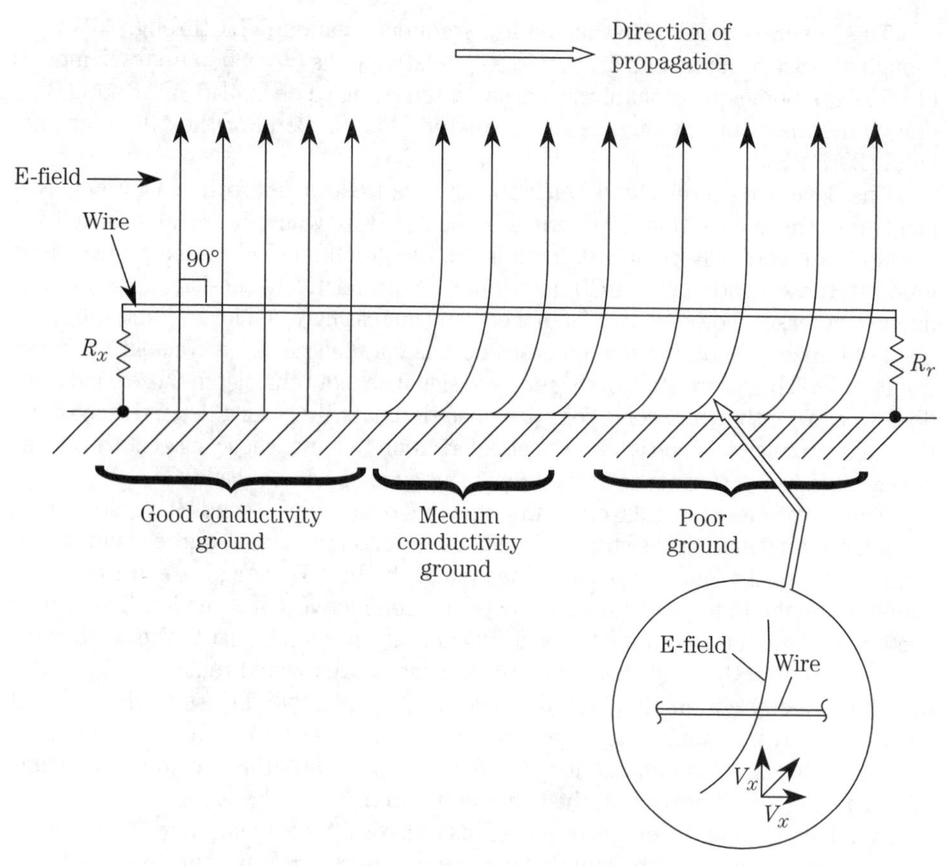

9-13 Performance of Beverage antenna over different values of ground conductance.

Like the longwire antenna, the Beverage needs a termination resistor that is connected to a good ground. This requirement might be harder to meet on Beverage antennas because they work best over lossy ground, which doesn't make a very good ground connection. As in the longwire case, insulated or bare wires, a quarter-wavelength long, make the best radials. However, a substantial improvement in the ground is possible using just bare wires measuring 15 to 20 feet long (which is much less than λ/4), buried in the soil just below the surface (far enough to prevent erosion from bringing it to the surface). Many articles and books on Beverages show ground rods of two or three feet long, which borders on the ridiculous. Poor soil requires longer ground rods, on the order of six to eight feet. Copper clad steel make the best rods.

In addition to the radials and ground rod, Misek also recommends using a wire connection between the ground connection at the termination resistor, and the ground connection at the receiver transformer (see again Fig. 9-12). According to Misek, this wire helps to stabilize the impedance variations at higher frequencies.

Installation of the Beverage antenna is not overly critical if certain rules are followed. The antenna should be installed at a height of six to ten feet off the ground, and it should be level with the ground over its entire length. If the ground is not flat enough to make a level installation possible, then try to use a height that is six to ten feet above the average terrain elevation along its run. A popular installation method is to erect 16-foot 4-×-4 lumber, such that three to four feet are buried in a concrete filled posthole. Use lumber that is treated for outdoor use (i.e., pressure treated lumber sold for decks and porches. The wire can be fastened to the 4-×-4 posts using either ceramic stand-off ("beehive") insulators, or electric fence insulators (which some people deem preferable). Try to use one contiguous length of wire for the antenna, if possible, in order to avoid soldered splices and joints.

One of the Beverage installation difficulties shared with the longwire is the need to slope down to a point where a termination resistor can be easily installed close to the ground. While the longwire can be sloped over a large portion of its length, the Beverage should only be sloped downwards over the last 60 feet or so.

Steerable notch Beverage antennas

A Beverage erected with two wires—parallel to each other, at the same height, spaced about 12 inches apart (Fig. 9-14), with a length that is a multiple of a half wavelength—is capable of *null steering*. That is, the rear null in the pattern can be steered over a range of 40 to 60 degrees. This feature allows strong, off-axis signals to be reduced in amplitude so that weaker signals in the main lobe of the pattern can be received. There are at least two varieties of the *steerable wave Beverage* (SWB).

If null steering behavior is desired, then a phase control circuit (PCC) will be required—consisting of a potentiometer, an inductance, and a variable capacitor in series with each other. Varying both the "pot" and the capacitor will steer the null. You can select the direction of reception, and the direction of the null, by using a switch to swap the receiver and the PCC between Port-A and Port-B.

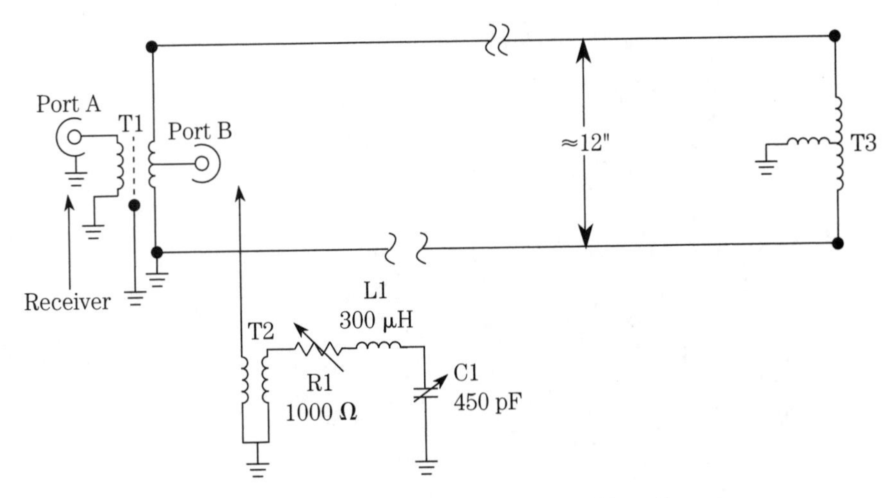

9-14 Beverage antenna with a steerable null.

10
CHAPTER

Hidden and limited-space antennas

ONE OF THE MOST SIGNIFICANT IMPEDIMENTS TO AMATEUR RADIO OPERATORS, CB operators, and shortwave listeners, is the space available for their antennas. In many thousands of other cases, the limitation is less one of space, but of regulators. More and more subdivisions are built with covenants on the deed that prohibit the buyer from installing outdoor antennas. Once limited to townhouse developments, where that breed of *contemptible vermin* (called the Homeowners Committee) routinely intruded on the affairs of people who mistakenly think they "own" their townhouse (ownership implies right of use, which is limited by the covenants). These onerous covenants are now routinely placed on single-family dwellings as well. In fact, it is the single most serious threat to amateur communications people in the country today. Other homeowners are no longer angered by the restriction on antennas because their television reception is now carried to them via cable systems in most parts of the country. Even where cable is not available, most users can install a moderately sized television antenna in their attic, or use rabbit ears. In other cases, the townhouse community will install a single master TV antenna and then distribute signals to each unit. The result is that the amateur radio operator, CB operator, and shortwave listener are left to fend for themselves without assistance from neighbors.

In this chapter we will examine some of the alternatives available to those readers who have either a limited space situation (such as a small city lot), or are unable to move out of a subdivision where there are stupid rules against outdoor antennas. The suggestions contained in this chapter are not universal, and indeed the author recommends that you adapt, as well as adopt, these recommendations . . . and come up with some of your own. Creativity within the constraints of the laws of physics governing radio antennas is encouraged.

Hidden antennas

A hidden antenna is one that is either completely shielded from view, or it is disguised as something else. Alternatively, you could also include, in this category, antennas that are in semi-open view, but which are not too obvious (except to the trained and diligent eye). Some people have opted for "hidden" longwires made of very fine wire (#26 enameled wire is popular). The user will install the wire in the open, high off of the ground (as in an apartment installation), and operate without anyone knowing the difference. One chap used #22 wire suspended between two 16-story apartment house buildings that were 100 yards apart. He had one of the best working longwires in town until a windy day when the whole thing came down. No one was injured; but had some been hurt, there might have been a lawsuit. Hidden antennas must be designed with an eye toward causing others no harm. It is neither ethical nor smart to place others at risk in enjoying our hobbies.

The dipole is a popular antenna with both shortwave listeners and amateur radio operators. Indeed, for the CB operator who wants to get on the air from an apartment or townhouse (or restricted single-family home), the dipole can represent a respectable alternative.

As you learned in chapter 6, the dipole is a horizontal wire (or pipe) antenna that is half wavelength long, and fed in the center (ideally) with 75-Ω coaxial cable. In the "townhouse" dipole, it is possible to build the antenna entirely inside the attic of the building. The length of the dipole is given approximately by:

$$L_{\text{ft}} = \frac{468}{F_{\text{MHz}}} \quad \textbf{[10.1]}$$

If you do some quick calculations you will find that antennas for the 10-, 13-, 15-, and (possibly) even 18- and 20-meter bands, will fit entirely inside the typical townhouse attic. This statement is also true of the 11-meter citizen's band antenna: it will fit inside the standard townhouse attic. But what about the lower frequency bands?

Figure 10-1 shows a possible solution to the use of the lower frequencies in the townhouse situation. The two quarter-wavelength arms of the dipole are ideally installed in line with each other, as was shown in the chapter on horizontal antennas. But in a sticky situation we can also install the dipole with the arms bent to accommodate the space available. In this example, only one of many possible methods for accomplishing this job is shown. Here each quarter-wavelength section is comprised of two legs, AB and CD, respectively. Ideally, segments B and C are the longest dimensions. Also, if possible, make segments A and D equal lengths. Another method is to reverse the direction of one end leg, say for example D, and run it to the other corner of the building over the peak of the roof.

The author is almost hesitant to offer only one drawing, despite space constraints, in the fear that readers will take the offered pattern as the only authorized version, or solution. In reality, you might not be able (for a variety of reasons) to use the exact pattern shown (so ad-lib a little bit).

How about performance? Will the constrained dipole of Fig. 10-1 work as well as a regular dipole installed a wavelength, or two, off the ground and away from ob-

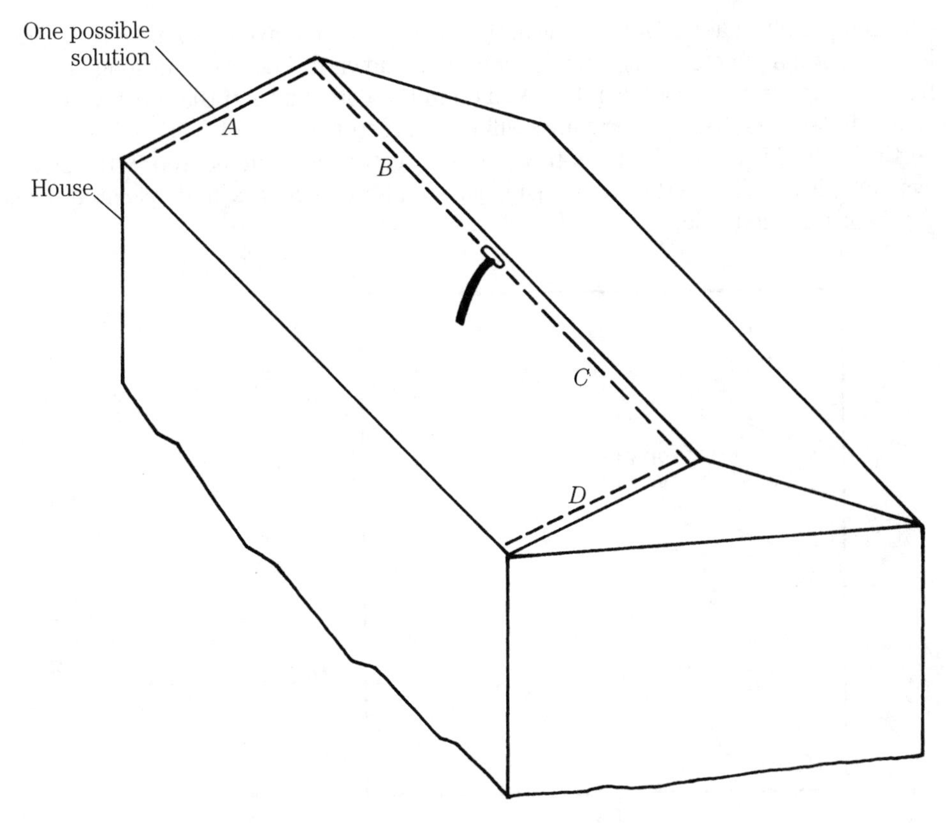

10-1 Installing dipole in attic.

jects? In a word: no. But that is not the problem being solved; getting on the air at all is the present problem. You will find that the pattern of the constrained dipole is distorted compared with that of the regular dipole. In addition, the feedpoint impedance is not going to be 73 Ω (except if by fluke), so you will be required to use an antenna tuner of some kind. You are well advised to read the chapter on antenna instruments and measurements so that you can be prepared to figure out any feed problems that crop up—as well they might.

The wire used in the constrained dipole (or other forms of attic antenna) should be mounted on TV-type screw-in stand-off insulators. Almost any outlet that sells TV antennas, or installation parts, will have them. In addition to electronics parts suppliers, these stand-off insulators are also available in many hardware stores, and department stores that sell TV antennas or "Harry and Harriet Homeowner" supplies. Do not simply tape the wire to the wooden underside of your roof. The reason is simple: in a poorly tuned antenna voltages can get high enough at the ends to produce *corona* effects . . . arcing could be a fire hazard. Also, use insulated wire to avoid accidental contact in the event that someone is in the attic when you are operating. Be aware that the insulation might affect the propagation velocity of signals in the wire, so it will slightly alter the required length.

Another alternative for the attic antenna, and one that avoids some of the problems of corona effects, is the nonresonant loop shown in Fig. 10-2. Although specified as a top view, the loop can be installed in the configuration that best uses the space. In fact, the best performance will be bidirectional when the loop is installed vertically. In this case, a large loop (as large as can be accommodated in the space available) is installed in the attic. Again, use stand-off insulators, and insulated wire, for the installation (Fig. 10-2).

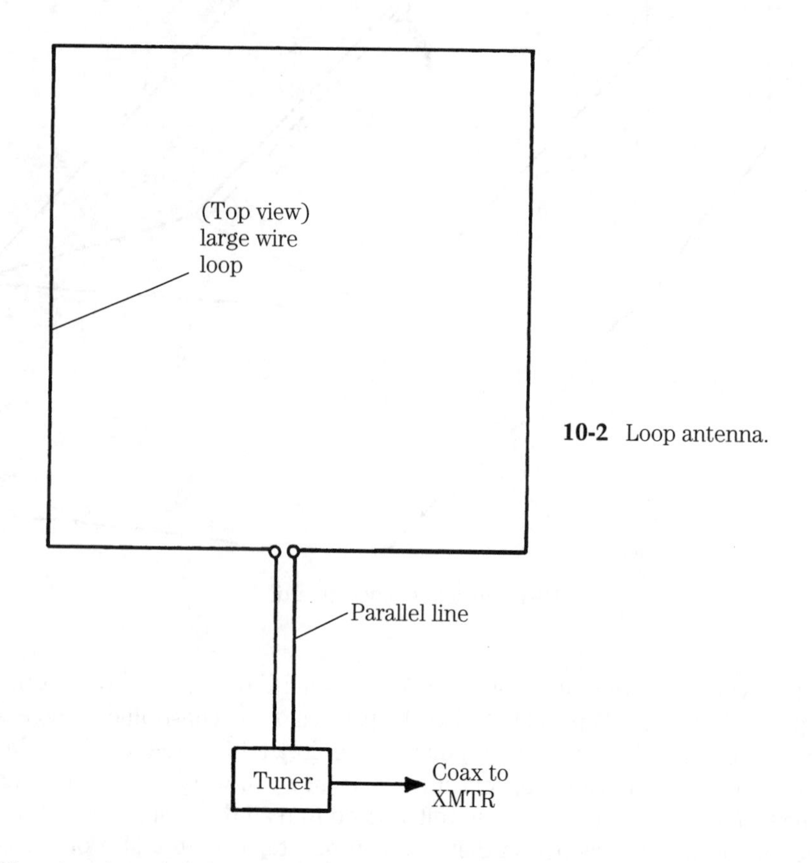

10-2 Loop antenna.

The giant loop is fed with parallel line, and is tuned with a balanced antenna tuning unit. As was true with the constrained dipole, the performance is not to be equated with the performance of more regular antennas; but, with the prospect of not being on the air at all . . .! Once again we have a compromise antenna for a compromising situation.

Another ploy is the old "flagpole trick" shown in Fig. 10-3. Some developments allow homeowners to vent their patriotism by installing flagpoles . . . and flagpoles can be disguised antennas. In the most obvious case, you can install a brass or aluminum flagpole and feed it directly from an antenna tuning unit. For single-band operation, especially on the higher frequencies, you can delta-feed the "flagpole" unobtrusively, and call your flagpole a vertical antenna. But that is not always the best solution.

Figure 10-3 shows two methods for creating a flagpole antenna, and both depend on using white PVC plumbing pipe as the pole. The heavier grades of PVC pipe are self-supporting to heights of 16 to 20 feet, although lighter grades are not self-supporting at all (hence, are not usable).

Figure 10-3A shows the use of a PVC flagpole in which a #12 (or #14) wire is hidden inside. This wire is the antenna radiator. For some frequencies, the wire will be resonant, and for others, it will surely be nonresonant. Because of this problem an

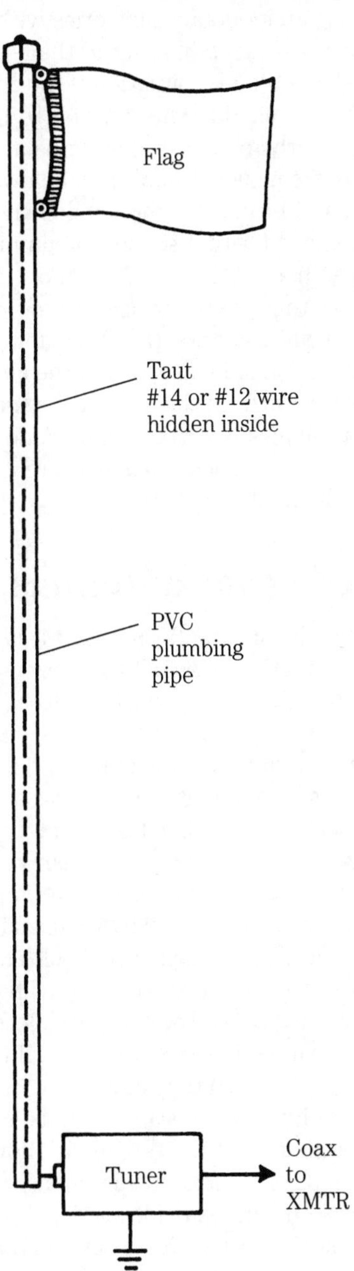

10-3A Flag pole antenna made from thick-walled PVC pipe.

antenna tuning unit is used either at the base of the antenna, or inside at the transmitter. If the wire is too long for resonance (as might happen in the higher bands), then place a capacitor in series with the wire. Various settings may be required, so use a multi-section transmitting variable that has a total capacitance selectable to more than 1000 pF. Alternatively, use a vacuum variable capacitor of the same range.

In cases where the antenna is too short for resonance, as will occur in the lower bands, insert an inductance in series with the line to "lengthen" it. Another alternative is to use an L-section tuner at the feedpoint.

A good compromise situation is the use of a 16-foot length of "flagpole" pipe with a 16-foot wire embedded inside. The 16-foot wire is resonant at 20 meters, so it will perform similarly to a vertical antenna at those frequencies. The tuner will then accommodate frequencies above and below 20 meters.

Another alternative is the version shown in Fig. 10-3B. In this case the wire radiator is replaced with a section of aluminum tubing. A wooden or plastic insert is fashioned with a drill and file to support the aluminum tubing inside of the PVC tubing. One way to make the support is to use a core bit in an electric drill to cut out a disk that fits snugly inside the PVC tubing. Rat out the center hole, left by the core bit pilot, to the outside diameter of the aluminum tubing. The support can be held in place with screws from the outside, or simply glued in place.

The problem of operating with a hidden antenna is a serious challenge. But with some of these guidelines and a little creativity, you can get on the air and enjoy your amateur radio hobby.

Limited-space antennas

Many people live in situations where it is permissible to install an outdoor antenna, but it is not practical to install a full size antenna. I once lived in a house that was 16 feet wide and 37 feet long, on a 33-×-100' lot. Very few full sized antennas could be installed on that lot because of the space constraints. Beam antennas were out because county laws required that the antenna not hang over the property line. Although vertical antennas were possible, there was a period of time when I used other antennas that were easier to install. In this chapter, you will examine some of the options open to those with limited space for amateur radio, CB, or SWL antennas.

Once again we return the simple dipole as the basis for our discussion. In Fig. 10-4, you see several alternatives for installing an outdoor antenna in a limited space. In Fig. 10-4A, the slanted dipole (or slipole) antenna uses the standard dipole configuration, but one end is connected to the high point of the building, while the other is anchored near the ground. The coaxial cable is connected to the mid-point of the antenna in the usual manner for regular dipoles. If the end of the dipole is within reach of people on the ground, then they may get a nasty RF burn if the antenna is touched while you are operating. Take precautions to keep people and pets away from this lower end.

Another method is the vee dipole shown in Fig. 10-4B. In a regular dipole installation, the ends of the antenna are along the same axis (in other words at an angle of 180 degrees). In the example of Fig. 10-4B, however, the angle between the elements is less than 180 degrees, but greater than 90 degrees. In some cases we might want to bend the elements, rather than install them in a vee-shape. Figure 10-4C

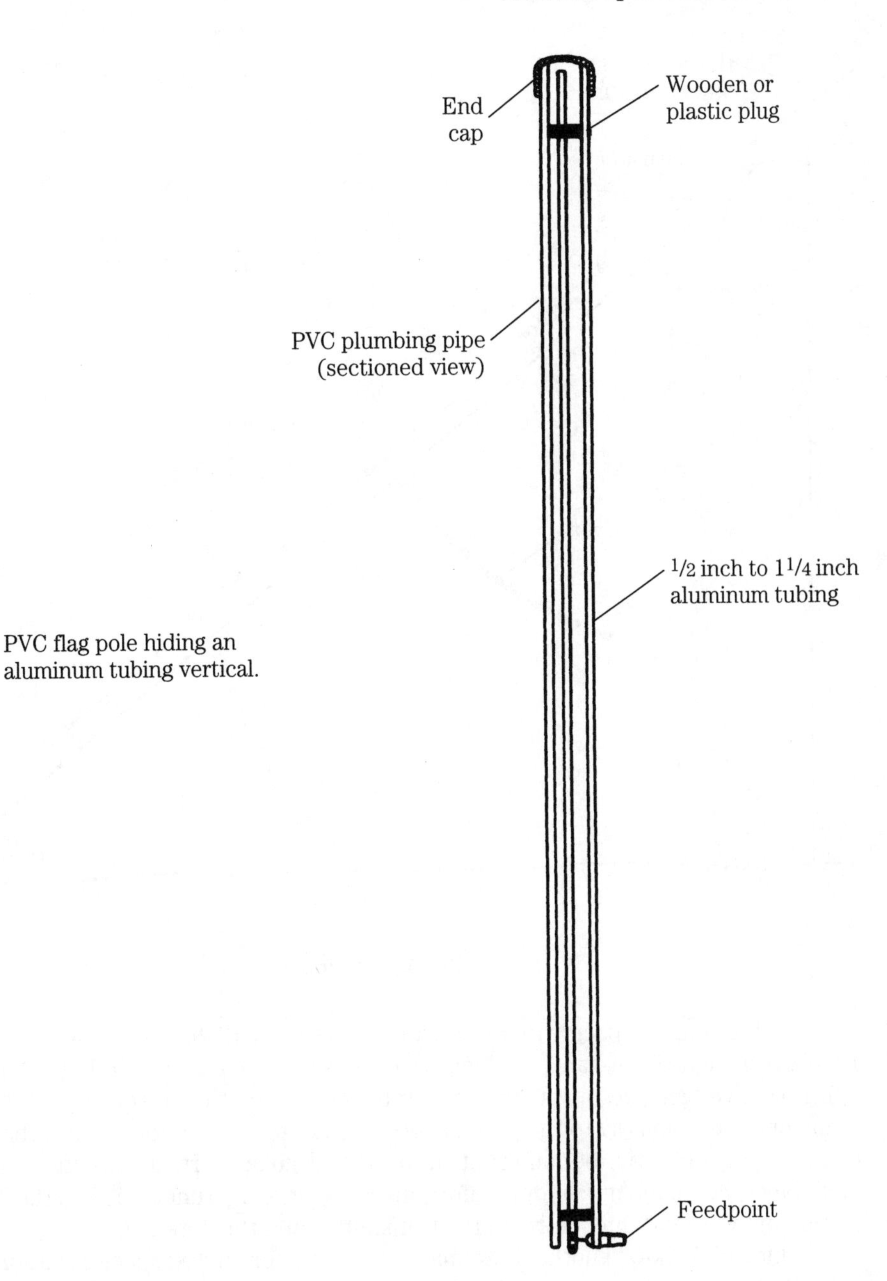

10-3B PVC flag pole hiding an aluminum tubing vertical.

shows an angled dipole with four segments. For the best performance (but not as good as for the regular dipole), make $A = D$ and $B = C$. In all three examples, Figs. 10-4A through 10-4C, you can expect to find the length needed for resonance varies somewhat from the standard $468/F_{MHz}$ value, and that the feedpoint impedance is other than 73 Ω. Also, the pattern will be distorted with respect to the regular dipole. Although these antennas do not work as well as a properly installed dipole, the performance is sufficient to allow successful operation.

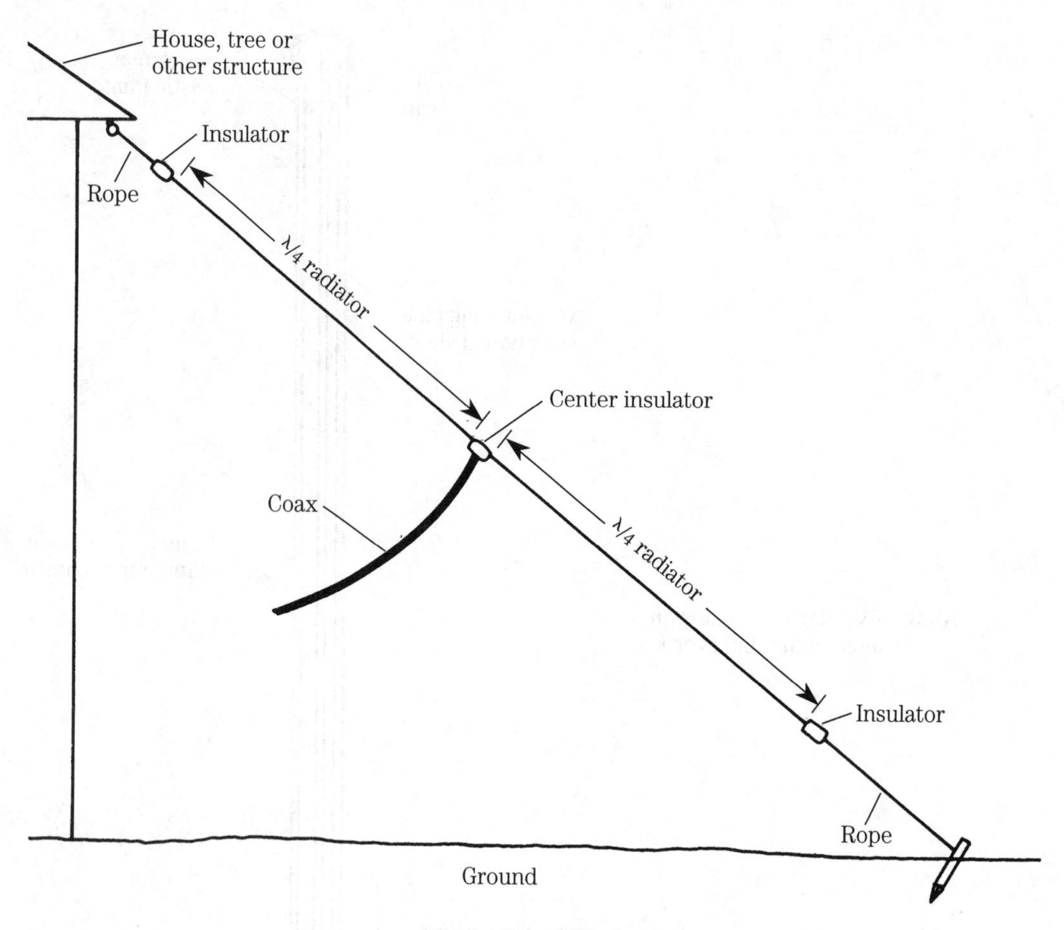

10-4A Sloper dipole.

Another limited space wire antenna is the (so called) half-slope shown in Fig. 10-4D. Although single-band versions are often seen, the example in Fig. 10-4D is a multi-band version. Resonant traps separate the different band segments. This antenna operates similarly to the vertical, but it is not omnidirectional. Also, the feed-point impedance will be different from the regular vertical situation. If the impedance varies too much for comfort, insert an antenna tuner, such as the Transmatch, in the coaxial line between the transmitter and the antenna.

Figure 10-5 shows another antenna that's useful for limited space situations. Although it is easily constructed from low-cost materials, the antenna is also sold by several companies under various rubrics including "cliff-dweller," "apartment house," "townhouse," or "traveler's" antennas. The antenna consists of a 4- to 16-foot section of aluminum or copper tubing. Some of the commercial antennas use a telescoping tubing that can be carried easily in luggage. As was true with the longwire, the window-sill antenna is tuned to resonance with an L-section coupler.

The L-section coupler must be tuned to produce the lowest possible VSWR; so either an RF power meter, or a VSWR meter, must be installed in the coaxial cable between the transmitter and the tuner. Again, a good ground (or radials system) will

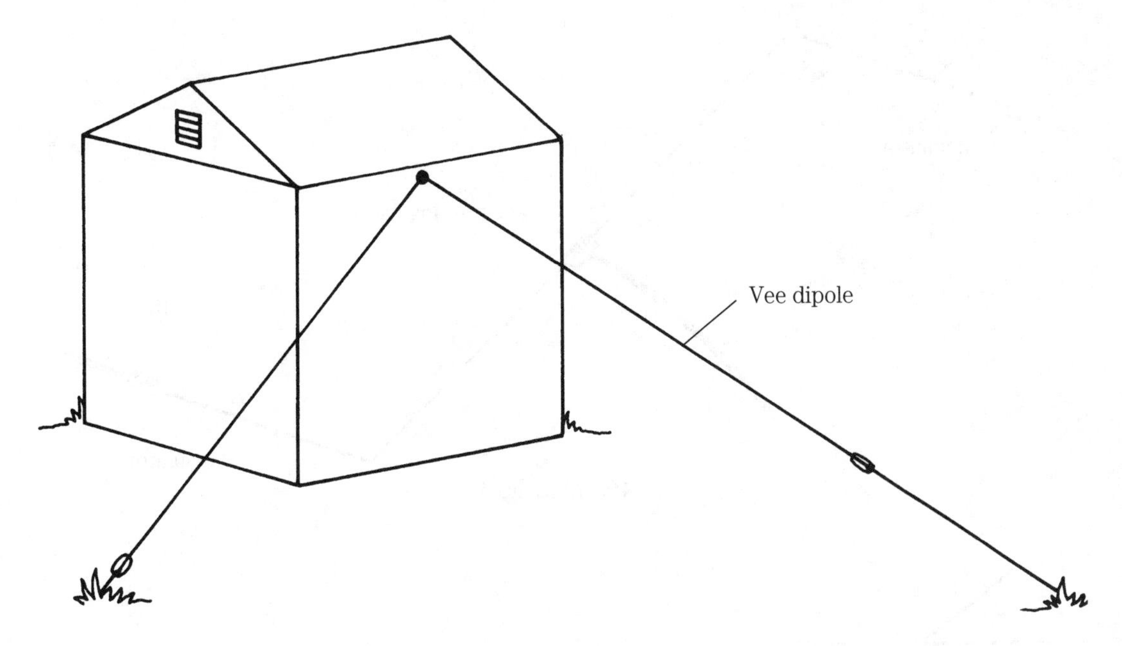

10-4B Vee dipole.

greatly improve the operation of this antenna. The performance should be considerably less than that of a good longwire, but it will work DX for you even on lower frequencies. Once again it is not the best antenna made, but it will get you on the air.

Figure 10-6 shows the use of a mobile antenna for a fixed or portable location. I used this type of antenna at one QTH to good effect. Mount a mobile antenna, such as the *Hustler*, on the window sill, or other convenient mount (at least one amateur radio operator uses one mounted on the roof of the house). Grounding is essential for this antenna, as was also true for the longwire and window-sill antenna (Fig. 10-5). The operation of this antenna is improved by installing at least two radials per band as a counterpoise ground. My installation worked well, even though only two radials were present.

A directional *rotatable* dipole is shown in Fig. 10-7. This antenna is made from a pair of mobile antennas connected "back-to-back" on a horizontal length of 1-×-2 lumber, and fed in the center with a coaxial cable. The end pieces of the mobile antenna set the resonance, and you must recognize that adjusting one requires a countervailing adjustment of the other as well.

Two examples of *helically* wound antennas are shown in Fig. 10-8. In this type of antenna, an insulating mast is wound with a half wavelength of antenna wire. The overall length of the antenna is considerably less than a half wavelength, except at the highest frequencies. In order to dissipate the high voltages that tend to build up at the ends of these antennas, a capacitance hat is used. These "hats" are either disks (pie tins work well), or rods of conductor about 16 to 24 inches long. The version shown in Fig. 10-8A is a vertical antenna, and like other verticals, it must be installed over either a good ground, or a counterpoise ground. The version shown in Fig. 10-8B is horizontally polarized.

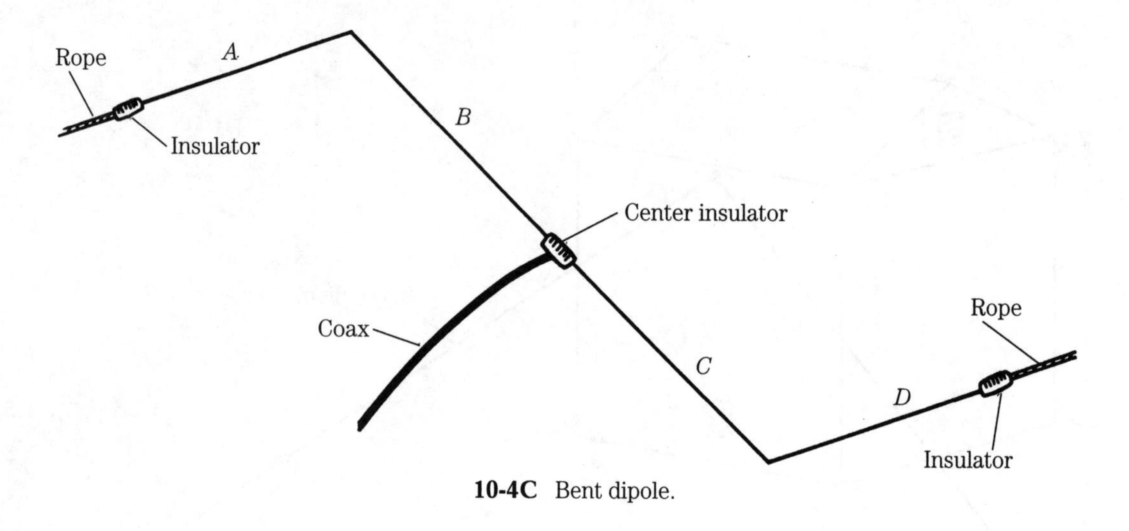

10-4C Bent dipole.

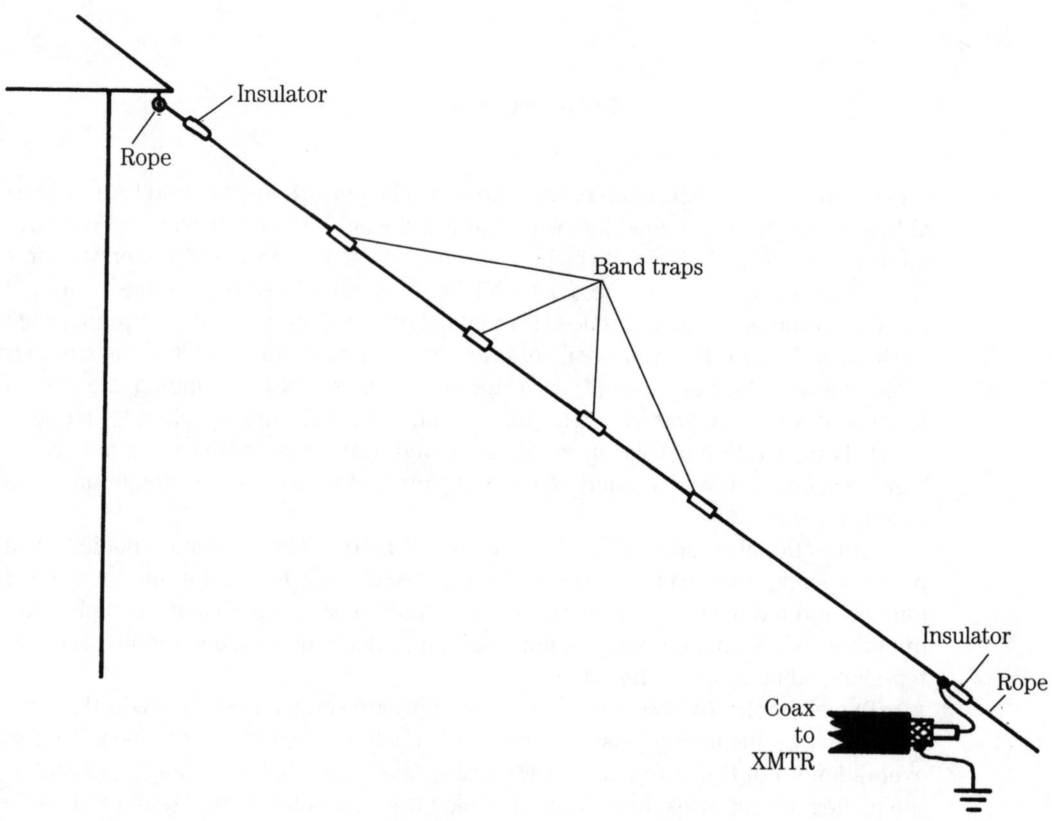

10-4D Sloper trap vertical.

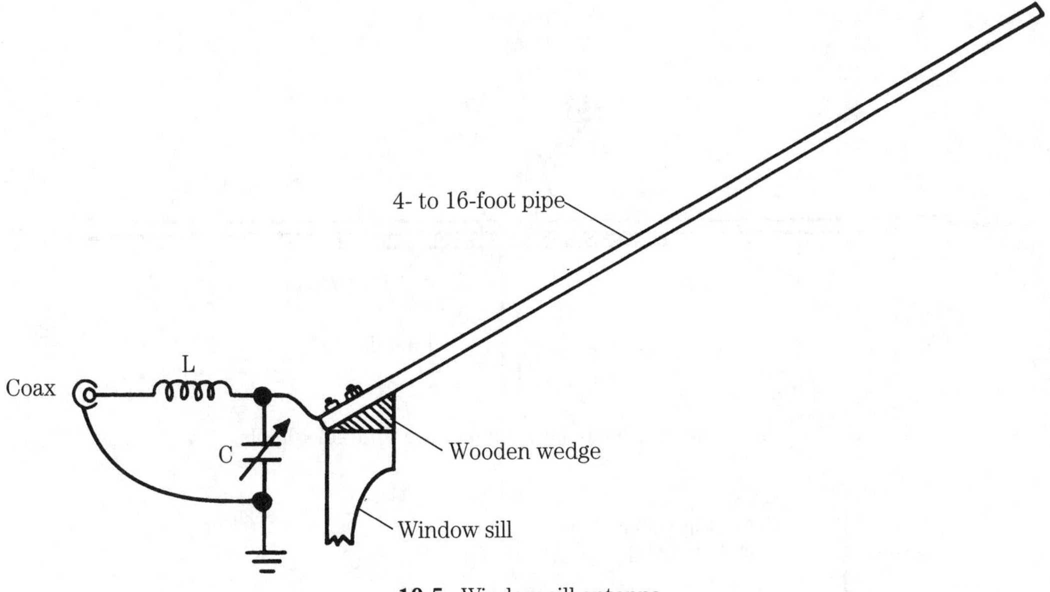

10-5 Window sill antenna.

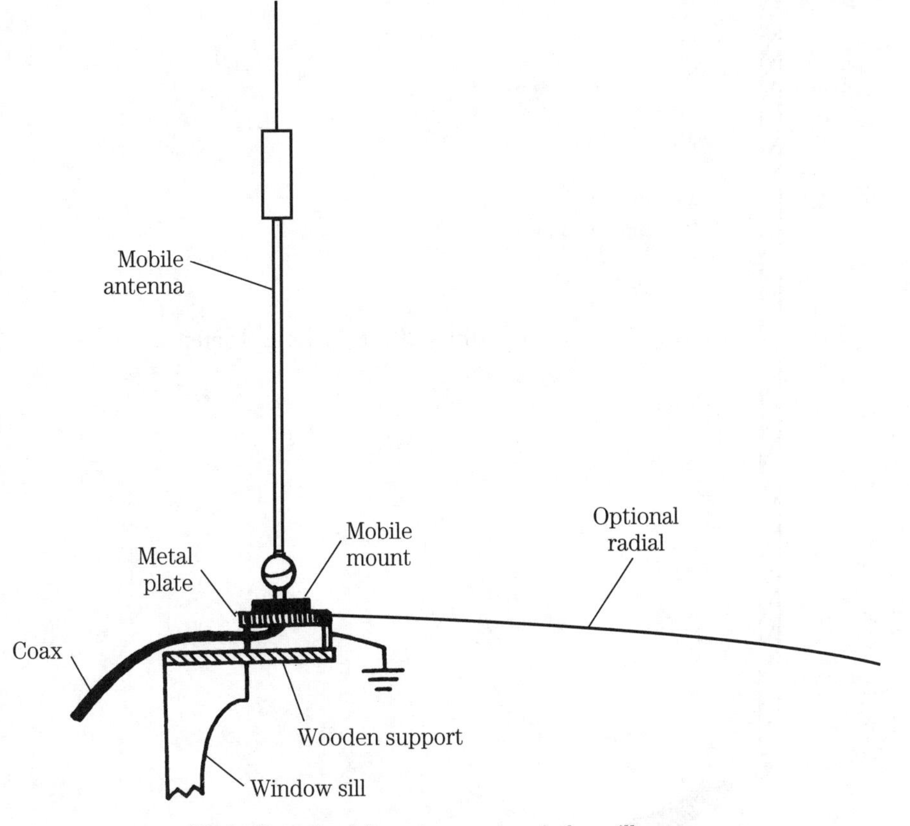

10-6 Use of mobile antenna as a window sill antenna.

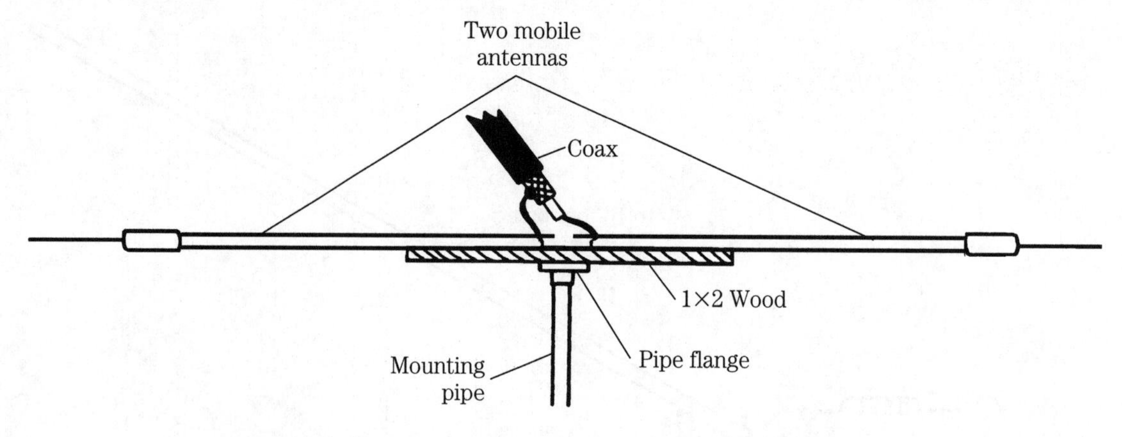

10-7 Use of two mobile antennas as a rotatable dipole.

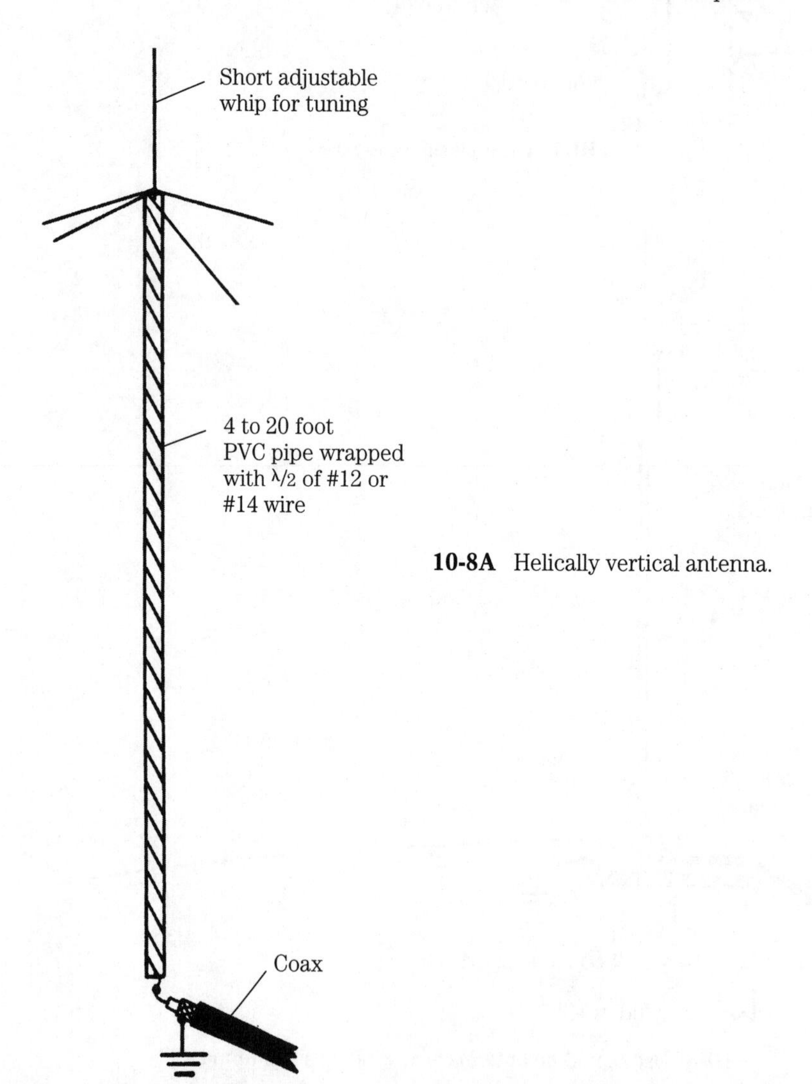

10-8A Helically vertical antenna.

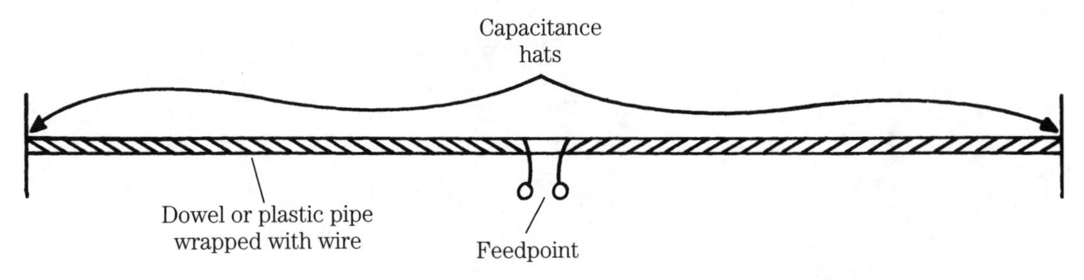

10-8B Helically wound dipole antenna.

11 CHAPTER

Directional phased vertical antennas

THE VERTICAL ANTENNA IS A PERENNIAL FAVORITE WITH RADIO COMMUNICATIONS users. The vertical is either praised, or cursed, depending upon the luck of the owner. "DXability" is usually the criterion for judging the antenna's quality. Some amateurs can't get out of their backyards with a vertical, and they let everyone within earshot know that such and such a brand is no good. Yet, another person routinely works New Zealand and Australia on 15 meters using exactly the same brand of vertical. The proper installation of vertical antennas is dealt with in another chapter, so, for the present, look at another problem attributed to vertical antennas.

The other problem with vertical antennas is that they are *omnidirectional* in the azimuth aspect; that is, they send out and receive equally well from all directions. Some people moan that this pattern dissipates their power, and gives them a weaker signal "out where it counts" (true): however, the main disadvantage of the omnidirectional pattern is noise (QRN and QRM). "*QRN*" is natural noise from thunderstorms and other sources. "*QRM*" is man-made noise, and can consist of other stations or the many assorted forms of electrical filth that clutter up the airwaves. All forms of noise, however, have one thing in common: they are directional with respect to the station. In other words, if you could null signals coming from the direction of the noise source (or undesired station), you would be able to hear desired stations much better.

Although most amateurs seem to think that the effective radiated power (ERP) increase that the directional antenna gives them is the real reason to own one, the main benefit is actually on receive. Think about it for a moment. With anywhere from 100 to 1500 watts available, the increase or decrease in signal strength (due to the directivity of the antenna) results in a minimal difference on the receive end . . . especially during good DX conditions. If we rotate the directional pattern, to null out interference, then we usually find that the change in our signal strength perceived by the other guy is small; the S-meter reading of the desired station is minimally affected; but the amplitude of the interference source is greatly attenuated! The overall effect is an apparent increase in the other guy's signal, even though the S-meter tells a slightly different story. The improvement of signal-to-noise ratio (SNR) is tremendously improved.

Directivity and phase shift

So, how does a vertical antenna owner get the benefit of directivity without the kilobuck investment that a beam or quad costs? The usual solution is to use phased verticals. AM broadcast stations, with more than one tower, are using this type of system (although for different reasons than hams). The idea is to place two or more antennas in close proximity and feed them at specific phase angles to produce a desired radiation pattern. A lot of material is available in the literature on phased vertical antenna systems, and it is far too much to be reproduced here. There are "standard patterns" dating from before World War II that are created with different spacings and different phase angles of feed current. In this chapter, we will consider only one system.

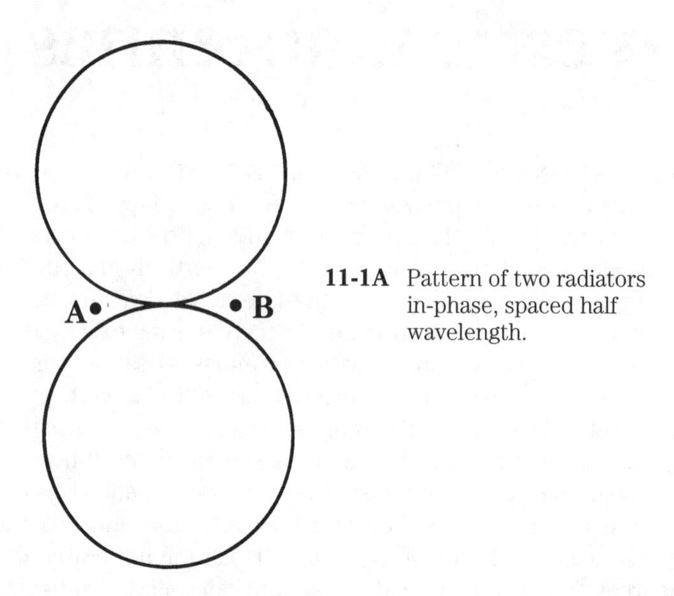

11-1A Pattern of two radiators in-phase, spaced half wavelength.

Figure 11-1 shows the patterns for a pair of quarter-wavelength vertical antennas spaced a half wavelength (180 degrees) apart. Without getting into complex phase shifting networks, there are basically two phasings that are easily obtained: 0 degrees (antennas in-phase) and 180 degrees (antennas out of phase with each other).

When the two antennas (A and B) are fed in-phase with equal currents, the radiation pattern (shown somewhat idealized here) is a bi-directional "figure 8" that is directionally perpendicular to the line of centers between the two antennas; this pattern is called a "broadside" pattern. A sharp null exists along the line of centers (A-B).

When the antennas are fed out of phase with each other by 180 degrees, the pattern rotates 90 degrees (a quarter way around the compass) and now exhibits directivity along the line of the centers (A-B); this is the "end fire" pattern. The interference cancelling null is now perpendicular to line A-B.

It should be apparent that you can select our directivity by selecting the phase angle of the feed currents in the two antennas. Figure 11-2 shows the two feeding

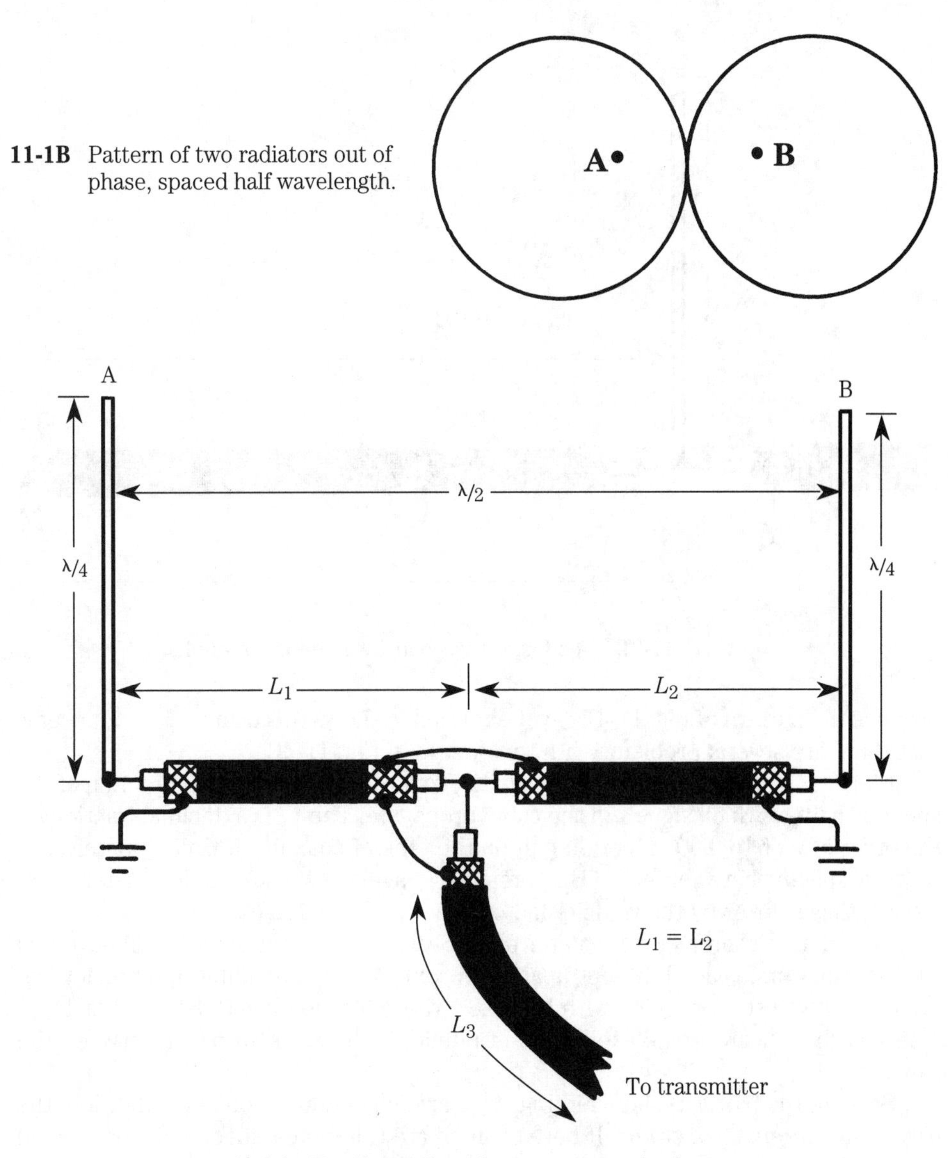

11-1B Pattern of two radiators out of phase, spaced half wavelength.

11-2A Feeding a phased array antenna in phase.

systems usually cited for in-phase (Fig. 11-2A) and out-of-phase (Fig. 11-2B) systems. Figure 11-2A shows the coax from the transmitter coming to a coax tee connector. From the connector to the antenna feedpoints are two lengths of coax (L1 and L2) that are equal to each other, and identical. Given the variation between coaxial cables, I suspect that it would work better if the two cables were not merely the same length ($L_1 = L_2$), but also that they came from the same roll.

The second variation, shown in Fig. 11-2B, supposedly produces a 180°-phase shift between antenna A and antenna B, when length L3 is an electrical half wavelength. According to a much-publicized theory, the system of Fig. 11-2B ought to

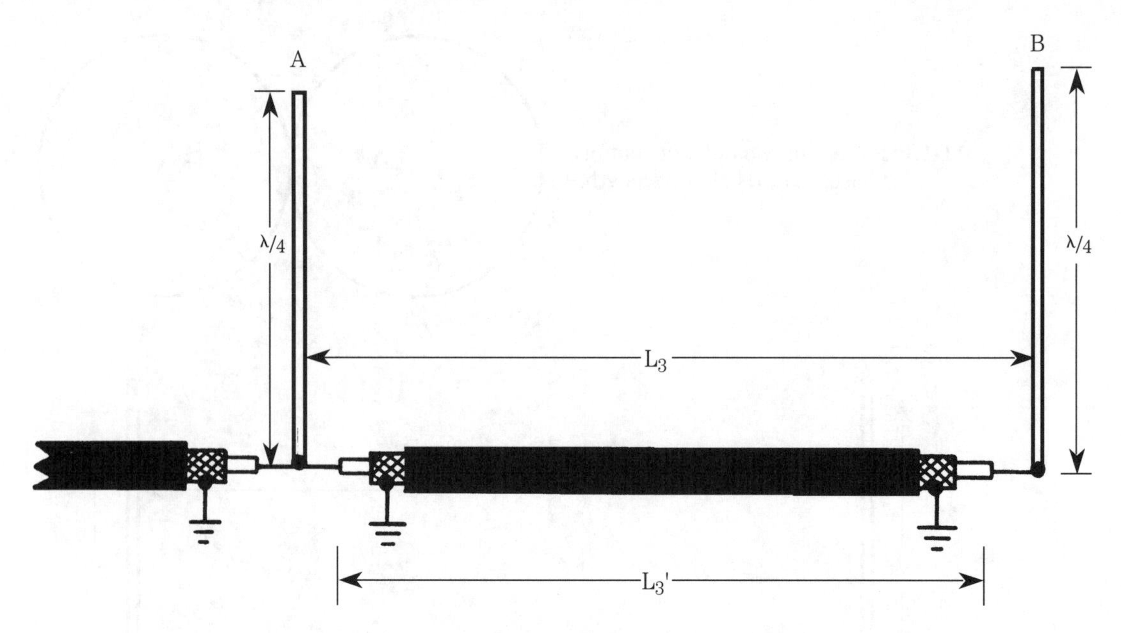

11-2B Feeding a phased array antenna out of phase.

produce the pattern of Fig. 11-1B—yet experience shows this claim is false. It seems that there are several problems with the system in Fig. 11-2B.

First, coax has a property called *velocity factor* (V_F), which is the fraction of the speed of light at which signals in the cable propagate. The V_F is a decimal fraction on the order of 0.66 to 0.90, depending upon the type of coax used. Unfortunately, the physical spacing between A and B is a real half wavelength ($L3 = 492/F$), but the cable length is shorter by the velocity factor ($L3' = ((V_F) \times 492/F)$.

Consider an example. A 15-meter phased vertical antenna system will have two 11-foot radiators, spaced 22 feet apart (approximately, depending upon exact frequency). If we use foam coax, with $V_F = 0.80$, the cable length is (0.8) × (22 feet), or 17.6 feet. In other words, despite lots of publicity, the cable won't fit between the towers!

Second, the patterns shown in Fig. 11-1 are dependent upon one condition: the *antenna currents are equal.* If both of them are the same impedance, and are fed from the same transmitter, then it is reasonable to assume that the currents are equal—right? No, wrong! What about coax loss? Because of normal coax loss, which increases at higher frequencies, the power available to antenna B in Fig. 11-1B is less than the power available to antenna A. Thus, the pattern will be somewhat distorted, because the current produced in B is less than the current in A, when they should be equal.

The first problem is sometimes fixed by using unequal lengths for cables L1 and L2 (Fig. 11-2A), and using it for the out of phase case. For example, if we make L1 a quarter wavelength, and L2 three quarter wavelength (Fig. 11-2C), antenna A is fed with a 90-degree phase lag (relative to the tee connector signal), while antenna B is fed with a 270° phase shift. The result is still a 180° phase difference. Unfortunately,

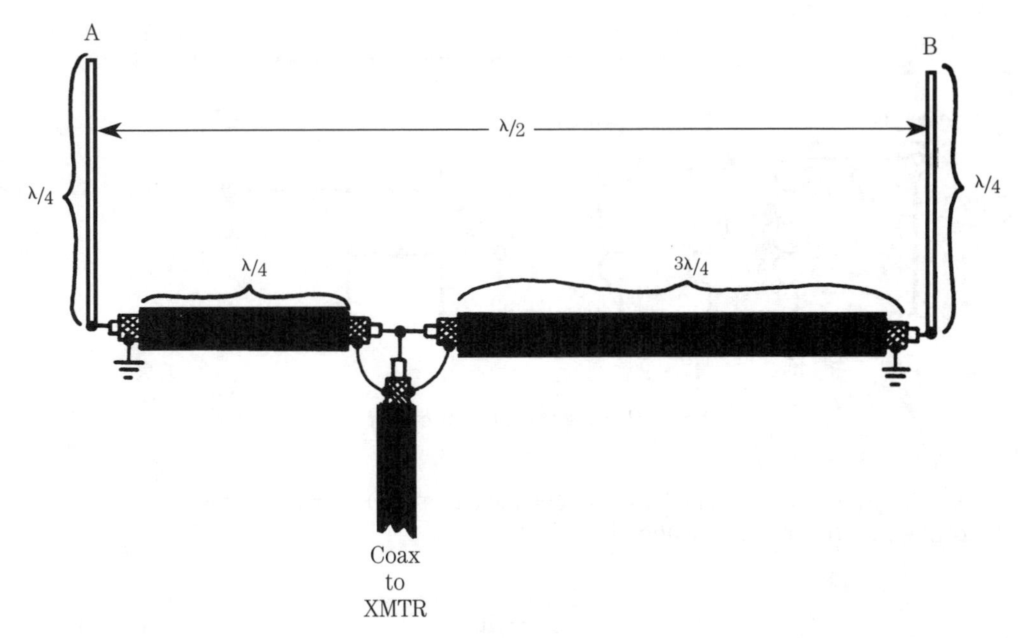

11-2C Corrected feed for a phased array antenna.

we have not solved the current level problem, and may have actually made it worse by adding still more lossy cable to the system.

There is still another problem that is generic to the whole class of phased verticals. Once installed, the pattern is fixed. This problem doesn't bother most point-to-point commercial stations, or broadcasters, because they tend to transmit in only one direction. But amateurs are likely to need a rotatable pattern. Neither Fig. 11-1A nor Fig. 11-1B is rotatable without a lot of effort—like changing the coax feeds, or physically digging up the verticals and repositioning them.

Fortunately, there is a single solution to all three problems. Figure 11-3 shows a two-port phasing transformer made from a toroidal BALUN kit. Use the kind of kit that makes a 1:1 BALUN transformer. Although we are not making a BALUN, we will need enough wire to make three windings, and that is the normal case for 1:1 BALUNs. *Amidon Associates* and others make toroidal BALUN kits.

Wind the three coils in trifilar style, according to the kit instructions. The dots in Fig. 11-3 show the "sense" of the coils, and they are important for correct phasing; call one end the "dot end," and the other end the "plain end," to keep them separate. If the dot end of the first coil is connected to J3 (and the transmitter), then connect the dot end of the second coil to the 0° output (J1, which goes to antenna A). The third coil is connected to a DPDT RF relay or switch. In the position shown, S1 causes the antennas to be 180-degrees out of phase. In the other position, the "sense" of the third coil is reversed, so the antennas are in-phase.

Another phasing method is shown in Fig. 11-4. In this scheme, two convenient, but equal, lengths of coaxial cable (L1 and L2) are used to carry RF power to the antennas. One segment (L1) is fed directly from the transmitter's coaxial cable (L3), while the other is fed from a phasing switch. The phasing switch is used to either by-

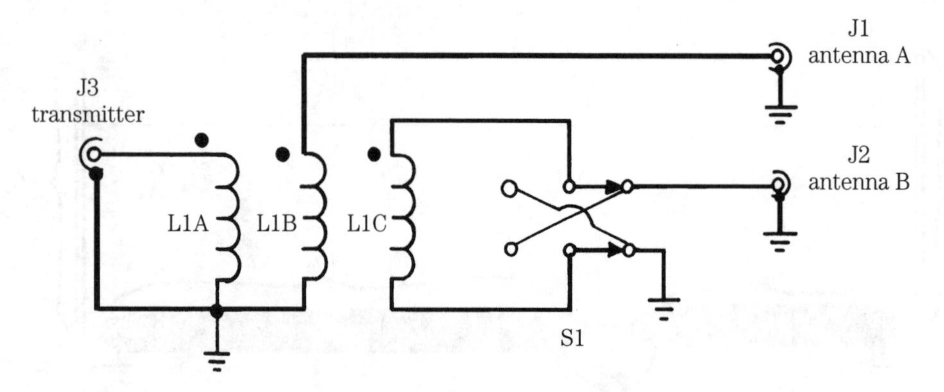

11-3A Phasing transformer circuit.

pass or insert a phase shifting length of coaxial cable (L4). For 180° phasing use the following equation to find the length (L4):

$$L = \frac{492\ V_F}{F_{\text{MHz}}} \text{ feet} \qquad [11.1]$$

Where:

L is the length of L4 in feet
V_F is the velocity factor (a decimal fraction)
F_{MHz} is the operating frequency in megahertz

Some people use a series of switches to select varying amounts of phase shift from 45° to 270°. Such a switch allows them to select any number of other patterns for special situations.

360 degree directional array

The phased vertical antennas concept can be used to provide 'round the compass control of the antenna pattern. Figure 11-5A shows how three quarter wavelength verticals (arranged in a triangle that is a half wavelength on each side) can be used to provide either end-fire, or broadside, patterns from any pair (A-B, A-C, or B-C). Any given antenna (A, B, or C) will be either grounded, fed at 0°, or fed with 180°. The table in Fig. 11-5B shows the relative phasing for each direction that was labelled in Fig. 11-5A. Either manual phase changing, or switch operated phase changing, can be used, although the latter is preferred for convenience.

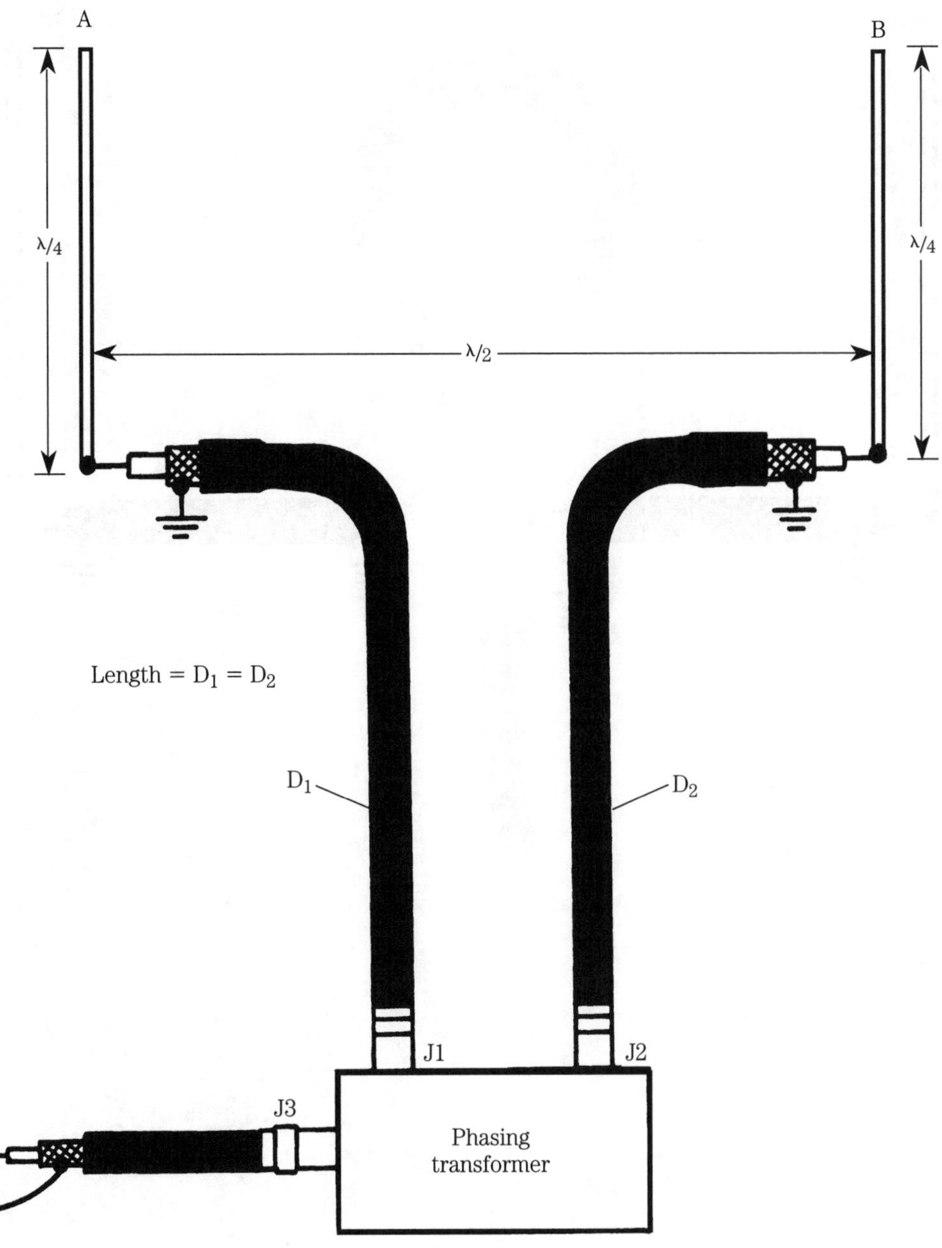

11-3B Connection to antennas.

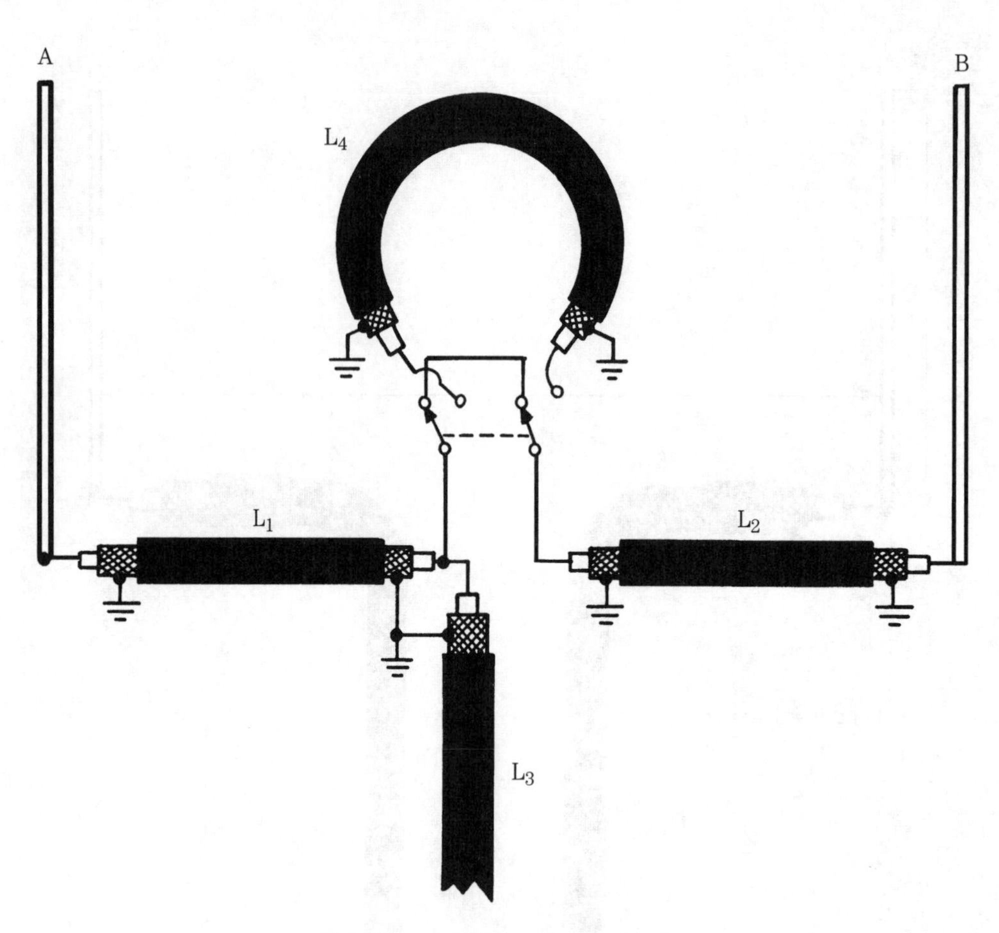

11-4 Phase shifting antenna circuit.

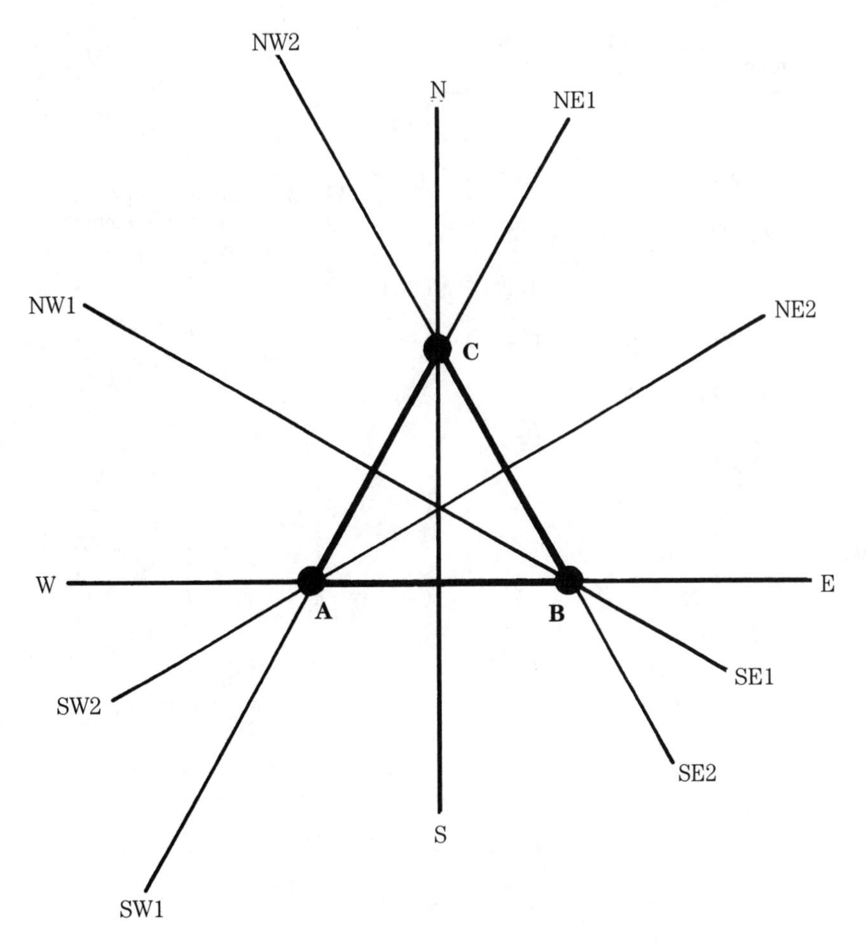

11-5A Three-element phased array.

Direction	Ant A	Ant B	Ant C
N-S	0°	0°	GND
NE1-SW1	0°	GND	180°
NE2-SW2	GND	0°	0°
E-W	0°	180°	GND
SE1-NW1	0°	GND	0°
SE2-NW2	GND	0°	180°

11-5B Table of feed phasing for the three-element array.

12 CHAPTER

Directional beam antennas

THE DIRECTIONAL BEAM ANTENNA DOES SEVERAL JOBS. FIRST, IT PROVIDES AN APPARENT increase in radiated power, because it focuses available transmitter power into a single (or at worst limited) direction. For this reason, a bidirectional dipole has a gain of a little less than 2 dB over an isotropic radiator. Add one or more additional elements, and the focusing becomes nearly unidirectional, which increases the *effective radiated power* (ERP) even more. Second, the beam increases the received signal available at the inputs of the receiver. Antennas are generally reciprocal, so will work for receiving as they do for transmitting. Finally, the directivity of the beam antenna allows the operator to null interfering stations. In fact, it is the latter attribute of the beam that is most useful on today's crowded bands. All in all if your funds are too little to provide both increased RF power, and a good antenna system, then spend what is available on the antenna—not on the power.

In this chapter we will focus on directional antennas that can be built relatively easily. It is assumed that most readers who want a tri-band multi-element Yagi will prefer to buy a commercial product, rather than build a homebrew model. The material herein concentrates on homebrew projects that are within the reach and capabilities of most readers. The first of these is not a beam at all, but rather a *rotatable dipole*.

Rotatable dipole

The *dipole* is a bidirectional antenna that has a figure pattern (when viewed from above). The dipole is a half wavelength, and is usually installed horizontally . . . although vertical half wavelength dipoles are known. Although the length of the dipole is too great for rotatability at the lower bands it is within reason for the higher band. For example, the size of the halfwave dipole is approximately 16 feet on 10 meters and 22 feet on 15 meters. Even the 33 foot length on 20 meters, is not unreasonable for amateur constructors. The length of the dipole is found from:

$$L = \frac{468}{F_{\text{MHz}}} \text{ feet} \qquad \textbf{[12.1]}$$

This length is approximate because of end effects and other phenomena, so some "cut and try" is required.

Example Find the length of a dipole antenna for a frequency of 24.930 MHz in the 12 meter amateur radio band.

Solution:

$$L = \frac{468}{F_{\text{MHz}}} \text{ feet}$$

$$L = \frac{468}{24.930 \text{ MHz}} = 18.77 \text{ feet}$$

The halfway dipole is fed in the center by coaxial cable. Each element of the dipole is one-half of the overall length (or, in the example given, about 9.4 feet).

Figure 12-1 shows a rotatable dipole that can be designed for use on 15 meters, 12 meters, and 10 meters. The radiator elements are made from 10-foot lengths of ¾-inch aluminmum tubing. The tubing is mounted on "beehive" standoff insulators, which in turn are mounted on a 4-foot length of 2-×-2 lumber. The lumber should be varnished against weathering. In a real pinch, the elements can be mounted directly to the lumber without the insulators, but this is not the recommended practice.

The mast is attached to the 2-×-2 lumber through any of several means. The preferred method is the use of a 1-inch pipe flange. These devices are available at hardware stores under the names "floor flange" and "right-angle flange."

The 10-foot lengths of pipe are the standard lengths available in hardware stores, so it was selected as being closest to the required 22 feet for 15 meters. A 0.14-μH loading coil is used at the center, between the elements, in order to make up for the short length. The dimensions of the coil are 4 to 5 turns, 0.5-inch diameter, 4 inch length. For low power levels, the coil can be made of #10 (or #12) solid wire—and, for higher levels, ⅛-inch copper tubing.

There are two basic ways to feed the antenna, and these are shown in details "A" and "B" in Fig. 12-1. The traditional method is to connect the coaxial cable (in parallel) across the inductor. This method is shown in Fig. 12-1, detail "A." A second method is to link couple the coil to the line through a 1 to 3 turn loop (as needed for impedance matching). This is the method that would be used is a toroidal inductor.

Lower frequencies can be accommodated by changing the dimensions of the coil. The coil cannot be scaled, simply because the relative length of the antenna changes as the frequency changes. But it is possible to "cut and try" by adding turns to the coil, one turn at a time, and remeasuring the resonant frequency. Adding inductance to the coil will make the antenna usable on 17 meters and 20 meters, as well as on 15 meters.

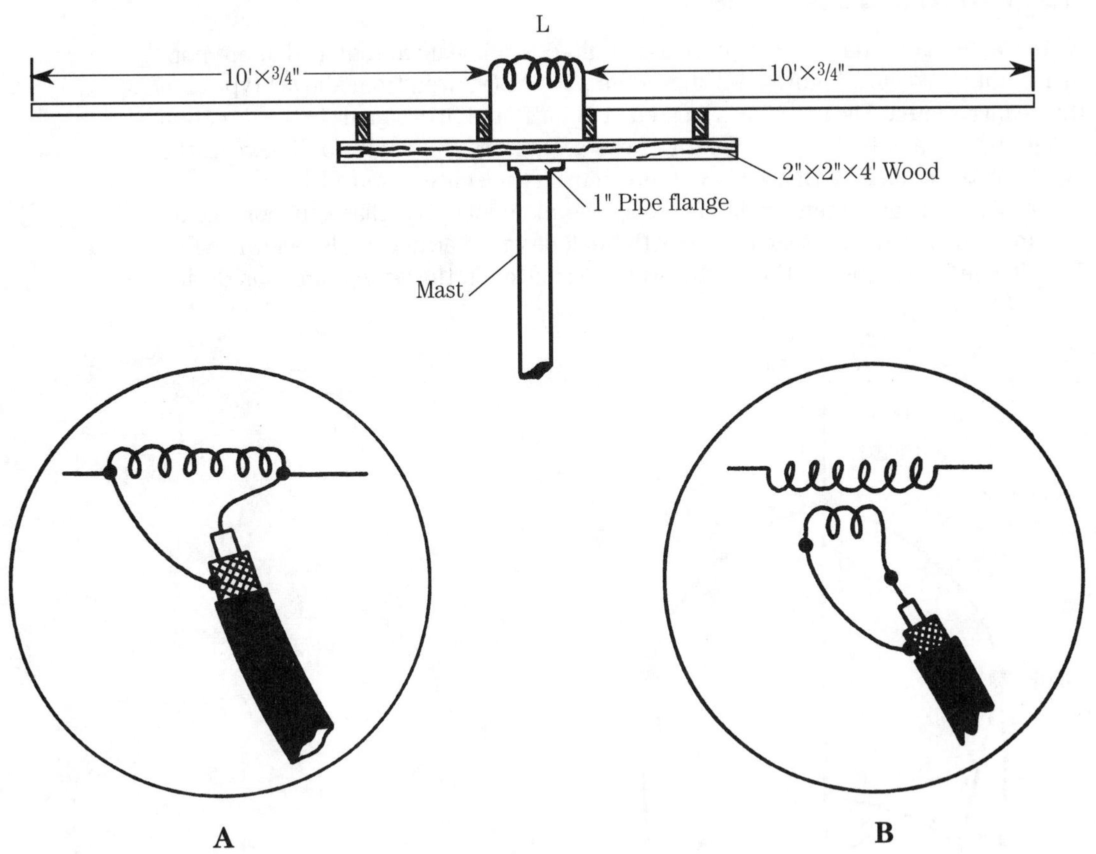

12-1 Rotatable dipole antenna. Inset A shows conventional feed; inset B shows transformer feed.

Another method for building a rotatable dipole for lower frequencies, is to increase the element lengths. On 17 meters, the overall length is approximately 27.4 feet, so each element length is 13.7 feet long. This length can be achieved in either of two methods. First, adjacent sizes of aluminum tubing are designed so that the smaller will be a slip-fit inside of the larger. What constitutes "adjacent sizes" depends on the wall thickness, but for one common brand, the ⅞-inch is adjacent to the ¾-inch size. You can use two smaller lengths to make the larger lengths of pipe, and cut it to size. This method is only available to those readers who have a commercial or industrial metals distributor nearby, because the 16-foot lengths are not generally available from hardware stores.

Bands higher than 15 meters (i.e., 12 meter and 10 meters can be accommodated using the 10-foot lengths of tubing, but without the inductor. The tubing is cut to the desired half-wavelength size and used directly.

Yagi beam antennas

A *Yagi* antenna is one member of a class of directional beam antennas that are popular in the HF bands. Figure 12-2 shows the pattern (viewed from above) typical of the beam antenna. The antenna is located at point P, and fires signals in the direction shown by the arrow. The beamwidth of the antenna is the angle (a) between the points on the main lobe that are –3 dB down from the center point ("C").

A perfect beam antenna will have only the main lobe, but that situation occurs only in dreams. All real antennas have both sidelobes and backlobes, also shown in Fig. 12-2. These lobes represent wasted power transmitted in the wrong direction during

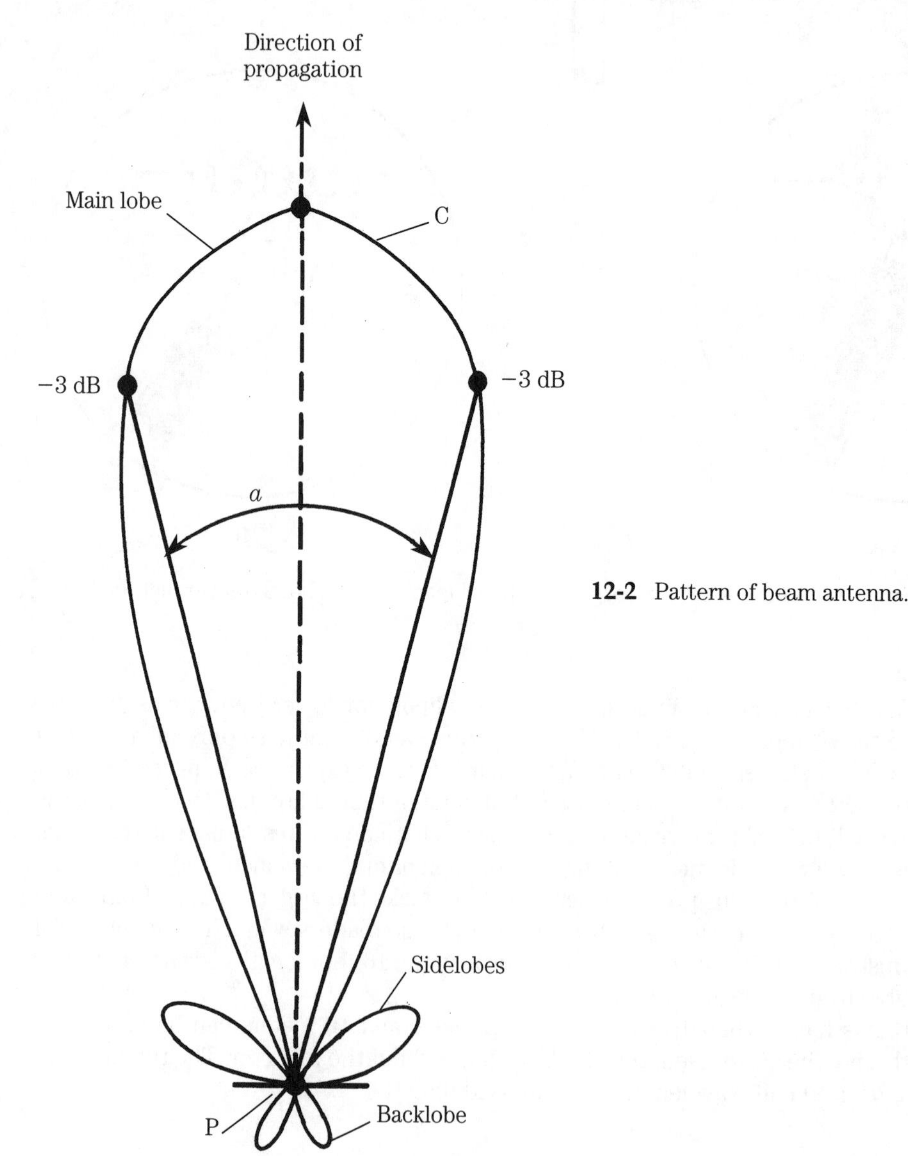

12-2 Pattern of beam antenna.

transmission and interference opportunities while receiving. The goal of the antenna designer is to increase the main lobe while decreasing the sidelobes and backlobes.

Figure 12-3 shows, schematically, the basic *Yagi-Uda* antenna (usually called simply *Yagi*). The driven element is a simple half-wavelength dipole fed in the center. There are two additional elements: *reflectors* and *directors*. These are called *parasitic elements* because they are not directly excited by RF, but rather, they receive energy radiated from the driven element and then re-radiate it. The reflector is placed behind the driven element, and is typically about 4-percent longer than the driven element. The director is placed in front of the driven element (relative to the direction of propagation). The director is typically about 4 percent shorter than the driven element. Although there is no fixed rule regarding the number of either reflectors or directors, it is common practice to use a single director and a driven element for two-element beams, and a single reflector and a single director, in addition to the driven element, for three element beams. Again, additional reflectors can be added for four and more element beams, but standard practice calls for addition of directors instead.

The length of the elements is given by:

$$L = \frac{K}{F_{\mathrm{MHz}}} \text{ feet} \qquad \textbf{[12.2]}$$

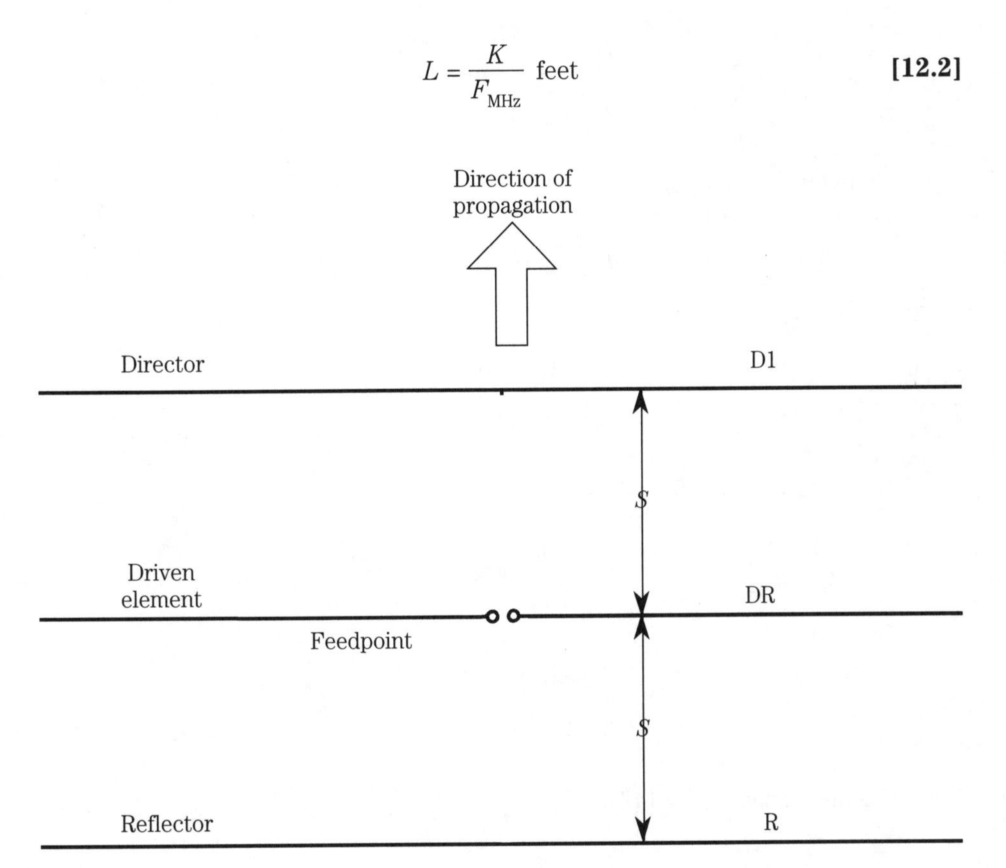

12-3 Basic Yagi-Uda antenna.

Where:

L is the length in feet
F_{MHz} is the frequency in megahertz
K is a factor obtained from a table

The spacing of the elements is typically from 0.15 to 0.308 wavelengths, although 0.2 and 0.25 are the most common values. The constant for K, in the case of element spacing, is given in the "S" column of the table.

Example Calculate the approximate element lengths for a three element 15-meter beam designed to operate on a frequency of 21.39 MHz.

Solution:

1. Driven Element ($K = 478$):

$$L = \frac{K}{F_{MHz}} \text{ feet}$$

$$L = \frac{478}{21.39 \text{ MHz}} \text{ feet}$$

$$L = 22.34 \text{ feet}$$

2. Reflector ($K = 492$):

$$L = \frac{K}{F_{MHz}} \text{ feet}$$

$$L = \frac{492}{21.39 \text{ MHz}} \text{ feet}$$

$$L = 23 \text{ feet}$$

3. Director ($L = 461.5$):

$$L = \frac{K}{F_{MHz}} \text{ feet}$$

$$L = \frac{461.5}{21.39 \text{ MHz}} \text{ feet}$$

$$L = 21.58 \text{ feet}$$

4. Element spacing (= 142):

$$L = \frac{K}{F_{MHz}} \text{ feet}$$

$$L = \frac{142}{21.39 \text{ MHz}} \text{ feet}$$

$$L = 6.64 \text{ feet}$$

The elements of a rotatable beam antenna can be built in a manner similar to the rotatable dipole described earlier. In the case of the beam, however, a boom is needed between the elements to support them. The boom can be made of metal or wood. In the case of a metal boom, however, the driven element must be insulated from the boom, even though the parasitic elements can be mounted directly to it. In general, however, it is usually better to use wood as a matter of convenience. Metal boom antennas can be obtained from commercial sources. The wood boom is easy to build and maintain, even though a little less durable than metal boom antennas.

The feedpoint impedance of a dipole is on the order of 72 Ω in free space, although the actual impedance will vary above, and below, that figure for antennas close to the earth's surface. In addition, adding parasitic elements reduces the impedance even more. The feedpoint impedance of the antenna is too low to be directly fed with coaxial cable, so some means of impedance matching is needed. Some people feed the antenna through an impedance matching BALUN transformer. Figure 12-4 shows the *gamma match* system. The driven element of the Vagi is not broken in the center, as in the case of the simple dipole. The outer conductor, or shield, of the coaxial cable is connected to the center point of the driven element. The center conductor is connected to the gamma match element. The dimensions of the gamma match are:

1. Gamma Match Length: $L/10$
2. Gamma Match director: $D/3$
3. Spacing of Gamma Match From Driven Element: $L/70$

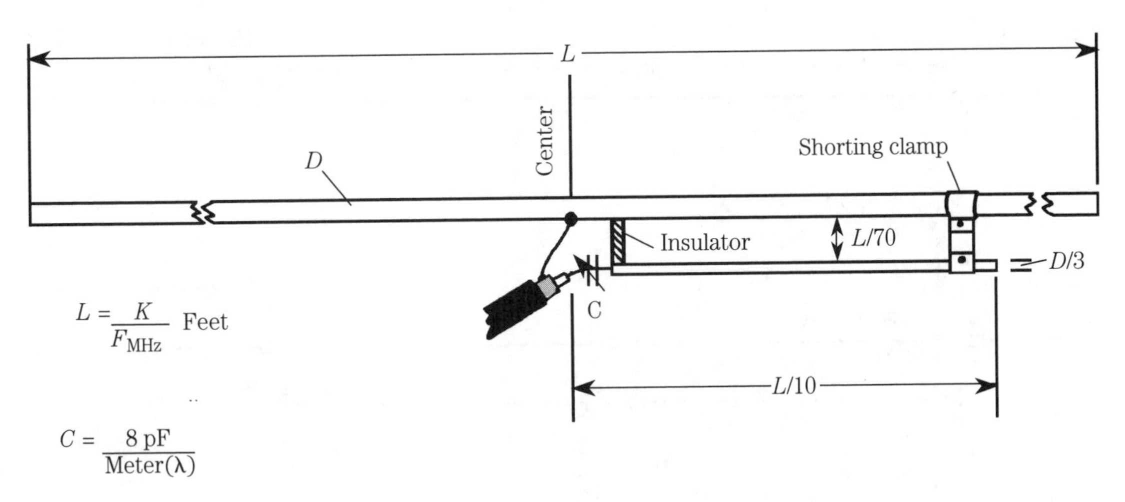

12-4 Gamma feed provides impedance matching.

Where:

L is the length of the driven element
D is the diameter of the driven element

The capacitor in series with the center conductor of the coaxial cable has a value of approximately 8 pF per meter of wavelength that the lowest frequency in the band of operation, or approximately:

$$C = \frac{2400}{F_{\text{MHz}}} \text{ picofarads} \quad \textbf{[12.3]}$$

The capacitor must be a high-voltage transmitting variable type. In general, the gamma match capacitors are either air or vacuum variables.

There are three aspects to the adjustment of the Yagi antenna. Resonance is determined by the length of the element. The length is increased or deceased in order to find the resonant point. This point can be determined by the use of a noise bridge, VSWR meter or other means. The capacitor, and the shorting bar/clamp, are adjusted to match the impedance of the antenna to the transmission line impedance. For the dimensions shown, the coaxial cable should be 52 Ω (RG-58 or RG-8).

It is not necessary to use tubing or pipes for the antenna elements in order to obtain the benefits of the Yagi beam antenna. An example of a wire beam is shown in Fig. 12-5. The wire beam is made as if it were two half-wavelength dipoles, installed parallel to, and about 0.2 to 0.25 wavelengths apart from each other. Although multi-element wire beams are possible, the two-element version is the most common. Perhaps the most frequent use of the wire beam antenna is on the lower bands (e.g., 40 meters and 75/80 meters) where rotatable beams are more difficult to build.

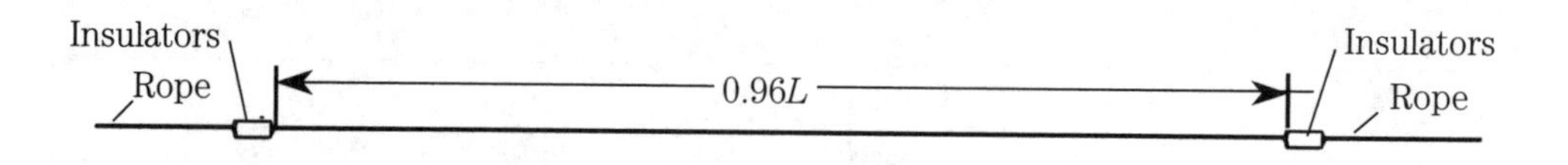

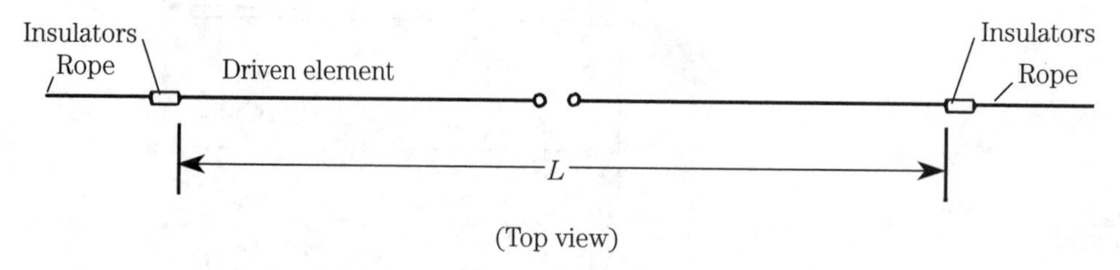

12-5 Wire beam useful for low frequencies.

20 meter ZL-special beam

The antenna shown in Fig. 12-6 is a close relative of the Yagi beam. It consists of a pair of folded dipoles, mounted approximately 0.12 wavelengths apart. The elements are 30.5 feet in length, and the spacing is 7.1 feet. The elements can be built from aluminum tubing if the antenna is to be rotatable. Alternatively, for a fixed antenna, the elements can be made of 300 Ω television type twin-lead. If the tubing type of construction is selected, then make the size of the tubing, and its spacing, sufficient for 300-Ω parallel transmission lines, according to:

$$Z_o = 276 \text{ LOG} \frac{2S}{d} \qquad [12.4]$$

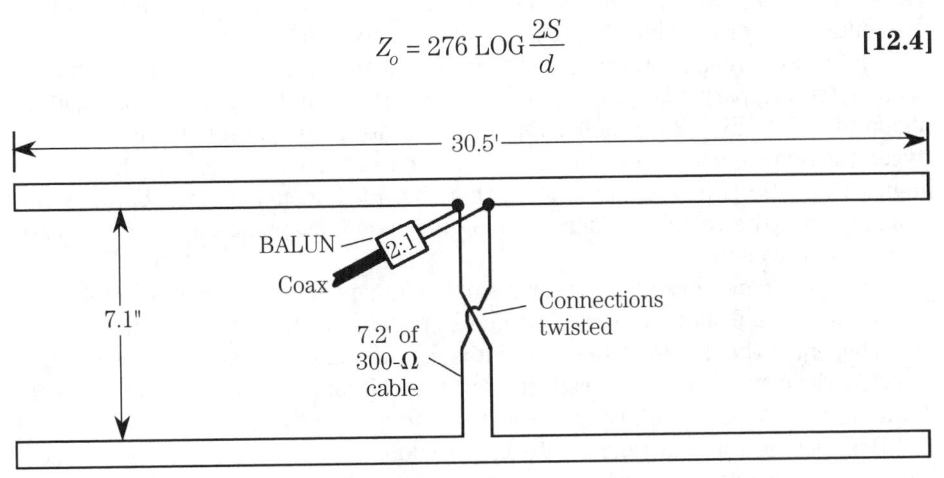

12-6 ZL-Special beam antenna.

Where:

Z_o is the impedance of the line (300 ohms)
S is the center to center spacing of the parallel conductors on the elements
d is the diameter of the conductors

The two half-wavelength elements of the *ZL-special* are fed 135 degrees out of phase with each other. The feedline is connected to one of the dipoles directly; and then to the other through a length of 300-Ω twinlead that has an electrical length of about 45 degrees (λ/8).

The feedpoint impedance is on the order of 100 to 150 Ω, so it will make a good match to either 52 Ω or 75 Ω coaxial cable if a 2:1 impedance matching transformer is used.

Cubical quad beam antenna

The *cubical quad* antenna is a one wavelength square wire loop. It was designed in the mid-1940s at radio station HCJB in Quito, Ecuador. HCJB is a Protestant missionary shortwave radio station with worldwide coverage. The location of the station

(Quito, Ecuador) is at a high altitude. This fact makes the Yagi antenna less useful than it is at lower altitudes. According to the story, HCJB originally used Yagi antennas. These antennas are fed in the center at a current node, so the ends are high voltage nodes. In the thin air of Quito, the high voltage at the ends caused corona arcing, and that arcing periodically destroyed the tips of the Yagi elements. Station engineer, Clarence Moore, designed the cubical quad antenna (Fig. 12-7) to solve this problem. Because it is a full-wavelength antenna, each side being a quarter wavelength, and fed at a current node in the center of one side, the voltage nodes occur in the middle of the adjacent sides . . . and that reduces or eliminates the arcing. The elements can be fed in the center of a horizontal side (Figs. 12-7 and 12-8A), in the center of a vertical side (Fig. 12-8B), or at the corner (Fig. 12-8C).

There is a running controversy regarding how the antenna compares with other beam antennas, particularly the Yagi. Some experts claim that the cubical quad has a gain of about 1.5 to 2 dB higher than a Yagi (with a comparable boom length between the two elements). In addition, some experts claim that the quad has a lower angle of radiation. Most experts agree that the quad seems to work better at low heights above the earth's surface, but the difference disappears at heights greater than a half wavelength.

The quad can be used as either a single element antenna, or in the form of a beam. Figure 12-9 shows a pair of elements spaced 0.13 to 0.22 wavelengths apart. One element is the driven element, and it is connected to the coaxial-cable feedline directly. The other element is a reflector, so it is a bit longer than the driven element. A tuning stub is used to adjust the reflector loop to resonance.

Because the wire is arranged into a square loop, one wavelength long, the actual length varies from the naturally resonant length by about 3 percent. The driven element is about 3 percent longer than the natural resonant point. The overall lengths of the wire elements are:

1. Driven element:

$$L = \frac{1005}{F_{\text{MHz}}} \text{ feet} \qquad \textbf{[12.5]}$$

2. Reflector:

$$L = \frac{1030}{F_{\text{MHz}}} \text{ feet} \qquad \textbf{[12.6]}$$

3. Director:

$$L = \frac{975}{F_{\text{MHz}}} \text{ feet} \qquad \textbf{[12.7]}$$

One method for the construction of the quad beam antenna is shown in Fig. 12-10. This particular scheme uses a 12-×-12 wooden plate at the center, bamboo (or fiberglass) spreaders, and a wooden (or metal) boom. The construction must be heavy duty in order to survive wind loads. For this reason, it is probably a better solution to buy a quad kit consisting of the spreaders and the center structural element.

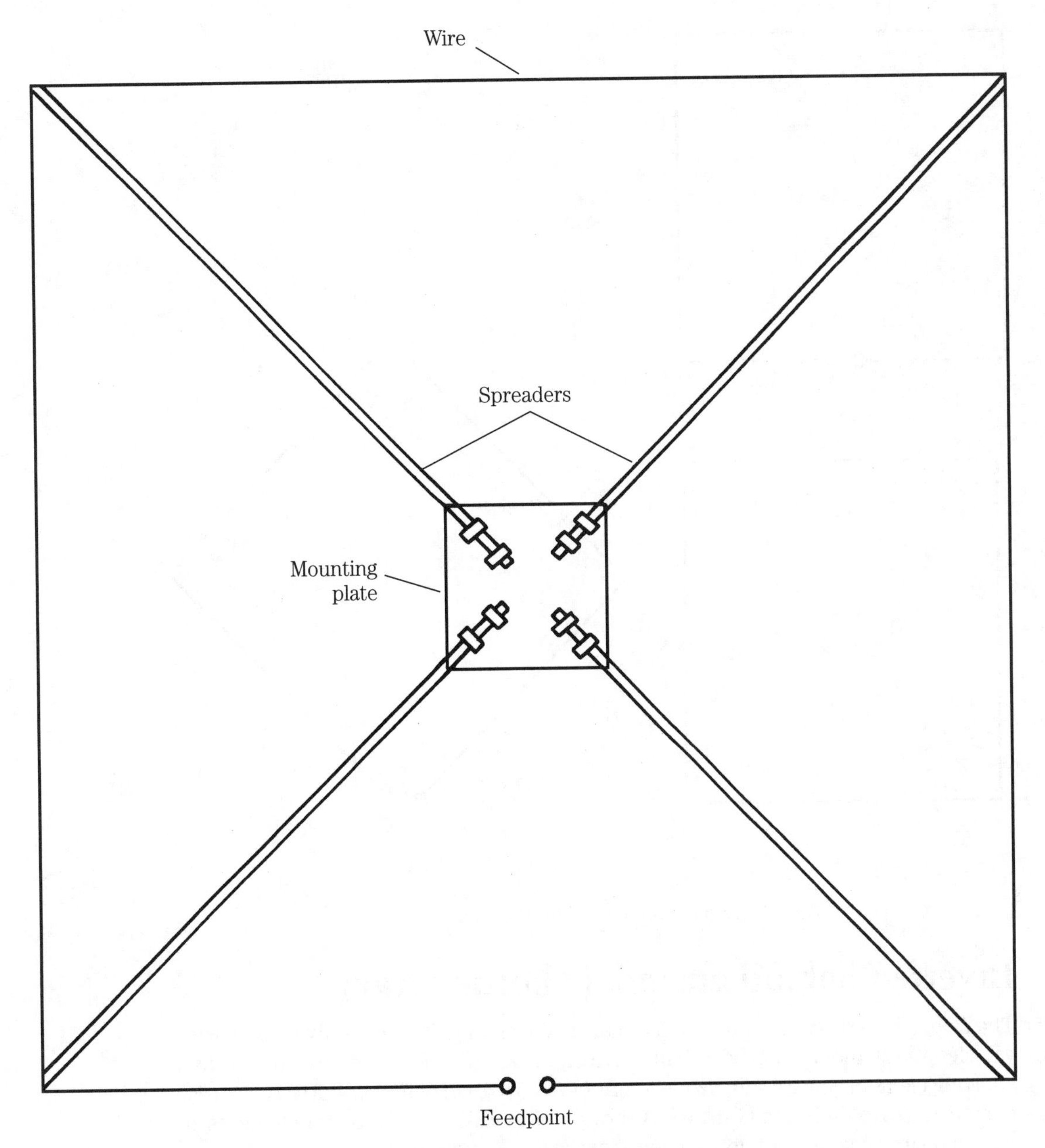

12-7 Quad loop antenna.

More than one band can be installed on a single set of spreaders. The size of the spreaders is set by the lowest band of operation, so higher frequency bands can be accommodated with shorter loops on the same set of spreaders.

This quad antenna is an example of a multi-element, large loop antenna. Additional information on large loops, but not in a beam antenna array, is found in chapter 14.

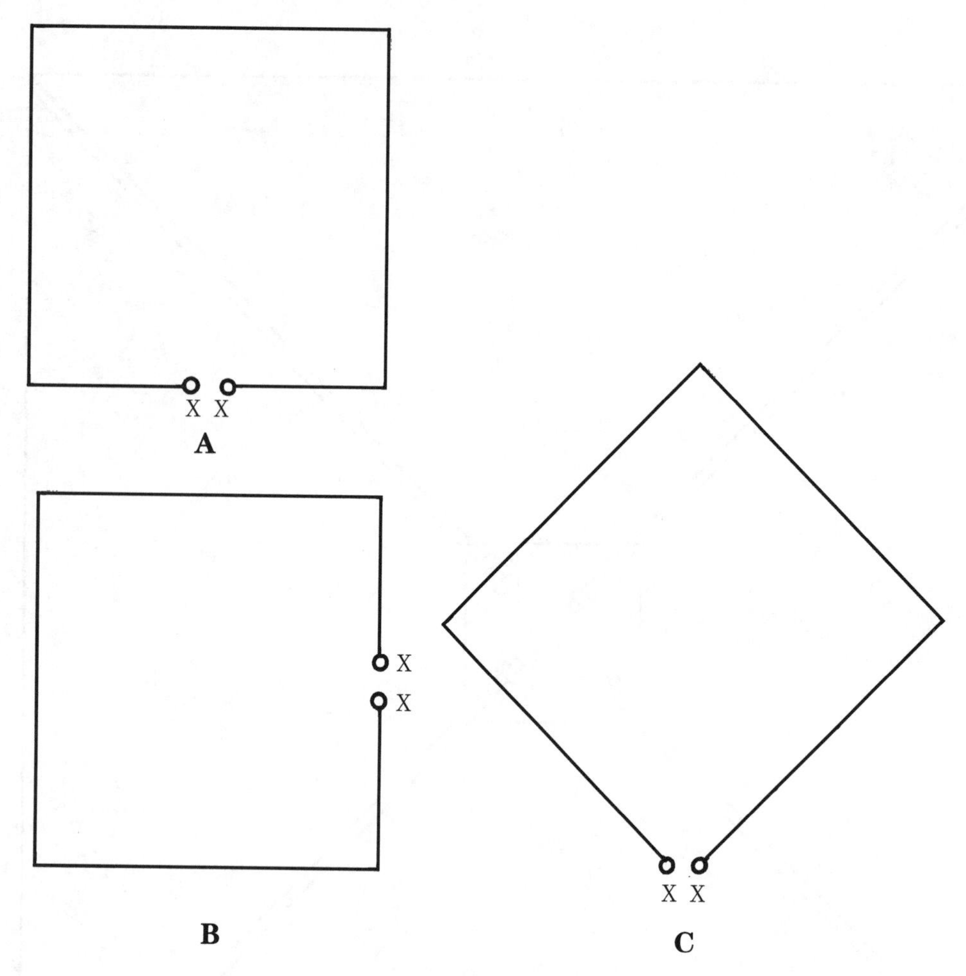

12-8 Feed options for the Quad.

Inverted bobtail curtain (Thorne array)

The *Bobtail Curtain* antenna is a fixed array consisting of three individual quarter-wavelength elements spaced a half wavelength apart, and fed from the top by a shorting element or wire. The *Inverted Bobtail Curtain*, or *Thorne Array*, consists of an *upside down* Bobtail Curtain as shown in Fig. 12-11. The radiator elements are each a quarter wavelength long. Their lengths are found from:

$$L = \frac{246}{F_{MHz}} \text{ feet} \qquad [12.8]$$

The lengths of spacing between the elements are exactly twice above the value or:

$$L = \frac{492}{F_{MHz}} \text{ feet} \qquad [12.9]$$

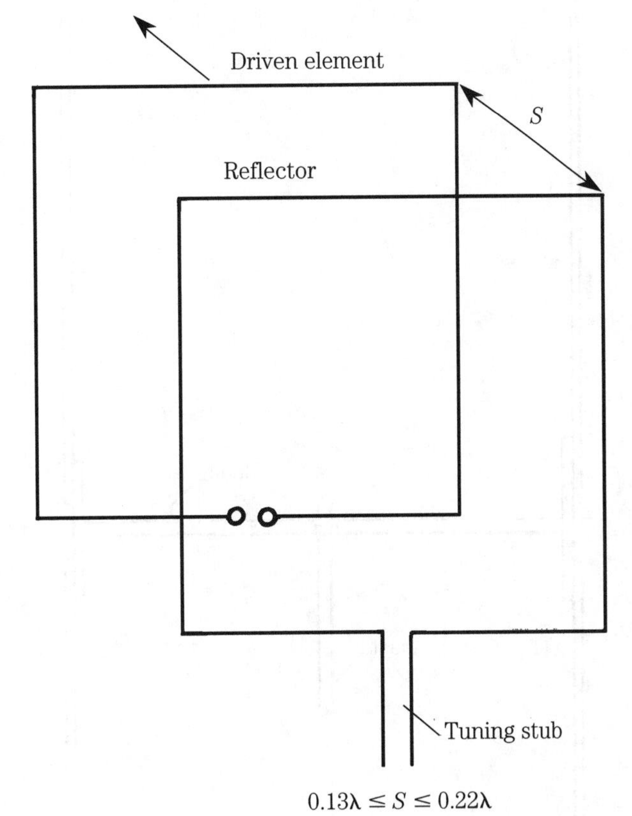

12-9 Quad beam antenna.

The antenna is fed at the base of the center element, through a parallel resonant tuner. The capacitor is a 100 to 200 pF transmitting variable, while the inductor is set to resonate at the band desired (with the capacitor at half to three-quarters full capacitance). A loop or link coupling scheme connects the tuner to the transmission line.

An alternate feed method, worked out by the late J.H. Thorne (K4NFU/5), Fig. 12-12 feeds the end elements from the sheild of the coaxial cable, and the center element of the array is fed from the center conductor of the coaxial cable. A coaxial impedance matching section is used between the cable transmitter, and the antenna feedpoint.

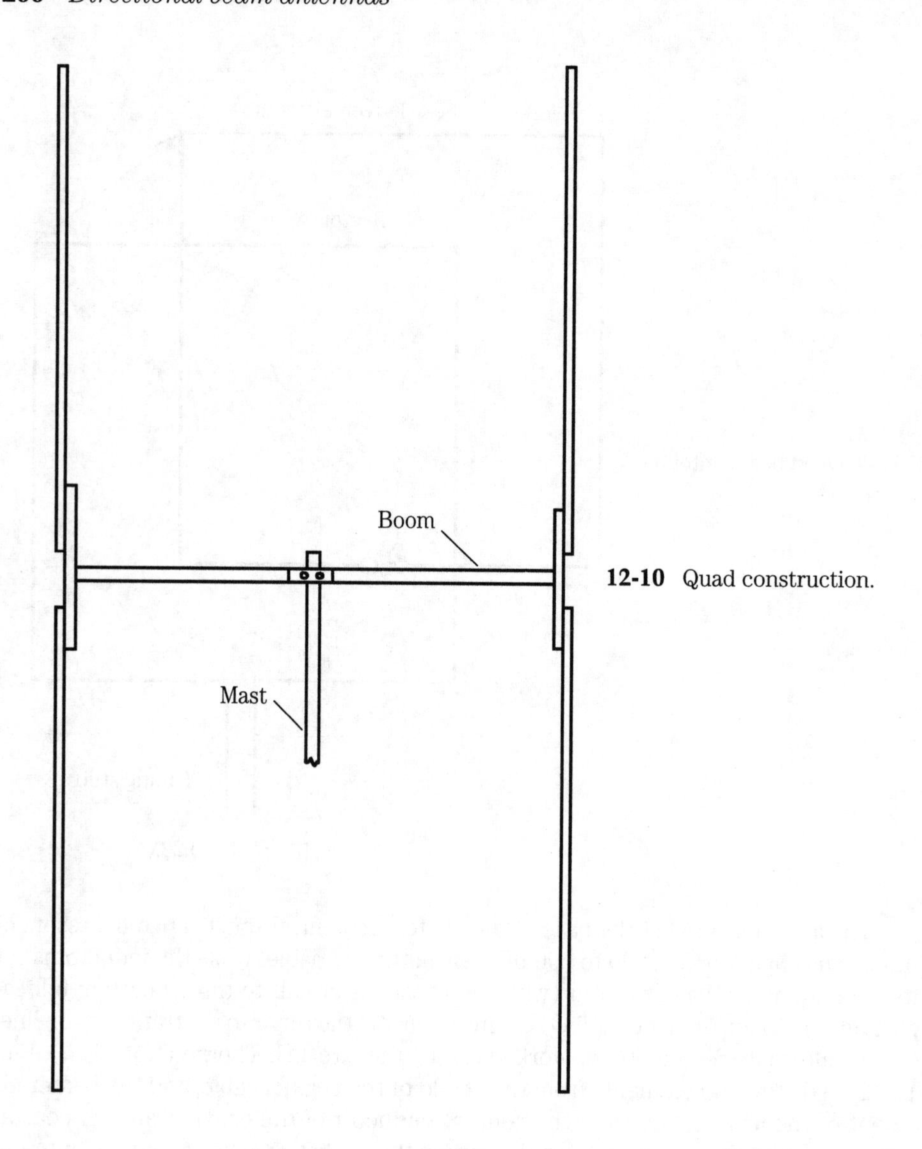

12-10 Quad construction.

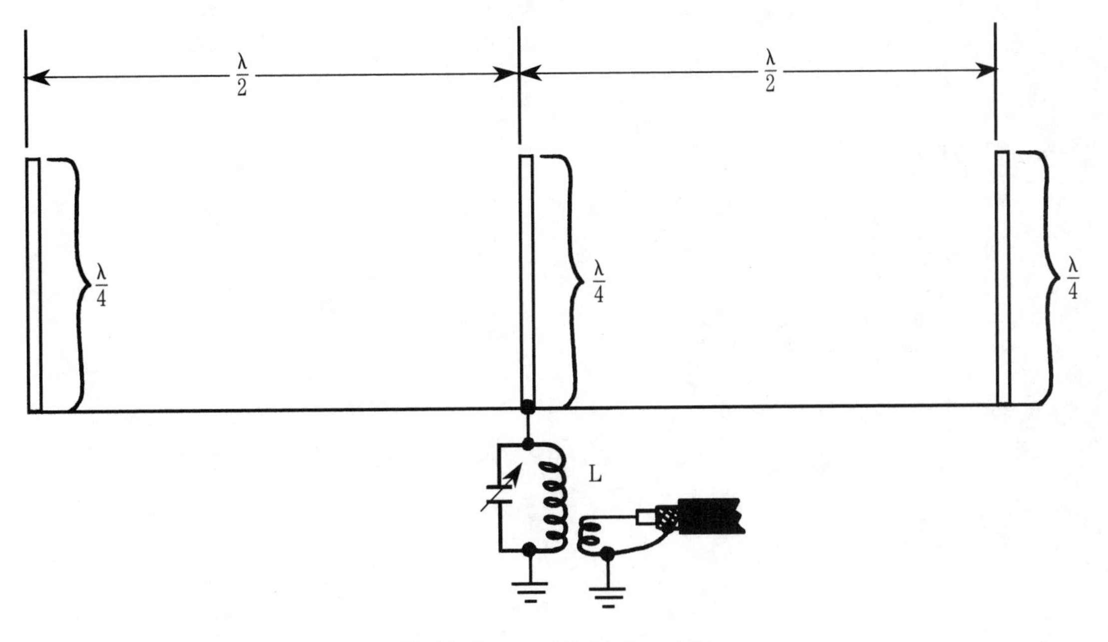

12-11 Inverted Bobtail curtain.

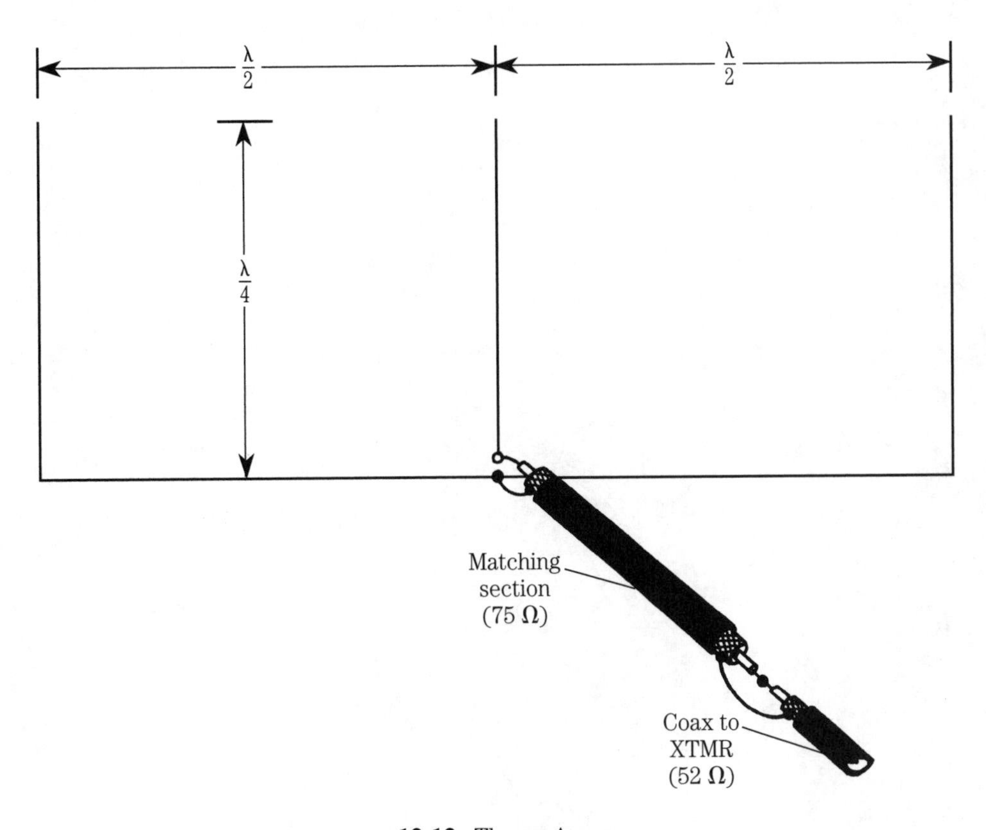

12-12 Thorne Array.

13
CHAPTER
Antennas for shortwave reception

ALTHOUGH MOST OF THIS BOOK ADDRESSES ANTENNAS FOR TRANSMITTING, THERE is a certain body of material that pertains purely to receiving antennas. This material also needs to be addressed, and that is the function of this chapter. There are two readers in mind for this chapter. First, and foremost, is the shortwave listener (SWL). Second, however, is the amateur radio operator who wants to either use a separate receiving antenna on the main station receiver, or use an ancilliary receiver (common amongst DXers).

The law of reciprocity

Antennas possess a property called *reciprocity*. That is a fancy way of saying that an antenna works as well on reception, as it does for transmission. Although articles occasionally appear in the literature claiming an HF or VHF design which violates the *Law of Reciprocity*, to date all have depended on either variable definitions, false premises or faulty measurements. There is even a school of thought that falsely argues against antenna reciprocity based on an ionospheric anomaly in which propagation depends on the *direction* traveled (see the end of chapter 2). The bottom line of reciprocity, for the SWL, is that every antenna described in this book can be also used with equal results on receive. For example, a half-wavelength dipole works equally well as a receiver antenna, or as a transmitter antenna.

Which properties are important?

Selecting a receiver antenna is a function of several factors. Assuming that you want more than a simple longwire (which we will deal with shortly), you will want to home-in on the properties desired for your particular monitoring application. Is the antenna to be fixed or rotatable? Do you want omnidirectional, or directional, reception? In which plane? What about gain?

What about "gain?" What is "gain," for that matter? The concept of antenna gain derives from the fact that directional antennas focus energy. Two kinds of gain figure are often quoted: gain referenced to a dipole antenna on the same frequency, and gain relative to a theoretical construct called an "isotropic radiator." In the isotropic case, the reference radiator to which the antenna is compared is a spherical point-source that radiates equally well in all directions. The dipole radiates equally well in all directions (in the horizontal plane), and not at all (or very little) in other directions. That is, the dipole exhibits a "figure-8" radiation and reception pattern.

The gain of an antenna is merely the ratio (usually expressed in decibels) of the power radiated in a given direction by the two antennas (i.e., the reference antenna and the test antenna). If an antenna gain is listed as "8 dB gain over isotropic," this means that, in the direction specified, the power radiated is 8 dB higher than the same total power applied to an isotropic radiator.

So, of what use is antenna gain? Two answers immediately present themselves. First, by accumulating more signal, the antenna essentially makes your receiver more sensitive. Note that the gain of the antenna does not create a higher powered signal, it merely increases the apparent signal power by focusing energy from a given direction. And, note well that *gain* implies *directivity*. Any antenna that claims to provide gain, but that is truly omnidirectional in all planes, is a fraud. The fundamental assumption is that *gain implies directivity*.

The concept of directivity (hence also of gain) is often taken to mean *horizontal directivity*, which is the case of a dipole antenna. But all forms of antenna radiate in three-dimensional space. Azimuth angle of radiation, and elevation angle of radiation, are both important. Certain 2-meter vertical antennas are listed as "gain antennas," yet the pattern in the horizontal direction is 360 degrees—implying omnidirectional behavior. In the vertical plane, however, lost energy is compressed into a smaller range of elevation angles, so gain occurs by refocussing energy that would have been radiated at a higher than useful angle.

The second application of directivity is in suppressing interfering signals. On the regular AM and FM broadcast bands, channelization permits receiver selectivity to overcome adjacent channel interference in most cases. But in the HF amateur radio and International Broadcast bands, channelization is either nonexistent, poorly defined, or ignored altogether. In these cases, interfering "adjacent" channel signals can wipe out a weaker, desired station. Similarly, with co-channel interference (i.e., when both stations are on the same frequency), two or more signals compete in a "dog fight" that neither will ever totally win. Consider Fig. 13-1A. Assume that two 9540-kHz signals, S1 and S2, arrive at the same omnidirectional vertical antenna. Either both signals will be heard, or the stronger signal will drown out the weaker.

Now, consider Fig. 13-1B. Here, a dipole is used as the receiving antenna, so a little directivity is obtained. The main lobes of the dipole are wide enough to provide decent reception of signal S1, even though the antenna is positioned such that S1 is not along the maxima line (dotted). But the positioning shown, places the interfering co-channel signal (S2) in the null off the ends of the dipole, so it weakens it considerably. The result will be enhanced reception of S1. In Fig. 13-1B, the idea is not to exploit the ability of the gain antenna to increase the level of S1. Indeed, by plac-

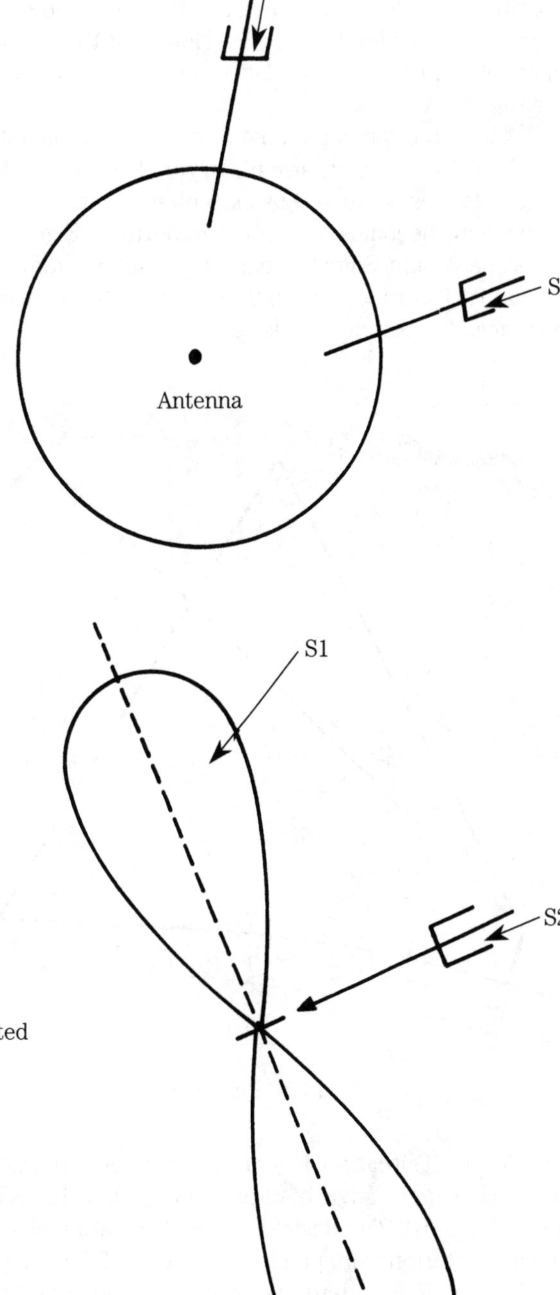

13-1A Omnidirectional antenna picks up co-channel interfering signals equally well.

13-1B Directional antenna can discriminate against unwanted signal.

ing the antenna as shown, we are not getting S1 levels as high as might otherwise be possible. The idea here is to place the unwanted signal (S2) into the "notch" in order to make it considerably weaker. Note that the notch is sharper than the peak of the main lobe. If the dipole is placed on a mast, with an antenna rotator, this ability is increased even more.

Another antenna parameter, of considerable interest, is *angle of radiation* (a), which (by reciprocity) also means angle of reception. Because HF propagation over long distances is created by skip phenomena, the angle at which the signal hits the ionosphere becomes extremely important. Figure 13-2 shows two situations from the same station. Signal S1 has a high angle of radiation (a_1), so its skip distance (D_1) is relatively short. On signal S_2, however, the angle of radiation (a_2) is low, so the skip distance (D_2) is much longer than D_1.

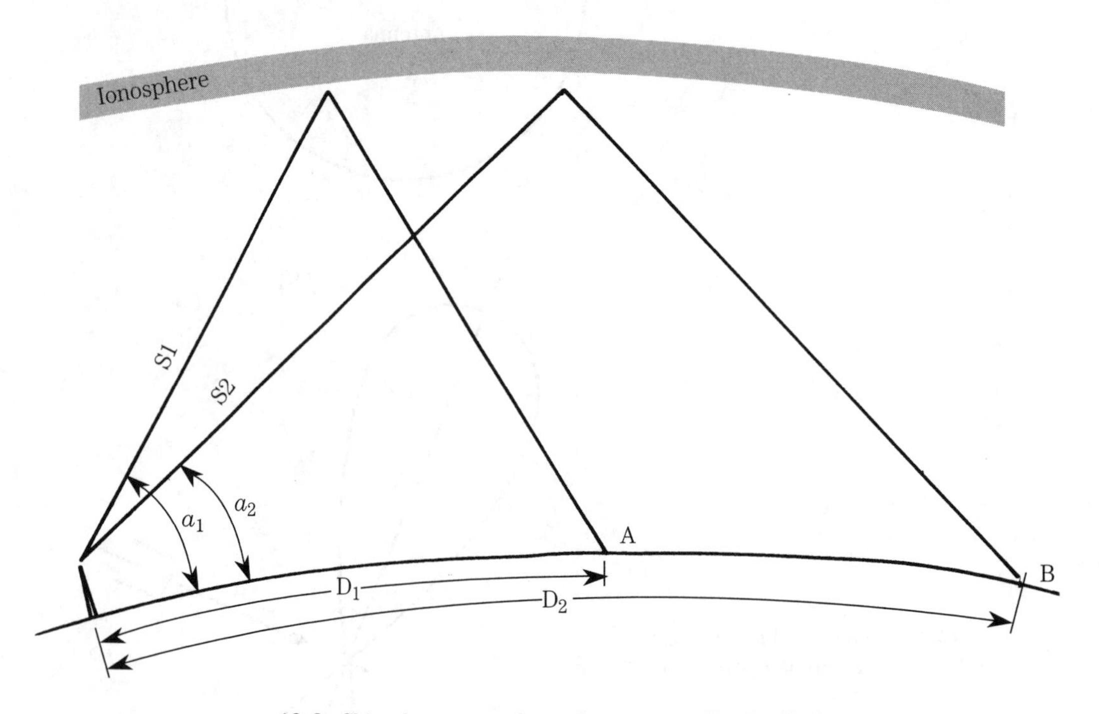

13-2 Skip phenomena dependence on angle of radiation.

So which situation do you want in your antenna? The impulsive answer would be the long distance angle of radiation (a_2), but that is often wrong. The correct answer is: "it depends!" The desired angle of radiation is a function of whether you want to receive a station from point "A" or point "B."

The angle of radiation of the antenna is fixed by its design, that is, by antenna physics. The desired angle is a function of the ionospheric properties at the time of interest, and the operating frequency. For this reason, some well-equipped radio hobbyists have several antennas, of differing properties, to enhance their listening.

Connection to the receiver

It's simply too naive to state, I suppose, but let's do it anyway: an antenna must be properly connected to the receiver before it can be effective. If your antenna uses coaxial cable, and the receiver accepts coax, then no discussion is needed: attach the proper coax connector and plug-in. But in other cases, non-coaxial cable antennas are used.

There are two major forms of antenna input connector used on shortwave receivers. One form uses two (or three) screws intended for either wrapped wire leads or spade lugs, while the other is one or more varieties of coaxial connector. This section covers how each type is connected to a single-wire antenna lead-in.

Consider first the screw-type connector (Fig. 13-3A). Depending upon the design, there will be either two or three screws. If only two screws are found, then one is for the antenna wire, and the other is for the ground wire. These screws will be marked something like "A/G" or "ANT/GND," or with the schematic symbols for antenna and ground.

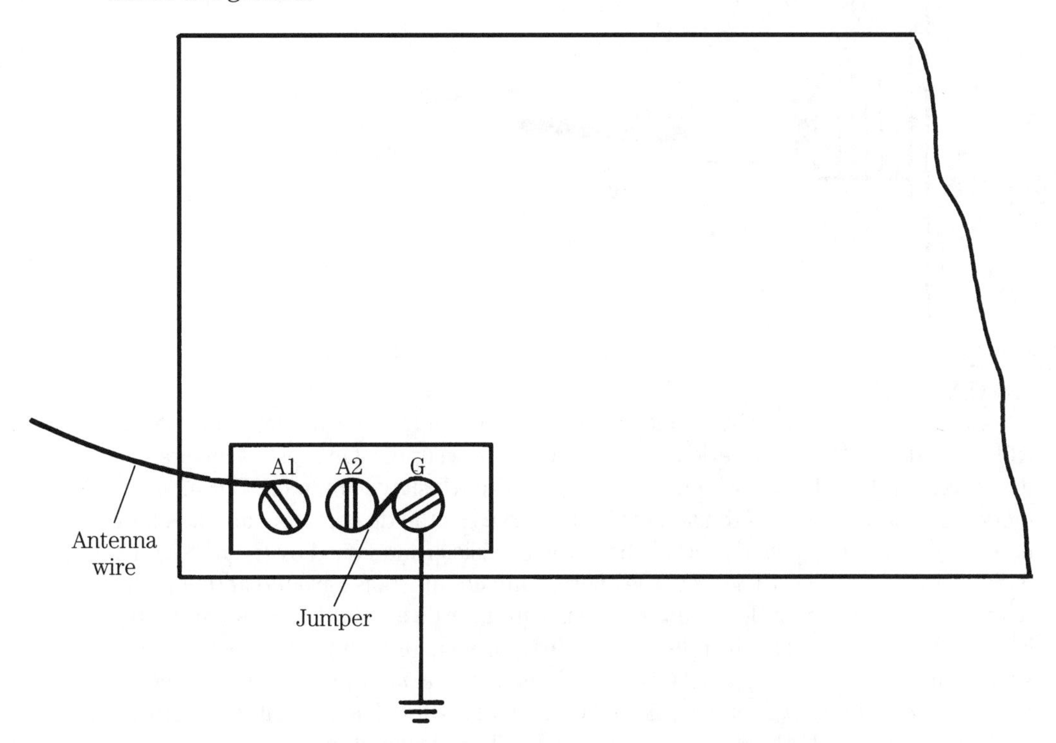

13-3A Connection of wire antenna to balanced antenna terminals on receiver.

Three-screw designs are intended to accommodate balanced transmission lines such as twin-lead, or parallel ladder line. Shortwave listeners can sometimes use ordinary ac line cord (called "zipcord") as an antenna transmission line. Zipcord has an impedance that approximates the 75-Ω impedance of a dipole. When parallel lines of any type are used, connect one lead to A1 and the other to A2. Of course, the ground terminal (G) is connected to the earth ground.

For single lead antenna lines connect a jumper wire or bar (i.e., a short piece of bare #22 solid hook-up wire) between A2 and G. This jumper converts the balanced input line to unbalanced. The A2/G terminal is connected to earth ground, while A1 is connected to the single-lead antenna wire.

On receivers that use an SO-239 coaxial connector, we can use either of two techniques to connect a single-lead wire. First, we can obtain the mating PL-259 plug, and solder the antenna lead to the center conductor pin. The PL-259 connector is then screwed into the mating SO-239 chassis connector. Regardless of the type of coaxial connector, however, the mate can be used for the antenna lead wire. But for SO-239 connectors another alternative is also available. Figure 13-3B shows a "banana plug" attached to the lead wire, and inserted into the receptacle of the SO-239.

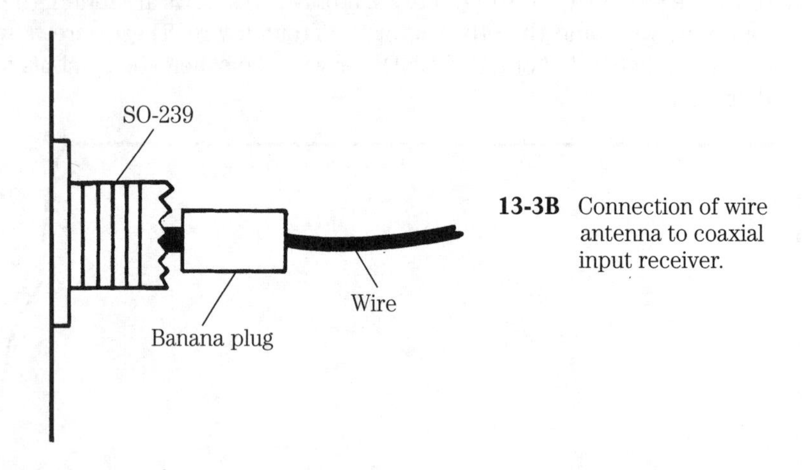

13-3B Connection of wire antenna to coaxial input receiver.

DANGER!

Certain low-cost receivers, especially older vacuum tube models, have a so-called "ac/dc" or "transformerless" internal dc power supply. On most receivers, the dc common is the chassis, which also serves as the RF signal common. But on ac/dc models the neutral wire of the ac power line serves as the dc common, and it is kept floating as a "counterpoise" ground above the chassis ground used by the RF signals. A capacitor (C1 in Fig. 13-4) sets the chassis and counterpoise ground at 0-volts RF potential, while keeping the counterpoise isolated for dc and 60-Hz ac. A danger exists if either the ac plug is installed backwards, or someone plugs the socket in the wall incorrectly (often happens!). Even if C1 is intact, *a nasty shock can be felt by touching the antenna ground* ("G" or "GND") *terminal.* The capacitive reactance of C1 is about 2.7 MΩ for 60-Hz ac, so at least a "bite" is going to happen. But if that capacitor is shorted, which is likely on older receivers, then the bite is considerably worse . . . and might even prove *fatal.* The problem, in that case, is that reversed ac line polarity will set the hot line from the ac socket on the ground lead. The least to expect is massive fireworks . . . and a possible fire hazard; the most to expect is a fire, and your possible electrocution.

The usual advice given to owners of such radios is to make sure that C1 is intact before using the radio. I prefer a better solution: buy, install, and use a 120:120 Vac isolation transformer to isolate your receiver from the ac power lines. Such a trans-

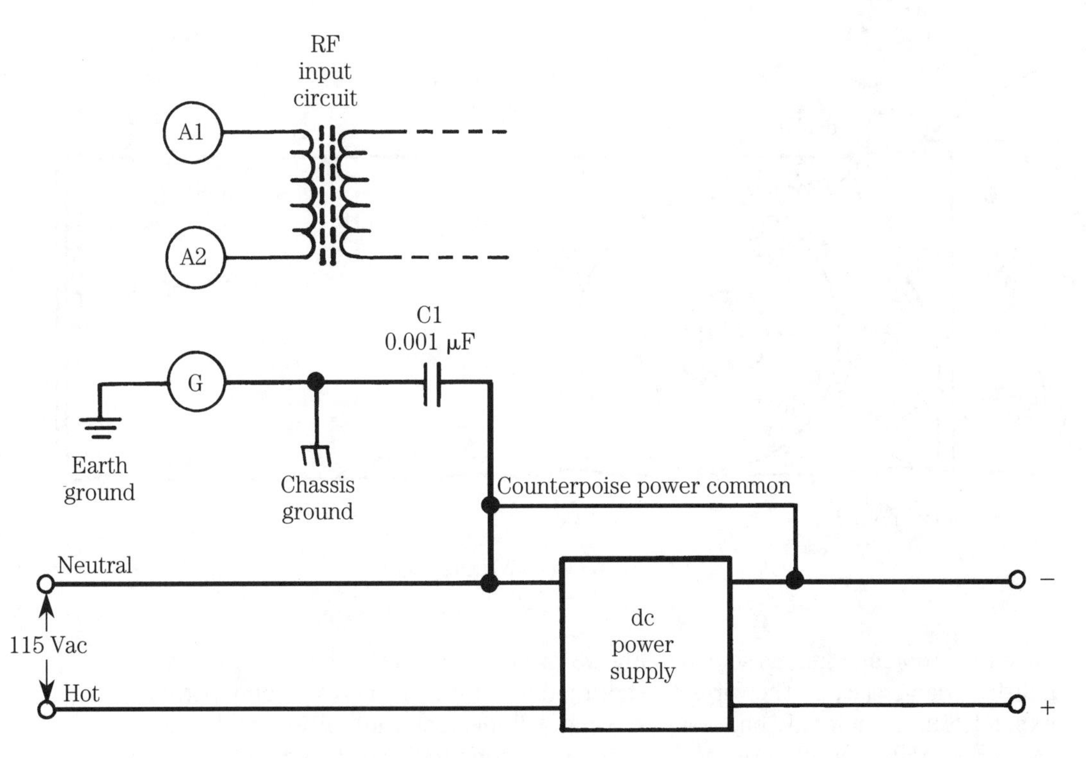

13-4 Dangerous form of antenna/receiver circuitry.

former is standard practice in repair shops, and it should also be standard practice in your house.

Wire antennas

This section reviews simple wire antennas that are suitable for the reception of shortwave signals, although not necessarily for transmitting. Once again, you are reminded that the Law of Reciprocity permits you to use any transmitting antenna found in other chapters for receiving also.

Figure 13-5 shows the common receiving longwire. The antenna element should be 30 to 150 feet in length. Although most texts show it horizontal to the ground (and indeed, a case can be made that performance is better that way), it is not strictly necessary. If you must slope the wire, then it is doubtful that you will notice any reception problems.

The far end of the wire is attached to a supporting structure through an insulator and a rope. The support structure can be another building, a tree, or a mast installed especially for this purpose. Chapter 23 deals with antenna construction practices.

Wind will cause motion in the antenna wire, and its supporting structure. Over time, the wind movement will fatigue the antenna wire and cause it to break. Also, if a big enough gust of wind (or a sustained storm) comes along, then even a new antenna can either sag badly, or break altogether. You can do either of two things to re-

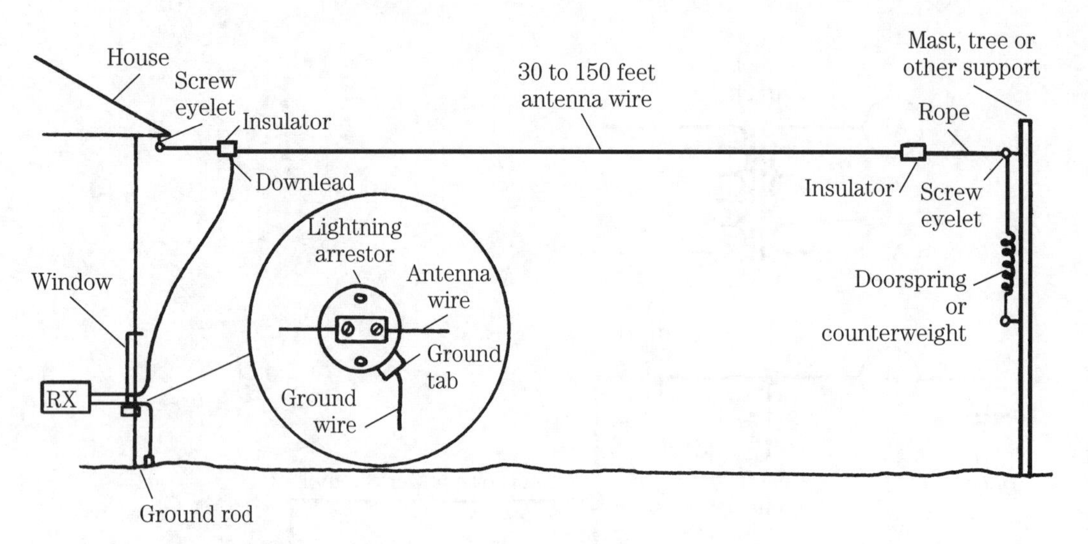

13-5 Longwire SWL antenna.

duce the problem. First, as shown in Fig. 13-5, a door spring can be used to provide a little variable slack in the wire. The spring tension is selected to be only partially expanded under normal conditions, so wind will increase the tension, and stretch the spring. Make sure that the spring is not too strong to be stretched by the action of wind on the antenna, or no good is accomplished.

Another tactic is to replace the spring with a counterweight that is heavy enough to keep the antenna nearly taut under normal conditions, but not so heavy that it fails to move under wind conditions. In other words, the antenna tension should exactly balance the counterweight under normal conditions, and not be too great that it stretches the antenna wire excessively.

The antenna wire should be either #12 or #14 hard drawn copper, or *Copperweld®* stranded wire. The latter is actually steel core wire, but has a copper coating on the outside. Because of "skin effect," RF signals only flow in the outer copper coating. Soft drawn copper wire will stretch and break prematurely, so it should be avoided.

The downlead of the antenna must be insulated, and it should also be stranded (which breaks less easily than solid wire). Again, #12 or #14 wire should be used, although #16 would be permissible. The point, where the downlead and antenna are joined, should be soldered to prevent corrosion of the joint. Mechanical strength is provided by proper splicing technique (see Fig. 13-6). Do not depend on the solder for mechanical strength, for it has none.

There are several ways to bring a downlead into the building. First, if you can tolerate a slight crack in the junction of the sash and sill, then run the wire underneath the sash and close the window. Alternatively, you can buy a flat strap connector to pass under the window. This method is electrically the same as running the lead, but is mechanically nicer. Chapter 23 deals with several methods, and should be consulted.

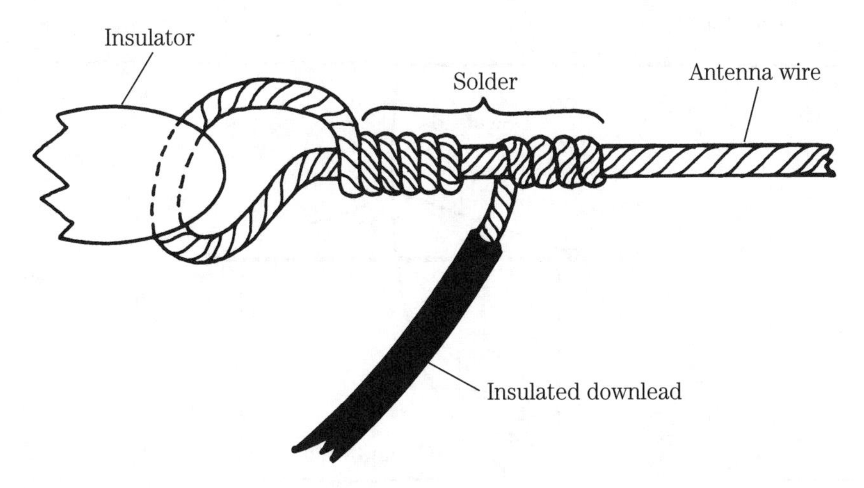

13-6 Construction details.

Grounding

The ground lead should be a heavy conductor, such as heavy wire or braid. The shield stripped from RG-8 or RG-11 coaxial cable is suitable for most applications. For reception purposes only, the ground may be a cold water pipe inside the house. Do not use the hot water pipes (which are not well grounded) or gas pipes (which are dangerous to use). Also, be aware that residential air conditioner liquid lines look like copper cold water pipes in some cases. Don't use them. Chapter 24 deals with grounding.

The lightning arrester is a safety precaution, and it *must* be used! Its purpose is to supply an alternate path to ground in the event of a lightning strike. Although at least one text calls the arrester optional, it is not. Besides the obvious safety reasons (which are reason enough), there are also legal and economic reasons for using the arrester. Your local government building and/or fire codes might require a lightning arrester for outdoor antennas. Also, your insurance company might not honor your homeowner's policy if the lightning arrester (required by local code) is not used. The antenna lightning arrester is *not optional*, so use it.

WARNING! DO NOT EVER ATTEMPT TO INSTALL AN ANTENNA BY CROSSING A POWER LINE! **EVER!** NO MATTER WHAT YOU BELIEVE OR WHAT YOUR FRIENDS TELL YOU, IT'S NEVER SAFE AND IT COULD KILL YOU.

The rest of this discussion touches on antennas, other than the receiver longwire. Because construction details are similar, we will not repeat them again. You are, by the way, encouraged to also read chapter 23 to glean more details on antenna construction methods.

The flattop antenna

The "flattop" antenna is shown in Fig. 13-7. This antenna is a relative of the longwire, with the exception that the downline is at the approximate center of the antenna

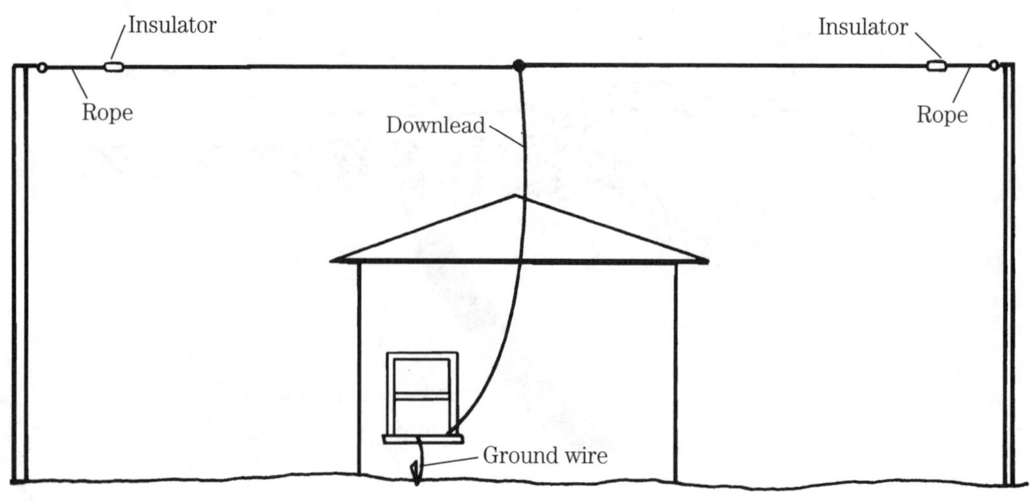

13-7 Flat top SWL antenna.

section. The flattop should be at least a half wavelength ($492/F_{MHz}$) at the lowest frequency of operation. The advantage of the flattop is that it allows the maximum use of space in the configuration shown.

Vertical antennas

It is also possible to build shortwave listener antennas in vertical polarization. Figure 13-8 shows one version of this type of antenna. A support (which could be a tree or building) with enough height to be at least a quarter wavelength on the lowest frequency of operation. The antenna is fed at the base with coaxial cable. The center conductor of the coax is connected to the antenna element, and the shield is connected to the ground rod at the base of the structure. You are encouraged to see chapter 10, which deals with limited space and hidden antennas, for a different version of this antenna. It is possible to install the wire (or multiple wires of different lengths) inside of a length of PVC plumbing pipe. The pipe serves as the support structure, and the conductors are placed inside.

Different lengths of conductor are required for different bands of operation. You can calculate the length (in feet) required for quarter wavelength vertical antennas from: $L_{ft} = 246/F_{MHz}$. Figure 13-9A shows how several bands are accommodated from the same feedline on the same support structure. In this particular case eight different antenna elements are supported from the same "tee-bar." Be sure to insulate them from each other, and from the support structure. Again, PVC piping can be used for the support. Another method for accommodating several different bands is to tie the upper ends of the wires to a sloping rope (as in Fig. 13-9B).

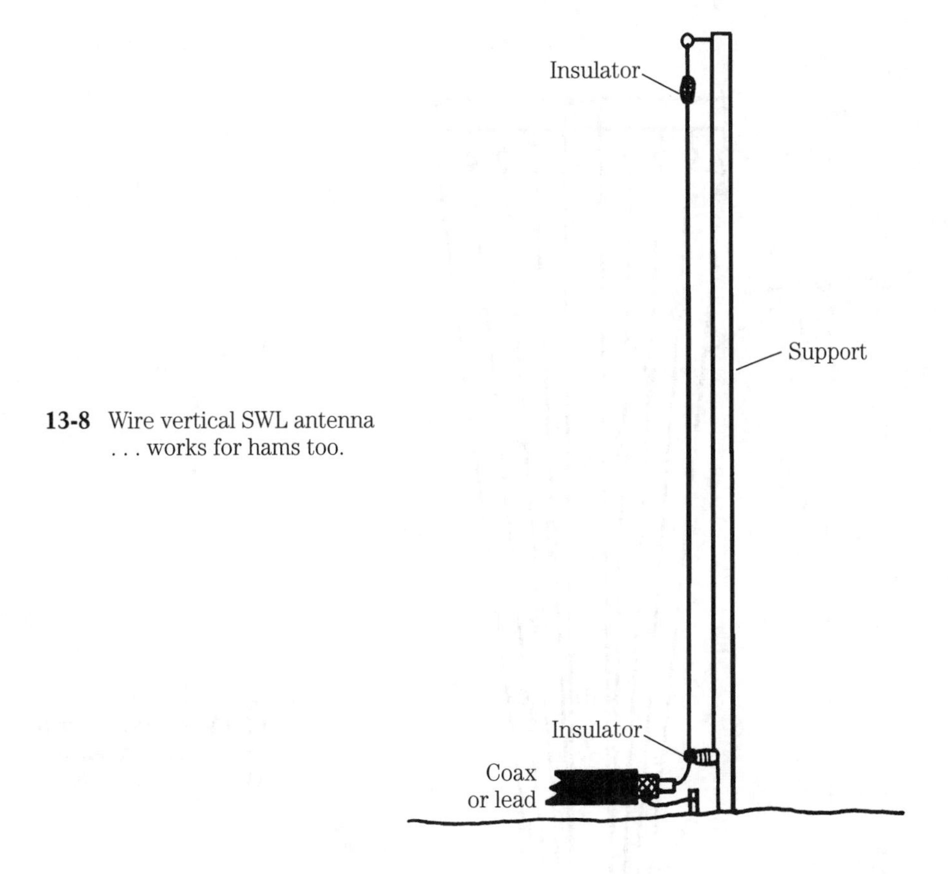

13-8 Wire vertical SWL antenna . . . works for hams too.

Wire directional antenna

A directional antenna has the ability to enhance reception of desired signals, while rejecting undesired signals arriving from slightly different directions. Although directivity normally means a beam antenna, or at least a rotatable dipole, there are certain types of antenna that allow fixed antennas to be both directive and variable. See chapter 7 for fixed, but variable directional antennas, and chapter 11 for fixed and not-variable directive arrays. Those antennas are transmitting antennas, but they work equally well for reception. This section shows a crude, but often effective, directional antenna that allows one to select the direction of reception with pin plugs or switches.

Consider Fig. 13-10. In this case, a number of quarter-wavelength radiators are fanned out from a common feedpoint at various angles from the building. At the near end of each element is a female banana jack. A pair of balanced feedlines from the receiver (300-Ω twin lead, or similar) are brought to the area where the antenna elements terminate. Each wire in the twin lead has a banana plug attached. By selecting which banana jack is plugged into which banana plug, you can select the directional pattern of the antenna. If the receiver is equipped with a balanced an-

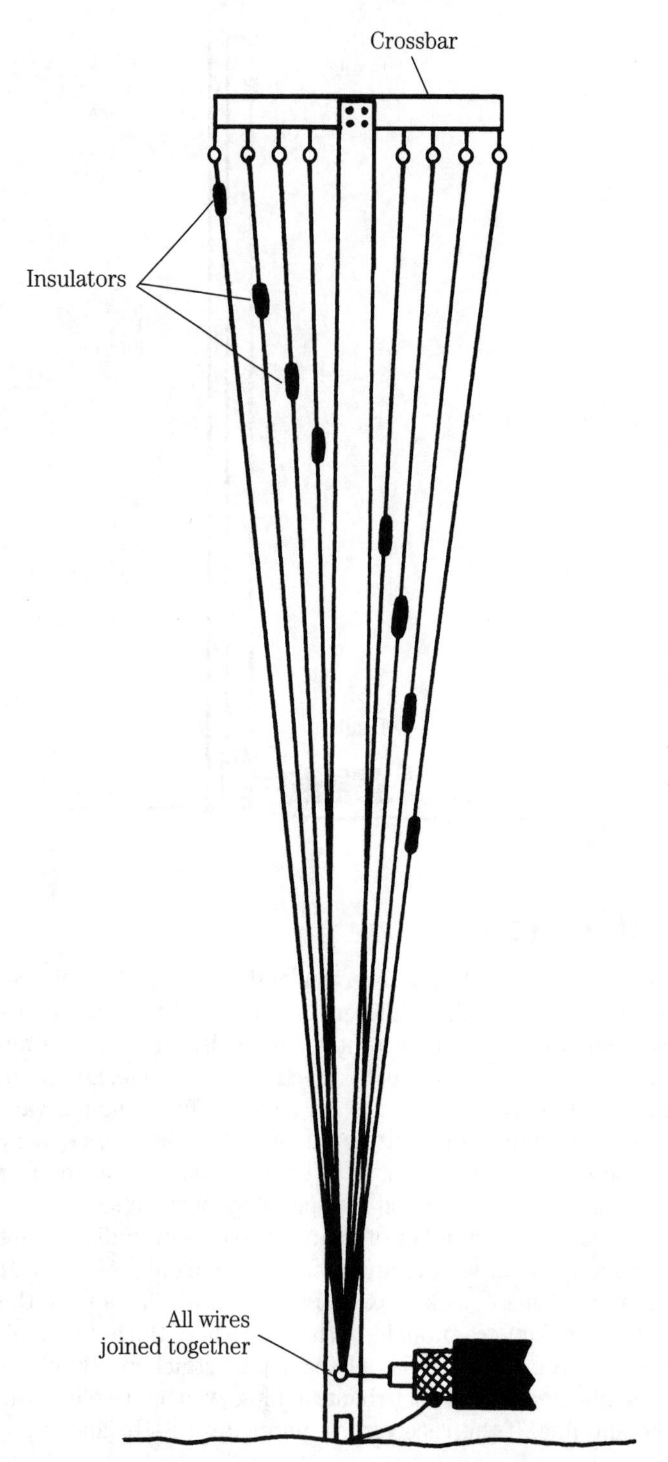

13-9A Multi-band wire vertical SWL antenna: A) T-bone style.

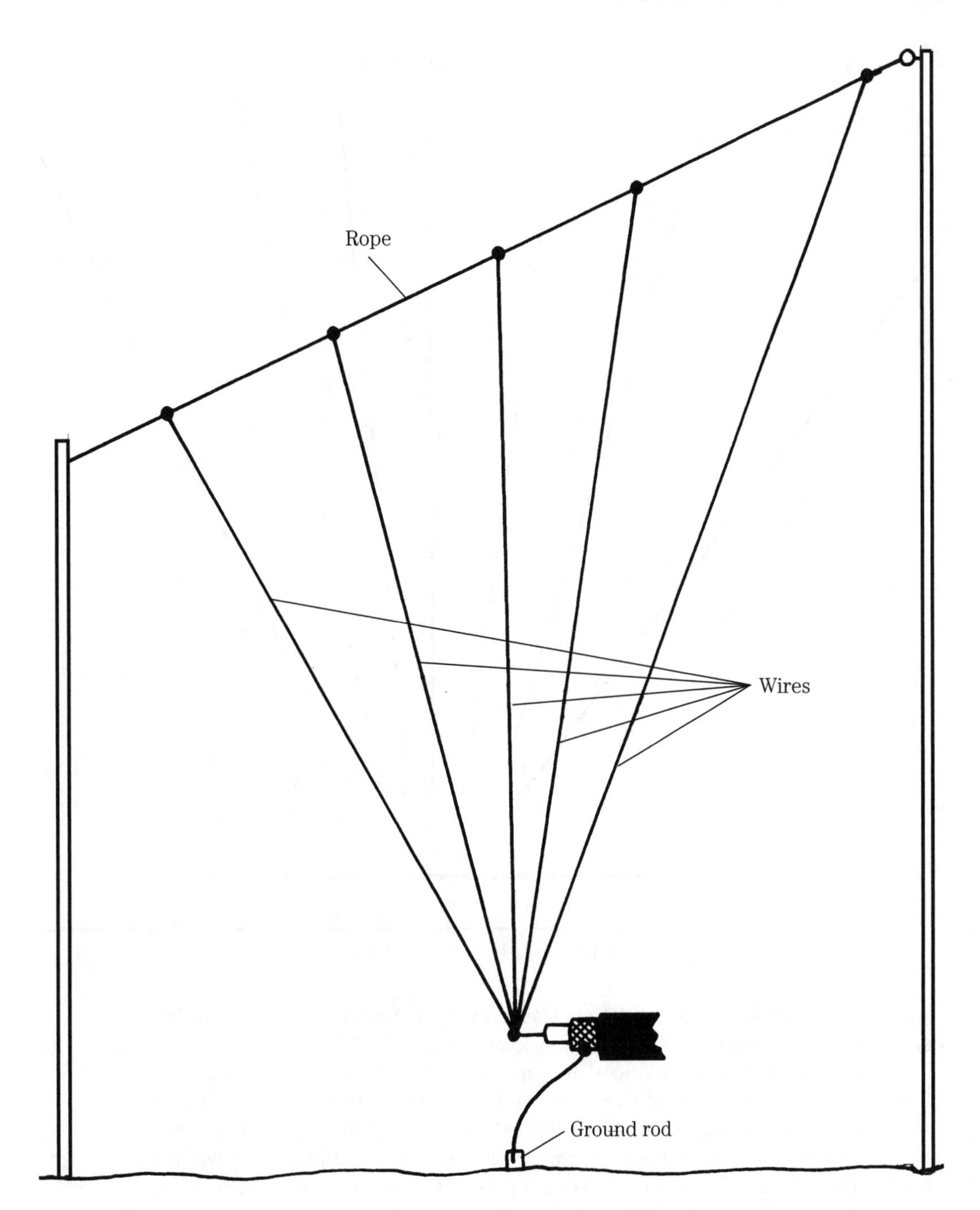

13-9B Multi-band wire vertical SWL antenna: B) fan style.

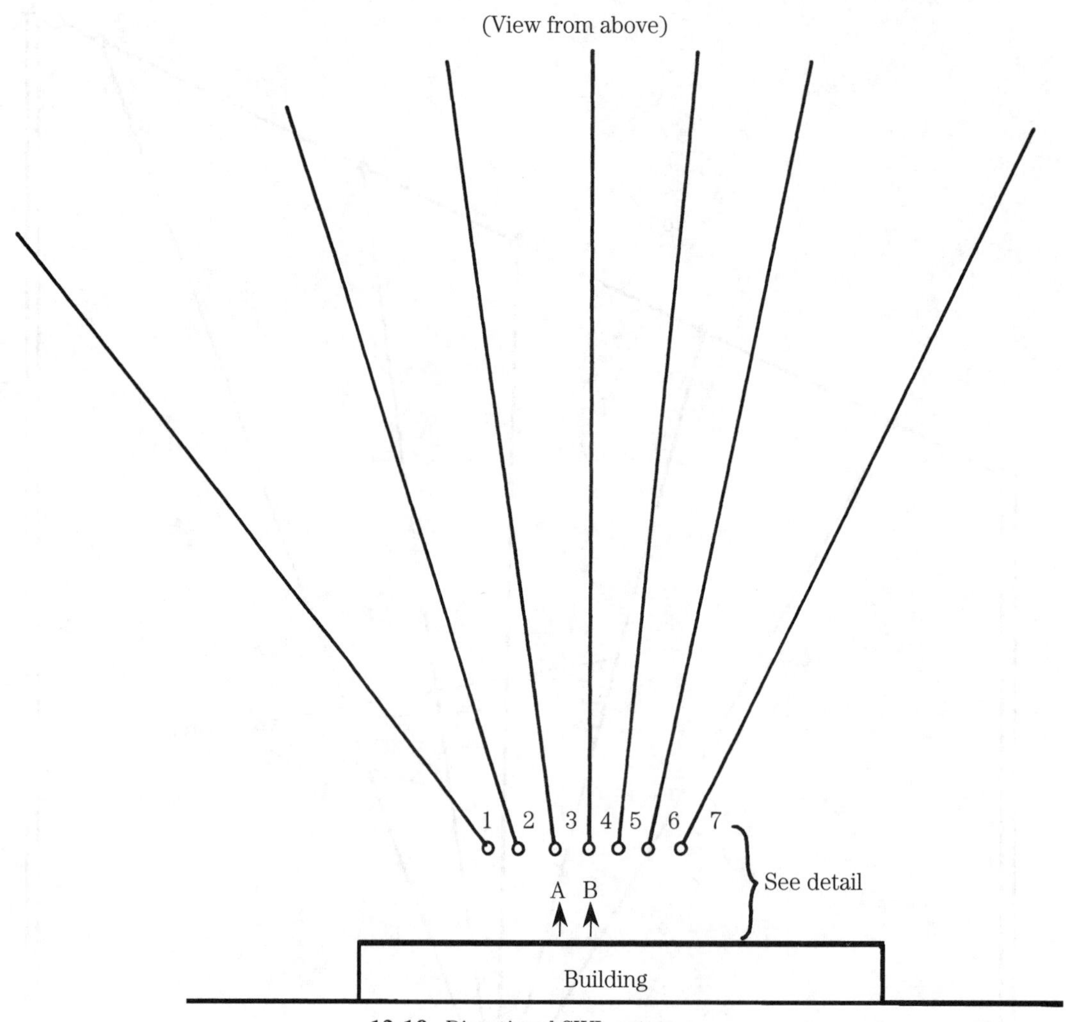

13-10 Directional SWL antenna.

tenna input, then simply connect the other end of the twin lead directly to the receiver. Otherwise, use one of the couplers shown in Fig. 13-11.

Figure 13-11A shows a balanced antenna coupler that is tuned to the frequency of reception. The coil is tuned to resonance by the interaction of the inductor and the capacitor. Antenna impedance is matched by selecting the taps on the inductor to which the feedline is attached. A simple RF broadband coupler is shown in Fig. 13-11B. This transformer is wound over a ferrite core, and consists of 12 to 24 turns of #26 enameled wire, with more turns being used for lower frequencies, and fewer for higher frequencies. Experiment with the number of turns in order to determine the correct value. Alternatively, use a 1:1 BALUN transformer instead of Fig. 13-11B; the type intended for amateur radio antennas is overkill powerwise, but it will work nicely.

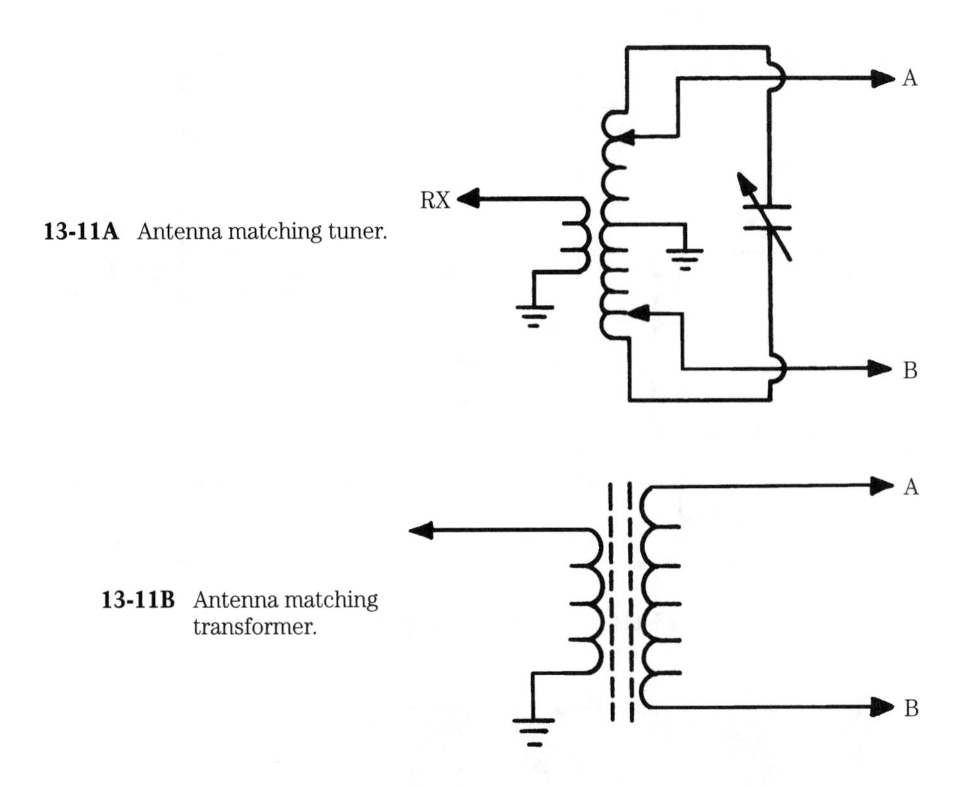

13-11A Antenna matching tuner.

13-11B Antenna matching transformer.

The antenna of Fig. 13-10 works by phasing the elements so as to null, or enhance, (as needed) certain directions. This operation becomes a little more flexible if you build a phasing transformer, as shown in Fig. 13-11C and 13-11D. Windings L1, L2, and L3 are wound "trifilar" style onto a ferrite core. Use 14 turns of #26 enameled wire for each winding. The idea in this circuit is to feed one element from coil L2 in the same way all of the time. This port becomes the 0° phase reference. The other port, B, is fed from a reversible winding, so it can either be in-phase, or 180° out of phase, with port A. Adjust the DPDT switch and the banana plugs of Fig. 13-10 for the best reception.

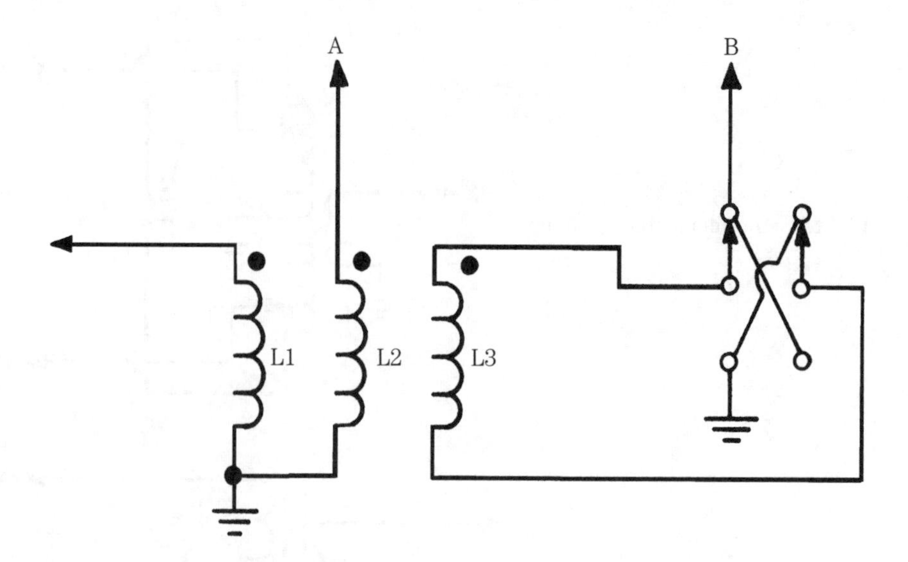

13-11C Phase switching antenna matcher.

13-11D Phasing box permits phasing antennas 0/180 degrees.

14 CHAPTER

Large wire loop antennas

THERE ARE TWO BASIC FORMS OF LOOP ANTENNAS: SMALL AND LARGE. THESE TWO types have different characteristics, work according to different principles, and have different purposes. Small loops are those in which the current flowing in the wire has the same phase and amplitude at every point in the loop (which fact implies a very short wire length, i.e., $<0.1\lambda$). Such loops respond to the magnetic field component of the electromagnetic radio wave. A large loop antenna has a wire length $>0.1\lambda$, with most being either $\lambda/2$, 1λ, or 2λ. The current in a large loop varies along the length of the wire in a manner similar to other wire antennas.

λ/2 large loops

The performance of large wire loop antennas depends in part on their size. Figure 14-1 shows a half-wavelength loop (i.e., one in which the four sides are each $\lambda/8$ long). There are two basic configurations for this antenna: *continuous* (S1 closed) and *open* (S1 open). In both cases, the feedpoint is at the midpoint of the side opposite the switch.

The direction of the main reception, or radiation lobe (i.e., the direction of maximum reception), depends on whether S1 is open or closed. With S1 closed, the main lobe is to the right (solid arrow); and with S1 open, it is to the left (broken arrow). Direction reversal can be achieved by using a switch (or relay) at S1, although some people opt for unidirectional operation by eliminating S1, and leaving the loop either open or closed.

The feedpoint impedance is considerably different in the two configurations. In the closed-loop situation (i.e., S1 closed), the antenna can be modeled as if it were a half-wavelength dipole bent into a square and fed at the ends. The feedpoint (X1–X2) impedance is on the order of 3 kΩ because it occurs at a voltage antinode (current

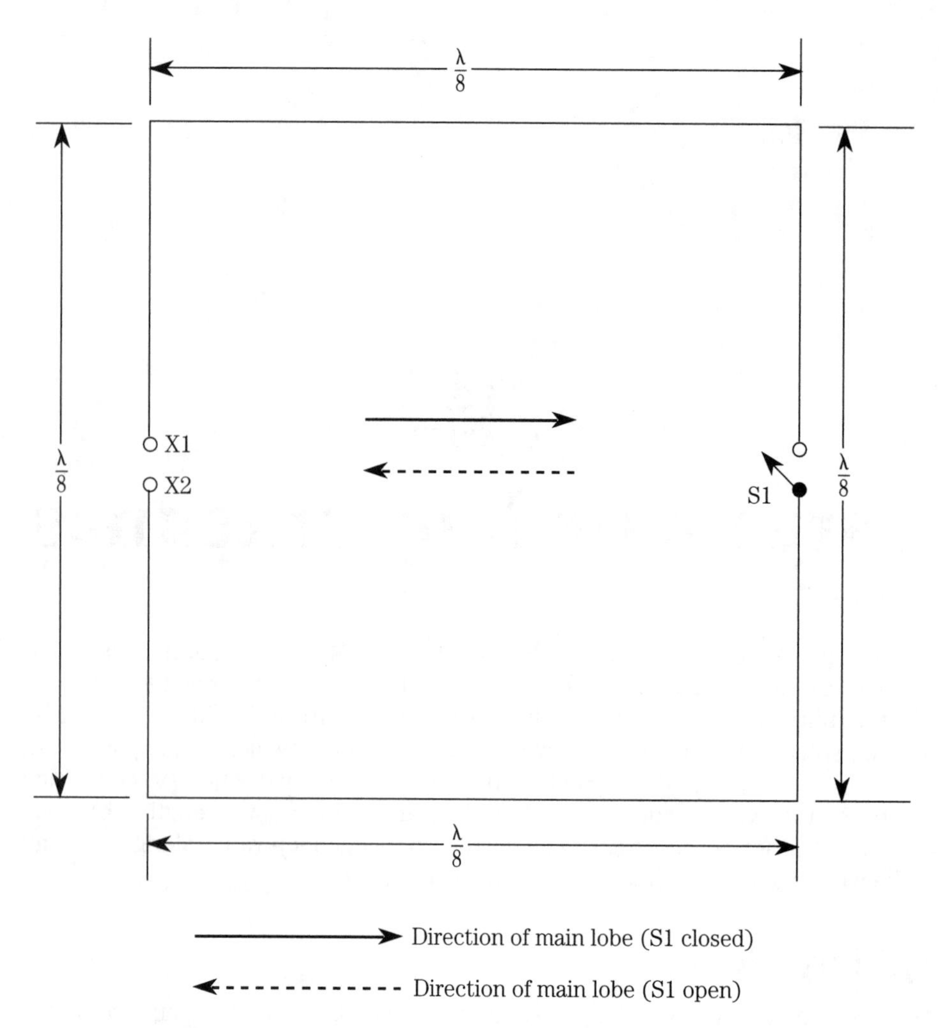

14-1 Half-wave square loop antenna with reversal switch.

node). The current antinode (i.e., I_{max}) is at S1, on the side opposite the feedpoint. An antenna tuning unit (ATU), or RF impedance transformer, must be used to match the lower impedance of the transmission lines needed to connect to receivers.

The feedpoint impedance of the open loop configuration (S1 open) is low because the current antinode occurs at X1–X2. Some texts list the impedance as "about 50 Ω," but my own measurements on several test loops were somewhat higher (about 70 Ω). In either case, the open loop is a reasonable match for either 52-Ω or 75-Ω coaxial cable.

Neither λ/2 loop configuration shows gain over a dipole. The figure usually quoted as –1 dB forward gain (i.e., a loss compared with a dipole), and about 6 dB of front-to-back ratio (FBR). Such low values of FBR indicate that there is no deep notch ("null") in the pattern.

A lossy antenna with a low FBR seems like a born loser, and in most cases it is.

But the λ/2 loop finds a niche where size must be constrained, for one reason or another. In those cases, the λ/2 loop can be an alternative. These antennas can be considered limited-space designs, and can be mounted in an attic, or other limited-access place, as appropriate.

A simple trick will change the gain, as well as the direction of radiation, of the closed version of the λ/2 loop. In Fig. 14-2, a pair of inductors (L1 and L2) are inserted into the circuit at the midpoints of the sides adjacent to the side containing the feedpoints. These inductors should have an inductive reactance (X_L) of about 360 Ω in the center of the band of operation. The inductance of the coil is:

$$L_{\mu H} = \frac{3.6 \times 10^8}{2\pi F_{Hz}} \quad [14.1]$$

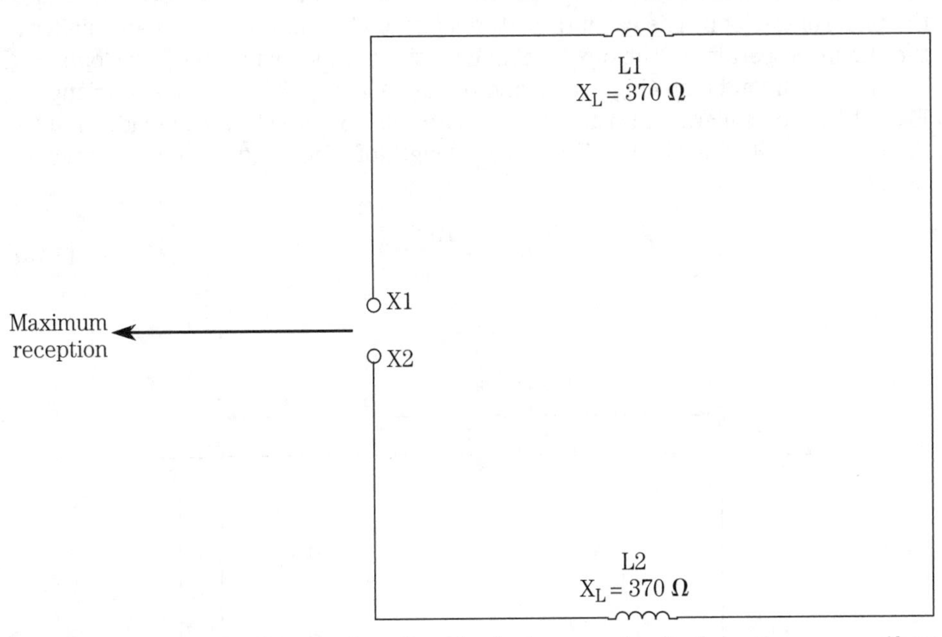

14-2 Use of inductive loading to reduce the side of antenna, and make the pattern more uniform.

Where:

$L_{\mu H}$ is the coil inductance in microhenrys (μH)
F_{Hz} is the mid-band frequency in hertz (Hz)

Example Find the inductance for the coils in a loaded half-wavelength closed-loop antenna that must operate in a band centered on 10.125 MHz.

Solution:

[Note: 10.125 MHz = 10,125,000 MHz]

$$L_{\mu H} = \frac{3.6 \times 10^8}{(2)(\pi)(10{,}125{,}000\ \text{Hz})} = 5.7\ \mu\text{H}$$

The coils force the current antinodes toward the feedpoint, reversing the direction of the main lobe, and creating a gain of about +1 dB over a half-wavelength dipole.

The currents flowing in the antenna can be quite high, so when making the coils, be sure to use a size that is sufficient for the power and current levels anticipated. The 2- to 3-inch B&W/*Air-Dux* style coils are sufficient for most amateur radio use. Smaller coils are available on the market, but their use is limited to low-power situations.

1λ large loops

If size is not forcing you to a λ/2 loop, then a 1λ loop might be just the ticket. It produces a gain of about +2 dB over a dipole in the directions that are perpendicular to the plane of the loop. The azimuth patterns formed by these antennas are similar to the "figure-8" pattern of the dipole. Three versions are shown: the square loop (Fig. 14-3), the diamond loop (Fig. 14-4), and the delta loop (a.k.a. D-loop and triangle—Fig. 14-5). The square and diamond loops are built with λ/4 on each side, and the delta loop is λ/3 on each side. The overall length of wire needed to build these antennas is:

$$L_{\text{feet}} = \frac{1005}{F_{\text{MHz}}} \quad [14.2]$$

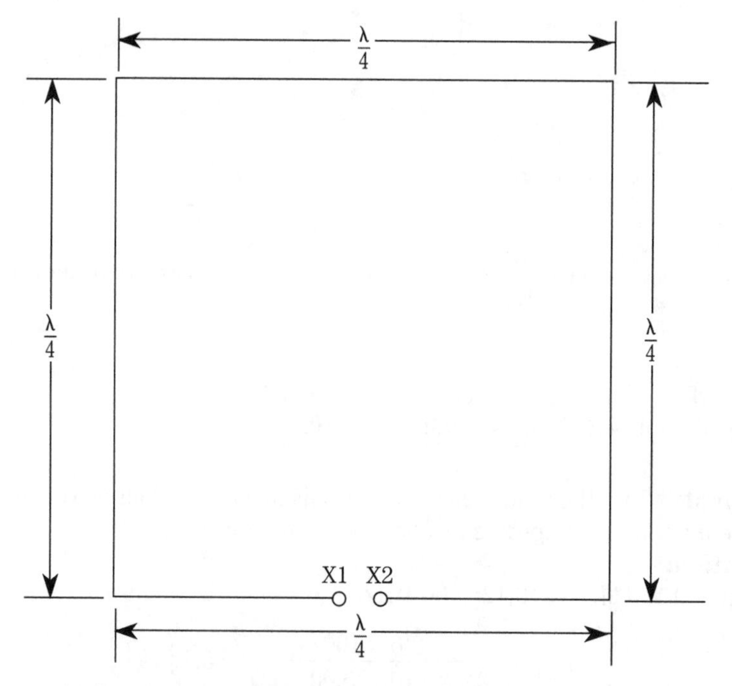

14-3 Quarter-wavelength square loop (single-element quad).

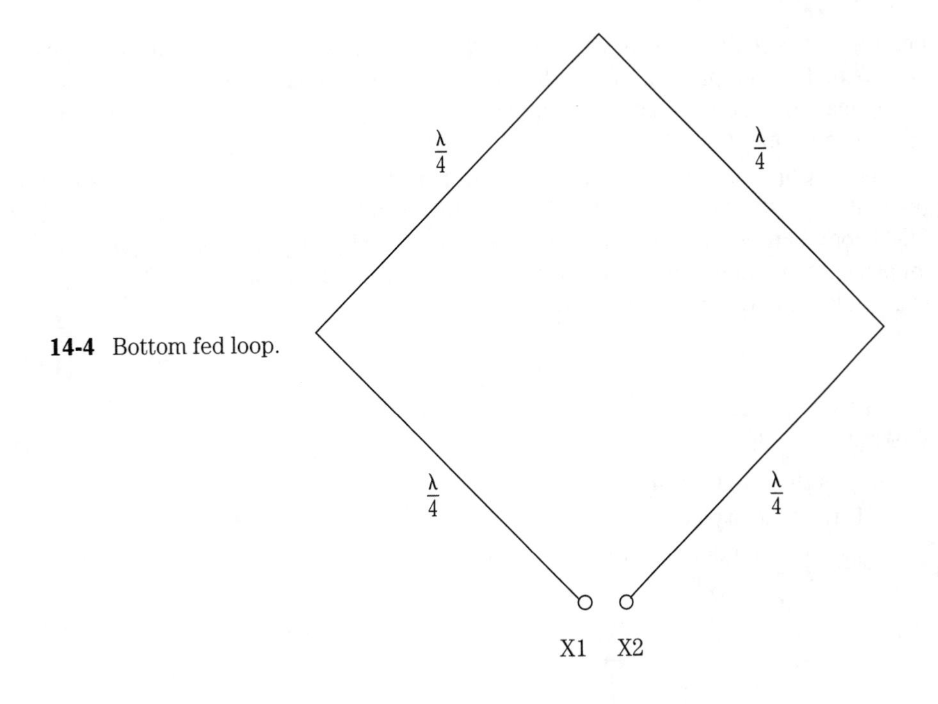

14-4 Bottom fed loop.

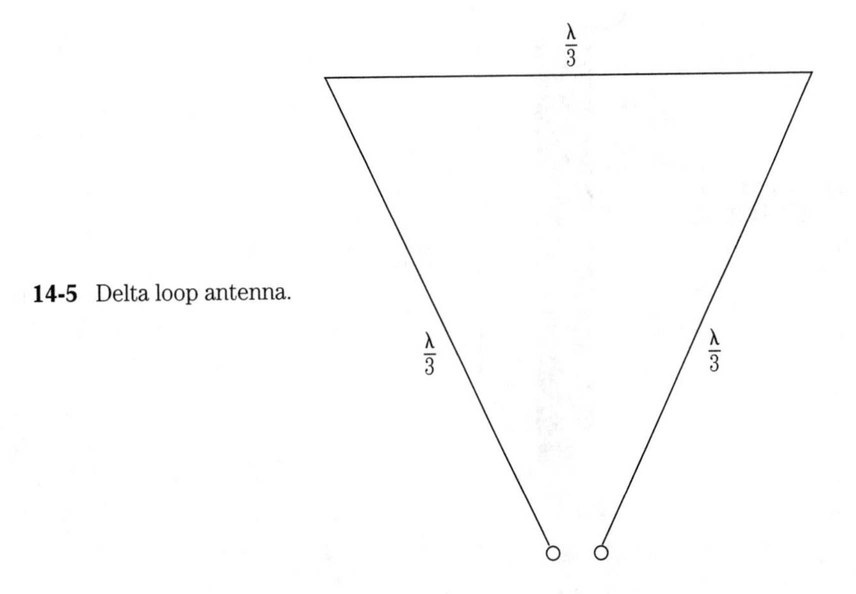

14-5 Delta loop antenna.

The polarization of the three loop antennas is horizontal, because of the location of the feedpoints. On the square loop, moving the feedpoint to the middle of either vertical side will provide vertical polarization. Similarly, on the diamond loop vertical polarization is realized by moving the feedpoint to either of the two adjacent apexes. On the delta loop, placing the feedpoint at either of the two other apexes produces a diagonal polarization that offers approximately equal vertical and horizontal polarization components.

The feedpoint impedance of the 1λ loop is around 100 W, so it provides a slight mismatch to 75-Ω coax, and a 2:1 mismatch to 52-Ω coax. A very good match to 52-Ω coax can be produced using the scheme of Fig. 14-6. Here, a quarter-wavelength coaxial cable matching section is made of 75-Ω coaxial cable. The length of this cable should be:

$$L_{feet} = \frac{246\,V}{F_{MHz}} \quad [14.3]$$

Where:

L_{feet} is the length in feet (ft)
V is the velocity factor of the coax
F_{MHz} is the frequency in megahertz (MHz)

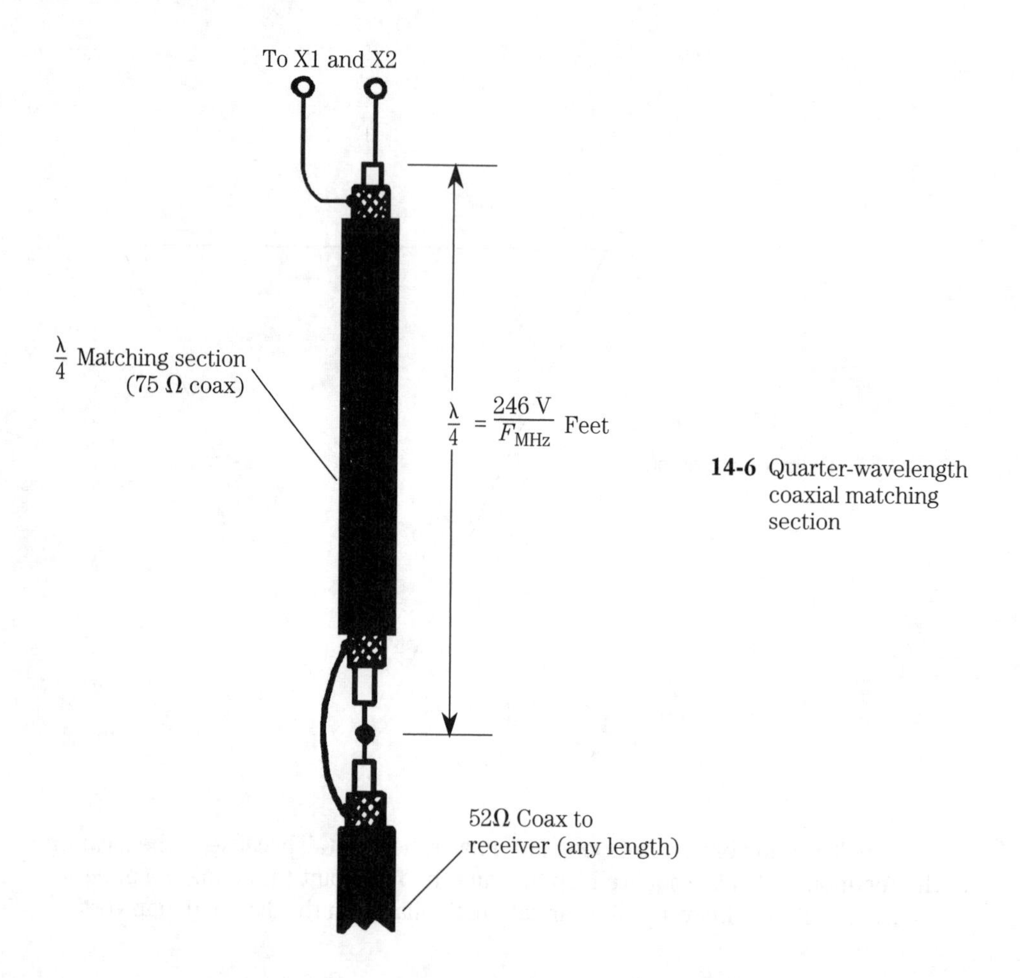

14-6 Quarter-wavelength coaxial matching section

The impedance (Z_o) of the cable used for the matching section should be:

$$Z_o = \sqrt{Z_L Z_s} \qquad \textbf{[14.4]}$$

Where:

Z_o is the characteristic impedance of the coax used in the matching section in ohms.
Z_L is the feedpoint impedance of the antenna in ohms
Z_s is the source impedance (i.e., the 52-Ω characteristic impedance of the line to the receiver in standard systems)

When Eq. 14.4 is applied to this system, where $Z_s = 52\ \Omega$, $Z_L = 100\ \Omega$:

$$Z_o = \sqrt{(100\Omega)(52\Omega)} = 72\Omega \qquad \textbf{[14.5]}$$

This is a very good match to 75-Ω coaxial cable.

Demi-quad loop antenna

The *demi-quad* is a single-element 1λ quad antenna. The length of the antenna is, like the cubical quad beam antenna (see chapter 12), one wavelength. Figure 14-7 shows a type of demi-quad based on the tee-cross type of mast.

The impedance matching section is a quarter-wavelength piece of 75-Ω coaxial cable (RG-58/U or RF-11/U). The length of the matching section is determined from:

$$L = \frac{246\ V}{F_{\text{MHz}}} \qquad \textbf{[14.6]}$$

Where:

L is the overall length in feet
F_{MHz} is the frequency in megahertz (MHz)
V is the velocity factor of the coaxial cable (typically 0.66, 0.70, or 0.80.

Delta loop

The *delta loop antenna*, like the Greek letter from which it draws its name, is triangle shaped (Fig. 14-8). The delta loop is a full wavelength, with elements approximately 2% longer than the natural wavelength (like the quad). The actual length will be a function of the proximity, and nature, of the underlying ground, so some experimentation is necessary. The approximate preadjustment lengths of the sides are found from:

$$L1 = \frac{437}{F_{\text{MHz}}}\ \text{feet} \qquad \textbf{[14.7]}$$

$$L2 = L3 = \frac{296}{F_{\text{MHz}}} \qquad \textbf{[14.8]}$$

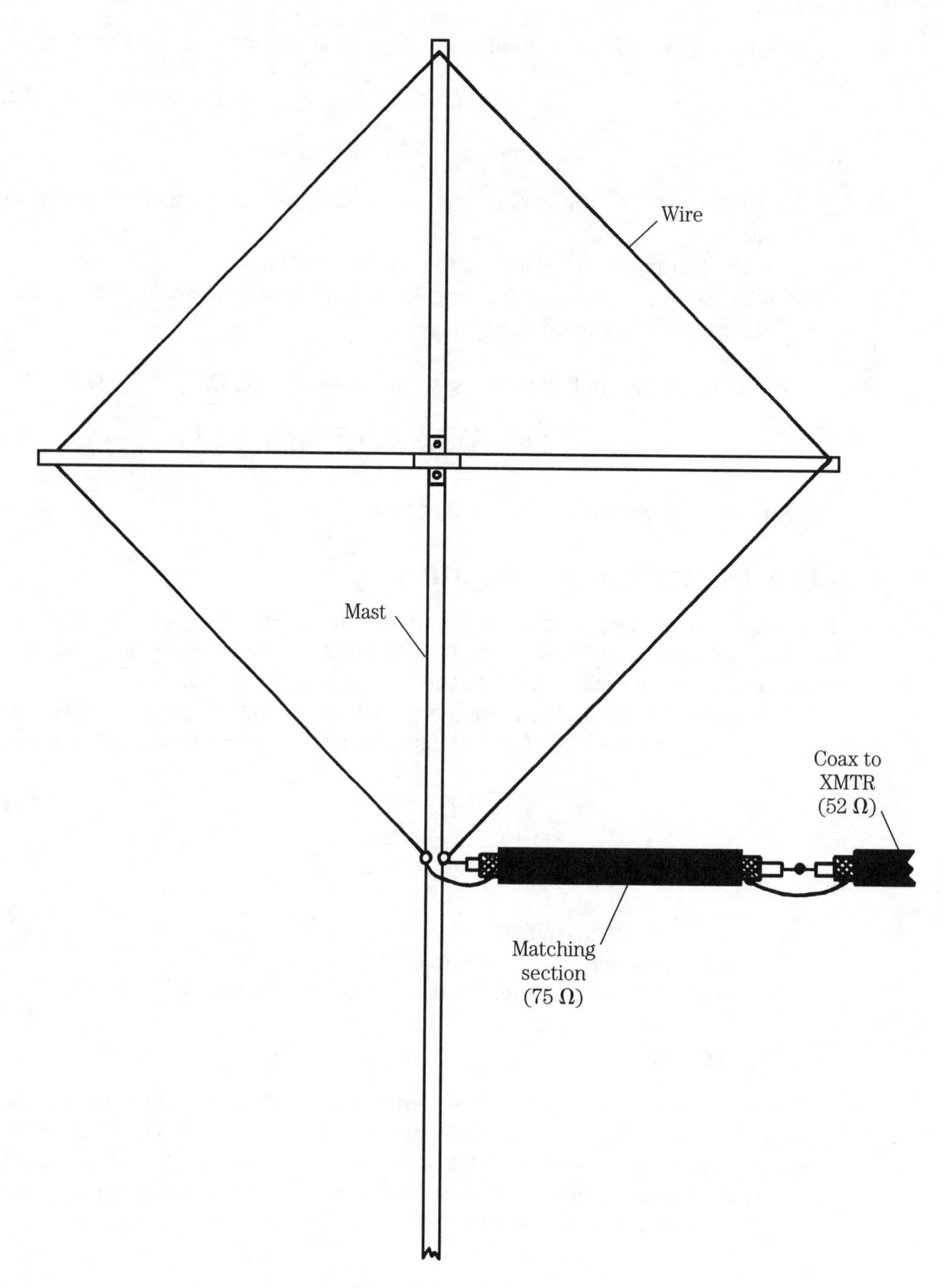

14-7 Demi-quad antenna.

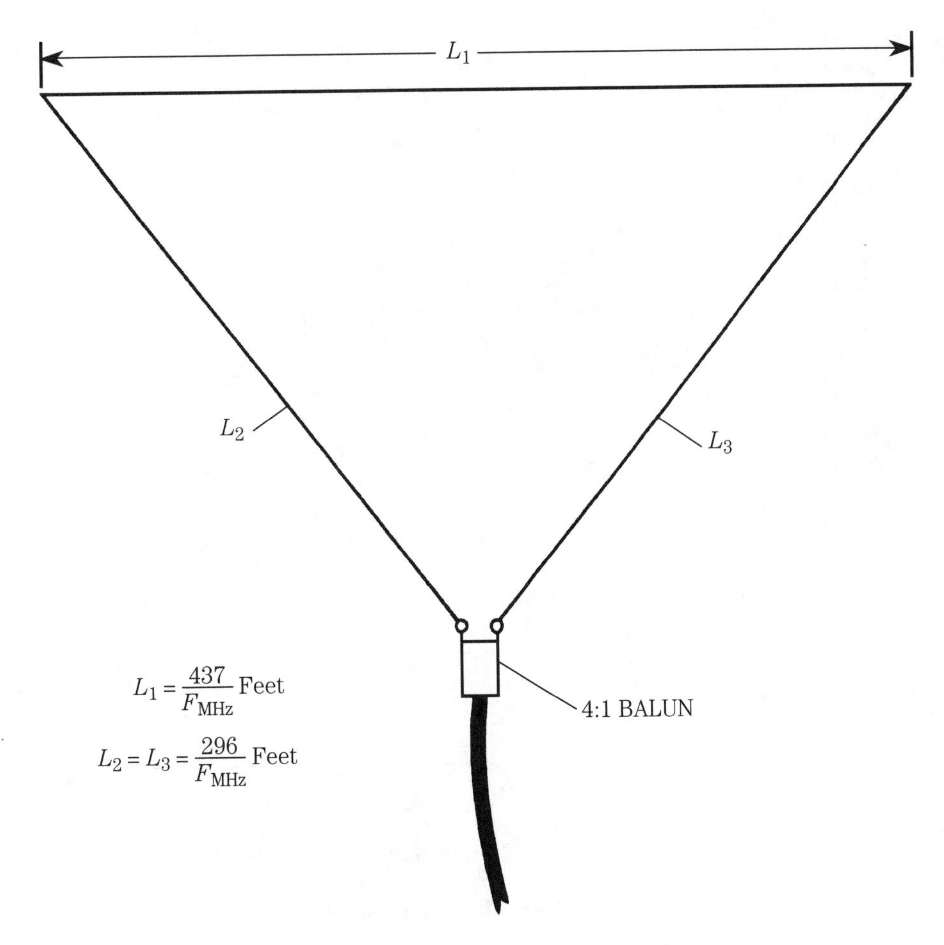

14-8 Delta loop antenna.

The delta loop antenna is fed from 52-Ω coaxial cable through a 4:1 BALUN transformer. The delta loop can be built in a fixed location, and will offer a bidirectional pattern.

Half-delta sloper (HDS)

The *half-delta sloper* (HDS) antenna (Fig. 14-9) is similar to the full delta loop, except that (like the quarter-wavelength vertical) half of the antenna is in the form of an "image" in the ground. Gains of 1.5 to 2 dB are achievable. The HDS antenna consists of two elements: a $\lambda/3$ wavelength sloping wire and a $\lambda/6$ vertical wire (on an insulated mast), or a $\lambda/6$ metal mast. Because the ground currents are very important, much like the vertical antenna, either an extensive radial system at both ends is needed, or a base ground return wire (buried) must be provided.

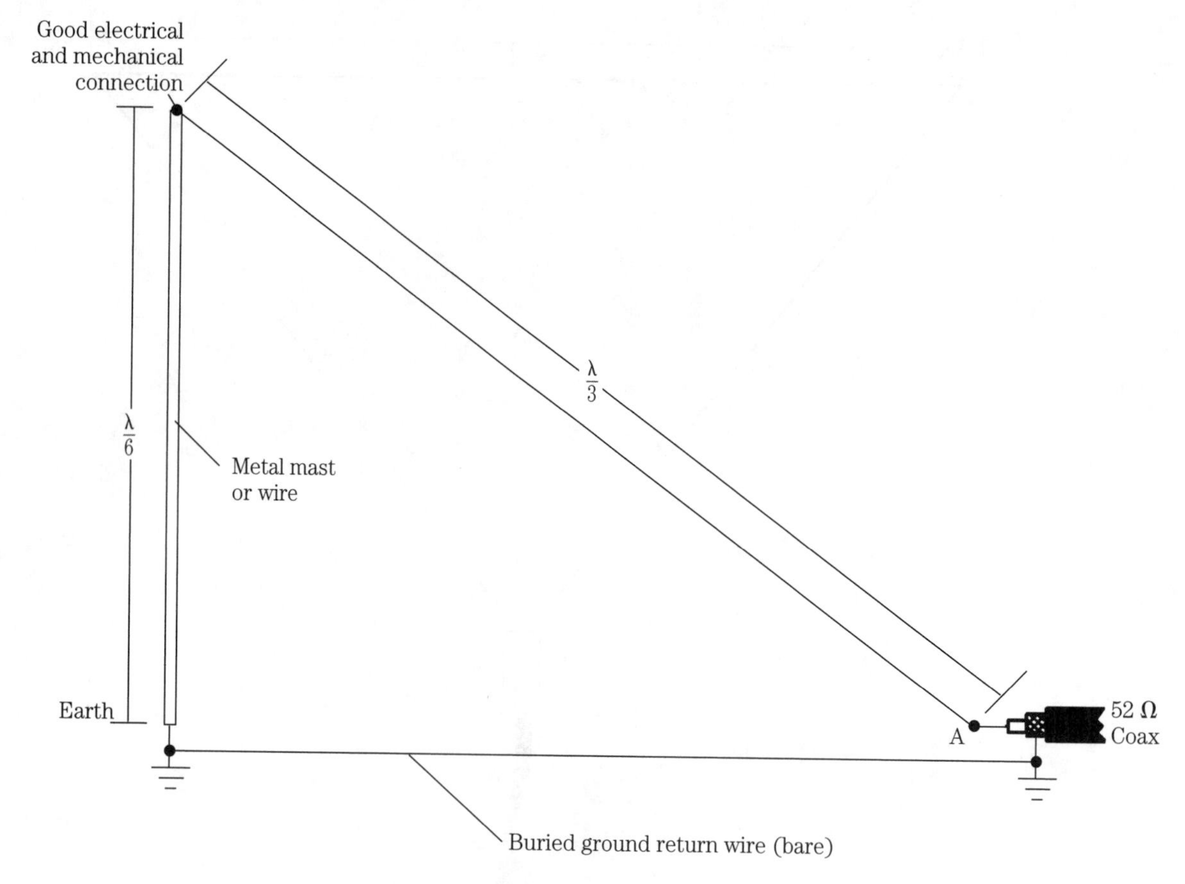

14-9 Half-delta loop antenna.

The HDS will work on its design frequency, plus harmonics of the design frequency. For a fundamental frequency of 5 MHz, a vertical segment of 33 feet and a sloping section of 66 feet is needed. The lengths for any frequency are found from:

$$\frac{\lambda}{3} = \frac{328}{F_{\text{MHz}}} \qquad \textbf{[14.9]}$$

and,

$$\frac{\lambda}{6} = \frac{164}{F_{\text{MHz}}} \qquad \textbf{[14.10]}$$

The HDS is fed at one corner, close to the ground. If only the fundamental frequency is desired, then you can feed it with 52-Ω coaxial cable. But at harmonics, the feedpoint impedance changes to as high as 1000 Ω. If harmonic operation is intended, then an antenna tuning unit (ATU) is needed at point "A" to match these impedances.

Bisquare loop antenna

The *bisquare antenna*, shown in Fig. 14-10, is similar to the other large loops, except that it is λ/2 on each side, making a total wire length of two wavelengths. This antenna is built like the diamond loop shown earlier (i.e., it is a large square loop fed at an apex that is set at the bottom of the assembly). In this case, the loop is fed either with an antenna tuning unit (to match a 1000-Ω impedance) or a quarter-wavelength matching section made of 300-Ω or 450-Ω twin-lead transmission line. A 1:1 BALUN transformer connects the 75-Ω coaxial cable to the matching section.

The bisquare antenna offers as much as 4 dB gain broadside to the plane of the antenna (i.e., in and out of the book page), in a figure-8 pattern, on the design frequency. It is horizontally polarized. When the frequency drops to one-half of the design frequency, the gain drops to about 2 dB, and the antenna works similarly to the diamond loop covered previously.

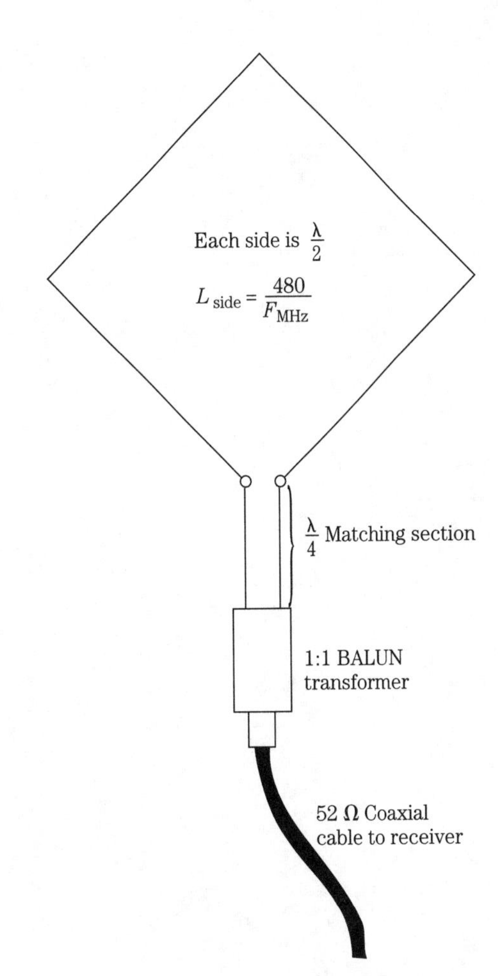

14-10 Bisquare 2λ square loop antenna.

15
CHAPTER

Small transmitting and receiving loops

AN ANTENNA STYLE THAT FINDS SOME USE, ESPECIALLY ON LOW FREQUENCY BANDS, IS the small loop. These antennas are fundamentally different from the large loop antennas covered in chapter 14. Small loops are those with an overall length that is $<0.18\lambda$, although some authorities say $<0.1\lambda$. Large loop antennas are sensitive primarily to the electric field of the electromagnetic radio wave, but small loop antennas are primarily sensitive to the magnetic field of the EM wave.

There are several general uses for small loop antennas. On transmit, they serve as a compact—although somewhat inefficient—antenna for both MW and HF frequencies. A square loop that is five or six feet to the side can be used on 75/80 meters for both transmitting and receiving. Although no one claims that the loop will work as well as a full-sized vertical or dipole, the fact is that they work very well for tightly limited spaces, such as the attic of a townhouse or apartment. These antennas are not nearly as efficient for transmitting (as are the dipoles and verticals), but they have the charming ability to get one on the air in circumstances where those other antennas are forbidden.

The best known small loop applications seem to be for receiving, although (as any amateur radio catalog will testify) they are used for transmitting as well. In the receiving capacity, the small loop antenna can be used in several different ways. First, it can be used as a *radio direction finder (RDF)*. The loop has distinct nulls that are useful for pinpointing the direction from which a signal arrives. Perform that operation from two separated locations, and the source of the signal will be at the point where the two tracks cross. Second, the small loop can be used to null out very strong interfering signals on the crowded low frequency bands. You will be surprised, when listening to 160-meters and 80-meters, how many DX stations are present when the kilowatt "loudenbooomer" trash clears up. In addition to this interference from other stations, there is also a lot of interference from man-made noise sources, and these tend to be directional, and thus can be nulled with the loop.

Form of the small loop

The small loop can be almost any regular shape—round, square, triangular, octagonal, hexagonal—but for the most part, the square loop is most often used. Perhaps its because of the ease of construction, but whatever the reason, the square small loop is very popular. Figure 15-1 shows the basic circuit and structure of the square small loop antenna. It consists of several turns of wire arranged in a square of side a, wound to a depth of b. A typical loop for the AM broadcast band through about 40-meters (7 MHz or so), will be on the order of 15 to 36 inches squared. Several loops that I built were 24 inches square. This size was selected because it is relatively easy to purchase wood for the frame stock in lengths of 24 inches, or integer multiples of 24 inches. For several receiving loops, I used 24" × 3" × 3⁄16" spruce purchased at

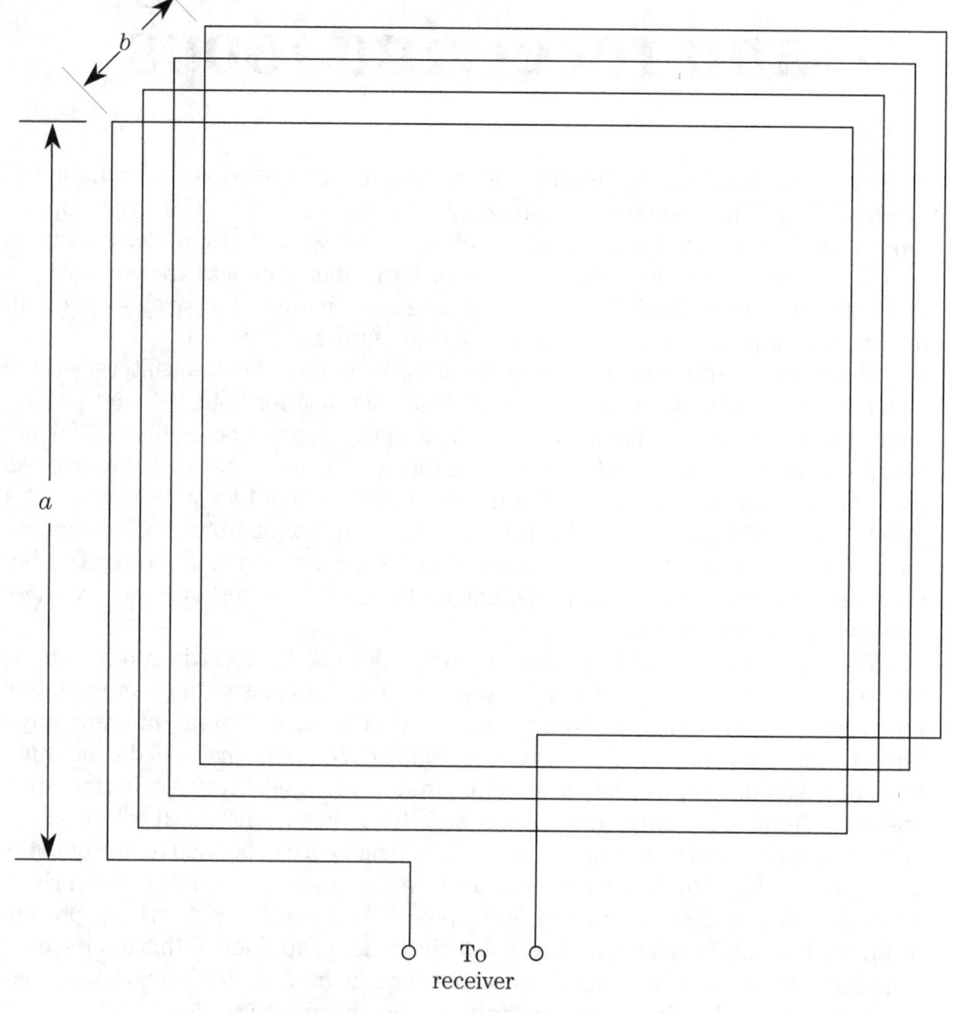

15-1 Small loop antenna (square configuration).

hobby and model-building shops. The spruce strips are usually sold in the same display as balsa wood strips.

It is remarkable how few turns of wire are needed to make a loop antenna work in the low frequency bands. On a 24-inch loop, I found decent AM BCB performance using 16 turns of wire. Even fewer turns are needed for 75-/80-meter loop antennas.

Small loop directivity

The directivity of the small loop is a figure-8 pattern, but its geometry is almost counterintuitive. Anyone who is familiar with dipole theory might suspect that the maximum response would be broadside to the plane of the loop, but that is incorrect. The loop directivity is 90° out from that direction! That is, the maxima of the figure-8 pattern are off the ends, and above and below, the loop (i.e., in the plane of the loop, while the nulls (i.e., minima) are broadside to the plane of the loop (Fig. 15-2). The null response can be as much as about –30 dB below the peak response of the maxima. Thus, tremendous improvement in signal-to-noise ratio is possible when an interfering station, or other signal source, is placed in the null.

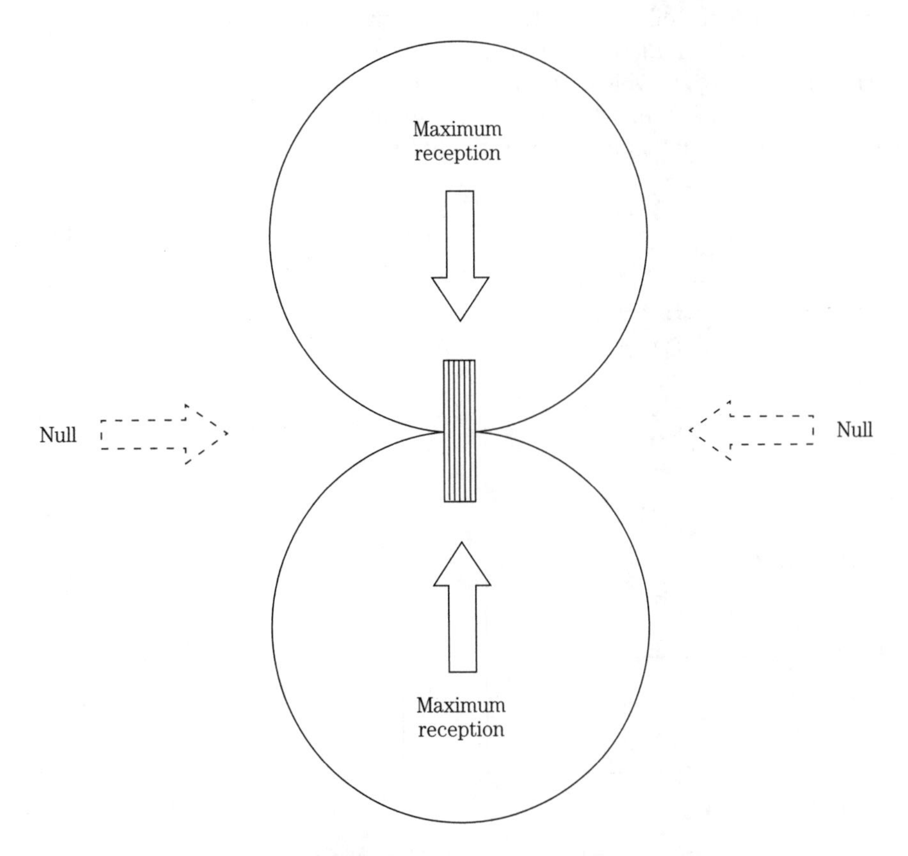

15-2 Small loop antenna pattern.

Small loop sensitivity

The small loop is sensitive to the magnetic field of the intercepted radio wave, not the electric field, as is the case with other antennas. When the magnetic field of the signal cuts across the turns of the loop, a very small current is induced in the conductors, and this gives rise to a very small signal voltage. Because this signal is so small, it is almost universally common practice to use either a tuned or untuned preamplifier between the output terminals of the receiving loop and the transmission line to the receiver. The preamplifier need not be very fancy, but it should have a gain of 15 to 30 dB, as needed.

Tuned small loop antennas

Another way to boost the signal level, available across the output terminals of the loop antenna, is to tune it to resonance on the frequency of the signal being received. A loop is an inductor, so it can be tuned to resonance using a capacitance (having a capacitive reactance that is the same value as the inductive reactance of the loop). The capacitance can be connected either in parallel (Fig. 15-3A), or in series (Fig. 15-3B), with the loop coil. In order to find out how much capacitance is needed, it is necessary to find the inductance of the loop coil.

The inductances of typical MW and HF loops are surprisingly high for so few turns of wire. Two different sets of equations are found in the technical literature for square-loop antenna inductance. Somerfield (1952) uses:

$$L = \frac{2\,a\mu_o n^2}{\pi}\,L_n\left(\frac{16\,a}{b}\right) \qquad [15.1]$$

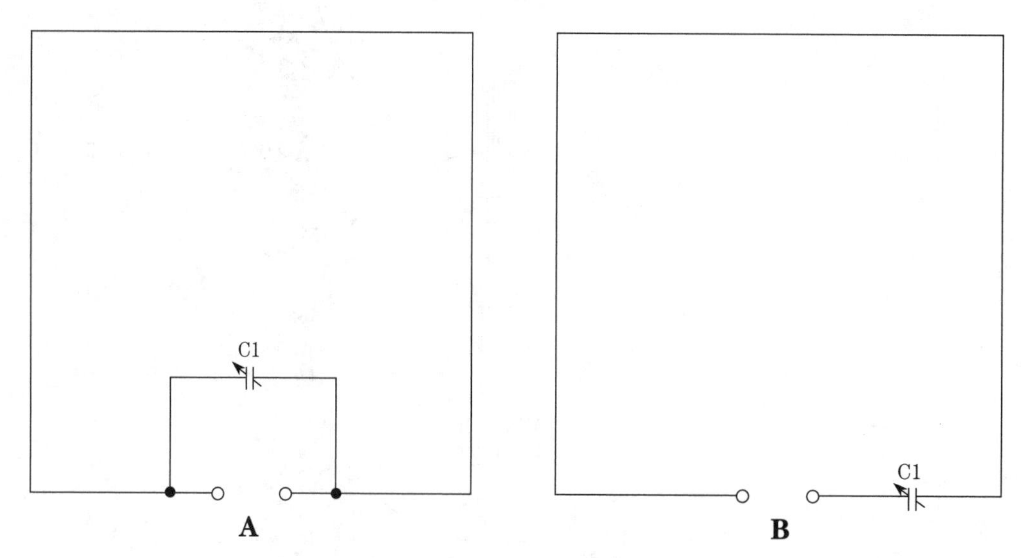

15-3 Resonating a small loop antenna: A) parallel version; B) series version.

Where:

$\mu_o = 4\pi \times 10^{-7}$
a, b, and n are parameters (see Eq. 15.2]
Turns are spaced . . . close-wound)

The Somerfield equations seem to work within an expected range of error, but Grover's equation (Grover, 1946) seems closer to the actual inductance measured in empirical tests:

$$L_{\mu H} = K_1 n^2 a \left(L_n \left(\frac{K_2 an}{(n+1)\,b} \right) + K_3 + \left(\frac{K_4\,(n+1)\,b}{an} \right) \right) \qquad \textbf{[15.2]}$$

Where:

$L_{\mu H}$ is the inductance in microhenrys (μH)
a is the length of a loop side in centimeters (cm)
b is the loop width in centimeters (cm)
n is the number of turns in the loop
K_1 through K_4 (shape constants) are given in Table 15-1
L_n is the natural log of this portion of the equation

Table 15-1. Shape constants

Shape	K_1	K_2	K_3	K_4
Triangle	0.006	1.155	0.655	0.135
Square	0.008	1.414	0.379	0.33
Hexagon	0.012	2.00	0.655	0.135
Octagon	0.016	2.613	0.7514	0.0715

This equation is a bit of a bear, but never fear, the BASIC program ("*Antlers*") will calculate it for you with ease. In fact, with that program, you can do iterative calculations to play mind games with a design, until you find one by computer cut-n-try that is practical for the given application.

People who are building a receive-only loop might wish to opt for using a voltage variable capacitance diode ("varactor") for tuning, rather than a regular capacitor. The reasons are two fold: first, air variable capacitors are hard to locate these days; and second, they are difficult to tune remotely (so you must be content with one channel, or spend a lot of time adjusting the capacitor). Figure 15-4 shows a means for using a varactor diode to tune the loop inductance to the correct frequency.

Transformer loop designs

The simple square small loop antenna can be connected directly to the transmission line, or to the preamplifier input. However, some people prefer to use a transformer coupling loop, such as shown in Fig. 15-5. In these antennas, the main loop will be exactly as described above, but it will have a one or two turn coupling loop in close

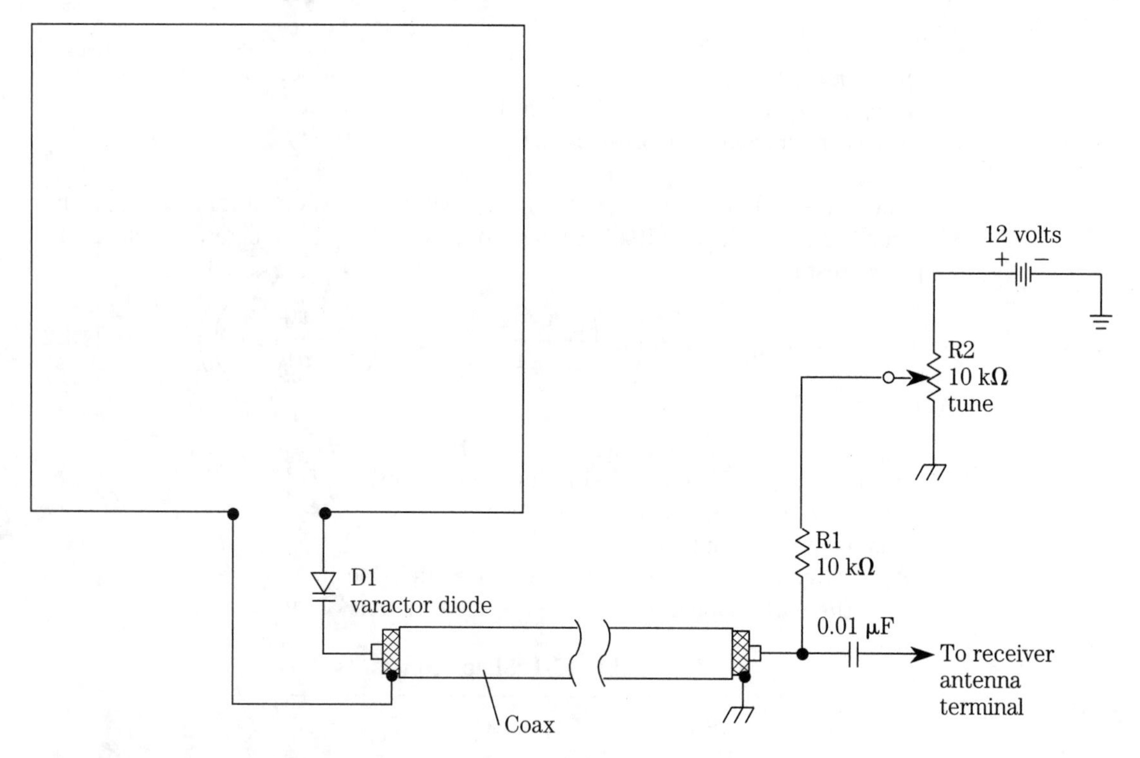

15-4 Remote tuning using a varactor diode instead of a capacitor. dc bias for the diode is fed through the coaxial cable.

proximity to the main loop. The main loop is often tuned to the operating frequency, and the coupling loop is usually untuned (although that is not a requirement, and both loops may be tuned).

Several implementations of the transmitting loop principle can be found in the amateur radio literature. Erwin (1991) discusses several variants on a design that is based on U.S. army work by K. Patterson—which was amateurized by Louis McCoy (W1ICP), Pat Hawker (G3VA), and Spenny (G6NA). An example of the loop is shown in Fig. 15-6, and a suitable 75-/80-meter coupling unit is shown in the inset to Fig. 15-6B. Alternatively, a coupling loop can be used (Fig. 15-7), along with its own special matching system.

Both Erwin (1991) and Koontz (1993) describe the use of small loop antennas in receiving arrays (i.e., two or more loop antennas used in conjunction with each other, as elements in a phased array). Koontz claims null depths of 25 to 45 dB on less than 100 feet of linear space.

Figure 15-8 shows the basic construction of a loop antenna based on Koontz (1993) and other sources. For low-frequency work, even down into the VLF bands, the antenna can be made of PVC plumbing pipe, with multiple turns of wire inside. One approach is to thread a multiconductor cable through the off-the-ground portion of the PVC pipe supports, and then cross-connect the wires to form a continuous loop.

For transmitting loops, the support portion can be made of PVC pipe, and the radiator element made of copper pipe cemented into the PVC supports.

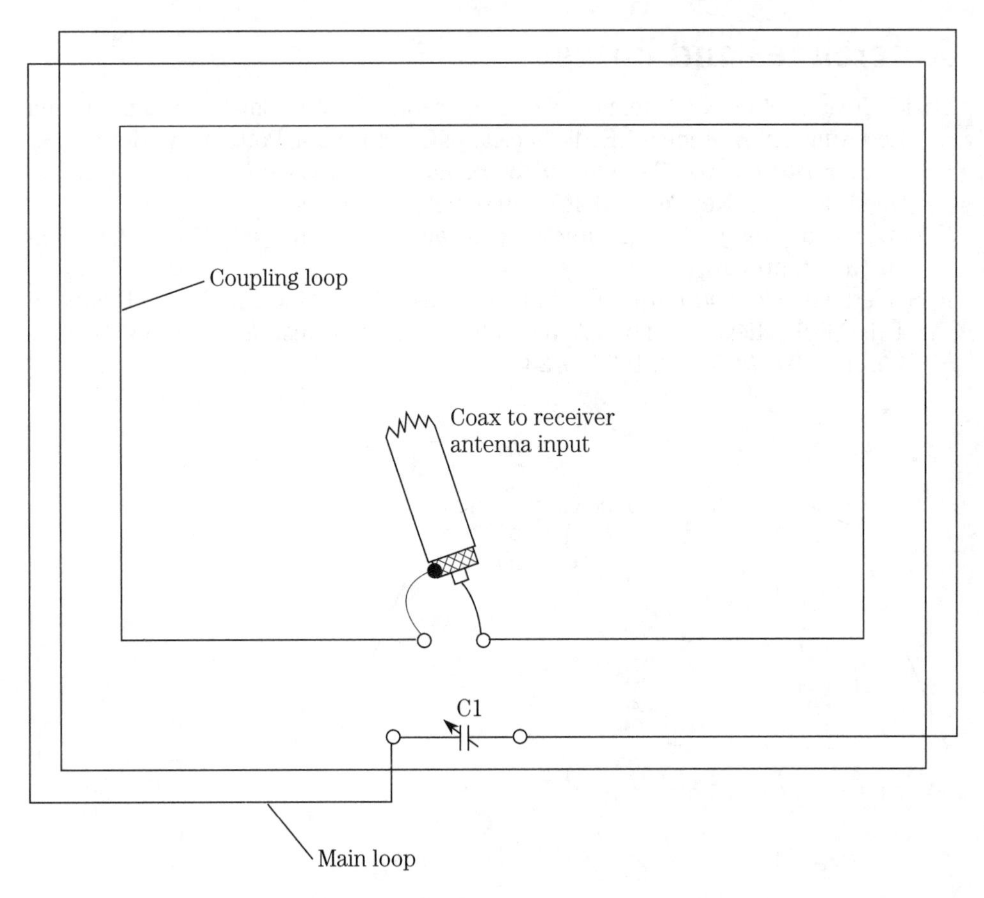

15-5 Use of a coupling loop with a small loop antenna.

When building transmitting loops, even when only low power is contemplated, be sure to use high-voltage "transmitting" variable capacitors for the matching units. Variable capacitors are needed because the transmitting loop is quite high-Q, and thus it tunes very narrowly. While this effect is of only marginal importance for receiving, it is critical for transmitting. The late Johnnie H. Thorne, K4NFU/5, was experimenting with 5-foot transmitting loops on 160 meters and 75/80 meters from his farmette near Austin, TX. When he showed me the set-up in the mid-1980s, the tuning capacitors were 1000-pF vacuum variables, with a motor unit on them, purchased from *Fair Radio Sales* in Lima, OH. The low voltage dc motor, that drove the capacitor's piston could be actuated from a small dc power supply inside the house.

Conclusion

Loop antennas solve some peculiar problems for low-frequency operators, especially those who have limited space, or those who are afflicted with tremendous amounts of QRN and QRM on the frequencies, where they plan to operate. And, for all of their usefulness, they are also quite simple to build and use.

References and notes

David, Erwin (1991), HF Antenna Collection, chapter 5, "Very Small Transmitting and Receiving Only Antennas," Radio Society of Great Britain, Potters Bar, Herts., UK.

F.W. Grover, *Inductance Calculation-Working Formulas and Tables*, D. VanNostrand Co., Inc. (New York, 1946). Cited in ARRL (op-cit).

Koontz, Floyd (1993), "A High-Directivity Receiving Antenna for 3.8 MHz," QST, August 1993, pp. 31-34.

A. Somerfield, *Electrodynamics*, Academic Press (New York, 1952), p.111; cited in T.H. O'Dell, "Resonant-Loop Antenna for Medium Waves," *Electronics World + Wireless World*, March 1992, p. 235.

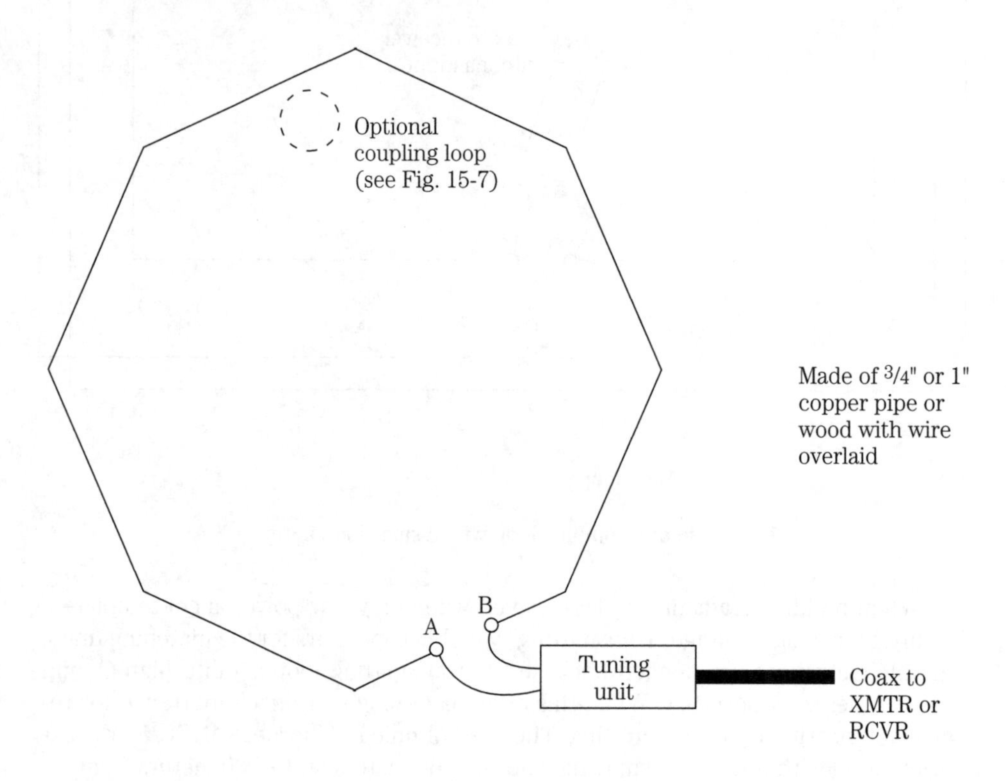

15-6A Octagonal small transmitting loop antenna. Inset shows the impedance-matching network.

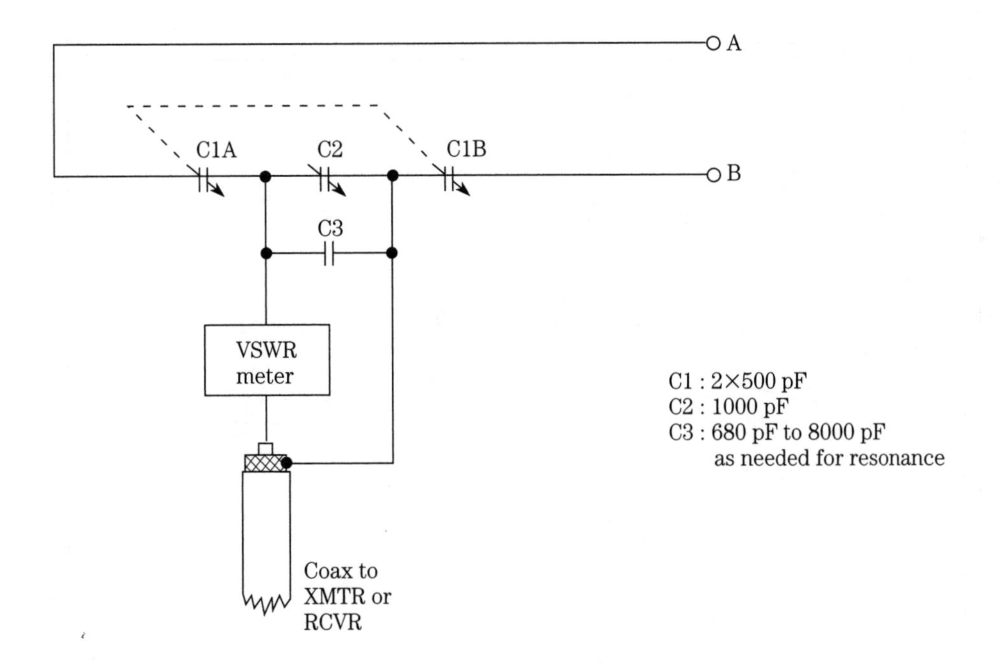

15-6B 80 meter coupling unit.

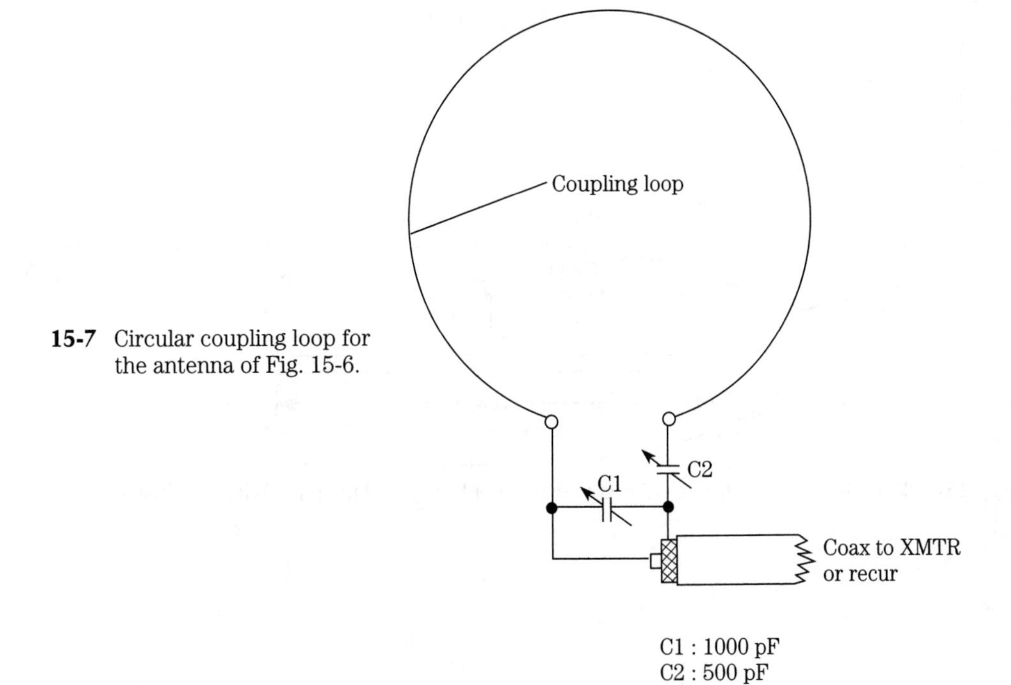

15-7 Circular coupling loop for the antenna of Fig. 15-6.

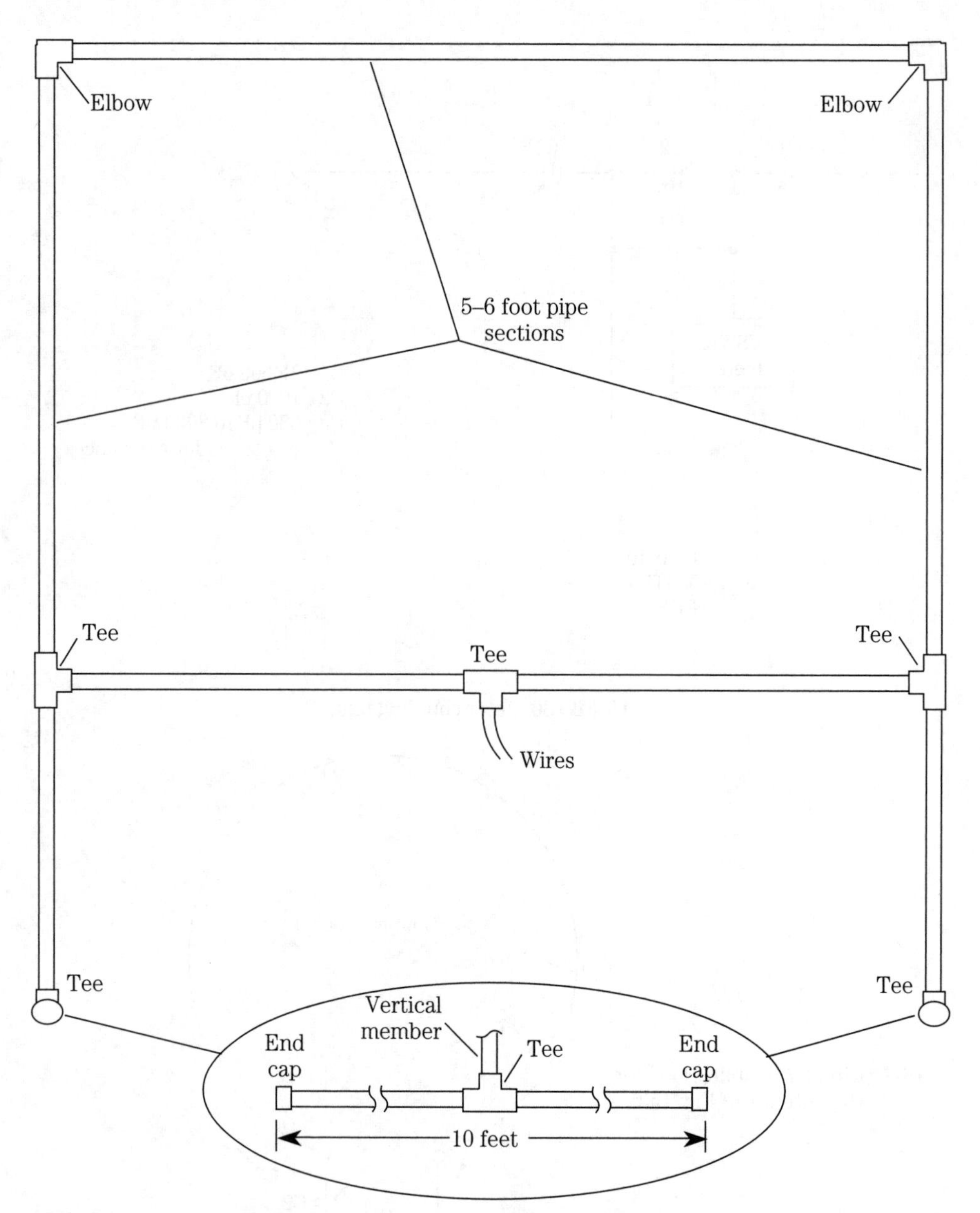

15-8 Small loop for low frequencies constructed of PVC plumbing pipe. Wires are inside the pipe.

16
CHAPTER

Other wire antennas

THIS CHAPTER CONTAINS SOME WIRE ANTENNAS THAT MANAGED TO ESCAPE PREVIOUS chapters, for one reason or another. Some are completely different types from those of other chapters, and others are unusual adaptations of antennas covered in those chapters. They all share one attribute, however, and that is that they are made of wire construction. No fancy metal-working skills are needed to construct these antennas. Basically, anyone who can successfully erect a dipole can also build the kinds of antennas covered in this chapter. So, unless you are one of those unfortunate souls who can't count their own fingers and come up with the same number twice in a row, these antennas are well within your technical reach, and may offer advantages over other types.

Off-centerfed fullwave doublet (OCFD) antennas

An antenna that superficially resembles the Windom, is the off-centerfed doublet (OCFD) antenna of Fig. 16-1. It is a single-band antenna, although at harmonics it will begin to act as a resonant, standing wave, longwire antenna. The overall length is one wavelength long:

$$L_{\text{feet}} = \frac{936}{F_{\text{MHz}}} \qquad \textbf{[16.1]}$$

This antenna works best at heights of at least $\lambda/2$ above ground, so practical considerations limit it to frequencies above about 10 MHz (i.e., 30-meter band). The feedpoint of the antenna is placed at a distance of $\lambda/4$ from one end, and is a good match for 75-Ω coaxial cable. A 1:1 BALUN transformer at the feedpoint is highly recommended. The pattern of a 1λ antenna is a four-lobe "cloverleaf," with the major lobes being about 53 degrees from the wire. The gain is about 1 dB.

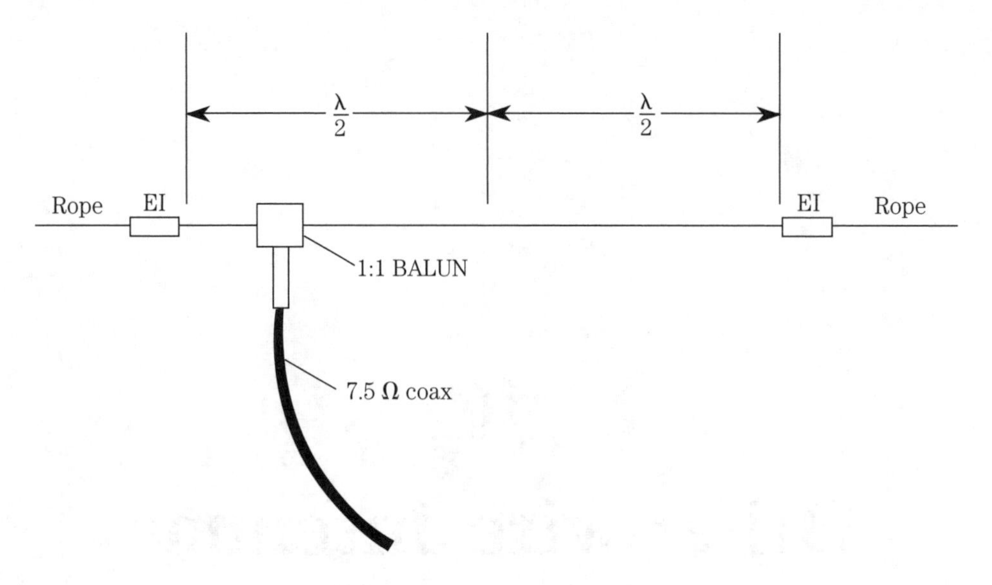

16-1 Off-centerfed 1-wire antenna.

Off-centerfed nonresonant sloper (OCFS)

Perhaps more viable for many people is the *nonresonant off-centerfed sloper* (OCFS) antenna of Fig. 16-2. This antenna consists of a wire radiator that must be longer than 3λ/2 at the lowest frequency of operation. The feedpoint is elevated at least λ/4 above ground at the lowest operating frequency. The antenna is fed with 75-Ω coaxial cable. The shield of the coax is connected to a λ/4 resonant radial (counterpoise ground). There should be at least one radial (more is better) per band of operation.

The far end of the radiator element is sloped to ground, where it is terminated in a 270-Ω noninductive resistor. The resistor should be able to dissipate up to one third of the power level applied by the transmitter.

Double extended Zepp antenna

The *double extended Zepp* antenna (Fig. 16-3) provides a gain of about 2 dB over a dipole at right angles to the antenna wire plane. It consists of two sections of wire, each one of a length:

$$L_{1\text{feet}} = \frac{600}{F_{\text{MHz}}} \qquad [16.2]$$

Typical lengths are: 20.7 feet on the 10-meter band, 28 feet on the 15-meter band, 42 feet on the 20-meter band, and 84 feet on the 40-meter band.

The double extended Zepp antenna can be fed directly with 450-Ω twin lead (see Appendix A for suppliers), especially if a balanced antenna tuner is available at the receiver. Alternatively, it can be fed from a quarter-wavelength matching section

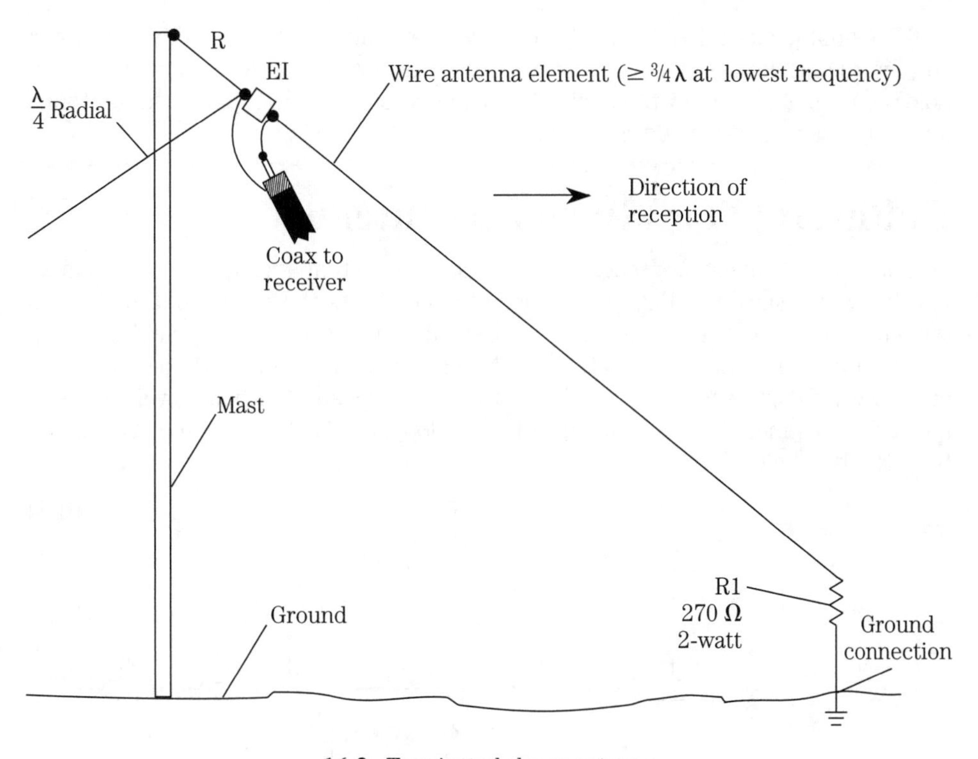

16-2 Terminated sloper antenna.

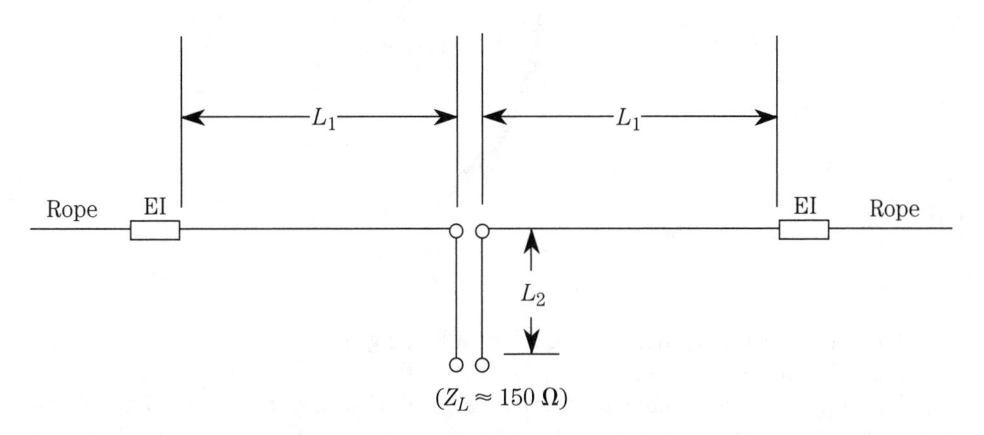

16-3 Double extended Zepp antenna.

(made of 450-Ω twin lead, or equivalent open air parallel line), as shown, and a BALUN if coax is preferred. The length of the matching section should be:

$$L_{2feet} = \frac{103}{F_{MHz}} \qquad \textbf{[16.3]}$$

The double extended Zepp will work on several different bands. For example, a 20-meter band double extended Zepp will work as a Zepp on the design band, a dipole on frequencies below the design band, and as a four-lobed "cloverleaf" antenna on frequencies above the design band.

Collinear "Franklin" array antenna

Perhaps the cheapest approach, to very serious antenna gain, is the Collinear Franklin array shown in Fig. 16-4. This antenna pushes the dipole and double extended Zepp concepts even farther. It consists of a half-wavelength dipole that is centerfed with a 4:1 BALUN and 75-Ω coaxial cable. At each end of the dipole, there is a quarter-wavelength *phase reversal stub* that endfeeds another half-wavelength element. Each element is a half-wavelength (λ/2) long, and its length can be calculated from Eq. 16.4 below:

$$\frac{\lambda}{2} = \frac{492}{F_{\mathrm{MHz}}} \qquad [16.4]$$

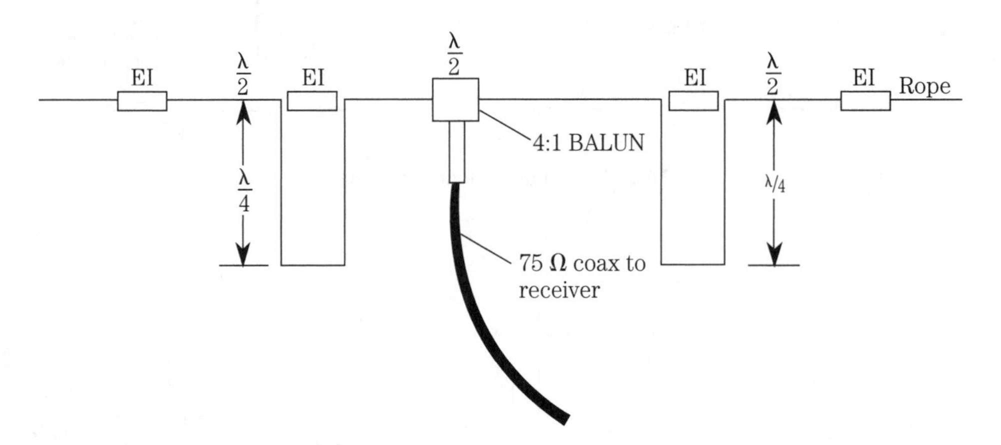

16-4 Wire collinear antenna.

The phase reversal stubs are a quarter-wavelength long, or one half the length calculated by Eq. 16.4 above.

The version of the "Collinear" shown in Fig. 16-4 has a gain of about 3 dB. There is no theoretical reason why you can't extend the design indefinitely, but there is a practical limit set by how much wire can be held by your supports, and how much real estate you own. A 4.5-dB version can be built by adding another half-wavelength section at each end, with an intervening quarter-wavelength phase reversal stub in between each new section, and the preceding section. Once you get longer than five half wavelengths, which provides the 4.5 dB gain, the physical size becomes a bit of a bother for most folks.

The TCFTFD dipole

The *tilted, center-fed, terminated, folded dipole* (*TCFTFD*, also called the T^2FD or *TTFD*) is an answer to both the noise pick-up and length problems that sometimes affect other antennas. For example, a random-length wire, even with antenna tuner, will pick up considerable amounts of noise. A dipole for 40-meters is 66 feet long.

This antenna was first described publicly in 1949 by Navy Captain C.L. Countryman, although the U.S. Navy tested it for a long period in California during World War II. The TCFTFD can offer claimed gains of 4 to 6 dB over a dipole, depending on the frequency and design, although 1 to 3 dB is probably closer to the mark in practice, and less than 1 dB will be obtained at some frequencies within its range (especially where the resistor has to absorb a substantial portion of the RF power). The main attraction of the TCFTFD is not its gain, but rather its *broadbandedness*.

In addition, the TCFTFD can also be used at higher frequencies than its design frequency. Some sources claim that the TCFTFD can be used over a 5 or 6:1 frequency range, although my own observations are that 4:1 is more likely. Nonetheless, a 40-meter antenna will work over a range of 7000 kHz to 25,000 kHz, with at least some decent performance up into the 11-meter Citizen's Band (27,000 kHz).

The basic TCFTFD (Fig. 16-5) resembles a folded dipole in that it has two parallel conductors, of length L, spaced a distance W apart, and shorted together at the ends. The feedpoint is the middle of one conductor, where a 4:1 BALUN coil and 75-Ω coaxial-cable transmission line to the transceiver are used. A noninductive, 390-Ω resistor is placed in the center of the other conductor. This resistor can be a carbon-composition (or metal-film) resistor, but it must not be a wirewound resistor. The resistor must be able to dissipate about one third of the applied RF power. The TCFTFD can be built from ordinary #14 stranded antenna wire.

For a TCFTFD antenna covering 40 through 11 meters, the spread between the conductors should be 19½ inches, while the length (L) is 27 feet. Note that length L includes one half of the 19-inch spread because it is measured from the center of the antenna element to the center of the end supports.

The TCFTFD is a sloping antenna, with the lower support being about 6 feet off the ground. The height of the upper support depends on the overall length of the antenna. For a 40-meter design, the height is on the order of 50 feet.

The parallel wires are kept apart by spreaders. At least one commercial TCFTFD antenna uses PVC spreaders, while others use ceramic. You can use wooden dowels of between 1-inch and ⅝-inch diameter; of course, a coating of varnish (or urethene spray) is recommended for weather protection. Drill two holes, of a size sufficient to pass the wire, that are the dimension W apart (19 inches for 40 meters). Once the spreaders are in place, take about a foot of spare antenna wire and make jumpers to hold the dowels in place. The jumper is wrapped around the antenna wire on either side of the dowel, and then soldered.

The two end supports can be made of 1"-x-2" wood treated with varnish or urethene spray. The wire is passed through screw eyes fastened to the supports. A support rope is passed through two holes on either end of the 1" × 2" and then tied off at an end insulator.

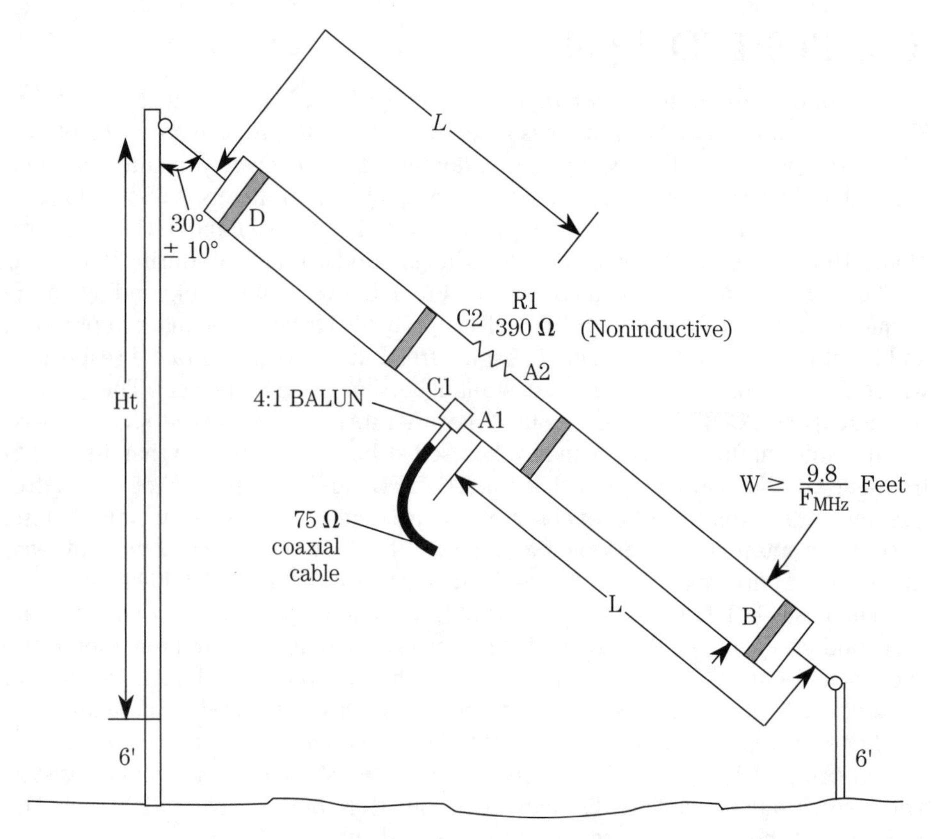

16-5 TTFD antenna.

The TCFTFD antenna is noticeably quieter than the random-length wire antenna, and somewhat quieter than the half-wavelength dipole. When the tilt angle is around 30 degrees, the pattern is close to omnidirectional. Although a little harder to build than dipoles, it offers some advantages that ought not to be overlooked. These dimensions will suffice when the "bottom end" frequency is the 40-meter band, and it will work well on higher bands.

Vee-sloper antenna

The *vee-sloper antenna* is shown in Fig. 16-6. It is related to the vee-beam (covered in chapter 9), but it is built like a sloper (i.e., with the feed end of the antenna high above ground). The supporting mast height should be about half (to three fourths) of the length of either antenna leg. The legs are sloped downward to terminating resistors at ground level. Each wire should be >1λ at the lowest operating frequency. The terminating resistors should be on the order of 270 Ω (about one half of the characteristic impedance of the antenna), with a power rating capable of dissipating one third of the transmitter power. Like other terminating resistors, these should be noninductive (carbon composition or metal film).

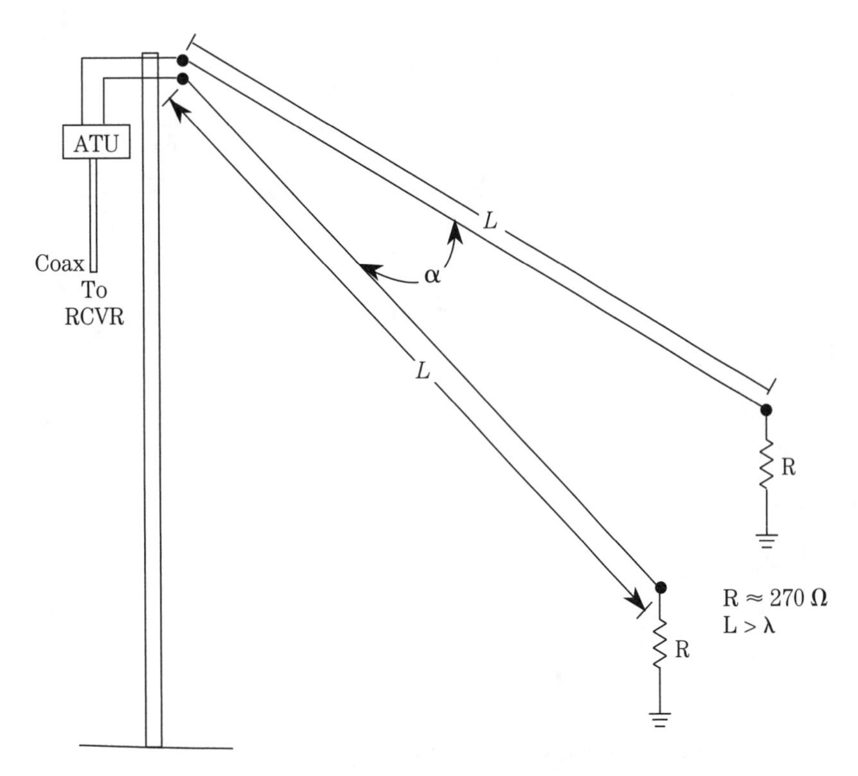

16-6 Sloper-vee antenna.

The advantage of this form of antenna over the vee-beam is that it is vertically polarized, and the resistors are close to the earth, so they are easily grounded.

Rhombic inverted-vee antenna

A variation on the theme is the vertically polarized rhombic of Fig. 16-7. Although sometimes called an *inverted-vee*—not to be confused with the dipole variant of the same name—this antenna is half a rhombic, with the missing half being "mirrored" in the ground (similar to a vertical). The angle at the top of the mast (Φ) is typically ≥ 90 degrees, and 120 to 145 degrees is more common. Each leg ("A") should be ≥λ, with the longer lengths being somewhat higher in gain, but harder to install for low frequencies.

Multi-band fan dipole

The basic half-wavelength dipole antenna is a very good performer, especially when cost is a factor. The dipole yields relatively good performance for practically no investment. A standard half-wavelength dipole offers a bidirectional, "figure-8" pattern on its basic band (i.e., where the length is a half wavelength), and a four lobe,

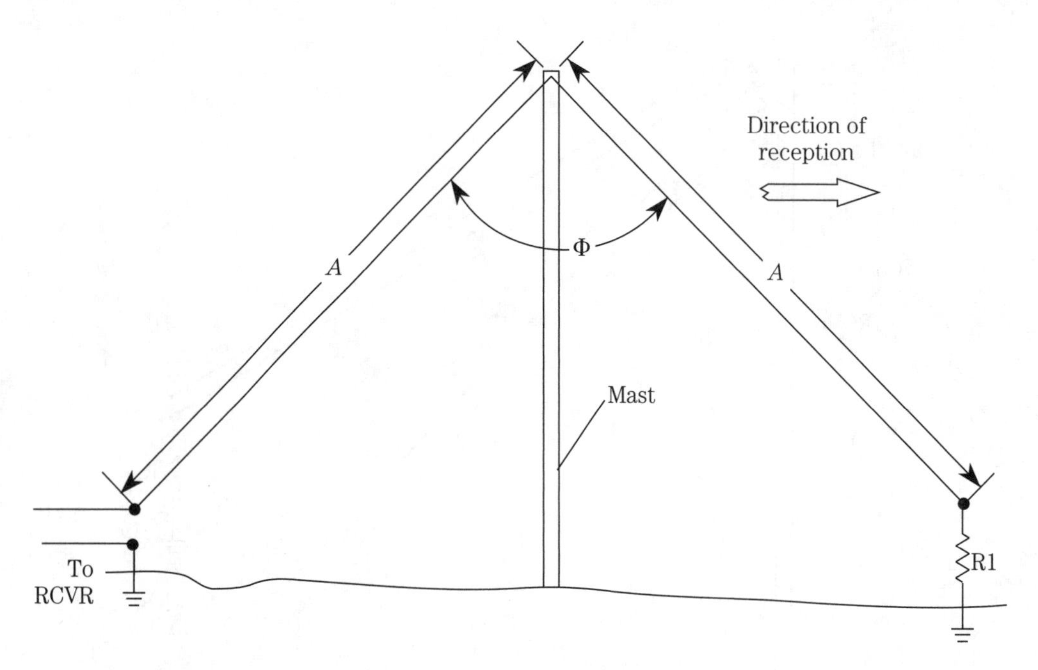

16-7 Inverted-vee antenna.

cloverleaf pattern at frequencies for which the physical length is $3\lambda/2$. Thus, a 40-meter half-wavelength dipole produces a bidirectional pattern on 40-meters, and a four lobe cloverleaf pattern on 15 meters.

The dipole is not easily multi-banded without resorting to traps (covered in chapter 8). One can, however, tie several dipoles to the same center insulator or BALUN transformer. Figure 16-8 shows three dipoles cut for different bands, operating from a common feedline and BALUN transformer: A1–A2, B1–B2, and C1–C2. Each of these antennas is a half wavelength (i.e., $L_{feet} = 468/F_{MHz}$).

There are two points to keep in mind when building this antenna. First, try to keep the ends spread a bit apart, and second, make sure that none of the antennas is cut as a half wavelength for a band for which another is $3\lambda/2$. For example, if you make A1–A2 cut for 40-meters, then don't cut any of the other three for 15-meters. If you do, the feedpoint impedance and the radiation pattern will be affected.

The counterpoise longwire

The *longwire antenna* is an end-fed wire $>2\lambda$ long. It provides considerable gain over a dipole, especially when a very long length can be accommodated. Although 75–80-meter, or even 40-meter, longwires are a bit difficult to erect at most locations, they are well within reason at the upper end of the HF spectrum. Low-VHF band operation is also practical. Indeed, I know one fellow who lived in far southwest Virginia as a teenager, and he was able to get his family television reception for very low cost by using a TV longwire (channel 6) on top of his mountain.

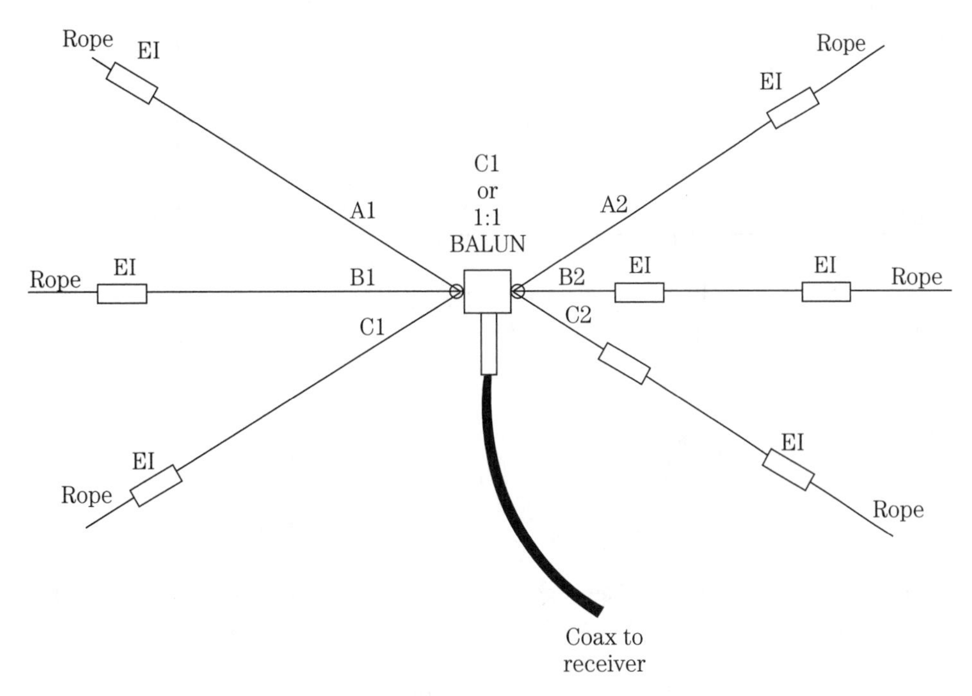

16-8 Multi-band dipole antenna.

There are some problems with longwires that are not often mentioned. Two problems seem to insulate themselves into the process. First, the Zepp feed is a bit cumbersome (not everyone is enamored of parallel transmission line). Second, how do you go about actually grounding that termination resistor? If it is above ground, then the wire to ground is long, and definitely not at ground potential for RF. If you want to avoid both the straight Zepp feed system employed by most such antennas, as well as the resistor-grounding problem, then you might want to consider the *counterpoise longwire antennas* shown in Fig. 16-9.

A counterpoise ground is a structure that acts like a ground, but is actually electrically floating above real ground (and it is not connected to ground). A ground plane of radials is sometimes used as a counterpoise ground for vertical antennas that are mounted above actual earth ground. In fact, these antennas are often called *ground-plane verticals*. In those antennas, the array of four (or more) radials from the shield of the coaxial cable are used as an artificial, or counterpoise, ground system.

In the counterpoise longwire of Fig. 16-9, there are two counterpoise grounds (although, for one reason or another, you might elect to use either, but not both). One counterpoise is at the feedpoint, where it connects to the "cold" side of the transmission line. The parallel line is then routed to an antenna tuning unit (ATU), and from there to the transmitter. The other counterpoise is from the cold end of the termination resistor to the support insulator. This second counterpoise makes it possible to eliminate the earth ground connection, and all the problems that might entail, especially in the higher end of the HF spectrum, where the wire to ground is of substantial length compared with 1λ of the operating frequency.

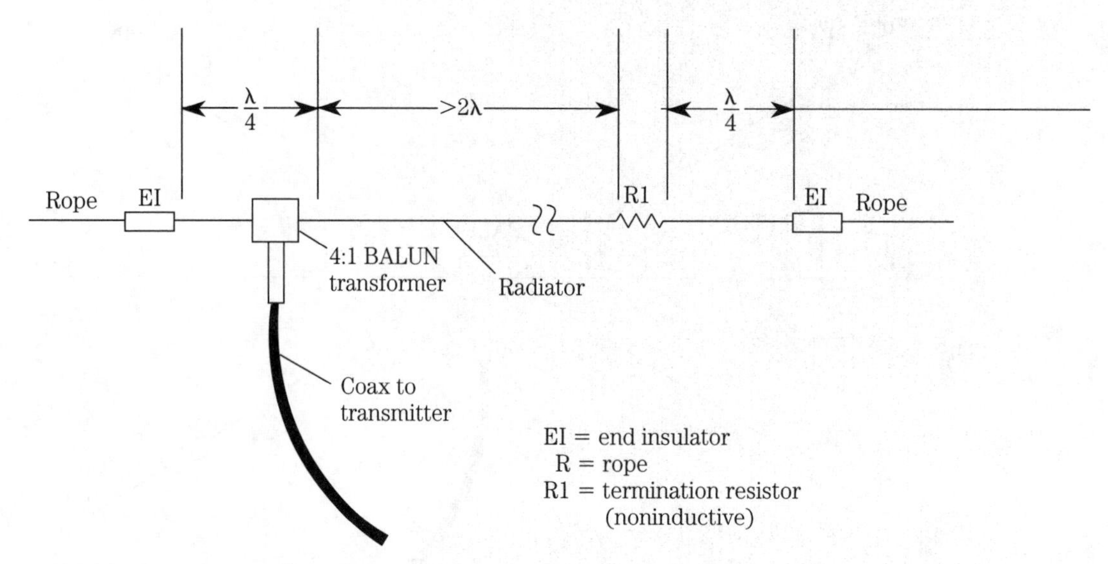

16-9A Longwire with λ/4 radials used as counterpoise ground at feedpoint and terminating resistor.

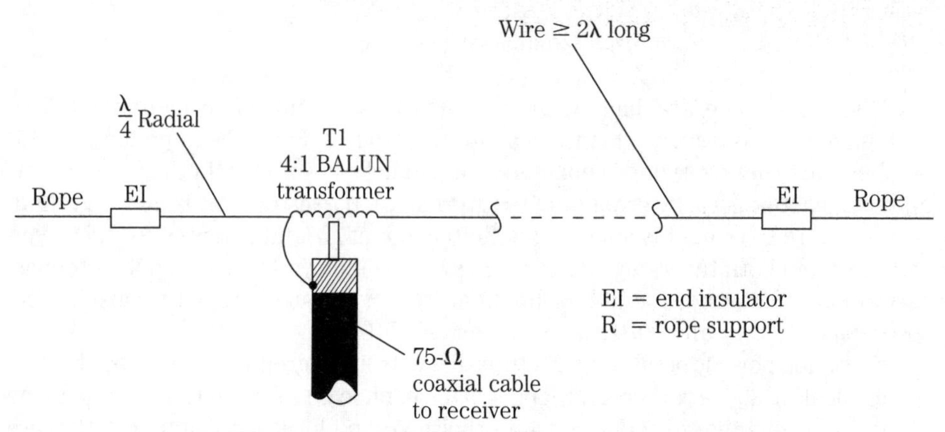

16-9B BALUN transformer feed for longwire antenna.

A slightly different scheme is used to adapt the antenna to coaxial cable is shown in Fig. 16-9B. In this case, the longwire is a resonant type (nonterminated). Normally, one would expect to find this antenna fed with 450-Ω parallel transmission line. But with a λ/4 radial acting as a counterpoise, a 4:1 BALUN transformer can be used to effect a reasonable match to 75-Ω coaxial cable. The radial is connected to the side of the BALUN that is also connected to the coaxial cable shield, and the other side of the BALUN is connected to the radiator element.

17
CHAPTER
VHF/UHF transmitting and receiving antennas

THE VHF/UHF SPECTRUM IS COMMONLY ACCEPTED TO RANGE FROM 30 MHz TO 900 MHz, although the upper breakpoint is open to some differences of opinion. The VHF spectrum is 30 MHz to 300 MHz, and the UHF spectrum is 300 MHz to 900 MHz. Above 900 MHz is the microwave spectrum. These bands are used principally for local "line-of-sight" communications, according to the standard wisdom. However, with the advent of OSCAR satellites, the possibility of long-distance direct communications is a reality for VHF/UHF operators. In addition, packet radio is becoming common; this means indirect long distance possibilities through networking. For the low end of the VHF spectrum (e.g., 6-meter amateur band), long-distance communications are a relatively common occurrence.

In many respects, the low-VHF region is much like the 10-meter amateur band and 11-meter Citizen's Band: skip is not an infrequent occurrence. Many years ago, I recall an event where such skip caused many a local police officer to skip a heart beat. In those days, our police department operated on 38.17 MHz, which is between the 6-meter and 10-meter amateur bands. They received an emergency broadcast concerning a bank robbery at a certain Wilson Boulevard address. After a race to the county line, they discovered that the reported address would be outside of the county . . . and in fact did not exist even in the neighboring county (a number was skipped). The problem was traced to a police department in a southwest city that also had a Wilson Boulevard, and for them the alarm was real.

The principal difference between the lower frequencies and the VHF/UHF spectrum is that the wavelengths are shorter in the VHF/UHF region. Consider the fact that the wavelengths for these bands range from 10 meters to 1 meter for the VHF region, and from 1 meter to 33 centimeters for the UHF region. Most antenna designs are based on wavelength, so that fact has some implications for VHF/UHF antenna design. For example, because bandwidth is a function of length/diameter ratio for many classes of antenna, broadbanding an antenna in the VHF/UHF region is relatively easy. If, say, 25-mm (i.e., 1-inch) aluminum tubing is used to make a quarter-wavelength vertical, then the approximate *L/D* ratio is 790 in the 8-meter band, and 20 in the 2-meter band. This feature is fortunate, because the VHF/UHF bands tend to be wider than the HF bands.

Another point to make is that many of the mechanical chores of antenna design and construction become easier for VHF/UHF antennas. One good example is the delta impedance matching scheme. At 80-meters, the delta-match dimensions are approximately 36 ×43 feet, and at 2 meters they are 9.5 × 12 *inches*. Clearly, delta matching is a bit more practical for most users at VHF than at HF.

Types of antennas usable for VHF/UHF

The concept "VHF/UHF antenna" is only partially valid because virtually all forms of antenna can be used at HF, MW, and VHF/UHF. The main limitations that distinguish supposedly VHF/UHF designs from others are mechanical: there are some things that are simply much easier to accomplish with small antennas. Besides the delta match mentioned previously, there is the ease of construction for multi-element antennas. A 14-element 20-meter beam would be a wonderful thing to have in a QRM-laden DX pile-up, but is simply too impractical for all but a few users because of its size. If you look on embassy rooftops around the world you will see many-element Yagi and log periodic HF antennas supported on massive towers . . . and some of them use a standard Size 25 tower (common for amateur use) as the *antenna boom*! A 14-element 80-/75-meter Yagi approaches impossibility. But at 2 meters, a 14-element Yagi beam antenna can be carried by one person, in one hand, unless the wind is acting up.

Safety note Large array beams, even at VHF/UHF, have a relatively high "windsail area," and even relatively modest winds can apply a lot of force to them. I once witnessed a large, strong technician blown off a ladder by wind acting on a modest, "suburban" sized, TV antenna. *It can happen to you, too. So always install antennas with a helper, and use hoists and other tools to actually handle the array.*

Lower band antennas on VHF/UHF

Between 1958 and 1962, the late Johnnie H. Thorne (K4NFU/5) and I had access to a radio club amateur radio station in a Red Cross Chapter House in Virginia. The "antenna farm" consisted of a 14-element 2-meter beam, a three-element tri-band HF beam (10-15-20 meters), and a five-band (80-10 meters) trap dipole. All of the coaxial cables came into the station through a wall; they were kept disconnected, and shorted out, when not in use because of the senior Red Cross official's concern over lightning.

One night, attempting to connect the 2-meter beam to the *Gonset* "gooney box" 2-meter AM transceiver, John accidentally used the cable from the five-band trap dipole instead. We worked a lot of stations that contest weekend, and scored lots of points. Later, we discovered the error, and asked a more technically competent adult (we were teenagers), "why the good reports?" He then gave us a lesson in longwire antenna theory. A good longwire is many wavelengths long. Consider that a half-wave antenna on 2-meters is (80 m/2 m), 40 wavelengths shorter than an 80-meter half-wave antenna. Thus, the 80-meter antenna, counting foreshortening of physical lengths because the traps, was on the order of 35 to 38 wavelengths long on 2-meters. We had a highly directional, but multi-lobed pattern.

Similarly, 40- to 10-meter and 80- to 10-meter trap verticals are often usable on VHF/UHF frequencies without any adjustments. Similarly, Citizens Band 11-meter antennas, many of which are ⅝-wavelength (18-feet high), will sometimes work on VHF frequencies. Check the VSWR of an HF antenna on 2 meters with a reliable VHF/UHF VSWR meter (or RF watt-meter) to discover the truth about any particular antenna. Always use the low-power setting on the transmitter to limit damage in cases where the specific antenna is not usable on a specific frequency.

The lesson to be learned is that antennas are often usable on much higher than the design frequency, even though useless on nearby bands. Care must be exercised when initially checking out the antenna, but that is not an inordinate difficulty.

VHF/UHF antenna impedance matching

The VHF/UHF antenna is no more or less immune from the need for impedance matching than lower frequency antennas. However, some methods are easier (coax BALUNs, delta match, etc.) and others become either difficult or impossible. An example of the latter case is the tuned LC impedance matching network. At 6 meters, and even to some limited extent 2 meters, inductor and capacitor LC networks can be used. But above 2 meters other methods are more reasonable. We can, however, mimmick the LC tuner by using stripline components, but that approach is not always suited to amateur needs.

The BALUN transformer makes an impedance transformation between BALanced and UNbalanced impedances. Although both 1:1 and 4:1 impedance ratios are possible, the 4:1 ratio is most commonly used for VHF/UHF antenna work. At lower frequencies it is easy to build broadband transformer BALUNs, but these become more of a problem at VHF and above.

For the VHF/UHF frequencies, a 4:1 impedance ratio coaxial BALUN (Fig. 17-1A) is normally used. Two sections of identical coaxial cable are needed. One section ("A") has a convenient length to reach between the antenna and the transmitter. Its characteristic impedance is Z_o. The other section ("B") is a half-wavelength long at the center of the frequency range of interest. The physical length is found from:

$$L = \frac{5904\ V}{F_{\text{MHz}}} \text{ inches} \qquad [17.1]$$

Where:

L is the cable length in inches
F_{MHz} is the operating frequency in megahertz
V is the velocity factor of the coaxial cable

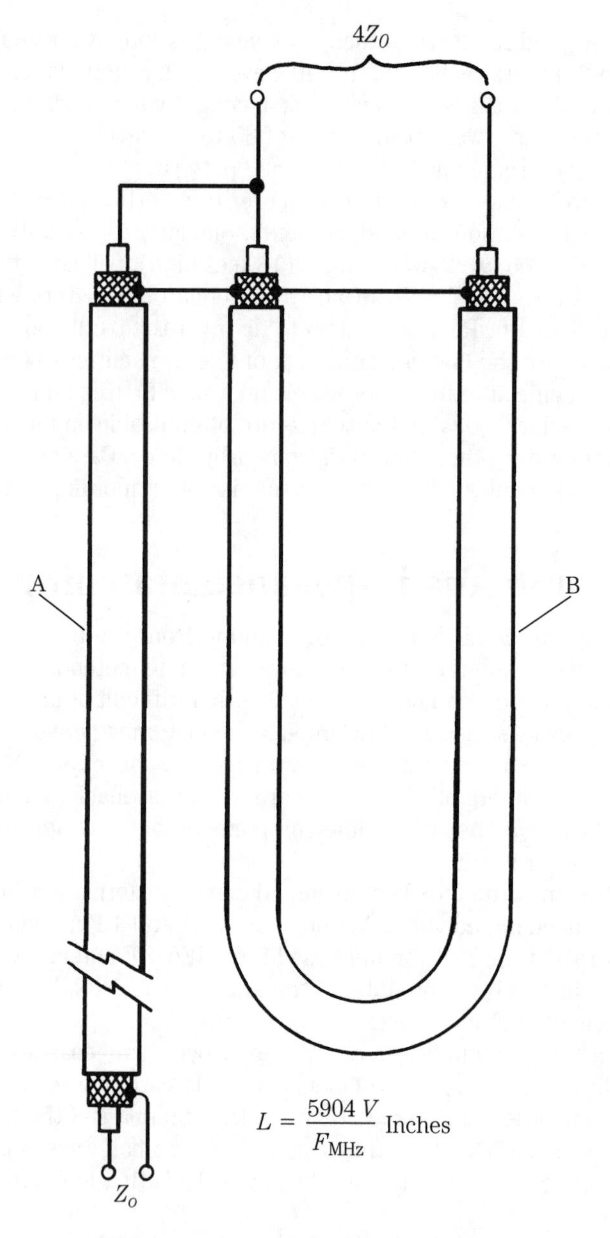

$$L = \frac{5904\,V}{F_{\text{MHz}}}\ \text{Inches}$$

17-1A Coaxial 4:1 BALUN transformer.

The velocity factors of common coaxial cables are shown in the table.

Coaxial cable velocity factors

Regular Polyethylene	0.66
Polyethylene Foam	0.80
Teflon	0.72

Example Calculate the physical length required of a 146 MHz 4:1 BALUN made of polyethylene foam coaxial cable.

Solution:

$$L = \frac{5904\ V}{F_{\mathrm{MHz}}}$$

$$L = \frac{(5904)\ (0.80)}{146\ \mathrm{MHz}}$$

$$L = \frac{4723.2}{146\ \mathrm{MHz}} = 32.4\ \text{inches}$$

A mechanical method of joining the coaxial cables is shown in Fig. 17-1B. In this example, three SO-239 coaxial receptacles are mounted on a metal plate. This arrangement has the effect of shorting together the shields of the three ends of coaxial cable. The center conductors are connected in the manner shown. This method is used especially where a mounting bracket is available on the antenna. The lengths of coaxial cable need PL-259 coax connectors installed in order to use this method.

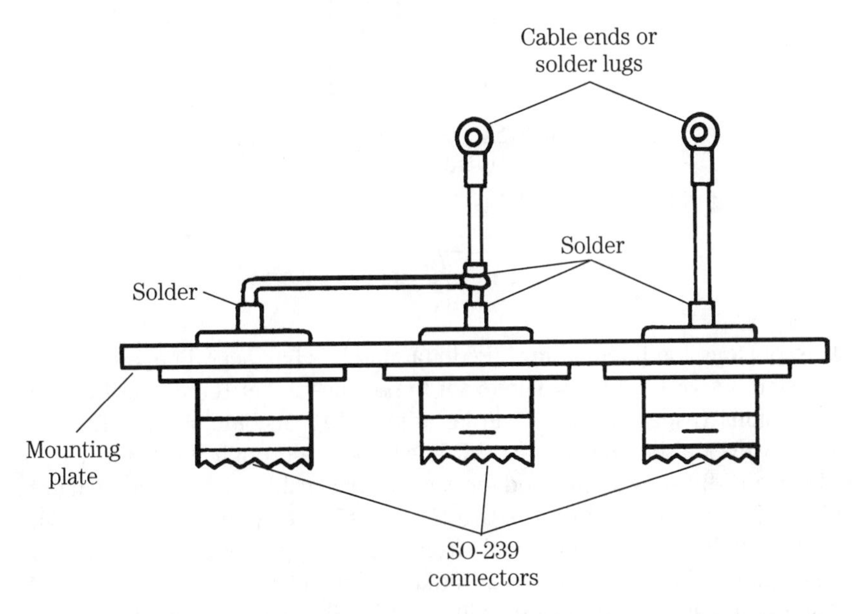

17-1B Practical implementation of 4:1 BALUN using connectors.

The "delta match" gets its name from the fact that the structure of the matching element has the shape of the Greek letter "delta," or a triangle. Figure 17-2A shows the basic delta match scheme. The matching element is attached to the driven element of the antenna (symmetrically, about the center point of the antenna). The width ("A") of the delta match is given by:

$$L = \frac{1416}{F_{\mathrm{MHz}}}\ \text{inches} \qquad [17.2]$$

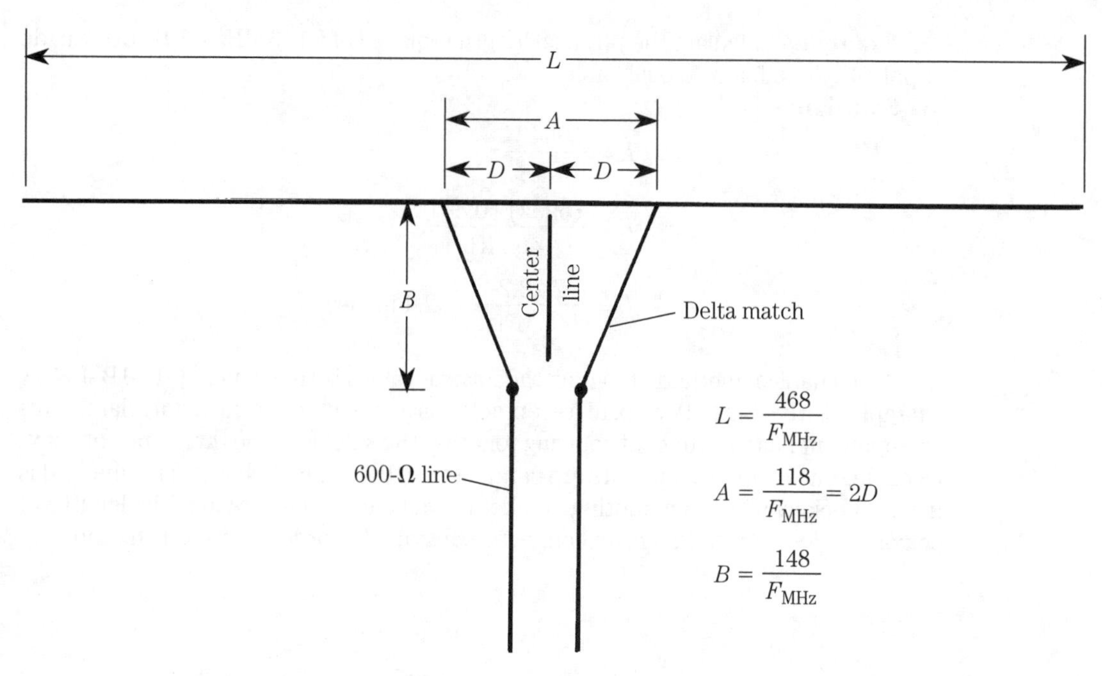

17-2A Delta feed matching system.

While the height of the match ("B") is:

$$L = \frac{1776}{F_{MHz}} \text{ inches} \qquad [17.3]$$

The transmission line feeding the delta match is balanced line, such as parallel transmission line or twin lead. The exact impedance is not terribly critical because the dimensions (especially "A") can be adjusted to accommodate differences. In general, however, either 450-Ω or 600-Ω line is used, although 300-Ω line can also be used. Figure 17-2B shows a method for using coaxial cable with the delta match. The impedance is transformed in a 4:1 BALUN transformer (see Fig. 17-1A). The elements of the delta match can be made from brass, copper, or aluminum tubing, or a bronze brazing rod bolted to the main radiator element.

A stub-matching system is shown in Fig. 17-3. In this case, the impedance transformation is accomplished although a half-wavelength shorted stub of transmission line. The exact impedance of the line is not very critical, and is found from:

$$Z_o = 276 \text{ LOG}_{10} \frac{2S}{d} \qquad [17.4]$$

The matching stub section is made from metal elements such as tubing, wire, or rods (all three are practical at VHF/UHF frequencies). For a ³⁄₁₆-inch rod, the spacing is approximately 2.56-inches to make a 450-Ω transmission line. A sliding short circuit is used to set the electrical length of the half-wave stub. The stub is tapped at a distance from the antenna feedpoint that matches the impedance of the transmis-

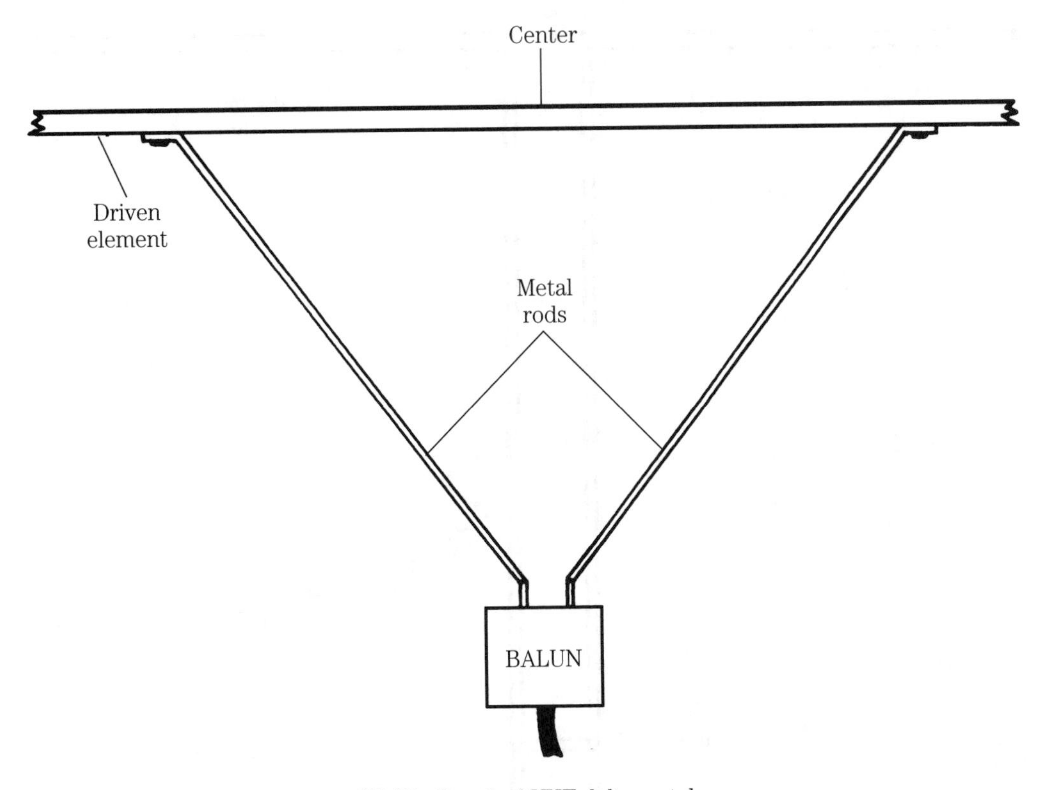

17-2B Practical VHF delta match.

sion line. In the example shown, the transmission line is coaxial cable, so a 4:1 BALUN transformer is used between the stub and the transmission line. The two adjustments to make in this system are: 1) the distance of the short from the feedpoint, and 2) the distance of the transmission line tap point from the feedpoint. Both are adjusted for minimum VSWR.

The gamma match is basically a half delta match, and operates according to similar principles (Fig. 17-4). The shield (outer conductor) of the coaxial cable is connected to the center point of the radiator element. The center conductor of the coaxial cable series feeds the gamma element through a variable capacitor.

VHF/UHF antenna examples

Although it is probably not necessary to reiterate the point, VHF/UHF antennas are not substantially different from HF antennas, especially those for the higher bands. However, for various practical reasons there are several forms that are specially suited, or at least popular, in the VHF/UHF bands. In this section we will take a look at some of them.

Coaxial vertical

The coaxial vertical is a quarter-wavelength vertically polarized antenna that is popular on VHF/UHF. There are two varieties. In Fig. 17-5A we see the coaxial antenna

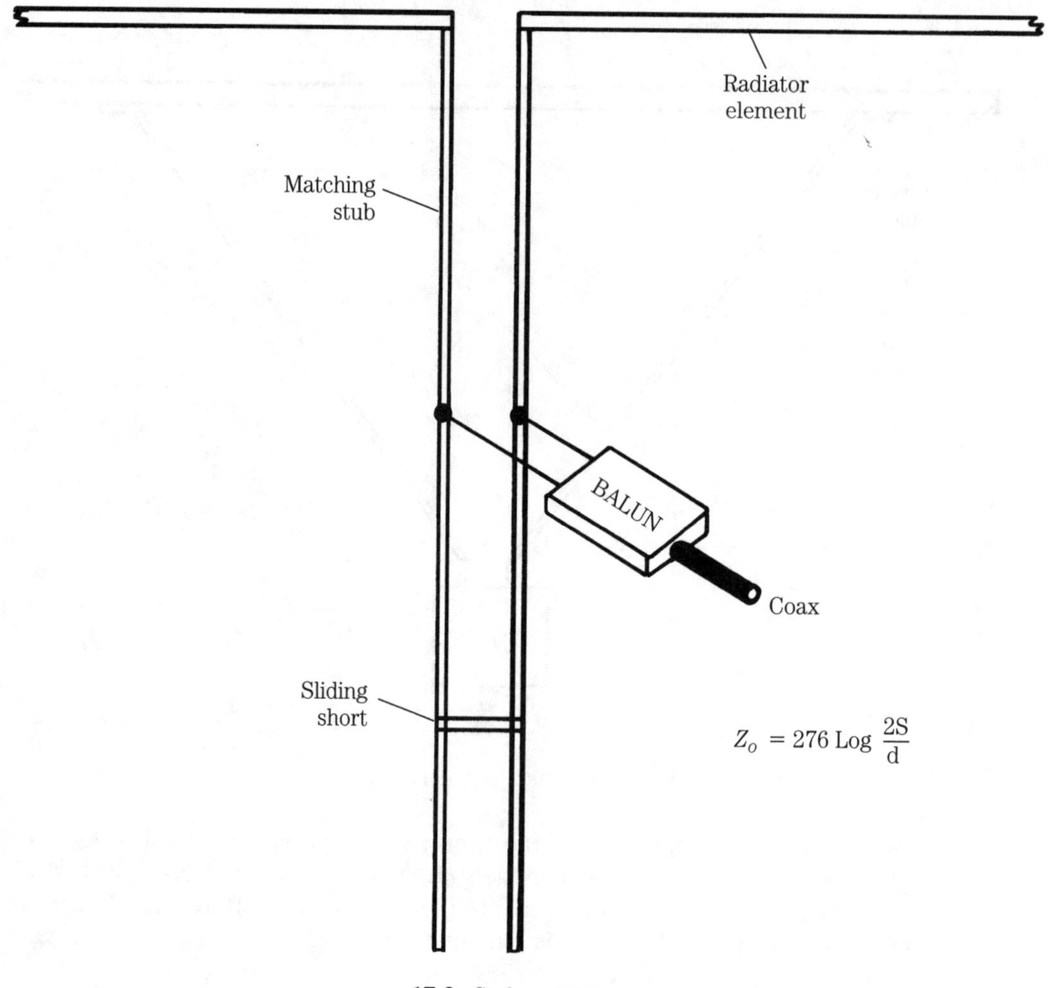

17-3 Stub matching.

made with coaxial cable. Although not terribly practical for long-term installation, the coax-coax antenna is very useful for short-term, portable or emergency applications. For example, a boater found himself adrift, and in dire trouble, after a storm damaged the boat. The mast-top VHF antenna was washed away, leaving only the end of the coaxial cable dangling loose. Fortunately, the boat operator was a two-way radio technician, and he knew how to strip back the coaxial cable to make an impromptu coaxial vertical.

The coax-coax antenna shown in Fig. 17-5A uses a quarter-wavelength radiator and a quarter-wavelength sleeve. The sleeve consists of the coax braid stripped back and folded down the length of the coax cable. The maximum length is found from the equation below (actual length is trimmed from this maximum):

$$L = \frac{2952\ V}{F_{\text{MHz}}} \text{ inches} \qquad \textbf{[17.5]}$$

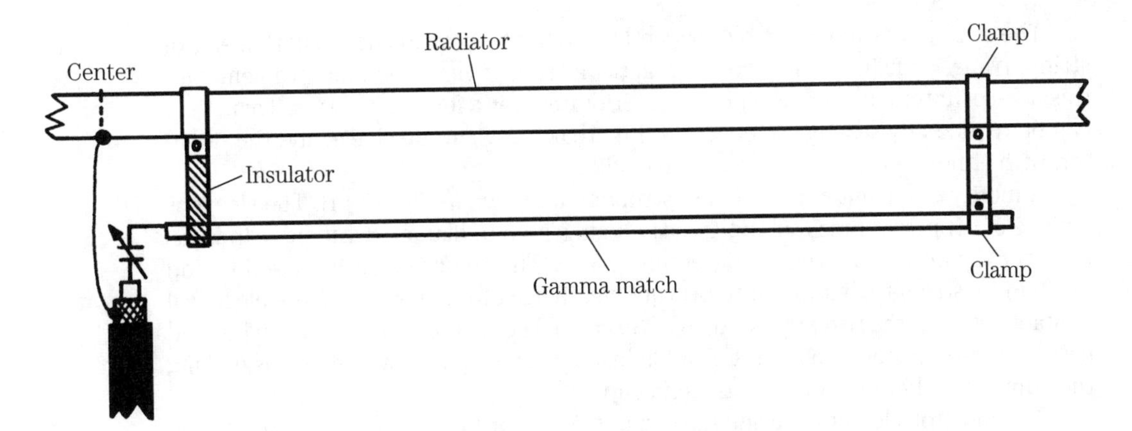

17-4 Gamma matching.

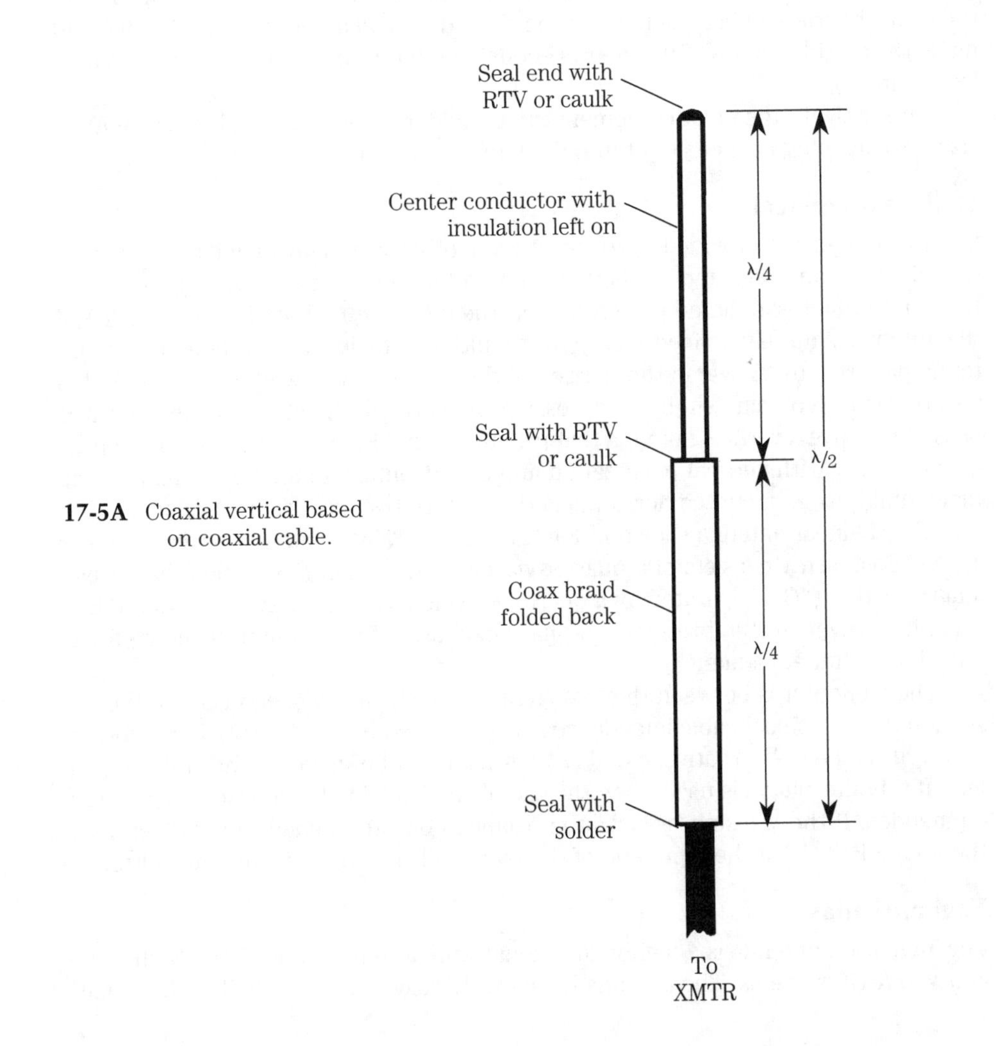

17-5A Coaxial vertical based on coaxial cable.

The antenna is mounted by suspending it from above using a short piece of string, twine, or fishing line. From a practical point of view, the only problem with this form of antenna is that it tends to deteriorate after a few rain storms. This effect can be reduced by sealing the end, and the break between the sleeve and the radiator, with either silicone *RTV* or bathtub caulk.

A more permanent method of construction is shown in Fig. 17-5B. The sleeve is a piece of copper or brass tubing (pipe) about 1-inch in diameter. An end cap is fitted over the end, and sweat soldered into place. The solder is not intended to add mechanical strength, but rather to prevent weathering from destroying the electrical contact between the two pieces. An SO-239 coaxial connector is mounted on the end cap. The coax is connected to the SO-239 inside of the pipe . . . which means making the connection before mounting the end cap.

The radiator element is a small piece of tubing (or brazing rod) soldered to the center conductor of a PL-259 coaxial connector. An insulator is used to prevent the rod from shorting to the outer shell of the PL-259 (note: an insulator salvaged from the smaller variety of banana plug can be shaved a small amount with a fine file and made to fit inside the PL-259. It allows enough center clearance for ⅛-inch or 3/16-inch brass tubing).

Alternatively, the radiator element can be soldered to a banana plug. The normal sized banana plug happens to fit into the female center conductor of the SO-239.

Collinear vertical

Gain in antennas is provided by *directivity*. In other words, by taking the power radiated by the antenna, and projecting it into a limited direction, we obtain the appearance of higher radiated power. In fact, the *effective radiated power* (ERP) of the antenna is merely its feedpoint power multiplied by its gain. Although most antenna patterns are shown in the horizontal dimension (as viewed from above), it is also possible to obtain gain by compressing the vertical aspect. In this manner it is possible to have a vertical antenna that produces gain. Figure 17-6 shows a collinear gain antenna, with vertical polarization and a horizontally omnidirectional pattern. Incidentally, when mounted horizontally the pattern becomes bidirectional.

The collinear antenna shown in Fig. 17-6 is basically a pair of stacked collinear arrays. Each array consists of a quarter-wavelength section ("A") and a half-wavelength section ("C") separated by a quarter-wavelength phase reversing stub ("B"). The phase reversal stub preserves in-phase excitation for the outer element (referenced to the inner element).

The feedpoint is between the two elements of the array (i.e., between the "A" sections). The coaxial-cable impedance is transformed by a 4:1 BALUN transformer (see Fig. 17-1A). Alternatively, 300-Ω twin lead can be used for the transmission line. If this alternative is used, then the use of UHF shielded twin lead is highly recommended. If the transmitter lacks the balanced output needed to feed twin lead, then use a BALUN at the input end of the twin lead (i.e., right at the transmitter).

Yagi antennas

The Yagi beam antenna is a highly directional gain antenna, and is used both in HF and VHF/UHF systems. The antenna is relatively easy to build at VHF/UHF. In fact,

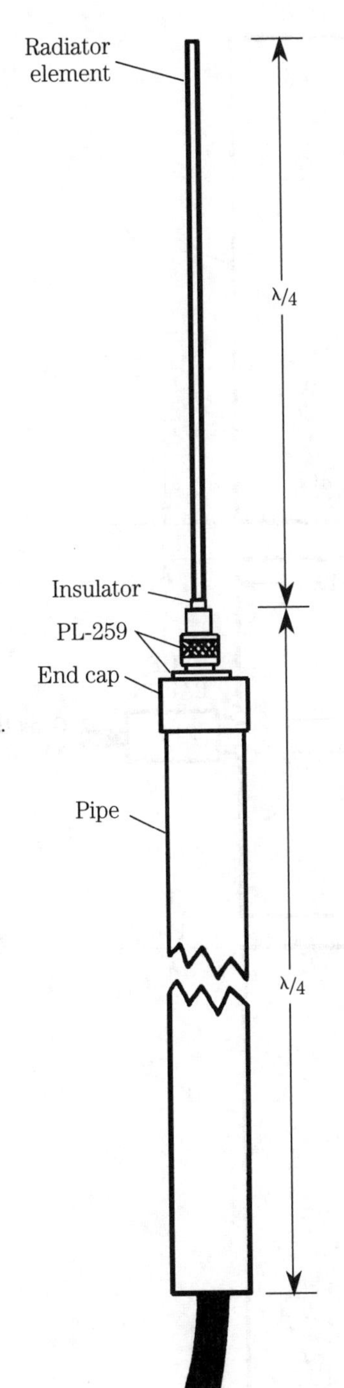

17-5B Tubing coaxial vertical.

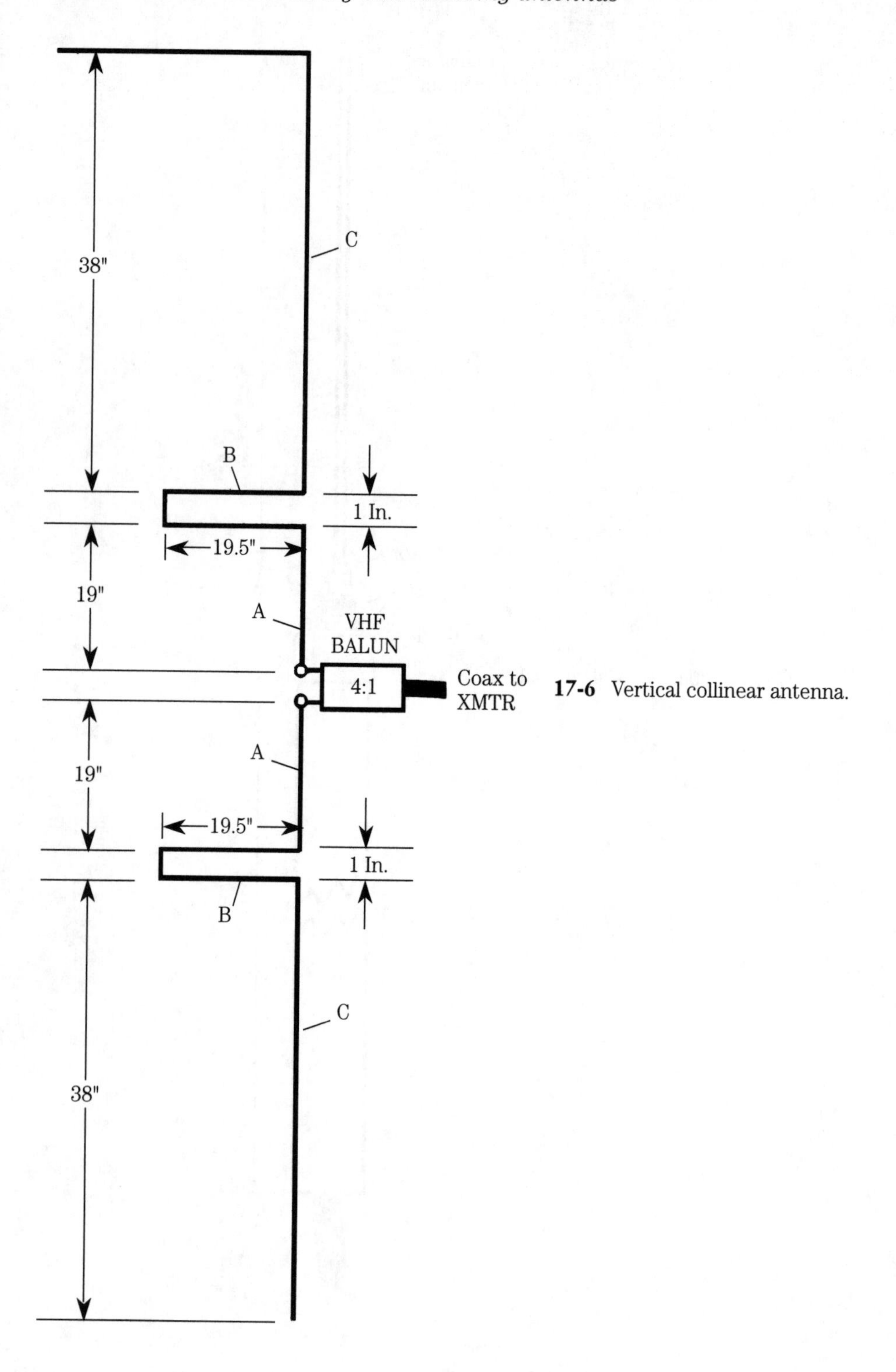

17-6 Vertical collinear antenna.

it is easier than for HF systems. The basic Yagi was covered in chapter 12, so we will only show examples of practical VHF devices. A 6-meter Yagi antenna is shown in Fig. 17-7. This particular antenna is a four element model. The reflector and directors can be mounted directly to a metallic boom, because they are merely parasitic. The driven element, however, must be insulated from the metal boom.

The driven element shown in Fig. 17-7 is a folded dipole. While this is common practice at VHF, because it tends to broadband the antenna, it is not strictly necessary. The dimensions of the driven element are found from Eq. 17.4. Set the equation equal to 300 Ω, select the diameter of the tubing from commercially available sources, and then calculate the spacing.

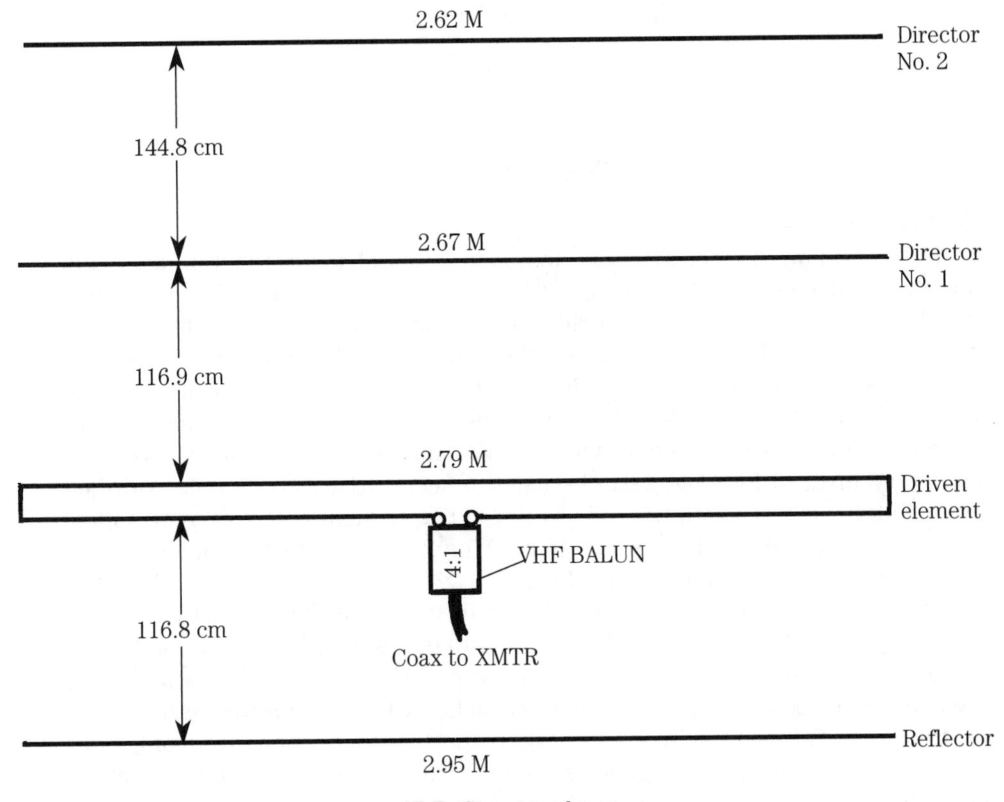

17-7 Six meter beam.

Example Calculate the spacing of a 300-Ω folded dipole when ¾-inch tubing is used in its construction.

Solution:

$$Z_o = 276 \text{ LOG}_{10} \frac{2S}{d}$$

$$300\ \Omega = 276 \text{ LOG}_{10} \frac{2S}{d}$$

$$\frac{300}{276} = \mathrm{LOG}_{10}\,\frac{2S}{d}$$

$$1.08 = \mathrm{LOG}_{10}\,\frac{2S}{d}$$

$$10^{1.08} = \frac{2S}{d}$$

Using 0.75-inch pipe results in:

$$10^{1.08} = \frac{2S}{0.75}$$

$$12.02 = 2.67\,S$$

so,

$$S = 4.5 \text{ inches}$$

Two-meter yagi

Figure 17-8 shows the construction details for a six-element 2-meter Yagi beam antenna. This antenna is built using a 2"-×–2"-wooden boom and elements made of either brass or copper rod. Threaded brass rod is particularly useful, but not strictly necessary. The job of securing the elements (other than the driven element) is easier when threaded rod is used, because it allows a pair of hex nuts, one on either side of the 2"-×–2"-boom, to be used to secure the element. Non-threaded elements can be secured with *RTV* sealing a press-fit. Alternatively, tie wires (see inset to Fig. 17-8) can be used to secure the rods. A hole is drilled through the 2" × 2" to admit the rod or tubing. The element is secure by wrapping a tie wire around the rod on either side of the 2"-×-2", and then soldering it in place. The tie wire is #14 to #10 solid wire.

Mounting of the antenna is accomplished by using a mast secured to the boom with an appropriate clamp. One alternative is to use an end-flange clamp, such as are sometimes used to support pole lamps, etc. The mast should be attached to the boom at the center of gravity, which is also known as the *balance point*. If you try to balance the antenna in one hand unsupported, there is one (and only one) point at which it is balanced (and won't fall). Attach the mast hardware at, or near, this point in order to prevent normal gravitational torques from tearing the mounting apart.

The antenna is fed with coaxial cable at the center of the driven element. Ordinarily, either a matching section of coax, or a gamma match, will be needed because the effect of parasitic elements on the driven element feedpoint impedance is to reduce it.

⅝-wavelength 2-meter antenna

The ⅝-wavelength antenna (Fig 17-9) is popular on 2-meters for mobile operation because it is easy to construct, and it provides a small amount of gain relative to a dipole. The radiator element is ⅝-wavelengths, so its physical length is found from:

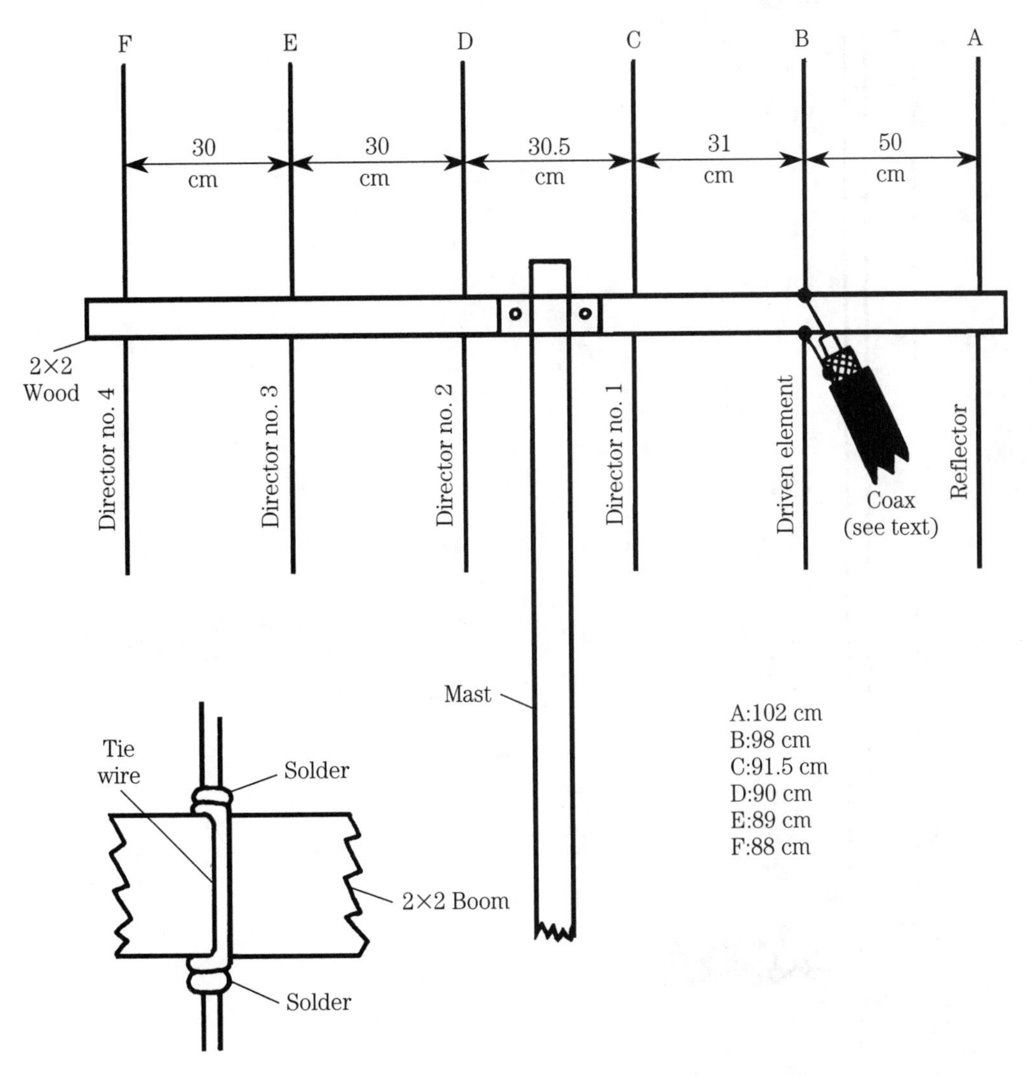

17-8 Two-meter vertical beam.

$$L = \frac{7380}{F_{\text{MHz}}} \text{ inches} \tag{17.6}$$

The ⅝-wavelength antenna is not a good match to any of the common forms of coaxial cables. Either a matching section of cable, or an inductor match, is normally used. In Fig. 17-9 an inductor match is used. The matching coil consists of 2 to 3 turns of #12 wire, wound over a ½-inch o.d. form, ½-inch long. The radiator element can be tubing, brazing rod, or a length of heavy "piano wire." Alternatively, for low-power systems, it can be a telescoping antenna that is bought as a replacement for portable radios or televisions. These antennas have the advantage of being capable of being adjusted to resonance without the need for cutting.

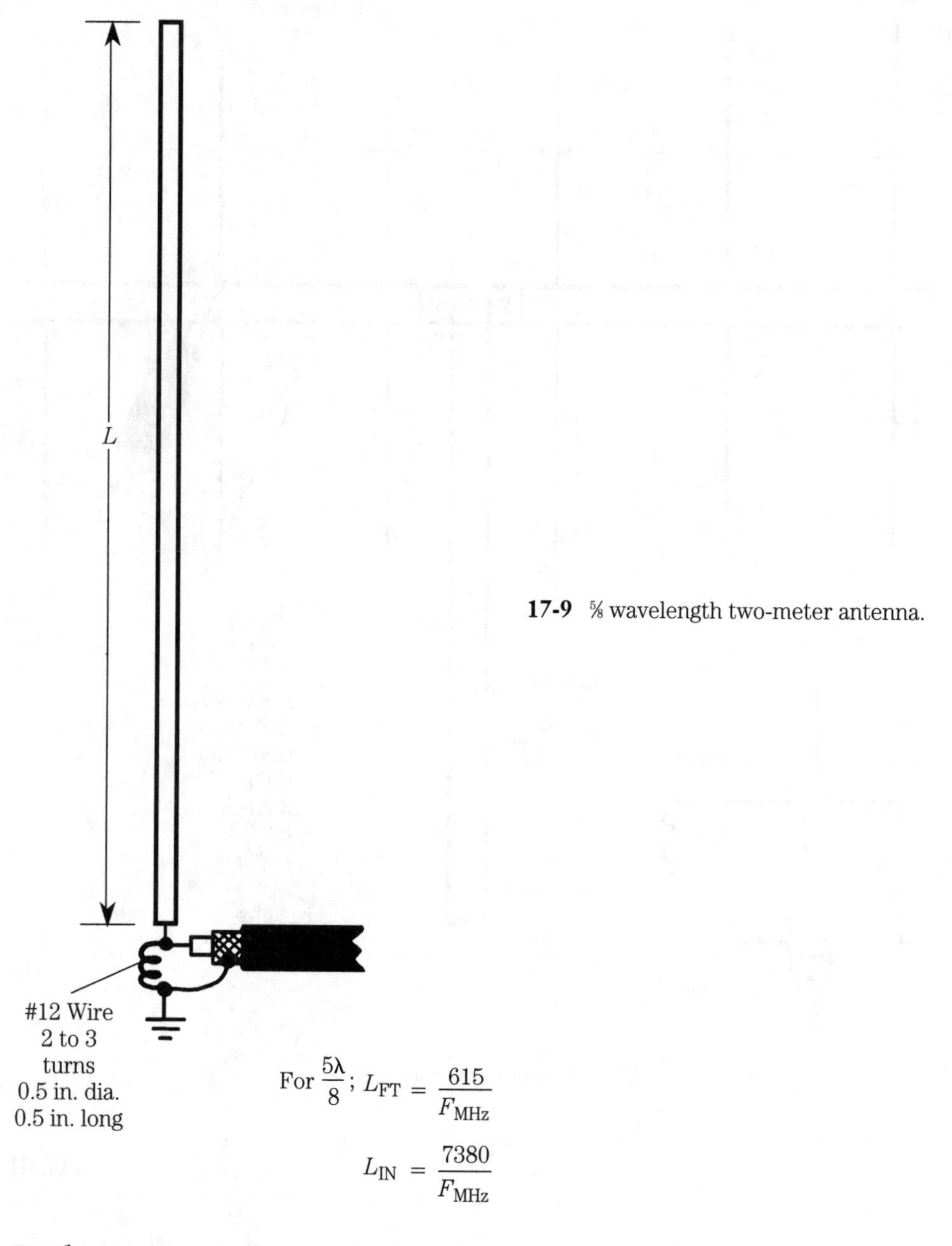

17-9 ⅝ wavelength two-meter antenna.

J-pole antennas

The *J-pole antenna* is another popular form of vertical on the VHF bands. It can be used at almost any frequency, although the example shown in Fig. 17-10 is for 2-meters. The antenna radiator is ¾-wavelength long, so its dimension is found from:

$$L = \frac{8838}{F_{MHz}} \text{ inches} \qquad [17.7]$$

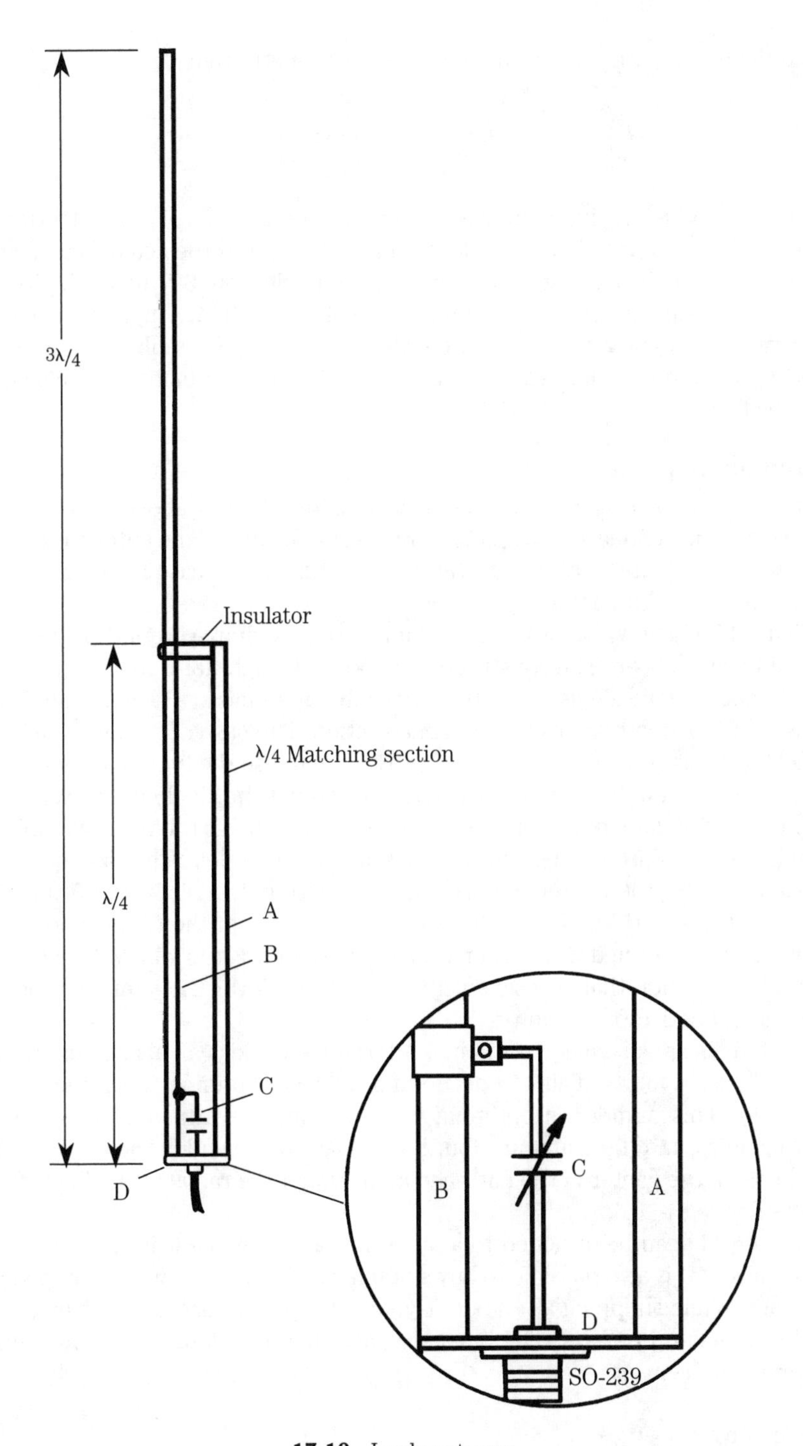

17-10 J-pole antenna.

. . . and the quarter-wavelength matching section length from:

$$L = \frac{2952}{F_{\text{MHz}}} \text{ inches} \qquad \textbf{[17.8]}$$

Taken together the matching section and the radiator form a parallel transmission line with a characteristic impedance that is four times the coaxial cable impedance. If 50-Ω coax is used, and the elements are made from 0.5 inch o.d. pipe, then a spacing of 1.5-inches will yield an impedance of about 200 Ω. Impedance matching is accomplished by a gamma match consisting of a 25-pF variable capacitor, connected by a clamp to the radiator, about 6 inches (experiment with placement) above the base.

Ground plane

The *ground plane antenna* is a vertical radiator situated above an artificial RF ground consisting of quarter-wavelength radiators. Ground plane antennas can be either ¼-wavelength or ⅝-wavelength (although for the latter case impedance matching is needed—see the previous example).

Figure 17-11 shows how to construct an extremely simple ground-plane antenna for 2-meters and above. The construction is too lightweight for 6-meter antennas (in general), because the element lengths on 6-meter antennas are long enough to make their weight too great for this type of construction. The base of the antenna is a single SO-239 chassis type coaxial connector. Be sure to use the type that requires four small machine screws to hold it to the chassis, and not the single nut variety.

The radiator element is a piece of $\frac{3}{16}$-inch or 4-mm brass tubing. This tubing can be bought at hobby stores that sell airplanes and other models. The sizes quoted just happen to fit over the center pin of an SO-239 with only a slight tap from a lightweight hammer—and I do mean *slight* tap. If the inside of the tubing and the connector pin are pre-tinned with solder, then sweat soldering the joint will make a good electrical connection that is resistant to weathering. Cover the joint with clear lacquer spray for added protection.

The radials are also made of tubing. Alternatively, rods can also be used for this purpose. At least four radials are needed for a proper antenna (only one shown in Fig. 17-11). This number is optimum because they are attached to the SO-239 mounting holes, and there are only four holes. Flatten one end of the radial, and drill a small hole in the center of the flattened area. Mount the radial to the SO-239 using small hardware (4-40, etc.).

The SO-239 can be attached to a metal L-bracket. While it is easy to fabricate such a bracket, it is also possible to buy suitable brackets in any well-equipped hardware store. While shopping at one do-it-yourself type of store, I found several reasonable candidate brackets. The bracket is attached to a length of 2"-×-2" lumber that serves as the mast.

Halo antennas

One of the more saintly antennas used on the VHF boards is the halo (Fig. 17-12). This antenna basically takes a half-wavelength dipole and bends it into a circle. The

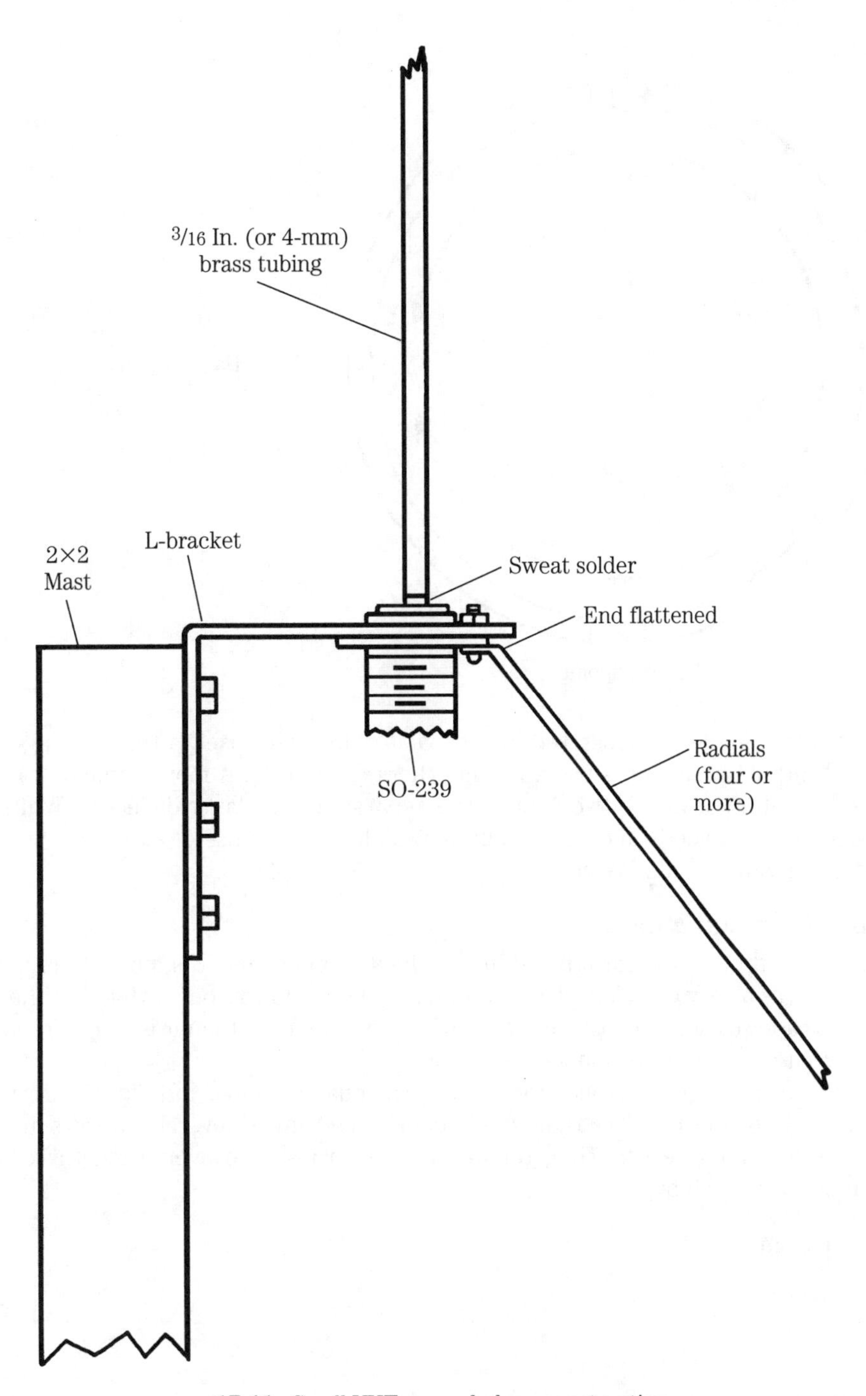

17-11 Small VHF ground plane construction.

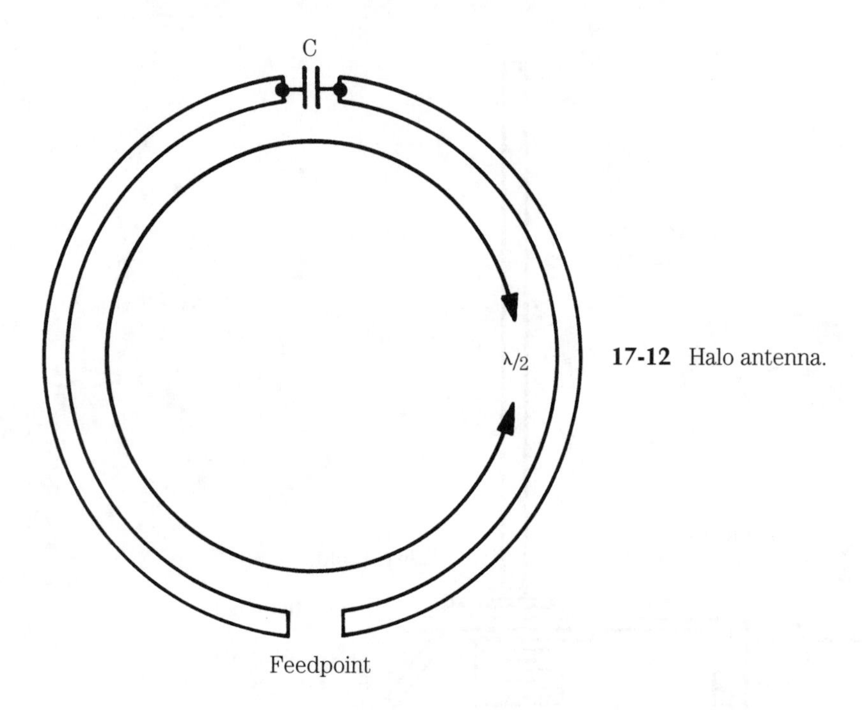

17-12 Halo antenna.

ends of the dipole are separated by a capacitor. In some cases, a transmitting-type mica "button" capacitor is used, but in others (and perhaps more commonly), the halo capacitor consists of two 3-inch disks separated by a plastic dielectric. While air also serves as a good (and perhaps better) dielectric, the use of plastic allows mechanical rigidity to the system.

Quad beam antennas

The quad antenna was introduced in the chapter on beams. It is, nonetheless, also emerging as a very good VHF/UHF antenna. It should go without saying that the antenna is a lot easier to construct at VHF/UHF frequencies, than it is at HF frequencies! Figure 17-13 shows a modest example.

There are several methods for building the quad antenna, and Fig. 17-13 represents only one of them. The radiator element can be any of several materials including heavy solid wire (#8–#12), tubing, or metal rods. The overall lengths of the elements are given by:

Driven element:

$$L = \frac{11826}{F_{\text{MHz}}} \text{ inches} \qquad \textbf{[17.9]}$$

Reflector:

$$L = \frac{12562}{F_{\text{MHz}}} \text{ inches} \qquad \textbf{[17.10]}$$

Overall lengths:

$$\text{Reflector: } L = \frac{12562}{F_{\text{MHz}}} \text{ Inches}$$

$$\text{Director: } L = \frac{11248}{F_{\text{MHz}}} \text{ Inches}$$

$$\text{Driven element: } L = \frac{11826}{F_{\text{MHz}}} \text{ Inches}$$

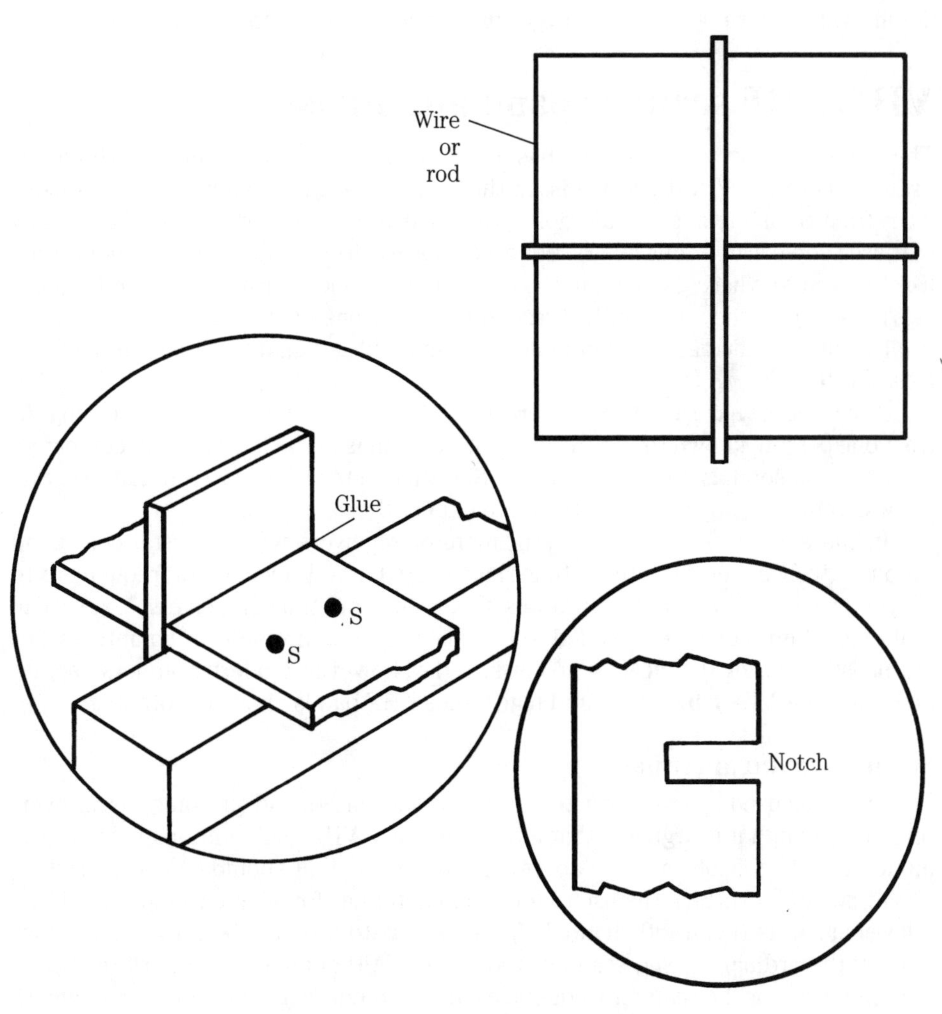

17-13 Quad loop construction on VHF.

Director:

$$L = \frac{11248}{F_{\text{MHz}}} \text{ inches} \qquad [17.11]$$

There are several alternatives for making the supports for the radiator. Because of the lightweight construction, almost any method can be adapted for this purpose. In the case shown in Fig. 17-13, the spreaders are made from either 1-inch firring strips, trim strips or (at above 2 meters) even wooden paint stirring sticks. The sticks are cut to length, and then half notched in the center (Fig. 17-13, detail "B"). The two spreaders for each element are joined together at right angles and glued (Fig. 17-13, detail "C"). The spreaders can be fastened to the wooden boom at points "S" in detail "C". The usual rules regarding element spacing (0.15 to 0.31 wavelength) are followed. See the information on quad antennas in chapter 12 for further details. Quads have been successfully built for all amateur bands up to 1296 MHz.

VHF/UHF scanner band antennas

The hobby of shortwave listening has always had a subset of adherents who listen exclusively to the VHF/UHF bands. In those bands, are found a rich variety of services from commercial two-way radios, to police and fire systems, and others. Many of these people are serious DXers, but others have a little of the voyeur in them (they like to listen to what's going on in the community by monitoring the police frequencies). As long as they don't either respond to the scene of a police or fire action, or use the information gained from monitoring in an illegal manner, they are perfectly free to listen in.

A few people have an unusually practical element to their VHF/UHF listening. At least one person, known to the author, routinely tunes in the local taxicab company's frequency as soon as she orders a cab. She then listens for her own address. She knows, from that, approximately when to expect the cab.

In the early 1960s, when all VHF monitor receivers where crystal controlled (and relatively expensive), the listener had a very limited selection of frequencies to choose from. Having worked in various shops that sold those receivers, I can recall that most of our customers tended to be police officers, firemen (or volunteers), or journalists covering the local crime news. Today, however, modern scanners operate on both major VHF bands (low and high), plus UHF bands and some others.

Scanner-vision antennas

The antennas used by scanner listeners are widely varied, and (in some cases) overpriced. Although it is arguable that a total coverage VHF/UHF antenna is worth the money, and it probably is, there are other possibilities that should be considered.

First, don't overlook the use of television antennas for scanner monitoring! The television bands (about 80 channels from 54 MHz to around 800 MHz) encompass most of the ordinarily used scanner frequencies. Although antenna performance is not optimized for the scanner frequencies, it is also not "zero" on those frequencies. If you already have an "all channel" TV antenna installed, then it is a simple matter to connect the antenna to the scanner receiver (Note: if the antenna uses 300-Ω twin lead, then install a 4:1 BALUN transformer that accepts 300 Ω in and produces 75 Ω out. These transformers are usually available at TV shops, video shops—including VCR type rental places—and Radio Shack stores.

The directional characteristic of the TV antenna makes it both an advantage, and a disadvantage, to the scanner user. If the antenna has a rotator, then there is no problem. Just rotate the antenna to the direction of interest. However, if the antenna is fixed, and the station of interest is elsewhere than where the antenna is pointed, then there is a bit of a problem. Nonetheless, some reception is possible for at least three reasons. One is that the main beam of the antenna is not infinitely thin, so the station of interest might well be within the beam—even if off the point a little bit. Second, there are always sidelobes on an antenna. These are areas outside of the main beam that offer reduced, but non-zero, reception characteristics. Finally, the sidelobes and main lobe of the antenna are optimized for the TV bands . . . and may not obtain the same directivity on certain scanner frequencies. It is, therefore, possible that a TV antenna will have an unusual lobe in the direction of interest for the scanner channel.

The TV antenna also offers other possibilities. The hardware and electronics, normally found useful on TV antenna systems, are also useful for scanner antennas. The use of the mounting hardware is obvious, but less obvious is the use of components such as multi-set couplers, impedance transformers, and wideband amplifiers. For example, Fig. 17-14 shows two scanner antennas joined together into a single transmission line using a two-set coupler. Although intended to allow two television sets to receive a signal from the same antenna, the device also works to combine two antennas into a single transmission line. Two popular TV uses include two antennas for different directions, or VHF and UHF antennas that share the same downlead. Be sure to buy a weatherproof model if you intend to mount the coupler on the antenna mast (the correct place for it) (Fig. 17-15).

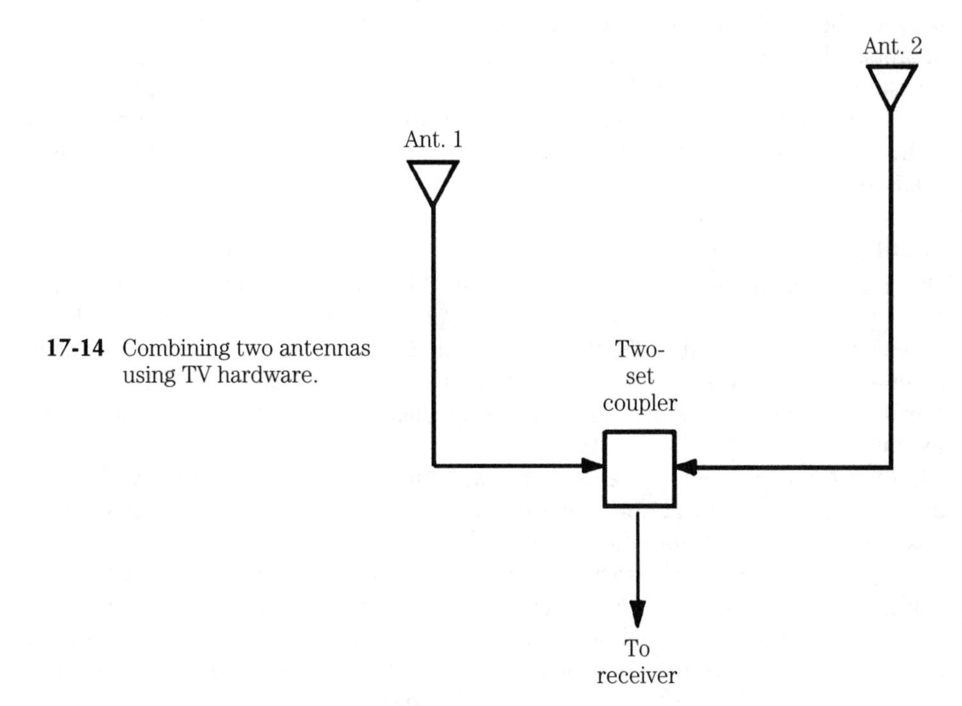

17-14 Combining two antennas using TV hardware.

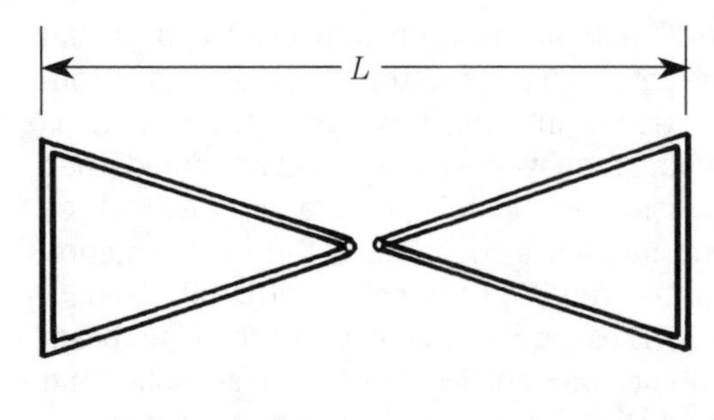

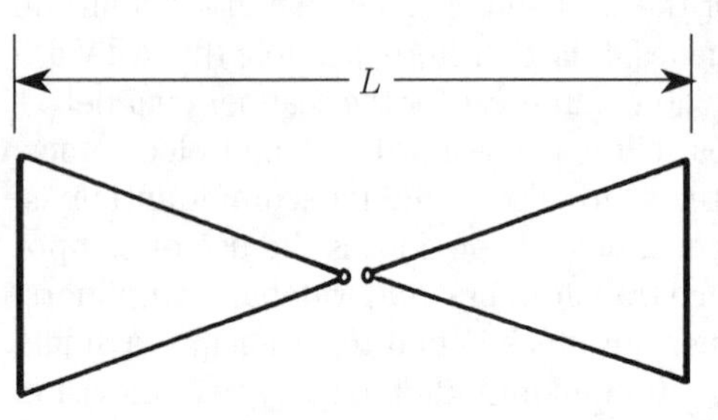

17-15 VHF/UHF "Bowie" dipoles for broadbanded reception.

TV antenna amplifiers are used to increase the weak signal from a distant station to a level compatible with the TV receiver. Because these amplifiers are wideband designs, they will also work well for the scanner bands that are inclusive within the 54- to 800-MHz TV bands. Some models are intended to be part of the antenna, while others are mounted indoors, some distance from the antenna. The basic rule of thumb is to place the amplifier as close as possible to the "head end" of the system where the antenna is located. This situation will permit signals to be built up to a level considerably above the noise level before being attenuated in the transmission line losses.

TV/FM receiver folded dipoles made of twin-lead transmission line also work well at other VHF/UHF frequencies. Figures 17-16A and 17-16B show the details of their construction.

Scanner skyhooks

Other antennas not to be overlooked are the CB, or amateur radio, high-frequency (shortwave) antennas. Many amateur radio operators use their 80-meter dipole antenna on 2 meters (144 to 148 MHz) and find that it has the gain and directivity characteristics of a longwire antenna. Similarly, with vertical "ham" and CB antennas, they have apparent gain on some VHF frequencies and will at least perform to some extent on others.

Even the lowly "random length" wire antenna used for your shortwave receiver ought to turn in decent performance as a VHF/UHF longwire antenna. These antennas are simply a 30- to 150-foot length of #14 wire attached to a distant support.

Figure 17-17 shows two variants to an antenna that is popular with VHF/UHF receiver operators: the *drooping dipole*. These antennas are similar to the inverted-vee dipole, in that it consists of two quarter-wavelength radiator elements. Unlike the ordinary dipole, the ends of the radiators, in the single-band drooping dipole of Fig. 17-7A, are lowered. The drooping dipole can be made using an SO-239 "UHF"

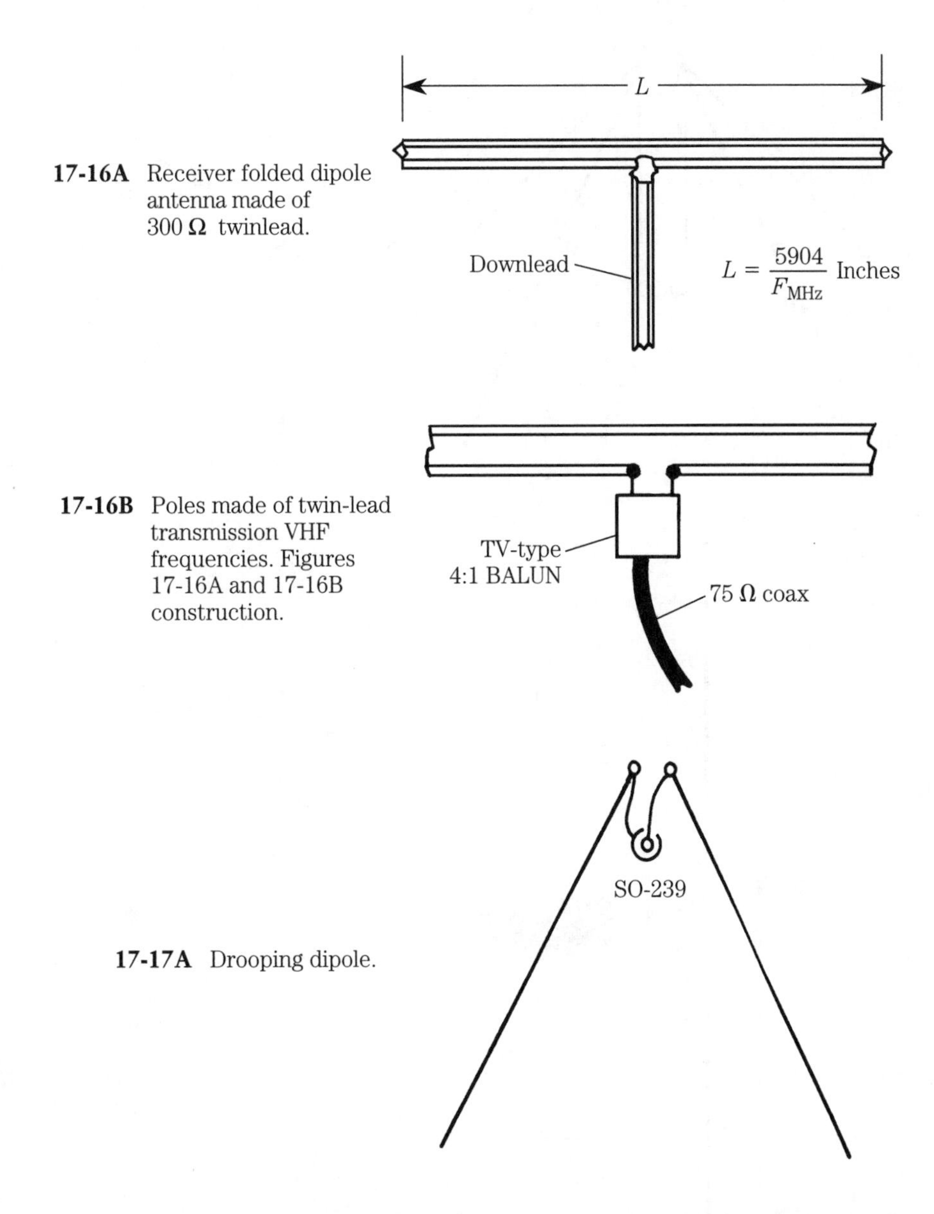

17-16A Receiver folded dipole antenna made of 300 Ω twinlead.

17-16B Poles made of twin-lead transmission VHF frequencies. Figures 17-16A and 17-16B construction.

17-17A Drooping dipole.

coaxial connector as the support for either stiff wire, brass hobbyist tubing or brazing rod elements cut to λ/4 each. The antenna can be made into a multiband, or very wideband, antenna by connecting several pairs of drooping dipoles in parallel, as in Fig. 17-17B.

The vertical dipole (Fig. 17-18A) also finds use in the VHF/UHF region. Each element is λ/4 long, and its physical length can be found from:

$$L = \frac{2832}{F_{MHz}} \text{ inches} \qquad \textbf{[17.12]}$$

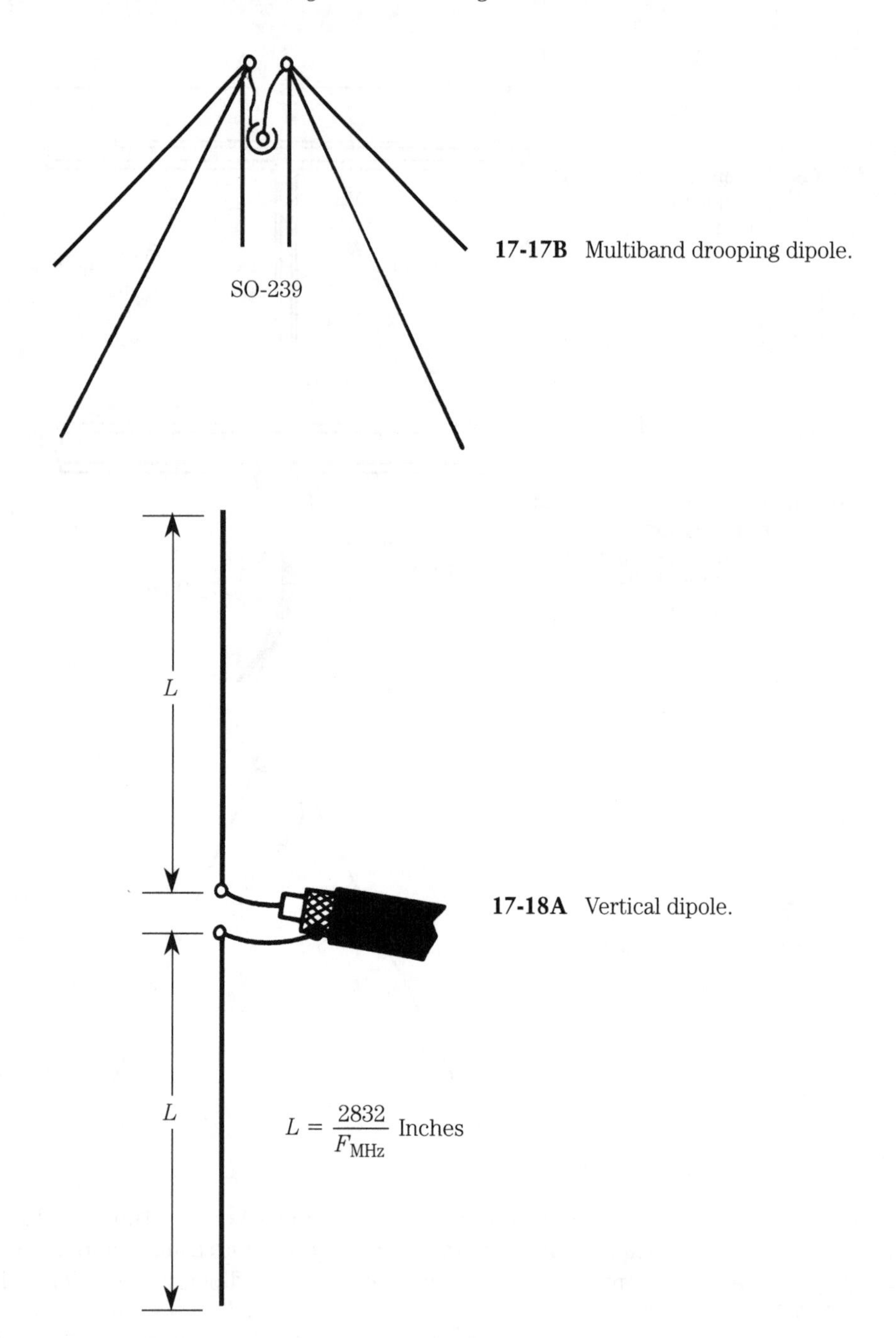

17-17B Multiband drooping dipole.

17-18A Vertical dipole.

This antenna can be built with the same types of materials as the drooping dipole. Although several different construction techniques are popular, the method of Fig. 17-18B is probably the most popular with homebrew builders.

A ground-plane antenna for VHF/UHF frequencies is shown in Fig. 17-19. The antenna sits on an L-bracket mounted to a 2"-×-4" wooden mast or support. The L-bracket

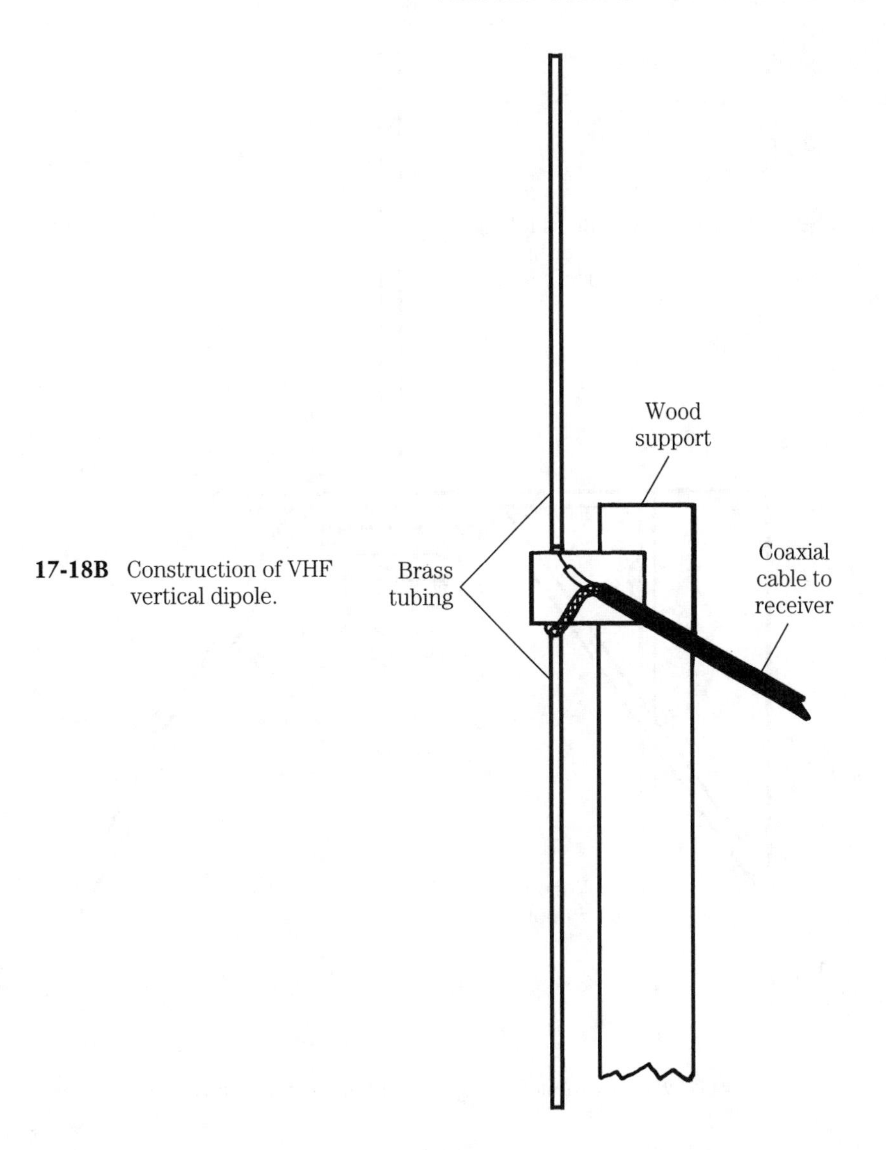

17-18B Construction of VHF vertical dipole.

can be manufactured for the purpose, or purchased at Happy Harry's hardware emporium. The base of the antenna is a chassis mount SO-239 UHF coaxial connector. A quarter-wave radiator element is made from ⅛-inch brass tubing, or brazing rod, soldered to the center conductor pin of the SO-239. Similarly, radials can be made from the same type of material, but is soldered directly to the body of the SO-239 (this action connects it to the coaxial cable shield). The four mounting holes of the SO-239 make reasonable anchors for the radials. Solder them for electrical integrity.

A coaxial vertical can be made from 1-inch brass tubing, or copper plumbing pipe, and either brazing rod or brass tubing. The sleeve is fitted with an end cap that has an SO-239 mounted on it (see inset). The radiator element can be made with a mate to the SO-239, namely the PL-259 coaxial plug. The radiator element is sol-

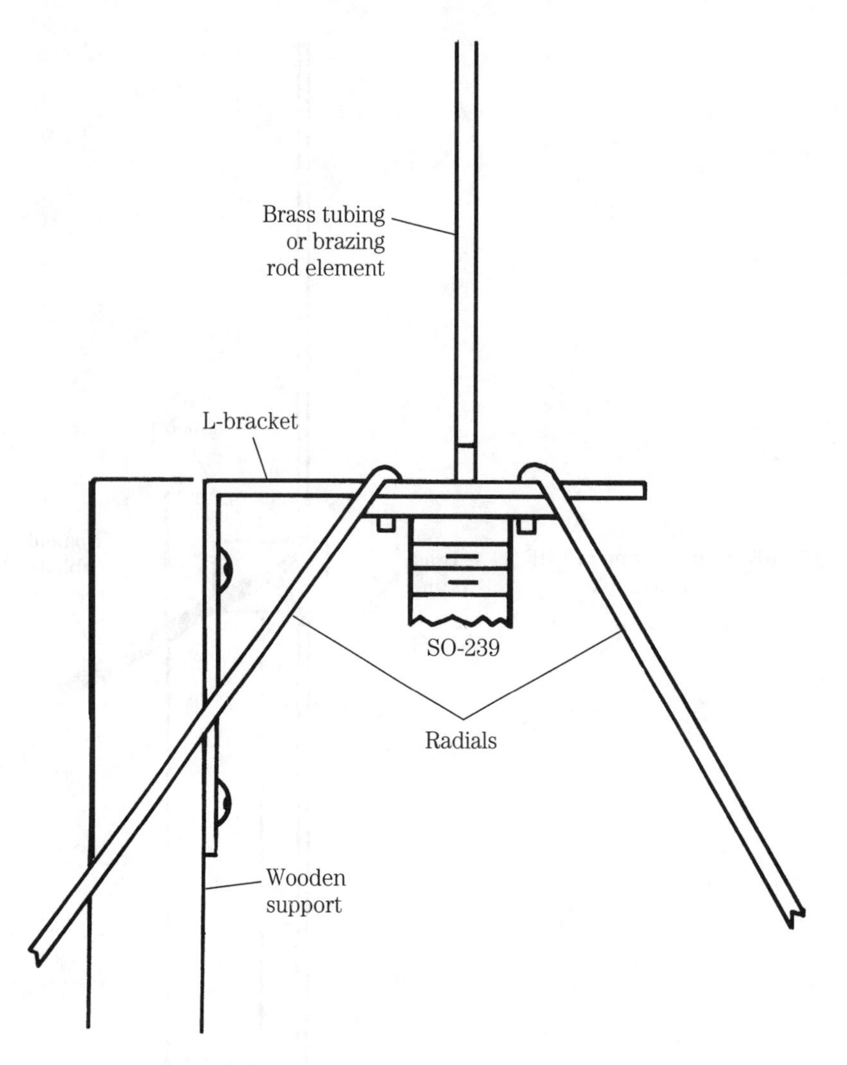

17-19 Construction of VHF ground plane for receivers.

dered to the center conductor, and it is insulated from the shield. You can buy coaxial whips with this type of construction, so may not have to actually build the radiator element.

Additional gain, about +3 dB, can be achieved by stacking VHF/UHF antennas together. Figure 17-20 shows a typical arrangement in which two half-wavelength dipole antennas are connected together through a quarter-wavelength harness of RG-59/U coaxial cable. This harness is shorter than a quarter wavelength by the velocity factor of the coaxial cable:

$$L = \frac{2832\ V}{F_{\text{MHz}}} \qquad \textbf{[17.13]}$$

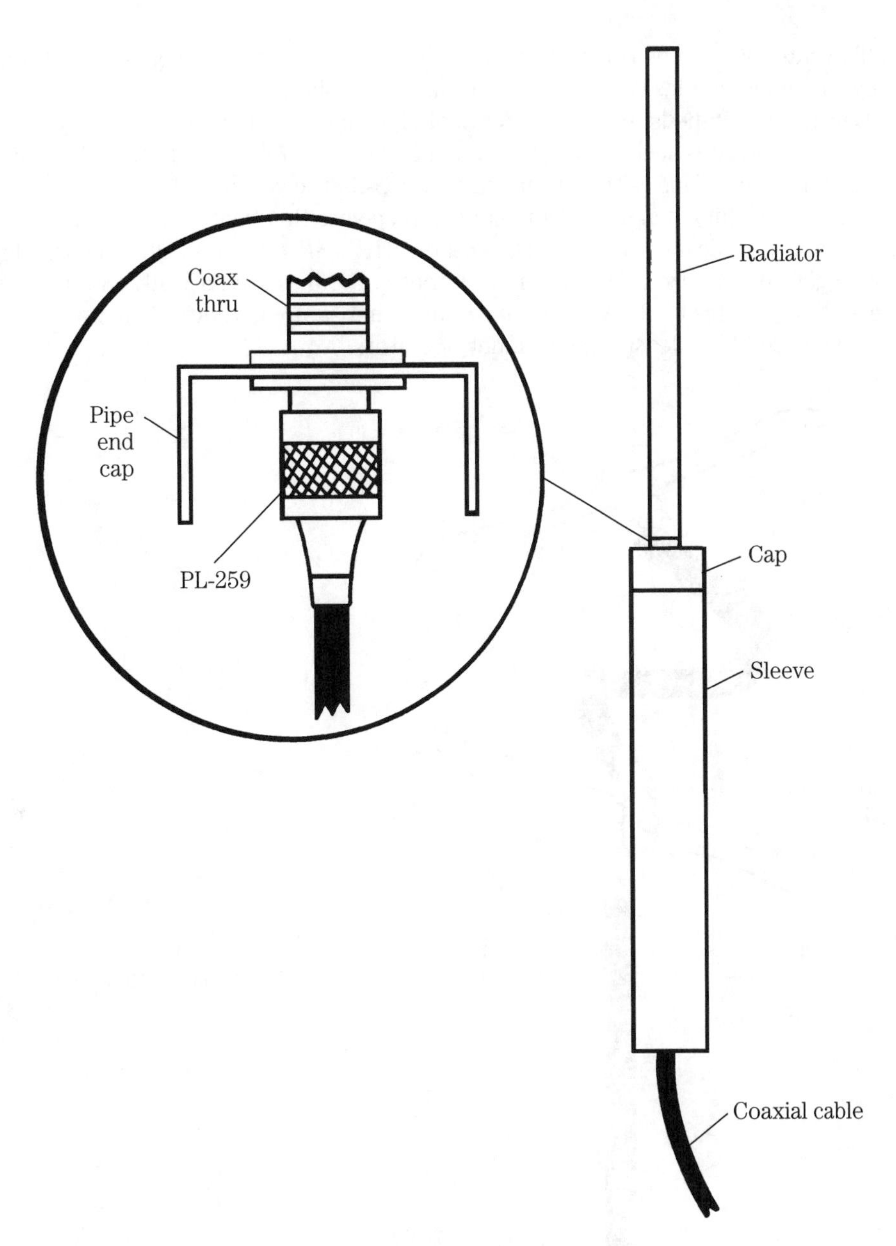

17-20 Coaxial dipole vertical.

Where:

L is the length in inches
V is the velocity factor (typically 0.66 or 0.80 for common coax)
F_{MHz} is the frequency in megahertz (MHz)

The antennas can be oriented in the same direction to increase gain, or orthogonally (as shown in Fig. 17-21) to make it more omnidirectional.

Because the impedance of two identical dipoles, fed in parallel, is one-half that of a single dipole, it is necessary to have an impedance matching section made of RG-58/U coaxial cable. This cable is then fed with RG-59/U coax from the receiver.

There is nothing magical about scanner receivers that require any form of antenna that is significantly different from other VHF/UHF antennas. Although the designs might be optimized for VHF or UHF, these antennas are basically the same as others shown in this book. As a matter of fact, almost any antenna, from any chapter, can be used by at least some scanner operators.

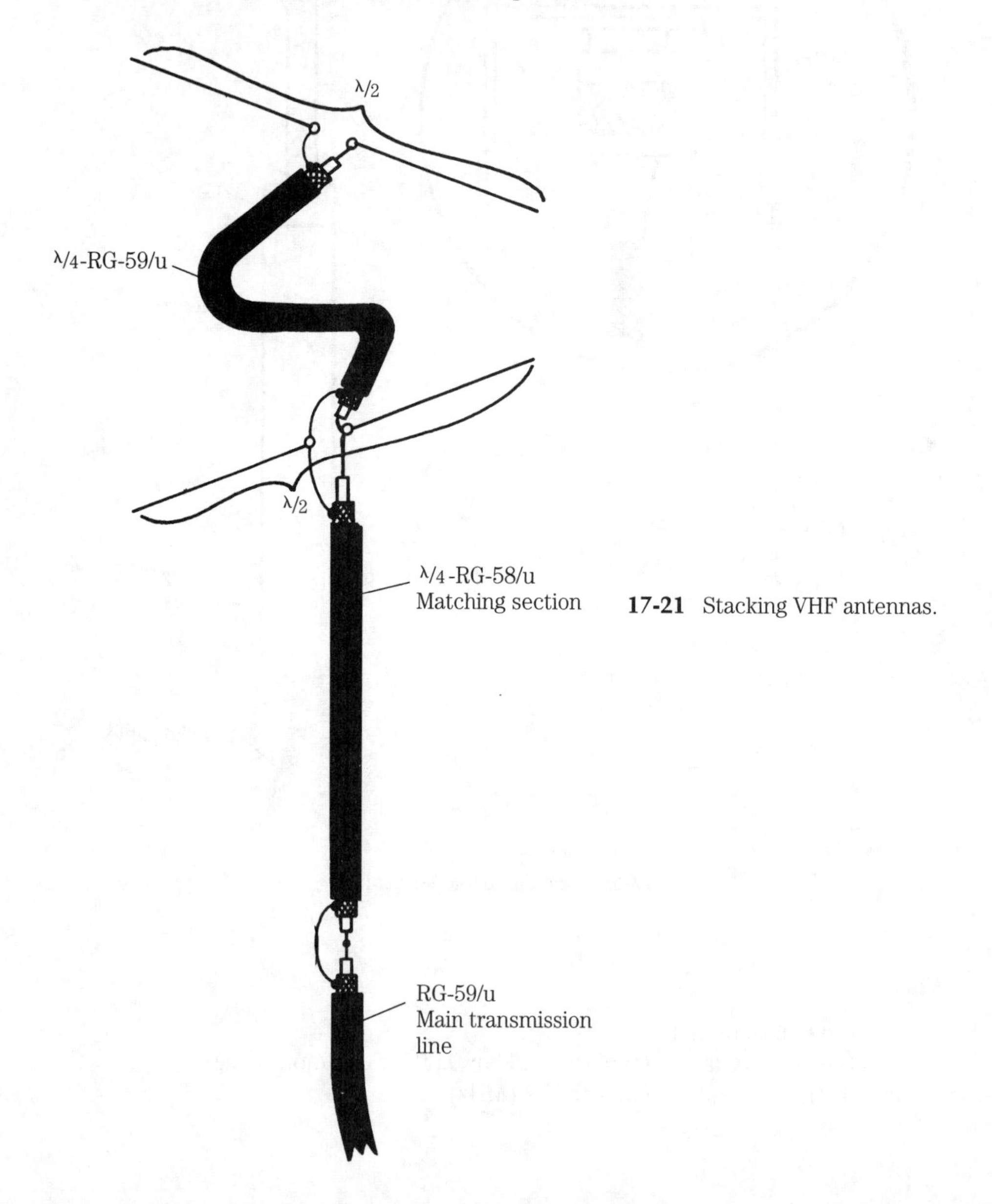

17-21 Stacking VHF antennas.

18
CHAPTER

Microwave waveguides and antennas

THE MICROWAVE PORTION OF THE RADIO SPECTRUM COVERS FREQUENCIES FROM ABOUT 900 MHz to 300 GHz, with wavelengths in free-space ranging from 33 cm down to 1 mm. Transmission lines can be used at frequencies from dc to about 50 or 60 GHz; although, above 5 GHz, only short runs are practical, because attenuation increases dramatically as frequency increases. There are three types of losses in conventional transmission lines: *ohmic*, *dielectric*, and *radiation*. The ohmic losses are caused by the current flowing in the resistance of the conductors making up the transmission lines. Because of the skin effect, which increases resistance at higher frequencies, these losses tend to increase in the microwave region. Dielectric losses are caused by the electric field acting on the molecules of the insulator and thereby causing heating through molecular agitation. Radiation losses represents loss of energy as an electromagnetic wave propagates away from the surface of the transmission line conductor.

Losses on long runs of coaxial transmission line (the type most commonly used) cause concern even as low as the 400-MHz region. Also, because of the increased losses, power handling capability decreases at high frequencies. Therefore, at higher microwave frequencies, or where long runs make coax attentuation losses unacceptable, or where high power levels would overheat the coax, *waveguides* are used in lieu of transmission lines.

What is a waveguide? Consider the "light pipe analogy" depicted in Fig. 18-1. A flashlight serves as our "RF source," which (given that light is also an electromagnetic wave) is not altogether unreasonable. In Fig. 18-1A the source radiates into free-space, and spreads out as a function of distance. The intensity per unit area, at the destination (a wall), falls off as a function of distance (D) according to the *inverse square law* ($1/D^2$).

But now consider the transmission scheme in Fig. 18-1B. The light wave still propagates over distance D, but is now confined to the interior of a mirrored pipe. Almost all of the energy (less small losses) coupled to the input end is delivered to

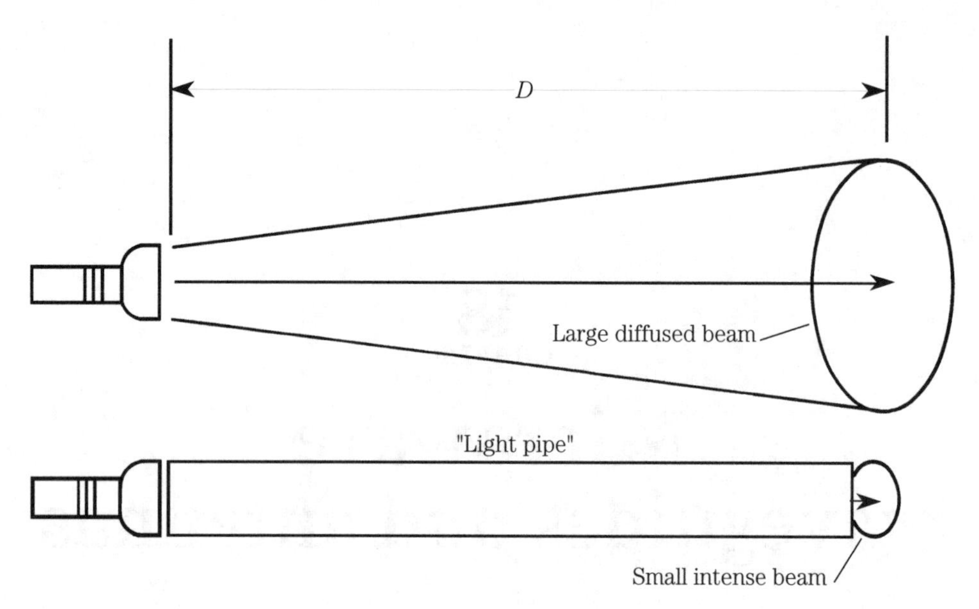

18-1 Waveguide analogy to light pipe.

the output end, where the intensity is practically undiminished. Although not perfect, the light pipe analogy neatly summarizes, on a simple level, the operation of microwave waveguides.

Thus, we can consider the waveguide as an "RF pipe" without seeming too serenely detached from reality. Similarly, fiber-optic technology is waveguide-like at optical (IR and visible) wavelengths. In fact, the analogy between fiber optics and waveguide can withstand more rigorous comparison than the simplistic light pipe analogy.

The internal walls of the waveguide are not mirrored surfaces, as in our optical analogy; but are, rather, electrical conductors. Most waveguides are made of either aluminum, brass, or copper. In order to reduce ohmic losses, some waveguides have their internal surfaces electroplated with either gold or silver, both of which have lower resistivities than the other metals mentioned above.

Waveguides are hollow metal pipes, and can have either circular or rectangular crossections (although the rectangular are, by far, the most common). Figure 18-2 shows an end view of the rectangular waveguide. The dimension a is the wider dimension, and b is the narrower. These letters are considered the standard form of notation for waveguide dimensions, and will be used in the equations developed in this chapter.

Development of the rectangular waveguide from parallel transmission lines

One way of visualizing how a waveguide works, is to develop the theory of waveguides from the theory of elementary parallel transmission lines (see chapter 4). Figure 18-3A shows the basic parallel transmission line which was introduced in chapter 3. The line

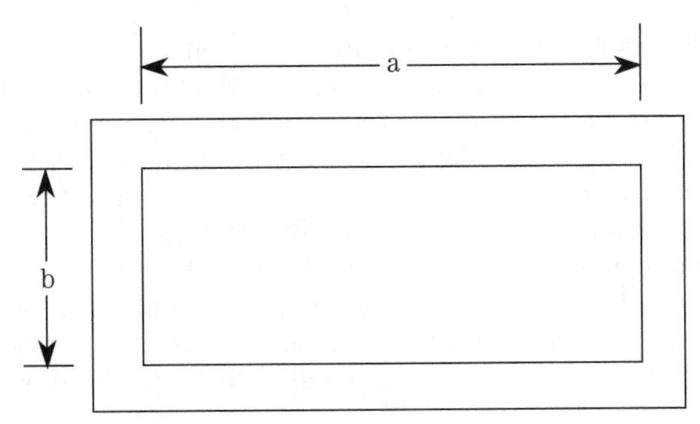

18-2 Rectangular waveguide (end view).

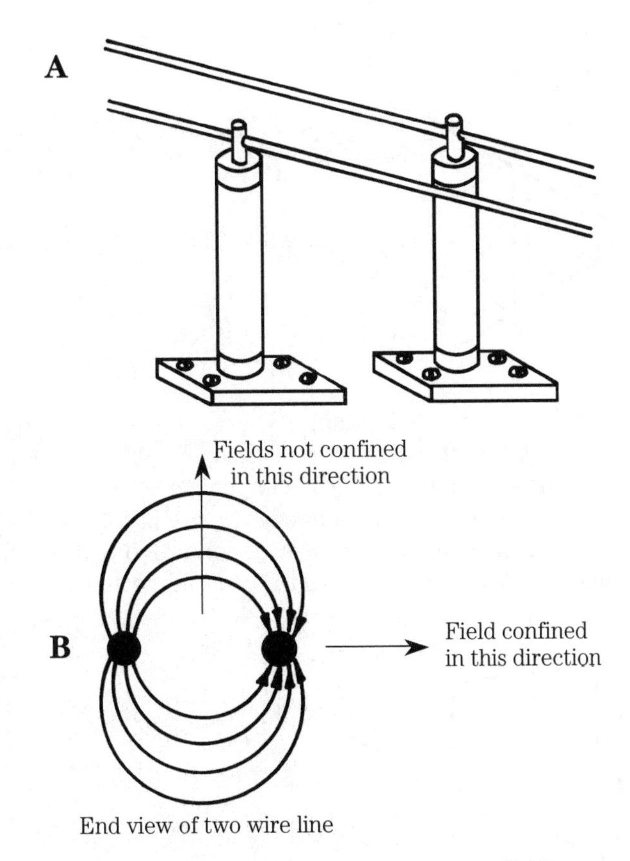

18-3 Development of waveguide from parallel line.

consists of two parallel conductors separated by an air dielectric. Because air won't support the conductors, ceramic or other material insulators are used as supports.

There are several reasons why the parallel transmission line *per se* is not used at microwave frequencies. Skin effect increases ohmic losses to a point that is unacceptable. Also, the insulators supporting the two conductors are significantly more lossy at microwave frequencies, than at lower frequencies. Finally, radiation losses increase dramatically. Figure 18-3B shows the electric fields surrounding the conductors. The fields add algebraically (either constructively or destructively) resulting in pinching of the resultant field along one axis, and bulging along the other. This geometry increases radiation losses at microwave frequencies.

Now let's consider the quarter-wavelength shorted stub. The "looking-in" impedance of such a stub is infinite. When placed in parallel across a transmission line (Fig. 18-4A) the stub acts like an insulator. In other words, at its resonant frequency, the stub is a *metallic insulator*, and can be used to physically support the transmission line.

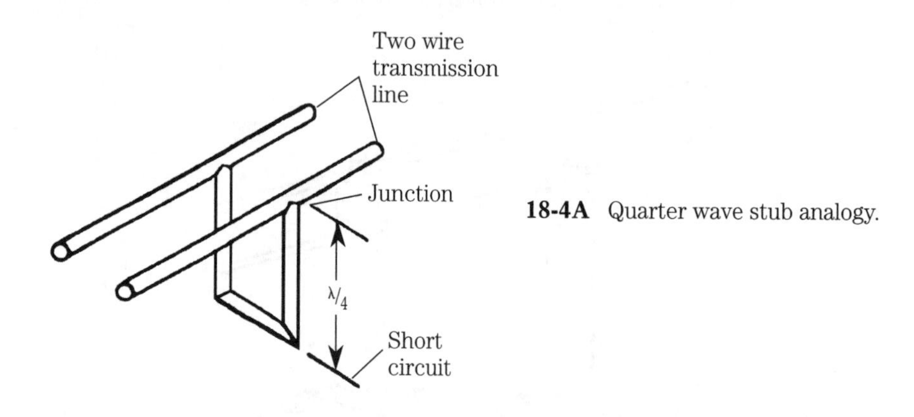

18-4A Quarter wave stub analogy.

Again, because the impedance is infinite we can connect two quarter-wavelength stubs in parallel with each other across the same points on the transmission line (Fig. 18-4B) without loading down the line impedance. This arrangement effectively forms a half-wavelength pair. The impedance is still infinite, so no harm is done. Likewise, we can parallel a large number of center-fed half-wavelength pairs along the line, as might be the case when a long line is supported at multiple points.

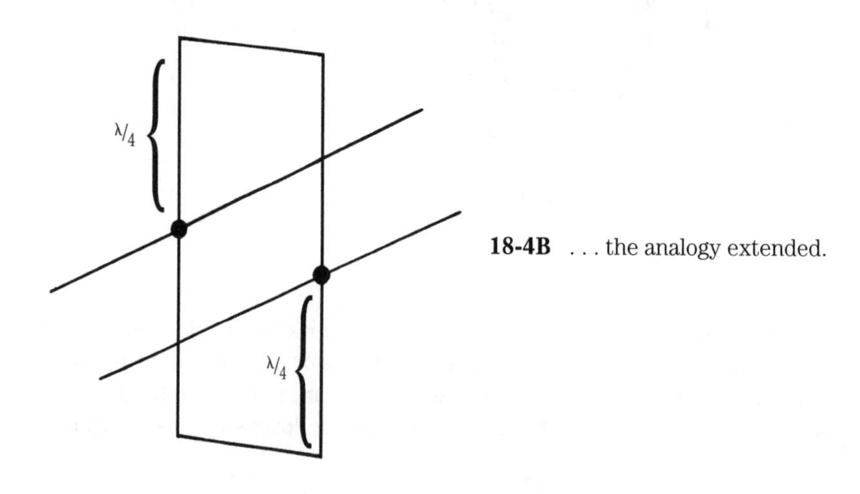

18-4B . . . the analogy extended.

The waveguide is analogous to an *infinite number of center-fed "half-wave pairs" of quarter-wave shorted stubs* connected across the line. The result is the continuous metal pipe structure of common rectangular waveguide (Fig. 18-4C).

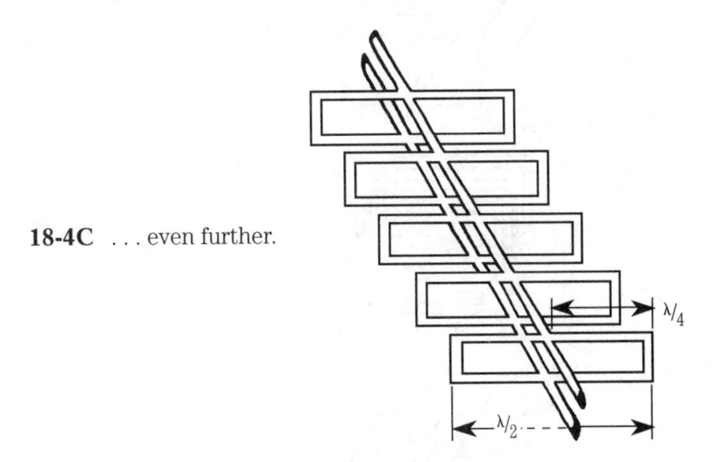

18-4C . . . even further.

On first glance, relating rectangular waveguide to quarter-wavelength shorted stubs seem to fall down, except at the exact resonant frequency. It turns out, however, that the analogy also holds up at other frequencies, so long as the frequency is higher than a certain minimum cut-off frequency. The waveguide thus acts like a high-pass filter. There is also a practical upper frequency limit. In general, waveguides support a bandwidth of 30 to 40 percent of cut-off frequency. As shown in Fig. 18-5, the center line of the waveguide (which represents the points where the conductors are in the parallel line analogy) becomes a "shorting bar" between segments, and that "bar" widens or narrows according to operating frequency. Thus, the active region is still a quarter-wavelength shorted stub.

Below the cut-off frequency, the structure disappears entirely, and the waveguide acts like a parallel transmission line with a low impedance inductive reactance shorted across the conductors. When modelled as a pair of quarter-wavelength stubs, the a dimension of the waveguide is half-wavelength long. The cut-off frequency is defined as the frequency at which the a dimension is less than half wavelength.

Propagation modes in waveguides

The signal in a microwave waveguide propagates as an electromagnetic wave, not as a current. Even in a transmission line, the signal propagates as a wave because the current in motion down the line gives rise to electric and magnetic fields, which behave as an electromagnetic field. The specific type of field found in transmission lines, however, is a *transverse electromagnetic* (*TEM*) *field*. The term *transverse* implies things at right angles to each other, so the electric and magnetic fields are perpendicular to the direction of travel. In addition to the word "transverse," these right angle waves are said to be "normal" or "orthogonal" to the direction of travel—three different ways of saying the same thing: right-angledness.

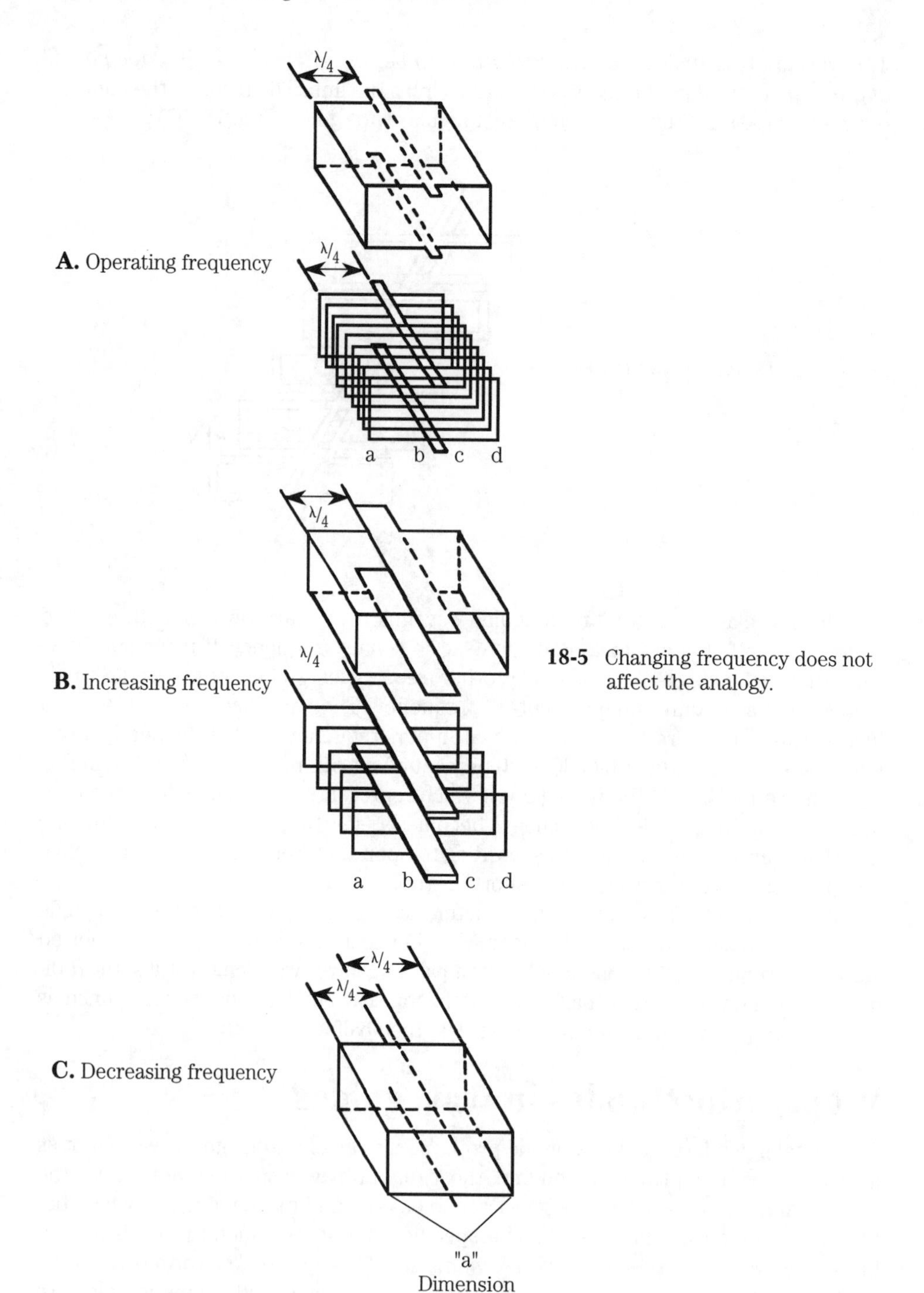

18-5 Changing frequency does not affect the analogy.

Boundary conditions

The TEM wave will not propagate in a waveguide because certain *boundary conditions* apply. Although the wave in the waveguide propagates through the air (or inert gas dielectric) in a manner similar to free-space propagation, the phenomena is bounded by the walls of the waveguide, and that implies certain conditions must be met. The boundary conditions for waveguides are:

1. *The electric field must be orthogonal to the conductor in order to exist at the surface of that conductor*; and
2. *The magnetic field must not be orthogonal to the surface of the waveguide.*

In order to satisfy these boundary conditions the waveguide gives rise to two types of propagation modes: *transverse electric mode* (TE-mode), and *transverse magnetic mode* (TM-mode). The TEM mode violates the boundary conditions because the magnetic field is not parallel to the surface, and so does not occur in waveguides.

The transverse electric field requirement means that the E-field must be perpendicular to the conductor wall of the waveguide. This requirement is met by use of a proper coupling scheme at the input end of the waveguide. A vertically polarized coupling radiator will provide the necessary transverse field.

One boundary condition requires that the magnetic (H) field must not be orthogonal to the conductor surface. Because it is at right angles to the E-field, it will meet this requirement (see Fig. 18-6). The planes formed by the magnetic field are parallel to both the direction of propagation and the wide dimension surface.

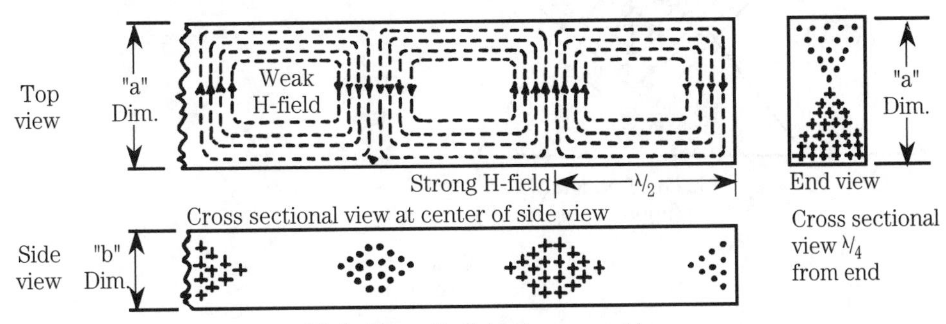

18-6 Magnetic fields in waveguide.

As the wave propagates away from the input radiator, it resolves into two components that are not along the axis of propagation, and are not orthogonal to the walls. The component along the waveguide axis violates the boundary conditions, so it is rapidly attenuated. For the sake of simplicity, only one component is shown in Fig. 18-7. Three cases are shown in Fig. 18-7: high, medium, and low frequency. Note that the angle of incidence with the waveguide wall increases as frequency drops. The angle rises toward 90° as the cut-off frequency is approached from above. Below the cut-off frequency the angle is 90° so the wave bounces back and forth between the walls without propagating.

Coordinate system and dominant mode in waveguides

Figure 18-8 shows the coordinate system used to denote dimensions and directions in microwave discussions. The a and b dimensions of the waveguide correspond to the X- and Y-axis of a Cartesian coordinate system, and the Z-axis is the direction of wave propagation.

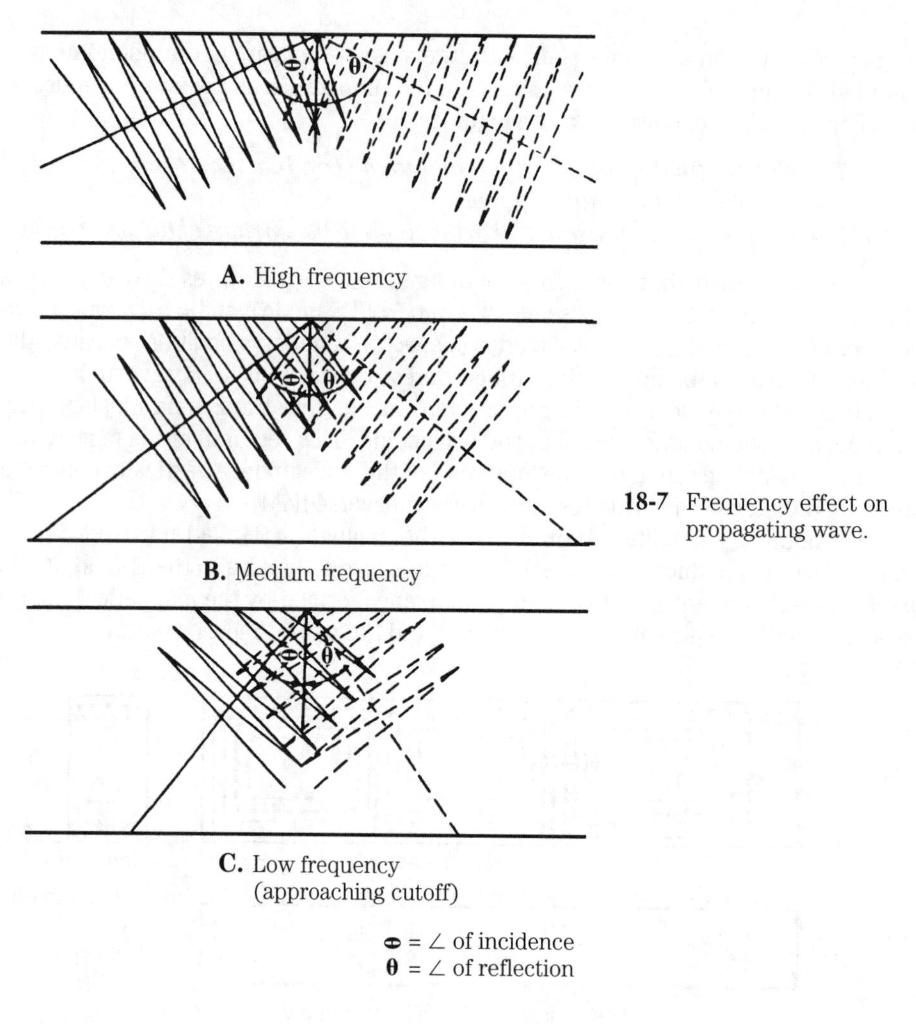

18-7 Frequency effect on propagating wave.

In describing the various modes of propagation, use a shorthand notation as follows:

$$Tx_{m,n}$$

Where:

- x is "E" for transverse electric mode, and "M" for transverse magnetic mode
- m is the number of half wavelengths along the X-axis (i.e., a dimension)
- n is the number of half wavelengths along the Y-axis (i.e., b dimension)

The TE_{10} mode is called the *dominant mode*, and is the best mode for low attenuation propagation in the Z-axis. The nomenclature "TE_{10}" indicates that there is 1 half wavelength in the a dimension, and zero half wavelengths in the b dimension. The dominant mode exists at the lowest frequency at which the waveguide is a half wavelength.

Velocity and wavelength in waveguides

Figure 18-9 shows the geometry for two wave components, simplified for the sake of illustration. There are three different wave velocities to consider with respect to

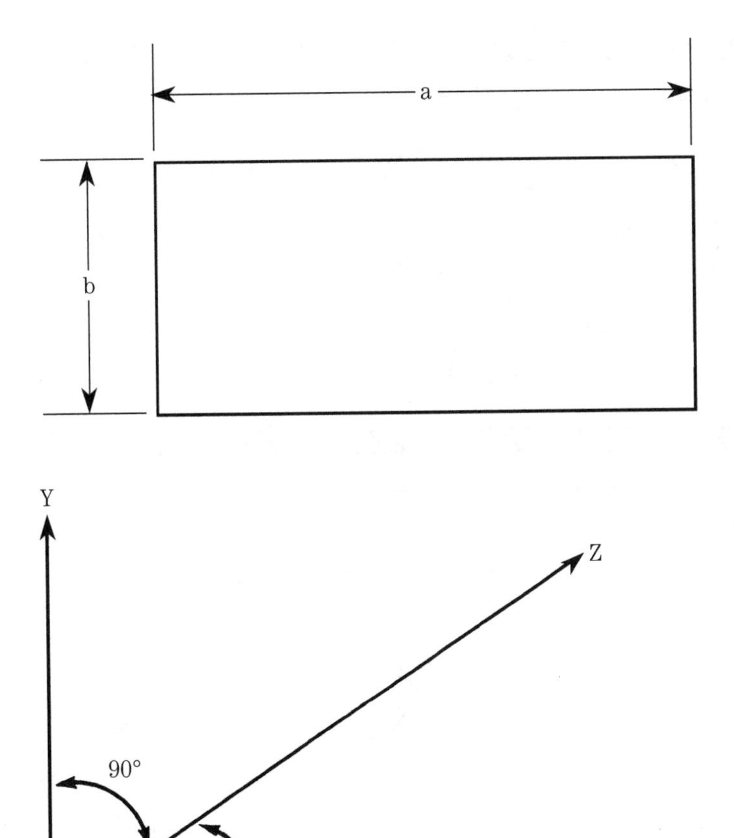

18-8 Rectangular waveguide coordinate system.

waveguides: *free space velocity* (*c*), *group velocity* (V_g), and *phase velocity* (V_p). The free-space velocity is the velocity of propagation in unbounded free space (i.e., the speed of light $c = 3 \times 10^8$ m/s).

The *group velocity* is the straight line velocity of propagation of the wave down the center-line (Z-axis) of the waveguides. The value of V_g is always less than *c*; because the actual path length taken, as the wave bounces back and forth, is longer than the straight line path (i.e., path *ABC* is longer than path *AC*). The relationship between *c* and V_g is:

$$V_g = c \sin a \qquad \textbf{[18.1]}$$

Where:

V_g is the group velocity in meters per second (m/s)
c is the free-space velocity (3×10^8 m/s)
a is the angle of incidence in the waveguide

The *phase velocity* is the velocity of propagation of the spot on the waveguide wall where the wave impinges (e.g., point "B" in Fig. 18-9B). This velocity, depend-

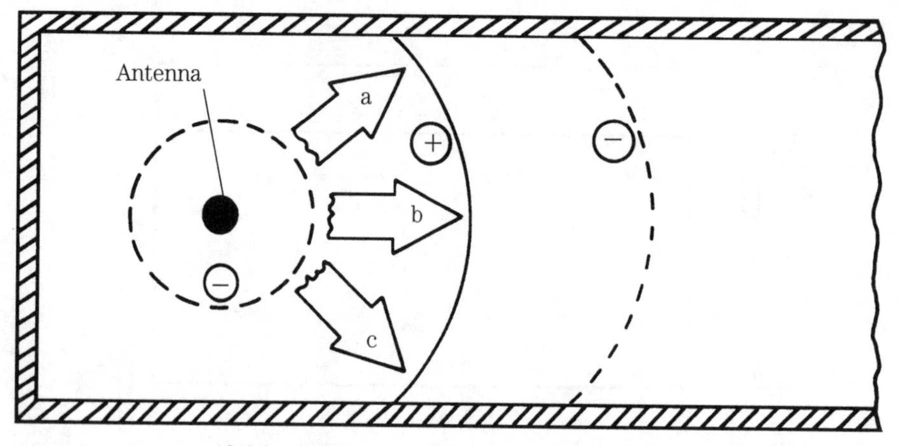

18-9A Antenna radiator in capped waveguide.

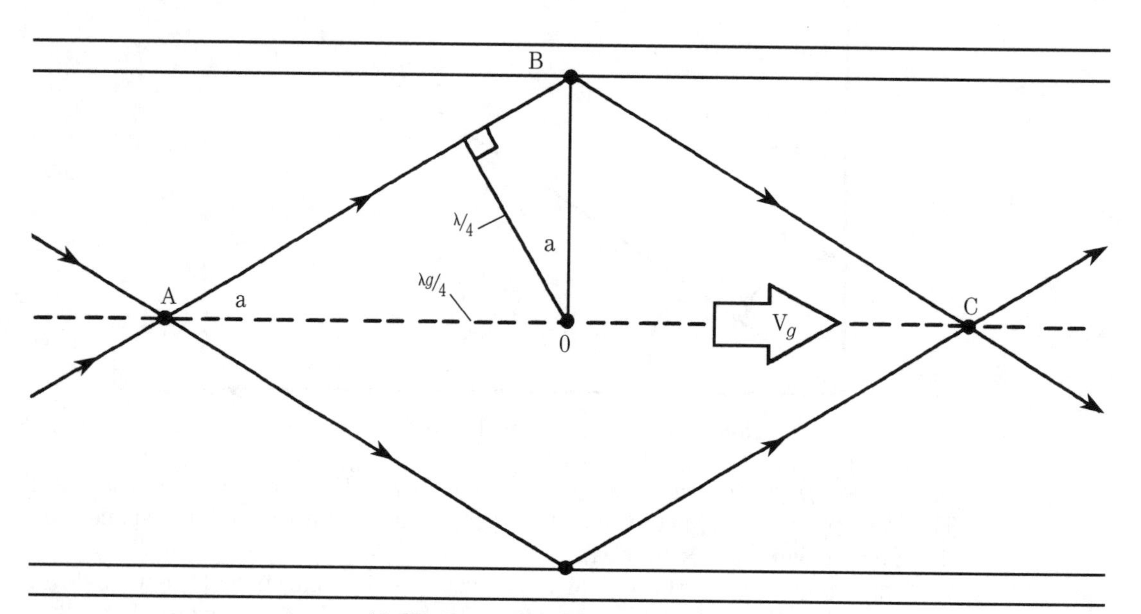

18-9B Wave propagation in waveguide.

ing upon the angle of incedence, can actually be faster than both the group velocity and the speed of light. The relationship between phase and group velocities can be seen in the "Beach analogy." Consider an ocean beach, on which the waves arrive from offshore at an angle other than 90°. In other words, the arriving wave fronts are not parallel to the shore. The arriving waves have a group velocity, V_g. But as a wave hits the shore, it will strike a point down the beach first, and the "point of strike" races up the beach at a much faster phase velocity, V_p, that is even faster than the group velocity. In a microwave waveguide, the phase velocity can be greater than c, as can be seen from Eq. 18.2:

$$V_p = \frac{c}{\sin a} \quad \textbf{[18.2]}$$

Example Calculate the group and phase velocities for an angle of incidence of 33°.
Solution:

a) *Group Velocity*

$$V_g = c \sin 2$$
$$V_g = (3 \times 10^8)(\sin 33)$$
$$V_g = (3 \times 10^8)(0.5446) = 1.6 \times 10^8 \text{ m/s}$$

b) *Phase Velocity*

$$V_p = c/\sin a$$
$$V_p = (3 \times 10^8 \text{ m/s})/\sin 33$$
$$V_p = (3 \times 10^8 \text{ m/s})/(0.5446)$$
$$V_p = 5.51 \times 10^8 \text{ m/s}$$

For this problem the solutions are:

$c = 3 \times 10^8$ m/s
$V_p = 5.51 \times 10^8$ m/s
$V_g = 1.6 \times 10^8$ m/s

We can also write a relationship between all three velocities by combining Eqs. 18.1 and 18.2, resulting in:

$$c = \sqrt{V_p V_g} \quad \textbf{[18.3]}$$

In any wave phenomena the product of *frequency* and *wavelength* is the *velocity*. Thus, for a TEM wave in unbounded free space we know that:

$$c = F \lambda_o \quad \textbf{[18.4]}$$

Because the frequency (F) is fixed by the generator, only the wavelength can change when the velocity changes. In a microwave waveguide we can relate phase velocity to wavelength as the wave is propagated in the waveguide (λ):

$$V_p = \frac{\lambda \times c}{\lambda_o} \quad \textbf{[18.5]}$$

Where:

Vp is the phase velocity in meters per second (m/s)
c is the free-space velocity (3×10^8 m/s)
λ is the wavelength in the waveguide, in meters (m)
λ_o is the wavelength in free-space (c/F), in meters (see Eq. 18.4)

Equation 18.5 can be re-arranged to find the wavelength in the waveguide:

$$\lambda = \frac{V_p \lambda_o}{c} \quad [18.6]$$

Example A 5.6-GHz microwave signal is propagated in a waveguide. Assume that the internal angle of incidence to the waveguide surfaces is 42 degrees. Calculate a) phase velocity, b) wavelength in unbounded free space, and c) wavelength of the signal in the waveguide.

Solution:

a) *Phase Velocity*

$$V_p = \frac{c}{\sin a}$$

$$V_p = \frac{3 \times 10^8 \text{ m/s}}{\sin 42}$$

$$V_p = \frac{3 \times 10^8 \text{ m/s}}{0.6991} = 4.5 \times 10^8 \text{ m/s}$$

b) *Wavelength in free space*

$$\lambda_o = c/F$$
$$\lambda_o = (3 \times 10^8 \text{ m/s})/(5.6 \times 10^9 \text{ Hz})$$
$$\lambda_o = 0.054 \text{ meters}$$

c) *Wavelength in waveguide*

$$\lambda = \frac{V_p \lambda_o}{c}$$

$$\lambda = \frac{(4.5 \times 10^8 \text{ m/s})(0.054 \text{ m})}{3 \times 10^8 \text{ m/s}} = 0.08 \text{ meters}$$

Comparing, we find that the free space wavelength is 0.054 meters, and wavelength inside of the waveguide increases to 0.08 meters.

Cut-off frequency (F_c)

The propagation of signals in a waveguide depends, in part, upon the operating frequency of the applied signal. As covered earlier, the angle of incidence made by the plane wave to the waveguide wall is a function of frequency. As the frequency drops, the angle of incidence increases towards 90 degrees.

The propagation of waves depends on the angle of incidence, and the associated reflection phenomena. Indeed, both phase and group velocities are functions of the angle of incidence. When the frequency drops to a point where the angle of incidence is 90 degrees, then group velocity is meaningless.

We can define a general mode equation based on our system of notation:

$$\frac{1}{[\lambda_c]^2} = \left[\frac{m}{2a}\right]^2 + \left[\frac{n}{2b}\right]^2 \qquad \textbf{[18.7]}$$

Where:

λ_c is the longest wavelength that will propagate
a, b are the waveguide dimensions (see Fig. 18-2)
m, n are integers that define the number of half wavelengths that will fit in the a and b dimensions, respectively

Evaluating Eq. 18.7 reveals that the longest TE-mode signal that will propagate in the dominant mode (TE_{10}) is given by:

$$\lambda_c = 2a \qquad \textbf{[18.8]}$$

From which, we can write an expression for the cut-off frequency:

$$F_c = \frac{c}{2a} \qquad \textbf{[18.9]}$$

Where:

F_c is the lowest frequency that will propagate, in Hertz
c is the speed of light (3×10^8 m/s)
a is the wide waveguide dimension

Example A rectangular waveguide has dimensions of 3 cm × 5 cm. Calculate the TE_{10} mode cut-off frequency.

Solution:

$$F_c = \frac{c}{2a}$$

$$F_c = \frac{(3 \times 108 \text{ m/s})}{(2)\left[5 \text{ cm} \times \frac{1 \text{ m}}{100 \text{ cm}}\right]}$$

$$F_c = \frac{3 \times 10^8 \text{ m/s}}{(2)(0.05 \text{ m})} = 3 \text{ GHz}$$

Equation 18.7 assumes that the dielectric inside the waveguide is air. A more generalized form, which can accommodate other dielectrics, is:

$$F_c = \frac{1}{2\sqrt{ue}}\sqrt{\left[\left[\frac{m}{a}\right]^2 + \left[\frac{n}{b}\right]^2\right]} \qquad \textbf{[18.10]}$$

Where:

e is the dielectric constant
u is the permeability constant

For air dielectrics, $u = u_o$ and $e = e_o$, from which:

$$c = \frac{1}{\sqrt{u_o e_o}} \quad \textbf{[18.11]}$$

To determine the cut-off wavelength, we can rearrange Eq. 18.10 to the form:

$$\lambda_c = \frac{2}{\sqrt{\left(\frac{m}{a}\right)^2 + \left(\frac{n}{b}\right)^2}} \quad \textbf{[18.12]}$$

One further expression for air-filled waveguide calculates the actual wavelength in the waveguide from a knowledge of the free-space wavelength and actual operating frequency:

$$\lambda_g = \frac{\lambda_o}{\sqrt{1 - \left[\frac{F_c}{F}\right]^2}} \quad \textbf{[18.13]}$$

Where:

λ_g is the wavelength in the waveguide
λ_o is the wavelength in free space
F_c is the waveguide cut-off frequency
F is the operating frequency

Example A waveguide with a 4.5-GHz cut-off frequency is excited with a 6.7-GHz signal. Find a) the wavelength in free space, and b) the wavelength in the waveguide.
Solution:

a)

$$\lambda_o = c/F$$

$$\lambda_o = \frac{3 \times 10^8 \text{ m/s}}{6.7 \text{ GHz} \times \frac{10^9 \text{ Hz}}{1 \text{ GHz}}}$$

$$\lambda_o = \frac{3 \times 10^8 \text{ m/s}}{6.7 \times 10^9 \text{ Hz}} = 0.0448 \text{ meters}$$

b)

$$\lambda_g = \frac{\lambda_o}{\sqrt{1 - \left[\frac{F_c}{F}\right]^2}}$$

$$\lambda_g = \frac{0.0448 \text{ meters}}{\sqrt{1 - \left[\frac{4.5 \text{ GHz}}{6.7 \text{ GHz}}\right]^2}}$$

$$\lambda_g = \frac{0.0448 \text{ meters}}{1 - 0.67}$$

$$\lambda_g = \frac{0.0448}{0.33} = 0.136 \text{ meters}$$

Transverse magnetic modes also propagate in waveguides, but the base TM_{10} mode is excluded by the boundary conditions. Thus, the TM_{11} mode is the lowest magnetic mode that will propagate.

Waveguide impedance

All forms of transmission line, including the waveguide, exhibit a characteristic impedance, although in the case of waveguide it is a little difficult to pin down conceptually. This concept was developed for ordinary transmission lines in chapter 4. For a waveguide, the characteristic impedance is approximately equal to the ratio of the electric and magnetic fields (E/H), and converges (as a function of frequency) to the intrinsic impedance of the dielectric (Fig. 18-10). The impedance of the waveguide is a function of waveguide characteristic impedance (Z_o) and the wavelength in the waveguide:

$$Z = \frac{Z_o \lambda_g}{\lambda_o} \qquad [18.14]$$

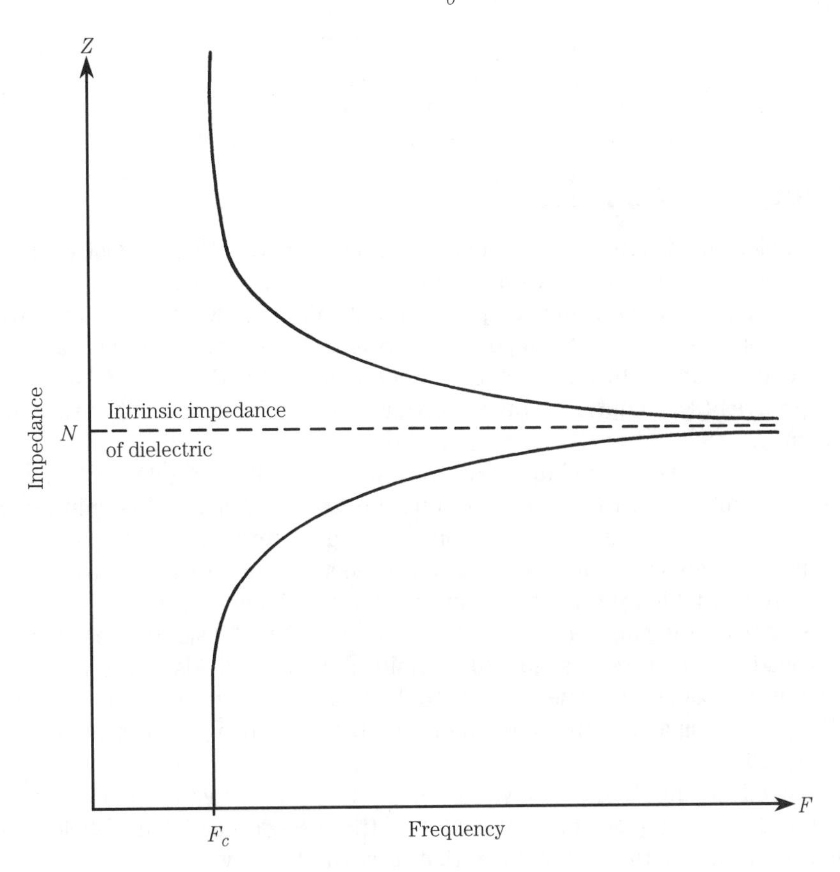

18-10 Impedance vs. frequency.

Or for rectangular waveguide, with constants taken into consideration:

$$Z = \frac{120\,\pi\,\lambda_g}{\lambda_o} \qquad [18.15]$$

The *propagation constant* (B) for rectangular waveguide is a function of both cut-off frequency and operating frequency:

$$B = W\sqrt{eu}\,\sqrt{1 - \left[\frac{F_c}{F}\right]^2} \qquad [18.16]$$

From which, we can express the TE-mode impedance:

$$Z_{TE} = \frac{\sqrt{[ue]}}{\sqrt{1 - \left[\frac{F_c}{F}\right]^2}} \qquad [18.17]$$

and the TM-mode impedance:

$$Z_M = 377\sqrt{1 - \left(\frac{F_c}{F}\right)^2} \qquad [18.18]$$

Waveguide terminations

When an electromagnetic wave propagates down a waveguide, it must eventually reach the end of the guide. If the end is open, then the wave will propagate into free-space. The horn radiator is an example of an unterminated waveguide. If the waveguide terminates in a metallic wall, then the wave reflects back down the waveguide, from whence it came. The interference between incident and reflected waves forms standing waves (see chapter 3). Such waves are stationary in space, but vary in the time domain.

In order to prevent standing waves, or more properly, the reflections that give rise to standing waves, the waveguide must be *terminated* in a matching impedance. When a properly designed antenna is used to terminate the waveguide, it forms the matched load required to prevent reflections. Otherwise, a *dummy load* must be provided. Figure 8-11 shows several types of dummy load.

The classic termination is shown in Fig. 18-11A. The "resistor" making up the dummy load is a mixture of sand and graphite. When the fields of the propagated wave enter the load, they cause currents to flow, which in turn causes heating. Thus, the RF power dissipates in the sand-graphite rather than being reflected back down the waveguide.

A second dummy load is shown in Fig. 18-11B. The resistor element is a carbonized rod critically placed at the center of the electric field. The E-field causes currents to flow, resulting in I^2R losses that dissipate the power.

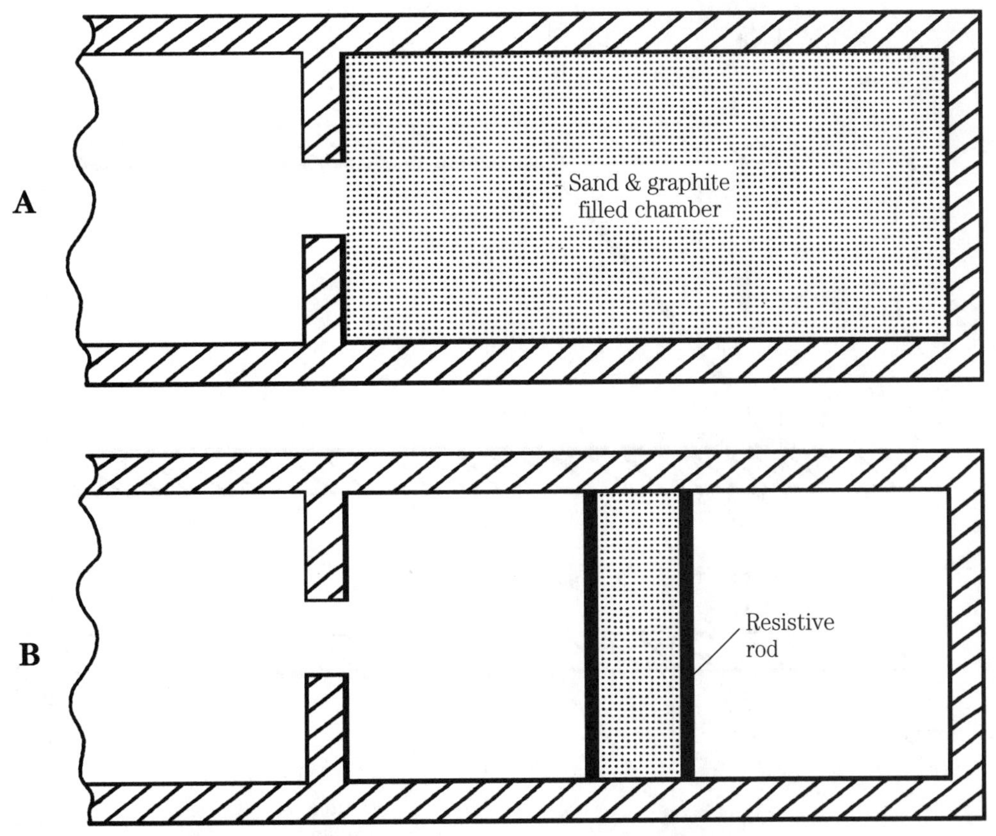

18-11 Dummy loads: A) sand/graphite chamber, B) resistive rod.

Bulk loads, similar to the graphite-sand chamber are shown in Figs. 18-11C, 18-11D, and 18-11E. Using bulk material such as graphite or a carbonized synthetic material, these loads are used in much the same way as the sand load (i.e., currents set up, and I^2R losses dissipate, the power).

The resistive vane load is shown in Fig. 18-11F. The plane of the element is orthogonal to the magnetic lines of force. When the magnetic lines cut across the vane, currents are induced, which gives rise to the I^2R losses. Very little RF energy reaches the metallic end of the waveguide, so there is little reflected energy and a low VSWR.

There are situations where it isn't desirable to terminate the waveguide in a dummy load. Several reflective terminations are shown in Fig. 18-12. Perhaps the simplest form is the permanent end plate shown in Fig. 18-12A. The metal cover must be welded or otherwise affixed through a very low-resistance joint. At the substantial power levels typically handled in transmitter waveguides, even small resistances can be important.

The end plate (shown in Fig. 18-12B) uses a quarter-wavelength cup to reduce the effect of joint resistances. The cup places the contact joint at a point that is a quarter wavelength from the end. This point is a minimum *current node*, so I^2R losses in the contact resistance becomes less important.

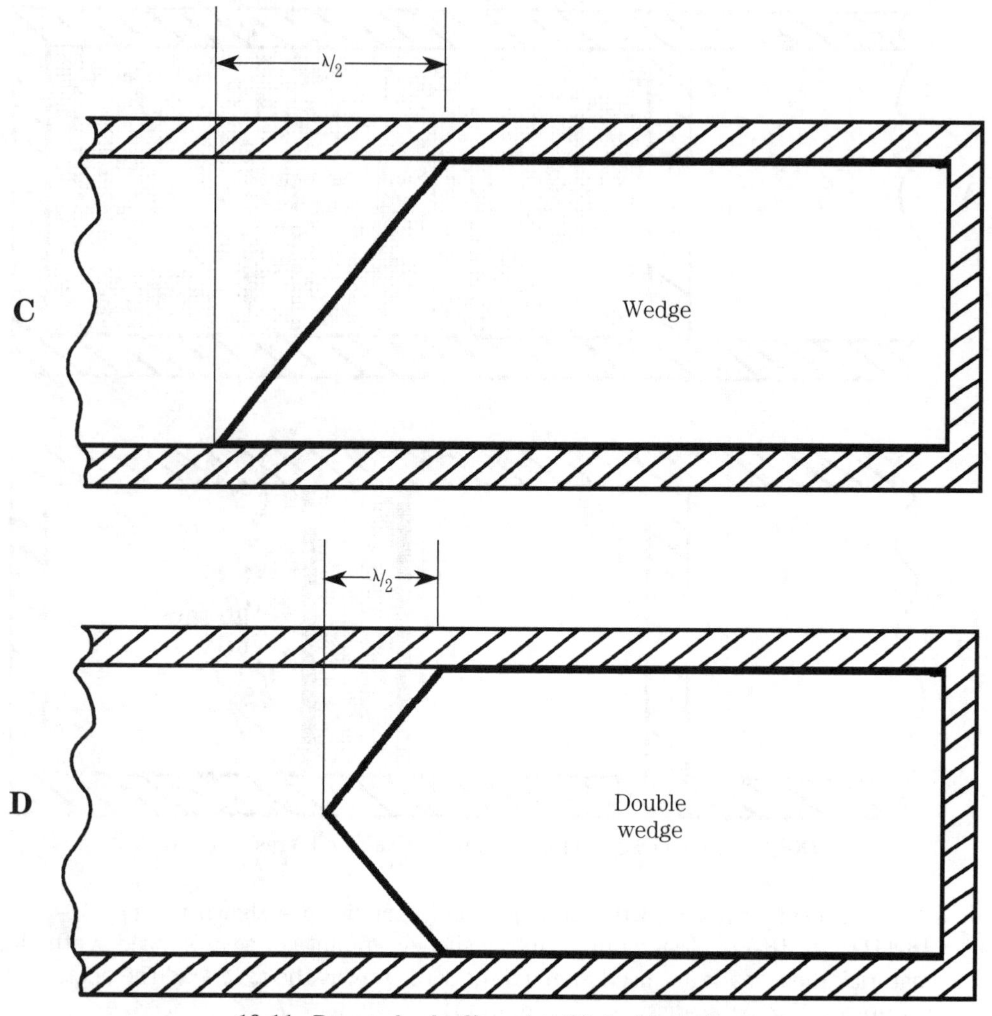

18-11 Dummy loads: C) wedge, D) double wedge.

The adjustable short circuit is shown in Fig. 18-12C. The walls of the waveguide and the surface of the plunger form a half-wavelength channel. Because the metallic end of the channel is a short circuit, the impedance reflected back to the front of the plunger is zero ohms, or nearly so. Thus, a *virtual short* exists at the points shown. By this means, the contact (or joint) resistance problem is overcome.

Waveguide joints and bends

Joints and bends in any form of transmission line or waveguide are seen as impedance discontinuities, and so are points at which disruptions occur. Thus, improperly formed bends and joints are substantial contributors to a poor VSWR. In general, bends, twists, joints or abrupt changes in waveguide dimension can deteriorate the VSWR by giving rise to reflections.

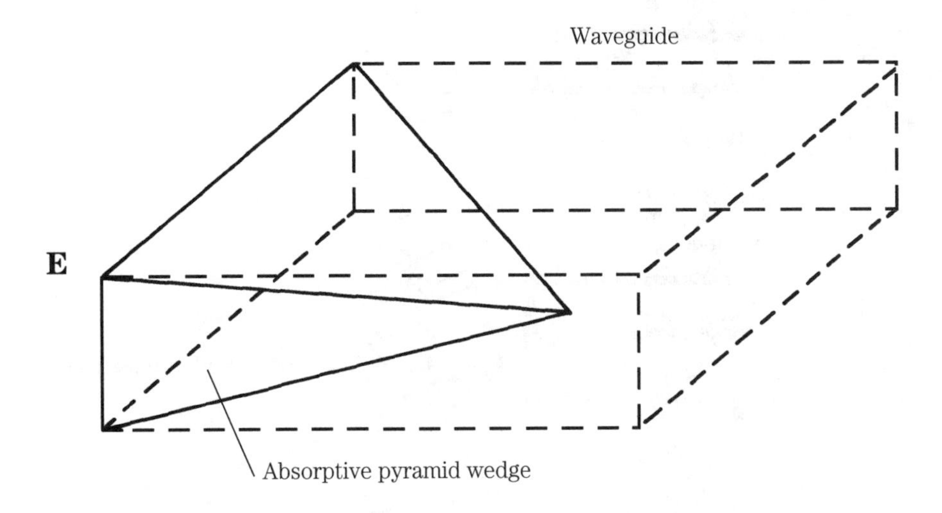

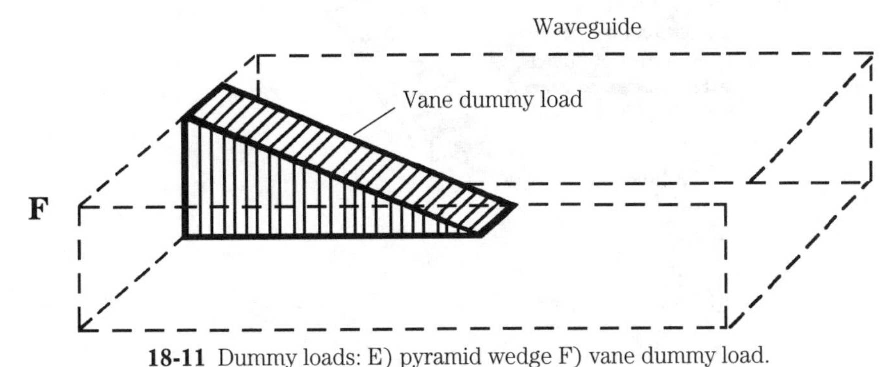

18-11 Dummy loads: E) pyramid wedge F) vane dummy load.

Extensive runs of waveguide are sometimes difficult to make in a straight line. Although some installations do permit a straight waveguide, many others require directional change. This possibility is especially likely on shipboard installations. Figure 18-13A shows the proper ways to bend a waveguide around a corner. In each case, the radius of the bend must be at least two wavelengths at the lowest frequency that will be propagated in the system.

The "twist" shown in Fig. 18-13B is used to rotate the polarity of the E- and H-fields by 90-degrees. This type of section is sometimes used in antenna arrays for phasing the elements. As in the case of the bend, the twist must be made over a distance of at least two wavelengths.

When an abrupt 90-degree transition is needed, it is better to use two successive 45-degree bends spaced one-quarter wavelength apart (see Fig. 18-13C). The theory (behind this kind of bend) is to cause the interference of the direct reflection of one bend, with the inverted reflection of the other end. The resultant relationship between the fields is reconstructed as if no reflections had taken place.

Joints are necessary in practical waveguides because it simply isn't possible to construct a single length of practical-size for all situations. Three types of common joints are used: *permanent*, *semi-permanent*, and *rotating*.

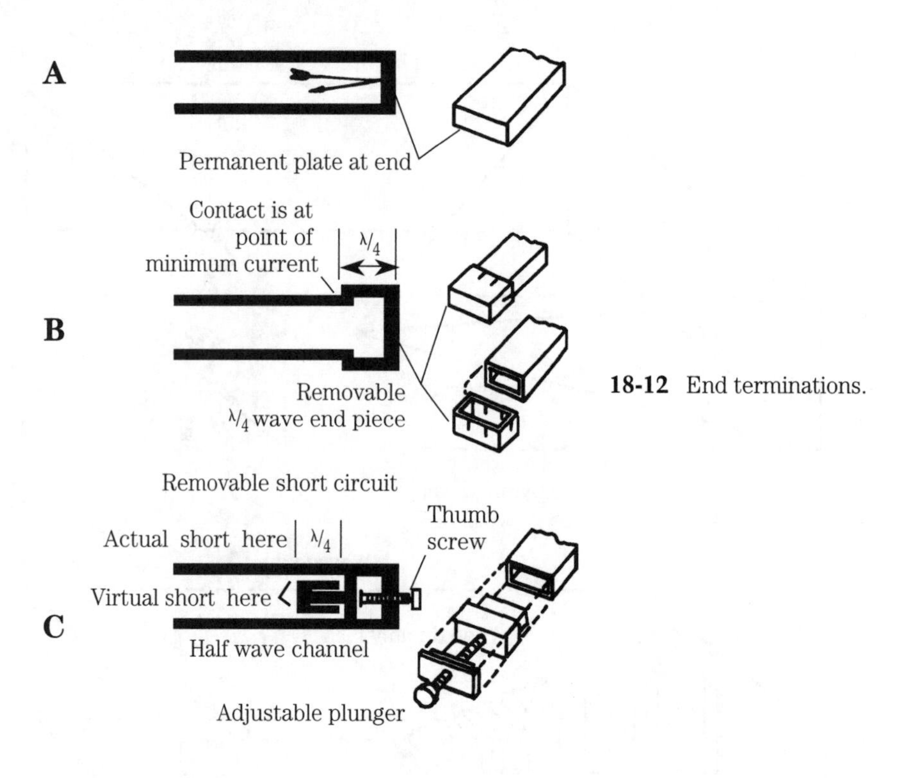

18-12 End terminations.

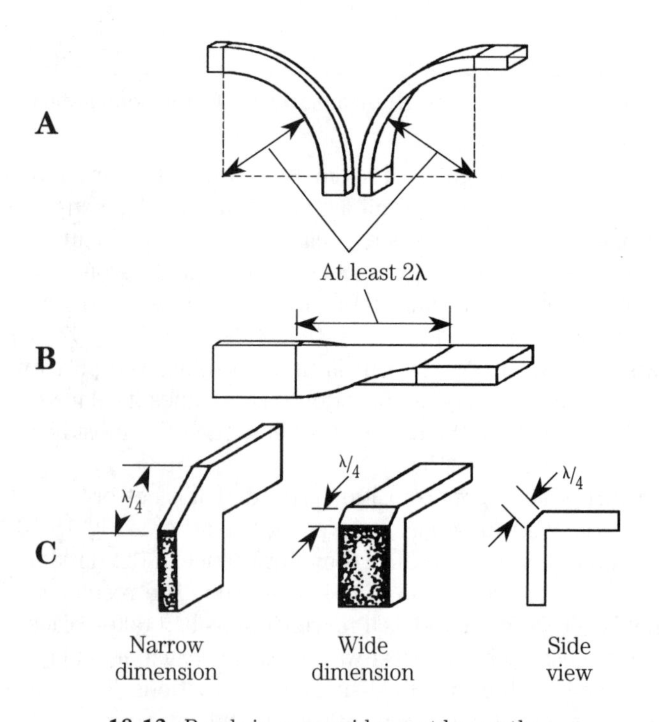

18-13 Bends in waveguide must be gentle.

To make a permanent joint the two waveguide ends must be machined extremely flat so that they can be butt-fitted together. A welded or brazed seam bonds the two sections together. Because such a surface represents a tremendous discontinuity, reflections and VSWR will result unless the interior surfaces are milled flat and then polished to a mirror-like finish.

A *semi-permanent joint* allows the joint to be disassembled for repair and maintenance, as well as allowing easier on-site assembly. The most common example of this class is the *choke joint* shown in Fig. 18-14.

One surface of the choke joint is machined flat, and is a simple butt-end planar flange. The other surface is the mate to the planar flange, but it has a quarter-wavelength circular slot cut at a distance of one quarter wavelength from the waveguide aperture. The two flanges are shown in side view in Fig. 18-14A, and the slotted end view is shown in Fig. 18-14B. The method for fitting the two ends together is shown in the oblique view in Fig. 18-14C.

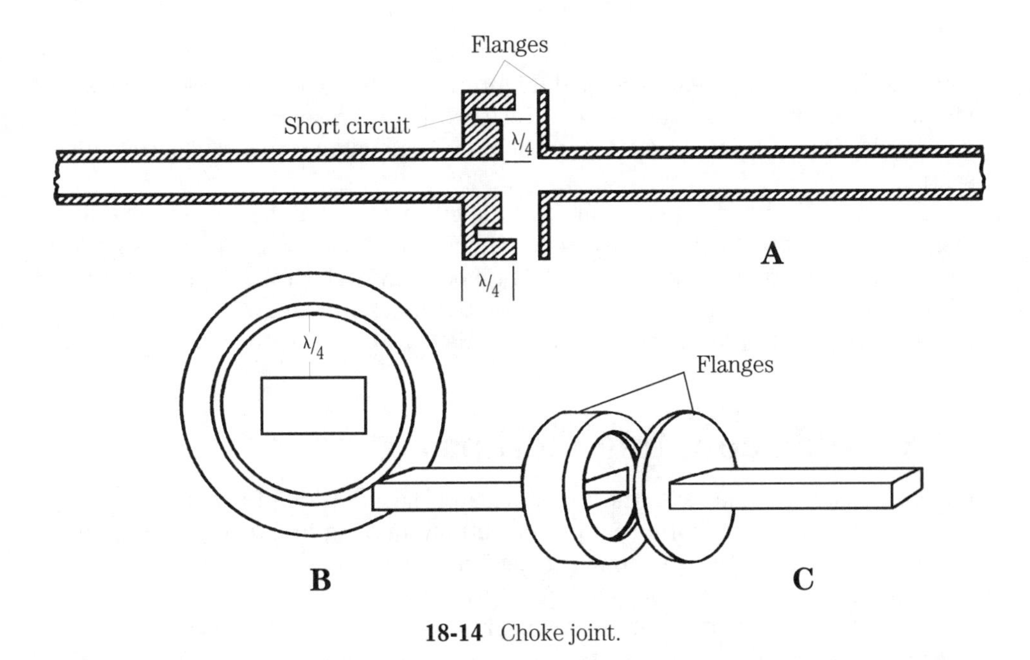

18-14 Choke joint.

Rotating joints are used in cases where the antenna has to point in different directions at different times. Perhaps the most common example of such an application is the radar antenna.

The simplest form of rotating joint is shown in Fig. 18-15. The key to its operation is that the selected mode is symmetrical about the rotating axis. For this reason, circular waveguide operating in the TM_{01} mode is selected. In this rotating choke joint, the actual waveguide rotates, but the internal fields do not (thereby minimizing reflections). Because most waveguide is rectangular, however, a some-

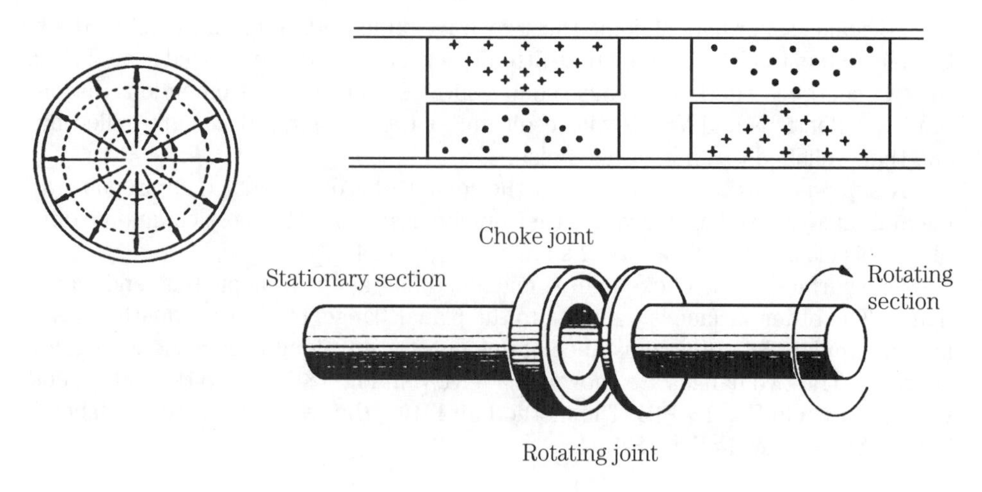

18-15 Basic rotating joint.

what more complex system is needed. Figure 18-16 shows a rotating joint consisting of two circular waveguide sections inserted between segments of rectangular waveguide. On each end of the joint, there is a rectangular-to-circular transition section.

In Fig. 18-16, the rectangular input waveguide operates in the TE_{10} mode that is most efficient for rectangular waveguide. The E-field lines of force couple with the circular segment, thereby setting up a TM_{01} mode wave. The TM_{01} mode has the required symmetry to permit coupling across the junction, where it meets another transition zone and is reconverted to TE_{10} mode.

Waveguide coupling methods

Except possibly for the case where an oscillator exists inside of a waveguide, it is necessary to have some form of input or output coupling in a waveguide system. There are three basic types of coupling used in a microwave waveguide: *capacitive* (or *probe*), *inductive* (or *loop*), and *aperture* (or *slot*).

Capacitive (also called probe) coupling is shown in Fig. 18-17. This type of coupling uses a vertical radiator inserted into one end of the waveguide. Typically, the probe is a quarter wavelength in a fixed-frequency system. The probe is analogous to the vertical antennas used at lower frequencies. A characteristic of this type of radiator is that the E-field is parallel to the waveguide top and bottom surfaces. This arrangement satisfies the first boundary condition for the dominant TE_{10} mode.

The radiator is placed at a point that is a quarter wavelength from the rear wall (Fig. 18-17B. By traversing the quarter wave distance (90° phase shift), being reflected from the rear wall (180° phase shift) and then re-traversing the quarter wavelength distance (another 90° phase shift), the wave undergoes a total phase shift of one complete cycle, or 360°. Thus, the reflected wave arrives back at the radiator in-phase to reinforce the outgoing wave. Hence, none of the excitation energy is lost.

Some waveguides have an adjustable end cap (Fig. 18-17C) in order to accom-

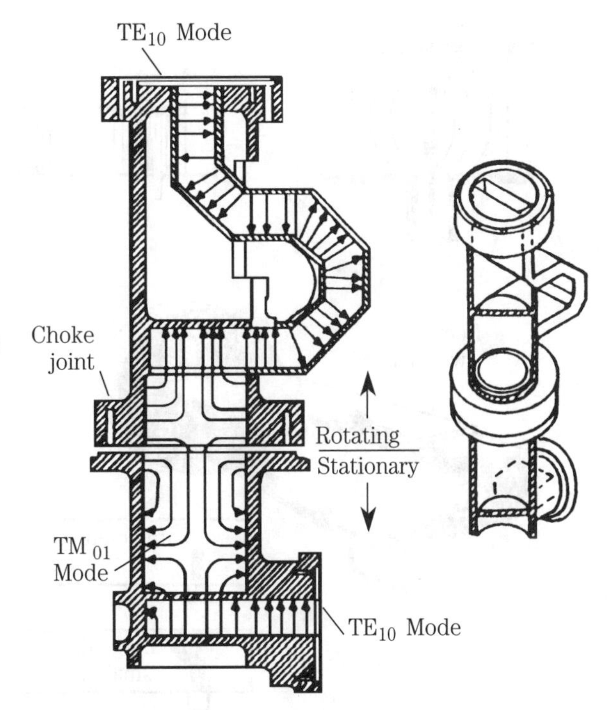

18-16 Representative practical rotating joint.

modate multiple frequencies. The end cap position is varied to accommodate the different wavelength signals.

Figure 18-17D shows high- and low-power broadband probes that are typically not a quarter wavelength except at one particular frequency. Broadbanding is accomplished by attention to the diameter-to-length ratio. The *degree-of-coupling* can be varied in any of several ways: the *length* of the probe can be varied; the *position* of the probe in the E-field can be changed; or *shielding* can be used to partially shade the radiator element.

Inductive or loop coupling is shown in Fig. 18-18. A small loop of wire (or other conductor) is placed such that the number of magnetic flux lines is maximized. This form of coupling is popular on microwave receiver antennas, in order to make a waveguide to coaxial cable transition. In some cases, the loop is formed by the pigtail lead of a detector diode that, when combined with a local oscillator, down-converts the microwave signal to an IF frequency in the 30- to 300-MHz region.

Aperture or slot coupling is shown in Fig. 18-19. This type of coupling is used to couple together two sections of waveguide, as on an antenna feed system. Slots can be designed to couple either electric, magnetic, or electromagnetic fields. In Fig. 18-19, slot "A" is placed at a point where the E-field peaks, so it allows electrical field coupling. Similarly, slot "B" is at a point where the H-field peaks, so it allows magnetic field coupling. Finally, we see slot "C" which allows electromagnetic field coupling.

Slots can also be characterized according to whether they are *radiating* or *nonradiating*. A nonradiating slot is cut at a point that does not interrupt the flow of currents in the waveguide walls. The radiating slot, on the other hand, does interrupt currents flowing in the walls. A radiating slot is the basis for several forms of antenna which are discussed at the end of this chapter.

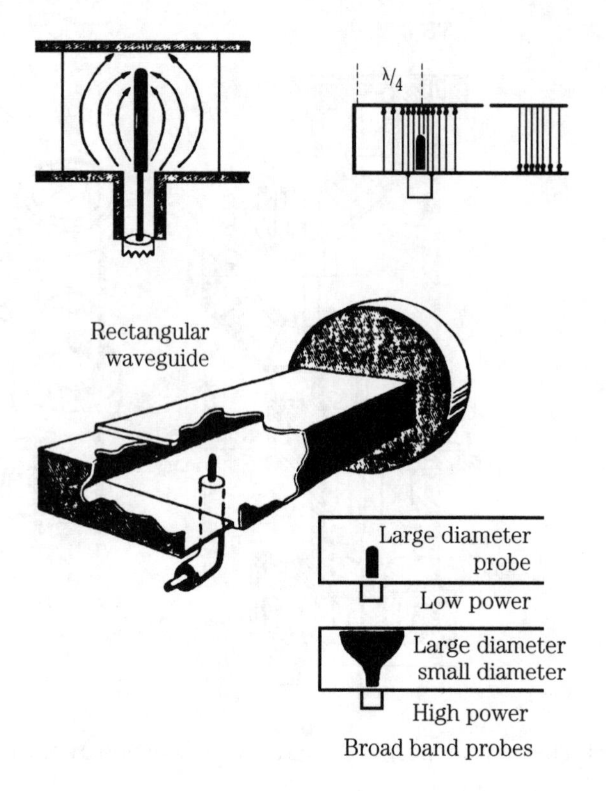

18-17 Probe (capacitive) coupling.

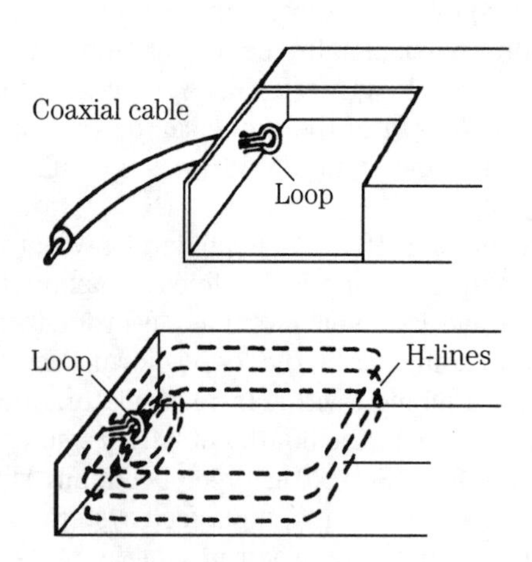

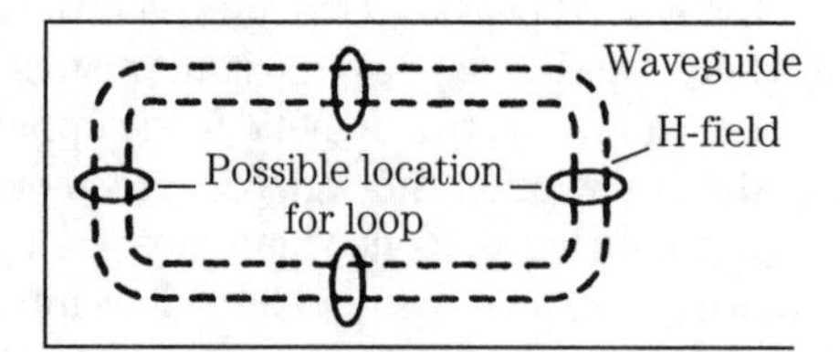

18-18 Loop (inductive) coupling.

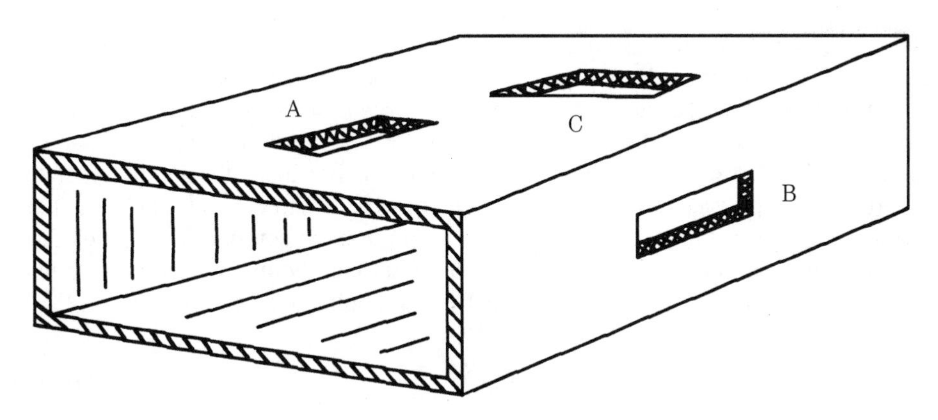

18-19 Slot coupling.

Microwave antennas

Antennas are used in communications, and radar systems, at frequencies from the very lowest to the very highest. In both theory and practice, antennas are used until frequencies reach infrared and visible light, at which point optics becomes more important. Microwaves are a transition region between ordinary "radio waves" and optical waves," so (as might be expected) microwave technology makes use of techniques from both worlds. For example, both dipoles and parabolic reflectors are used in microwave systems.

The purpose of an antenna is to act as a *transducer* between either electrical oscillations or propagated "guided waves" (i.e., in transmission lines or waveguides), and a propagating electromagnetic wave in free space. A principal function of the antenna is to act as an *impedance matcher* between the waveguide, or transmission line, impedance and the impedance of free space.

Antennas can be used equally well for both receiving and transmitting signals because they obey the *Law of Reciprocity*. That is, the same antenna can be used to both receive and transmit with equal success. Although there might be practical or mechanical reasons to prefer specific antennas for one or the other modes, electrically they are the same.

In the transmit mode, the antenna must radiate electromagnetic energy. For this job, the important property is *gain*, *G*. In the receive mode, the job of the antenna is to gather energy from impinging electromagnetic waves in free space. The important property for receiver antennas is the *effective aperture*, A_e, which is a function of the antenna's physical area. Because of reciprocity, a large gain usually infers a large effective aperture and vice versa. *Effective aperture* is defined as the area of the impinging radio wavefront that contains the same power as is delivered to a matched resistive load across the feedpoint terminals.

The isotropic "antenna"

Antenna definitions and specifications can become useless unless a means is provided for putting everything on a common footing. Although a variety of systems exist for describing antenna behavior, the most common system compares a specific antenna with a theoretical construct, called the *isotropic radiator*.

An isotropic radiator is a spherical point source that radiates equally well in all directions. By definition, the directivity of the isotropic antenna is unity (1), and all antenna gains are measured against this standard. Because the geometry of the sphere and the physics of radiation are well known, we can calculate field strength and power density at any point. These figures can then be compared with the actual values from an antenna being tested. From spherical geometry, we can calculate isotropic power density at any distance R from the point source:

$$P_d = \frac{P}{4\pi R^2} \qquad \textbf{[18.19]}$$

Where:

P_d is the power density in W/m^2
P is the power in watts input to the isotropic radiator
R is the radius in meters at which point power density is measured

Example Calculate the power density in W/m^2 at a distance of 1 kilometer (1000 m) from a 1000-watt isotropic source.

Solution:

$$P_d = \frac{P}{4\pi R^2}$$

$$P_d = \frac{(1000 \text{ Watts})}{(4)\,\pi\,(1000 \text{ m})^2}$$

$$P_d = 7.95 \times 10^{-5} \text{ W/m}^2$$

The rest of this chapter covers antenna gains and directivities that are relative to isotropic radiators.

Near field and far field

Antennas are defined in terms of *gain* and *directivity*, both of which are measured by looking at the radiated field of the antenna. There are two fields to consider: *near field* and *far field*. The patterns published for an antenna tend to reflect far field performance. The far field for most antennas falls off according to the *inverse square law*. That is, the intensity falls off according to the square of the distance ($1/R^2$), as in Eq. 18.19.

The near field of the antenna contains more energy than the far field because of the electric and magnetic fields close to the antenna radiator element. The near field tends to diminish rapidly according to a $1/R^4$ function. The minimum distance to the edge of the near field is a function of both the wavelength of the radiated signals and the antenna dimensions:

$$r_{\min} \frac{2\,d^2}{\lambda} \qquad \textbf{[18.20]}$$

Where:

r_{min} is the near field distance
d is the largest antenna dimension
λ is the wavelength of the radiated signal (all factors in the same units)

Example An antenna with a length of 6 cm radiates a 12-cm wavelength signal. Calculate the near field distance.

Solution:

$$r_{min} = \frac{2\,d^2}{\lambda}$$

$$r_{min} = \frac{(2)\ (6\text{ cm})2}{(12\text{ cm})}$$

$$r_{min} = 2/12 = 6\text{ cm}$$

Antenna impedance

Impedance represents the total opposition to the flow of alternating current (e.g., RF), and includes both resistive and reactive components. The reactive components can be either capacitive or inductive, or a combination of both. Impedance can be expressed in either of two notations:

$$Z = \sqrt{[R^2 + (X_L - X_c)^2]} \qquad \textbf{[18.21]}$$

or

$$Z = R \pm jX \qquad \textbf{[18.22]}$$

Of these, Eq. 18.22 is perhaps the most commonly used in RF applications. The reactive part of antenna impedance results from the magnetic and electrical fields (close to the radiator) returning energy to the antenna radiator during every cycle. The resistive part of impedance consists of two elements: *ohmic losses* (R_o) and *radiation resistance* (R_r). The ohmic losses are due to heating of the antenna conductor elements by RF current passing through, as when current passes through any conductor.

The radiation resistance relates to the radiated energy. An efficiency factor (k) compares the loss and radiation resistances:

$$k = \frac{R_r}{R_r + R_o} \qquad \textbf{[18.23]}$$

The goal of the antenna designer is to reduce R_o to a minimum. The value of R_r is set by the antenna design and installation, and is defined as the quotient of the voltage over the current at the feedpoint, less losses.

Dipole antenna elements

The *di-pole* is a two-pole antenna (Fig. 18-20) that can be modelled as either a single radiator fed at the center (Fig. 18-20A), or a pair of radiators fed back-to-back

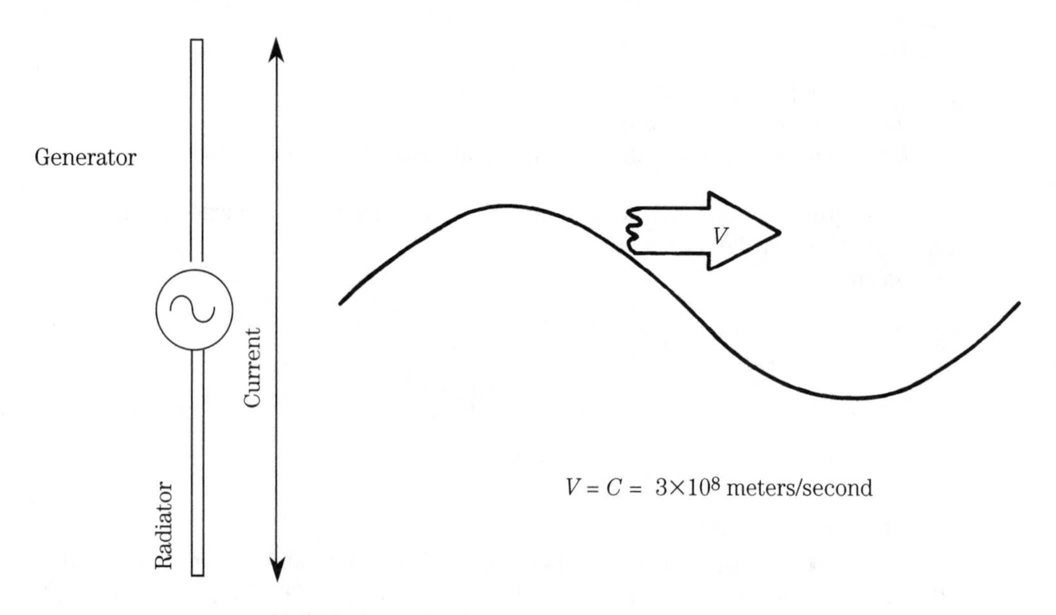

18-20A Basic dipole antenna showing propagation

(Fig. 18-20B). RF current from the source oscillates back and forth in the radiator element, causing an electromagnetic wave to propagate in a direction *perpendicular* to the radiator element. The polarity of any electromagnetic field is the direction of the electrical field vector (see chapter 2). In the dipole, the polarization is parallel to the radiator element: a horizontal element produces a horizontally polarized signal, while a vertical element produces a vertically polarized signal.

Figure 18-21 shows the radiator patterns for the dipole viewed from two perspectives. Figure 18-21A shows the pattern of a horizontal half-wavelength dipole as viewed from above. This plot shows the *directivity* of the dipole: maximum radia-

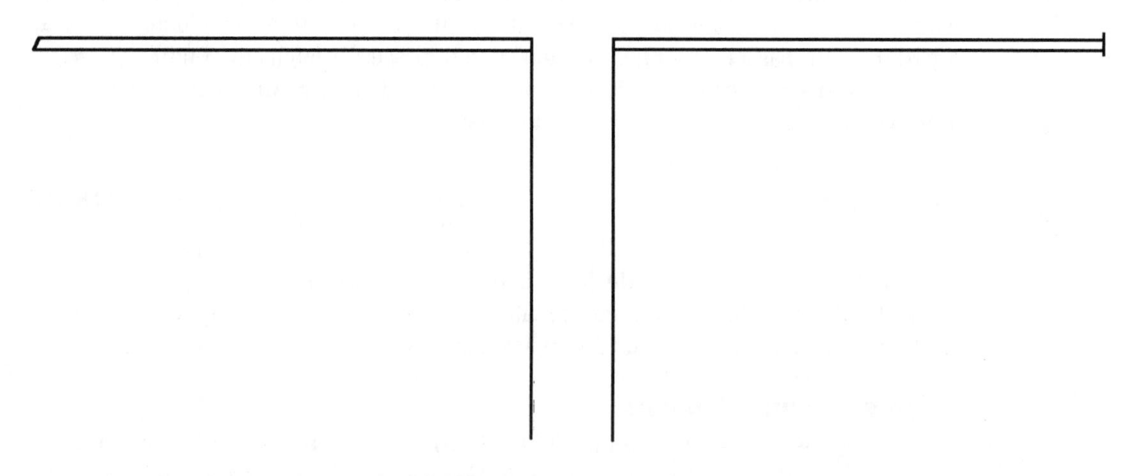

18-20B Basic dipole antenna.

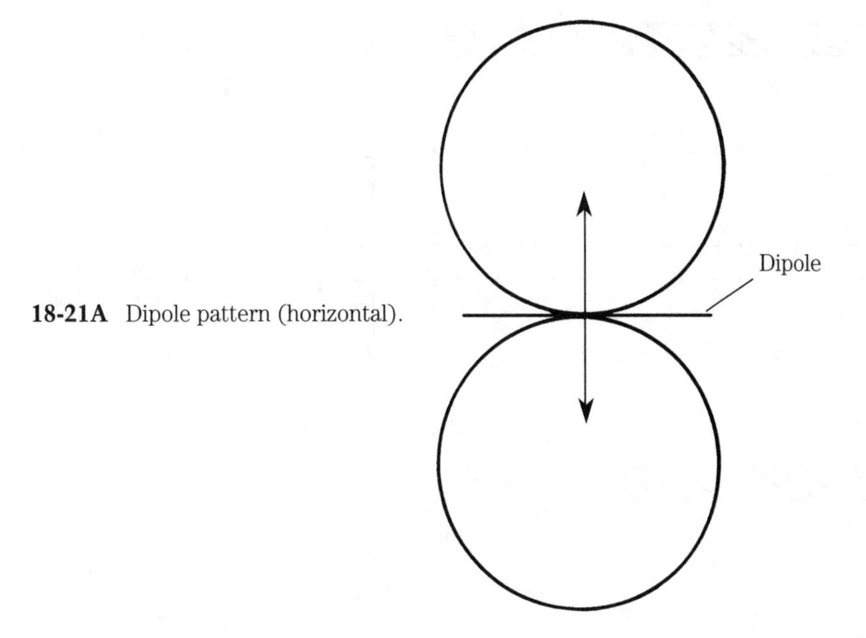

18-21A Dipole pattern (horizontal).

tion is found in two lobes perpendicular to the radiator length. The plot in Fig. 18-21B shows the end-on pattern of the dipole. This omnidirectional pattern serves for a vertically polarized dipole viewed from above. The end-on pattern of a horizontal dipole would be similar, except that it is distorted by ground effects unless the antenna is a very large number of wavelengths above the ground.

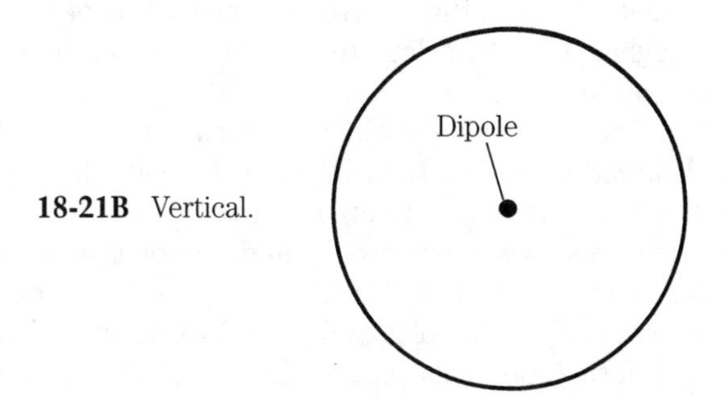

18-21B Vertical.

A microwave dipole is shown in Fig. 18-22. The antenna radiator element consists of a short conductor at the end of a section of waveguide. Although most low frequency dipoles are a half wavelength, microwave dipoles might be either a half wavelength, less than a half wavelength, or greater than a half wavelength, depending upon application. For example, because most microwave dipoles are used to illuminate a reflector antenna of some sort, the length of the dipole depends upon the exact illumination function required for proper operation of the reflector. Most, however, will be a half wavelength.

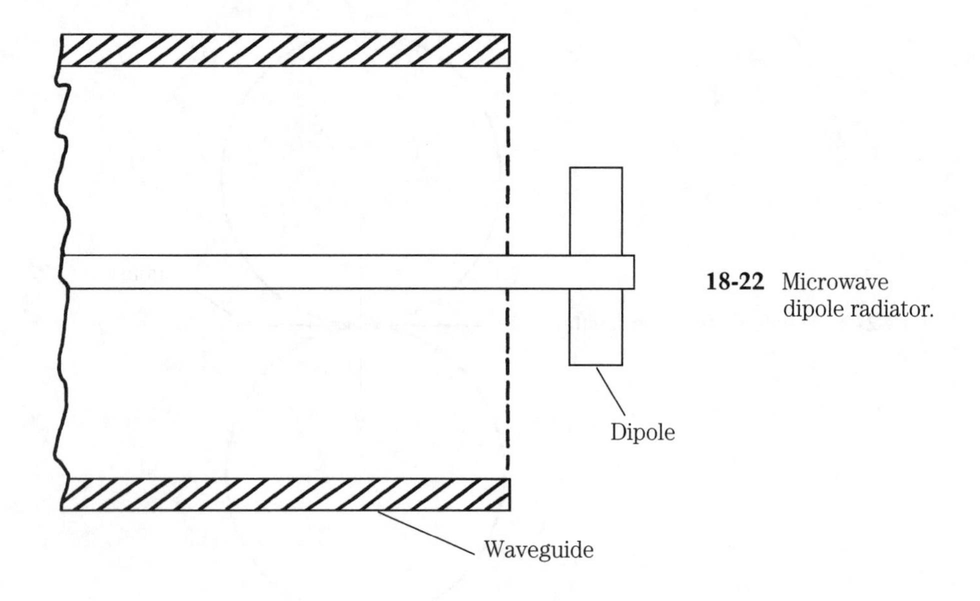

18-22 Microwave dipole radiator.

Antenna directivity and gain

The dipole discussed illustrated a fundamental property of the type of antennas generally used at microwave frequencies: *directivity* and *gain*. These two concepts are different but so interrelated that they are usually discussed at the same time. Because of the directivity, the antenna focuses energy in only two directions, which means that all of the energy is found in those directions (Fig. 18-21A), rather than being distributed over a spherical surface. Thus, the dipole has a gain approximately 2.1 dB greater than isotropic. In other words, the measured power density at any point will be 2.1 dB higher than the calculated isotropic power density for the same RF input power to the antenna.

Directivity The directivity of an antenna is a measure of its ability to direct RF energy in a limited direction, rather than in all (spherical) directions equally. As shown in Fig. 18-21A, the horizontal directivity of the dipole forms a bidirectional "figure-8" pattern. Two methods for showing unidirectional antenna patterns are shown in Fig. 18-23. The method of Fig. 18-23A is a polar plot viewed from above. The main lobe is centered on 0°. The plot of Fig. 18-23B is a rectangular method for displaying the same information. This pattern follows a Sin X/X function (or [Sin $x/x]^2$ for power).

Directivity (D) is a measure of relative power densities:

$$D = \frac{P_{max}}{P_{av}} \qquad [18.24]$$

Or, referenced to isotropic:

$$D = \frac{4\pi}{\Phi} \qquad [18.25]$$

Where:

D is the directivity
P_{max} is the maximum power
P_{av} is the average power
Φ is the solid angle subtended by the main lobe

The term Φ is a solid angle, which emphasizes the fact that antenna patterns must be examined in at least two extents: horizontal and vertical.

A common method for specifying antenna directivity is *beamwidth* (B_W). The definition of B_W is the angular displacement between points on the main lobe (see Figs. 18-23A and 18-23B), where the power density drops to one-half (–3 dB) of its maximum main lobe power density. This angle is shown in Fig. 18-23A as *a*.

In an ideal antenna system 100 percent of the radiated power is in the main lobe, and there are no other lobes. But in real antennas certain design and installation anomalies cause additional minor lobes, such as the *sidelobes* and *backlobe* shown in Fig. 18-23A. Several problems derive from the minor lobes. First is the loss of usable power. For a given power density required at a distant receiver site the transmitter must supply whatever additional power is needed to make up for the minor lobe losses.

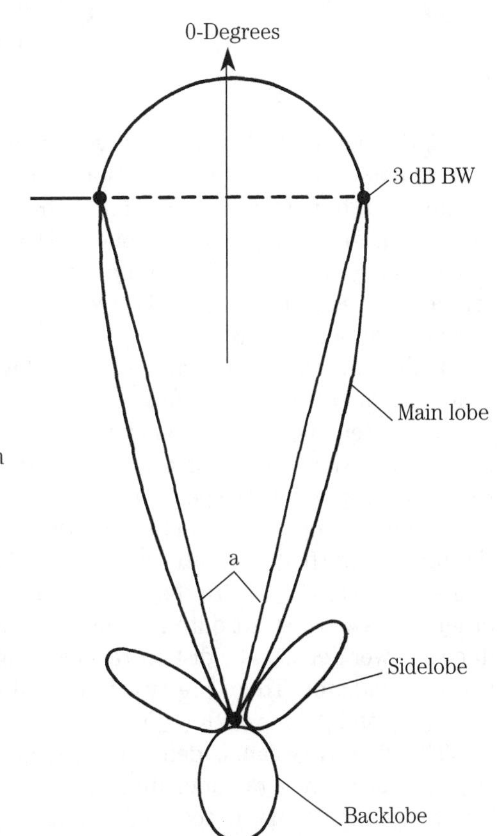

18-23A Directional antenna pattern (top view).

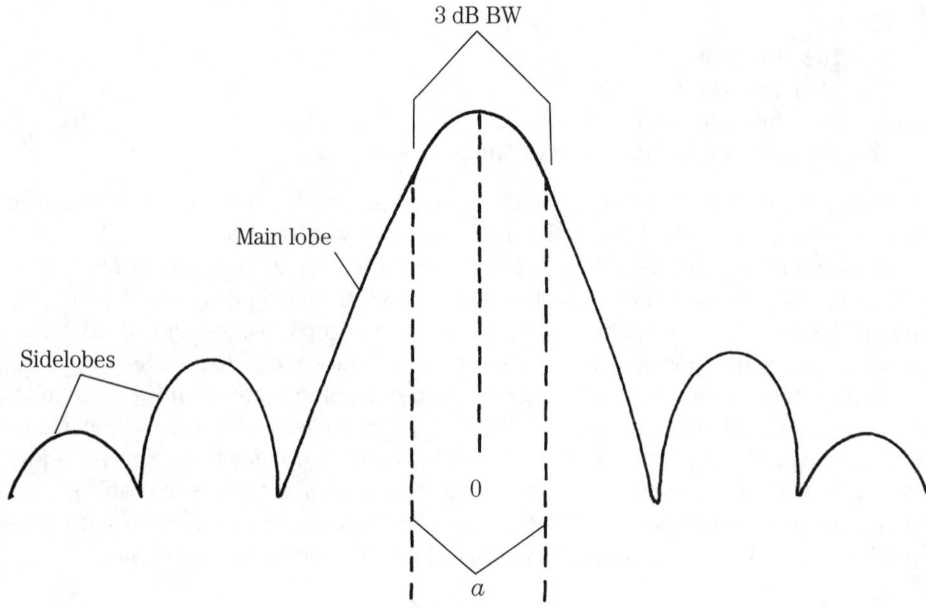

18-23B Graphically presented pattern.

The second problem is intersystem interference. A major application of directional antennas is the prevention of mutual interference between nearby cochannel stations. In radar systems, high sidelobes translate to errors in detected targets. If, for example, a sidelobe is large enough to detect a target, then the radar display will show this off-axis target, as if it was in the main lobe of the antenna. The result is an azimuth error that could be important in terms of marine and aeronautical navigation.

Gain Antenna gain derives from the fact that energy is squeezed into a limited space instead of being distributed over a spherical surface. The term *gain* implies that the antenna creates a higher power; when, in fact, it merely concentrates the power into a single direction that would otherwise be spread out over a larger area. Even so, it is possible to speak of an apparent increase in power. Antenna-transmitter systems are often rated in terms of *effective radiated power* (ERP). The ERP is the product of the transmitter power and the antenna gain. For example, if an antenna has a gain of +3 dB, the ERP will be twice the transmitter output power. In other words, a 100-watt output transmitter connected to a +3-dB antenna will produce a power density at a distant receiver equal to a 200-watt transmitter feeding an isotropic radiator. There are two interrelated gains to be considered: *directivity gain* (G_d) and *power gain* (G_p).

The directivity gain is defined as the quotient of the maximum radiation intensity over the average radiation intensity (note the similarity to directivity definition). This measure of gain is based on the shape of the antenna radiation pattern, and can be calculated with respect to an isotropic radiator ($D = 1$) from:

$$G_d = \frac{4\pi P_a}{P_r} \quad \textbf{[18.26]}$$

Where:

G_d is the directivity gain
P_a is the maximum power radiated per unit of solid angle
P_r is the total power radiated by the antenna

The power gain is similar in nature, but slightly different from directivity gain; it includes dissipative losses in the antenna. Not included, in the power gain, are losses caused by cross polarization or impedance mismatch between the waveguide (or transmission line) and the antenna. There are two commonly used means for determining power gain:

$$G_p = \frac{4\,\pi\,P_a}{P_n} \quad \textbf{[18.27]}$$

and,

$$G_p = \frac{P_{ai}}{P_i} \quad \textbf{[18.28]}$$

(Equation assumes equal power to antenna and comparison isotropic source.)

Where:

P_a is the maximum radiated per unit solid angle
P_n is the net power accepted by the antenna (i.e., less mismatch losses)
P_{ai} is the average intensity at a distant point
P_i is the intensity at the same point from an isotropic radiator fed the same RF power level as the antenna

Provided that ohmic losses are kept negligible, the relationship between directivity gain, and power gain, is given by:

$$G_p = \frac{P_r G_d}{P_n} \quad \textbf{[18.29]}$$

(All terms as previously defined)

Relationship of gain and aperture Antennas obey the Law of Reciprocity, which means that any given antenna will work as well to receive as to transmit. The function of the receiver antenna is to gather energy from the electromagnetic field radiated by the transmitter antenna. The aperture is related to, and often closely approximates, the physical area of the antenna. But in some designs the effective aperture (A_e) is less than the physical area (A), so there is an effectiveness factor (n)

that must be applied. In general, however, a high gain transmitter antenna also exhibits a high receiving aperture, and the relationship can be expressed as:

$$G = \frac{4\pi A_e n}{\lambda^2} \qquad [18.30]$$

Where:

- A_e is the effective aperture
- n is the aperture effectiveness (n = 1 for a perfect, lossless antenna)
- λ is the wavelength of the signal

Horn antenna radiators

The horn radiator is a tapered termination of a length of waveguide (see Figs. 18-24A through 18-24C) that provides the impedance transformation between the waveguide impedance and the free-space impedance. Horn radiators are used both as antennas in their own right, and as illuminators for reflector antennas. Horn antennas are not a perfect match to the waveguide, although standing wave ratios of 1.5:1 or less are achievable.

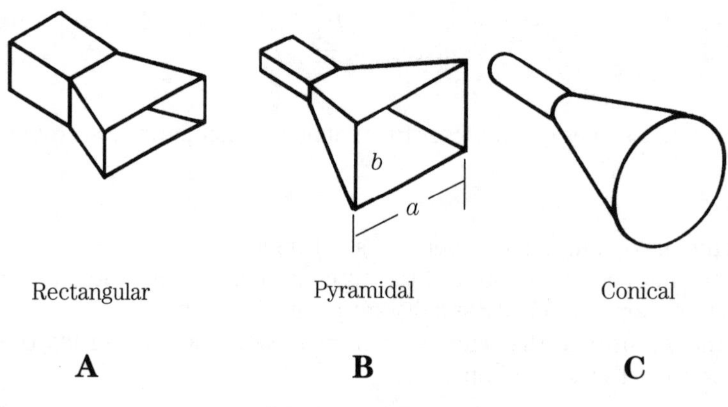

18-24 Horn radiators.

The gain of a horn radiator is proportional to the area (A) of the flared open flange ($A = ab$ in Fig. 18-24B), and inversely proportional to the square of the wavelength:

$$G = \frac{10\,A}{\lambda^2} \qquad [18.31]$$

Where:

- A is the flange area
- λ is the wavelength (both in same units)

The –3 dB beamwidth for vertical and horizontal extents can be approximated from:

Vertical

$$\Phi_v = \frac{51\,\lambda}{b} \text{ degrees} \qquad \textbf{[18.32]}$$

Horizontal

$$\Phi_h = \frac{0\,\lambda}{a} \text{ degrees} \qquad \textbf{[18.33]}$$

Where:

Φ_v is the vertical beamwidth in degrees
Φ_h is the horizontal beamwidth in degrees
a, b are dimensions of the flared flange
λ is the wavelength

A form of antenna, related to the horn, is the cavity antenna of Fig. 18-25. In this type of antenna, a quarter-wavelength radiating element extends from the waveguide (or transmission line connector) into a resonant cavity. The radiator element is placed one-quarter wavelength into a resonant cavity; and is spaced a quarter wavelength from the rear wall of the cavity. A tuning disk is used to alter cavity dimensions in order to provide a limited tuning range for the antenna. Gains to about 6 dB are possible with this arrangement.

Reflector antennas

At microwave frequencies, it becomes possible to use *reflector antennas* because of the short wavelengths involved. Reflectors are theoretically possible at lower frequencies, but because of the longer wavelengths, the antennas would be so large

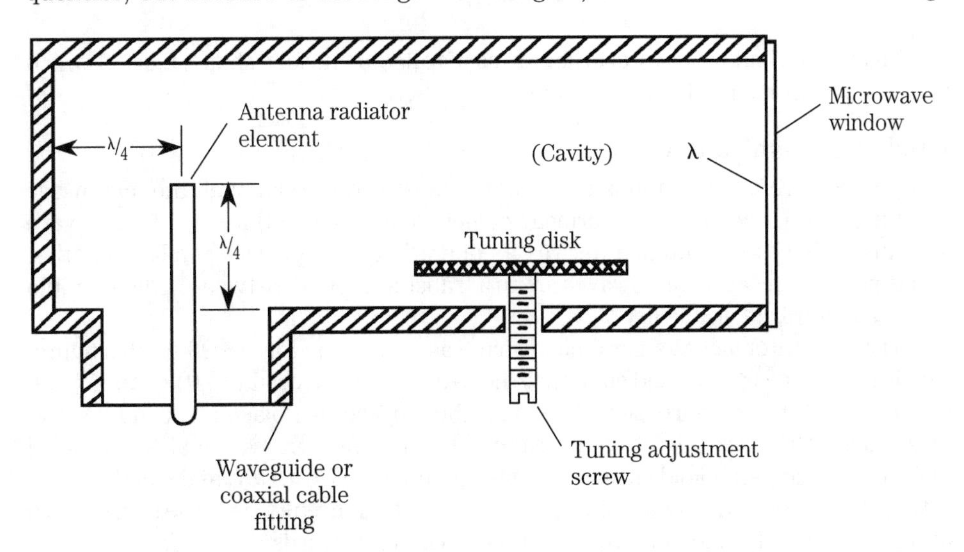

18-25 Cavity antenna.

that they become impractical. Several forms of reflector are used (Figs. 18-26 and 18-27. In Fig. 18-26 we see the *corner reflector* antenna, which is used primarily in the high-UHF and low-microwave region. A dipole element is placed at the "focal point" of the corner reflector, so it receives (in-phase) the reflected wavefronts from the surface. Either solid metallic reflector surfaces or wire mesh may be used. When mesh is used, however, the holes in the mesh must be ½ wavelength or smaller.

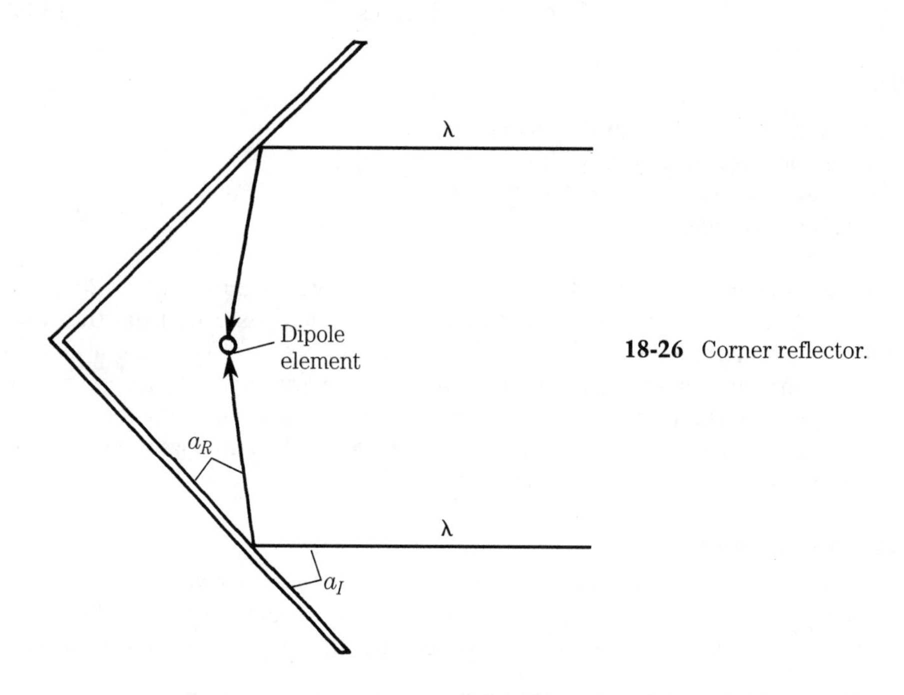

18-26 Corner reflector.

Figure 18-27 shows several other forms of reflector surface shape, most of which are used in assorted radar applications.

Parabolic "dish" antennas

The parabolic reflector antenna is one of the most widespread of all the microwave antennas, and is the type that normally comes to mind when thinking of microwave systems. This type of antenna derives its operation from physics similar to optics, and is possible because microwaves are in a transition region between ordinary radio waves and infrared/visible light.

The dish antenna has a paraboloid shape as defined by Fig. 18-28. In this figure, the dish surface is positioned such that the center is at the origin (0,0) of an *X-Y* coordinate system. For purposes of defining the surface, we place a second vertical axis (called the *directrix*) (Y') a distance behind the surface equal to the focal length (u). The paraboloid surface follows the function $Y^2 = 4uX$, and has the property that a line from the focal point (F) to any point on the surface is the same length as a line from that same point to the directrix (in other words, $MN = MF$).

If a radiator element is placed at the focal point (F), then it will illuminate the reflector surface, causing wavefronts to be propagated away from the surface in-

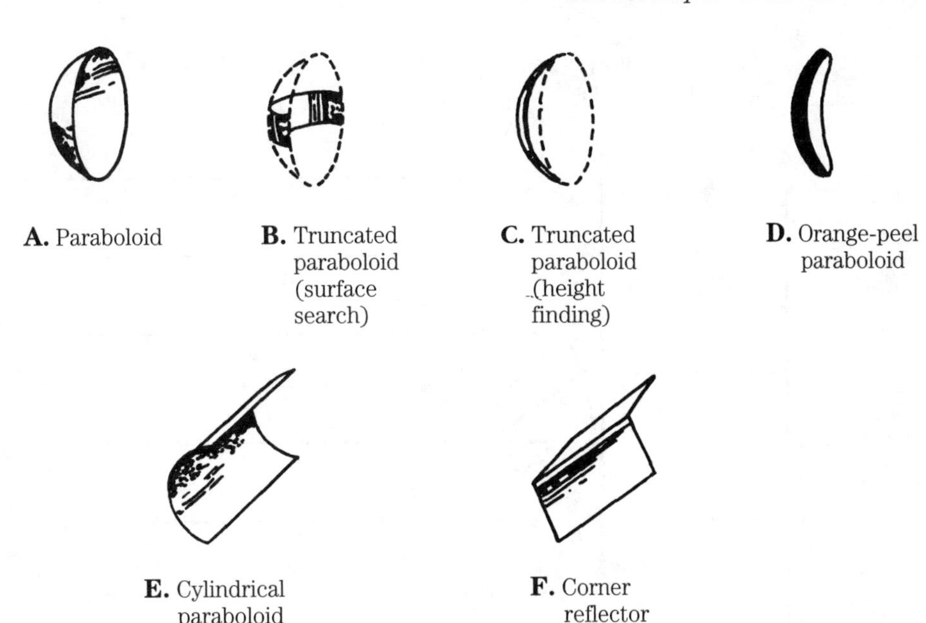

18-27 Reflector antennas.

phase. Similarly, wavefronts, intercepted by the reflector surface, are reflected to the focal point.

Gain The gain of a parabolic antenna is a function of several factors: dish diameter, feed illumination, and surface accuracy. The dish diameter (D) should be large compared with its depth. Surface accuracy refers to the degree of surface irregularities. For commercial antennas, ⅛ wavelength surface accuracy is usually sufficient, although on certain radar antennas the surface accuracy specification must be tighter.

The feed illumination refers to how evenly the feed element radiates to the reflector surface. For circular parabolic dishes, a circular waveguide feed produces optimum illumination, and rectangular waveguides are less than optimum. The TE_{11} mode is desired. For best performance, the illumination should drop off evenly from the center to the edge, with the edge being –10 dB down from the center. The diameter, length, and beamwidth of the radiator element (or horn) must be optimized for the specific F/d ratio of the dish. The cut-off frequency is approximated from:

$$f_{cutoff} = \frac{175698}{d_{mm}} \quad [18.34]$$

Where:

F_{cutoff} is the cutoff frequency
d is the inside diameter of the circular feedhorn

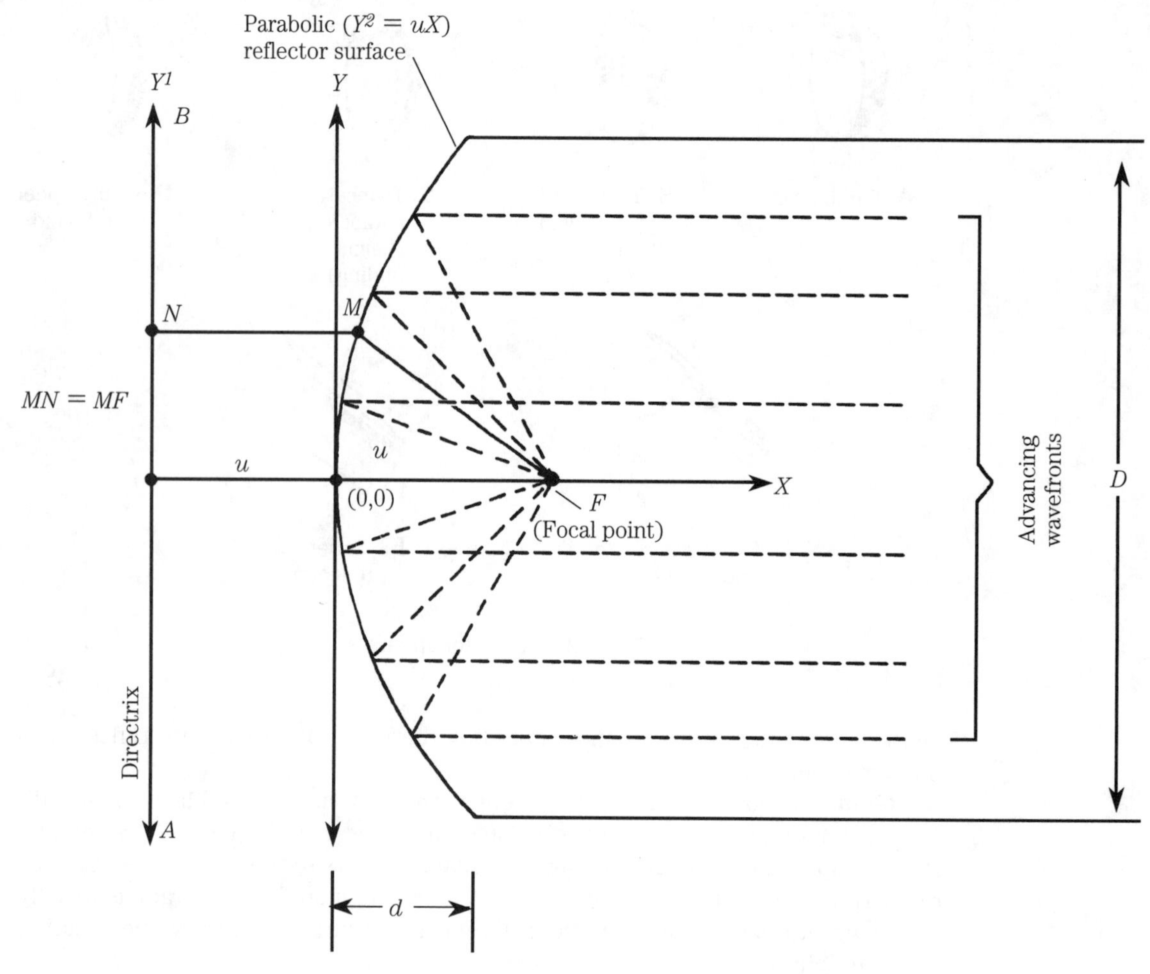

18-28 Ray tracing shows operation of parabolic antenna.

The gain of the parabolic dish antenna is found from:

$$G = \frac{k\,(\pi D)^2}{\lambda^2} \quad \textbf{[18.35]}$$

Where:

G is the gain over isotropic
D is the diameter
λ is the wavelength (same units as D)
k is the reflection efficiency (0.4 to 0.7, with 0.55 being most common)

The –3 dB beamwidth of the parabolic dish antenna is approximated by:

$$BW = \frac{70\,\lambda}{D} \quad \textbf{[18.36]}$$

and the focal length by:

$$f = \frac{D^2}{16\,d} \qquad \textbf{[18.37]}$$

For receiving applications, the effective aperture is the relevant specification, and is found from:

$$A_e = k\,\pi\,(D/2)^2 \qquad \textbf{[18.38]}$$

The antenna pattern radiated by the antenna is similar to Fig. 18-23B. With horn illumination, the sidelobes tend to be 23 to 28 dB below the main lobe, or 10 to 15 dB below isotropic. It is found that 50 percent of the energy radiated by the parabolic dish is within the –3 dB beamwidth, and 90 percent is between the first nulls on either side of the main lobe.

If a dipole element is used for the feed device, then a *splash plate* is placed ¼ wavelength behind the dipole in order to improve illumination. The splash plate must be several wavelengths in diameter, and is used to reflect the backlobe back toward the reflector surface. When added to the ½-wave phase reversal inherent in the reflection process, the two-way ¼ wavelength adds another ½ wavelength and thereby permits the backwave to move out in-phase with the front lobe wave.

Parabolic dish feed geometries Figure 18-29 shows two methods for feeding parabolic dish antennas, regardless of which form of radiator (horn, dipole, etc.) is used. In Fig. 18-29A we see the method in which the radiator element is placed at the focal point, and either a waveguide (or transmission line) is routed to it. This method is used in low-cost installations such as home satellite TV receive-only (TVRO) antennas.

Figure 18-29B shows the *Cassegrain feed* system. This system is modelled after the Cassegrain optical telescope. The radiator element is placed at an opening at the center of the dish. A hyperbolic sub-reflector is placed at the focal point, and it is used to reflect the wavefronts to the radiator element. The Cassegrain system results in lower noise operation because of several factors: less transmission line length, lower sidelobes, and the fact that the open horn sees sky instead of earth (which has lower temperature); on the negative side, galactic and solar noise might be slightly higher on a Cassegrain dish.

Figure 18-30A shows the *monopulse* feed geometry. In this system, a pair of radiator elements are placed at the focal point, and fed to a power splitter network that outputs both *sum* (Fig. 18-30B) and *difference* (Fig. 18-30C) signals. When these are combined, the resultant beam shape has improved –3 dB beamwidth as a result of the algebraic summation of the two.

Array antennas

When antenna radiators are arranged in a precision array, an increase in gain occurs. An array might be a series of dipole elements, as in the broadside array of Fig. 18-31 (which is used in the UHF region), or a series of slots, horns, or other radiators. The overall gain of an array antenna is proportional to the number of elements, as well as

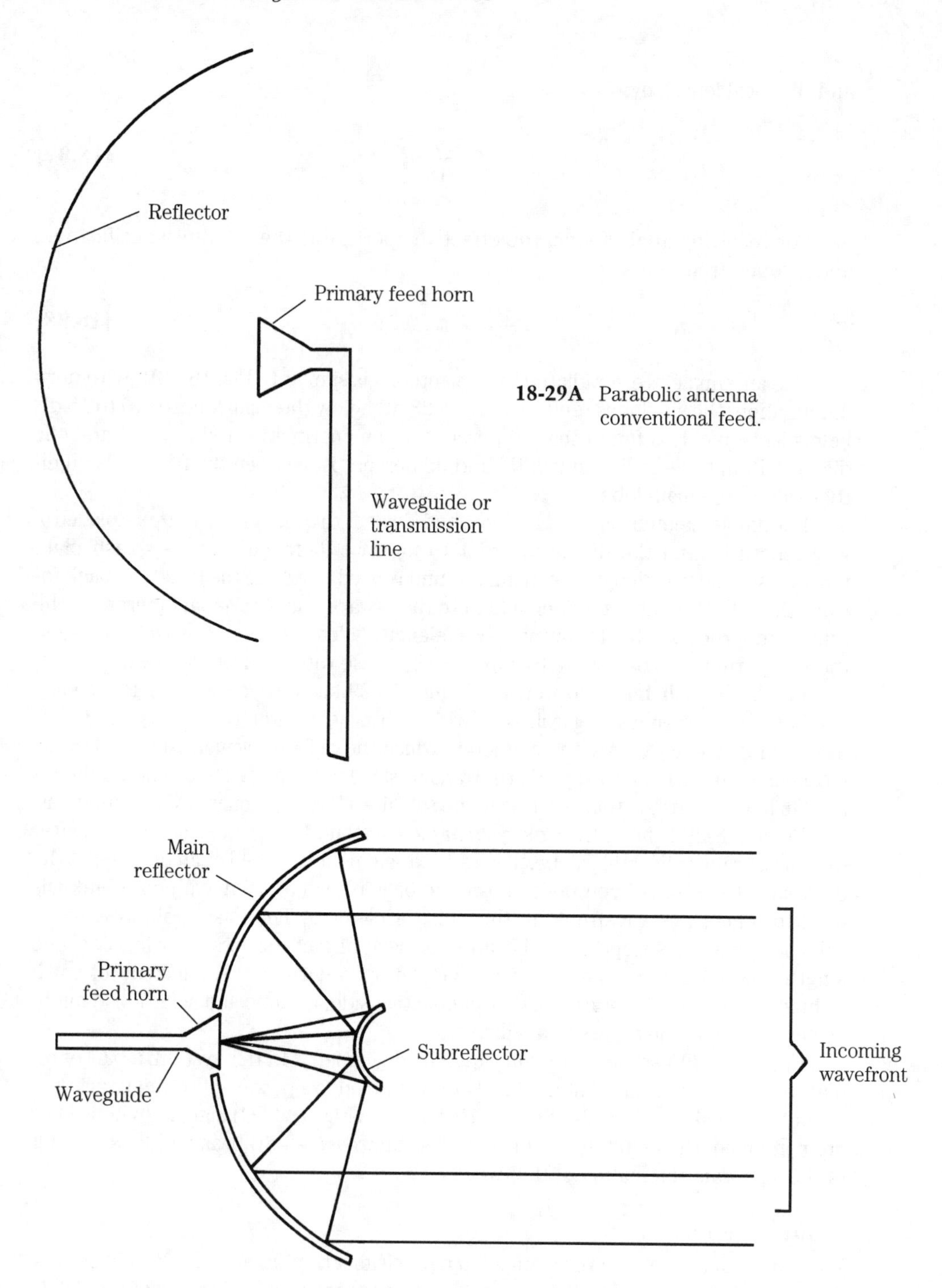

18-29A Parabolic antenna conventional feed.

18-29B Parabolic antenna cassegrain feed.

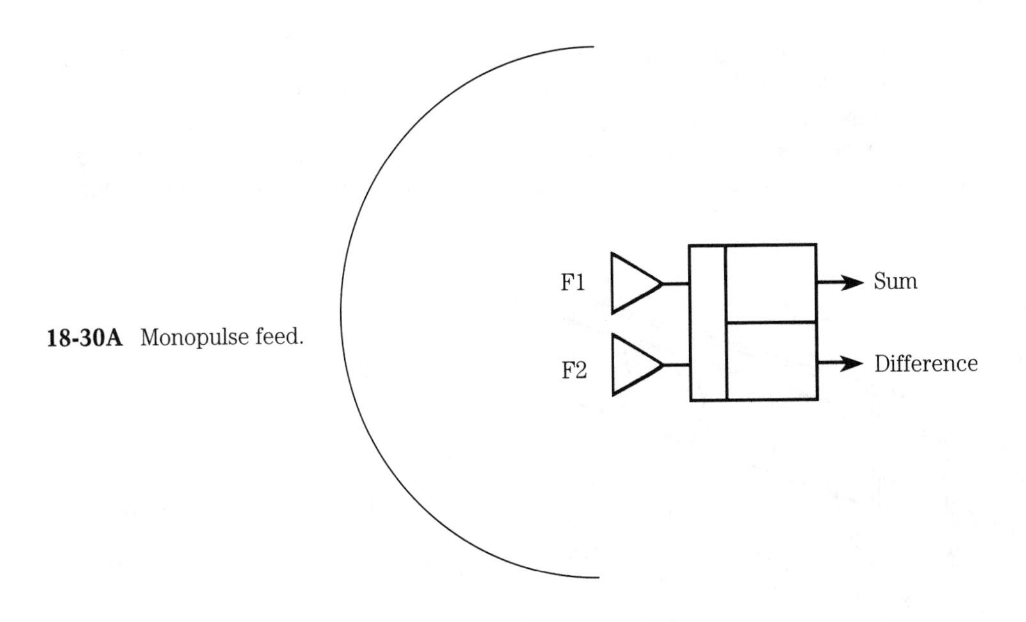

18-30A Monopulse feed.

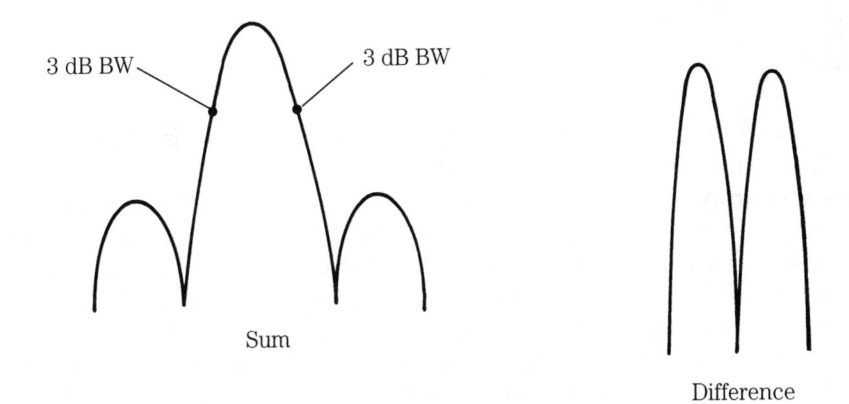

18-30B,C, D Monopulse patterns.

the details of their spacing. In this, and other antennas, a method of *phase shifting* is needed. In Fig. 18-31, the phase shifting is caused by the crossed feeding of the elements, but in more modern arrays, other forms of phase shifter are used.

Two methods of feeding an array are shown in Fig. 18-32. The *corporate feed* method connects all elements, and their phase shifters, in parallel with the source.

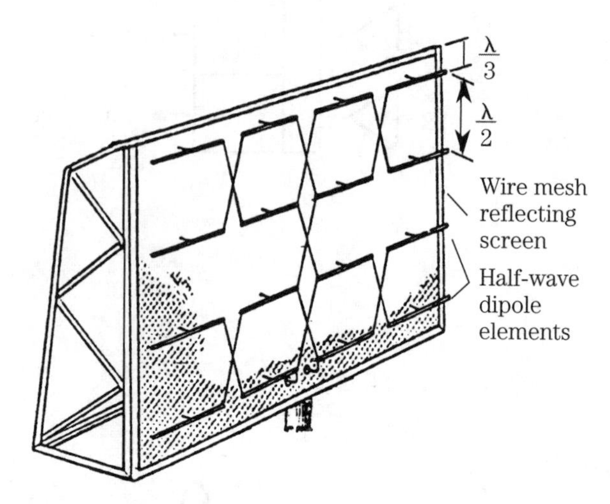

18-31 Reflector array antenna.

The *branch feed* method breaks the waveguide network into two (or more) separate paths; both branch and corporate feed are shown in Fig. 18-32.

Solid-state array antennas Some modern radar sets use solid-state array antennas consisting of a large number of elements, each of which are capable of shifting the phase of a microwave input signal. Two forms are known: *passive* (Fig. 18-33A) and *active* (Fig. 18-33B). In the passive type of element, a ferrite (or PIN diode) phase shifter is placed in the transmission path between the RF input and the radiator element (usually a slot). By changing the phase of the RF signal selectively, it is possible to form and steer the beam at will. A 3-bit (i.e., three discrete state) phase shifter allows the phase to shift in 45-degree increments, while a 4-bit phase shifter allows 22.5° increments of phase shift.

The active element contains a phase shifter in addition to a transmit power amplifier (1 or 2 watts) and a low noise amplifier for receiving. A pair of transmit/receive (T/R) switches select to which path the RF signal is directed. The total output power of this antenna is the sum of all output powers from all elements in the array. For example, an array of 1000 2-watt elements makes a 2000-watt system.

Slot array antennas

A resonant slot (cut into a wall of a section of waveguide) somewhat analogous, if not identical, to a dipole. By cutting several slots in the waveguide, we obtain the advantages of an array antenna in which the elements are several slot radiators. Slot array antennas are used for marine navigation radars, telemetry systems and for the re-

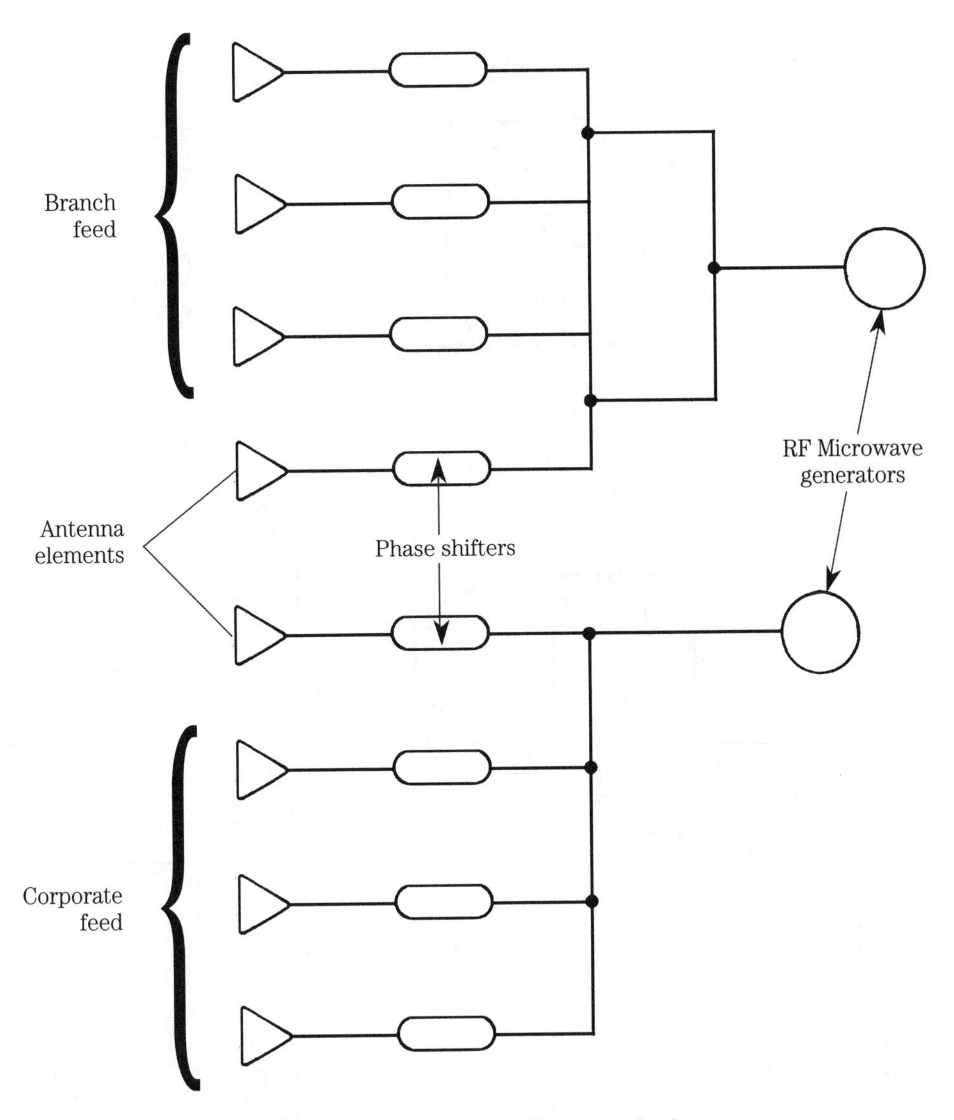

18-32 Branch feed and Corporate feed.

ception of microwave television signals in the Multipoint Distribution Service (MDS) on 2.145 GHz.

Figure 18-34 shows a simple slot antenna used in telemetry applications. A slotted section of rectangular waveguide is mounted to a right-angle waveguide flange. An internal wedge (not shown) is placed at the top of the waveguide, and serves as a matching impedance termination to prevent internal reflected waves. Directivity is enhanced by attaching flanges to the slotted section of waveguide parallel to the direction of propagation (see and view in Fig. 18-34).

Figure 18-35 shows two forms of *flatplate array* antennas constructed from slotted waveguide radiator elements (shown as insets). Figure 18-35A shows the

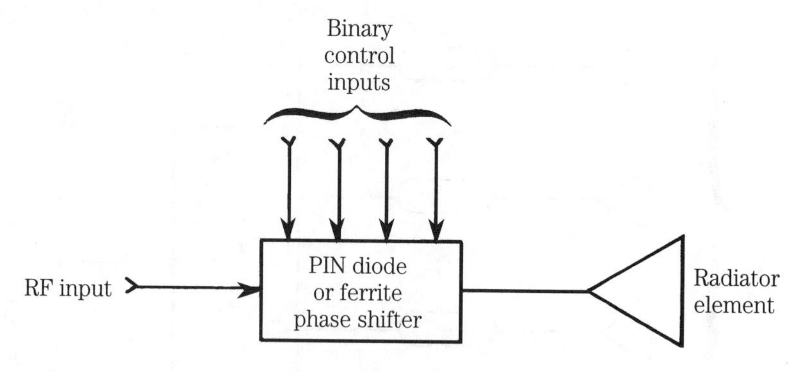

18-33A Phase shifter.

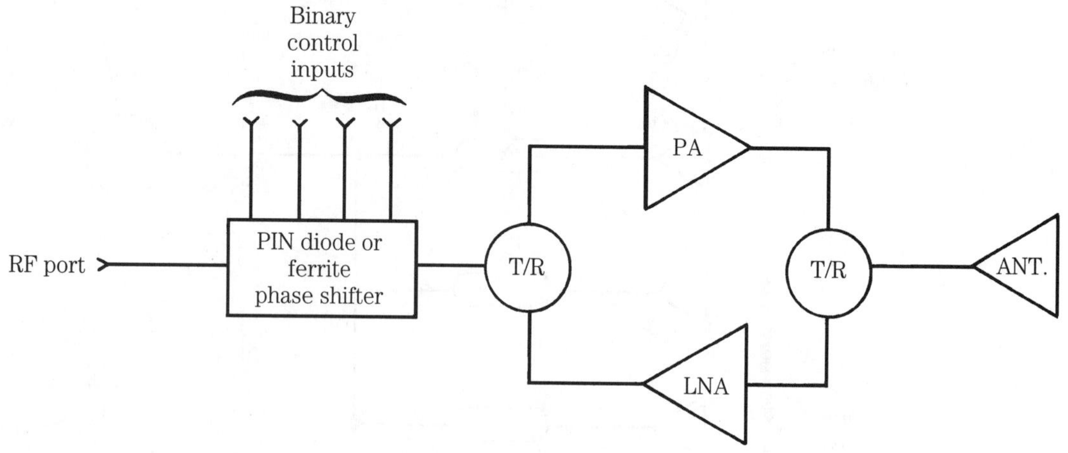

18-33B Phase shift T/R element.

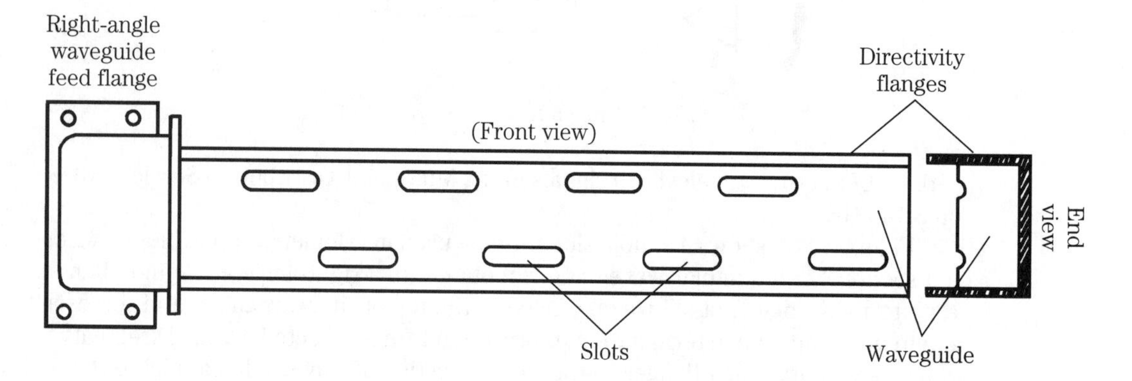

18-34 Slot antenna.

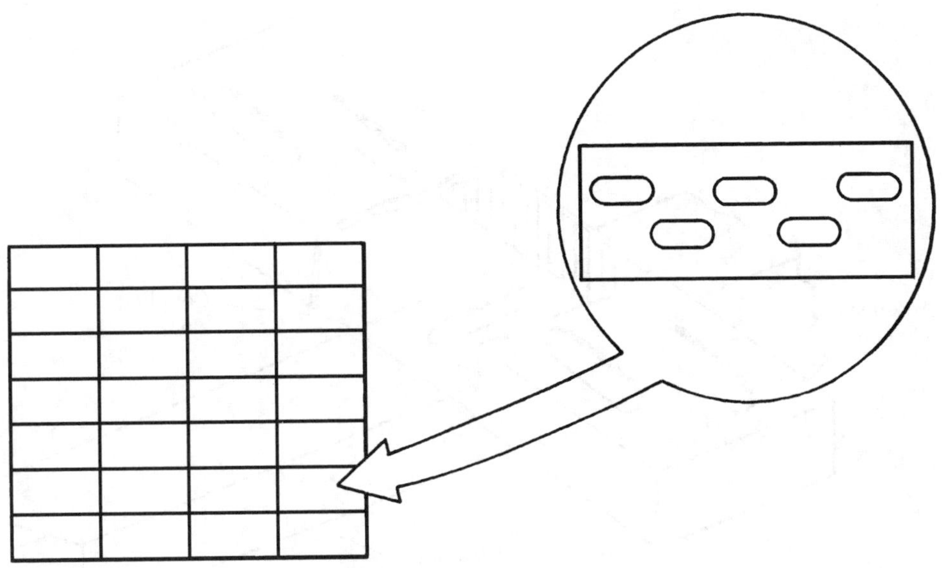

18-35A Flat plate slot array.

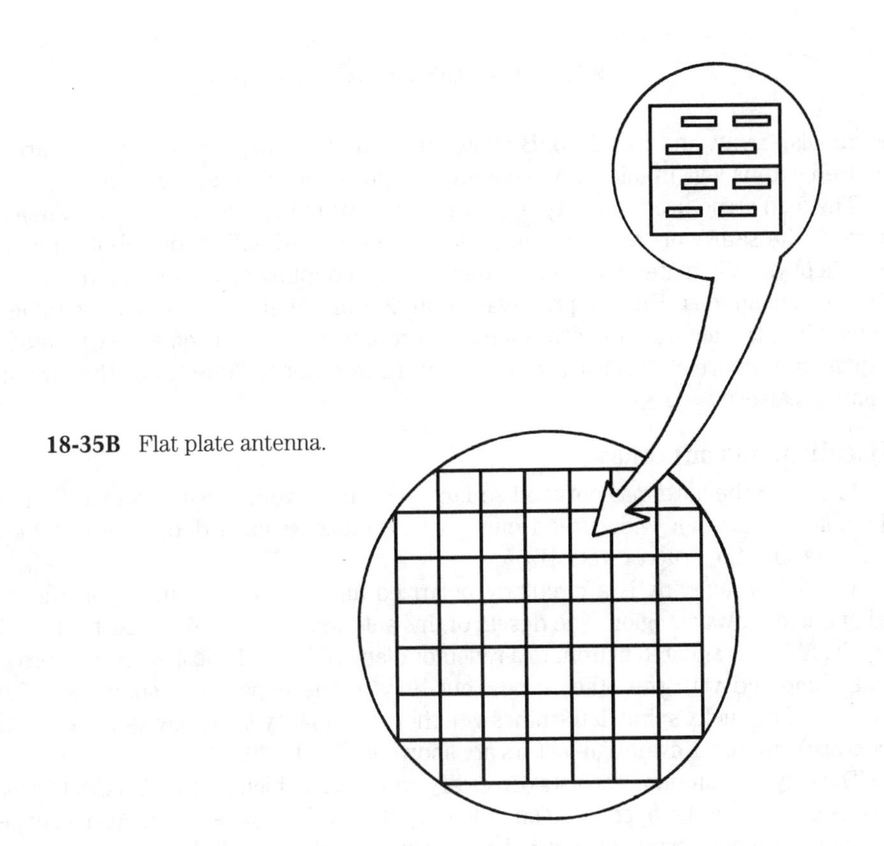

18-35B Flat plate antenna.

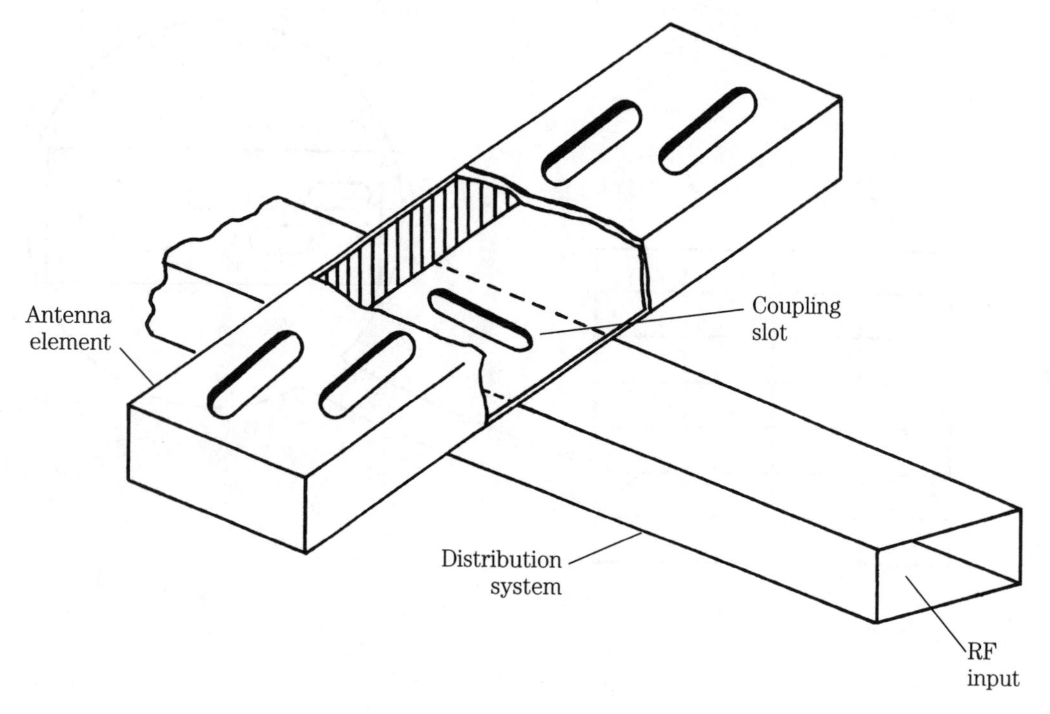

18-35C Flat plate antenna feed coupling.

rectangular array, and Fig. 18-35B shows the circular array. These flatplate arrays are used extensively in microwave communications and radar applications.

The feed structure for a flatplate array is shown in Fig. 18-35C. The antenna element is the same sort as shown in Figs. 18-35A and 18-35B. A distribution waveguide is physically mated with the element, and a coupling slot is provided between the two waveguides. Energy propagating in the distribution system waveguide is coupled into the antenna radiator element through this slot. In some cases metallic, or dielectric, phase-shifting stubs are also used in order to "fine tune" the antenna radiation pattern.

Miscellaneous antennas

In addition to the antennas covered so far, there are several designs which find application, but are not widespread enough to be discussed in detail here: *helical*, *polycone*, and the *ring* or *disk Yagi*.

The helical antenna is a circularly polarized antenna used primarily in the low end of the microwave region. The design of this antenna (Fig. 18-36) consists of a helical coil of N turns situated in front of a reflector plane (G). The helical antenna is broadband, compared with the other types, but only at the expense of some gain. The direction of the helix spiral determines whether the polarity is clockwise (or counterclockwise) circular. Key design factors are shown in Fig. 18-36.

The polycone antenna consists of a cavity antenna in which a cone of dielectric material is inserted in the open end (i.e., the output iris). The dielectric material shapes the beam into a narrower beamwidth than is otherwise obtained. Polycone antennas are

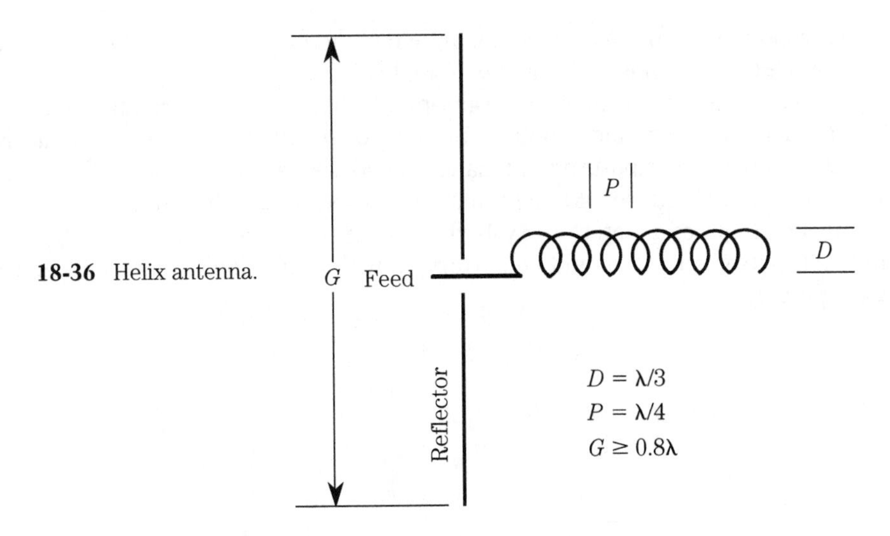

18-36 Helix antenna.

used in various low-power applications such as microwave signaling, control system sensors (including doppler traffic control systems) and in some police CW speed radar guns.

The classical Yagi-Uda array (commonly called "Yagi") is used extensively in HF, VHF, UHF, and low microwave bands; modified versions are used in the mid-microwave region (several GHz). The Yagi antenna consists of a half-wavelength centerfed dipole "driven element" and one or more parasitic elements (microwave versions tend to use ten or more elements). The parasitic elements are placed parallel to the driven element about 0.1 to 0.3 wavelengths away from it or each other. Two sorts of parasitic elements are used: *reflectors* and *directors*. The reflectors are slightly longer than (approximately 4 percent), and are placed behind, the driven element; directors are slightly longer, and placed in front of the driven element. The main lobe is in the direction of the director elements. Yagis are commonly used to frequencies of 800 or 900 MHz. Modified versions of this antenna use rings or disks for the elements, and are useful to frequencies in the 3- or 4-GHz region.

Microwave antenna safety note

Microwave RF energy is dangerous to your health. Anything that can cook a roast beef can also cook you! The U.S. government sets a safety limit for microwave exposure of 10 mW/cm^2 averaged over six minutes; some other countries use a level one-tenth of the U.S. standard. The principal problem is tissue heating, and eyes seem especially sensitive to microwave energy. Some authorities believe that cataracts form from prolonged exposure. Some authorities also believe that genetic damage to offspring is possible as well as other long-term effects as a result of cumulative exposure.

Because of their relatively high gain, microwave antennas can produce hazardous field strengths in close proximity—even at *relatively low RF input power levels*. At least one technician in a TV satellite earth station suffered abdominal adhesions, solid matter in the urine and genital disfunction after servicing a 45-meter diameter 3.5-GHz antenna with RF power applied.

Be very careful around microwave antennas. Do not service a radiating antenna. When servicing nonradiating antennas, be sure to stow them in a position that prevents the inadvertent exposure of humans, should power accidentally be applied. A *Radiation Hazard* sign should be prominently displayed on the antenna. Good design practice requires an interlock system that prevents radiation in such situations. "Hot" transmitter service should be performed with a shielded dummy load replacing the antenna.

19

CHAPTER

Impedance matching in antenna systems

ONE OF THE FIRST THINGS THAT YOU LEARN IN RADIO COMMUNICATIONS AND broadcasting is that antenna impedance must be matched to the transmission line impedance, and that the transmission line impedance must be matched to the output impedance of the transmitter. The reason for this requirement is that maximum power transfer, between a source and a load, always occurs when the system impedances are matched. In other words, more power is transmitted from the system when the load impedance (the antenna), the transmission line impedance and the transmitter output impedance are all matched to each other.

Of course, the trivial case is where all three sections of our system have the same impedance. For example, we could have an antenna with a simple 75-Ω resistive feedpoint impedance (typical of a half-wave dipole in free space), and a transmitter with an output impedance that will match 75 Ω. In that case, we need only connect a standard-impedance 75-Ω length of coaxial cable between the transmitter and the antenna. Job done! Or so it seems.

But there are other cases where the job is not so simple. In the case of our "standard" antenna, for example, the feedpoint impedance is rarely what the books say it should be. That ubiquitous dipole, for example, is nominally rated at 75 Ω, but even the simplest antenna books tell us that value is an approximation of the theoretical free-space impedance. At locations closer to the earth's surface, the impedance could vary over the approximation range of 30 Ω to 130 Ω and it might have a substantial reactive component.

But there is a way out of this situation. We can construct an impedance-matching system that will marry the source impedance to the load impedance. This chapter examines several matching systems that might prove useful in a number of antenna situations.

Impedance matching approaches

Antenna impedance can contain both reactive and resistive components. In most practical applications, we are searching for a purely resistive impedance ($Z = R$), but that ideal is rarely achieved. A dipole antenna, for example, has a theoretical free-space impedance of 73 Ω at resonance. As the frequency applied to the dipole is varied away from resonance, however, a reactive component appears. When the frequency is greater than resonance, then the antenna tends to look like an inductive reactance, so the impedance is $Z = R + jX$. Similarly, when the frequency is less than the resonance frequency, the antenna looks like a capacitive reactance, so the impedance is $Z = R - jX$. Also, at distances closer to the earth's surface the resistive component may not be exactly 73 Ω, but may vary from about 30 Ω to 130 Ω. Clearly, whatever impedance coaxial cable is selected to feed the dipole, it stands a good chance of being wrong.

The method used for matching a complex load impedance (such as an antenna) to a resistive source (the most frequently encountered situation in practical radio work), is to interpose a matching network between the load and the source (Fig. 19-1). The matching network must have an impedance that is the *complex conjugate* of the complex load impedance. For example, if the load impedance is $R + jX$, then the matching network must have an impedance of $R - jX$; similarly, if the load is $R - jX$ then the matching network must be $R + jX$. The sections to follow look at some of the more popular networks that accomplish this job.

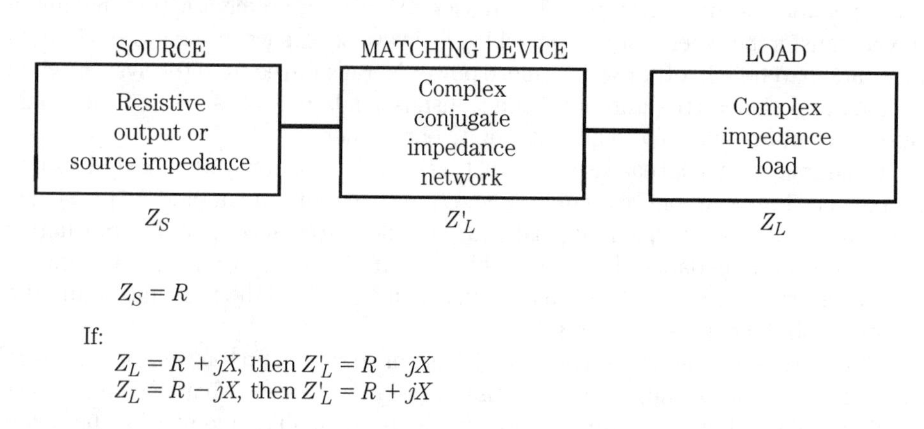

19-1 Transmitter, matcher and antenna system.

L-section network

The L-section network is one of the most used, or at least the most published, antenna matching networks in existance: it rivals even the pi-network. A circuit for the L-section network is shown in Fig. 19-2A. The two resistors represent the source (R_1) and load (R_2) impedances. The elementary assumption of this network is that $R_1 < R_2$. The design equations are:

$$R_1 < R_2, \text{ and } 1 < Q < 5 \qquad [19.1]$$

$$X_L = 6.28FL = Q \times R_1 \quad [19.2]$$

$$X_C = \frac{1}{6.28FC} \quad [19.3]$$

$$Q = \sqrt{\frac{R_2}{R_1} - 1} \quad [19.4]$$

also,

$$Q = \frac{X_L}{R_1} = \frac{R_2}{X_C} \quad [19.5]$$

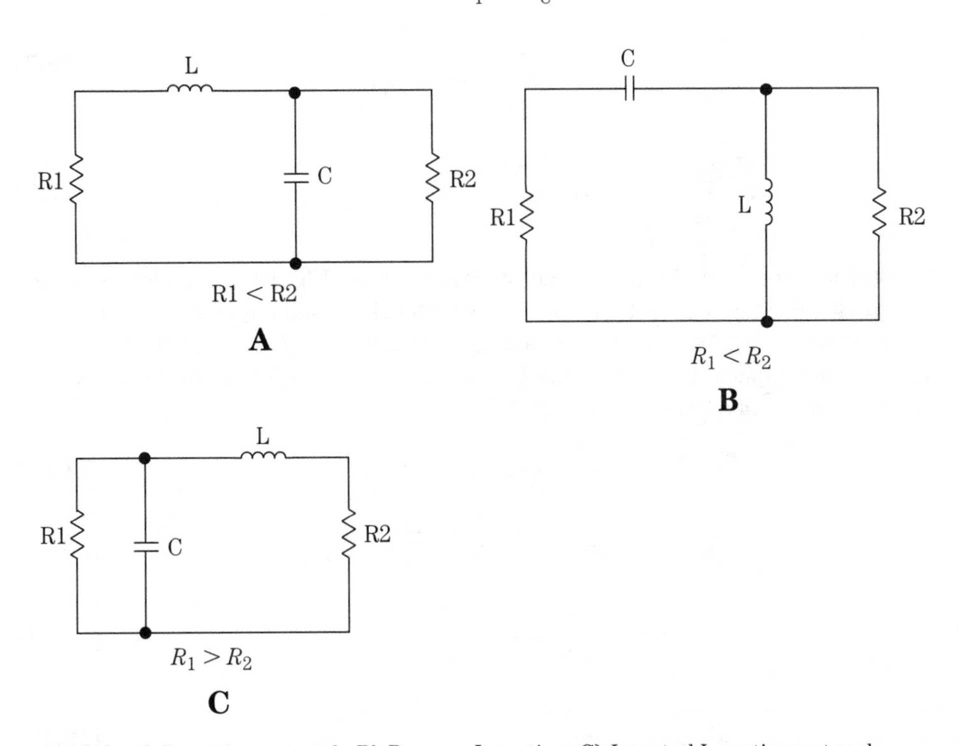

19-2 A) L-section network, B) Reverse L-section, C) Inverted L-section network.

You will most often see this network published in conjunction with less-than-quarter-wavelength longwire antennas. One common fault of those books and articles is that they typically call for a "good ground" in order to make the antennas work properly. But they don't tell you, 1) what a "good ground" is, and 2) how you can obtain it. Unfortunately, at most locations a good ground means burying a lot of copper conductor (see chapter 24)—something that most of us cannot afford. In addition, it is often the case that the person who is forced to use a longwire, instead of a better antenna, cannot construct a "good ground" under any circumstances because of landlords (and/or logistics) problems. The very factors that prompt the use of a long-wire antenna in the first place also prohibit any form of practically obtainable "good ground." But there is a way out: *radicals*. A good ground can be simulated with a counterpoise ground constructed of quarter-wavelength radials. These radials have a length in feet equal to $246/F_{MHz}$, and as few as two of them will work wonders. See chapter 24 for a more complete discussion of grounding problems.

Another form of L-section network is shown in Fig. 19-2B. This circuit differs from the previous circuit in that the roles of the L and C components are reversed. As you might suspect, this role reversal brings about a reversal of the impedance relationships: in this circuit the assumption is that the driving source impedance R_1 is larger than the load impedance R_2 (i.e., $R_1 > R_2$). The equations are shown below:

$$R_2 > R_1 \qquad [19.6]$$

$$X_L = R_2 \sqrt{\frac{R_1}{(R_2 - R_1)}} \qquad [19.7]$$

$$X_C = \frac{(R_1 R_2)}{X_L} \qquad [19.8]$$

Still another form of L-section network is shown in Fig. 19-2C. Again, we are assuming that driving source impedance R_1 is larger than load impedance R_2 (i.e., $R_1 > R_2$). In this circuit, the elements are arranged similar to Fig. 19-2A, with the exception that the capacitor is at the input rather than the output of the network. The equations governing this network are:

$$R_1 > R_2 \text{ and } 1 < Q < 5 \qquad [19.9]$$

$$X_L = 6.28FL = \sqrt{(R_1 R_2) - (R_2)^2} \qquad [19.10]$$

$$X_C = \frac{1}{6.28FC} = \frac{R_1 R_2}{X_L} \qquad [19.11]$$

$$C = \frac{1}{6.28FX_C} \qquad [19.12]$$

$$L = \frac{X_L}{6.28F} \qquad [19.13]$$

Thus far, we have considered only matching networks that are based on inductor and capacitor circuits. But there is also a possibility of using transmission line segments as impedance matching devices. Two basic forms are available to us: quarter-wave sections, and the series matching section.

Pi-networks

The pi-network shown in Fig. 19-3 is used to match a high source impedance to a low load impedance. These circuits are typically used in vacuum tube RF power ampifiers that need to match low antenna impedances. The name of the circuit comes from its resemblance to the Greek letter "pi." The equations for the pi-network are:

$$R_1 > R_2 \text{ and } 5 < Q < 15 \qquad [19.14]$$

$$Q > \sqrt{\frac{R_1}{R_2} - 1} \qquad [19.15]$$

$$X_{C2} = \frac{R_2}{\sqrt{\left(\frac{R_2}{R_1}\right)(1 + Q^2) - 1}} \qquad [19.16]$$

$$X_{C1} = \frac{R_1}{Q}$$

$$X_L = \frac{(R_1 (Q + (R_2/X_{C2})))}{Q^2 + 1} \qquad [19.17]$$

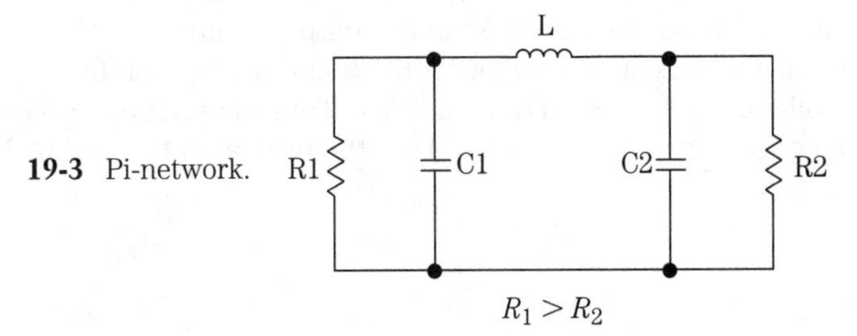

19-3 Pi-network.

Split-capacitor network

The split capacitor network shown in Fig. 19-4 is used to transform a source impedance that is less than the load impedance. In addition to matching antennas, this circuit is also used for interstage impedance matching inside communications equipment. The equations for design are:

$$R_1 < R_2 \qquad [19.18]$$

$$Q > \sqrt{\frac{R_1}{R_2} - 1} \qquad [19.19]$$

$$X_L = \frac{R_2}{Q} \qquad [19.20]$$

$$X_{C1B} = \sqrt{\frac{R_1 (Q_2 + 1)^2}{}} - 1 \qquad [19.21]$$

$$X_{C1A} = \frac{R_2 Q}{Q^2 + 1} \left[1 - \frac{R_1}{Q X_{C1A}} \right] \qquad [19.22]$$

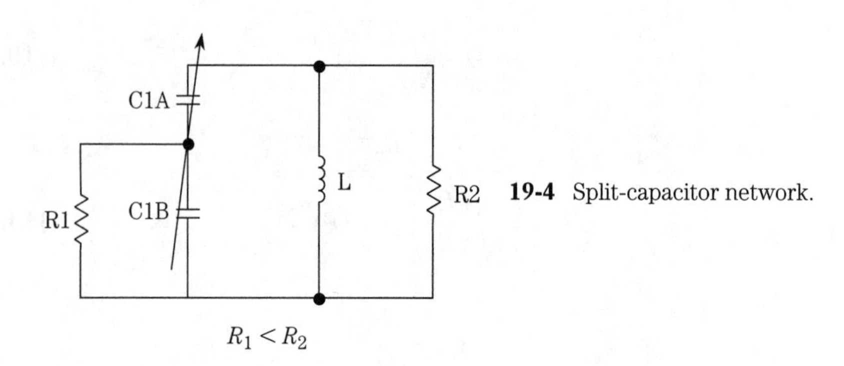

19-4 Split-capacitor network.

Transmatch circuit

One version of the *Transmatch* is shown in Fig. 19-5. This circuit is basically a combination of the split-capacitor network, and an output tuning capacitor (C2). For the HF bands, the capacitors are on the order of 150 pF per section for C1, and 250 pF for C2. The collar inductor should be 28 μH. The Transmatch is essentially a coax-to-coax impedance matcher, and is used to trim the mismatch from a line before it affects the transmitter.

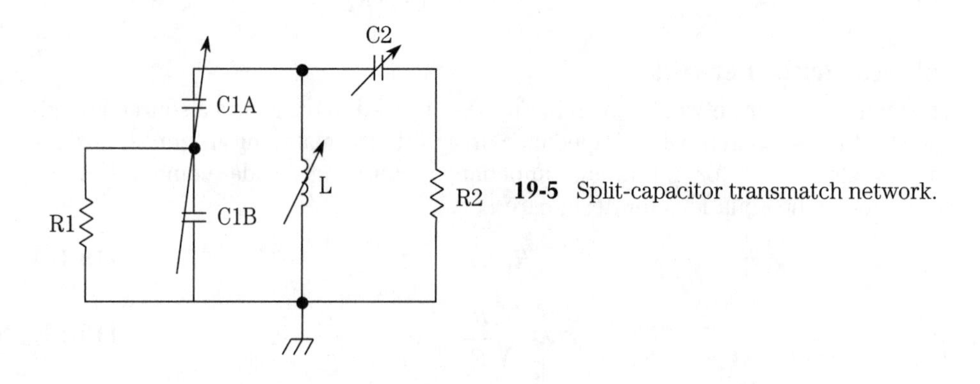

19-5 Split-capacitor transmatch network.

Perhaps the most common form of Transmatch circuit is the tee-network shown in Fig. 19-6. This network is lower in cost than some of the others, but suffers a problem. Although it does, in fact, match impedance (and thereby, in a naive sense, "tunes out" VSWR on coaxial lines), it also suffers a "high-pass" characteristic. The network, therefore, does not reduce the harmonic output of the transmitter. The simple tee-network does not serve one of the main purposes of the antenna tuner: harmonic reduction.

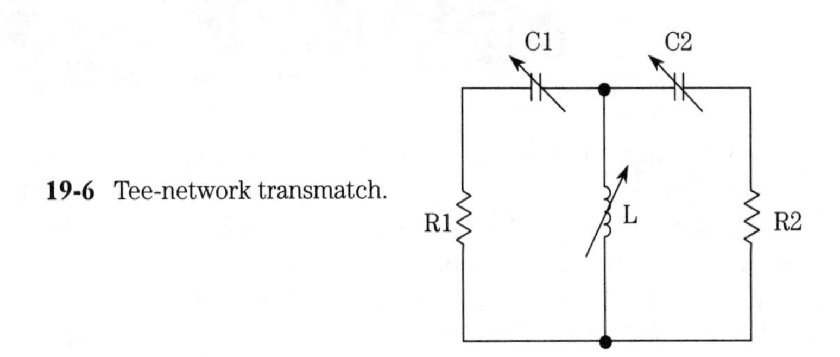

19-6 Tee-network transmatch.

An alternative network, called the *SPC Transmatch*, is shown in Fig. 19-7. This version of the circuit offers harmonic attenuation, as well as matching impedance.

Figures 19-8A/B/C shows commercially available antenna tuners based on this transmatch design. The unit shown in Fig. 19-8A is manufactured by MFJ Electronics, Inc. It contains the usual three tuning controls, here labelled *Transmitter, Antenna, and Inductor*. Included in this instrument is an antenna selector switch that allows the operator to select a coax antenna through the tuner, to connect input to output (coax) without regard to the tuner, select a balanced antenna, or an internal dummy load. The instrument also contains a multi-function meter that can measure 200 watts or 2000 watts (full scale), in either forward or reverse directions. In addition, the meter operates as a VSWR meter.

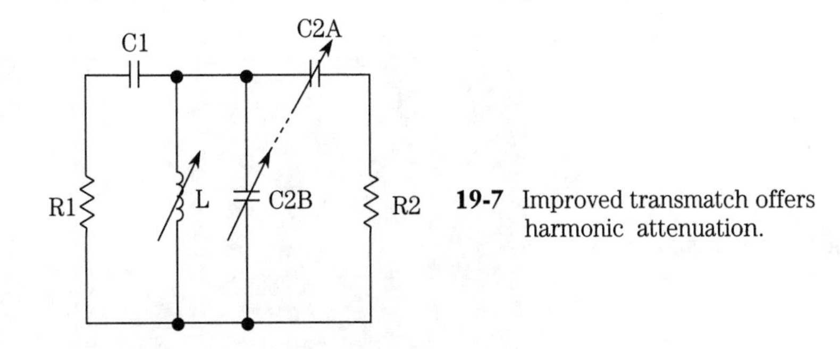

19-7 Improved transmatch offers harmonic attenuation.

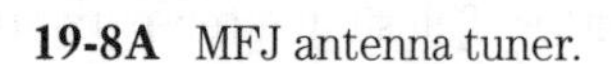

19-8A MFJ antenna tuner.

Figures 19-8B through 19-8D show an imported tuner from the United Kingdom. This instrument, called the *Nevada* model, is a low-cost model, but contains the three basic controls. For proper operation, an external RF power meter, or VSWR meter, is required. The tuner is shown in Fig. 19-8B; also shown is a Heathkit transmatch antenna tuner. The rear panels of the instruments are shown in Fig. 19-8C. There are SO-239 coaxial connectors for input and unbalanced output, along with a pair of posts for the parallel line output. A three-post panel is used to select which antenna the RF goes to: unbalanced (coax) or parallel. The internal circuitry of the *Nevada* is shown in Fig. 19-9D. The roller inductor is in the center and allows the user to set the tuner to a wide range of impedances over the entire 3- to 30-MHz HF band.

19-8B Author's Heathkit antenna tuner and the Nevada tuner from the United Kingdom.

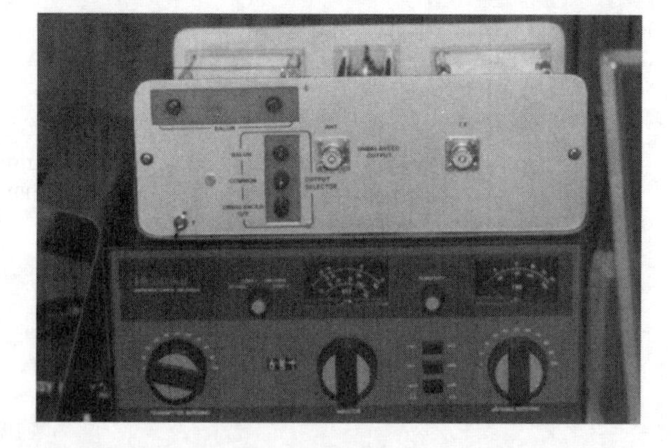

19-8C Rear of Nevada tuner.

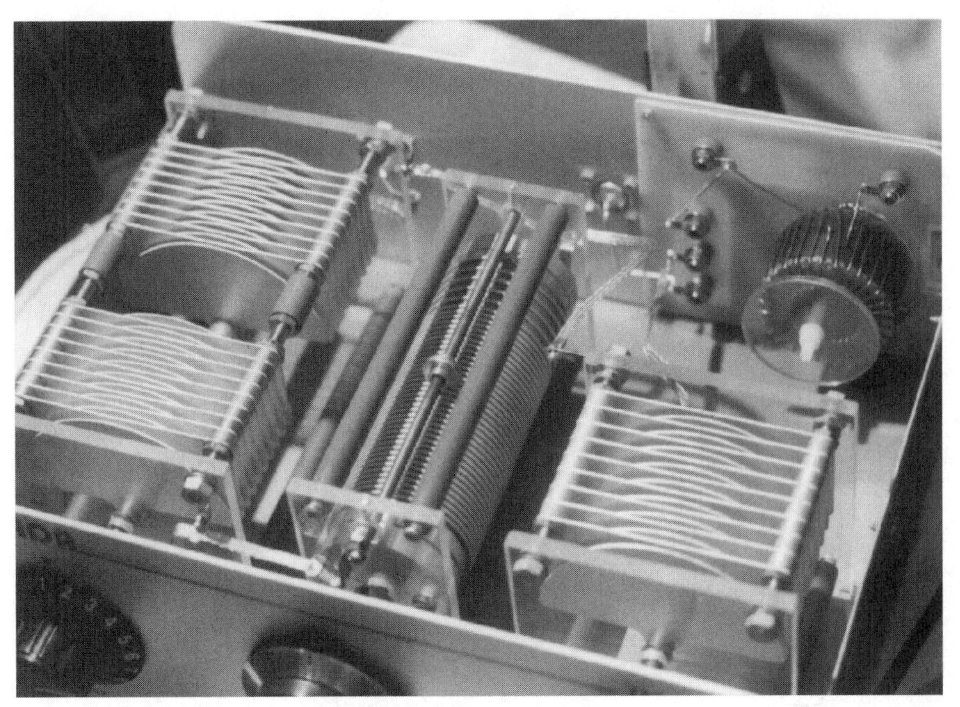

19-8D Insides of Nevada tuner.

Coaxial cable BALUNS

A BALUN is a transformer that matches an "UNbalanced" resistive source impedance (such as a coaxial cable), and a "BALanced" load (such as a dipole antenna). With the circuit of Fig. 19-9, we can make a BALUN that will transform impedance at a 4:1 ratio, with $R_2 = 4 \times R_1$. The length of the BALUN section of coaxial cable is:

$$L_{\text{ft}} = \frac{492\ V}{F_{\text{MHz}}} \qquad [19.23]$$

Where:

L_{ft} is the length in feet
V is the velocity factor of the coaxial cable (a decimal fraction)
F_{MHz} is the operating frequency in megahertz

Matching stubs

A shorted stub can be built to produce almost any value of reactance. This fact can be used to make an impedance matching device that cancels the reactive portion of a complex impedance. If we have an impedance of, $Z = R + j30\ \Omega$ we need to make a stub with a reactance of $-j30\ \Omega$ to match it. Two forms of matching stub are shown in Figs. 19-10A and 19-10B. These stubs are connected exactly at the feedpoint of the complex load impedance, although they are sometimes placed further back on the line at a (perhaps) more convenient point. In that case, however, the reactance required will be transformed by the transmission line between the load and the stub.

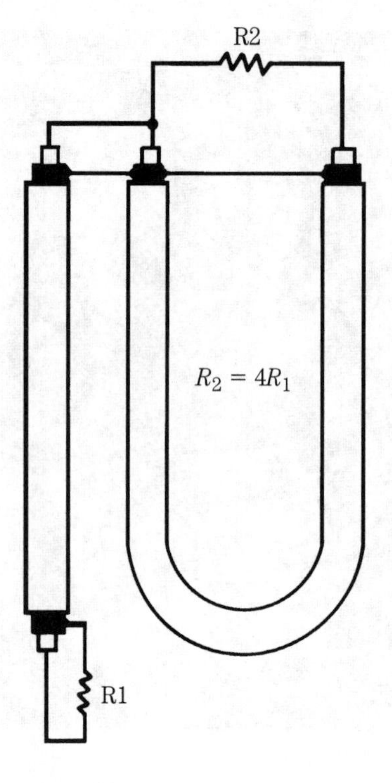

19-9 Coaxial BALUN transformer.

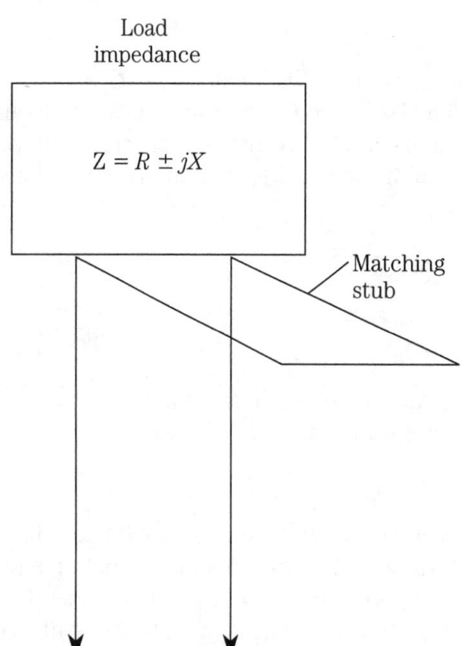

19-10A Stub match scheme.

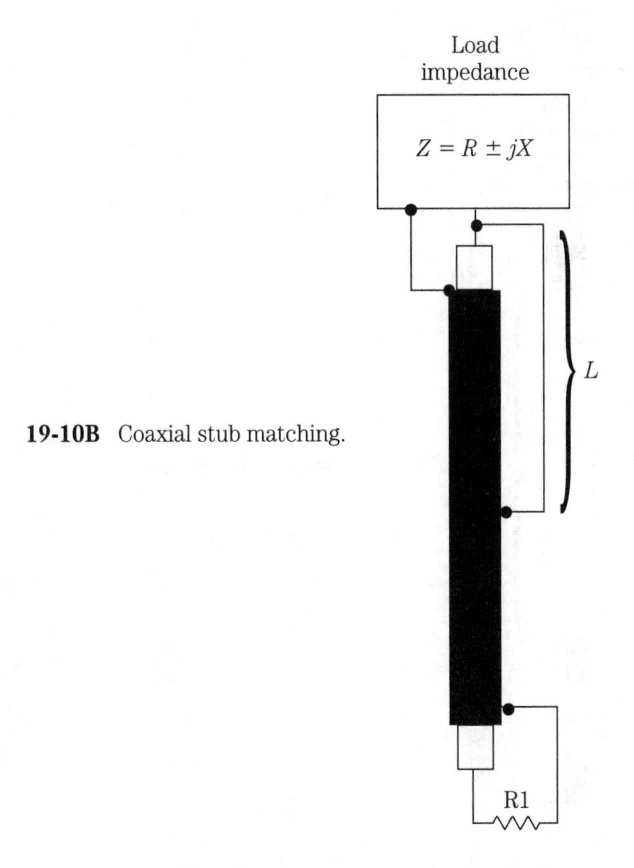

19-10B Coaxial stub matching.

Quarter-wave matching sections

Figure 19-11 shows the elementary quarter-wavelength transformer section connected between the transmission line and the antenna load. This transformer is also sometimes called a *Q-section*. When designed correctly, this transmission line transformer is capable of matching the normal feedline impedance (Z_s) to the antenna feedpoint impedance (Z_R). The key factor is to have available a piece of transmission line that has an impedance Z_o of:

$$Z_o = \sqrt{Z_s Z_R} \qquad \textbf{[19.24]}$$

Most texts shows this circuit for use with coaxial cable. Although it is certainly possible, and even practical in some cases, for the most part there is a serious flaw in using coax for this project. It seems that the normal range of antenna feedpoint impedances, coupled with the rigidly fixed values of coaxial-cable surge impedance available on the market, combines to yield unavailable values of Z_o. Although there are certainly situations that yield to this requirement, many times the quarter-wave section is not useable on coaxial-cable antenna systems using standard impedance values.

On parallel transmission-line systems, on the other hand, it is quite easy to achieve the correct impedance for the matching section. We use the equation above to find a value for Z_o, and then calculate the dimensions of the parallel feeders. Be-

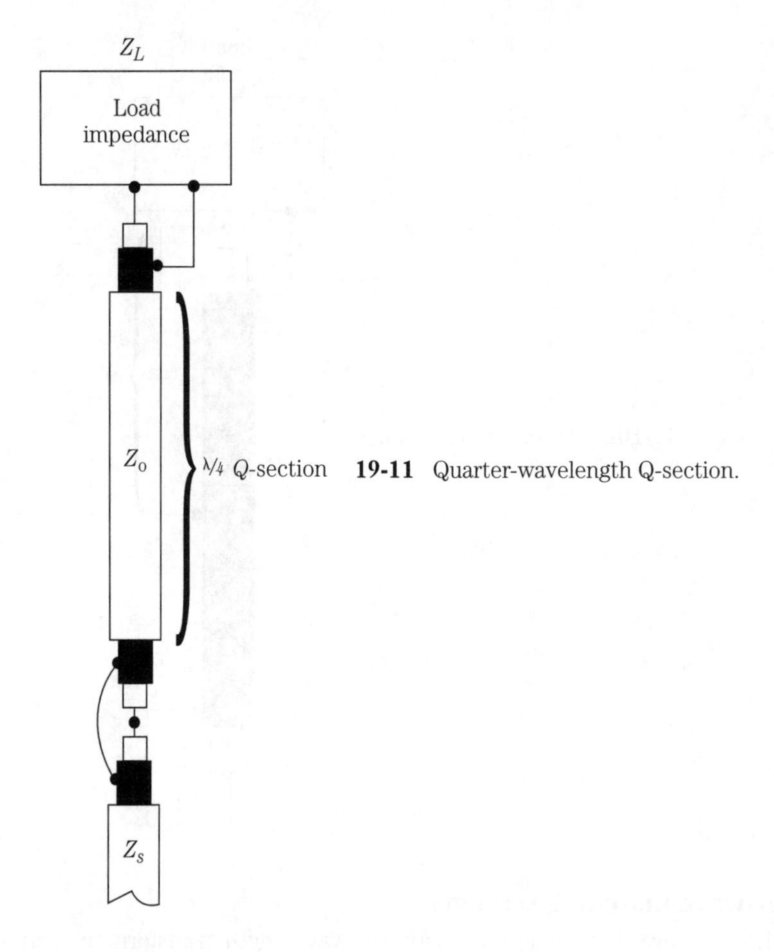

19-11 Quarter-wavelength Q-section.

cause we know the impedance, and can more often than not select the conductor diameter from available wire supplies, you can use the equation below to calculate conductor spacing:

$$S = D10^{(z/276)} \qquad \textbf{[19.25]}$$

Where:

> S is the spacing, D is the conductor diameter (D and S in the same units), and Z is the desired surge impedance. From there you can calculate the length of the quarter-wave section from the familiar $246/F_{\text{MHz}}$.

Series matching section

The quarter-wavelength section, covered in the preceding section, suffers from several drawbacks: it must be a quarter wavelength, and it must use a specified (often non-standard) value of impedance. The series matching section is a generalized case of the same idea, and it permits us to build an impedance transformer that overcomes most of these faults. According to *The ARRL Antenna Book*, this form of transformer is capable of matching any load resistance between about 5 Ω and 1200 Ω. In addition, the transformer section is not located at the antenna feedpoint.

Figure 19-12 shows the basic form of the series matching section. There are three lengths of coaxial cable: L1, L2, and the line to the transmitter. Length L1 and the line to the transmitter (which is any convenient length) have the same characteristic impedance, usually 75 Ω. Section L2 has a different impedance from L1 and the line to the transmitter. Note that only standard, easily obtainable values of impedance are used here.

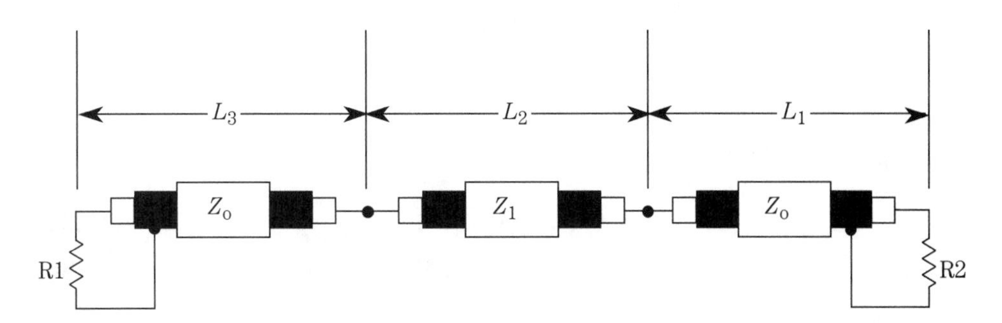

19-12 Odd-length Q-section.

The design of this transformer consists of finding the correct lengths for L1 and L2. You must know the characteristic impedance of the two lines (50 Ω and 75 Ω given as examples), and the complex antenna impedance. In the case where the antenna is non-resonant, this impedance is of the form $Z = R \pm jX$, where R is the resistive portion, X is the reactive portion (inductive or capacitive) and j is the so-called "imaginary" operator (i.e., square root of minus one). If the antenna is resonant, then X=0, and the impedance is simply R.

The first chore in designing the transformer is to *normalize* the impedances:

$$N = \frac{Z_{L1}}{Z_o} \quad \textbf{[19.26]}$$

$$R = \frac{R_L}{Z_o} \quad \textbf{[19.27]}$$

$$X = \frac{X_L}{Z_o} \quad \textbf{[19.28]}$$

The lengths are determined in electrical degrees, and from that determination we can find length in feet or meters. If we adopt ARRL notation and define A = TAN (LI), and B = TAN (L2), then the following equations can be written:

$$\text{If: } Z_L = R \pm jX_L$$

$$\text{TAN } L_2 = B = \frac{(r-1)^2 + X^2}{r(N - (1/N)^2 - (r-1)^2 - X^2} \quad \textbf{[19.29]}$$

$$\text{TAN}\, L_1 = A = \frac{(N - (r/R)) \times B) + X}{r + (XNB) - 1} \quad \textbf{[19.30]}$$

Where:

$$N = \frac{Z_1}{Z_o}$$

$$r = \frac{R_L}{Z_o}$$

$$Z = \frac{X_L}{Z_o}$$

Constraints:

$$Z_1 > Z_o \sqrt{\text{VSWR}} \quad \textbf{[19.31]}$$

$$Z_1 < Z_o \sqrt{\text{VSWR}} \quad \textbf{[19.32]}$$

If $L_1 < 0$, then add 180 degrees
If $B < 0$, then Z_1 is too close to Z_o
Z_1 not equal to Z_o
$Z_o \sqrt{\text{VSWR}} < Z_1 < Z_o \sqrt{\text{VSWR}}$
Physical length in feet:

$$L_1 = \frac{L_1 \lambda}{360} \quad \textbf{[19.33]}$$

$$L_2 = \frac{L_2 \lambda}{360} \quad \textbf{[19.34]}$$

Where:

$$\lambda = \frac{984 \times \textit{Velocity Factor}}{\textit{Frequency in megahertz}} \quad \textbf{[19.35]}$$

The physical length is determined from ARCTAN (*A*) and ARCTAN (*B*), divided by 360, and multiplied by the wavelength along the line and the velocity factor.

Although the sign of B can be selected as either – or +, the use of + is preferred because a shorter section is obtained. In the event that the sign of *A* turns out negative, add 180 degrees to the result.

There are constraints on the design of this transformer. For one thing, the impedance of the two sections (L1 and L2) cannot be too close together. In general, the following relationships must be observed:

Either,

$$Z_{L1} > Z_o \times SWR \quad \textbf{[19.36]}$$

or,

$$\frac{Z_{L1} < Z_o}{SWR} \qquad [19.37]$$

Ferrite core inductors

The word "ferrite" refers to any of several examples of a class of materials that behave similarly to powdered iron compounds, and are used in radio equipment in the form of inductors and transformers. Although the original materials were of powdered iron, and indeed the name "*ferr*ite" still implies iron, many modern materials are composed of other compounds. According to the Amidon Associates literature, ferrites with a permeability of 800 to 5000 mu are generally of the manganese-zinc type of material, and cores with permeabilities of 20 to 800 are of nickel-zinc. The latter are useful in the 0.5- to 100-MHz range.

Toroid cores

A *toroid* is a "doughnut shaped" object, so one can reasonably expect a toroidal core to be an inductor (or transformer) form, made of a ferrite material, in the general shape of a doughnut. The type of core must be known for application purposes, and is given by the type number. The number will be of the form: $FT - xx - nn$, where FT means "ferrite toroid" and describes the shape, "xx" indicates the size and "nn" indicates the material type. The "F" in "FT" is sometimes deleted in parts lists, and the core defined as a "T-xx-nn."

A chart is provided by Amidon that gives the dimensions, and a description of the properties of the different types of material, along with a lot of other physical data. Some of these data are also available in the annual ARRL publication *The ARRL Handbook for the Radio Amateur* (the same material has appeared in earlier editions also).

Tables are derived in part from both Amidon and ARRL sources. From these tables, you can see the sizes and properties of various popular toroids. These tables, incidentally, are not exhaustive of either the variety of toroids available, or all of the properties of the toroids mentioned. Using the nomenclature mentioned, a T-50-2 core refers to a core that is useful from 1 to 30 MHz, has a permeability of 10, is painted red and has the following dimensions: o.d. = 0.500 inches (1.27 cm), i.d. = 0.281 inches (0.714 cm) and a height (i.e., thickness) of 0.188 inches (0.448 cm).

Toroidal transformers

A magazine reader once asked a question of this author concerning the winding protocol for toroidal transformers as seen in textbooks and magazine articles. My correspondent included a partial circuit (Fig. 19-13A) as typical of the dilemma. The question was "how do you wind it," and a couple of alternative methods were proposed. At first I thought it was a silly question because the answer was "obvious," and then I realized that perhaps I was wrong, and to many people the answer was not at all that obvious.

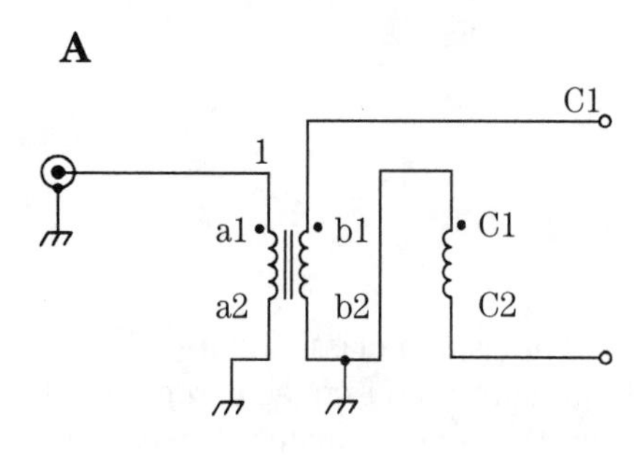

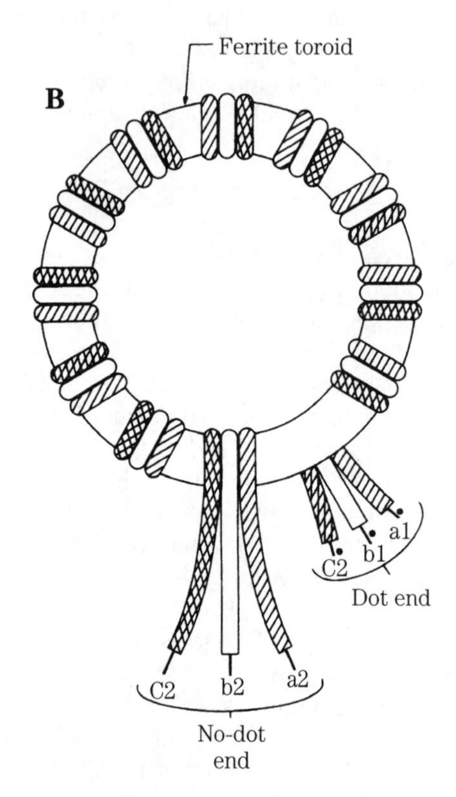

19-13 Broadband RF transformer (from author's column in *Ham Radio* magazine).

The answer to the question is that all windings are wound together in a "multifilar" manner. Because there are three windings, in this case we are talking about "trifilar" windings. Figure 19-13B shows the trifilar winding method. For the sake of clarity, I have patterned all three wires differently so that you can follow it. This practice is also a good idea for practical situations. Since most small construction projects use #26, #28 or #30 enameled wire to wind coils, I keep three colors of each size on hand, and wind each winding with a different color [Note: for transmitting antenna transformers use #16, #14, #12, or #10 wire]. Otherwise, label the ends with adhesive labels.

The dots in the schematic, and on the pictorial, are provided to identify one end of the coil windings. Thus, the "dot" and "no-dot" ends are different from each other, and it usually makes a difference to circuit operation which way the ends are connected into the circuit (the issue is signal phasing).

Figure 19-14 shows two accepted methods for winding a multifilar coil on a toroidal core. Figure 19-14A is the same method as in Fig. 19-13B, but on an actual toroidal instead of a pictorial representation. The wires are laid down parallel to each other as shown previously. The method in Fig. 19-14B uses twisted wires. The three wires are "chucked up" in a drill and twisted together before being wound on the core. With one end of the three wires secured in the drill chuck, anchor the other end of three wires in something that will hold it taut. Some people use a bench vise for this purpose. Turn on the drill on slow speed and allow the wires to twist together until the desired pitch is achieved.

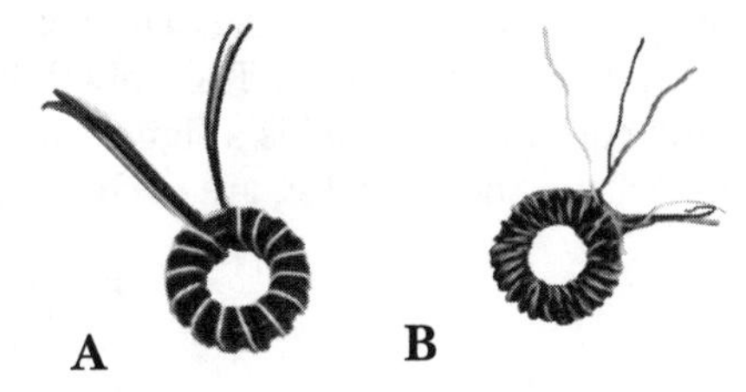

19-14 Winding a toroidal transformer A) parallel wound, B) twist round.

Be very careful when performing this operation. If you don't have a variable speed electric drill (so that it can be run at very low speed), then use an old-fashioned manual hand drill. If you use an electric drill, then wear eye protection. If the wire breaks, or gets loose from its mooring at the end opposite the drill, it will whip around wildly until the drill stops. That whipping wire will cause painful welts on the skin, and can easily damage eyes permanently.

Of the two methods for winding toroids, the method of Figs. 19-13B and 19-14A is preferred. When winding toroids, at least those of relatively few windings, pass the wire through the "doughnut hole" until the toroid is about in the middle of the length of wire. Then, loop the wire over the outside surface of the toroid, and pass it through the hole again. Repeat this process until the correct number of turns are wound onto the core. Be sure to press the wire against the toroid form, and keep it taut as you wind the coils.

Enameled wire is usually used for toroid transformers and inductors and that type of wire can lead to a problem. The enamel can chip and cause the copper conductor to contact the core. On larger cores, such as those for antenna matching tranformers (and baluns used at kilowatt power levels), the practical solution is to wrap the bare toroid core layer of fiberglass packing tape. Wrap the tape exactly as if it was wire, but overlap the turns slightly to ensure covering the entire circumference of the core.

On some projects, especially those in which the coils and transformers used very fine wire (e.g., #30), you may experience a tendency for the wire windings to unravel after the winding is completed. This problem is also easily curable. At the ends of the windings, place a tiny dab of rubber cement or *RTV* silicone sealer.

Mounting toroid cores Now that you have a properly wound toroidal inductor or transformer, it is time to actually mount it in a circuit being constructed. There are three easy ways to do this job. First, you might be able to ignore it. If the wire is heavy enough, then just use the wire connections to the circuit board or terminal strip to support the component. But if this is not satisfactory, and in mobile equipment (or wherever else vibration is a factor) it won't be, then try laying the toroid flat on the board and cementing it in place with silicone seal or rubber cement. The third method, is to drill a hole in the wiring board and use a screw and nut to secure the toroid. Do not use metallic hardware for mounting the toroid! Metallic fasteners will alter the inductance of the component and possibly render it unusable. Use nylon hardware for mounting the inductor, or transformer.

How many turns? Three factors must be taken into consideration when making toroid transformers or inductors: *toroid size, core material,* and *number of turns of wire.* The toroid size is selected as a function of power-handling capability, or for convenience of handling. The core material is selected according to the frequency range of the circuit. The only thing left to vary is the number of turns. The size and core material yields a figure called the A_L factor. The required value of inductance, and the A_L factor, are used in the following equation:

$$N = 100\sqrt{\frac{L_{\mu H}}{A_L}} \quad [19.38]$$

Where:

N is the number of turns
$L_{\mu H}$ is the inductance in microhenrys
A_L is the core factor in microhenrys per 100 turns

Example Calculate the number of turns required to make a 5μH inductor on a T-50-6 core. The A_L factor is 40.

Solution:

$$N = 100\sqrt{\frac{L_{\mu H}}{A_L}}$$

$$N = 100\sqrt{\frac{5_{\mu H}}{40}}$$

$$N = 100\sqrt{0.125}$$

$$N = (100)\,(0.35) = 35$$

Don't take the equation value too seriously, however, because it is my experience that a wide tolerance exists on amateur-grade ferrite cores. Although it isn't too much of a problem when building transformers, it can be critical when making inductors for a tuned circuit. When you find that the tuned circuit takes considerably more (or less) capacitance than called for in the standard equation, and all of the stray capacitance is properly taken into consideration, then it may be that the actual A_L value of your particular core is different from the table value.

Ferrite rods

Another form of ferrite core, available on the market, is the rod, as shown in Fig. 19-15A. This type of core is used to make RF chokes, such as the RFC used in the vacuum tube filament lines of a linear amplifier power tube (Fig. 19-15B). They are also used for BALUNs, by some people. The two windings are wound in a bifilar manner over the ferrite rod. Of course, the wires used must be heavy enough to carry the filament current of the tube. As was true in the toroidal transformer, I used two different wire colors in order to make it easy to identify windings.

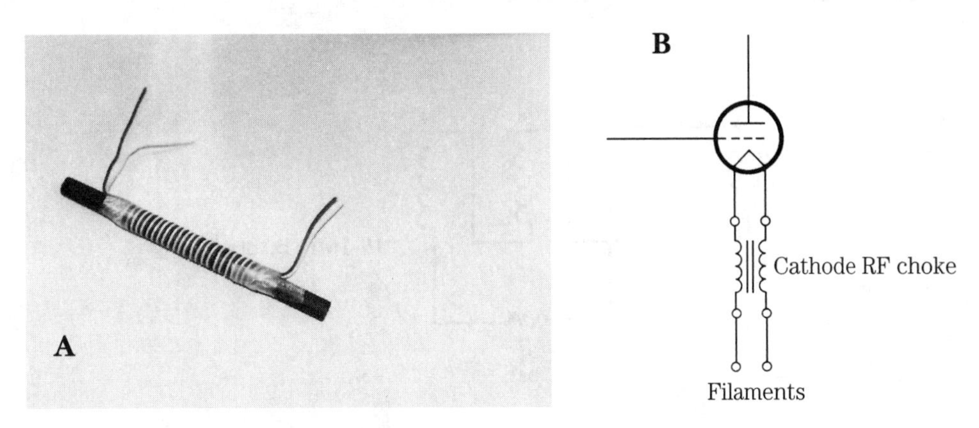

19-15 Ferrite rod inductor A) construction, B) use in amplifier.

Ferrite rods are also used in receiving antennas. Although the amateur use is not extensive, there are places where a ferrite rod antenna (or "loopstick") is used. For example, in radio direction-finding antennas, it is common to see the ferrite loop. Also, some amateurs report that they use a loopstick receiving antenna when operating on crowded bands, such as 40 and 75 meters. The small loopstick has an extremely directional characteristic, so it is capable of nulling out interfering signals. Of course, you would not want to use the loopstick for transmitting, so some means must be found for transferring the antennas over between the transmit and receive functions.

Mounting ferrite rods Ferrite rods can be mounted in several manners, two of which are analagous to the methods used on toroids. We can, for example, mount the rod using either its own wires for support or by using a dab of cement or silicone sealer to fasten it to the board. Although we cannot use simple nylon screws the way we can on toroids, we can use insulating cable clamps to secure the ends of the rod to the board.

Toroid broadbanded impedance matching transformers

The toroidal transformer forms a broadbanded means for matching antenna impedance to the transmission line, or matching the transmission line to the transmitter. The other matching methods (shown thus far) are frequency sensitive, and must be readjusted whenever the operating frequency is changed even a small amount. Although this problem is of no great concern to fixed-frequency radio stations, it is of critical importance to stations that operate on a variety of frequencies.

Figure 19-16A shows a trifilar transformer that provides a 1:1 impedance ratio, but it will transform an unbalanced transmission line (e.g., coaxial cable) to a balanced signal required to feed a dipole antenna. Although it *provides no impedance transformation*, it does tend to balance the feed currents in the two halves of the antenna. This fact makes it possible to obtain a more accurate "figure-8" dipole radiation pattern in the horizontal plane. Many station owners make it standard practice to use a BALUN at the antenna feedpoint.

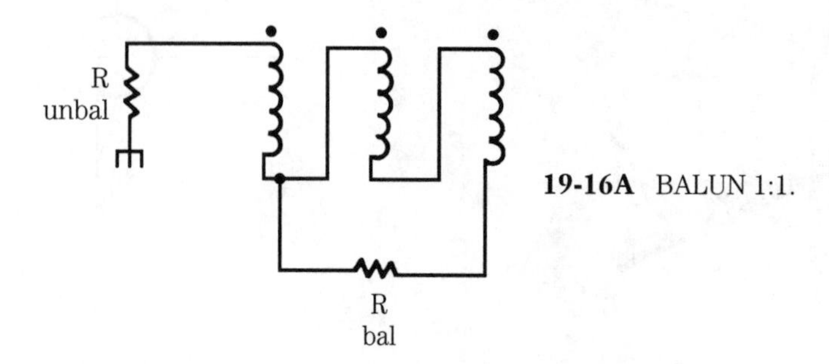

19-16A BALUN 1:1.

The BALUN shown in Fig. 19-16B is designed to provide the unbalanced to balanced transformation, while also providing a 4:1 impedance ratio. Thus, a 300 Ω folded dipole feedpoint impedance will be transformed to 75 Ω unbalanced. This type of BALUN is often included inside antenna tuners, including all three models shown in Figs. 19-8A and 19-8B. A variable, (or at least "settable") broadbanded transformer is shown in Fig. 19-16C. In this case the output winding is tapped, and the operator selects the correct tap needed to provide the desired impedance ratio. The usual turns ratio criterion applies.

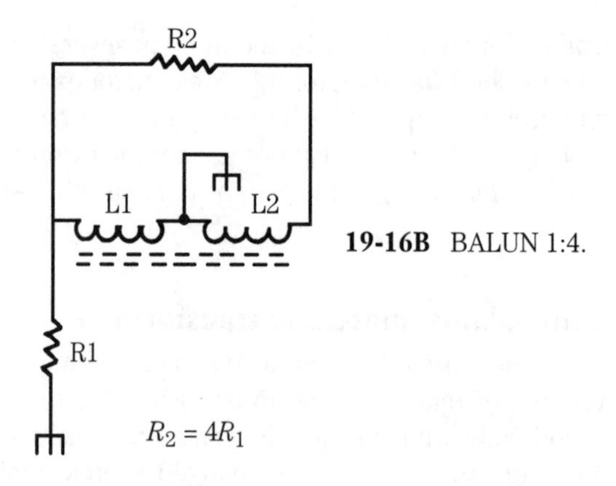

19-16B BALUN 1:4.

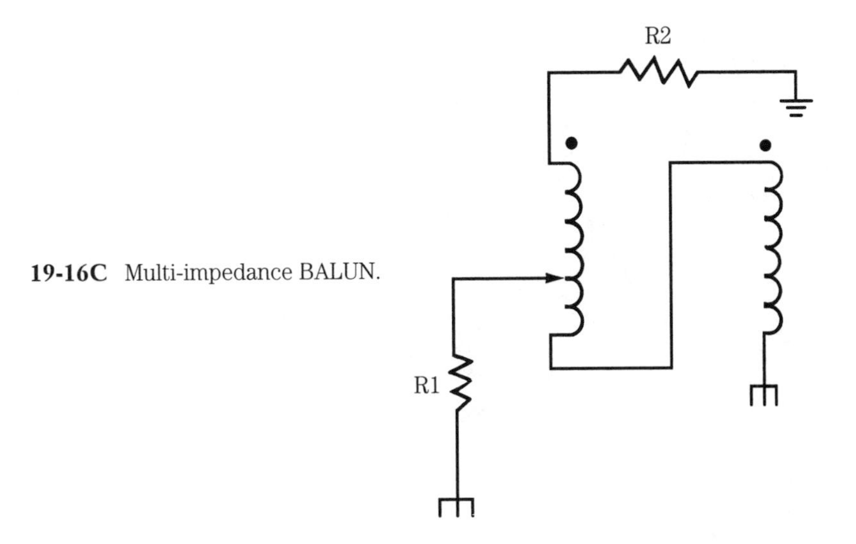

19-16C Multi-impedance BALUN.

Another multiple impedance transformer is shown in Fig. 19-16D. In this case, the operator can select impedance transformation ratios of 1.5:1, 4:1, 9:1 or 16:1. A commercial version of this type of transformer is shown in Fig. 19-16E. It is manufactured by Palomar Engineers, and is intended for vertical antenna feeding in the HF bands. It will, however, work well on antennas other than simple verticals.

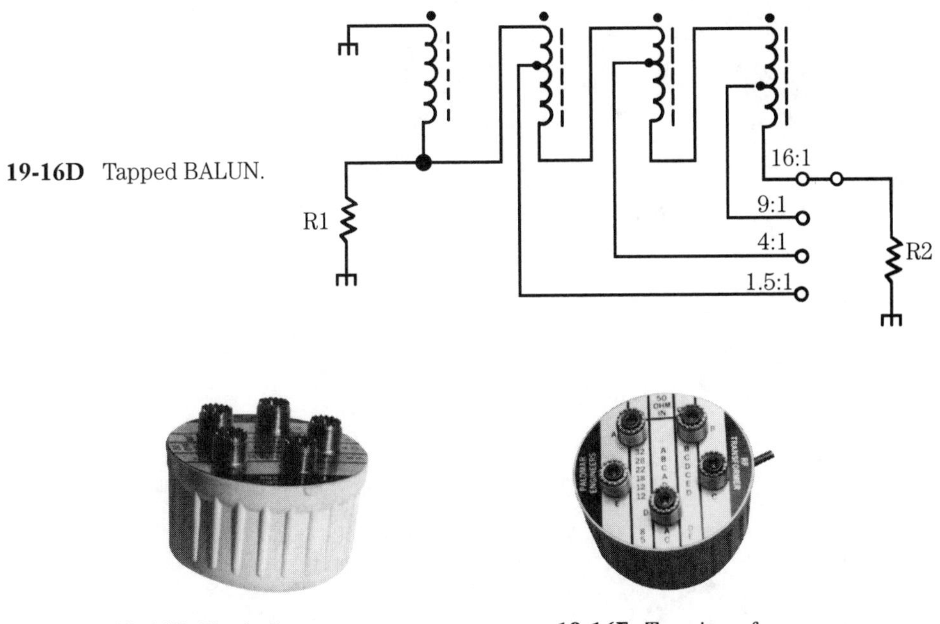

19-16D Tapped BALUN.

19-16E Vertical antenna feeding transformer.

19-16F Top view of Fig. 19-16E.

20
CHAPTER

Mobile, emergency, and marine antennas

MOBILE OPERATION OF RADIO COMMUNICATIONS EQUIPMENT DATES BACK TO ONLY A little later than "base station" operation. From its earliest times, radio buffs have attempted to place radio communications equipment in vehicles. Unfortunately, two-way radio was not terribly practical until the 1930s, when the earliest applications were amateur radio and police radio (which used frequencies in the 1.7- to 2.0-MHz region). Over the years, the landmobile and amateur radio mobile operation has moved progressively higher in frequency because of certain practical considerations. The higher the frequency, for example, the shorter the wavelength; and, therefore, the shorter a full-size antenna. On the 11-meter citizen's band, for example, a quarter-wavelength whip antenna is 102 inches long, and for the 10-meter amateur band only 96 inches long. At VHF frequencies, antennas become even shorter. As a result, much mobile activity takes place in the VHF and UHF region.

The amateur 144-MHz, 220-MHz, and 440-MHz bands are popular because of several factors, not the least of which is the ease of making ¼- and ⅝-wavelength antennas. Because low-cost commercial antennas are available for these frequencies, however, we will not examine such antennas in this chapter. Rather, we will concentrate on high-frequency antennas.

Mobile HF antennas

High frequency (HF) mobile operation requires substantially different antennas than VHF or UHF. Quarter-wavelength antennas are only feasible on the 11 and 10 meter bands, with some argument in favor of 13 meters as well. But by the time the frequency drops to the 21-MHz (15 meter) band, however, the antenna size must be approximately 11 feet long, and that is too long for practical mobile operation. Because of practical considerations, you must limit antenna size to 8 or 9 feet; in the amateur radio bands the 8-foot antenna is most popular because it is resonant on 10 meters. At all lower frequencies, the 8-foot whip becomes capacitive, and therefore requires an equal inductive reactance to cancel the capacitive reactance of the antenna.

Figure 20-1 shows three basic configurations of coil-loaded HF antennas for frequencies lower than the natural resonant frequency of the antenna. In each case, the antennas are series fed with coaxial cable from the base; point "A" is connected to the coaxial cable center conductor, and point "B" is connected to the shield, and the car body (which serves as ground). The system shown in Fig. 20-1A is base loaded. Although convenient, there is some evidence that the current distribution is less than optimum. The version shown in Fig. 20-1B is center-loaded, and results in an improved current distribution. This configuration is probably the most common amongst commercially available HF mobile antennas, although the coil is often located slightly above the center point. Finally, we see the top-loaded coil system. In all three cases, the point is to use the inductance of the coil to cancel the capacitive reactance of the antenna.

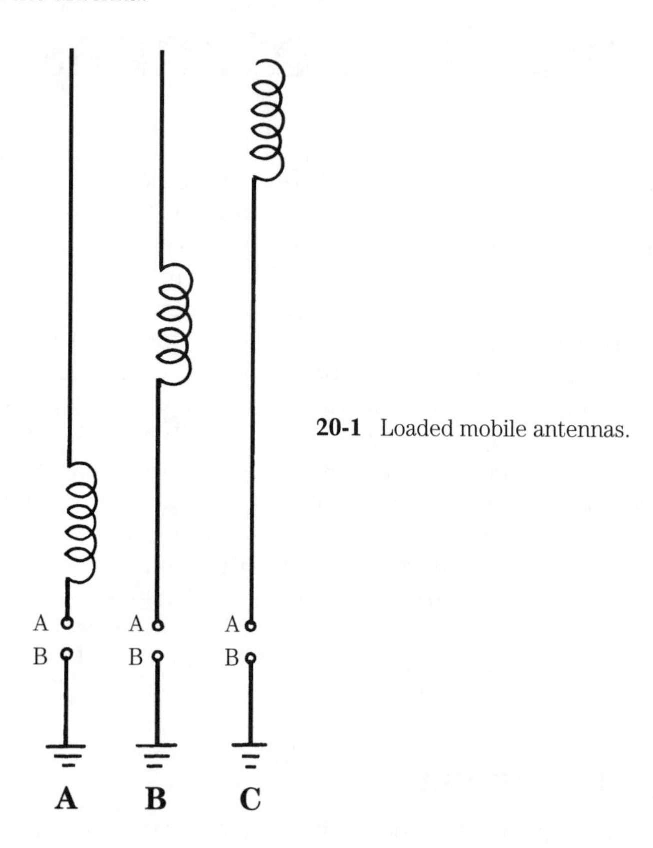

20-1 Loaded mobile antennas.

A modified version of the coil-loaded HF mobile antenna is shown in Fig. 20-2A. In this configuration the loading inductance is divided between two individual coils, L1 and L2. Coil L1 is adjustable with respect to the antenna, while L2 is fixed; coil L2 is tapped, however, in order to match the impedance of the antenna to the characteristic impedance of the coaxial-cable transmission line. When tuning this antenna, two instruments are needed: a field strength meter and a VSWR meter. The field strength meter gives a relative indication of the amount of power radiated from the antenna, while the VSWR meter helps determine the state of the impedance match. A pair of relative field strength meter projects are discussed later in this chapter.

Variations on the theme are shown in Figs. 20-2B and 20-2C. A helical whip is shown in Fig. 20-2B. In this configuration, the inductor is distributed along the length of a fiberglass antenna rod. The conductor is a wire that is helically wound on the fiberglass shaft. An adjustable tip sets the antenna to resonance. The length of this tip can be set by a local field strength meter. Fixed versions of this type of antenna are very popular on the Citizens Band where they may be as short as 30 inches, or as long as 48 inches. There are also some amateur radio commercial antennas based on this concept.

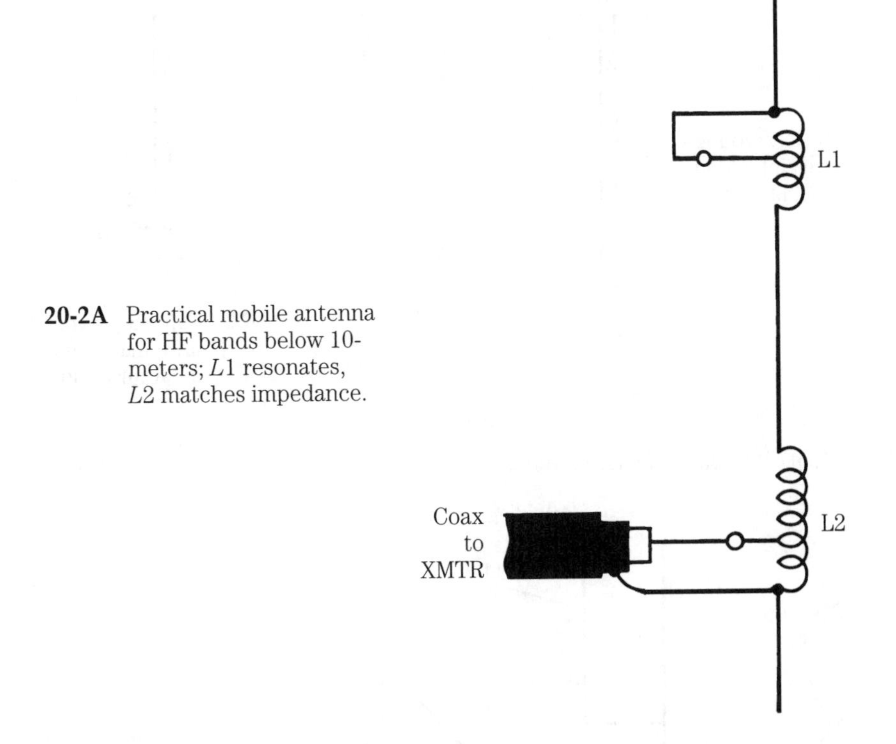

20-2A Practical mobile antenna for HF bands below 10-meters; *L*1 resonates, *L*2 matches impedance.

The other variant is shown in Fig. 20-2C. In this case, the lower end of the radiator consists of a metal tube topped with a loading coil. An adjustable shaft at the top end is used to tune the antenna to resonance. This form of antenna is popular among commercial makers of HF mobile antennas, such as the *Hustler*. The fixed shaft tends to be universal for all bands, while the coil and adjustable shaft form a separate "resonator" for each band. Multiband operation of this form of antenna can be accomplished by using a bracket such as shown in Fig. 20-2D. Although early versions of this scheme were homebrewed, several manufacturers currently make factory-built versions. The idea in Fig. 20-2D is to mount two or more resonators to a common fixed shaft.

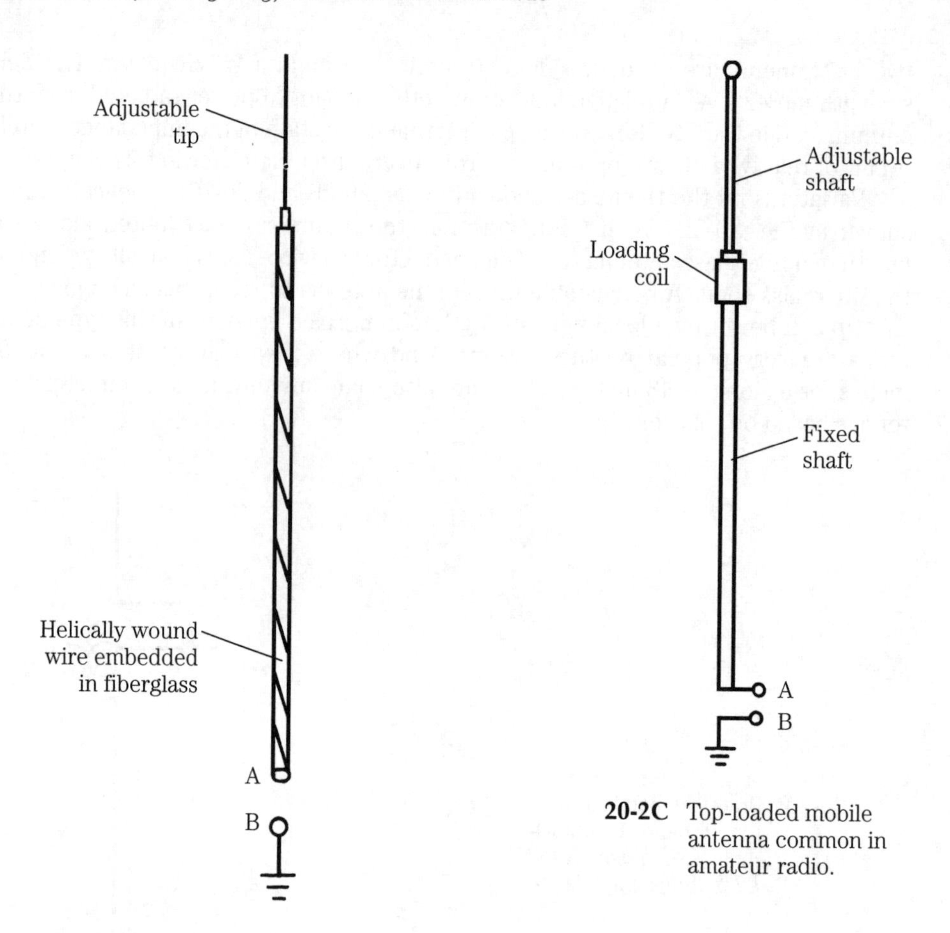

20-2C Top-loaded mobile antenna common in amateur radio.

20-2B Helically wound mobile antenna.

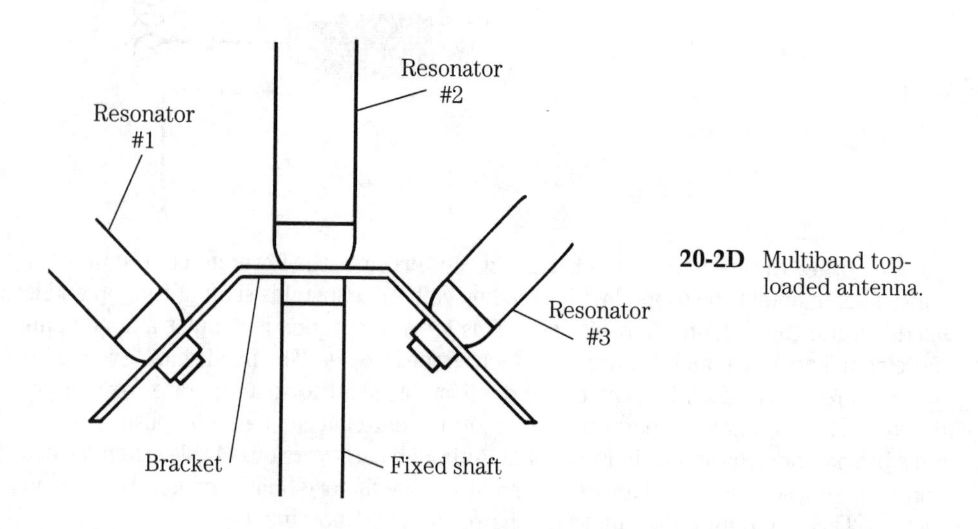

20-2D Multiband top-loaded antenna.

A common problem, with all coil-loaded mobile antennas, is that they tend to be very high *Q* antennas. In other words, they are very sharply tuned. The VSWR tends to rise rapidly as the operating frequency departs from the frequency to which the antenna is tuned; as little as 25 kHz change of operating frequency will detune the antenna significantly. Although an antenna tuner at the output of the transmitter will reduce the VSWR to a point that allows the transmitter to operate; that type of tuner is merely a line flattener that does not fully address the problem. The actual problem is that the antenna is not resonant. The efficiency of the antenna drops off rapidly as the frequency changes. The only cure for this problem is to readjust the resonator's adjustable shaft as the band segment is changed (not merely the band, but the *band segment*). Unfortunately, this solution requires tools, and the operator getting out of the vehicle in order to do a good job of retuning. Another solution is to use a motor driven variable inductor for the loading coil. Several manufacturers offer both base-loaded and center-loaded coils that are either motor driven, or relay selectable, to permit frequency changing from the operator's seat.

Tuning HF mobile antennas

Although the procedures for tuning certain specific antennas might be different from those given below, the basic principles are the same, and can easily be adapted to any given situation. There are two situations to consider. First is the antenna in which there is either a single adjustable (or tap selectable) loading coil, or a fixed loading coil and an adjustable shaft resonator. Second is the case (per Fig. 20-2A) where there is an adjustable or tapped coil in series with a tapped impedance-matching coil at the base.

Case no. 1. In this case, we assume that the antenna has a single adjustment, either a shaft resonator or adjustable coil. In this case we need to use a field strength meter (FSM) to measure the relative field strength of the radiated signal. The antenna resonator is adjusted until the radiated field strength is maximum. Of course, this procedure must be done in steps, keying the transmitter after each adjustment to see what happened. Alternatively, a VSWR meter can be used to set the resonator to a minimum VSWR.

Case no. 2. Here, we have a coil at the base that is fixed with respect to the antenna radiator element, but is tapped with respect to the coaxial cable from the transmitter. A second coil is also used in the antenna (see Fig. 20-2A), which can be either in the center, or at the top. This coil is adjustable for setting resonance. Alternatively, this coil may be fixed, and an adjustable resonator shaft is used to set operating frequency. In this type of antenna the upper coil, or resonator shaft, is adjusted to resonance using a field strength meter. The lower coil is adjusted for minimum VSWR. Both adjustments are needed to complete the job.

Field strength meters

A *field strength meter* (FSM) is an instrument that measures the radiated field from an antenna. Commercial engineering-grade instruments are calibrated in terms of either watts/cm^2 or volts/meter, and are used for jobs such as broadcast station "proof-of-performance" tests and other professional jobs. For adjusting antennas, however, a considerably different instrument is sufficient. This section describes two simple, passive (which means no dc power is required) field strength meters usable for adjusting HF radio antennas, including both CB and amateur radio antennas.

Two forms are shown in Fig. 20-3; both are basically variations on the old-fashioned "crystal set" theme. Figure 20-3A shows the simplest form of untuned FSM. In this circuit a small whip antenna (used for signal pick-up) is connected to one end of a grounded RF choke (RFC). The RF voltage developed across the RFC is applied to a germanium diode detector; either IN34 or IN60 can be used. Silicon diodes are normally preferred in signal applications, but in this case we need the lower contact potential of germanium diodes in order to improve sensitivity (V_g is 0.2 to 0.3 volts for Ge, and 0.6 to 0.7. volts for Si). A potentiometer is used both as the load for the diode, and as a sensitivity control to set the meter reading to a convenient level.

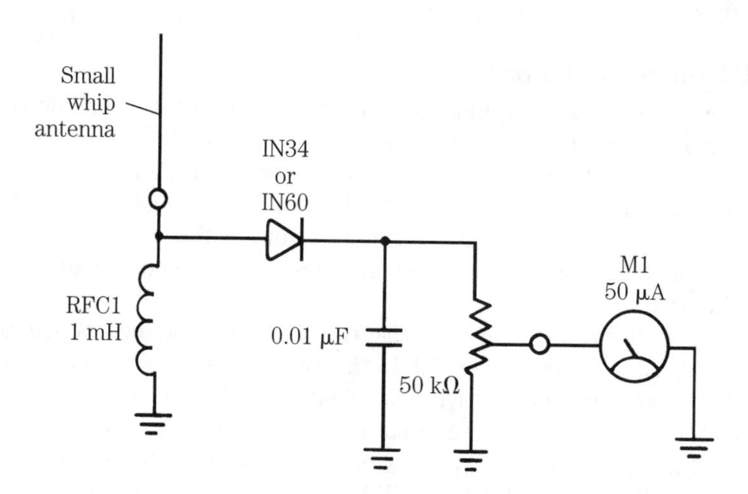

20-3A Simple field strength meters.

The untuned version of the FSM is usable even at low powers, but suffers from a lack of sensitivity. Only a certain amount of signal can be developed across the RFC, so this limits the sensitivity. Also, the RFC does not make a good impedance match to the detector diode (D1). An improvement is possible by adding a tuned circuit and an impedance matching scheme, as shown in Fig. 20-3B. In this case a variable capacitor (C3) is used as a TUNE control. The tuning capacitor is parallel resonant with inductor L1. A tapped capacitive voltage divider (C1/C2) is used to provide impedance matching to the diode.

Operation of the tunable FSM is simple and straightforward. Set the sensitivity control to approximately half-scale, and then key the transmitter; adjust the tuning control (C3) for maximum deflection of the meter pointer (readjustment of the sensitivity control may be needed). After these adjustments are made, the tunable FSM works just like any other FSM.

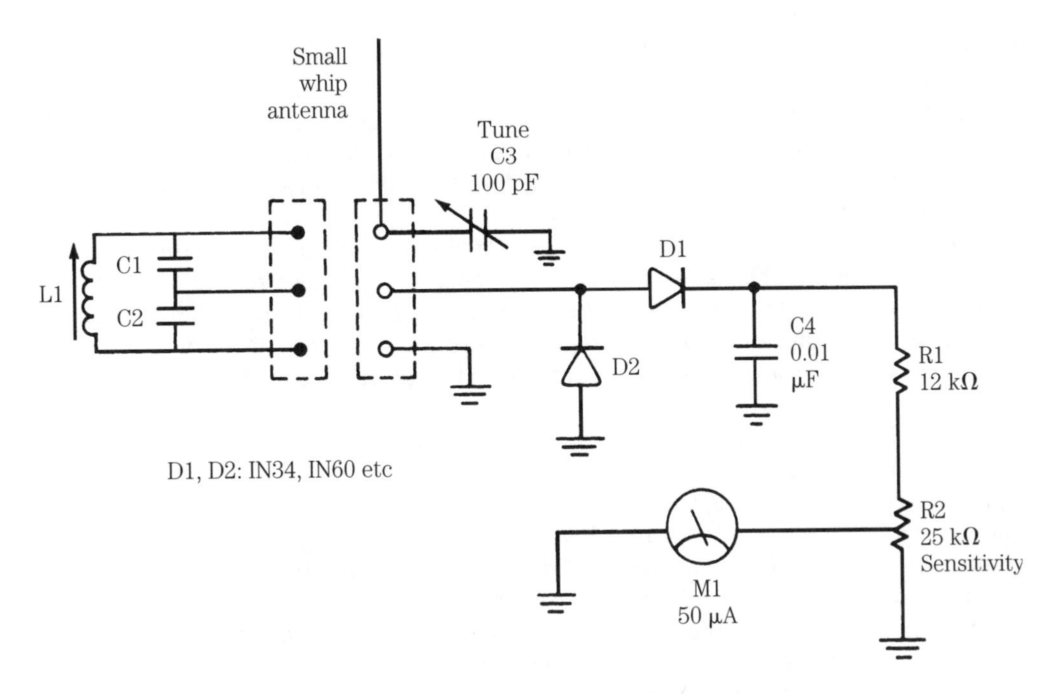

20-3B Tuned field strength meters.

20-3C Values chart for field strength meters.

Band (meters)	L1 (μH)	C1 (pF)	C2 (pF)
75/80	27	25	100
40	10	15	68
30	4.4	15	68
20	2.25	10	68
17	1.7	10	56
15	1.3	10	47
13	0.8	10	39
10	0.5	10	39

Antennas for emergency operations

Some time ago, I met an interesting character at a convention. As a medical doctor, he was a medical missionary working at a relief station in Sudan. Because of his unique business address, he was able to discuss mobile and portable antennas for communications from the boondocks. His *bona fides* for this knowledge includes the fact that he is licensed to operate on both the amateur radio bands, and as a land-mobile (or point-to-point) station in the 6.2-MHz band. The desert where he travels is among the worst in the world. The path they euphemistically call a "road" is occasionally littered with camel corpses because of the harsh conditions. The doctor's organization requires him to check in twice daily on either 6.2 MHz or 3.885 MHz (which some missionary hams in Africa use as an unofficial calling frequency). If he misses two check-ins in a row, then the search and rescue planes are sent up. As a result of his unique "housecalls," he does a lot of mobile and portable operating in the lower-HF region of the spectrum. His problem is this: how do you reliably get through the QRM and tropical QRN with only 200 watts PEP and a standard loaded mobile antenna?

Another fellow I once met, works in Alaska for a government agency. He faces many of the same problems as the doctor in Sudan, but at close to 100 degrees colder. He frequently takes his 100-watt mobile rig into the boondocks with him in a four-wheel drive vehicle. Again, with only 100 watts into a poor-efficiency loaded mobile antenna, how does one reliably cut through the interference to be heard back at the homestead?

An earthquake, or hurricane, strikes your community. Antenna towers collapse, tribanders become tangled masses of aluminum tubing, dipoles are snarled globs of #14 Copperweld, and the rig and linear amplifier are smashed under the rubble of one corner of your house. All that remains is the 100-watt HF rig in your car. How do you reliably establish communications in "kilowatt alley" with a 100-watt mobile driving a 75-watt loaded whip? Of course, you always got through one way or another before, but now communications are not for fun—they are deadly serious. Somehow, the distant problems of a Sudanese missionary doctor, and a KL7 government forester, don't seem too very far away.

For these operators, communications often means life or death for someone, perhaps themselves. Given the inefficiency of the loaded whips typically used as mobile antennas in the low-HF region, the generally low power levels used in available mobile rigs, and the crowded band conditions on the 80-, 75-, and 40-meter bands, it becomes a matter of more than academic interest how you might increase the signal strength from your portable (or mobile) emergency station. Anything we can do, easily and cheaply, to improve the signal is like having money in the bank. Fortunately, there are several tricks of the trade that will help us out in a pinch.

Figure 20-4 shows a typical mobile antenna for a low-HF band. Because quarter-wavelength antennas on these frequencies are 30 to 70 feet high, full-size vertical whip antennas are not practical. In fact, at frequencies below 10 meters, full-size whips are not generally used. Short antennas exhibit capacitive reactance, so we add a loading coil (inductor L in Fig. 20-4) to the radiator to make up the difference (its inductive reactance cancels the capacitive reactance of the antenna). The inductor in such a loaded antenna can be placed almost anywhere along the radiator, although base, center, and top-loaded designs predominate. The actual inductance needed varies somewhat with coil placement, as does antenna performance. The "resonators" used on commercial low-HF mobile antennas are loading coils encapsulated in a weather-tight housing.

The mobile configuration is inefficient by its nature, and little can be done to improve matters. Of course, an antenna-matching device or tuner will help in optimizing the power transfer to the antenna, and should always be used in any event (especially with solid-state final amplifiers that don't tolerate VSWR as easily as do tube finals). In the portable configuration, however, we can both improve the performance of "mobile" antennas, and look into antenna options not open to the mobile

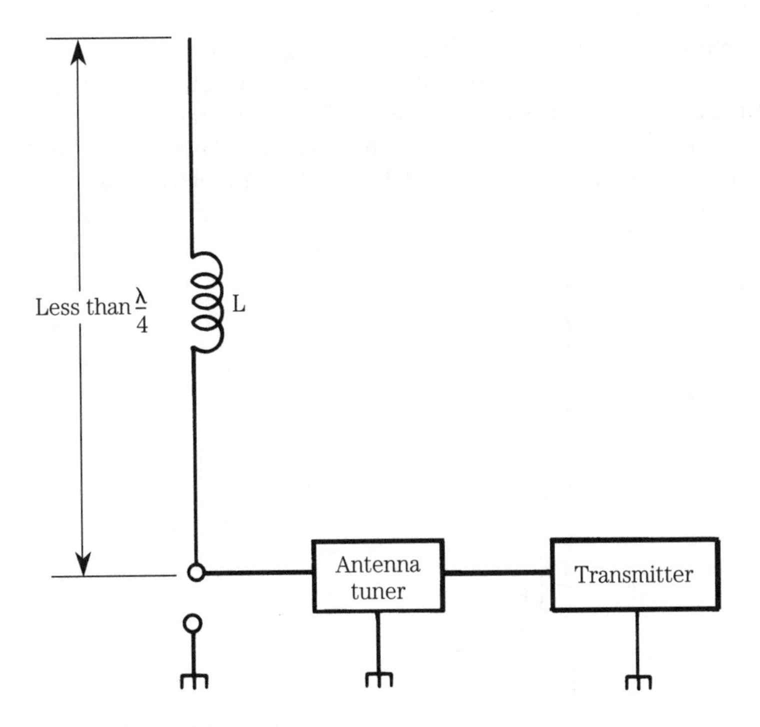

20-4 Basic mobile antenna system often proves unsatisfactory in marginal cases.

operator. We can make a basic assumption for these improvements: the operator needing emergency or urgent communications is not driving down the interstate somewhere, but is stopped in the back country and needs to call out.

In cases of emergencies on most highways, we are in range of some repeater, so we would use a VHF band (probably 2-meters) to contact police or other emergency services through a repeater autopatch. In fact, with the wide availability of repeaters around the country it behooves any amateur, backpacking or four-wheeling into remote areas, to be familiar with repeater locations and frequencies. From hilltops, especially, it is often possible to hit repeaters from a surprisingly long distance. I have seen handheld 2-meter rigs trigger mountain top repeaters from the mountains of southwestern Virginia, and presume that other areas of the country are as well off repeater-wise. This chapter deals with HF rigs, especially those operating in the lower end of the HF spectrum, in situations where a temporary antenna must be erected.

Most amateurs, who have low-HF mobile rigs, will testify that getting out is a pain in the mike button. Antenna efficiencies are simply too poor. One factor in this dismal equation is that the vehicle body makes a very poor antenna ground system. One could, I suppose, carry a supply of 6- or 8-foot copper-clad steel ground rods, and a 9-pound sledge hammer to drive them into the earth wherever a portable station is needed. Somehow, that solution doesn't seem too appealing at this point! Besides, have you ever tried to remove an effective 8-foot ground rod from the ground? Forget it, it's too much pain for too little gain. A better solution is to provide a counterpoise ground plane, as shown in Fig. 20-5. In Fig. 20-5A we see the electrical situation and connection, while in Fig. 20-5B we see the mechanical scheme for a specific situation: a ground plane consists of two or more (even *one* helps) quarter-wavelength radials connected to the antenna ground point (i.e., where the coaxial cable shield connects to the vehicle body). The radials are made of #14 wire, so are relatively easy to stow.

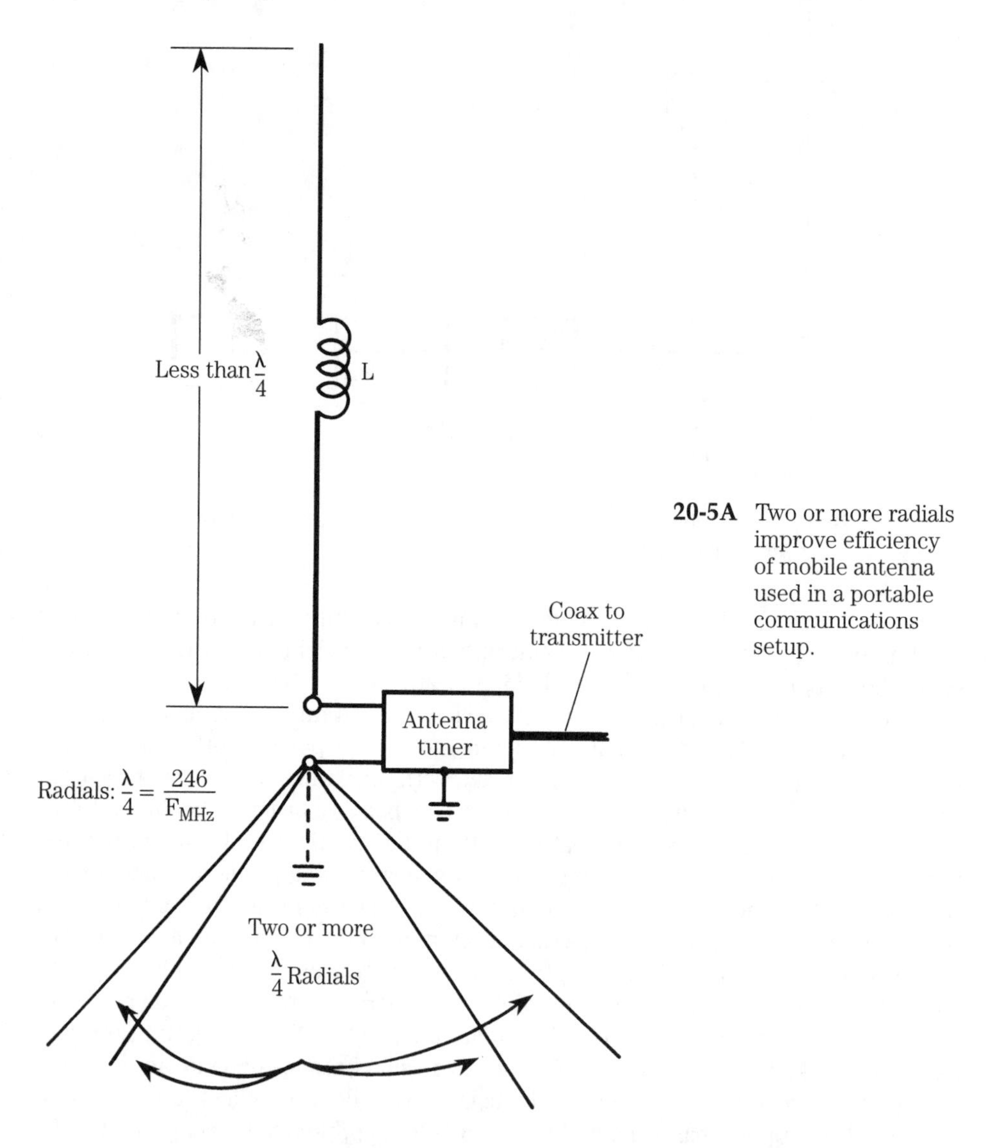

20-5A Two or more radials improve efficiency of mobile antenna used in a portable communications setup.

Figure 20-5B shows a workable system that will improve the performance of a mobile rig in stationary situations. The mobile antenna uses the normal basemount attached to the rear quarter panel of the car adjacent to the trunk lid. An all-metal grounding-type binding post is installed through an extra hole drilled in the base insu-

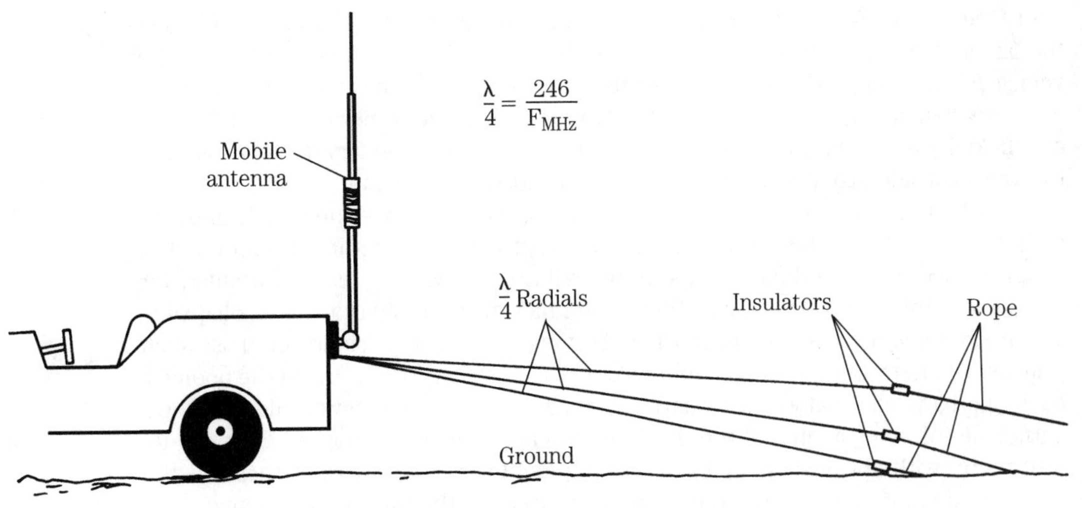

20-5B Multiple radials are even better.

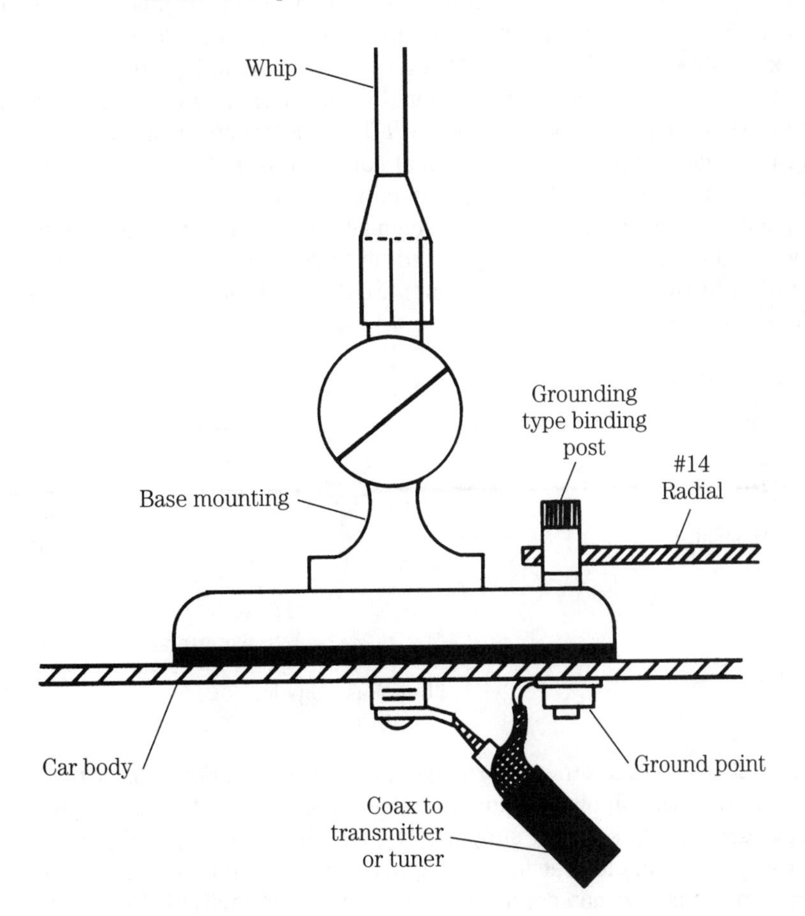

20-5C Connection of radials to mobile antenna.

lator (see Fig. 20-5). Radials for portable operation are attached at this point. Although the binding post is small, it easily accommodates two #14 radials. The owner used the vehicle for camping, and the radials were used to improve antenna efficiency. I don't believe the dramatic accounts enthusiastically reported by the owner of the car, but my own field day experience (and the testimony of the Sudan missionary doctor) leads me to expect considerable improvement over the unaided loaded whip.

A potential solution to the inefficiency problem is to replace the mobile antenna with a more efficient, but stowable, antenna that can be brought out and erected when needed. One candidate is a surplus military HF whip antenna. Intended for jeeps and communications trucks, these antenna/tuner combinations are collapsible and are as efficient as any on the market. Although my Alaskan friend could use such a surplus find, my Sudan friend could not. He told me that Americans in the boonies, even (perhaps especially) missionaries, are always suspected by pin-headed local authorities of being agents of the dreaded "Cuban Invasion Authority." Showing up "out there" with obviously military radio equipment serves only to seal that belief!

The common dipole is often looked down upon by the owners of massive array antennas, but those old-fashioned antennas are capable of turning in some impressive results. The dipole is made by connecting two quarter-wavelength pieces of wire to a coaxial-cable transmission line. One length is connected to the center conductor, and the other is connected to the shield of the coax. In a pinch, zip (i.e., lamp) cord and twisted pairs of hook-up wire will do for a transmission line. Figure 20-6 shows the common dipole, and the normal equation for determining *approximate* length. The actual length is found by trimming length until the VSWR drops to its lowest point . . . a nicety that might not seem altogether important in an emergency. The ends of the dipole must be supported on trees, masts or some other elevated structure. Unfortunately, mounting points aren't easily found. My Sudanese friend doesn't see many trees!

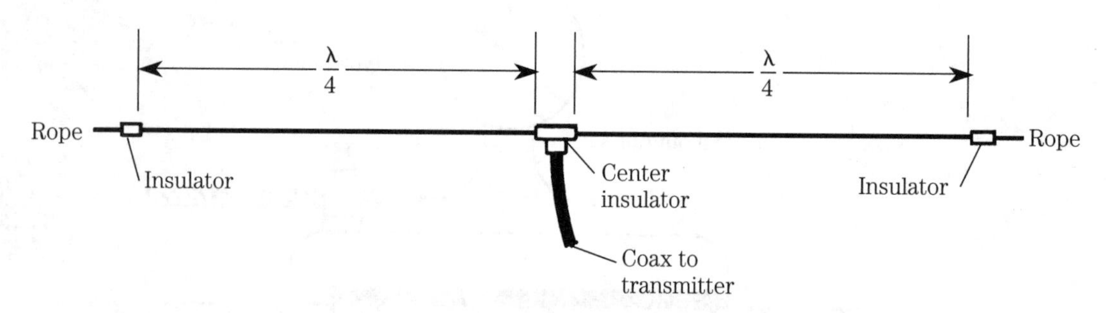

20-6 Basic dipole.

Figure 20-7 shows another alternative antenna that works well for portable operation. Based on the dipole, this inverted-vee dipole does not require two end supports but rather a single center support. The length of each leg is 6 percent longer than for a nominal dipole. Because the application is both emergency related, and temporary in nature, we can get away with construction methods that would be unthinkable in more permanent installations.

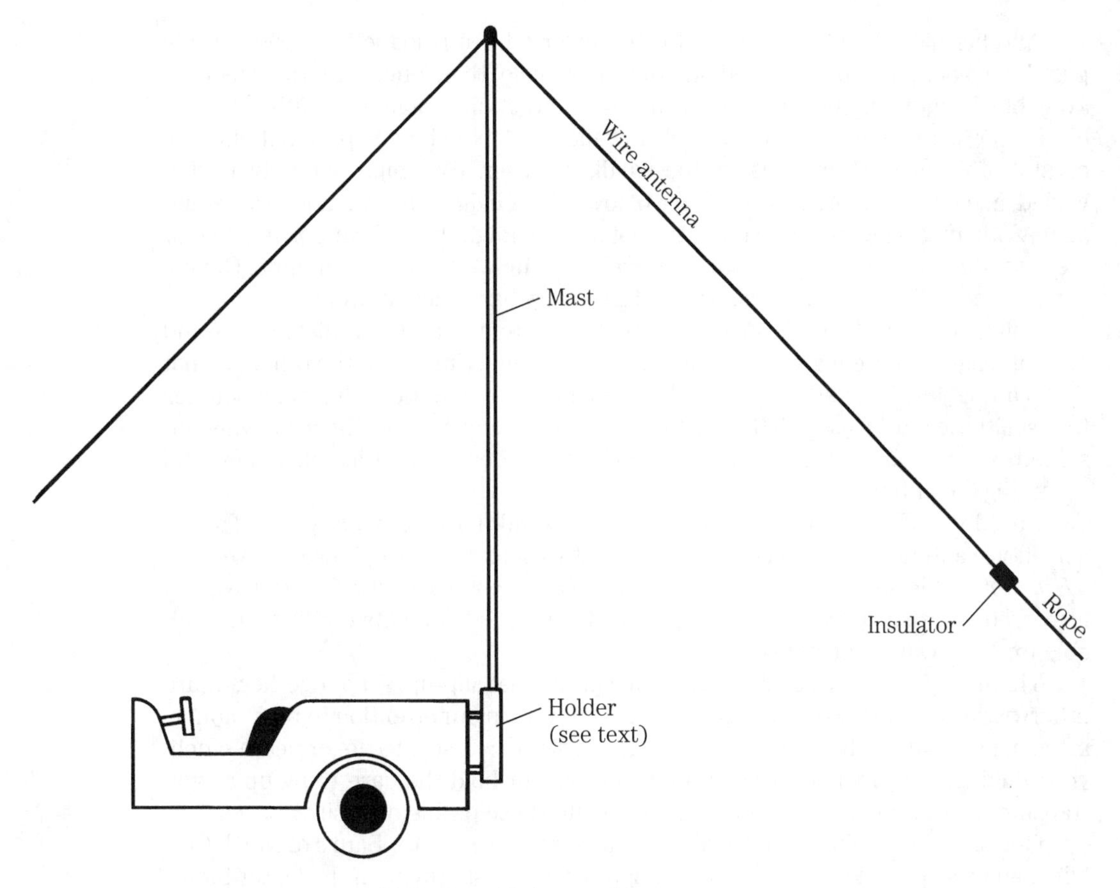

20-7 Portable inverted-vee antenna.

Three problems present themselves. First, the antenna must be portable for backpackers, or stowable in the case of those people who motor into the back country. Size and weight are major constraints in both cases, even though the vehicular case is a lot easier to work with than the backpacker. Second, what materials and means of construction are needed for the mast? Third, how is the mast supported? Because the answer to the third question depends in part on the answer selected for the second, we will deal with it first. The first question is avoided at this point because I have selected the vehicular case.

One alternative is to use a telescoping TV antenna mast to support your antenna. We can either mount a mobile whip and its associated radials at the top, or install an inverted-vee per Fig. 20-7. These masts collapse to 6 or 8 feet in length, but can be slipped up to heights of 18 feet, 25 feet, 30 feet, 40 feet or 50 feet, depending upon the type selected. Keep in mind, when shopping for these masts, that the larger models are considerably heavier than shorter models, and they require two or more people to install them. Erecting a 40- or 50-foot telescoping mast is not a single-person job, despite that you might know someone who has done it. I've done it, and won't do it again! Even the 30-foot models are a bit hairy to install alone.

Another mast that can be used for the inverted vee is made from PVC plastic plumbing pipe. If it can be carried on or in the vehicle, then lengths up to 10 feet are available. Longer lengths are made at the site of erection by joining together 10-foot or shorter sections with couplings (also available at plumbing supply outlets). Be careful of using PVC pipe that is too small, however. PVC pipe is relatively thin walled, and it is therefore flexible. Sizes below 1.5-inch diameter will not easily stand alone without guying. While a single 10-foot section might be self-supporting, two or more sections together will not support itself and the weight of the antenna. Guying can be accomplished with ropes, or on a temporary basis, heavy twine.

Still another alternative is to carry steel TV antenna masts. Available in 5- and 10-foot lengths, these masts are flared on one end and crimped on the other, so that they can be joined together to form longer lengths. In addition, the same sources that sell these masts also sell the guy wire rings that help support the mast when installed. There is also a variety of roof-top mounting devices that aid also in ground mounting the antenna.

I used the TV mast solution on Field Day expeditions for many years. The inverted-vee antenna is, of course, a natural for this solution. Once, however, we used a *Hustler* mobile antenna mounted at the top of a 20-foot mast, with four radials, and it worked surprisingly well with the 30-watt limit for extra points credit then available for Field Day contestants.

There are, then, several alternatives for masts that slip-up, snap together, or are otherwise easily unstowed and installed. Let's now turn our attention to base mounting schemes that can be used to support the mast. Of course, in a life or death pinch, you could always just pile rocks around the base, or hold the darn thing up as you transmit. But a little forethought could eliminate those problems easily.

One of the first solutions I'll deal with was seen on the Outer Banks of North Carolina, and was provided to me by a CB operator. Surf fishermen on the Outer Banks use four-wheel drive vehicles to get out on the beach (to the surf, where the big sea bass lurk). Welded to either the front or rear bumper attachments are steel tubes (see Fig. 20-7) used for mounting the very, very long surf casting rods they use to catch fish. The CB operator had a 20-foot mast consisting of two 10-foot TV mast sections mounted in one of the rod holders. At the upper end of the mast was his 11-meter ground plane antenna. The same method of mounting would also support similar amateur antennas, inverted-vee dipoles, or VHF/UHF antennas.

Given that the antenna installation will be temporary, lasting only a few hours or a couple of days in the worst case, we need not worry about long-term integrity, or the practicality of the installation. Mounting the mast to the back of a four-wheeler, or pick-up truck, with a pair of U-bolts is not terribly practical if you must move the vehicle, but works nicely if you plan to camp (or are stranded) for a few days.

For lightweight masts, up to about 25 feet, the base support could be an X-shaped base made of 2"-×-4"-lumber (Fig. 20-8A), or even a Christmas tree holder. Alternatively, a TV antenna rooftop tripod mount (Fig. 20-8B) is easily adopted for use on the ground. None of these three alternatives can be depended upon for self-supporting installations, and must be guyed even if used for only a short period. Again, because of the temporary nature of the installation, aluminum or wooden tent pegs can be used to anchor the guy wires. Although they are insufficient for long-term installations, they work fine for the short run.

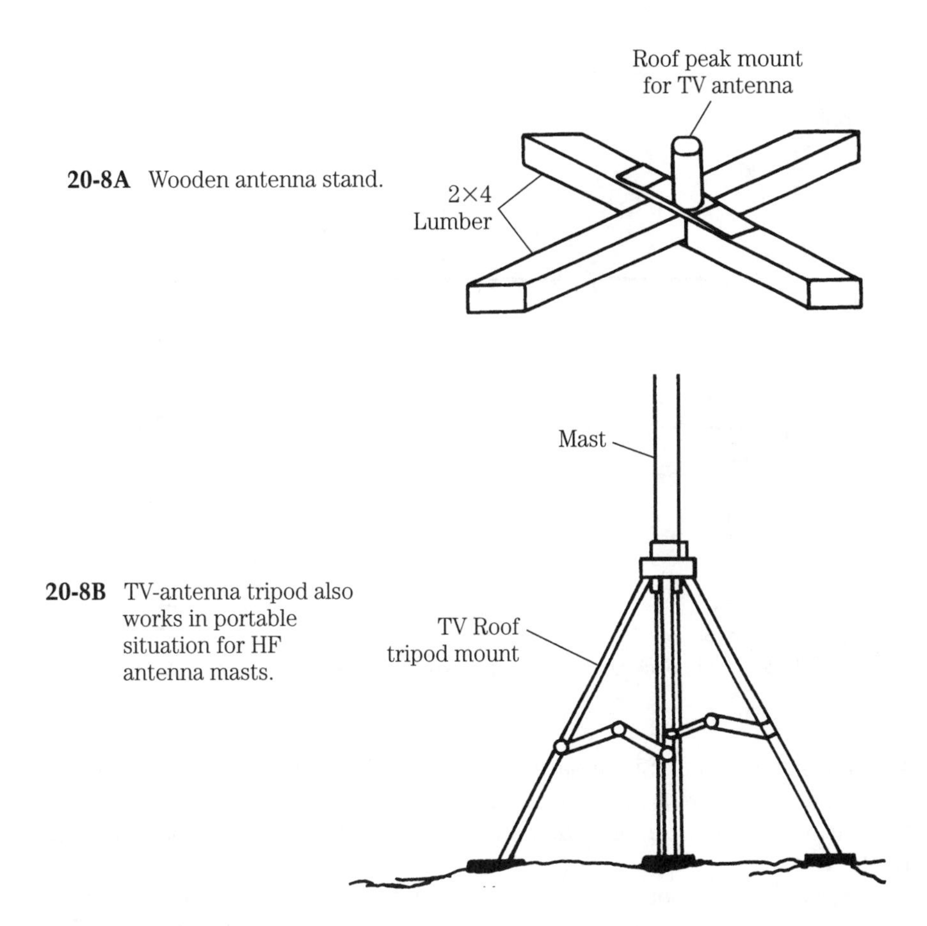

20-8A Wooden antenna stand.

20-8B TV-antenna tripod also works in portable situation for HF antenna masts.

Another problem

Field Day is a reasonably good training ground for people who anticipate operating radio equipment under primitive conditions. If you pay attention to the sometimes humorous foul-ups seen while getting on the air from atop High Peak Mountain in Goose Bump National Park, then you might avoid some of the same mistakes when "foul-ups" mean more than a quick drive to the nearest town. You will learn, for example, which simple tools are a must for constructing and repairing simple antennas (including the 12-Vdc soldering iron that runs off the vehicle battery). You will also learn a bit about electricity needed to run the rig.

Two alternatives present themselves. First, you could use 12 Vdc from the vehicle electrical system . . . which makes sense with the current crop of 12 Vdc HF rigs on the market. Or, you could operate from 120 Vac generated by a light plant generator.

When boondocking in a four-wheeler (or other vehicle), it is wise to use a dual battery system such as shown in Fig. 20-9. Two separate 12-Vdc auto batteries (preferably high ampere-hour capacity) are connected essentially in parallel with the alternative charging system. Diodes D1 and D2 are rated at 100 A, 50 V PIV, and are used to isolate the two batteries from each other. Diode assemblies such as this can be built, or purchased from van conversion and recreational vehicle shops. Their

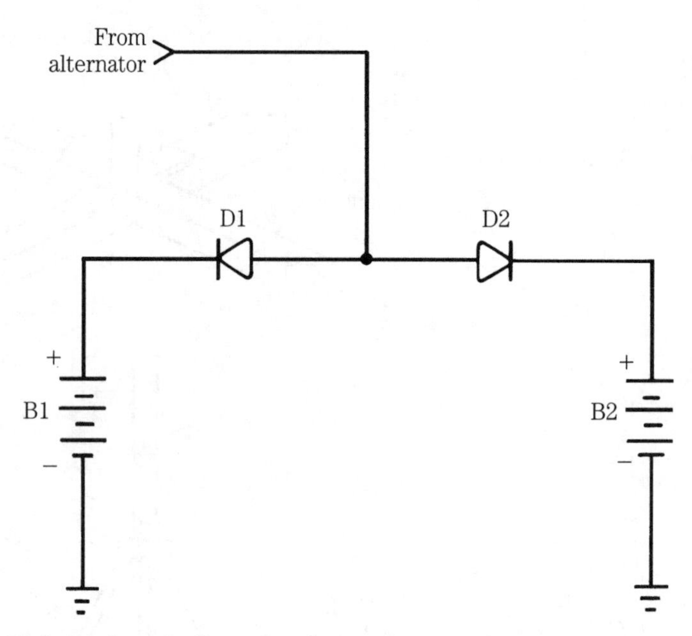

20-9 Diode pack allows charging two batteries from system charger.

use is the same as yours: it's a darn shame to run down the battery needed to start the vehicle just by operating your rig. You not only cannot start the vehicle in such a case, but cannot muster enough power to call for the wilderness equivalent of "Triple-A."

One experienced boondocker I met in southern Arizona totes either a 450-watt Kawasaki, or 500-watt Honda, mini light plant generator in the back of his Bronco. These models are surprisingly lightweight and quiet operating. One of the generators has a 12-Vdc, 8-A outlet that can be used to charge a dead battery. The other has only the 110-Vac outlet, but can be used for battery charging if a small charger is provided. Simple 10-A battery chargers can usually be bought at auto parts, or accessories stores, for about $40.

Operating radio communications equipment successfully under primitive conditions depends upon two major factors: available electrical power and a proper, efficient, antenna system. Although dealing in detail with the means of obtaining power in remote locations is beyond the scope of this book, we have provided some ideas from which you can start planning your own "survival" radio system.

Marine radio antennas

Radio communications considerably lessened the dangers inherent in sea travel. So much so, in fact, that the maritime industry took to the new "wireless telegraphy" earlier than any other segment of society. In the early days of radio, a number of exciting rescues occurred because of wireless. Even the infamous *S.S. Titanic* sinking might have cost less human life if the wireless operator aboard a nearby ship had been on duty. Today, shore stations and ships maintain 24-hour surveillance, and vessels can be equipped with autoalarm devices to wake up those who are not on duty. Even small pleasure boats are now equipped with radio communications, and these are monitored around the clock by the U.S. Coast Guard.

The small shipboard operator can have a selection of either HF single-sideband communications or VHF-FM communications. The general rule is that a station must have the VHF-FM, and can be licensed for the HF SSB mode only if VHF-FM is also aboard. The VHF-FM radio is used in coastal waters, in-land waters and in harbors. These radios are equipped with a high/low power switch that permits the use of low power (1-watt typically) in the harbor, but higher power when underway. The HF SSB radios are more powerful (100 watts typically), and are used for off-shore long-distance communications beyond the "line-of-radio-sight" capability of the VHF-FM band.

The problems of antennas on boats are the same as for shore installations, but it is aggravated by certain factors. Space, for one thing, is less on a boat so most HF antennas must be compensation types. Also, grounds are harder to come by on a fiberglass or wooden boat, so external grounds must be provided.

A typical powerboat example is shown in Fig. 20-10. The radio is connected to a whip antenna through a transmission line (and a tuner on HF), while also being grounded to an externally provided ground plate. Over the years, radio grounds have taken a lot of different forms. For example, the ground might be copper or aluminum foil cemented to the boat hull. Alternatively, it might be a bronze plate or hollow bronze tube along the center line of the boat, or along the sides just below the water line. The ground will also be connected to the engine. Careful attention must be paid

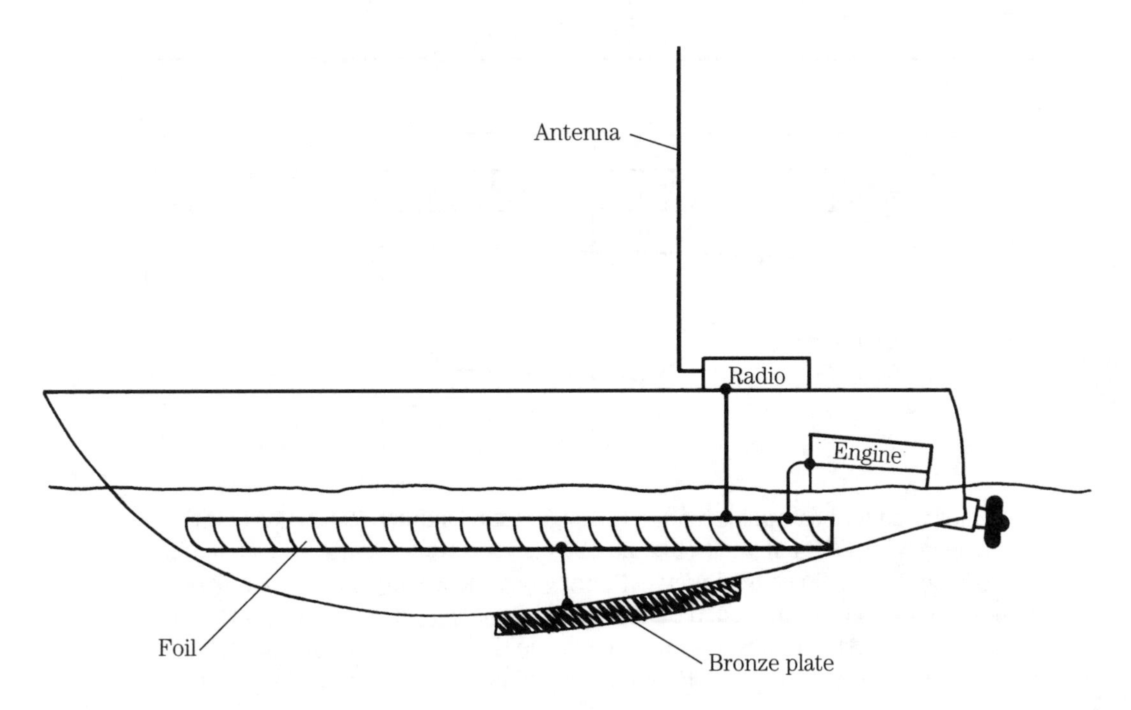

20-10 Motorboat grounding system.

to the electrical system of the boat when creating external grounds to prevent electrolytic corrosion from inadvertent current flows.

On sailboats the whip antenna (especially VHF-FM) might be mast-mounted as shown in Fig. 20-11. The same grounding scheme applies with the addition of a metal keel or metal foil over a nonmetallic keel.

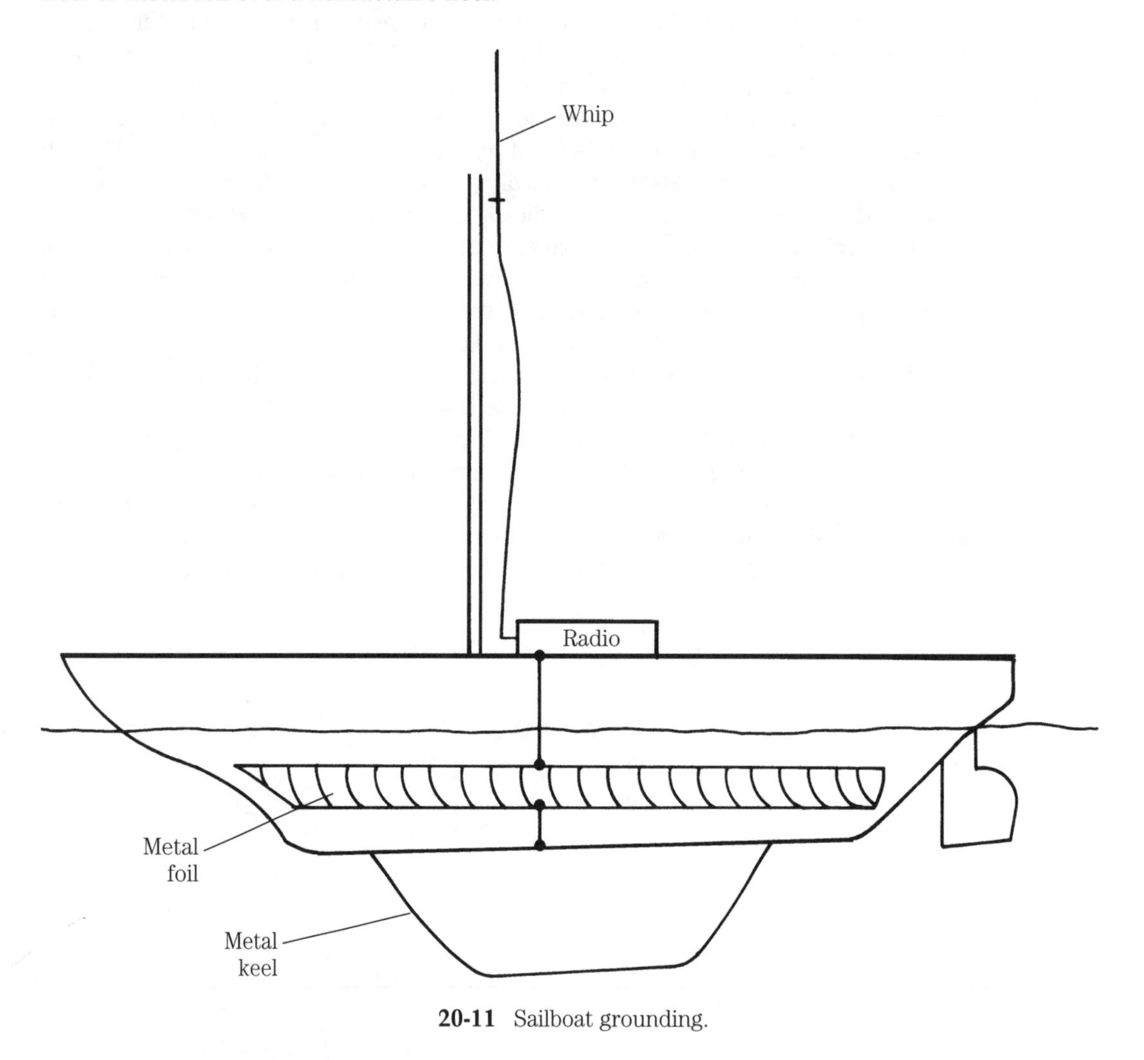

20-11 Sailboat grounding.

The whips used for boat radios tend to be longer than landmobile antennas for the same frequency. The VHF-FM whip (Fig. 20-12) can be several quarter wavelengths and take advantage of gain characteristics thereby obtained. Whips for the HF bands tend to be 10 to 30 feet in length, and often look like trolling rods on power boats. You will also see Citizen's Band radios on-board boats, and the CB antennas are also whips. It is not a good idea to rely solely on CB for boat communications, because it is a lot less

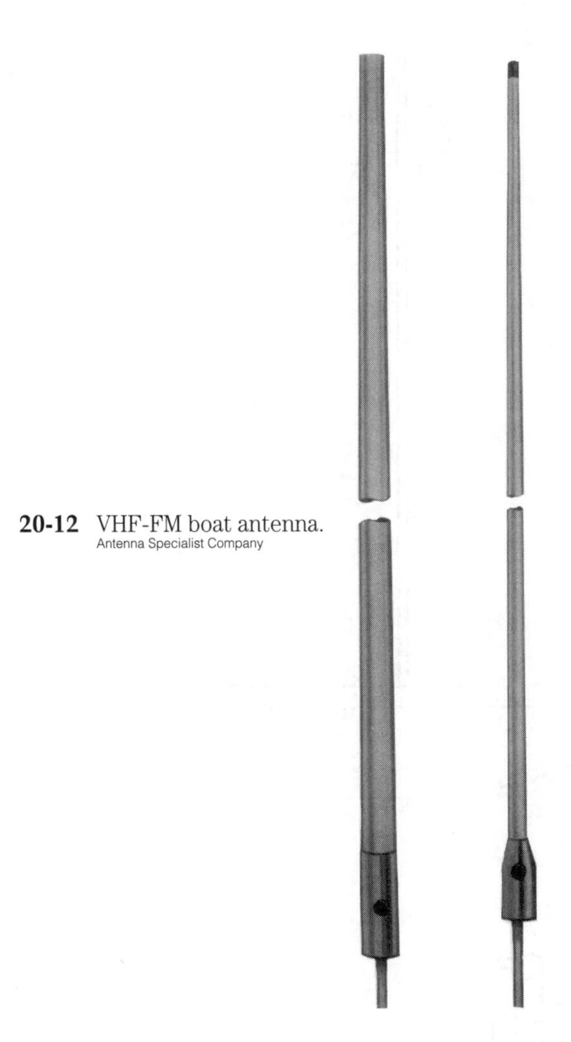

20-12 VHF-FM boat antenna.
Antenna Specialist Company

likely to be useful in emergencies. Also, some boaters use amateur radio sets, often illegally, for emergency communications in-lieu of a proper HF-SSB unit.

Longwire antennas also find use in marine service. Figure 20-13 shows two installations. The antenna in Fig. 20-13A shows a wire stretched between the stern and bow by way of the mast. The antenna is end-fed from an antenna tuner of "line flattener." The longwire shown in Fig. 20-13B is similar in concept, but runs from the bottom to the top of the mast. Again, a tuner is needed to match the antenna to the radio transmission line. Notice that the antennas used in this manner are actually not "longwires" in the truly rigorous sense of the term, but rather "random length" antennas.

The format of the tuner can be any of several designs, shown in Fig. 20-14. The reversed L-section coupler shown in Fig. 20-14A is used when the antenna radiator element is less than a quarter wavelength. Similarly, when the antenna is greater than a quarter wavelength, the circuit of Fig. 20-14B is the tuner of choice. This circuit is a modified L-section coupler that uses two variable capacitors and the inductor.

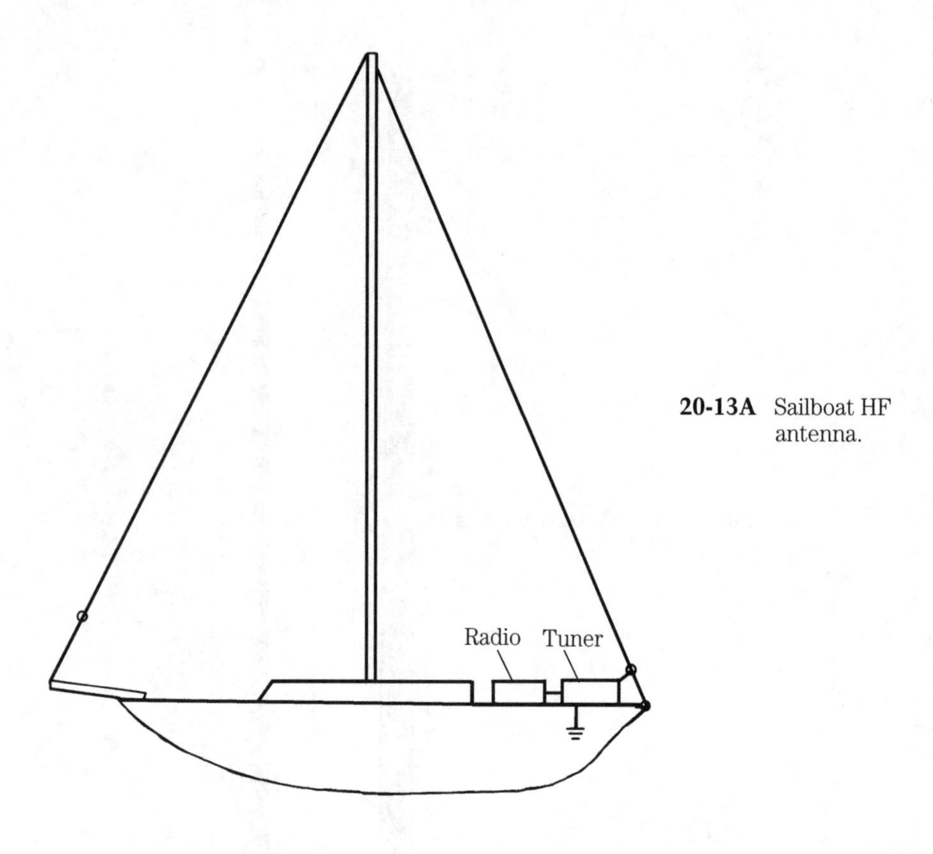

20-13A Sailboat HF antenna.

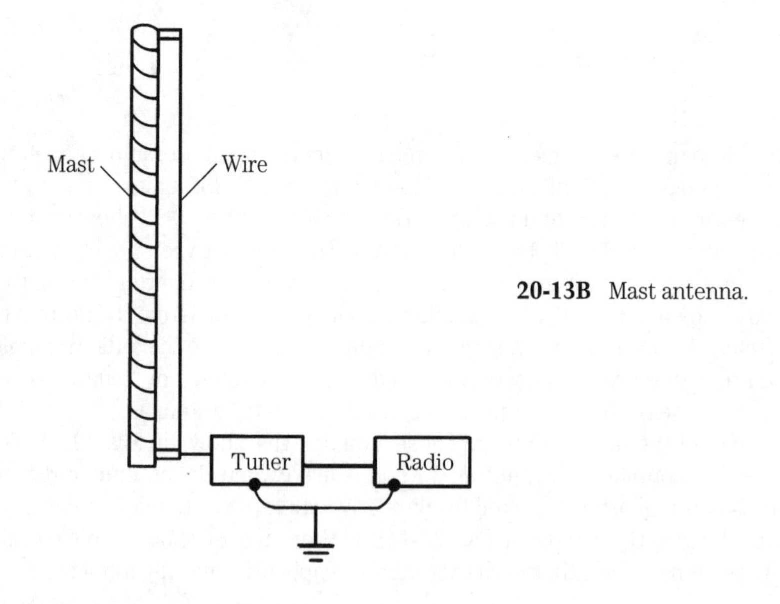

20-13B Mast antenna.

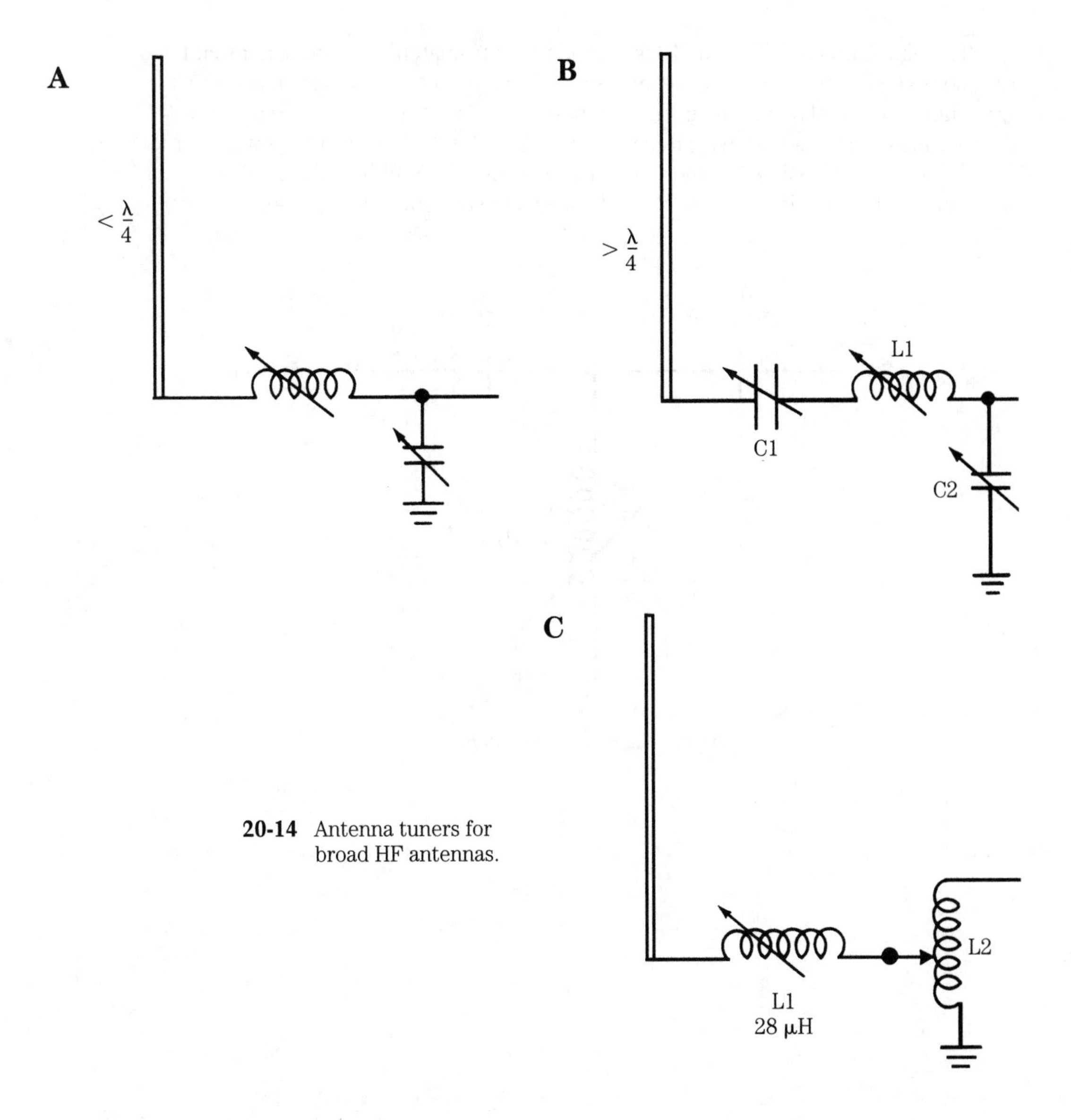

20-14 Antenna tuners for broad HF antennas.

Finally, we see the coupler used on many radios for random-length antennas. Two variable inductors are used; L1 is used to resonate the antenna, and L2 is used to match the impedance looking back to the transmitter to the system impedance. In some designs, the inductors are not actually variable, but rather use switch-selected taps on the coils. The correct coil taps are selected when the operator selects a channel. This approach is less frequently encountered today, when frequency synthesizers give the owner a selection of channels to use. In those cases, either the antenna must be tuned every time the frequency is changed, or an automatic (or motor driven) preselected tuner is used.

The "line flattener" (Fig. 20-15) is a standard "transmatch" antenna tuner that flattens coaxial-cable transmission line for VSWR. This type of tuner is especially useful for transmitters with solid-state finals that are not "happy" with high VSWR. Some of those designs incorporate shutdown circuits that reduce (and then cutoff) power as the VSWR increases. The line flattener basically tunes out the VSWR at the transmitter. It does nothing to tune the antenna, but only makes the transmitter operable.

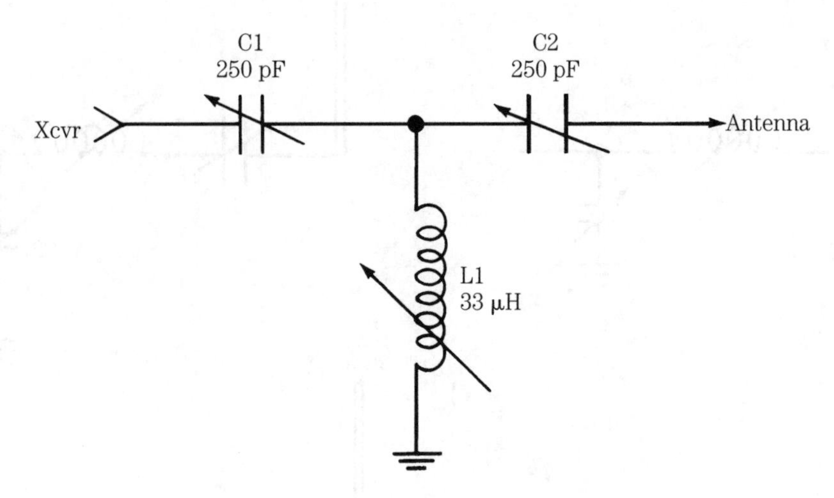

20-15 "Line flattener" tuner.

21

CHAPTER

Antennas for low-frequency operation

LOW-FREQUENCY OPERATION (E.G., 160- AND 75/80-METER BAND) POSES CERTAIN difficulties for the antenna. Of course, the first thing that springs to mind is the large size of those antennas. A half-wavelength dipole is between 117 and 133 feet long on 75/80 meters, and on 160 meters it is about twice that length. On my own suburban lot, I cannot erect a half-wave 75/80-meter band antenna and stay within the property lines.

A similar situation is seen with vertical antennas. Although a 40-meter vertical (33 feet high) is not an unreasonable mechanical job, the 66 foot 75/80-meter vertical (never mind the 120-foot 160-meter vertical) is a nightmare. In addition, the local authorities might not require any special inspections or permits (check!) on the 33-foot antenna, yet impose rigid and very exacting requirements on the higher structure. On suburban or urban lots, a typical 40-meter antenna might well be able to fall over and still not cross the property line—or come in close proximity to power lines. A longer antenna, however, almost inevitably suffers one—or the other—problem when it falls.

Grounding

Still another problem involves grounding. The ground system of a higher-frequency vertical antenna can be installed using $\lambda/4$ radials buried a few inches below the surface. But that same idea becomes problematic when the radials are 66 or 125 feet long. Burying half a dozen 125-foot radials might remind your neighbors of the Galloping Gung-Ho Gopher . . . and not be to their amusement!

There are, however, some solutions to the problem—without having you buy a farm in the flatter regions of the midwest. For example, the length problem can be solved by using one of the bent dipoles covered in chapter 10. No, they are not as effective as an antenna that is correctly installed. But they will serve to get you on the air. Some of them work surprisingly well.

The ground radial problem can be solved by bending the radial system around your property (Fig. 21-1). You will never have to cross your neighbor's line. And, as for the gopher track appearance, it is not necessary to work sloppily, and you can install radials so that nary an eye, practiced or otherwise, can see their location.

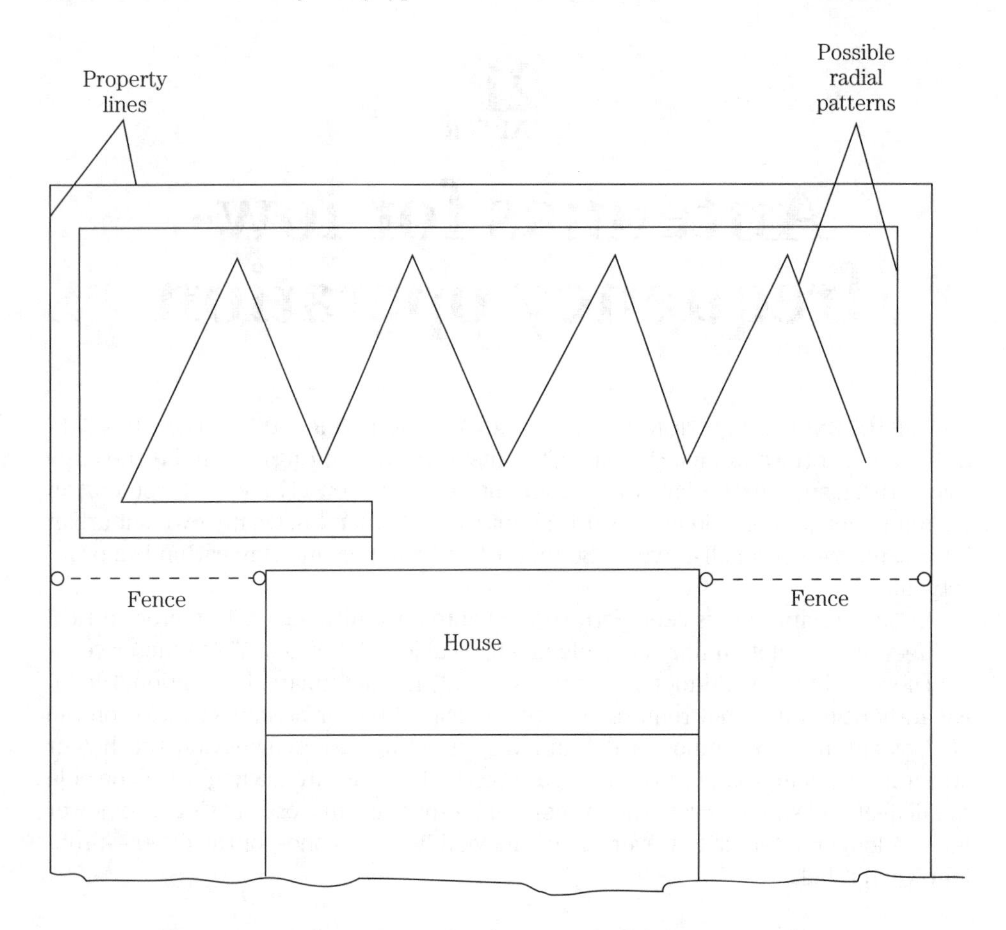

21-1 Methods for laying out low frequency radials in a cramped space.

Shortened vertical antennas

The biggest problem for most low-frequency DXers, as you have seen, is the excessive size of antennas for those frequencies; it is "not for nothing" that those AM broadcast band (<1.6 MHz) towers are usually hundreds of feet tall. But there are ways to shorten an antenna—not for free, because the TANSTAAFL[1] principle still

---

[1]TANSTAAFL = There Ain't No Such Thing As A Free Lunch!

applies—to a point where it becomes mechanically possible. Let me reiterate once again that these compensation antennas will not work as well as a properly installed full-sized antenna, but they will serve to get you on the air on frequencies where it is otherwise utterly impossible. Several different compensation configurations are popular, and these are shown in Fig. 21-2.

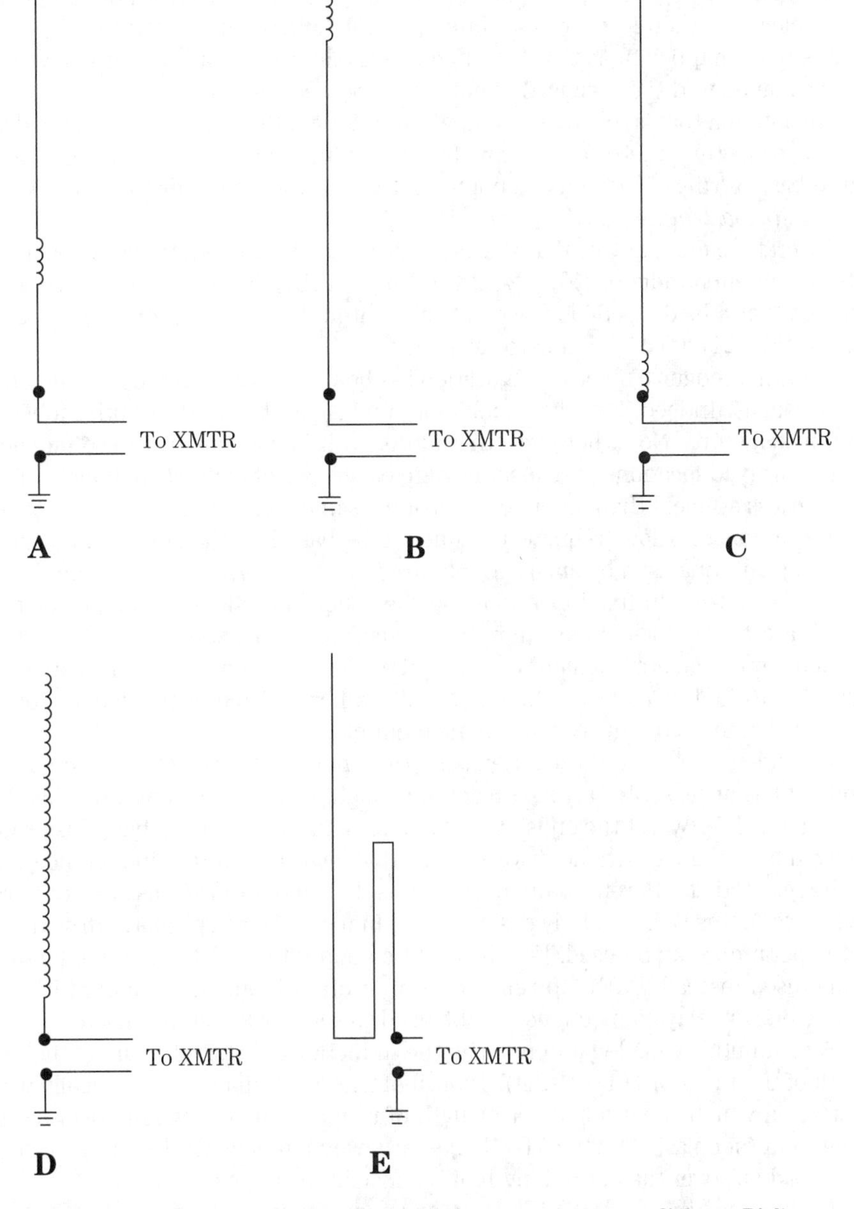

21-2 Inductance loading of vertical antennas A) center; B) top; C) base; D) linear or length loaded; E) linear hairpin loaded.

The basic foundation for these antennas is a very short vertical antenna. The "standard" vertical antenna is a quarter wavelength ($\lambda/4$) (i.e., 90° electrical length), and it is unbalanced with respect to ground.

Recall that a vertical antenna that is too short for its operating frequency (i.e., less than $\lambda/4$) will exhibit capacitive reactance. In order to resonate that antenna, it is necessary to cancel the capacitive reactance with an equivalent inductive reactance, such as $|X_L| = |X_C|$. By placing an inductance in series with the antenna radiator element, therefore, we can effectively "lengthen" it electrically. Of course, what is really happening is that the effects of the lower operating frequency are being accommodated (i.e., "cancelled out") of a too-short antenna.

An antenna that is reactance compensated for a different frequency is said to be *loaded*, or in the case of a very low frequency antenna it is *inductively loaded*. Three basic forms of loading are popular: *discrete loading, continuous loading,* and *linear loading*.

Discrete loading means that there is a discrete, or lumped, inductance in series with the antenna radiator (Figs. 21-2A through 21-2C). These antennas are so constructed that a loading coil is placed at the center (Fig. 21-2A), top (Fig. 21-2B) or bottom (Fig. 21-2C) of the radiator element.

You will recognize these configurations as being the same as those found on mobile antennas. Indeed, low-band mobile antennas can be used in both mobile and fixed installations. Note, however, that although it is convenient to use mobile antennas for fixed locations (because they are easily available in "store bought" form); they are less efficient than other versions of the same concept. The reason is that the mobile antenna, for low frequencies, tends to be based on the standard 96- to 102-inch whip antenna used by amateur operators on 10 meters or Citizens Band operators on 11 meters. In fixed locations, on the other hand, longer radiator elements (which are more efficient) are more easily handled. For example, a 16- to 30-foot high aluminum radiator element can easily be constructed of readily available materials. The 16-foot element can be bought in the form of one or two lengths of 1- to 1.5-inch aluminum tubing at do-it-yourself outlets.

A problem seen with these antennas is that they tend to be rather high-Q, so the bandwidth is necessarily narrow. An antenna might work in the center of a band, but present a high VSWR at the ends of the band . . . and thus be unusable. This problem is solved by making the inductor variable so that slightly different inductance values can be selected at different frequencies across the band. For the base loaded version (Fig. 21-2C), this is particularly easy: a rotary inductor (perhaps motor driven for remote operation) can be used. For the other configurations, a tapped fixed inductor can be used instead. Each tap represents a different inductance value. Either clip connectors, or relay connections, can be used to select which tap is used.

A continuously loaded antenna has the inductance distributed along the entire length of the radiator (Fig. 21-2D). Typical of these antennas is the helically wound verticals in which about a half wavelength of insulated wire is wound over an insulating form (such as a length of PVC pipe, or a wooden dowel); the turns of the coil are spread out over the entire length of the insulated support.

Linear loading (Fig. 21-2E) is an arrangement whereby a section of the antenna is folded back on itself in a manner similar to a stub. Antennas of this sort have been successfully built from the same type of aluminum tubing as regular verticals. For 75/80-meter (3.75 MHz) operation a length of 30 feet for the radiator represents 41 degrees, while a normal $\lambda/4$ vertical is 90 degrees. The difference between 90 and 41 degrees of electrical length is made up by the "hairpin" structure at the base.

Shortened horizontal antennas

The same strategies that worked for vertical antennas, also work for horizontal antennas, although in the horizontal case we are simulating a half wavelength (180 degree) balanced antenna—rather than a $\lambda/4$ unbalanced antenna. Figure 21-3 shows several different shortened, low frequency antennas that are based on the same methods as the verticals shown previously.

Figures 21-3A through 21-3C show discretely loaded dipole antennas. Figures 21-3A and 21-3B are center-loaded versions, and Fig. 21-3C is a center (of each element) loaded version. In each case, it is assumed that the radiator elements are the same physical length. In the example of Fig. 21-3A, the coil is tapped to provide a match to either 52- or 75-Ω transmission line. In some variants, the shield of the coaxial-cable transmission line is connected directly to the junction of the coil and one radiator element, and the coax center conductor is connected to the tap that best matches the impedance of the line.

The impedance-matching problem is solved a little differently in Fig. 21-3B. In this type of antenna, the coil is used to center load the dipole, but the transmission line is connected to a link wound on the same form as the loading inductor. The turns ratio between the loading inductor and the coupling link determines the impedance match.

A "hairpin" linear loading scheme is shown in Fig. 21-3D. This design is basically the same as for the vertical case, but it is balanced out of respect for the design of the dipole. This type of design is used for both the driven and parasitic elements on some commercial 40-meter beam antennas. In addition, some "add-on" 40-meter dipoles (designed for beam antennas that were intended for operation at frequencies in the 20-meter and higher bands) use this method. Such antennas are essentially "rotatable dipoles."

The continuously loaded dipole of Fig. 21-3E is constructed in a manner similar to the helical vertical (i.e., about half wavelength of insulated wire is wound over the entire length of an insulated rod or pipe of some sort). The winding can be either broken at its center point to accommodate the feedline, or link coupled to the transmitter, or receiver, as shown in Fig. 21-3F. Of the two feed methods, the most popular appears to be the type that breaks the winding into two pieces.

Another form of continuously loaded dipole is shown in Fig. 21-3E. This antenna has a combination of the two methods of feed system. The distributed loading coil is broken into two sections, as is often done on continuously loaded dipoles. But in this case, a portion of the overall inductance is made by using a discrete inductor that is part of a toroidal transformer (T1). The length of the wire used for the continuous loading coil is somewhat shorter (find by experimentation) than otherwise would be the case because of the inductance of the transformer secondary. Transformer T1 can be wound using the same sort of toroidal ferrite, or powdered iron cores, as are used for BALUN transformers.

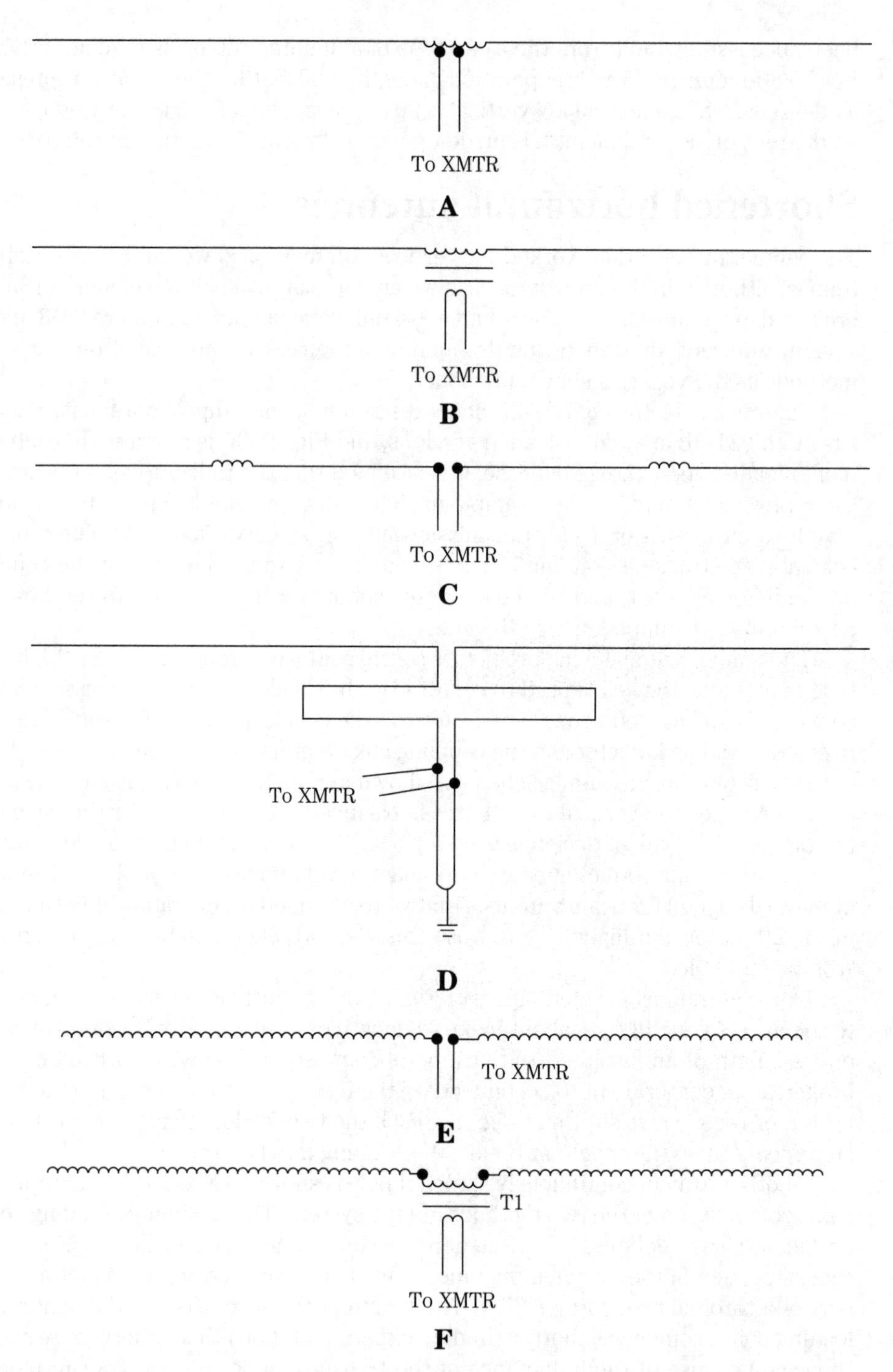

21-3 Inductance loading a dipole: A) center loading with impedance tapped coil; B) center loading with transformer coupling; C) Loading at center of elements; D) linear loading (hairpin); E) linear or length loading; F) transformer coupling.

Loaded tower designs

Many amateur radio stations and other services, use a rotatable beam antenna on top of a tower. Typical towers are from 30 to 120 feet in height. In my locality, towers to 30 feet, or less than 3 feet above the roof line whichever is taller, can be erected without a permit, as long as they are attached to the house and will not fall onto a power line, or across the property line, in a catastrophic failure. Higher towers must be erected under a mechanical permit, and properly inspected by the county. These towers typically support a two or more element (three is popular) Yagi /Uda beam antenna, or a quad, or some other highly directional antenna. But they can also be treated as a vertical antenna under the right circumstances.

If the tower is close to 66 feet high (which is a popular height for amateur radio towers), then the tower (already $\lambda/4$) can be used as a resonant vertical on 75/80 meters. The same tower can also be used on lower frequencies if proper antenna tuning unit (ATU) components are provided.

Figure 21-4 shows a situation in which an 80- to 110-foot tower is insulated from ground, and is considered a random length vertically polarized Marconi antenna. Like other such antennas, it is fed using a simple L-section, or Reverse L-section, ATU. Note the approximate values of the inductor and capacitor in each configuration, and also their configuration. On the 40- and 75/80-meter versions, the tower is too long, so a series capacitance and shunt inductance are used for the ATU (i.e., Reverse L-section coupler). On the other hand, the tower is too short for 160 meters, so the inductor and capacitor are reversed.

There is only one small problem with the design of Fig. 21-4: compared with grounded towers, insulated towers are expensive to buy and install. In addition, lightning protection is probably better in the grounded tower. So how do we work a grounded tower? See Fig. 21-5. In this antenna, the tower is grounded at its base. If the tower is mounted to a concrete pedestal (the usual arrangement), then a separate ground rod and ground wire adjacent to the pedestal must be provided. Concrete is not a good insulator, but it is also not a good conductor. A *delta feed* system is used with this antenna. That is, a single wire from the ATU is connected to a point on the antenna where an impedance match can be achieved. It is important that the ATU be spaced away from the tower base (as shown), and the wire must be run straight to the feedpoint on the antenna. This system is an example of several possible *shunt feeding* systems.

Random length Marconi

One of the old standbys for all bands is the random-length Marconi antenna. Such antennas consist of a length of wire, typically (but not always) less a $\lambda/4$, and fed at one end with coaxial cable and an L-section coupler (Fig. 21-6). The antenna radiator element can be angled in any direction as needed, but it works best if the radiator is either as horizontal, or vertical, as possible (for pure polarization). The usual situation, however, is to run the wire at an inclined angle, or with about equal portions horizontal and vertical (see the inverted-L) which yields complex polarization.

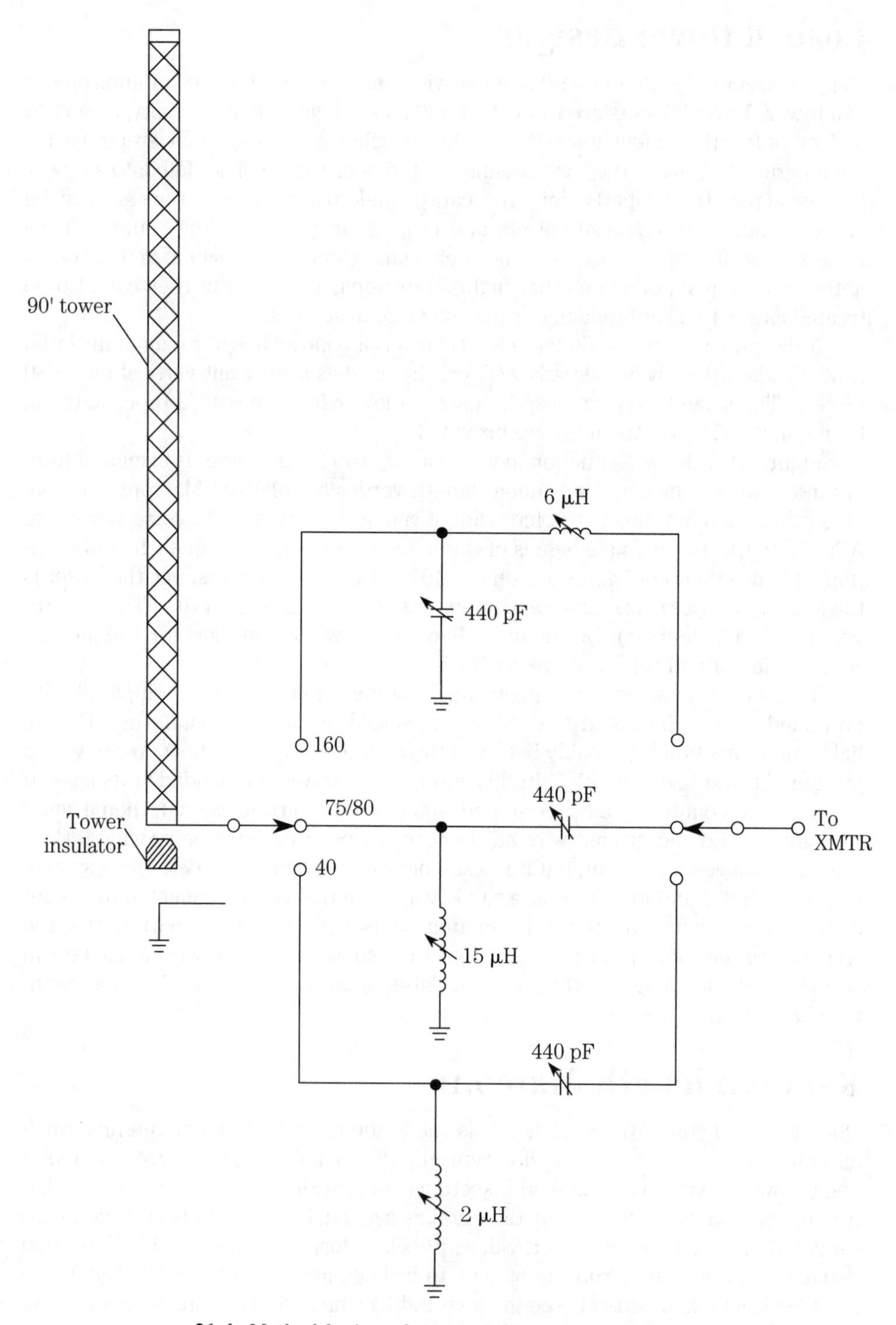

21-4 Method for impedance matching a tower antenna.

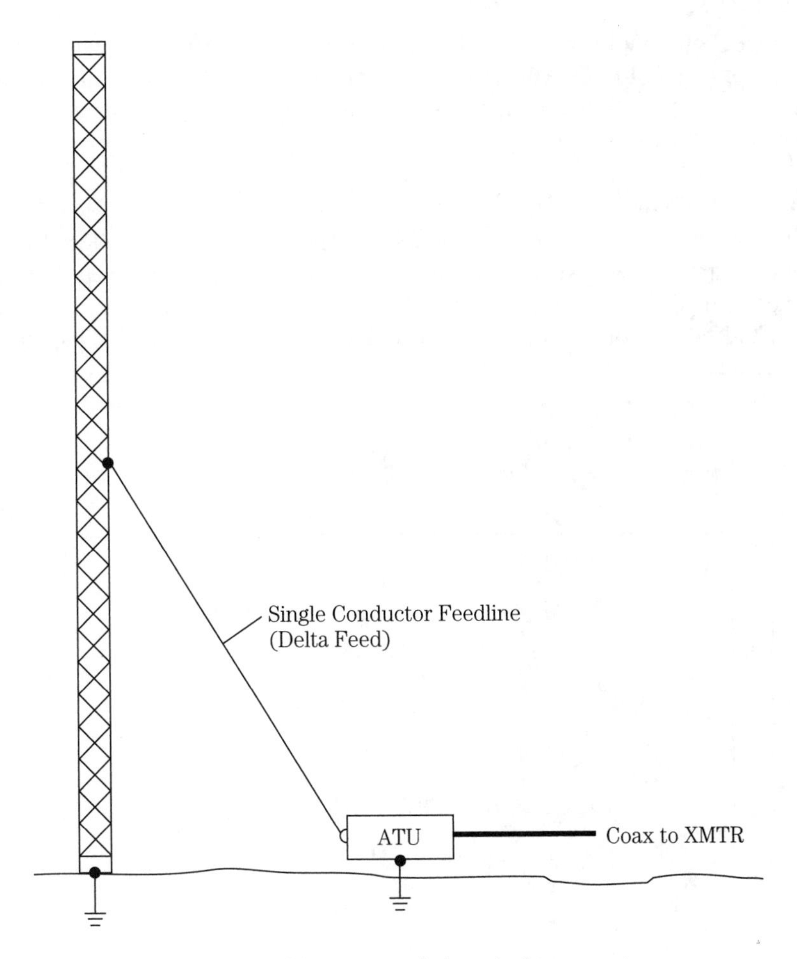

21-5 Delta matching a grounded vertical tower antenna.

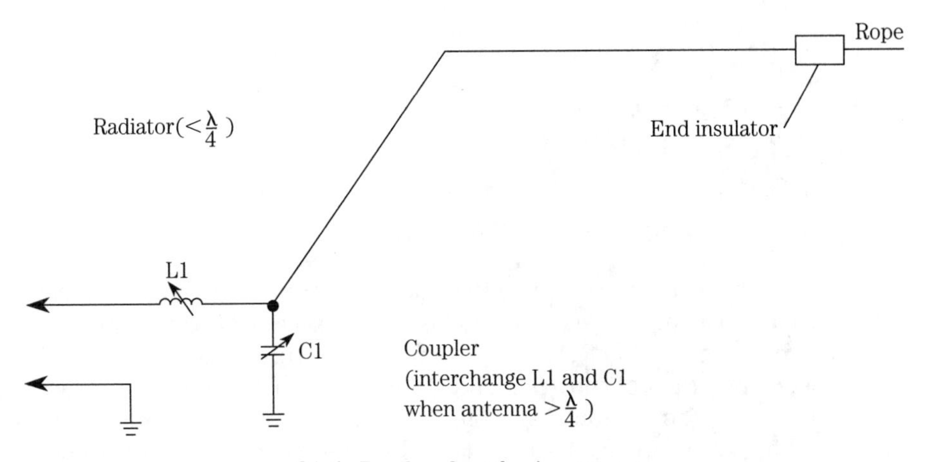

21-6 Random-length wire antenna.

The L-section coupler shown in Fig. 21-6 is set up for the case where the antenna radiator element is less than λ/4 (i.e., a series inductor and a shunt capacitor). If the antenna is longer than λ/4 on some frequency, then reverse the positions of the capacitor and inductor.

Inverted-L antennas

Another popular antenna for low frequencies is the *λ/4 inverted-L* (Fig. 21-7). In this type of antenna, two sections are erected at a 90° angle with respect to each other; one vertical and the other horizontal. One way to think of this antenna is bent λ/4 vertical, although some people liken it to a top-loaded vertical. The feedline can be 52- or 75-Ω coaxial cable.

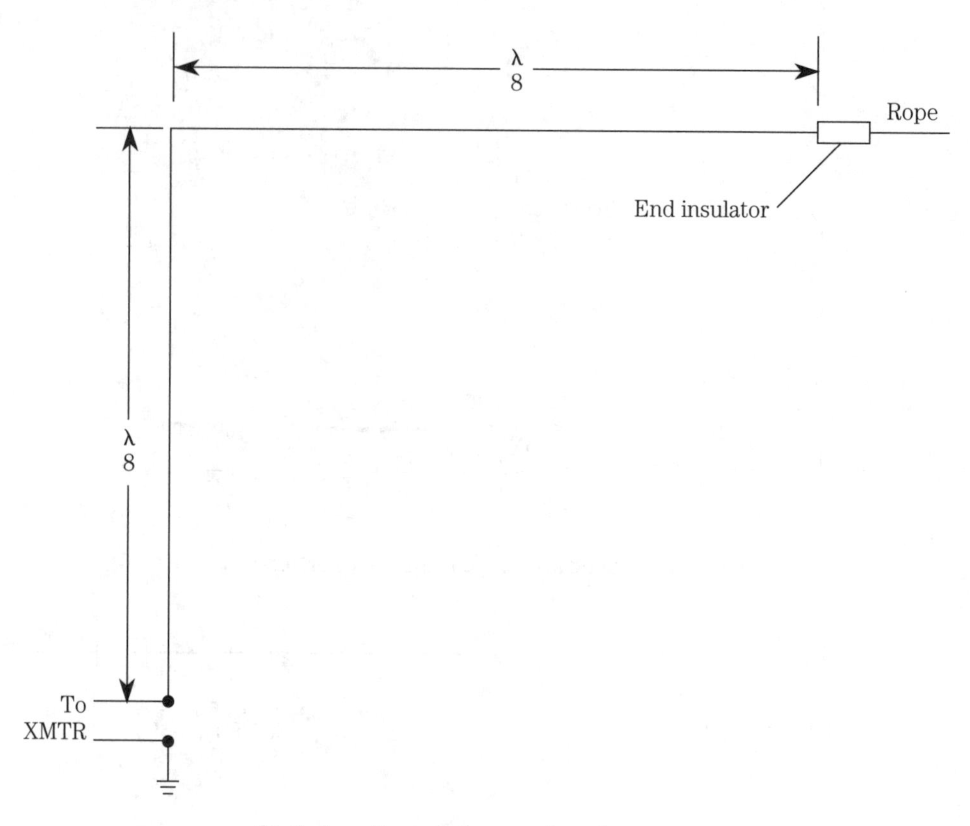

21-7 L-section quarter-wavelength antenna.

It is generally the case that the sections of the inverted-L are equal length (λ/8 each), but that is not strictly necessary. As the vertical section becomes longer (with overall length remaining at λ/4), the angle of radiation depresses.

One popular method of construction is to use a tower for the vertical section, and a run of wire for the horizontal section. If you already have a 60-foot tower to accommodate the beam antenna used on higher frequencies, then it is relatively easy to build an inverted-L antenna for 160-meters.

The linearly loaded "tee" antenna

A "tee" antenna consists of a horizontal radiator element fed at the center with a single conductor wire (not coax) that is a quarter-wavelength long, and goes away from the radiator at a right angle. Such antennas were popular, prior to World War II, in the United States, although popularity fell off in the late 1930s as other types (and reasonably priced transmission lines) became available. The Tee is possibly one of the oldest forms of radio transmitting antennas. Like other antennas at low frequencies, however, it is far too long for easy use in most locations.

We can, however, linearly load the Tee by folding the ends of the radiator back on themselves to form the zig-zag pattern of Fig. 21-8. Popular in Europe for some time, this antenna can provide reasonable performance on the lower frequencies without using too much horizontal space.

The radiator element consists of three sections that are each about $\lambda/6$ long ($L_1 = 164/F_{MHz}$), and spaced 8 to 12 inches apart; they are parallel to each other. The quarter-wavelength transmission line is connected to the center point of the middle section of the radiator element.

Figure 21-9 shows how the linearly loaded Tee antenna could be built using wire for the radiator elements. Spreaders made of 24-inch long 1"-x-2" lumber (treated against the weather), plastic, or some other synthetic material can be used. Each spreader has three holes drilled in it spaced about 10 inches apart. Spreaders A and E are end spreaders, and are identical—except for being upside down with respect to each other. In each case, one of the radiator element conductor sections is terminated, while the other passes through to the back of the spreader to join the center radiator element. The other spreaders are used either for center support, as in C (note the transmission line attachment), or for interim support between the center insulator and the end insulators.

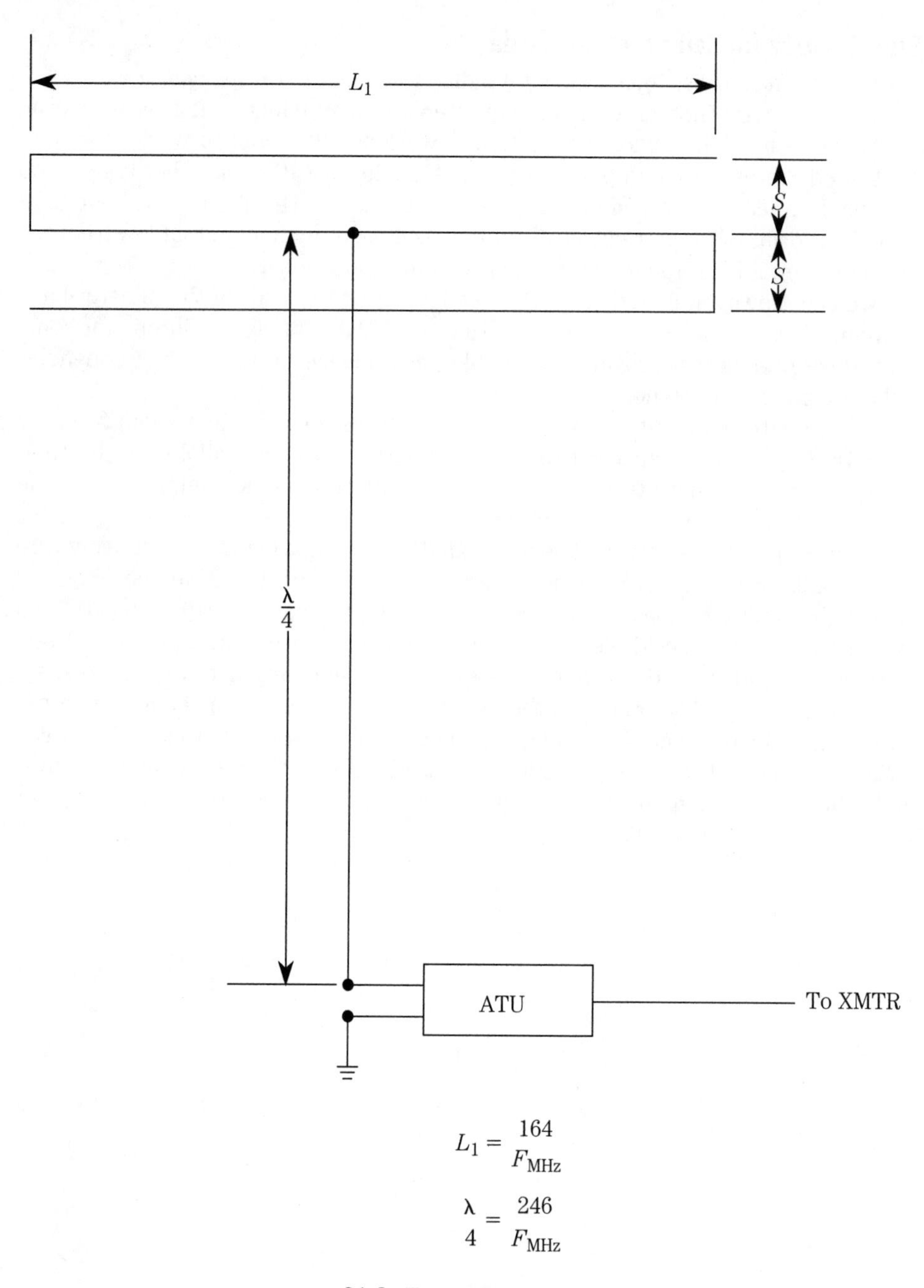

$$L_1 = \frac{164}{F_{MHz}}$$

$$\frac{\lambda}{4} = \frac{246}{F_{MHz}}$$

21-8 Tee-loaded antenna.

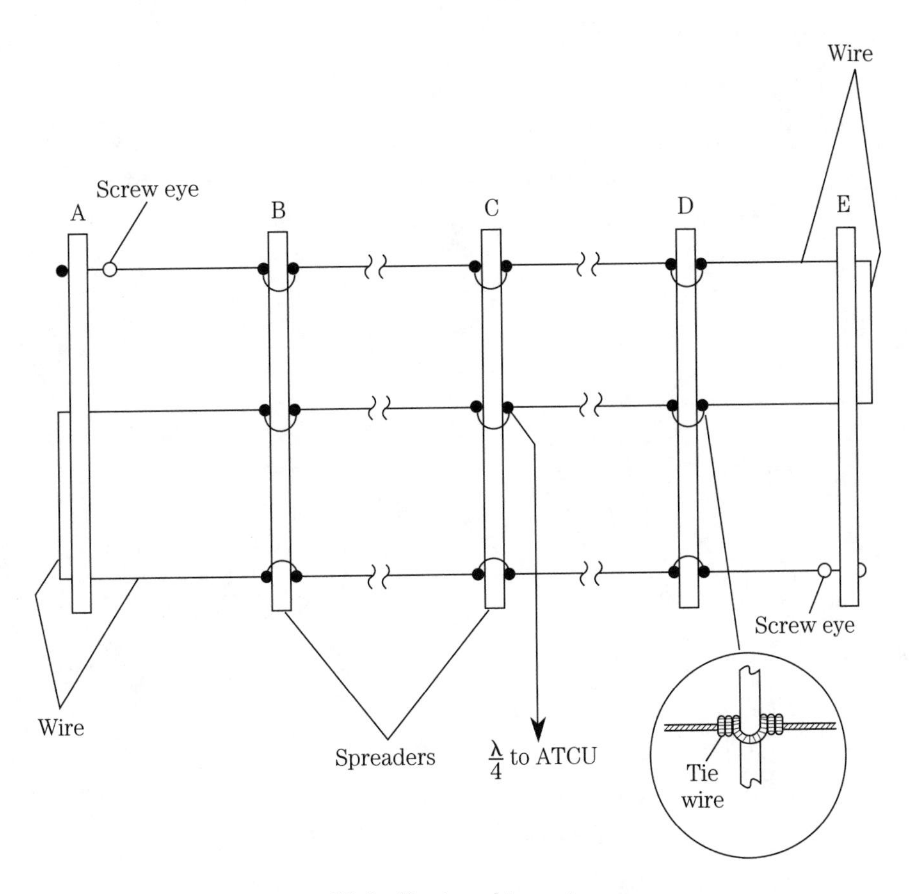

21-9 Shortened tee antenna.

22 CHAPTER

Measurements and adjustment techniques

THIS CHAPTER EXAMINES SOME OF THE INSTRUMENTS AND TECHNIQUES FOR TESTING antenna systems, whether brand new installations, or in troubleshooting situations on older antenna. The basic radio system is shown in Fig. 22-1. We have several elements in the system: transmitter, low-pass filter, impedance matching unit, a coaxial relay (RLY) if the receiver is separate from the transmitter, and the antenna. Connecting these elements are lengths of transmission line. In most modern radio systems below the microwaves, the transmission line is coaxial cable.

The low-pass filter and matching unit may be considered optional by some, but some designers regard them as essentially standard equipment, especially for HF and low-band VHF systems. The low-pass filter has the job of removing harmonics from the output signal that could interfere with other radio systems or radio and TV broadcast reception. It will pass only those frequencies below a certain cut-off point. The impedance matching unit is used to "tune out" impedance mismatches that cause the transmitter to see an excessive VSWR. These units are common in VLF, MW, and HF communications stations, as well as AM broadcasting stations. The matching unit also provides additional attenuation of the harmonics, so it makes the output even cleaner than is possible with the low-pass filter alone.

Another reason to use the matching unit is to allow the radio transmitter to put out the maximum allowable RF power. Modern solid-state final amplifiers are not tolerant of VSWR. In addition, these units use fixed-tuned low-pass filters for each band rather than the wide-range pi-networks common on vacuum tube transmitters. Solid-state transmitters usually include a feedback *Automatic Load Control* (ALC) that reduces output power when a high VSWR is sensed. The VSWR cut-in knee begins around 1.5:1, and completely shuts off the transmitter when the VSWR gets high (typically above 2.5:1 or 3:1). The external T/R relay is not used on most modern systems because the receiver and transmitter are typically housed inside of the same box. The coaxial relay was used in days past, when a

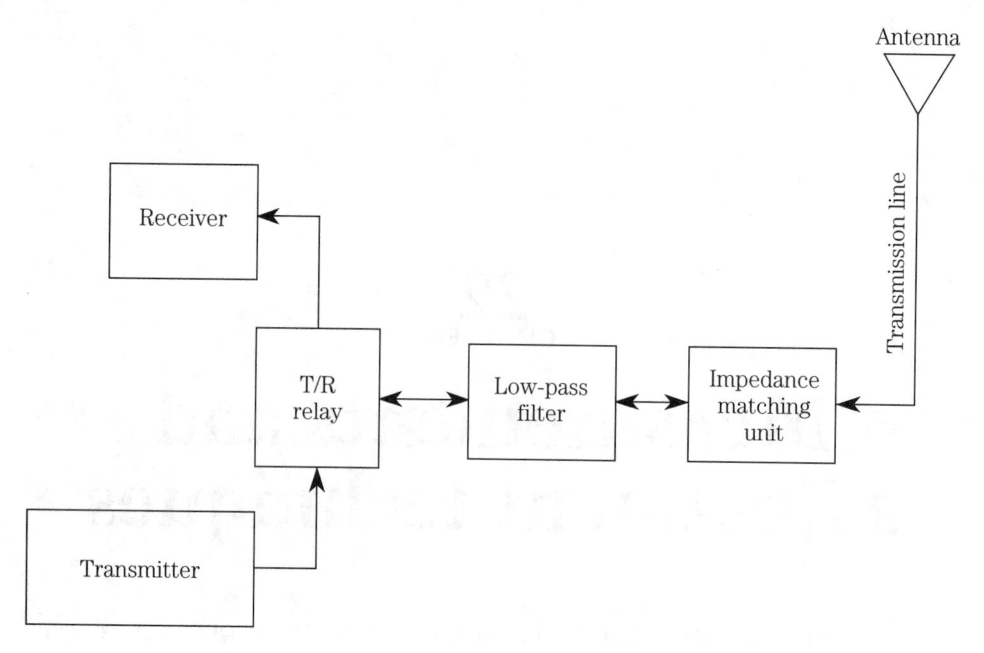

22-1 Basic communications radio station setup.

separate receiver needed to use the same antenna. In addition, many modern solid-state equipments use PIN diode T/R switching.

Transmission lines

The transmission line is not merely a wire that carries RF power to the antenna. It is actually a complex circuit that simulates an infinite LC network. There is a characteristic impedance (Z_o), also called surge impedance, which describes each transmission line. This impedance is the square root of the ratio of the capacitance and inductance per unit of length. When a load having a "resistive only" impedance equal to the surge impedance of the transmission line is connected, then we will see maximum transfer of power between the line and the antenna.

We cannot deal extensively with transmission-line theory here, and refer the reader instead to previous chapters. We must, however, at least have some idea of what the circuit looks like. Figure 22-2 shows a model of a transmission line in which Z_o is the surge impedance of the line, R_2 is the load impedance of the antenna, and R_2 is the output impedance of the transmitter. In a properly designed system all three impedances will be either equal ($Z_o = R_1 = R_2$), or a matching network will make them equal.

We must consider the electrical situation along the transmission line in order to understand the readings that we see on our instruments. Figure 22-3 shows several possible situations. These graphs are of the RF voltage along the line, with voltage on the vertical axis and transmission line length (expressed in wavelengths of the RF signal) along the horizontal axis. When the system is matched ($Z_o = R_1$), the voltage

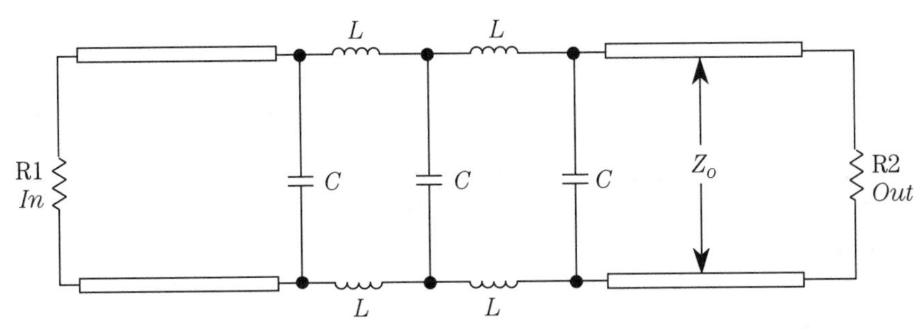

22-2 Equivalent circuit of transmission line.

is the same everywhere along the line (Fig. 22-3A). This line is said to be "flat." But when Z_o and R_2 are not equal, then the voltage varies along the line with wavelength. In mismatched systems, not all of the power is radiated by the antenna, but rather is reflected back to the transmitter. The forward and reflected waves combine algebraically at each point along the line to form standing waves (Fig. 22-3B). We can plot the voltage maxima (V_{max}) and minima (V_{min}). Keep this graph in mind for a few minutes because you will refer again to it when we deal with VSWR.

Two special situations occur in transmission line and antenna systems that yield similar results. The entire forward power is reflected back to the transmitter (none radiated) if the load (i.e., antenna) end of the transmission line is either open or shorted. The voltage plot for an open transmission line (R_2 is infinite) is shown in Fig. 22-3C, and that for the shorted line is shown in Fig. 22-3D. Note that they are very similar to each other except for where the minima ($V_{min} = 0$) occur. The minima are offset from each other by 90 degrees (i.e., quarter wavelength).

Calculating standing wave ratio

The *VSWR* can be calculated from any of several bits of knowledge. Even if you don't have a VSWR meter, therefore, it is possible to determine *VSWR*. If the antenna load impedance (R_2) is not equal to Z_o, then we can calculate *VSWR* from one of the following:

If Z_o is greater than R_2:

$$VSWR = \frac{Z_o}{R_2} \qquad [22.1]$$

If Z_o is less than R_2:

$$VSWR = \frac{R_2}{Z_o} \qquad [22.2]$$

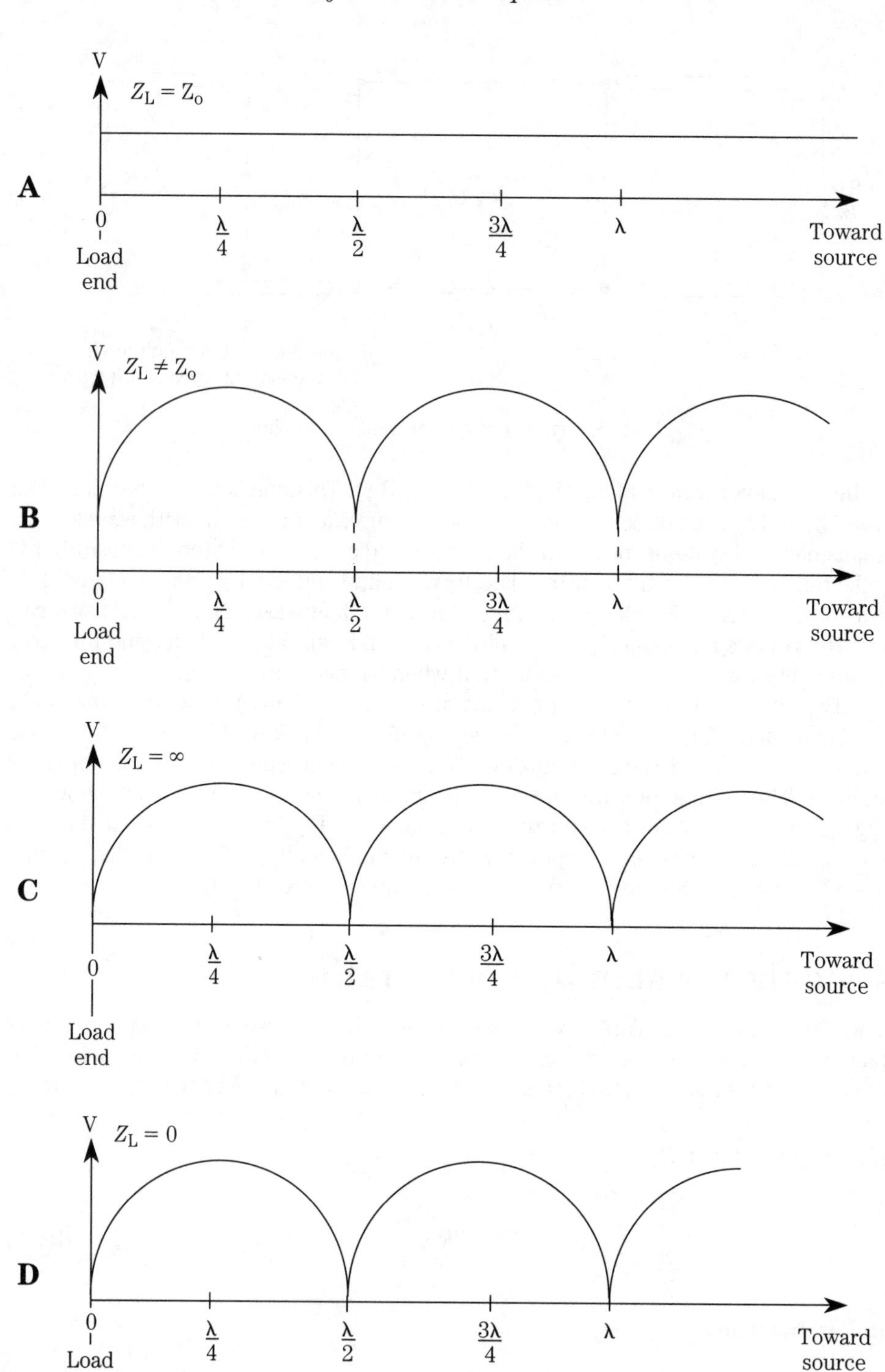

22-3 Voltage along line vs. line electrical length: A) flat line with matched impedance, B) unequal impedances, C) open load, D) shorted load.

We can also measure the forward and reflected power, and calculate the VSWR from those readings:

$$VSWR = \frac{1 + \sqrt{(P_r/P_f)}}{1 - \sqrt{(P_r/P_f)}} \quad \textbf{[22.3]}$$

Where:

$VSWR$ is the voltage standing wave ratio
P_r is the reflected power
P_f is the forward power

If we can measure either the voltage maxima and minima, or the current maxima and minima, then we can calculate *SWR*:

$$VSWR = \frac{V_{max}}{V_{min}} \quad \textbf{[22.4]}$$

Finally, if the forward and reflected voltage components at any given point on the transmission line can be measured, then we can calculate the *VSWR* from:

$$VSWR = \frac{V_f + V_r}{V_f - V_r} \quad \textbf{[22.5]}$$

Where:

V_f is the forward voltage component
V_r is the reflected voltage

The latter equation, based on the forward and reflected voltages, is the basis for many modern VSWR and RF power meters.

Impedance bridges

We can make antenna impedance measurements using a variant of the old-fashioned Wheatstone bridge. Figure 22-4A shows the basic form of the bridge in its most generalized form. The current flowing in the meter will be zero when $(Z_1/Z_2) = (Z_3/Z_4)$. If one arm of the bridge is the antenna impedance,then we can adjust the others to make the bridge null to make the measurement. A typical example is shown in Fig. 22-4B. The antenna connected to J2 is one arm of the bridge, while R2 is a second. The value of R2 should be 50 Ω or 75 Ω, depending upon the value of the expected antenna impedance. The choice of 68 Ω is a good compromise for meters to operate on both types of antennas. The other two arms of the bridge are the reactances of C1A and C1B, which is a single differential capacitor. Tune C1 until the meter is nulled, and then read the antenna from the dial. At least one instrument allowed the technician to plug in a resistor element equal to system impedance.

Calibrating the instrument is simple. A series of noninductive, carbon composition resistors having standard values from 10 to 1000 Ω are connected across J2. The meter is then nulled, and the value of the load resistor is inscribed on the dial at the point.

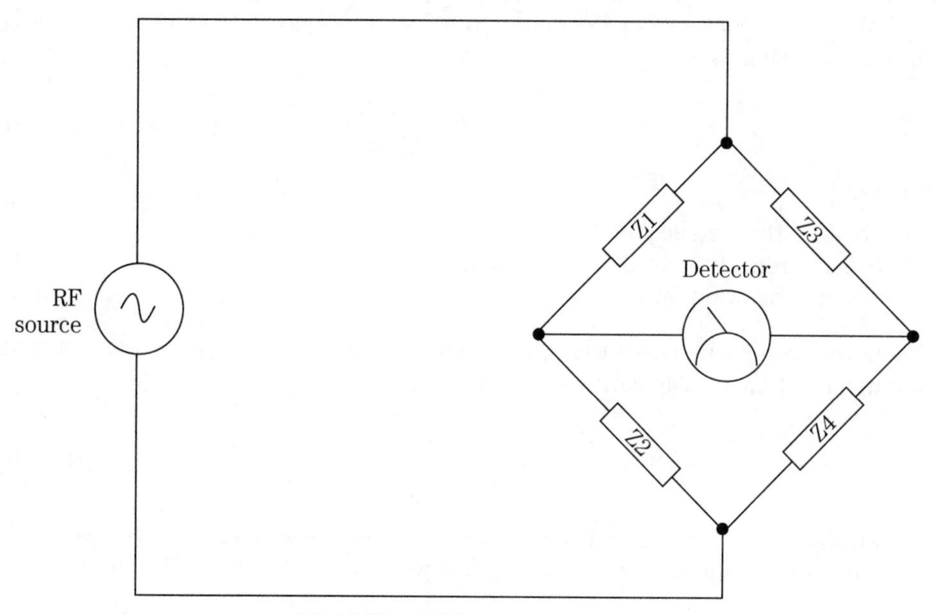

22-4A Basic Wheatstone bridge.

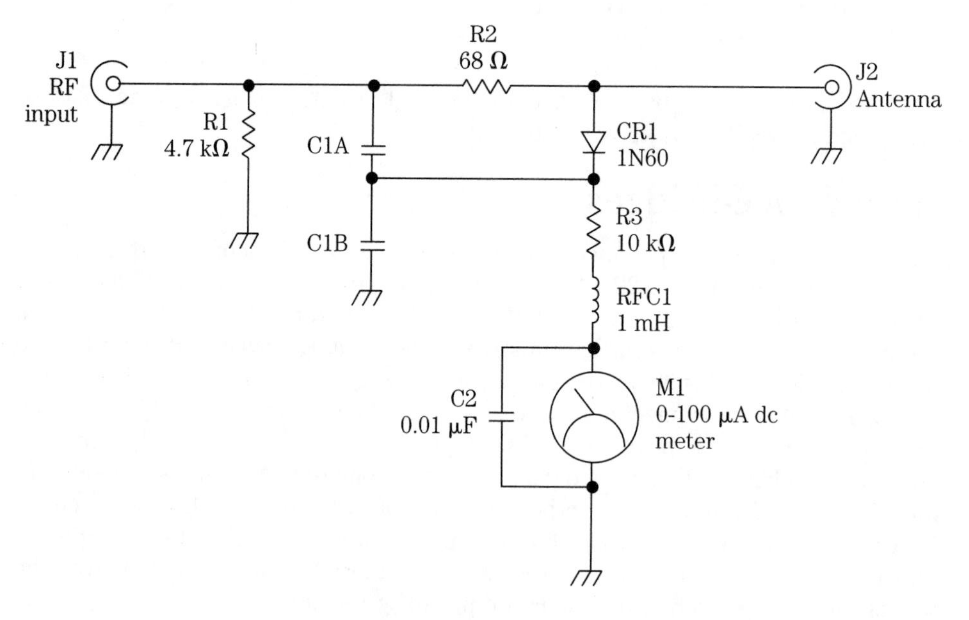

22-4B SWR bridge.

The basic circuit of Fig. 22-4B is useful only to measure the resistive component of impedance. We can modify the circuit as shown in Fig. 22-4C to account for the reactive component. An example of a commercial, handheld radio antenna impedance bridge is shown in Fig. 22-4D.

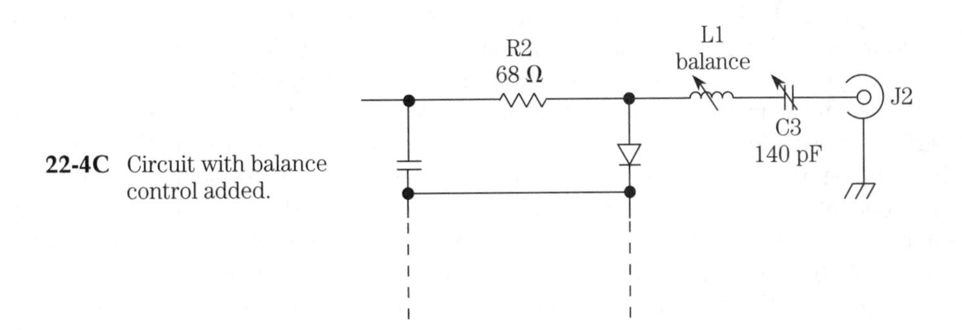

22-4C Circuit with balance control added.

The RF noise bridge

This section explores a device that was once only associated with engineering laboratories, but turns out to have applications in general communications servicing as well: the *RF noise bridge*. It is one of the most useful, low-cost, and often overlooked test instruments in our armamentarium.

Several companies have produced low-cost noise bridges: Omega-T, Palomar Engineers, The Heath Company, and, most recently, MFJ. The Omega-T and the Palomar Engineers models are shown in Fig. 22-5. The Omega-T device is a small cube with minimal dials, and a pair of BNC coax connectors (ANTENNA and RECEIVER). The dial is calibrated in ohms, and measures only the resistive component of impedance. The Palomar Engineers device is a little less eye-appealing, but does everything the Omega-T does, plus, it allows you to make a rough measurement of the reactive component of impedance. The Heath Company added their Model HD-1422 to the line-up. And, the latest entry is the MFJ-204B antenna bridge.

Over the years, some people have found the noise bridge very useful for a vari-

22-5 Commercially available amateur noise bridges.

ety of test and measurement applications, especially in the HF and low VHF regions. Those applications are not limited to the testing of antennas (which is the main job of the noise bridge). In fact, although the two-way technician (including CB) or amateur radio operator, will measure antennas, tuned circuits, and resonant cavities with the device, consumer electronics technicians will find other applications.

Figure 22-6 shows the circuit of a noise bridge instrument. The bridge consists of four arms. The inductive arms (L1b and L1c) form a trifilar wound transformer over a ferrite core with L11a, so signal applied to L1a is injected into the bridge circuit. The measurement consists of a series circuit of a 200-Ω potentiometer and a 120-pF variable capacitor. The potentiometer sets the range (0 to 200 Ω) of the resistive component of measured impedance, while the capacitor sets the reactance component. Capacitor C2 in the UNKNOWN arm of the bridge is used to balance the measurement capacitor. With C2 in the circuit, the bridge is balanced when C is approximately in the center of its range. This arrangement accommodates both inductive and capacitive reactances, which appear on either side of the "zero" point (i.e., the mid-range capacitance of C). When the bridge is in balance, the settings of R and C reveal the impedance across the UNKNOWN terminal.

A reverse-biased zener diode (zeners normally operate in the reverse bias mode) produces a large amount of noise because of the avalanche process inherent in zener operation. Although this noise is a problem in many other applications, in a noise bridge it is highly desirable: the richer the noise spectrum the better the performance. The spec-

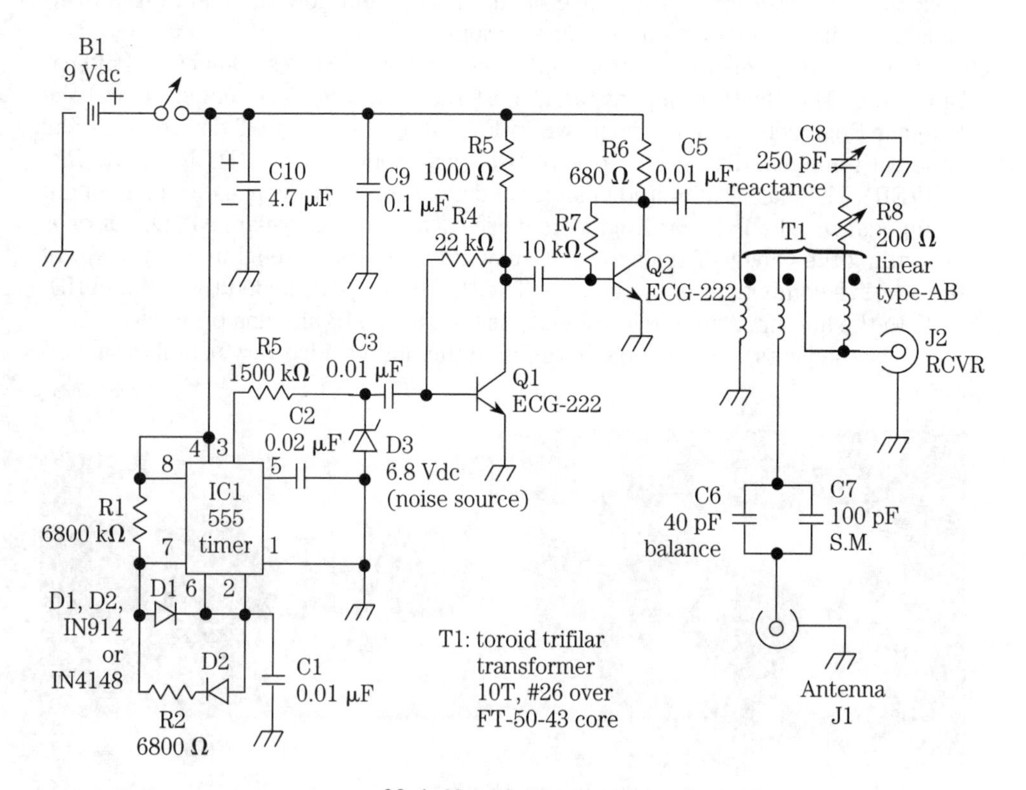

22-6 Noise bridge circuit.

trum is enhanced because of the 1-kHz square-wave modulator that chops the noise signal. An amplifier boosts the noise signal to the level needed in the bridge circuit.

The detector used in the noise bridge is a tunable receiver covering the frequencies of interest. The preferable receiver uses an AM demodulator, but both CW (morse code) and SSB receivers will do in a pinch. The quality of the receiver depends entirely on the precision with which you need to know the operating frequency of the device under test.

Adjusting antennas

Perhaps the most common use for the antenna noise bridge is finding the impedance and resonant points of an HF antenna. Connect the RECEIVER terminal of the bridge to the ANTENNA input of the HF receiver through a short length of coaxial cable as shown in Fig. 22-7. The length should be as short as possible, and the characteristic impedance should match that of the antenna feedline. Next, connect the coaxial feedline from the antenna to the ANTENNA terminals on the bridge. You are now ready to test the antenna.

Set the noise bridge resistance control to the antenna feedline impedance (usually 50 to 75 Ω for most common antennas). Set the reactance control to mid-range (zero). Next, tune the receiver to the *expected* resonant frequency (F_{exp}) of the antenna. Turn the noise bridge *on*, and look for a noise signal of about S9 (will vary on different receivers, and if—in the unlikely event that the antenna is resonant on the expected frequency—you will find yourself right in the middle of the null.

Adjust the *Resistance* control (R) on the bridge for a null, i.e., minimum noise as indicated by the S-meter. Next, adjust the *Resistance* control (C) for a null. Repeat the adjustments of the R and C controls for the deepest possible null, as indicated by the lowest noise output on the S-meter (there is some interaction between the two controls).

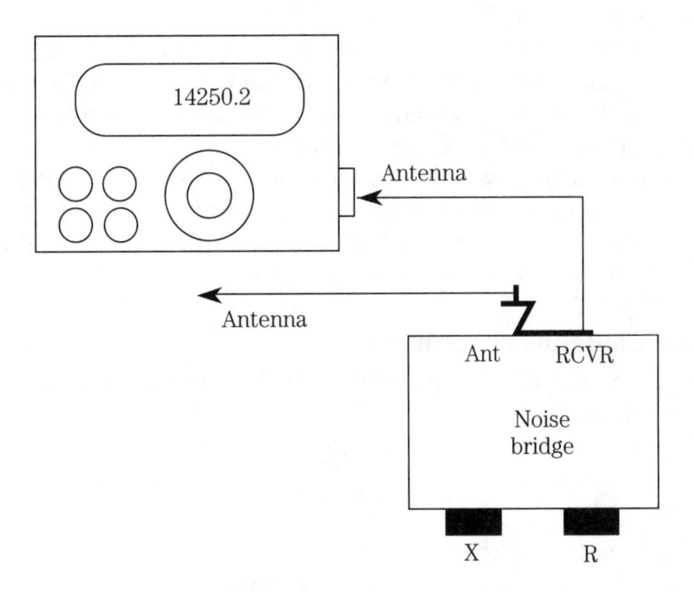

22-7 Connection of noise bridge.

A perfectly resonant antenna will have a reactance reading of zero ohms, and a resistance of 50 to 75 Ω. Real antennas might have some reactance (the less the better), and a resistance that is somewhat different from 50 to 75 Ω. Impedance-matching methods can be used to transform the actual resistive component to the 50- or 75-Ω characteristic impedance of the transmission line. The results to expect:

1. If the resistance is close to zero, then suspect that there is a short circuit on the transmission line; and an open circuit, if the resistance is close to 200 Ω.
2. A reactance reading on the X_L side of zero indicates that the antenna is too long, while a reading on the X_C side of zero indicates an antenna that is too short.

An antenna that is too long or too short should be adjusted to the correct length. To determine the correct length, we must find the actual resonant frequency, F_r. To do this, reset the *Reactance* control to zero, and then *slowly* tune the receiver in the proper direction—down-band for too-long and up-band for too-short—until the null is found. On a high *Q* antenna, the null is easy to miss if you tune too fast. Don't be surprised if that null is out of band by quite a bit. The percentage of change is given by dividing the expected resonant frequency (F_{exp}), by the actual resonant frequency (F_r), and multiply by 100: Change =

$$\frac{(F_{exp} \times 100\%)}{F_r} \qquad \textbf{[22.6]}$$

Connect the antenna, noise bridge and the receiver in the same manner as above. Set the receiver to the expected resonant frequency (i.e., approximately 468/*F* for half-wavelength types, and 234/*F* for quarter-wavelength types). Set the resistance control to 50 Ω or 75 Ω, as appropriate for the normal antenna impedance and the transmission line impedance. Set the reactance control to zero. Turn the bridge on and listen for the noise signal.

Slowly rock the reactance control back and forth to find on which side of zero the null appears. Once the direction of the null is determined, set the reactance control to zero, and then tune the receiver toward the null direction (downband if null is on X_L side, and upband if it is on the X_C side of zero).

A less-than-ideal antenna will not have exactly 50 or 75 Ω impedance, so some adjustment of R and C to find the deepest null is in order. You will be surprised how far off some dipoles and other forms of antennas can be if they are not in "free space," (i.e., if they are close to the earth's surface).

Nonresonant antenna adjustment

We can operate antennas on frequencies other than their resonant frequency if we know the impedance:

$$X_C = X = \frac{55}{68 - C} - 2340 \qquad \textbf{[22.7]}$$

or,

$$X_L = X = 2340 - \frac{159155}{68 + C} \qquad \textbf{[22.8]}$$

Now, plug "X" calculated from Eq. 22.7 or 22.8 into $X_f = X/F$ where F is the desired frequency in MHz.

Other jobs

The noise bridge can be used for a variety of jobs. We can find the values of capacitors and inductors, the characteristics of series- and parallel-tuned resonant circuits, and for adjusting transmission lines.

Transmission line measurements Some antennas and non-noise measurements require antenna feedlines that are either a quarter wavelength or half wavelength at some specific frequency. In other cases, a piece of coaxial cable of specified length is required for other purposes: for instance the dummy load used to service depth sounders is nothing but a long piece of shorted coax that returns the echo at a time interval that corresponds to a specific depth. We can use the bridge to find these lengths as follows:

1. Connect a short-circuit across the UNKNOWN terminals and adjust R and X for the best null at the frequency of interest (note: both will be near zero);
2. Remove the short circuit:
3. Connect the length of transmission line to the UNKNOWN terminal—it should be longer than the expected length;
4. For quarter-wavelength lines, shorten the line until the null is very close to the desired frequency. For half-wavelength lines, do the same thing, except that the line must be shorted at the far end for each trial length.

The velocity factor of a transmission line (usually designated by the letter V in equations) is a decimal fraction that tells us how fast the radio wave propagates along the line relative to the speed of light in free space. For example, foam dielectric coaxial cable is said to have a velocity factor of $V = 0.80$. This number means that the signals in the line travel at a speed 0.80 (or 80%) of the speed of light.

Because all radio wavelength formulas are based on the velocity of light, you need the V value to calculate the physical length needed to equal any given electrical length. For example, a half-wavelength piece of coax has a physical length of $(492 \times V)/F_{\text{MHz}}$ feet. Unfortunately, the real value of V is often a bit different from the published value. You can use the noise bridge to find the actual value of V for any sample of coaxial cable as follows:

1. Select a convenient length of the coax more than 12 feet in length and install a PL-259 RF connector (or other connector compatible with your instrument) on one end, and short-circuit the other end.
2. Accurately measure the physical length of the coax in feet; convert the "remainder" inches to a decimal fraction of one foot by dividing by 12 (e.g., 32' 8" = 32.67' because 8"/12" = 0.67). Alternatively, cut off the cable to the nearest foot and reconnect the short circuit.
3. Set the bridge RESISTANCE and REACTANCE controls to zero.
4. Adjust the monitor receiver for deepest null. Use the null frequency to find the velocity factor $V = FL/492$, where V is the velocity factor (a decimal fraction); F is the frequency in MHz; and L is the cable length in feet.

Tuned circuit measurements An inductor/capacitor (LC) tuned "tank" circuit is the circuit equivalent of a resonant antenna, so there is some similarity between the two measurements. You can measure resonant frequency with the noise bridge to within ±20 percent (or better if care is taken). This accuracy might seem poor, but it is better than you can usually get with low-cost signal generators, dip meters, absorption wavemeters and the like.

A *series tuned circuit* exhibits a low impedance at the resonant frequency, and a high impedance at all other frequencies. Start the measurement by connecting the series tuned circuit under test across the UNKNOWN terminals of the dip meter. Set the RESISTANCE control to a low resistance value, close to zero ohms. Set the REACTANCE control at mid-scale (zero mark). Next, tune the receiver to the expected null frequency, and then tune for the null. Make sure that the null is at its deepest point by rocking the R and X controls for best null. At this point, the receiver frequency is the resonant frequency of the tank circuit.

A *parallel resonant circuit* exhibits a high impedance at resonance, and a low impedance at all other frequencies. The measurement is made in exactly the same manner as for the series resonant circuits, except that the connection is different. Figure 22-8 shows a two-turn link coupling that is needed to inject the noise signal into the parallel resonant tank circuit. If the inductor is the toroidal type, then the link must go through the hole in the doughnut-shaped core and then connects to the UNKNOWN terminals on the bridge. After this, proceed exactly as you would for the series tuned tank measurement.

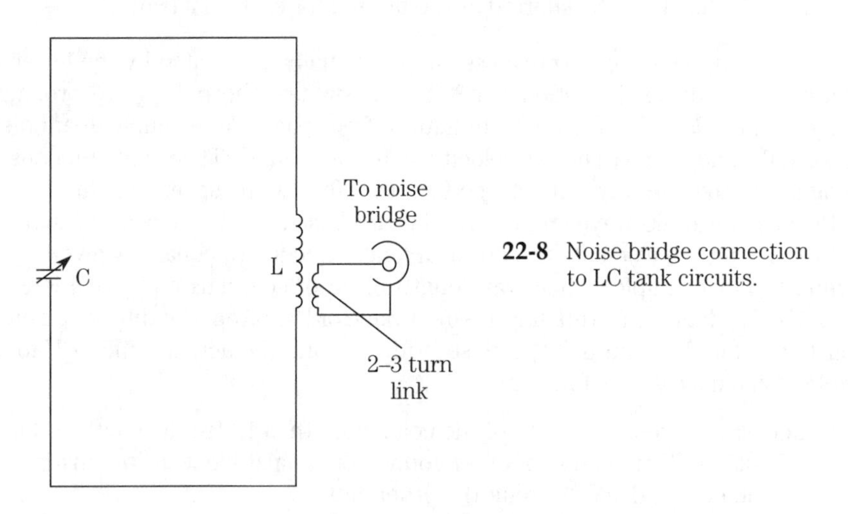

22-8 Noise bridge connection to LC tank circuits.

Capacitance and inductance measurements The bridge requires a 100-pF silver mica test capacitor and a 4.7-μH test inductor, which are used to measure inductance and capacitance, respectively. The idea is to use the test components to form a series-tuned resonant circuit with an unknown component. If you find a resonant frequency, then you can calculate the unknown value. In both cases, the series tuned circuit is connected across the UNKNOWN terminals of the dip meter, and the series-tuned procedure is followed.

To measure inductance, connect the 100-pF capacitor in series with the unknown coil across the UNKNOWN terminals of the dip meter. When the null frequency is found, find the inductance from:

$$L = \frac{253}{F^2} \quad \textbf{[22.9]}$$

Where:

L is the inductance in microhenry (μH)
F is the frequency in megahertz (MHz)

Connect the test inductor across the UNKNOWN terminals in series with the unknown capacitance. Set the RESISTANCE control to zero, tune the receiver to 2 MHz, and readjust the REACTANCE control for null. Without readjusting the noise bridge control, connect the test inductor in series with the unknown capacitance and retune the receiver for a null. Capacitance can now be calculated from:

$$C = \frac{5389}{F^2} \quad \textbf{[22.10]}$$

Where:

C is in picofarads (pF)
F is in megahertz (MHz)

Dip oscillators

One of the most common instruments for determining the resonant frequency of an antenna is the so-called "dip oscillator," or "dip meter." Originally called the *grid dip-meter*, the basis for this instrument is the fact that its output energy can be absorbed by a nearby resonant circuit (or antenna, which is a electrically resonant LC tank circuit). When the inductor of the dip oscillator (see Fig. 22-9) is brought into

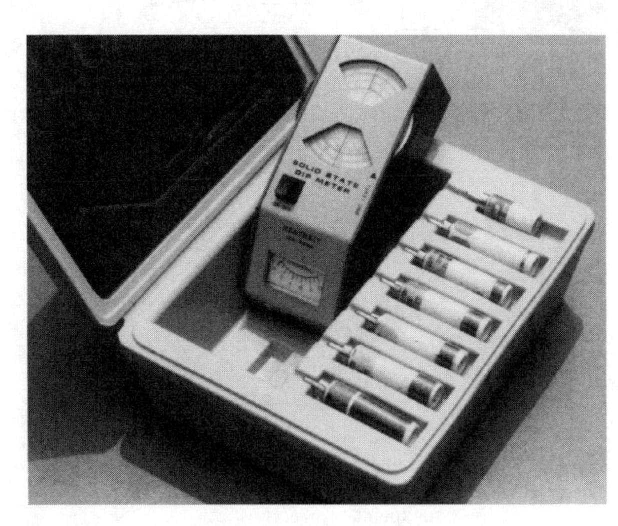

22-9 Dip meter.

close proximity to a resonant tank circuit, and the oscillator is operating on the resonant frequency, then a small amount of energy is transferred. This energy loss shows up on the meter pointer as a slight "dipping" action. The dip is extremely sharp, and is easily missed if the meter frequency dial is tuned too rapidly.

Antennas are resonant circuits, and can be treated in a manner similar to LC tank circuits. Figure 22-10A shows one way to couple the dip oscillator to a vertical antenna radiator. The inductor of the dipper is brought into close proximity to the base of the radiator. Figure 22-10B shows the means for coupling the dip oscillator

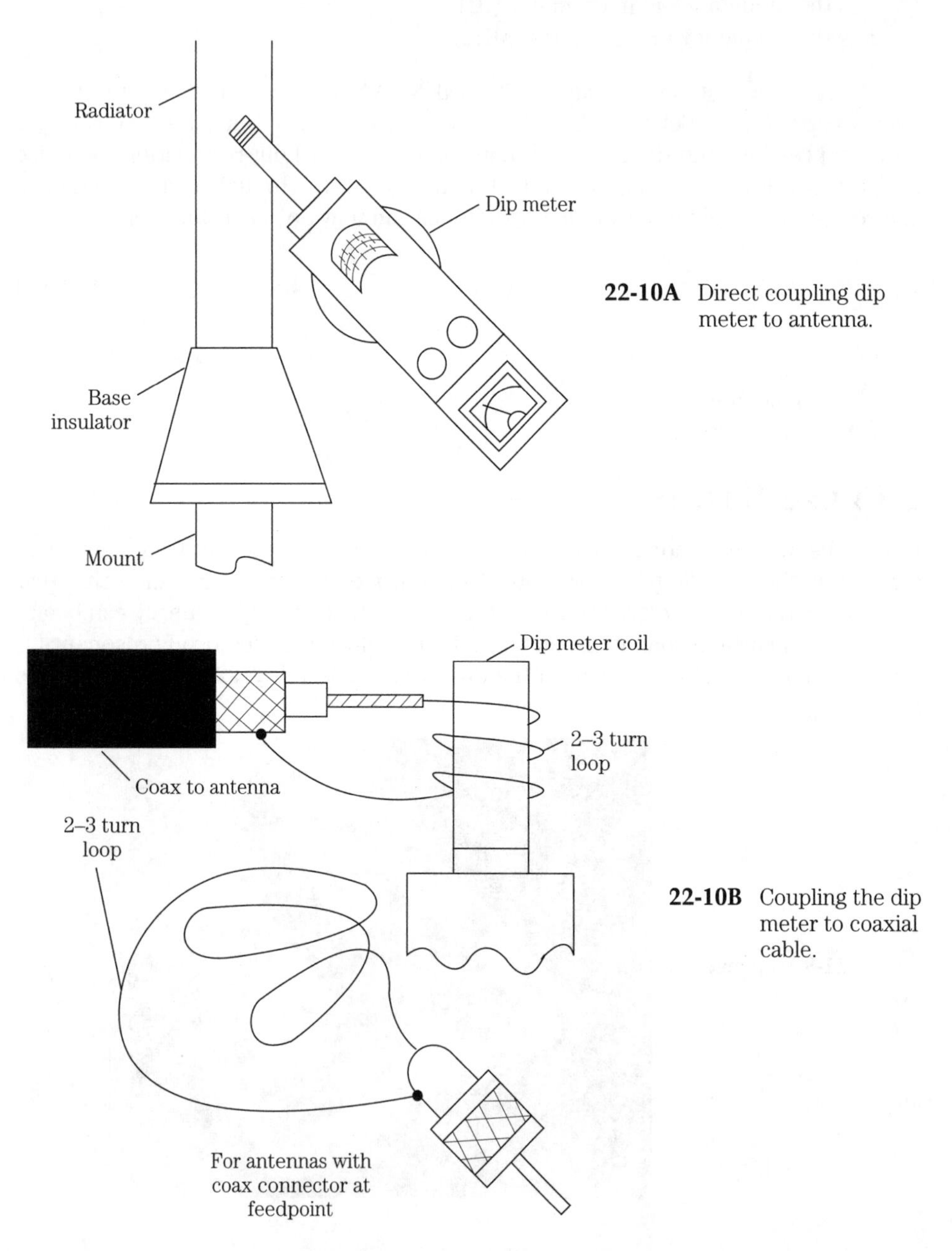

22-10A Direct coupling dip meter to antenna.

22-10B Coupling the dip meter to coaxial cable.

to systems where the radiator is not easily accessed (as when the antenna is still erected). We connect a small two or three turn loop to the transmitter end of the transmission line, and then bring the inductor of the dipper close to it. A better way is to connect the loop directly to the antenna feedpoint.

There are two problems with dip meters that must be recognized in order to best use the instrument. First, the dip is very sharp. It is easy to tune past the dip and not even see it. To make matters worse, it is normal for the meter reading to drop off gradually from one end of the tuning range to the other. But if you tune very slowly, then you will notice a very sharp dip when the resonant point is reached.

The second problem is the dial calibration. The dial gradations of inexpensive dip meters are too close together, and are often erroneous. It is better to monitor the output of the dip oscillator on a receiver, and depend upon the calibration of the receiver for data.

Selecting and using RF wattmeters and antenna VSWR meters

A key instrument required in checking the performance of, or troubleshooting, radio transmitters is the RF power meter (or "wattmeter"). These instruments measure the output power of the transmitter, and display the result in watts, or some related unit. Closely related to RF wattmeters is the antenna VSWR meter. These instruments also examine the output of the transmitter, and give a relative indication of output power. They can be calibrated to display the dimensionless units of *voltage standing wave ratio* (VSWR). Many modern instruments, a couple of which will be covered as examples in this chapter, combine both RF power and VSWR measurement capabilities.

Measuring RF power

Measuring RF power has traditionally been notoriously difficult, except perhaps in the singular case of continuous wave (CW) sources that produce nice, well-behaved sine waves. Even in that limited case, however, some measurement methods are distinctly better than others.

The peak voltage of a waveform is 100 V (i.e., peak-to-peak 200 V). Given that the CW waveform is sinusoidal, we know that the RMS voltage is 0.707 V. The output power is related to the RMS voltage across the load by:

$$P = \frac{(V_{rms})^2}{Z_o} \qquad \textbf{[22.11]}$$

Where:

P is the power in Watts
V_{rms} is the RMS potential in Volts
Z_o is the load impedance in ohms

If we assume a load impedance of 50 Ω, then we can state that the power in our hypothetical illustration waveform is 100 W.

We can measure power on unmodulated sinusoidal waveforms, by measuring either the RMS or peak values of either voltage or current, assuming that a constant value resistance load is present. But the problem becomes more complex on modulated signals. The various power readings on a Bird Model 4311 peak power meter, the peak (PEP) and average powers, vary markedly with modulation type.

One of the earliest forms of practical RF power measurement was the thermocouple RF ammeter (see Fig. 22-11). This instrument works by dissipating a small amount of power in a small resistance inside the meter, and then measuring the heat generated with a thermocouple. A dc current meter monitors the output of the thermocouple device, and indicates the level of current flowing in the heating element. Because it works on the basis of the power dissipated heating a resistance, *a thermocouple RF ammeter is inherently an RMS-reading device*. Because of this feature it is very useful for making average power measurements. If we know the RMS current and the resistive component of the load impedance, and if the reactive component is zero or very low, then we can determine RF power from the familiar expression:

$$P = I^2 \times R_i \qquad [22.12]$$

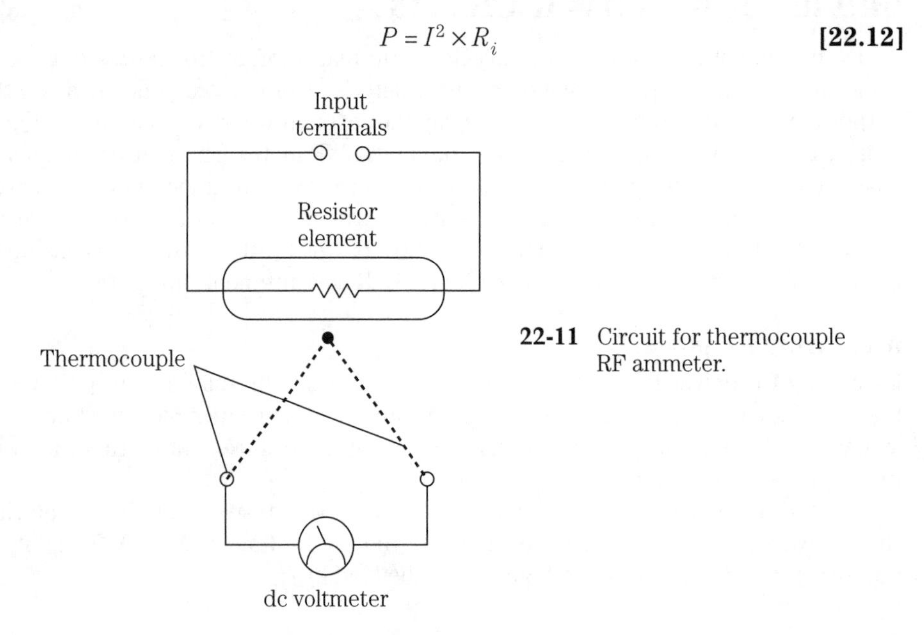

22-11 Circuit for thermocouple RF ammeter.

There is, however, a significant problem that keeps thermocouple RF ammeters from being universally used in RF power measurement: those instruments are highly frequency dependent. Even at low frequencies, it is recommended that the meters be mounted on insulating material with at least ⅝-inch spacing between the meter and its metal cabinet. Even with that precaution, however, there is a strong frequency dependence that renders the meter less useful at higher frequencies. Some meters are advertised to operate into the low-VHF region, but a note of caution is necessary. That recommendation requires a copy of the calibrated frequency response curve for that specific meter, so that a correction factor can be added (or subtracted) from the reading. At 10 MHz and higher, the readings of the thermocouple RF ammeter must be taken with a certain amount of skepticism unless the original calibration chart is available.

We can also measure RF power by measuring the voltage across the load resistance (see Fig. 22-12). In the circuit of Fig. 22-12, the RF voltage appearing across the load is scaled downward to a level compatible with the voltmeter by the resistor voltage divider (R2/R3). The output of this divider is rectified by CR1, and filtered to dc by the action of capacitor C2.

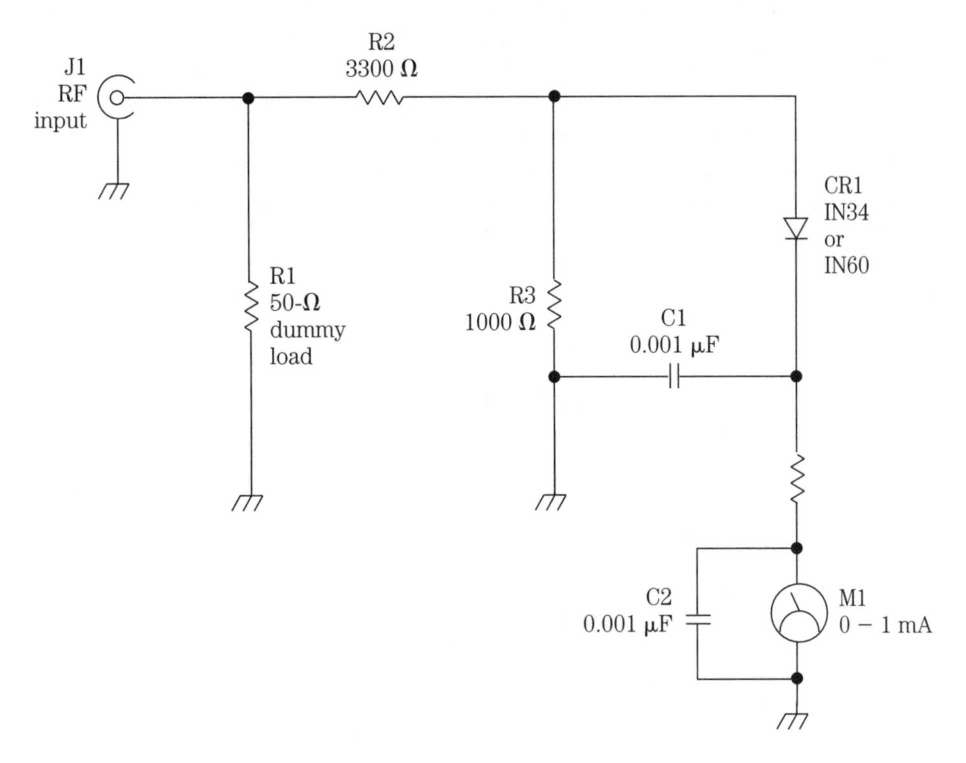

22-12 RF voltmeter "wattmeter."

The method of measuring the voltage in a simple diode voltmeter is valid only if the RF signal is unmodulated and has a sinusoidal waveshape. While these criteria are met in many transmitters, they are not universal. If the voltmeter circuit is peak reading, as in Fig. 22-12, then the peak power is:

$$P = \frac{(V_o)^2}{\text{R1}} \qquad \textbf{[22.13]}$$

The average power is then found by multiplying the peak power by 0.707. Some meter circuits include voltage dividers that precede the meter and thereby convert the reading to RMS, and thus convert the power to average power. Again, it must be stressed that terms like "RMS," "average," and "peak" have meaning only when the input RF signal is both unmodulated and sinusoidal. Otherwise, the readings are meaningless unless calibrated against some other source.

It is also possible to use various bridge methods for measurement of RF power. Figure 22-13 shows a bridge set up to measure both forward and reverse power. This circuit was once popular for VSWR meters. There are four elements in this quasi-Wheatstone bridge circuit. R1, R2, R3, and the antenna impedance (connected to the bridge at J2). If R_{ant} is the antenna resistance, then we know that the bridge is in balance (i.e., the null condition) when the ratios R_1/R_2 and R_3/R_{ant} are equal. In an ideal situation, resistor R3 will have a resistance equal to R_{ant}, but that might overly limit the usefulness of the bridge. In some cases, therefore, the bridge will use a compromise value such as 67 Ω for R3. Such a resistor will be usable on both 50-Ω and 75-Ω antenna systems with only small errors. Typically, these meters are designed to read relative power level, rather than the actual power.

An advantage of this type of meter is that we can get an accurate measurement of VSWR by proper calibration. With the switch in the FORWARD position, and RF power applied to J1 ("XMTR"), potentiometer R6 is adjusted to produce a full-scale deflection on meter M1. When the switch is then set to the REVERSE position, the meter will read reverse power relative to the VSWR. An appropriate "VSWR" scale is provided.

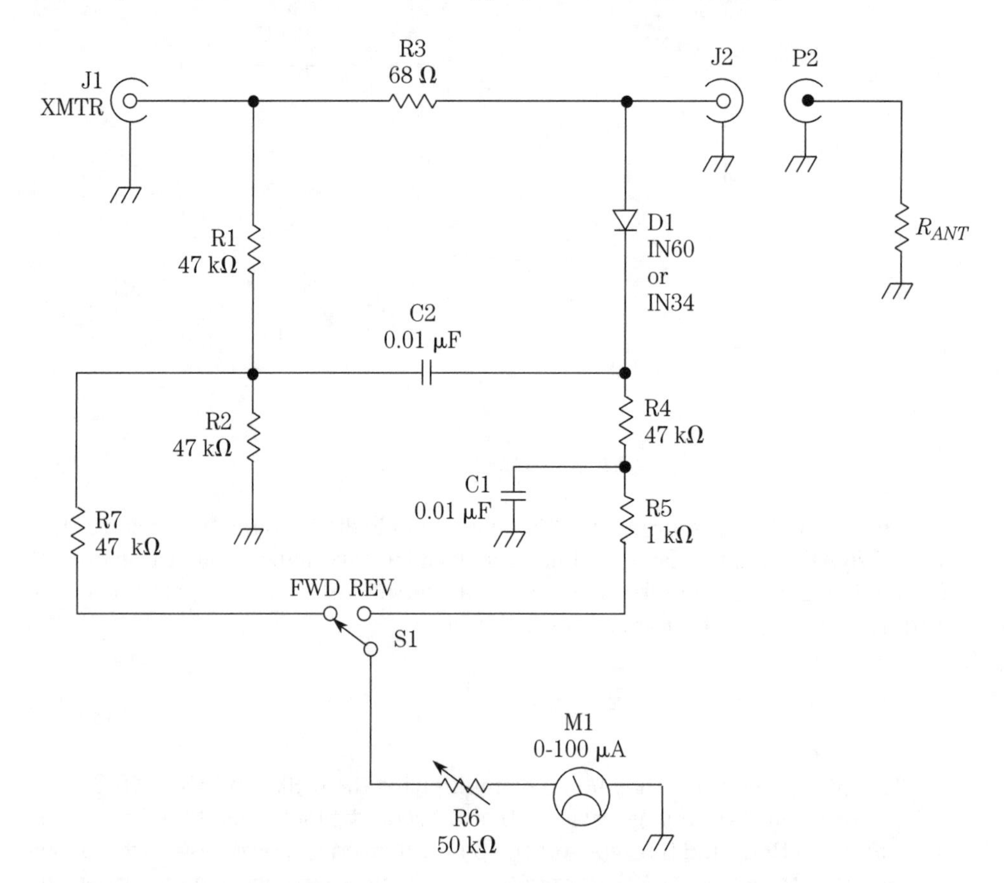

22-13 A bridge arrangement to measure both forward and reverse power.

A significant problem with the bridge of Fig. 22-13 is that it cannot be left in the circuit while transmitting because it dissipates a considerable amount of RF power in the internal resistances. These meters, during the time when they were popular, were provided with switches that bypassed the bridge when transmitting. The bridge was only in the circuit when making a measurement.

An improved bridge circuit is the capacitor/resistor bridge in Fig. 22-14; this circuit is called the "Micromatch" bridge. Immediately, we see that the Micromatch is improved over the conventional bridge because it uses only 1 Ω in series with the line (R_i). This resistor dissipates considerably less power than the resistance used in the previous example. Because of this low-value resistance, we can leave the Micromatch in the line while transmitting. Recall that the ratios of the bridge arms must be equal for the null condition to occur. In this case, the capacitive reactance ratio of C_1/C_2 must match the resistance ratio R_1/R_{ant}. For a 50-Ω antenna, the ratio is 1/50, and for 75-Ω antennas it is 1/75 (or, for the compromise situation, 1/68. The small-value trimmer capacitor (C2) must be adjusted for a reactance ratio with C1 of 1/50, 1/75, or 1/68, depending upon how the bridge is set up.

The sensitivity control can be used to calibrate the meter. In one version of the Micromatch, there are three power ranges (10 watts, 100 watts, and 1000 watts). Each range has its own sensitivity control, and these are switched in and out of the circuit as needed.

The Monomatch bridge circuit in Fig. 22-15 is the instrument of choice for HF and low-VHF applications. In the Monomatch design, the transmission line is seg-

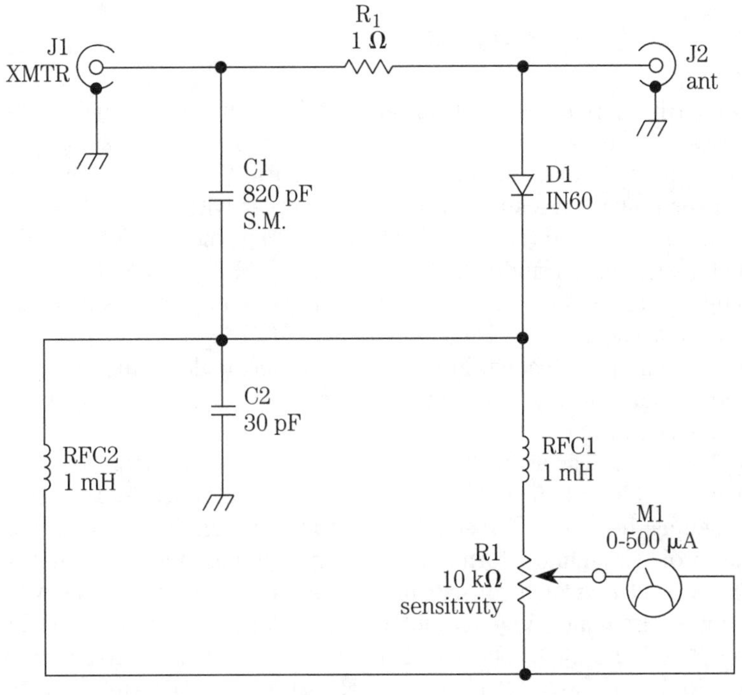

22-14 Micromatch wattmeter.

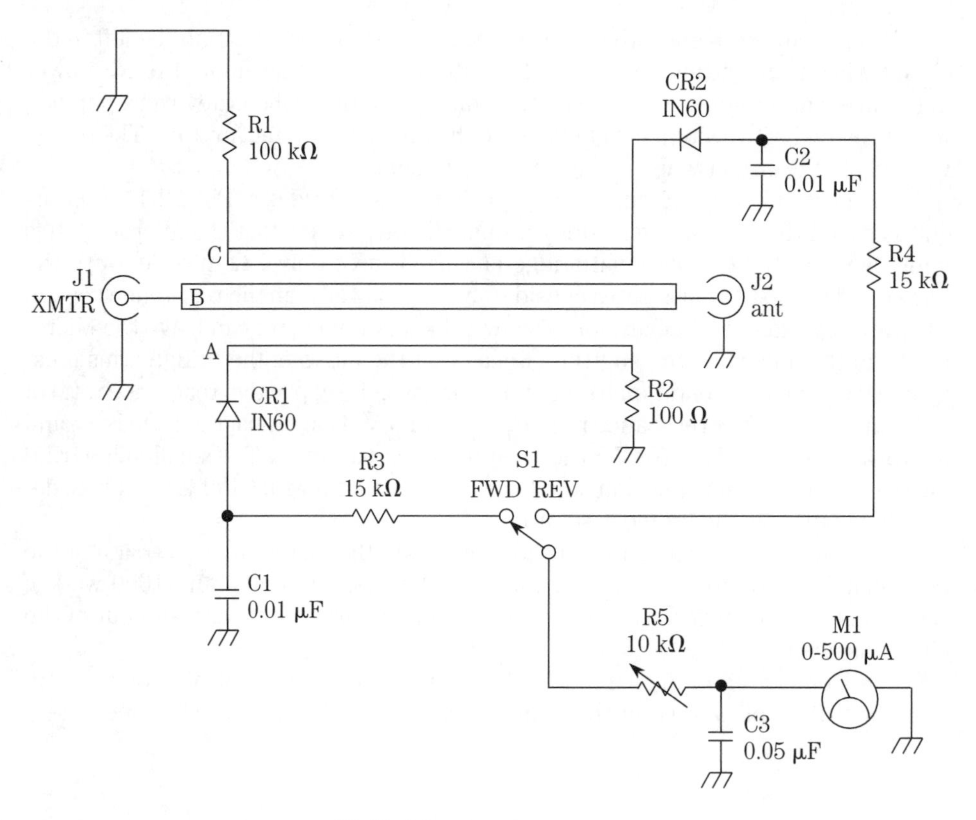

22-15 Monomatch wattmeter.

ment B, while RF sampling elements are formed by segments A and C. Although the original designs were based on a coaxial cable sensor, later versions used either printed circuit foil transmission line segments or parallel brass rods for A, B, and C.

The sensor unit is basically a directional coupler with a detector element for both forward and reverse directions. For best accuracy, diodes CR1 and CR2 should be matched, as should R1 and R2. The resistance of R1 and R2 should match the transmission line surge impedance, although in many instruments a 68-Ω compromise resistance is used.

The particular circuit shown in Fig. 22-15 uses a single dc meter movement to monitor the output power. Many modern designs use two meters (one each for forward and reverse power).

One of the latest designs in VSWR meter sensors is the current transformer assembly shown in Fig. 22-16. In this instrument, a single-turn ferrite toroid transformer is used as the directional sensor. The transmission line passing through the hole in the toroid "doughnut" forms the primary winding of a broadband RF transformer. The secondary, which consists of 10 to 40 turns of small enamel wire, is connected to a measurement bridge circuit (C_1 + C_2 + *load*) with a rectified dc output.

Figures 22-17 and 22-18 show instruments based on the current transformer technique. Shown in Fig. 22-17 is the Heath model HM-102 high-frequency VSWR/Power meter. The sensor is a variant on the current transformer method. This

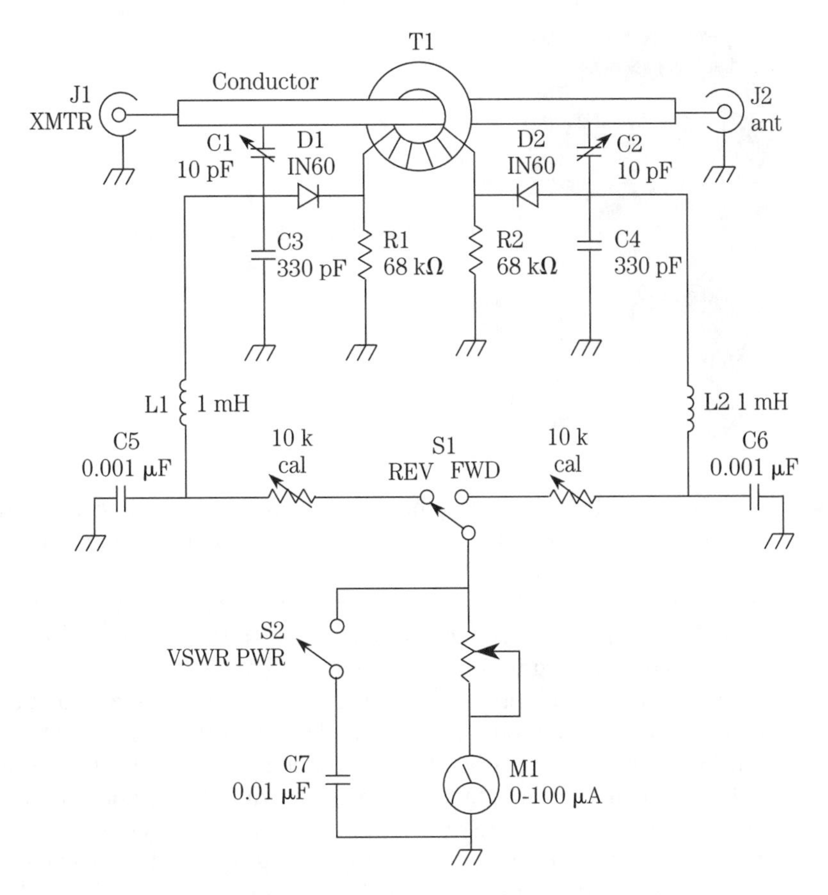

22-16 Current transformer wattmeter.

22-17 Amateur radio RF wattmeter.

instrument measures both forward and reflected power, and it can be calibrated to measure VSWR.

The Bird Model 43 Thruline RF wattmeter shown in Fig. 22-18 has for years been one of the industry standards in communications service work. Although it is slightly more expensive than lesser instruments, it is also versatile, and it is accurate and rugged. The Thruline meter can be inserted into the transmission line of an an-

22-18 Bird Model 43 RF wattmeter.

tenna system with so little loss, that it may be left permanently in the line during normal operations. The Model 43 Thruline is popular with landmobile and marine radio technicians.

The heart of the Thruline meter is the directional coupler transmission line assembly shown in Fig. 22-19A; it is connected in series with the antenna, or dummy load, transmission line. The plug-in directional element can be rotated 180 degrees to measure both forward and reverse power levels. A sampling loop and diode detector are contained within each plug-in element. The main RF barrel is actually a special coaxial line segment with a 50-Ω characteristic impedance. The Thruline sensor works due to the mutual inductance between the sample loop and center conductor of the coaxial element. Figure 22-19B shows an equivalent circuit. The output voltage from the sampler (e) is the sum of two voltages, e_r and e_m. Voltage e_r is created by the voltage divider action of R and C on transmission line voltage E. If R is much less then X_C, then we may write the expression for e_r below:

$$e_r = \frac{RE}{X_C} = RE(j\,\omega\,\mathrm{C}) \qquad \textbf{[22.14]}$$

Voltage e_m, on the other hand, is due to mutual inductance, and is expressed by:

$$e_m = I(j\,\omega) \pm \mathrm{M} \qquad \textbf{[22.15]}$$

We now have the expression for both factors that contribute to the total voltage e. We know that:

$$e = e_r + e_m \qquad \textbf{[22.16]}$$

so, by substitution,

$$e = j\omega M(\left(\frac{E}{Z_o}\right) \pm I) \qquad \textbf{[22.17]}$$

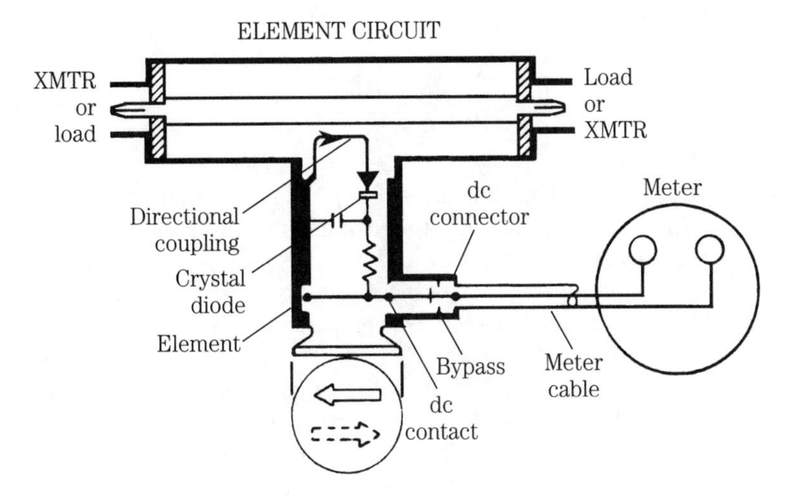

22-19A Thruline sensor circuit.

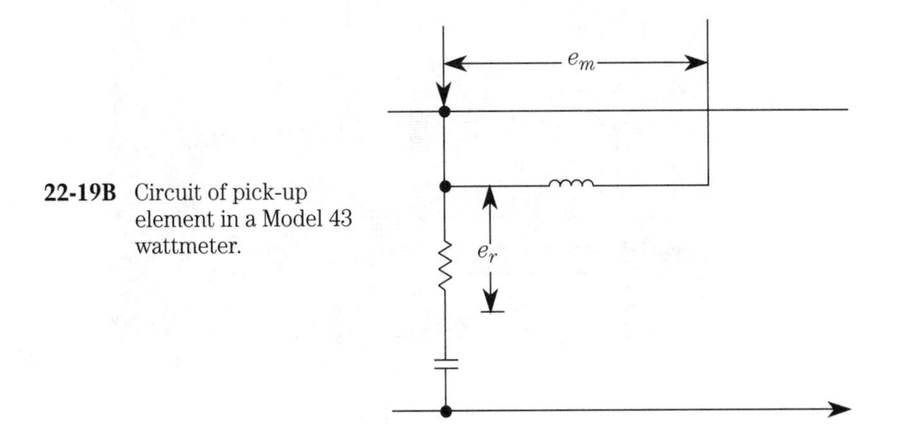

22-19B Circuit of pick-up element in a Model 43 wattmeter.

By recognizing that, at any given point in a transmission line, E is the sum of the forward (E_f) and reflected (E_r) voltages, and that the line current is equal to:

$$I = \frac{E_f}{Z_o} - \frac{E_r}{Z_o} \qquad \textbf{[22.18]}$$

Where Z_o is the transmission line impedance.

We may specify e in the forms:

$$e = \frac{j\,\omega M(2E_f)}{Z_o} \qquad \textbf{[22.19]}$$

and

$$e = \frac{j\,\omega\,M(2E_r)}{Z_o} \qquad [22.20]$$

The output voltage e of the coupler then, is proportional to the mutual inductance and frequency (by virtue of $j\omega M$). But the manufacturer terminates R in a capacitive reactance, so the frequency dependence is lessened (see Fig. 22-19C). Each element is custom calibrated, therefore, for a specific frequency and power range. Beyond the specified range for any given element, however, performance is not guaranteed. There are a large number of elements available that cover most commercial applications. The Thruline meter is not a VSWR meter, but rather a power meter. VSWR can be determined from the formula, or by using the nomograph in Fig. 22-20. Some of the Thruline series intended for very high power (Fig. 22-19D) applications use an in-line coaxial cable coupler (for broadcast style hardline) and a remote indicator.

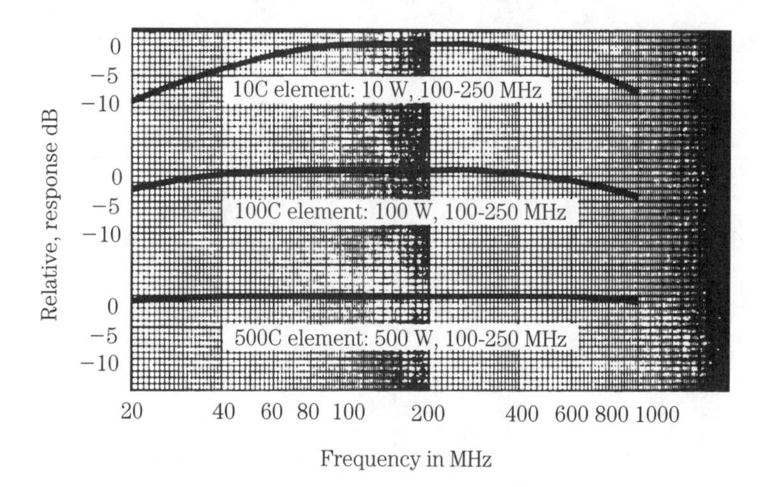

22-19C Power-frequency calibration.

22-19D High-power RF wattmeters.

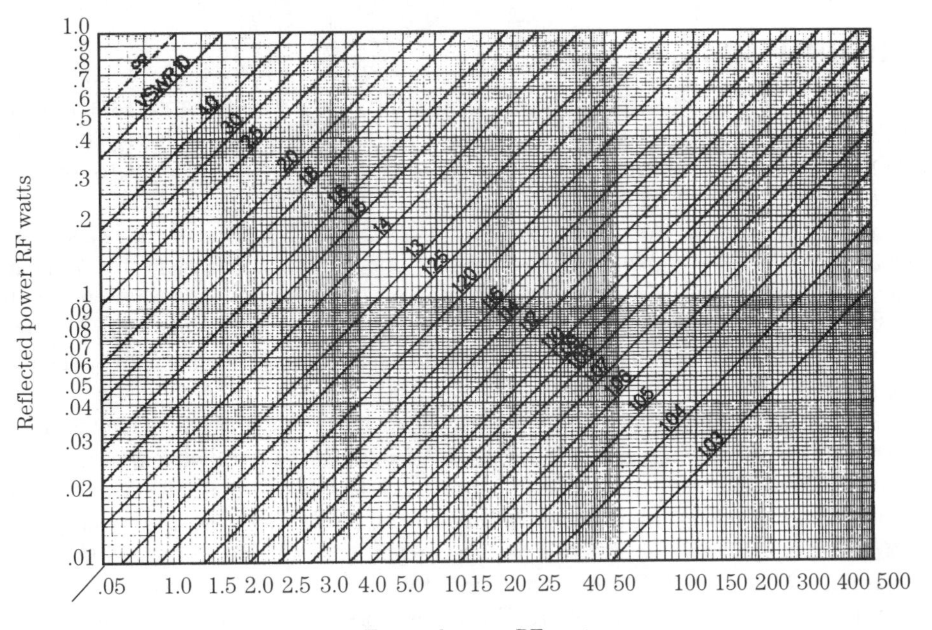

A

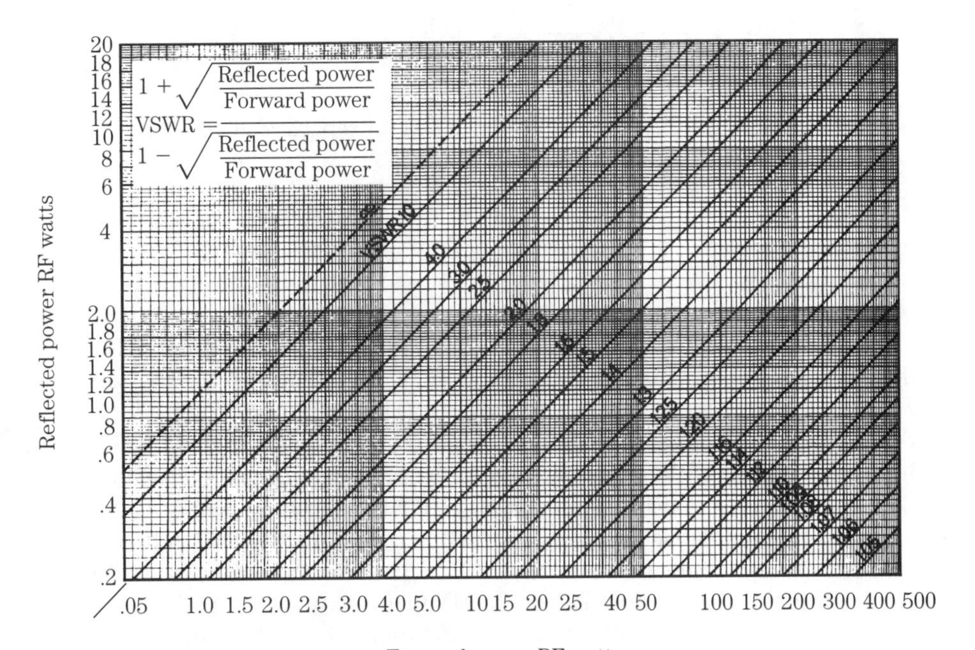

B

22-20B Forward versus reflected power graphs.

Dummy loads

A *dummy load* is a substitute antenna for making measurements and tests. In fact, British radio engineers often refer to dummy loads as "artificial aerials." There are several uses for these devices. Radio operators should use dummy loads to tune up on crowded channels, and (only then) transfer to the live antenna.

The other use is in troubleshooting antenna systems. Suppose we have a system in which the VSWR is high enough to affect the operation of the transmitter. You can disconnect each successive element and connect the dummy load to its output. If the VSWR goes down to the normal range, then the difficulty is downstream (i.e., toward the antenna). You will eventually find the bad element (which is usually the antenna itself).

Figure 22-21 shows the most elementary form of dummy load, which consists of one or more resistors connected in parallel, series, or series-parallel so that the total resistance is equal to the desired load impedance. The power dissipation is the sum of the individual power dissipations. *It is essential that noninductive resistors be used for this application.* For this reason, carbon-composition or metal-film resistors are used. For very low frequency work, it is permissible to use special counterwound wire low-inductance resistors. These resistors, however, cannot be used over a few hundred kilohertz.

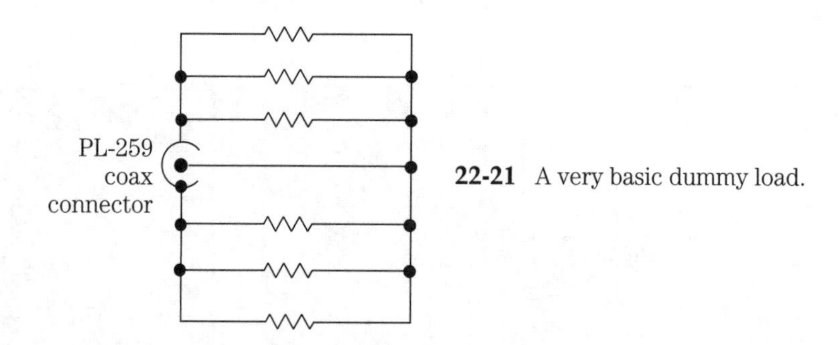

22-21 A very basic dummy load.

Several commercial dummy loads are shown in Figs. 22-22 through 22-24. The load in Fig. 22-22A is a 5-W model, and is typically used in Citizen's Band servicing. The resistor is mounted directly on a PL-259 coaxial connector. These loads typically work to about 300 MHz, although many are not really useful over about 150 MHz. A higher power version of the same type is shown in Fig. 22-22B. This device works to the low VHF region, and dissipates up to 50 W. I have used this dummy load for servicing high VHF landmobile rigs, VHF-FM marine rigs, and low-VHF landmobile rigs. Very high power loads are shown in Fig. 22-23. These devices are *Bird Electronics* "coaxial resistors," and operate to power levels up to 10 kW, 30 kW, and 40 kW. These high-power loads are cooled by flowing water through the body of the resistor, and then exhausting the heat in an air-cooled radiator.

Our final dummy load resistor is shown in Fig. 22-24. The actual resistor is shown in Fig. 22-24A, and a schematic view is shown in Fig. 22-24B. The long, high-power noninductive resistor element is rated at 50 Ω, and can dissipate 1000 W for

22-22A CB-style 5-W dummy load.

22-22B 15-W dummy load.

22-23 Termaline RF wattmeter.

22-24A Drake DL-1000 dummy load.

several minutes. If longer times, or higher powers, are anticipated, then forced air cooling is applied by adding a blower to one end of the cage. The device is modified by adding the BNC sampling jack. This jack is connected internally to either a two-turn loop, made of #22 insulated hook-up wire, or brass rods that are positioned alongside the resistor element. It will therefore pick up a sample of the signal so that it can be viewed on an oscilloscope, or used for other instrumentation purposes.

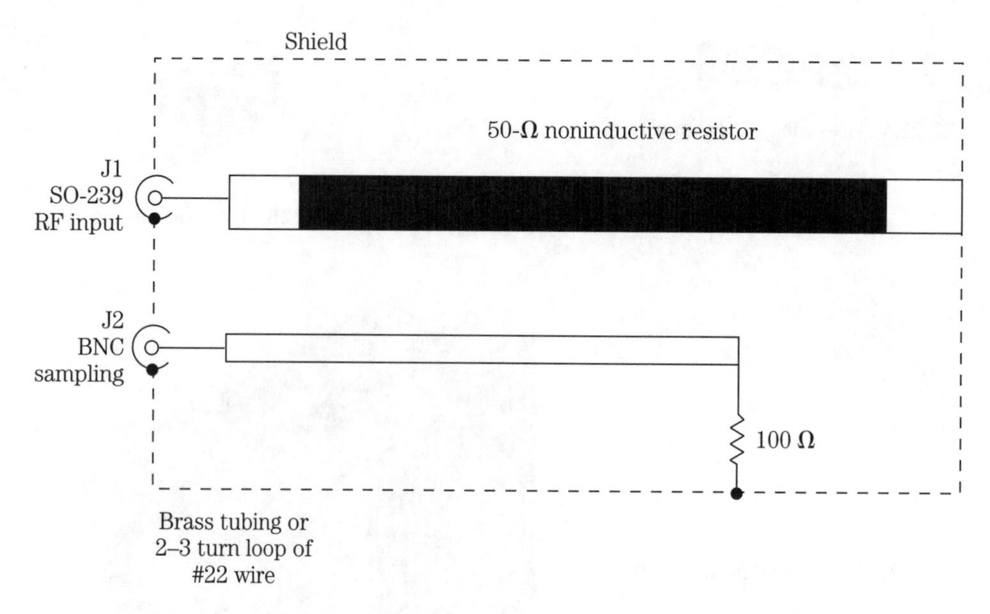

22-24B HF dummy load with pick-up loop.

23
CHAPTER

Antenna construction techniques

ANTENNA INSTALLATION CHORES VARY IN COMPLEXITY FROM THE SIMPLE, THAT A teenager, with no assistance, can properly and safely execute, all the way up to large-scale projects that are best left to professional antenna riggers. If your only antenna installation chore is a dipole hanging between the peak of the house roof and a nearby tree, then little need be said outside of what was said in the chapter on that form of antenna (*NOTE: Do read the electrical safety note below, however.*)

The purpose of this chapter is to cover the intermediate installation job that can be done by amateurs if instructed properly. Keep in mind when planning an installation, however, that the information given herein is merely informal guidance. Always have your plans reviewed by a professional mechanical or structural engineer licensed to practice in your state. In many locales, there are good insurance reasons for this step. Also, learn and abide by the local mechanical and building codes. It's simply not worth the pain, and risk, to skirt around the local laws.

Installing vertical antennas

Despite continuous (and mostly untrue) stories of very poor performance, the four- or five-band commercially manufactured trap vertical antenna remains one of the most popular amateur radio "antlers" in the HF bands. They are economical, compared with directional rotating beam antennas; easy to erect; and don't have to occupy a lot of real estate (the "footprint" of the vertical can be very small, especially if you bury the radials). Unfortunately, a misinstalled vertical is both dangerous and a very poor performer. This section covers both problems in order to give you a good chance of success.

Safety first

Before dealing with the radio and performance issues, let's first deal with safety matters . . . you don't want to be hurt either during installation, or during the next wind storm. Two problems present themselves: *reliable mechanical installation* and *electrical safety*.

Electrical safety note

Every year we read sad news in the magazines of a colleague being electrocuted while installing, or working on, an antenna. In all of these tragic cases, the antenna somehow came into contact with the electrical power lines. Keep in mind one dictum and make it an absolute:

There is never a time or situation when it it safe to let an antenna contact the electrical power lines!

This advice includes dipoles and longwires "thrown over" supposedly insulated lines, as well as antennas built from aluminum tubing. The excuse that the lines are insulated is hogwash. Old insulation crumbles on contact with even a thin wire antenna. *Don't do it*! The operant word is *never*!

Consider a typical scenario involving a four-band trap vertical. It will be 18 to 26 feet tall (judging from ads in magazines), and will be mounted on a roof, or mast, 12 to 30 feet off the ground. At my home in Virginia, a 25-foot tall trap vertical is installed atop a 15-foot Radio Shack telescoping TV antenna mast (Fig. 23-1). The total height above ground is the sum of the two heights: 15 feet + 25 feet = 40 feet. The tip of the vertical is 40 feet above the ground. I had to select a location, on the side of my house, at which a 40-foot aluminum pole could fall. Although that requirement limited the selection of locations for the antenna, neither my father-in-law (who helped install the thing) or myself was injured during the work session. Neither will a wind storm cause a shorted or downed power line if that antenna falls over.

In some jurisdictions, there might be another limitation on antenna location. Some local governments have a requirement that the antenna be able to fall over and land entirely on your own property. About 25 years ago, a Maryland county required the antenna to be installed at double its own length plus 50 feet from the nearest property line—until a local ham club sued, on grounds that about 99.9 percent of the rooftop TV antennas in the county were noncompliant. Before installing the antenna, check local building codes.

When installing a trap vertical, especially one that is not ground mounted, make sure that you have help. *It takes at least two people to safely install the standard HF vertical antenna*, and more may be needed for especially large models. If you are alone, then go find some friends to lend an arm or two. Wrenched backs, smashed antennas, (and house parts) and other calamities simply don't happen as often on a well-organized work party that has sufficient "hands" to do the job safely.

Mechanical integrity

The second issue in installing vertical antennas is old-fashioned mechanical integrity. Two problems are seen. First, you must comply with local building regulations and inspections. Even though the courts seem to forbid local governments from prohibiting

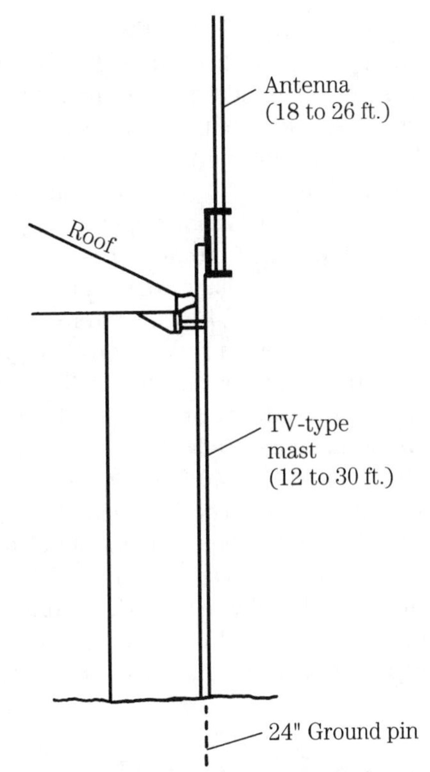

23-1 Mast-mounted vertical antenna.

amateur radio activity (on grounds that it is federal prerogative), the local government has a reasonable interest, and absolute right, to impose reasonable engineering standards on the mechanical installation of radio antennas. The second issue is that, it is in your own best interest to make the installation as good as possible. View local regulations as the *minimum* acceptable standard, not the maximum; go them one better. In other words, build the antenna installation like a brick outhouse.

Both of these mechanical integrity issues become extremely important if a problem develops. For example, suppose a wind or snow storm wrecks the antenna, plus a part of your house. The insurance company will not pay off (in most cases) if your local government requires inspections, and you failed to get same. Make sure the mechanical and/or electrical inspector (as required by law) leaves a certificate or receipt proving that the final inspection was done. It could come in handy when disputing with the insurance company over damage.

A quality installation starts with the selection of good hardware for the installation. Any radio-TV parts distributor who sells TV antenna hardware will have what you need. I used Radio Shack standoff brackets, ground pin, and a 19-foot telescoping mast. Wherever you buy, select the best quality, strongest material that you can find. Opt for steel masts and brackets over aluminum, no matter what the salesperson behind the counter tells you. Keep in mind that, although sales people can be knowledgeable and helpful, you, not they, are responsible for the integrity of the installation. In my own case, I found that the 19-foot mast was considerably sturdier at

15 feet, so I opted to use less than the full length because the installation is unguyed. Because I have never trusted the little cotter pin method of securing the mast at the slipup height, I drilled a single hole through both bottom and slip-up segments (which telescope together), and secured the antenna mast with a 5⁄16-inch steel bolt. The bolt was "double nutted" in order to ensure that it did not come loose over time.

The TV mast is set on a ground mounting pin/plate that is set into a 30-inch deep (local frost line regulations required only 28 inches) fencepost hole filled with concrete. The top end of the mast was secured to the roof overhang of the house (see Fig. 23-1). That overhang was beefed up with 2"-×-8" kiln-dried lumber that was bolted between two, 24-inch center, roof rafters. I felt it necessary to do that because the roof is only plywood, and the gutter guard is only 1"-×-6" lumber (and is old), and the soffits are aluminum. There wasn't enough strength to support a 40-foot leverarm, whipping around in a 35-knot wind.

Wind can be a terrible force, especially when acting on the "sail area" of the antenna through a 25- to 40-foot leverarm. A shabby installation will tear apart in wind, causing the antenna to be damaged, damage to the house, and destruction of the installation. That's why I recommend "brick outhouse" construction methods. Over the 33 years I have been in amateur radio, I have seen a lot of verticals toppled over. Except for a few shabby models that were so poorly built that they should not have been on the market in the first place, all of these failed installations were caused by either poor installation design or cheapskate materials.

Electrical installation

Figure 23-2 shows the usual form of multi-band trap vertical antenna. Each trap (TR1-TR3) is a parallel resonant LC tank circuit that blocks a certain frequency, but passes all others. In Fig. 23-2, TR1 is the 10-meter trap, TR2 is the 15-meter trap and TR3 is the 20-meter trap. No 40-meter trap is needed because the antenna resonates the entire length of the tubing on 40 meters. Each section (except perhaps the 10-meter section) is actually a little shorter than might be expected from the standard quarter-wavelength formulas. That is because the traps tend to act inductively, and so lessen the length required to resonate on any given band.

The vertical manufacturer might give suggested lengths for the various segments between traps. *Do not make the mistake* of assuming that these are absolute numbers. They are only recommended starting points, even though the literature packed with the antenna may suggest otherwise. Loosely (meaning don't tighten the clamps too much), but safely install the antenna and then adjust each segment for resonance. Start with the 10-meter band, and then work each lower frequency band in succession: 10-15-20-40, etc. After each lower band is adjusted, recheck the higher bands to make sure that nothing has shifted, because there might be a little interaction between bands. Once the antenna is properly resonant, tighten the clamps, and make the final installation. I know this is a pain in the neck, and means putting the antenna up, and taking it down, a couple of times, but it pays dividends in the end. I failed to do this once, and found that the 15-meter band was useless: it resonated at 19.2 MHz.

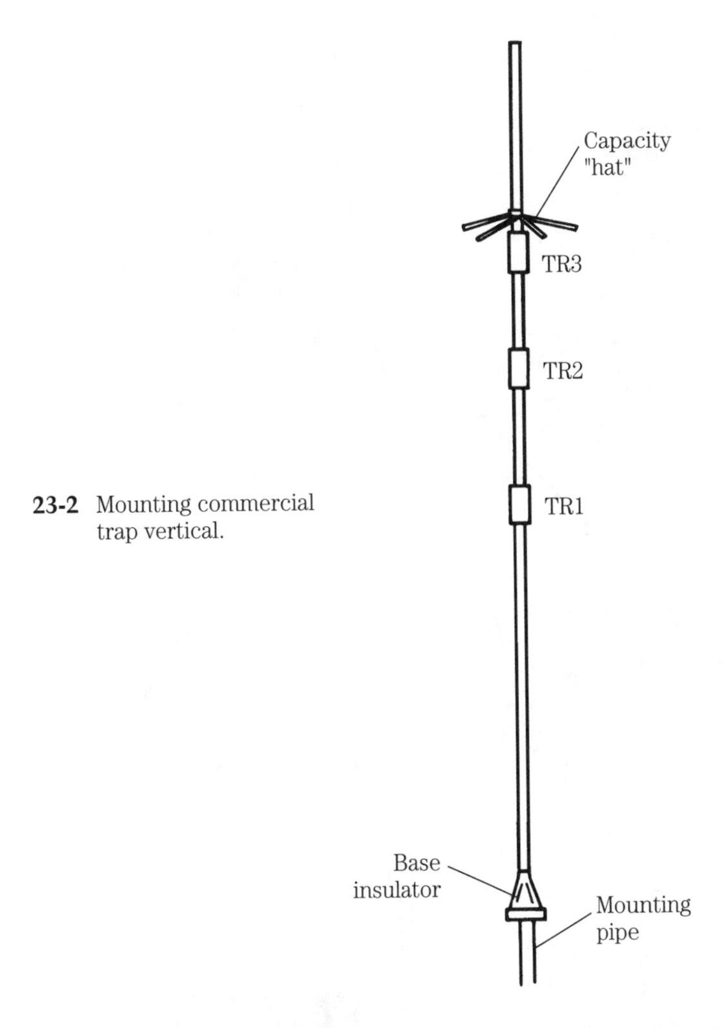

23-2 Mounting commercial trap vertical.

Radials make or break a vertical antenna; they form the ground plane for the antenna. AM broadcast stations typically use vertical antennas, and must have up to 120 radials for the ground plane. Antenna reference books usually contain a graph plotted to show the effectiveness of radials, and demonstrate a decreasing return on investment after about 32 radials. For amateur work, I recommend not less than two radials per band, and preferably four, arranged so that they are equally spaced around the antenna. If you can't space them correctly, never fear: they'll work anyway. On a four-band antenna, that means 16 radials, which really isn't a lot. Of course, the general rule is that the more radials the better the antenna, at least up to the point where diminishing returns are realized.

The radials are made of #14 wire, and must be a quarter wavelength ($246/F_{MHZ}$). Figure 23-3 shows how to mount the radials on a mastmounted (or roofmounted) installation, and Fig. 23-4 shows radials in a ground mounted situation. I do not recommend radials on the surface for ground-mounted antennas. It is too easy for a guest, or even a trespasser, to trip over the radials and that could land you in court.

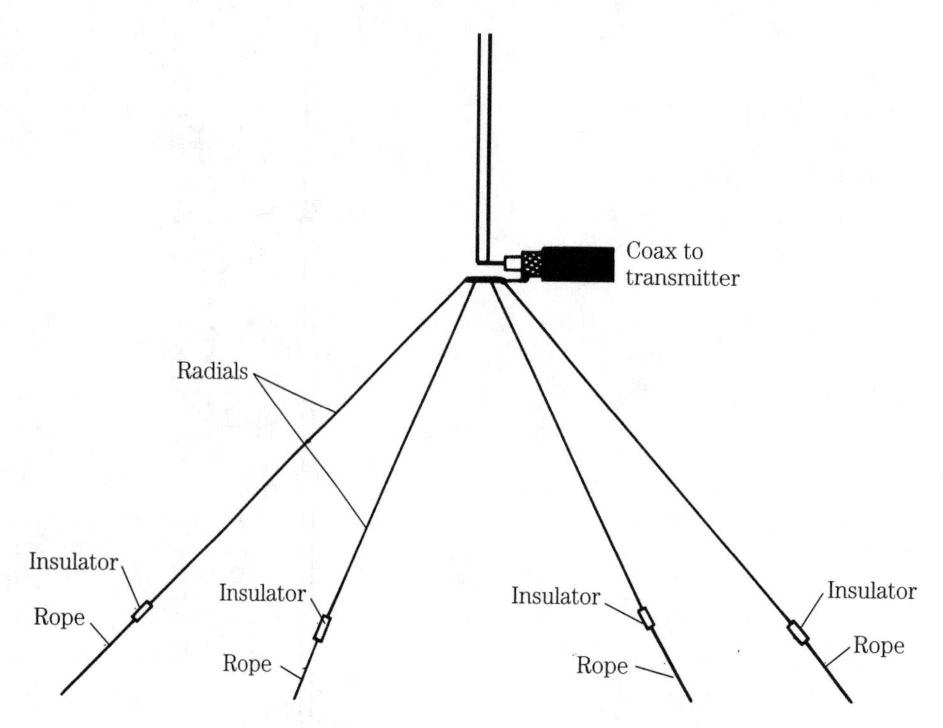

23-3 Feeding the mast-mounted vertical antenna.

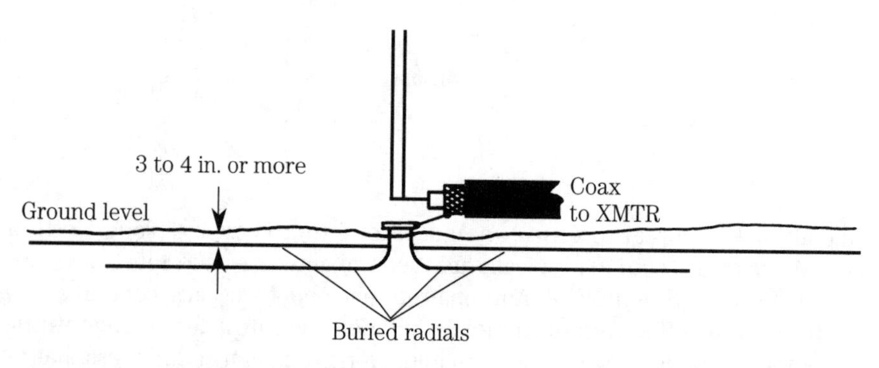

23-4 Feeding the ground-mounted vertical antenna.

Wire radials can be buried using a spade tip to cut a 3 to 4 inch slit and they even make for a better ground.

Ground mounting verticals and masts

In the previous section, the essential assumption was that the antenna is mounted above ground, on a mast or other support. In this section, you will take a look at the methods of ground mounting the antenna, or its mast. The same mast ideas are also useful for erecting supports for dipoles and other wire antennas.

Figure 23-5 shows a standard chain link fence post, used as a support for either a vertical antenna or a mast (which could be metal, or PVC, or wood). The typical fence post is 1.25 to 2.00 inches outside diameter, and made of thick-wall galvanized steel. Thus, such a pipe makes a tough installation. Some authorities believe that the pipe will last longer if painted its entire length (especially the underground portion) with a rust inhibiting paint such as Rustoleum.

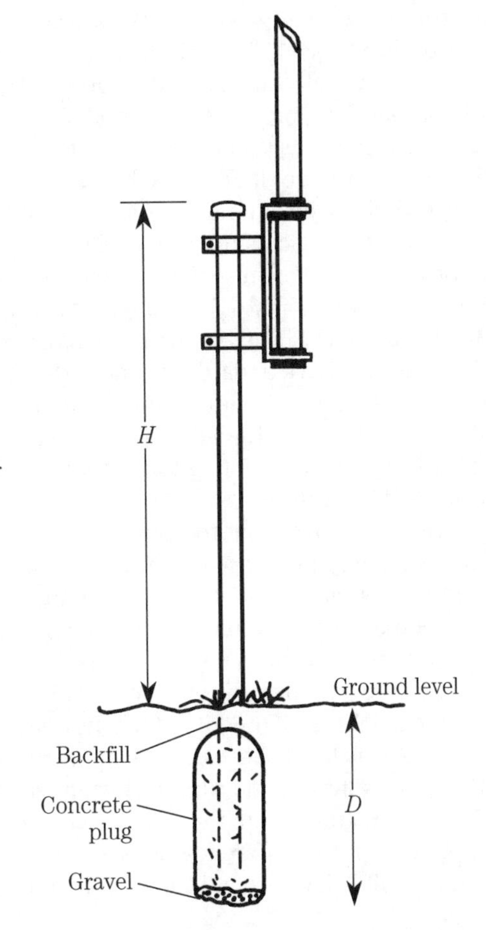

23-5 Fence post mounting.

The beauty of the fence post as an antenna support is that it is relatively easy to install, and supplies are easily obtained from hardware stores. The post is mounted in a concrete plug set at depth *D* in the ground (see again Fig. 23-5). The depth (*D*) is a function of local climate and local building codes. In the absence of local regulations to the contrary, make the hole at least 24 inches deep (or more). The depth is determined by the 100-year depth of the frost line in your area, a figure well familiar to local building experts. Keep in mind, however, that shallower depths may be legal for fence posts but installation of a vertical antenna (or mast) on top of the fence post, changes the mechanical situation considerably.

The hole can be dug with a post-hole digger tool, or an "earth auger bit" tool. The latter can often be rented in either gasoline engine, electrical motor, or manual versions from tool rental stores.

Once the hole is dug, place about 4 inches (or local requirement) of gravel at the bottom of the hole. This gravel can be bought in bags from hardware stores for small quantity jobs. The post is installed at this point, and made "plumb," (i.e., it is made to sit vertical in the hole). Gravel can be used to force the post to remain upright as the job is finished. On top of the gravel, is a concrete plug that fills the hole to at least 4 inches of the surface. The rest of the hole is backfilled with dirt, unless you want the plug to extend all the way to the surface. Don't put anything on top of the post for at least four days, or whatever period of time the concrete manufacturer recommends. Concrete needs time to cure, and four to seven days seems to be the recommended period of time. It is probably a good idea to install the post one weekend, and then complete the antenna installation the following weekend.

Chain-link fence posts tend to be about four-feet high, and that is sufficient for installation of masts for inverted-vee and dipole support. However, for vertical installation, make the height (H) of the post about 18-inches (otherwise, the antenna tries to act differently from the instructions).

If the antenna is a standard vertical, with the usual form of offset base (as shown in Fig. 23-5), then use either the mounting brackets that come with the antenna or U-bolts fastened around the fence post. Never use a single mounting point; always use at least two-point mounting to prevent the antenna from pivoting and shearing off the mounting hardware.

Figure 23-6 shows how to mount a 2-×-4 mast to a fence post. Good-quality 2-×-4 lumber can be purchased in lengths up to 20 feet, although, 8-foot, 10-foot, 12-foot, and 16-foot lengths are the most commonly available (in my area, 16-foot and 20-foot lengths have to be bought at professional "contractor" lumberyards, rather than those that serve homeowners). The 2-×-4 selected should be kiln-dried and pressure treated. It should also be painted in order to minimize the effects of weathering on the mechanical integrity of the installation. Good quality 2-×-4 lumber is relatively inexpensive, but the less costly material (which is only marginally less costly than the best) will warp and ruin the installation within one year.

The 2-×-4 is attached to the fence post using U-bolts. A 2-×-4 scrap is used as a wedge to take up the difference between the post and the mast. In some installations, the U-bolt will go around the perimeter of the 2-×-4; while in others, a pair of holes will be drilled in the 2-×-4 to admit the U-bolt arms. The U-bolt must be at least ¼-inch × 20 thread, and either a 5⁄16-inch bolt, or a ⅜-inch bolt, are highly recommend if the mast supports any significant weight.

Some installers like to place a brick at the base of the 2-×-4. This support bears the static and dynamic loads (i.e., gravity and wind) that would eventually cause the mast to list because of the eccentric loading on the footer. The auxiliary purpose is to keep the butt end of the 2-×-4 off the ground, and thereby prevent rot. Some additional mechanical stability might be afforded by replacing the brick with a cinder block that is filled with concrete.

Another approach is to install a pressure-treated 4-×-4 of the sort used to support patios and decks on the back of houses (see Fig. 23-7). Hardware stores and

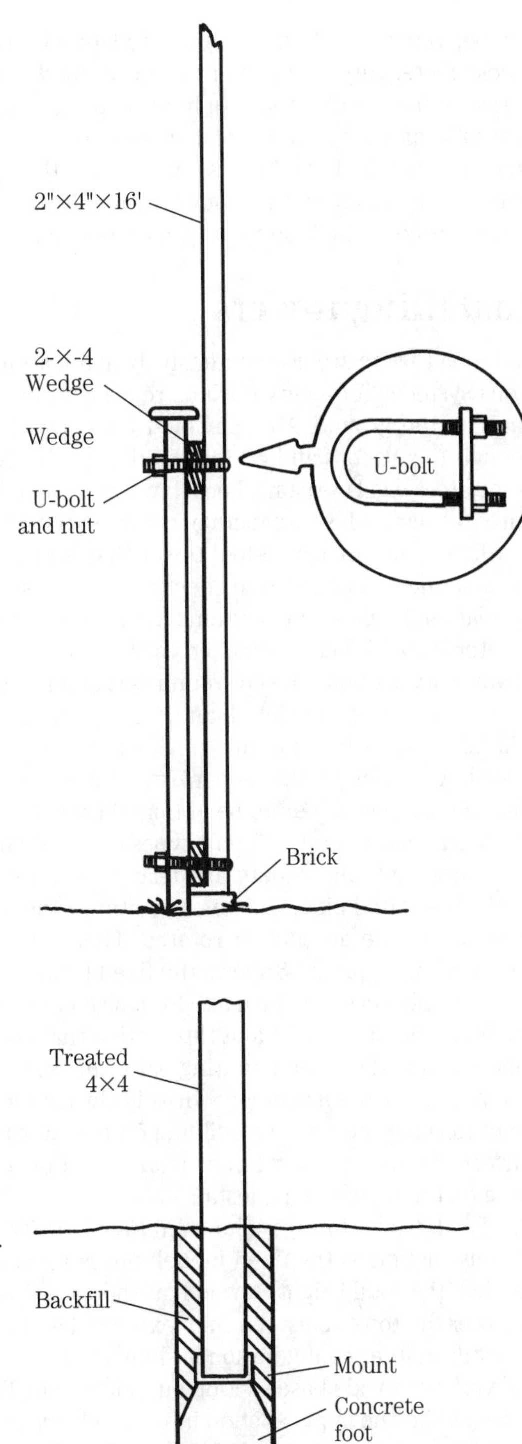

23-6 Making a mast mounted to a fence post.

23-7 4-×-4 antenna support.

lumber yards sell the specially treated 4-×-4 posts, and the concrete for the footers. These footers are dropped onto a gravel bed at the bottom of a hole (see the previous section for a discussion of how deep the hole must be). A typical deck footer has a plastic or heavy-duty metal mount to support the 4-×-4. After the footer and mount is installed, the hole is backfilled with soil. Some people prefer to backfill, at least to a point above the mount, with builder's gravel (of the same sort as used beneath the concrete footer); or, concrete can be used for maximum stability.

Installing towers

Antenna towers are used extensively in amateur and commercial radio communications systems, for a very obvious reason: antennas tend to work better "at altitude" than on the ground. Amateur operators tend to install 30- to 100-foot towers, although the 40-foot and 50-foot models probably predominate for practical (and in some areas legal) reasons. Local laws vary considerably, from one area to another. Be sure to consult the bureaucracy before erecting a tower.

In my own area, a 30-foot tower that is physically attached to a two-story house does not need a permit or inspection (but must be constructed according to certain regulations), although anything mounted away from the house (or at a height greater than 30 feet, if attached to the house), requires a permit and the usual mechanical inspection. The law requires that an inspection be performed when the hole for the footing is completed, and 4 inches of gravel is on the bottom. The inspector will take a yardstick and measure both the depth of the hole and the depth of the gravel. A final inspection is performed when the antenna is installed. Prior to all of that mess, a permit had to be obtained (which means drawings submitted.)

There are several different types of towers used by amateurs. Some are square-base designed, and others are triangular base designs. At least one commercial model is a round pipe, and has a rotator at the base. In this model, the entire tower turns when the antenna is rotated. There are several tower configurations commonly seen. Figure 23-8 shows the fixed form of tower. This type of tower is the lowest cost, and a case can be made for it also being the most stable. It is made of 10-foot sections, and an 8- to 12-foot top section that contains the thrust bearing for the antenna mast and rotator mounting platform. Some fixed towers are self-supporting up to a point, but the use of guy wires is always recommended unless a firm mounting point (against the side of a building for example) is provided about 18 feet above the surface. The fixed tower can be made to tilt over at the base by using a hinged base-plate on the mounting pedestal.

The tilt-over tower is shown in Fig. 23-9. This type of tower uses the same sort of construction as the fixed model, but is hinged at a point between one third and one half the total height. A winch at the base drives a steel wire that is used to raise or lower the tower between fully extended and fold-over positions.

Still another configuration is shown in Fig. 23-10. This type of tower is the "slip-up" variety. A wide base section supports a smaller movable upper section. A winch is used to lift the upper section into its fully extended position. Typical examples will raise from a low position of 20 feet, to a height of 40 feet or so when fully extended (or 20/50 feet). When the tower is fully extended it is locked into place at the bottom

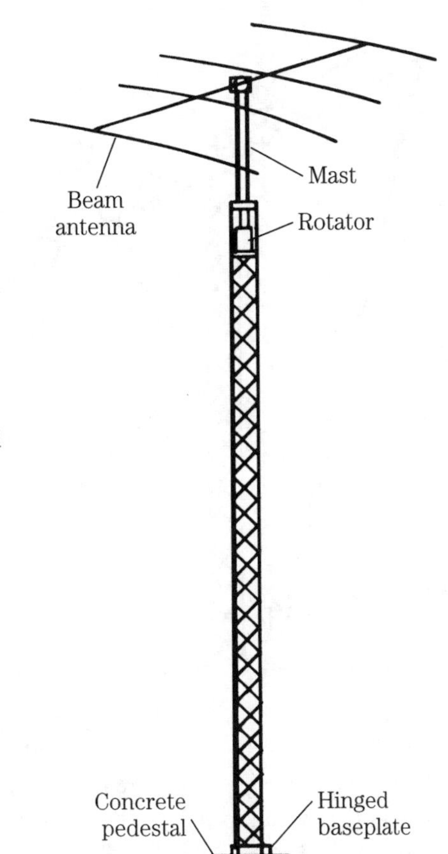

23-8 Antenna tower on hinged baseplate.

of the upper section with steel bolts. A lot of shear force is applied to these bolts, so it is wise to use several very hard stainless steel bolts for the lockdown.

There is a serious danger inherent in the design of the slip-up tower: the *guillotine effect*. If the upper section comes loose while you are working on it, it will come plummeting down the shaft formed by the lower section and shear off any arms, legs, or other body parts that get in the way. Whenever you work on this form of tower, use steel fence posts (or similar pieces of metal) as safety measure (see Fig. 23-11); at least two should be used and both should be securely attached at both ends with rope. The purpose of the rope is to prevent the safety pipes from working loose and falling out. These pipes are used in addition to, not instead of, the normal bolt fasteners that keep the antenna tower erect.

The antenna tower, regardless of its configuration, is typically installed on a buried pedestal made of concrete (Fig. 23-12). The construction must follow local laws, but there are some general considerations to keep in mind. The surface area of the pedestal is 24^2 inches to 36^2 inches. The top surface is fitted with either a fixed or hinged baseplate (which must be installed before the concrete sets). The depth of the pedestal, and the depth of the gravel base tends to reflect the local frost line).

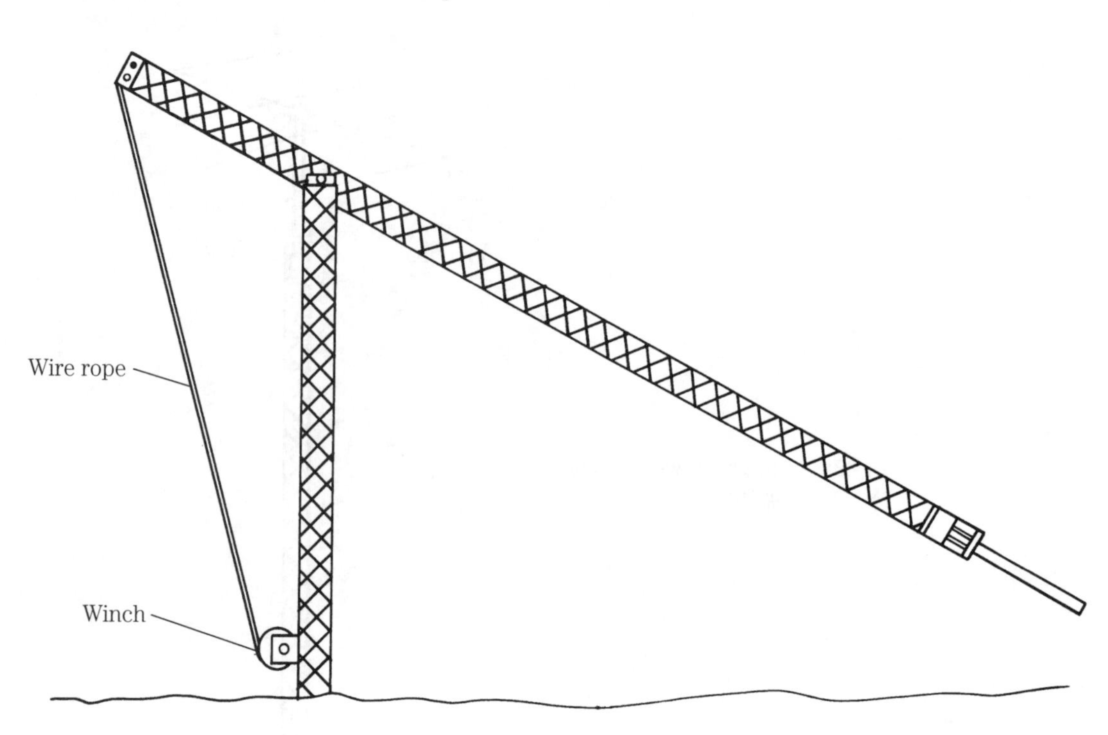

23-9 Fold-over tower.

Regardless of how general local law is, good advice is to make the pedestal at least 28 inches deep—even if local law permits less.

Raising the tower can be better by either of two methods, although in my opinion, only one of them is safe. Figure 23-13 shows a method by which the tower is laid out on the ground, and then raised with a heavy rope over a high support (such as the peak of the house roof). This method is dangerous unless certain precautions are taken. For example, a support must be placed beneath the antenna as it is raised. This requirement means that a constantly increasing height of the tower means a constantly increasing lower support.

The "end-over-end" method of raising a tower is shown in Fig. 23-14. A "gin pole" is required for this job. The gin pole is a length of pipe, fitted with a pulley at the top and a pair of clamps at the bottom. The gin pole is clamped onto the top end of the highest section, so that it can be used to raise the next 10 foot section into place where it can be bolted down. The gin pole is then raised to the top of the newly installed section where the operation is repeated again. The end-over-end method is preferred by tower manufacturers, and is the method recommended by most of them.

The antenna tower will be most stable if guy wires are installed at appropriate places determined by the tower manufacturer. Although many people are inclined to install towers "free standing", and some towers are advertised as such, it is always good practice to use guy wires attached to at least one point (or more, for tall towers). Use at least four guy wires on a square tower or three in a triangle tower (one in each direction). Figure 23-15 shows two methods for securing the guy wires to the ground. One approach is to bury a heavy concrete or metal "deadman" weight un-

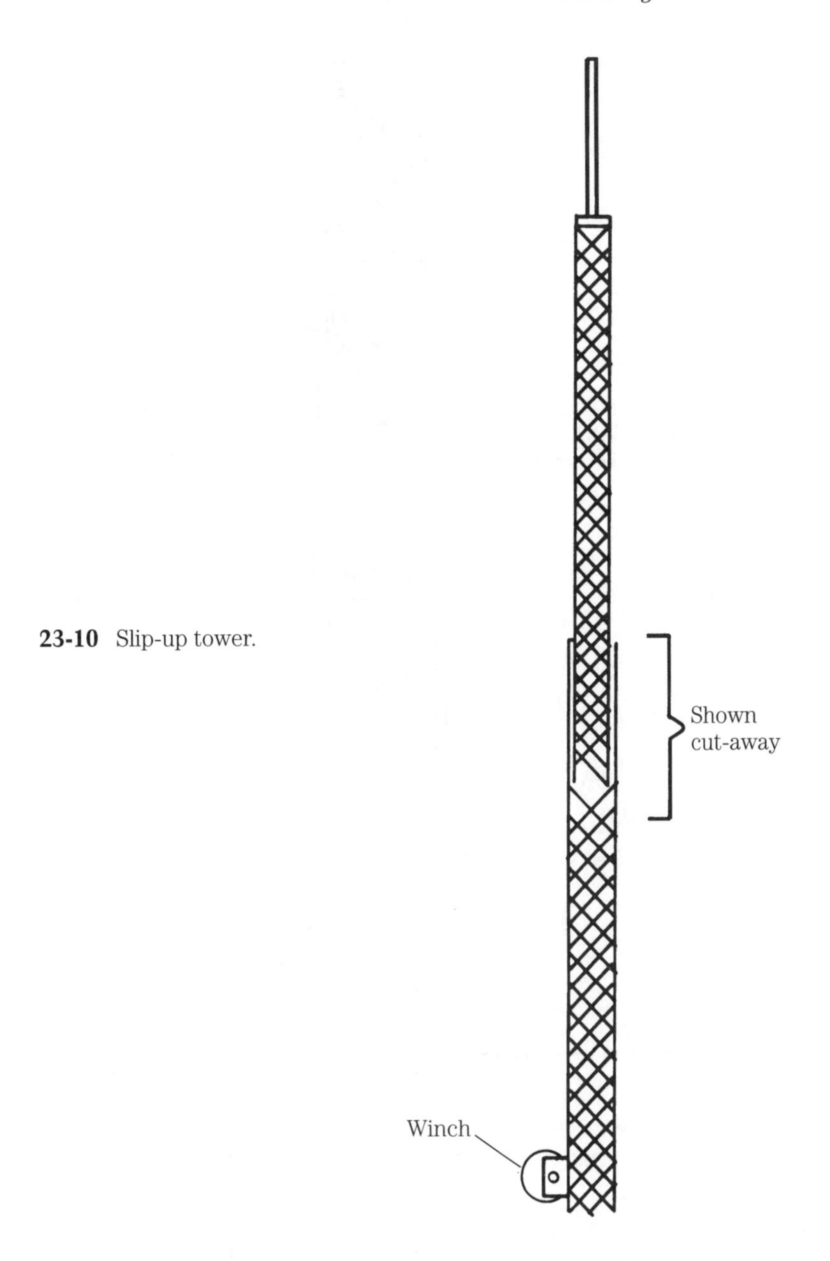

23-10 Slip-up tower.

derground. Alternatively, use at least one 4-foot to 6-foot pipe (1 to 1.5-inch diameter) buried up to a length of 12 to 24 inches (18 inches is shown in Fig. 23-15).

Note: It is absolutely essential to use safety belts when climbing a tower. Use two belts that are completely independent of each other so that no "single-point failure" can eliminate the belt's function. Always keep one belt fastened to the tower; when climbing unhook one belt and move it up on the tower, and then disconnect the other only when the first is reattached. Be sure to inspect the belts prior to use . . . leather and metal wear out and break . . . and you don't want to find that failure at 50 feet up. Also, don't buy cheap belts.

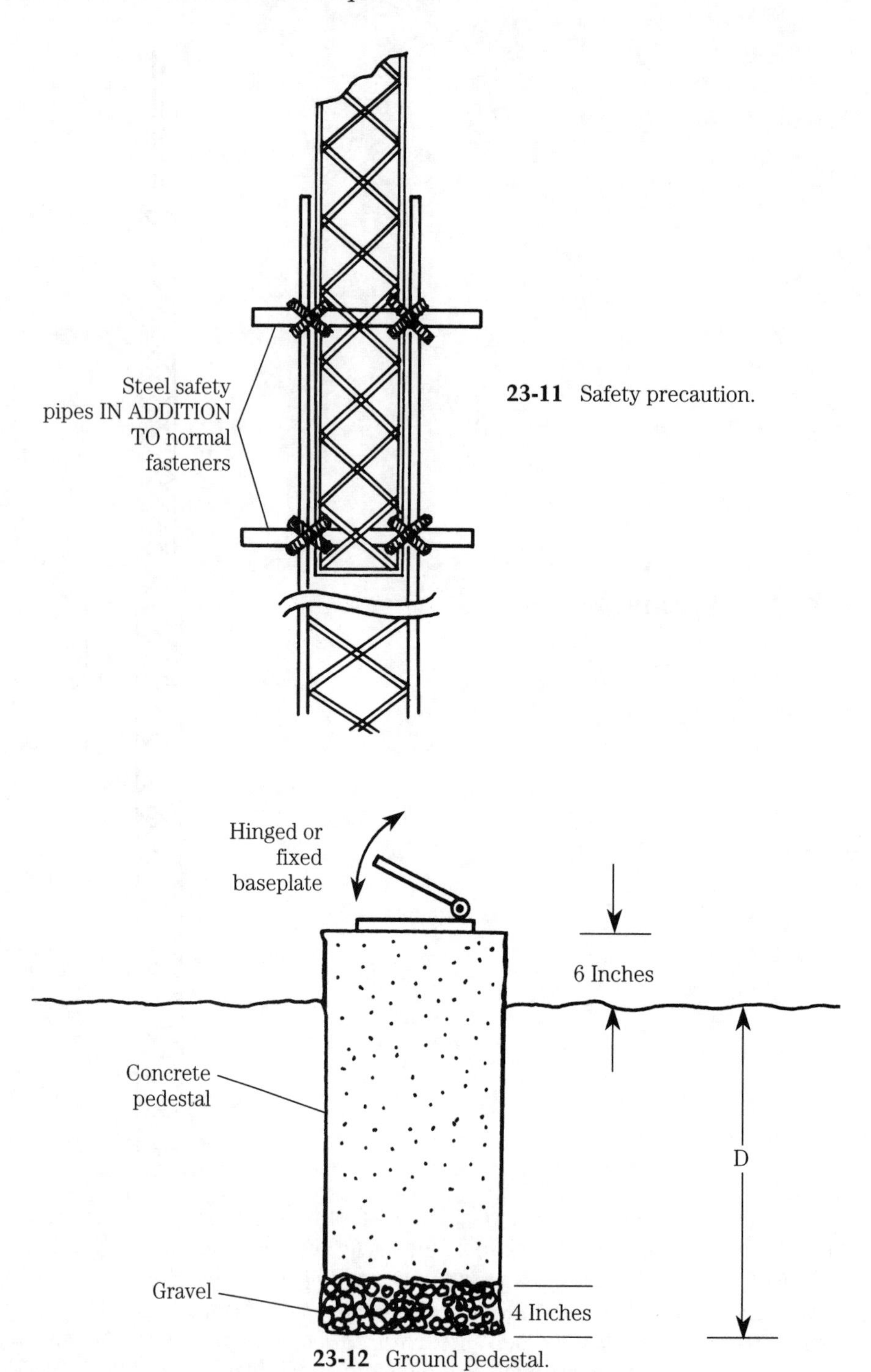

23-11 Safety precaution.

23-12 Ground pedestal.

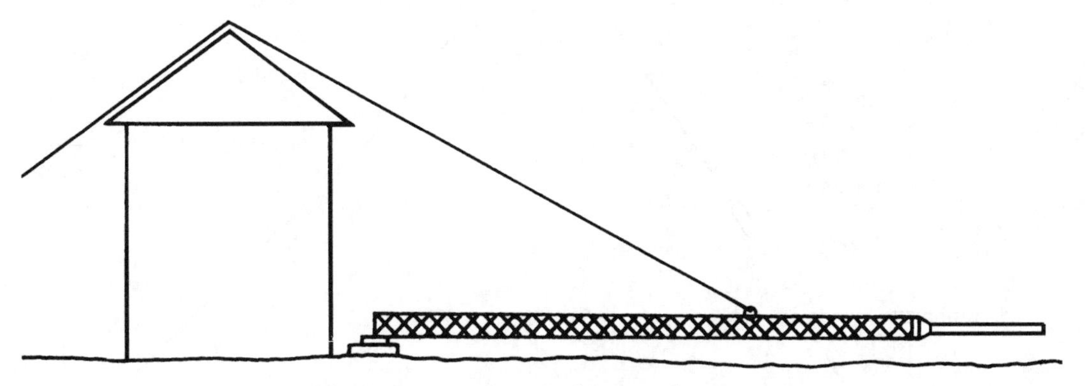

23-13 Dangerous method for erecting tower.

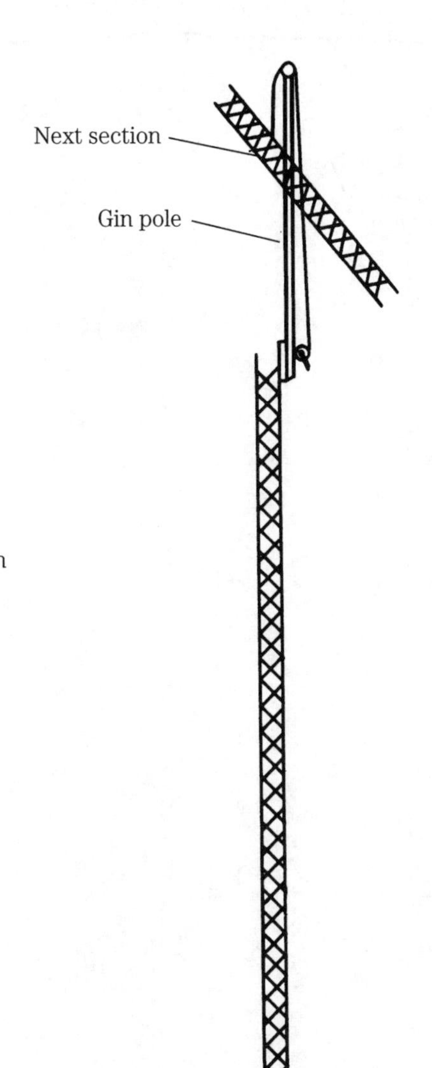

23-14 Proper tower installation using gin pole.

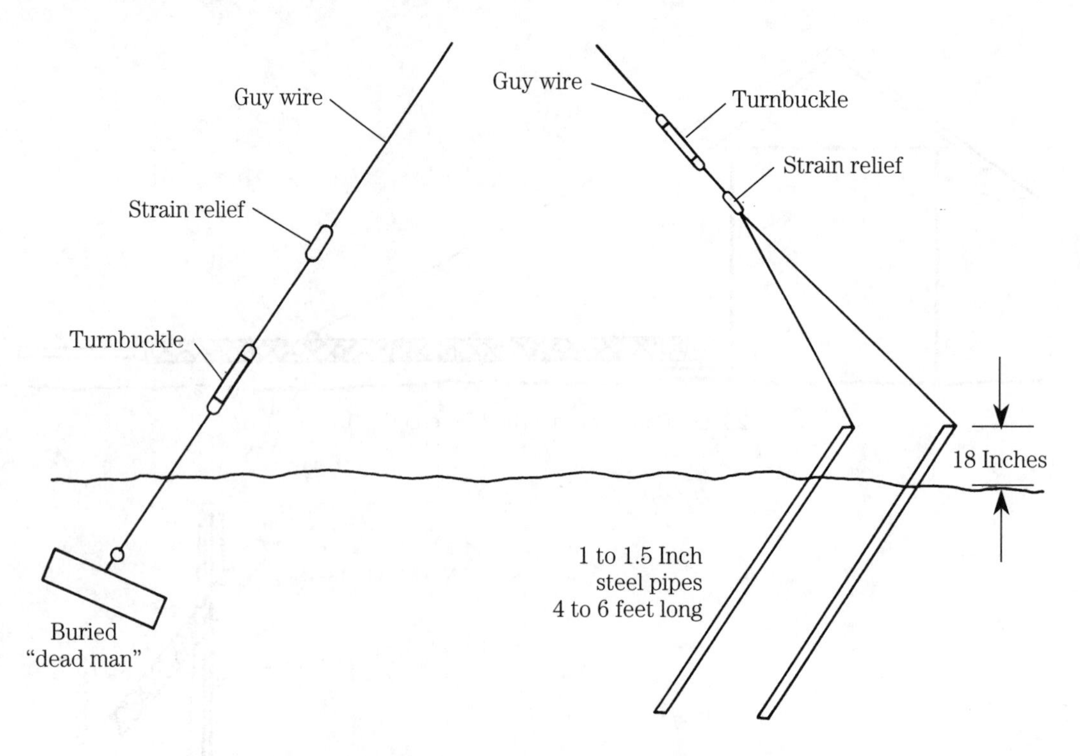

23-15 Ground stakes for guy wires.

24 CHAPTER

Grounding the antenna: What is a good ground

THE SUCCESS OR FAILURE OF A RADIO ANTENNA SYSTEM OFTEN (PERHAPS USUALLY) hangs on whether or not it has a good RF ground. Poor grounds cause most antennas to operate at less than best efficiency. In fact, it is possible to burn up between 50 and 90 percent of your RF power heating the ground losses under the antenna, instead of propagating into the air. Ground resistances can vary from very low values of 5 Ω, up to more than 100 Ω (5 to 30 Ω is a frequently quoted range). RF power is dissipated in the ground resistance. The factors that affect the ground resistance include the conductivity of the ground, its composition, and the water content. The ideal ground depth is rarely right on the surface, and depending on local water table level might be a couple meters or so below the surface.

It is common practice among some amateur radio operators to use the building electrical ground wiring for the RF antenna ground of their station. Neglecting to install an outdoor ground that will properly do the job, they opt instead for a single connection to the grounded "third wire" in a nearby electrical outlet. Besides being dangerous to work with, unless you know what you are doing, it is also a very poor RF ground. It is too long for even the low RF bands, and radiates RF around the house in large quantity. Stations that use the household electrical wiring as the radio ground tend to cause TVI, BCI, and other electromagnetic interference both in their own building and in nearby buildings.

Fortunately, there are some fixes that will help your situation. We can reduce the ground resistance by either altering the composition of the earth surrounding the ground point, or by using a large surface area conductor as the ground point.

Figure 24-1 shows the traditional ground rod used on small radio stations, including amateur stations. Use either a copper (or copper-clad steel) rod at least 6 feet long (8 feet preferred). Electrical supply houses, as well as amateur radio and communications equipment suppliers, also sell these ground rods. Do not use the

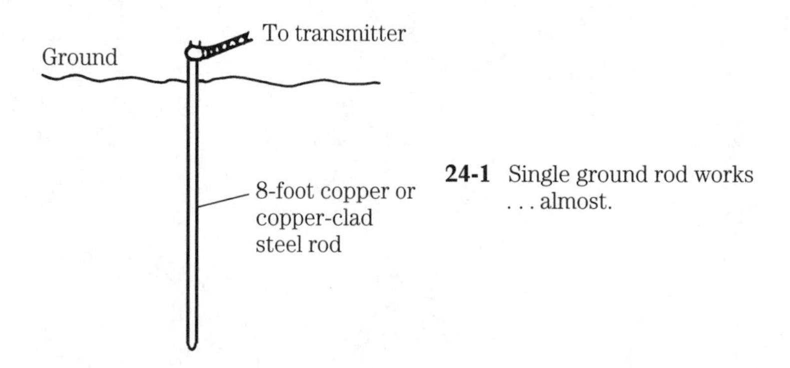

24-1 Single ground rod works . . . almost.

nonclad steel types sold by some electrical supply houses. They are usable by electricians when making a service entrance ground on your home or workplace, but RF applications require the low skin resistance of the copper-clad variety. The rod need not be all-copper, because of skin effect forcing the RF current to flow only on the outer surface of the rod. Try to use an 8-foot rod if at all possible, because it will work better than the shorter kind. Do not bother with the small TV-antenna 4-foot ground rods; they are next to useless for HF radio stations. Drive the ground rod into the earth until only 6 inches or so remains above the surface. Connect a ground wire from your station to the ground rod.

The ground wire should be as short as possible. Furthermore, it should be a low-inductance conductor. Use either heavy braid (or the outer conductor stripped from RG-8 or RG-11 coaxial cable), or sheet copper. You can buy rolls of sheet copper from metal distributors in widths from 4 inches up to about 18 inches. Some amateurs prefer to use 7-inch-wide foil that is rated at a weight of 1 lb/linear-foot. Sweat solder the ground wire to the rod. You can get away with using mechanical connections like the electricians use, but will eventually have to service the installation when corrosion takes its toll. I prefer to use soldered connections, and then cover the joint with either petroleum jelly or acrylic spray lacquer.

Another alternative is to use a copper plumbing pipe as the ground rod. The pipe can be purchased in 8-foot through 16-foot lengths from plumbing supply shops or hardware stores. The pipe selected should be ¾ inch or larger. Some people report using up to 2-inch pipe for this application. The surface area of the hollow pipe is greater than that of a solid rod of the same diameter. Because of certain current flow geometries in the system, however, the ground resistance is not half the resistance of a rod the same diameter, but is nonetheless lower.

Driving a long pipe into the ground is not easily done. Unlike the copper-clad steel rod, the pipe has no compression strength and will deform when hit with a hammer or other driving tool. To overcome this problem (see Fig. 24-2) you can use a garden hose as a water drill. Sweat solder a tee-joint on the top end of the pipe, and then sweat solder a faucet fitting that matches the garden hose end on one end of the tee-joint. Cap off the other port of the tee-joint. Use the tee-joint

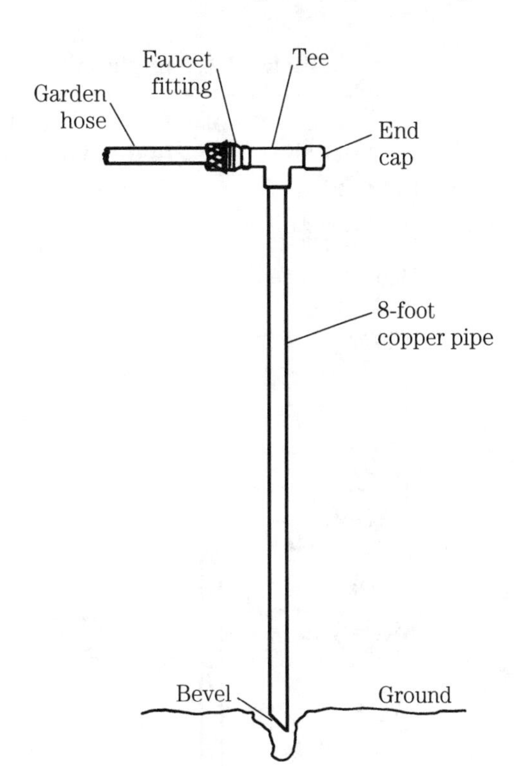

24-2 Hose method of installing ground pipe.

as a handle to drive the pipe into the ground. When water pressure is applied, the pipe will sink into the ground as you apply a downward pressure on the tee-handle. In some cases, the pipe will slip into the ground easily, requiring only a few minutes. In other cases, where the soil is hard or has a heavy clay content, it will take considerably greater effort and more time. When you finish the task, turn off the water and remove the garden hose. Some people also recommend unsoldering and removing the tee-joint.

Altering soil conductivity

The conductivity of the soil determines how well, or how poorly, it conducts electrical current (Table 24-1). Moist soil over a brackish water dome conducts best (southern swamps make better radio station locations), and the sand of the western deserts make the worst. Figure 24-3 shows a method for reducing the soil electrical resistance by treating with one of two chemicals: copper sulphate or common rocksalt. The rocksalt is one of several salt materials used for snow and ice melting in snow-prone areas of the country. If you cannot locate rocksalt in a hardware store, then look for a store that sells ice cream making supplies. Rocksalt is a principal ingredient in the process (but not the product).

Table 24-1. Sample soil conductivity values

Type of Soil	Dielectric constant	Conductivity (Siemans/meter)	Relative quality
Salt water	81	5	Best
Fresh water	80	0.001	Very poor
Pastoral hills	14–20	0.03–0.01	Very good
Marshy, wooded	12	0.0075	Average/poor
Rocky hills	12–14	10	Poor
Sandy	10	0.002	Poor
Cities	3–5	0.001	Very poor

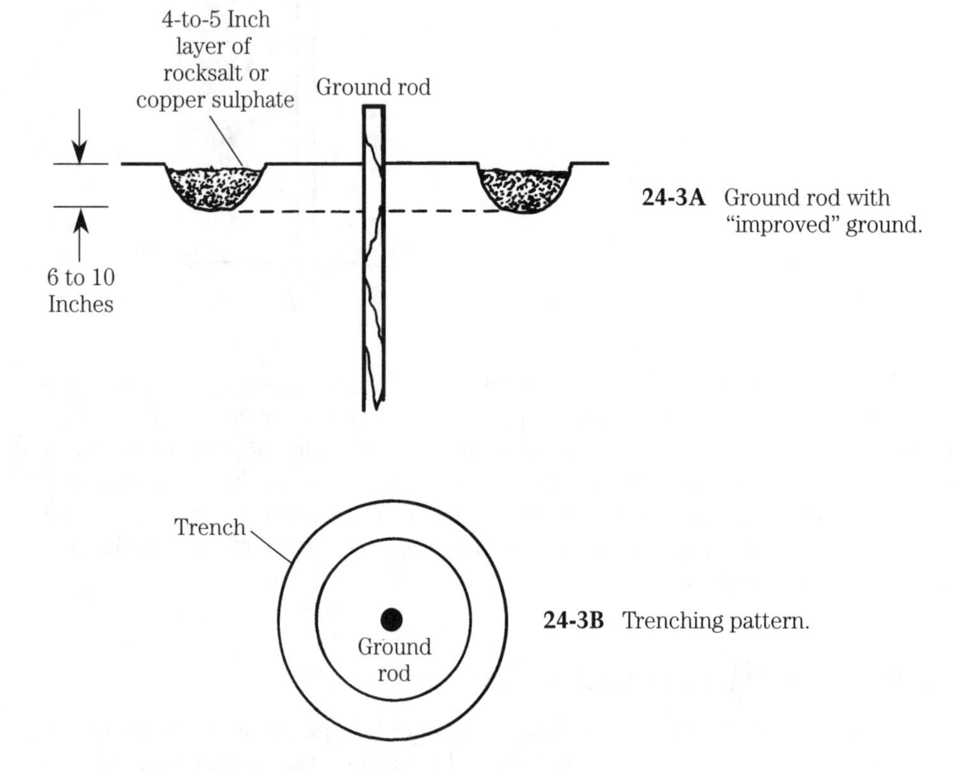

24-3A Ground rod with "improved" ground.

24-3B Trenching pattern.

Figures 24-3A and 24-3B show side and top views (respectively) of a slit trench method of applying the chemical treatment. Dig a 6- to 10-inch-deep trench, about 12 to 24 inches from the ground rod. Fill that trench with a 4- or 5-inch layer of rocksalt or copper sulphate. Cover the remaining depth with soil removed when you dug the trench. Water the trench well for about 15 minutes. The treatment will need to be repeated every 12 to 36 months, depending upon local rainfall and soil composition.

Figure 24-3C shows an alternative method: the saltpipe. Use either copper or PVC plumbing pipe, up to 4 inches diameter (although 2 to 3 inches is easier to work). The overall length of the pipe should be at least 18 to 24 inches, although

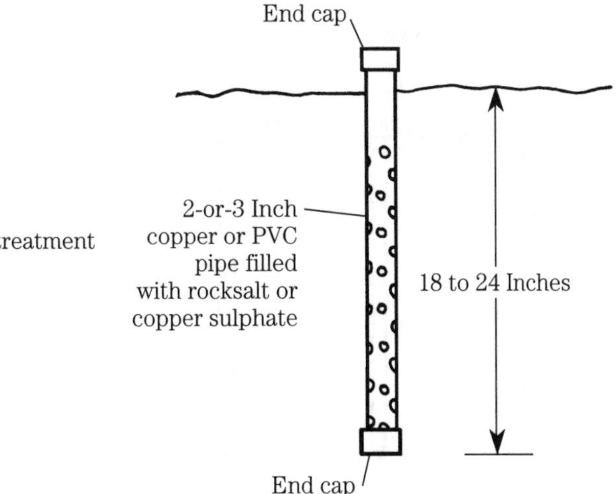

24-3C Slow release ground treatment pipe.

longer is useful. Drill a large number of small holes in the pipe (no hole over ⅜ inch), sparing only the end that will be above the surface. Cap off both ends.

Install several saltpipes around the ground rod. Installation is best accomplished using a fencepost hole digger. Drop the saltpipe into the hole and backfill with water. Remove the top endcap and hose down for about 15 minutes. Refill the pipe occasionally to account for the salt leaching out.

Multiple ground rods

The key to a low resistance ground is the surface area in contact with soil. One means for gaining surface area, and thereby reducing ground resistance, is to use multiple ground rods. Although multiple 4-foot rods are better than a single 4-foot ground rod, the use of 6- or 8-foot rods is recommended even in cases where multiple rods are used.

Figure 24-4A shows the use of three ground rods in the same system. The 8-foot rods are placed 12 to 18 inches apart for low and medium power levels, and perhaps 30 inches for higher amateur power levels. An electrical connection is made between the rods on the surface using either copper stripping or copper braid. The connections are sweat soldered in the manner described above, with the feedpoint at the center rod.

Figure 24-4B shows a more complex system with a larger number of rods laid out in an array. The rods are arrayed about 24 inches apart. As in the previous case, the rods are electrically connected (by braid or foil) on the surface. Again, connection to the system is made from the rod at the geometric center of the array.

Radials/counterpoise grounds

The effectiveness of the ground system is enhanced substantially by the use of radials either above ground or buried under the surface. Figure 24-5 shows a vertical an-

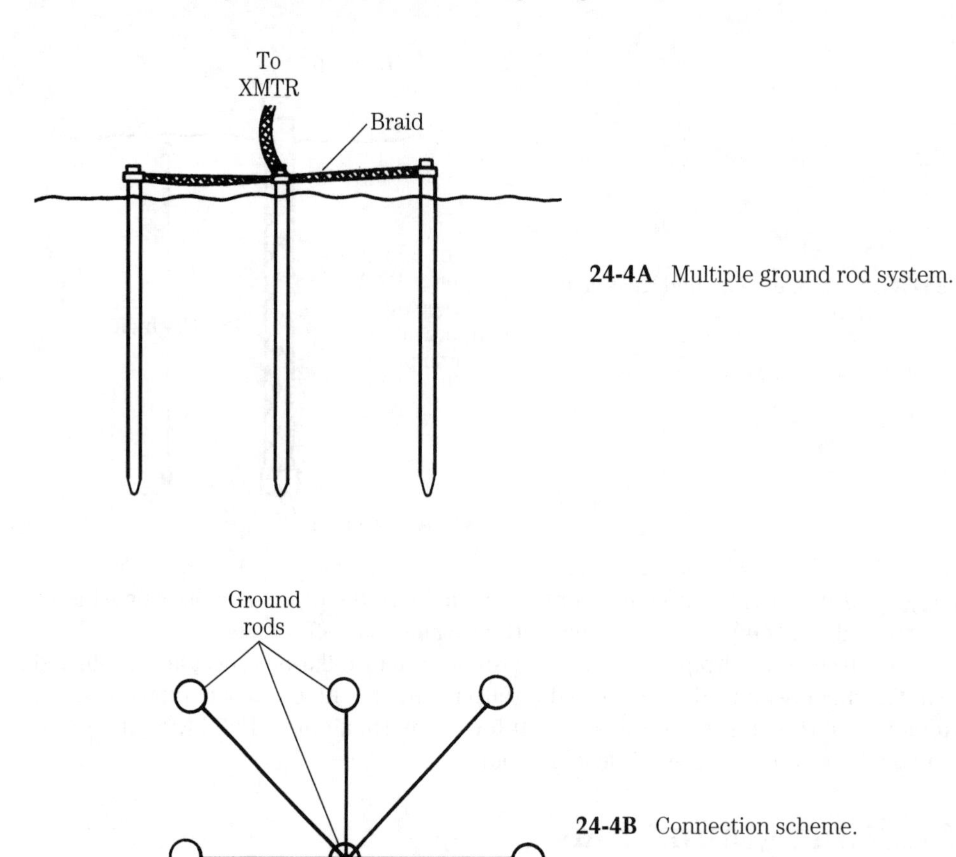

24-4A Multiple ground rod system.

24-4B Connection scheme.

tenna with three different forms of radials: above ground, subsurface, and ground rod. It is not unreasonable to use both radials and a ground rod. Note (from chapter 7) that vertical antennas are relatively ineffective unless provided with a good ground system, and for most installations that requirement is best met through a system of ground radials.

An effective system of radials requires a large number of radials. Although as few as two quarter-wavelength resonant radials will provide an improvement, the best performance is to use more. Broadcasters in the AM band (550 to 1640 kHz) are advised to use 120 half-wavelength radials. Installing more than 120 radials is both expensive and time consuming, but does not provide any substantial improvement. For amateur and small commercial stations, use a minimum of 16 quarter-wavelength radials.

Above ground, the use of insulated wire is recommended, but not required. Below ground, noninsulated wire is preferred. Although some sources claim that any

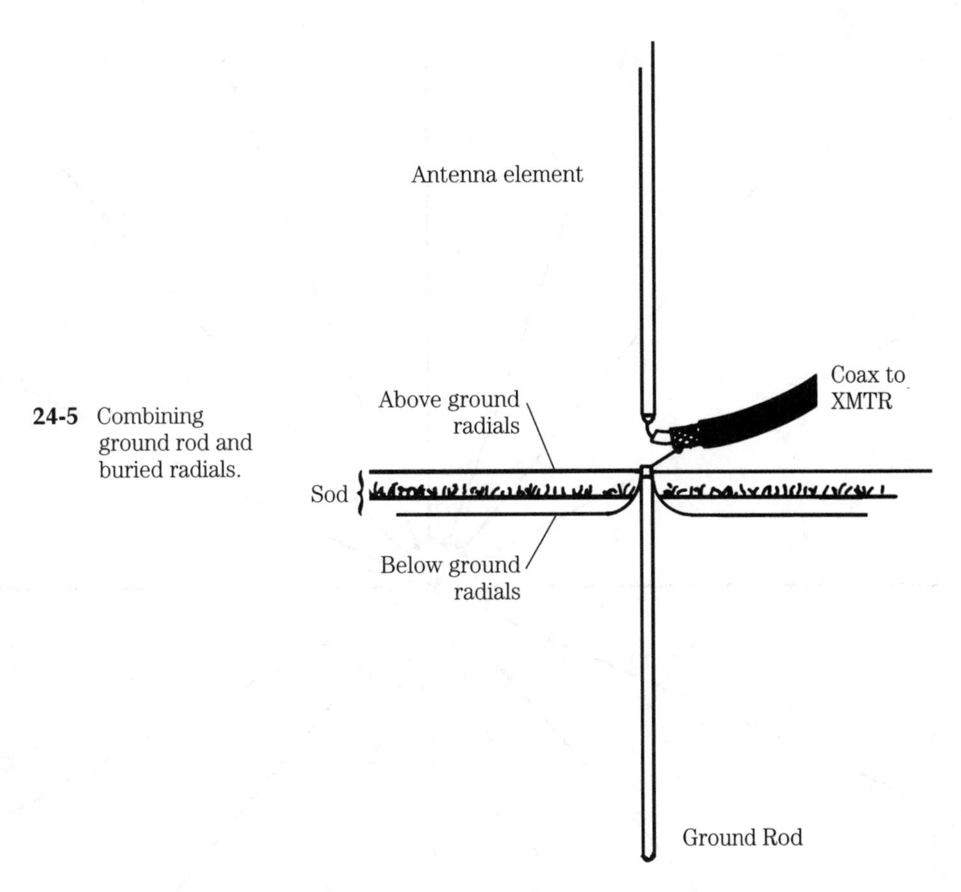

24-5 Combining ground rod and buried radials.

size wire from #26 up to #10 can be used, it is best to use larger sizes in that range (i.e., #14 through #10). Either solid or stranded wire can be used.

The layout for a system of radials in a vertical antenna system is depicted in a view from above in Fig. 24-6. Here, the radials are laid out in a uniform pattern around the antenna element. This coverage provides both the lowest resistance and the best radiation pattern for the antenna. Solder all radials together at a common point, which might be the ground or mounting rod used to support the vertical antenna.

Station grounding

It does no good to provide a top-flight ground system, such as those shown earlier in this chapter, if the connection between the station equipment and the ground system is substandard. Figure 24-7 shows a method used by the author to good effect. On the back of the operating position is a sheet of copper, 7-inches wide, running the length of the equipment platform. This form of copper, in the 1-lb/ft weight, is used on older houses for roofing flashing. Each piece of equipment is connected to the ground sheet through a short length of braid harvested from RG-8 or RG-11 coaxial cable shield. RF accessories, such as the low-pass TVI filter (if used—it should be) are mounted directly to the copper sheet. In one installation, the author was able to

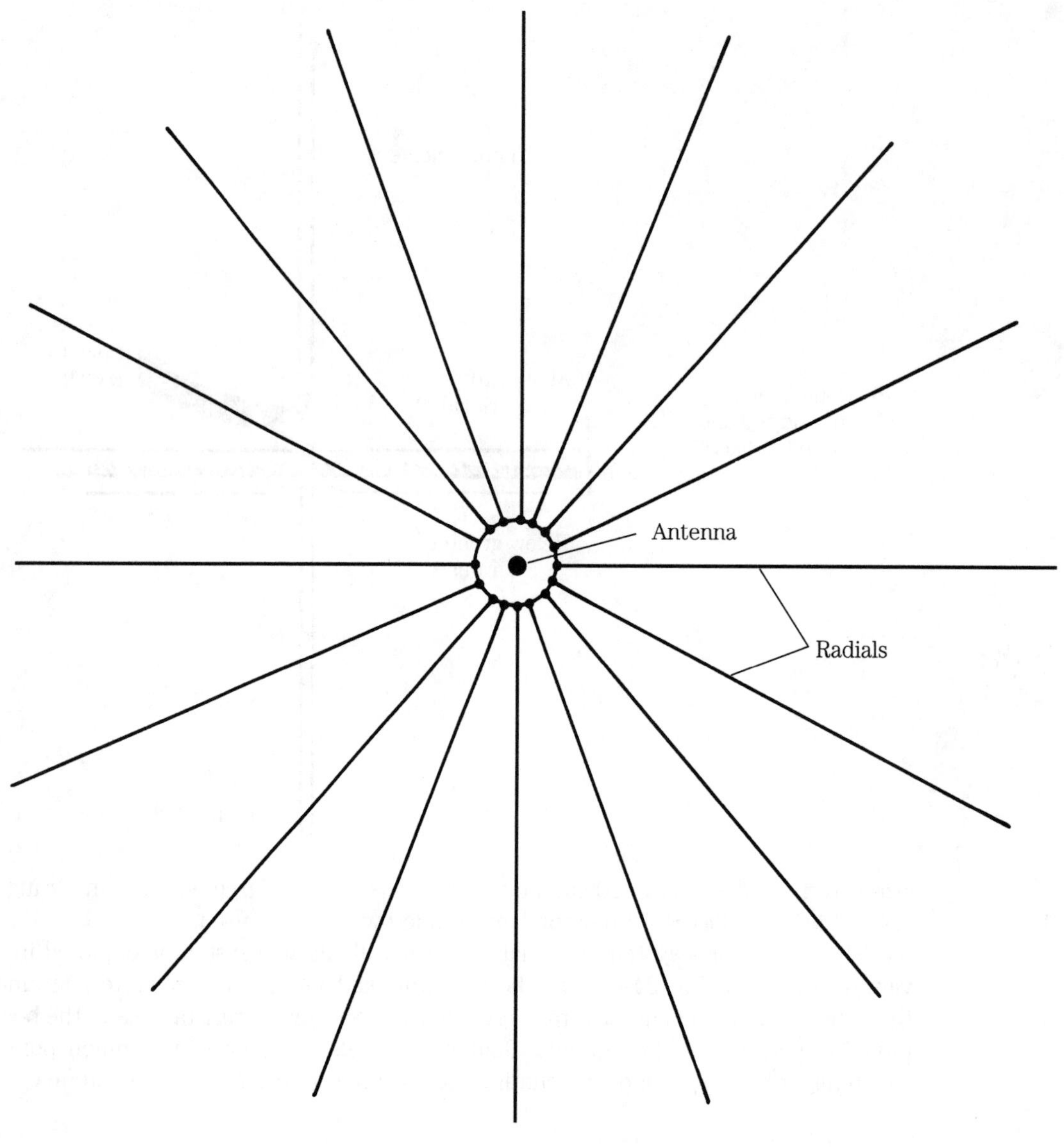

24-6 Radial pattern.

drop the copper sheet down from the table to connect directly to the ground system outside the building. The run was less than 40 inches. But in other cases, a short length of braid wire will be more practical.

Tuning the ground wire

An alternative that some operators use is the ground wire tuner. These instruments insert an inductor or capacitor (or a LC network) in series with the ground line. You

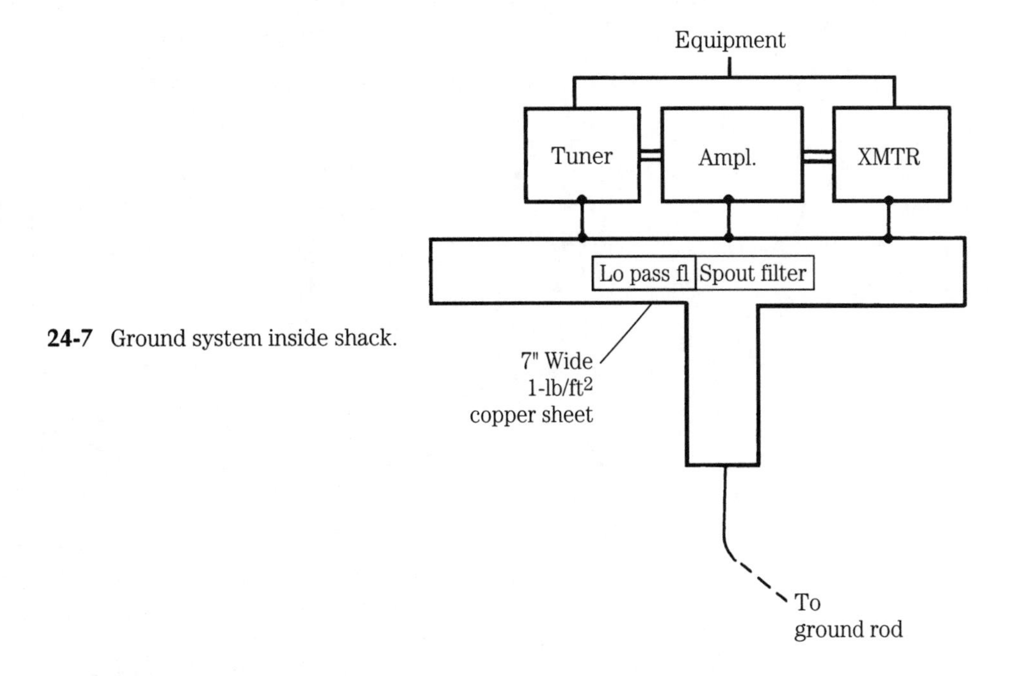

24-7 Ground system inside shack.

adjust the ground line tuner for maximum ground current at the operating frequency. MFJ Electronics, Inc., Mississippi State, MS, makes one of these devices.

Conclusion

A high-quality, low-resistance ground might seem costly to install, but in reality, it pays rich dividends in the functioning of your antenna. Don't overlook the quality of the ground, or you might be in the position of being penny-wise and pound-foolish.

Conclusion

Appendix A
Sources of supply

Radiokit
POB 973
Pelham, NH 03076
(603) 635-2235

Baker & Williamson
10 Canal Street
Bristol, PA 19007
(215) 788-5581

Unadilla/Antennas Etc.
POB 215 BV
Andover, MA 01810-0814
(616)475-7831

Van Gorden Engineering
POB 21305
South Euclid, OH 44121

Fair Radio Sales
POB 1105
Lima, OH 45802
(419)223-2196
227-6573
FAX: (419) 227-1313

SPI-RO Manufacturing, Inc.
POB 1538
Hendersonville, NC 28793

Alpha Delta Communications, Inc.
POB 571
Centerville, OH 45459
(513) 435-4772

Kelvin Electronics
7 Fairchild Avenue
Plainview, NY 11803
(800) 645-9212
(516) 349-7620
FAX: (516) 349-7830

Appendix B
Decibels

THE SUBJECT OF "DECIBELS" ALMOST ALWAYS CONFUSES THE NEWCOMER TO communications, and even many an old-timer seems to have occasional memory lapses regarding the subject. For the benefit of both groups, and because the subject is so vitally important to understanding communications systems, we will examine decibel notation.

The decibel measurement originated with the telephone industry, and was named after telephone inventer Alexander Graham Bell. The original unit was the "bel." The prefix "deci" means 1/10, so the "decibel" is one-tenth of a bel. The bel is too large for most common applications, so it is rarely if ever used. Thus, we will concentrate only on the more familiar decibel (dB).

The decibel is nothing more than a means of expressing a ratio between two signal levels; for example, the "output-over-input" ratio of an amplifier. Because the decibel is a ratio, it is also dimensionless . . . despite the fact that "dB" looks like a dimension. Consider the voltage amplifier as an example of dimensionless gain; its gain is expressed as the output voltage over the input voltage (V_o/V_{in}).

Example A voltage amplifier outputs 6 Volts when the input signal has a potential of 0.5 Volts. Find the gain (A_v).

$$A_v = V_o/V_{in}$$
$$A_v = (6 \text{ Volts})/(0.5 \text{ Volts})$$
$$A_v = 12$$

Note above that the "Volts" units appeared in both the numerator and denominator, and so "canceled out" leaving only a dimensionless "12" behind.

In order to analyze systems using simple addition and subtraction, rather than multiplication and division, a little math trick is used on the ratio. We take the base-

10 logarithm of the ratio, and then multiply it by a scaling factor (either 10 or 20). For voltage systems, such as our voltage amplifier, the expression becomes:

$$dB = 20 \text{ LOG } (V_1/V_2)$$

In the example given earlier we had a voltage amplifier with a gain of 12 because 0.5 volts input produced a 6 volt output. How is the this gain (i.e., V_o/V_{in} ratio) expressed in decibels?

$$dB = 20 \text{ LOG } (V_o/V_{in})$$

$$dB = 20 \text{ LOG } (6/0.5)$$

$$dB = 20 \text{ LOG } (12)$$

$$dB = 21.6$$

Despite the fact that we have massaged the ratio by converting it to a logarithm, the decibel is nonetheless nothing more than a means for expressing a ratio. Thus, a voltage gain of 12 can also be expressed as a gain of 21.6 dB.

A similar expression can be used for current amplifiers, where the gain ratio is (I_o/I_{in}):

$$dB = 20 \text{ LOG } (I_o/I_{in})$$

For power measurements, we need a modified expression to account for the fact that power is proportional to the square of the voltage or current:

$$dB = 10 \text{ LOG } (P_1/P_2)$$

We now have three basic equations for calculating decibels.

Adding it all up

Why bother converting seemingly easy to handle, dimensionless numbers like voltage or power gains, to a logarithmic number like decibels? Fair question. The answer is that it makes calculating signal strengths in a system easier. To see this effect, let's consider the multistage system in Fig. B-1. Here we have a hypothetical electronic circuit in which there are three amplifier stages and an attenuator pad. The stage gains are as follows:

$$A_1 = V_1/V_{in} = 0.2/0.010 = 20$$

$$Atten = V_2/V_1 = 0.1/0.2 = 0.5$$

$$A_2 = V_3/V_2 = 1.5/0.1 = 15$$

$$A_3 = V_o/V_3 = 6/1.5 = 4$$

The overall gain is the *product* of the stage gains in the system:

$$A_v = A_1 \times Atten \times A_2 \times A_3$$

$$A_v = (20) \times (0.5) \times (15) \times (4)$$

$$A_v = 600$$

When converted to dB, the gains are expressed as:

$$A_1 = 26.02$$

$$Atten = -6.02$$

$$A_2 = 23.52$$

$$A_3 = 12.04$$

The overall gain of the system (in dB) is the *sum* of these numbers:

$$A_v(\text{dB}) = A_1 + Attn + A_2 + A_3$$

$$A_v(\text{dB}) = (26.02) + (-6.02) + (23.52) + (12.04)$$

$$A_v(\text{dB}) = 55.56 \text{ dB}$$

The system gain calculated earlier was 600, and this number should be the same as above:

$$A_{\text{dB}} = 20 \text{ LOG } (600)$$

$$A_{\text{dB}} = 55.56 \text{ dB}$$

They're the same.

One convenience, of the decibel scheme, is that gains are expressed as positive numbers, and losses are negative numbers. Conceptually it seems easier to understand a loss of "–6.02 dB," than a loss represented as a "gain" of +0.50.

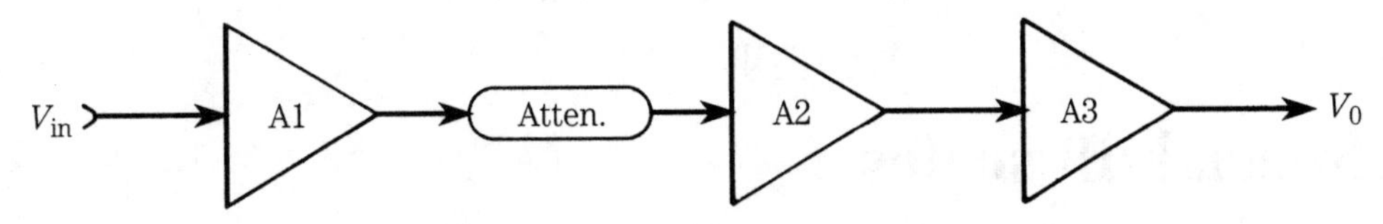

Fig. B-1 Cascade amplifier system.

Converting between dB notation and gain notation

We sometimes face situations where gain is expressed in dB, and we want to calculate the gain in terms of the output/input ratio. For example, suppose we have a +20 dB amplifier with a 1-millivolt (1 mV = 0.001 V) input signal. What is the expected output voltage? It's 20 dB higher than 0.001 volt, right? Yes, that's true, but your meter or oscilloscope is probably not calibrated in decibels but rather in volts. By using a little algebra, we can re-arrange the expression ($dB = 20 \text{ LOG } [V_o/V_{in}]$) to solve for output voltage, V_o. The new expression is:

$$V_o = V_{in}\, 10^{(dB/20)}$$

Which is also sometimes written in the alternate form:

$$V_o = V_{in}\, EXP\ (dB/20)$$

In the example above we want to calculate V_o if the gain in dB and the input signal voltage are known. We can calculate it from the above equations. Using the values above (20 dB and 1 mV):

$$V_o = V_{in}\, EXP\ (dB/20)$$

$$V_o = (0.001)\ EXP\ (20/20)$$

$$V_o = (0.001\ EXP\ (1)$$

$$V_o = (0.001)\ (10)$$

$$V_o = 0.001 \text{ volts}$$

Again we see the convenience of decibel scales over gain ratios. If we want to calculate system gain of a circuit that has a gain of 10,000, and an attenuation of 1/1000 in series, then we can do it either way:

$$A_v = 10{,}000 \times 0.001$$

$$A_v = 10$$

Or,

$$A_v = (+80 \text{ dB}) + (-60 \text{ dB})$$

$$A_v = +20 \text{ dB}$$

Special dB scales

Various user groups have defined special dB-based scales that meet their own needs. They make a special scale by defining a certain signal level as "0 dB," and referencing all other signal levels to the defined 0 dB point. In the dimensionless dB scale, 0 dB corresponds to a gain of unity. But if we define "0 dB" as a particular signal level, then we obtain one of the special scales. Below are listed several such scales commonly used in electronics:

1. *dBm*. Used in RF measurements, defines 0 dBm as 1 milliwatt of RF dissipated in a 50-Ω resistive load.
2. *Volume Units (VU)*. The VU scale is used in audio work, defines 0 VU as 1 milliwatt of 1000 Hertz audio signal dissipated in a 600 Ω resistive load.
3. *dB (now obsolete)*. Defined 0 dB as 6 milliwatts of 1000 Hertz audio signal dissipated in a 500-Ω load (once used in telephone work).
4. *dBmv*. Used in television antenna coaxial cable systems with a 75-Ω resistive impedance, the dBmV system uses 1000 microvolts (1 mV) across a 75-Ω resistive load as the 0 dBmv reference point.

Consider the case of the RF signal generator. In RF systems using standard 50-Ω input and output impedances, all power levels are referenced to 0 dBm being 1 mW (i.e., 0.001 Watt). To write signal levels in *dBm*, we used the modified power dB expression:

$$dBm = 10 \text{ LOG } (P/1 \text{ mW})$$

Example What is the signal level 9 milliwatts as expressed in *dBm*?

$$dBm = 10 \text{ LOG } (P/1 \text{ mW})$$

$$dBm = 10 \text{ LOG } (9/1)$$

$$dBm = 9.54 \ dBm$$

Thus, when we refer to a signal level of 9.54 *dBm*, we mean an RF power of 9 mW dissipated in a 50-Ω load. Signal levels less than 1 mW show up as negative dBm. For example, 0.02 mW is also written as –17 dBm.

Converting dBm to volts

Signal generator output controls and level meters are frequently calibrated in microvolts or millivolts (although some are also calibrated in dBm). How do we convert dBm to volts, or volts to dBm?

1. *Microvolts to dBm*. Use the expression $P = V^2/R = V^2/50$ to find milliwatts, and then use the dBm expression above.

Example Express a signal level of 800 µV (i.e., 0.8 mV) RMS in *dBm*.

$$P = V^2/50$$

$$P = (0.8)^2/50$$

$$P = 0.65/50 = 0.0128 \text{ mW}$$

$$dBm = 10 \text{ LOG } (P/1 \text{ mW})$$

$$dBm = 10 \text{ LOG } (0.0128 \text{ mW}/1 \text{ mW})$$

$$dBm = -18.9$$

2. *Converting dBm to Microvolts or Millivolts.* Find the power level represented by the dBm level, and then calculate the voltage using 50 Ω as the load.

Example What voltage exists across a 50-Ω resistive load when –6 dBm is dissipated in the load?

$$P = (1 \text{ mW}) \times (10^{(\text{dBm}/10)})$$

$$P = (1 \text{ mW}) \times (10^{(-6 \text{ dBm}/10)})$$
$$P = (1 \text{ mW}) \times (10^{-0.6})$$

$$P = (1 \text{ mW}) \times (0.25) = 0.25 \text{ mW}$$

If $P = V^2/50$,

then,

$$V = \sqrt{50P} = 0.707 \left(\sqrt{P}\right)$$

so:

$$V = (7.07) \times \sqrt{P}$$

$$V = (7.07) \times (0.25^{1/2})$$

$$V = 3.535 \text{ millivolts}$$

(Note: Because power is expressed in milliwatts, the resulting answer is in millivolts. To convert to microvolts, multiply result by 1,000.)

Conclusion

Decibels are both easy to understand and easy to use. They make chores (such as audio or RF system design) an easier task.

Appendix C
BASIC programs for antenna system design

THIS APPENDIX FEATURES TWO PROGRAMS FOR ANTENNA DESIGN. ALTHOUGH THEY ARE not sophisticated, they will suffice to allow you to find the correct starting dimensions for your selected antenna. These programs were written in GW-BASIC, so they should run on any MS-DOS (IBM-compatible) machine that includes BASICA or GW-BASIC.

The antenna design program is called *Antlers*, and it will allow you to design a variety of dipoles, verticals, miscellaneous wire antennas, as well as loops. For the loop antennas, it will permit you to find the inductance of the loop, and calculate the capacitance needed to resonate it to a particular frequency. These functions work over the frequency range of 10 kHz to 50 MHz. Graphics permit you to see the basic form of certain of the antennas.

The second program permits you to design an impedance matching "Antenna Tuning Unit" network over the same range as *Antlers*. This program is called *Zmatch*. The graphics will show you the basic form of the ATU network circuit, although the text of this book also contains the same information . . . as well as the math equations on which *Zmatch* is based.

Owners of later model IBM-compatible computers might find that they do not have a version of either BASICA or GW-BASIC on their operating systems. These machines typically have a different dialect, but in most cases can import BASICA or GW-BASIC programs if the programs was saved in ASCII format. Those readers might want to type in the program listing on a word processor such as WordPerfect or Word, using zero margins, and then store the text in ASCII (called *DOS Text* in *WordPerfect* and *Text File* in *Word*). When you save the text to disk, besides using the ASCII format, you must also use a ".BAS" extender on the filename. For example, if you wish to import *Antlers* to a modern *Windows* or other BASIC dialect, then save it in ASCII as "ANTLERS.BAS." The use of ASCII, and the .BAS extender permits *QuickBASIC 4.5* and the lesser variants to read the file from disk.

These two programs are very long, so the author will make an executable form of the program at a nominal cost for those who don't care to "fingerbone" the code into their computer. The code was compiled into an executable ".EXE" file by *QuickBASIC 4.5*. Contact the author at:

Joseph J. Carr
P.O. Box 1099
Falls Church, VA 22041

Antlers listing (BASIC language)

```
100 'ANTLERS (Version 1.2, 17 December 1992)
110 'Copyright 1992 Joseph J. Carr
120 'P.O. Box 1099, Falls Church, VA 22041
130 'SET UP COMPUTER PARAMETERS
140     CLS:KEY OFF:CLEAR:SCREEN 9
150 'TABLE OF DIMENSIONS AND CONSTANTS
160     PI=3.141593:PI#=PI
170 'EXECUTION PORTION OF PROGRAM
180    GOSUB 620:'Get opening screen (logo)
190    GOSUB 730:'Get opening announcement
200      COLOR 1,0:COLOR 14
210 'Advertisement for my books
220       GOSUB 4640
230    GOSUB 1210:'Get main menu
240      IF MENU = 8 THEN 100
250      IF MENU = 9 THEN 310
260      IF MENU = 6 THEN GOSUB 4480
270      IF MENU = 6 THEN GOTO 230
280    GOSUB 4840:'Select operating frequency (FREQ=MHz,FREQHZ=Hz)
290 ON MENU GOSUB 1630,10330,6540,9420,7760,4480,310,100,310
300    GOTO 230
310 'END OF PROGRAM SUBROUTINE
320    COLOR 1,0:COLOR 14
330    LINE (406,242)-(202,135),3,BF:'Draw box
340    LOCATE 11,27:PRINT "                        "
350    LOCATE 12,27:PRINT "     Joseph J. Carr     "
360    LOCATE 13,27:PRINT "         K4IPV          "
370    LOCATE 14,27:PRINT "                        "
380    LOCATE 15,27:PRINT "       POB 1099         "
390    LOCATE 16,27:PRINT " Falls Church, VA 22041 "
400    LOCATE 17,27:PRINT "                        "
410 N = .2:DIT = .75:DASH = 2.8
420 SOUND (1000),DASH
430 CODE=TIMER:WHILE TIMER<CODE+N:WEND
440 SOUND (1000),DASH
450 CODE=TIMER:WHILE TIMER<CODE+N:WEND
460 SOUND (1000),DIT
470 CODE=TIMER:WHILE TIMER<CODE+N/2:WEND
480 SOUND (1000),DIT
490 CODE=TIMER:WHILE TIMER<CODE+N/2:WEND
500 SOUND (1000),DIT
510 CODE=TIMER:WHILE TIMER<CODE+3*N:WEND:'end first character
520 SOUND (1000),DIT
530 CODE=TIMER:WHILE TIMER<CODE+N/2:WEND
540 SOUND (1000),DIT
550 CODE=TIMER:WHILE TIMER<CODE+N/2:WEND
560 SOUND (1000),DIT
570 CODE=TIMER:WHILE TIMER<CODE+N/2:WEND
580 SOUND (1000),DASH
```

```
590 CODE=TIMER:WHILE TIMER<CODE+N:WEND
600 SOUND (1000),DASH
610 CLS:SCREEN 0:END
620 ' OPENING SCREEN SUBROUTINE (LOGO)
630     NTE(1)=523.25:NTE(2)=493.88:NTE(3)=523.25:NTE(4)=587.33:NTE(5)=659.26
640     NTE(6)=698.46:NTE(7)=783.99:NTE(8)=880:NTE(9)=987.77:NTE(10)=1046.5
650     CLS:SCREEN 9:XXX1=400:XXX2=100:YYY1=50:YYY2=200:M=10:COLOR 15
660     LINE (XXX1,YYY1)-(XXX2,YYY2),,B:SOUND NTE(M),10
670     M=M-1:IF M = 0 THEN 690 ELSE 680
680     XXX1=XXX1+10:XXX2=XXX2+10:YYY1=YYY1+10:YYY2=YYY2+10:GOTO 660
690     COLOR 14:LOCATE 12,34:PRINT " ANTLERS ":COLOR 15
700     LOCATE 14,25:PRINT " Copyright 1992 J.J. Carr "
710   TIMELOOP=TIMER:WHILE TIMER < TIMELOOP + 2:WEND:'End of open logo
720 CLS:RETURN:'End of opening logo subroutine.
730 'NAME OF PROGRAM SCREEN SUBROUTINE
740    LINE (450,240)-(200,240),1:'Draw box
750    LINE (449,239)-(201,239),1
760    LINE (450,240)-(450,100),1
770    LINE (449,240)-(449,100),1
780    LINE (450,100)-(200,100),9
790    LINE (200,240)-(200,100),9
800    LINE (444,236)-(204,104),1,BF
810    LOCATE 9,28:PRINT  "                              "
820    LOCATE 10,28:PRINT "  ANTLERS is a computer       "
830    LOCATE 11,28:PRINT "  program that calculates     "
840    LOCATE 12,28:PRINT "  the lengths of various      "
850    LOCATE 13,28:PRINT "  types of shortwave and      "
860    LOCATE 14,28:PRINT "  VHF and UHF receiver        "
870    LOCATE 15,28:PRINT "  antennas.                   "
880    LOCATE 16,28:PRINT "                              "
890  TIMELOOP=TIMER:WHILE TIMER < TIMELOOP+5:WEND
900    COLOR 7,1:COLOR 14:'Reset screen colors
910 'Draw antenna symbol
920    CLS
930    LINE (130,50)-(74,50)
940    LINE (102,85)-(74,50)
950    LINE (130,50)-(102,85)
960    LINE (102,200)-(102,85)
970    LINE (140,200)-(102,200)
980    LINE (140,210)-(102,210)
990    LINE (102,230)-(102,210)
1000    LINE (114,230)-(90,230)
1010    LINE (108,235)-(96,235)
1020    LINE (104,240)-(100,240)
1030    LINE (200,220)-(140,190),,B
1040    LINE (325,220)-(230,190),,B
1050    LINE (230,205)-(200,205)
1060    LINE (230,204)-(200,204)
1070    LINE (375,205)-(325,205)
1080    LINE (375,205)-(370,200)
1090    LINE (375,205)-(370,210)
1100    LINE (150,131)-(105,131)
1110    LINE (105,131)-(112,136):'Draw arrow
1120    LINE (105,131)-(112,125)
1130    LOCATE 15,20:PRINT " ATU "
1140    LOCATE 15,31:PRINT " RECEIVER "
1150    LOCATE 15,48:PRINT " YOU! "
1160    LOCATE 3,10:PRINT " ANTLERS "
1170    LOCATE 10,20:PRINT " DOWNLEAD/TRANSMISSION LINE "
1180    LOCATE 19,10:PRINT " GROUND "
1190    TIMELOOP=TIMER:WHILE TIMER<TIMELOOP+3:WEND
1200 BEEP:CLS:RETURN:'End of subroutine
1210 'MAIN MENU SUBROUTINE
1220    CLS
```

```
1230    LINE (455,330)-(190,100),3,BF
1240    LOCATE 9,28:PRINT  "                                 "
1250    LOCATE 10,28:PRINT " MAIN MENU - SELECT CHOICE "
1260    LOCATE 11,28:PRINT "                                 "
1270    LOCATE 12,28:PRINT " (D)ipoles                       "
1280    LOCATE 13,28:PRINT " (O)ther Wire Antennas           "
1290    LOCATE 14,28:PRINT " (V)erticals                     "
1300    LOCATE 15,28:PRINT " (B)eams                         "
1310    LOCATE 16,28:PRINT " (L)oops (small RDF)             "
1320    LOCATE 17,28:PRINT " (G)ateway                       "
1330    LOCATE 19,28:PRINT " (R)estart Program               "
1340    LOCATE 20,28:PRINT " (E)nd Program                   "
1350    LOCATE 21,28:PRINT "                                 "
1360    LOCATE 22,28:PRINT "                                 "
1370    LOCATE 23,28:PRINT " Make Selection:                 ";
1380    MENU$ = INPUT$(1)
1390 'Test input value
1400    IF MENU$ = "D" THEN MENU = 1
1410    IF MENU$ = "d" THEN MENU = 1
1420    IF MENU$ = "O" THEN MENU = 2
1430    IF MENU$ = "o" THEN MENU = 2
1440    IF MENU$ = "V" THEN MENU = 3
1450    IF MENU$ = "v" THEN MENU = 3
1460    IF MENU$ = "B" THEN MENU = 4
1470    IF MENU$ = "b" THEN MENU = 4
1480    IF MENU$ = "L" THEN MENU = 5
1490    IF MENU$ = "l" THEN MENU = 5
1500    IF MENU$ = "G" THEN MENU = 6
1510    IF MENU$ = "g" THEN MENU = 6
1520    IF MENU$ = "I" THEN MENU = 7
1530    IF MENU$ = "i" THEN MENU = 7
1540    IF MENU$ = "R" THEN MENU = 8
1550    IF MENU$ = "r" THEN MENU = 8
1560    IF MENU$ = "E" THEN MENU = 9
1570    IF MENU$ = "e" THEN MENU = 9
1580      IF MENU < 0 THEN 1210
1590      IF MENU = 0 THEN 1210
1600      IF MENU > 9 THEN 1210
1610      IF INT(MENU)=MENU THEN 1620 ELSE 1210
1620 CLS:RETURN:'End subroutine
1630 'SELECT TYPE OF DIPOLE
1640    CLS
1650    LINE (456,243)-(162,105),3,BF
1660    LOCATE  9,22:PRINT "            DIPOLE MENU            "
1670    LOCATE 10,22:PRINT "                                   "
1680    LOCATE 11,22:PRINT " (R)egular Half wavelength Dipole  "
1690    LOCATE 12,22:PRINT " (I)nverted-Vee Dipole             "
1700    LOCATE 13,22:PRINT " (T)hree-Quarter Wavelength Dipole "
1710    LOCATE 14,22:PRINT " (S)hortened Coil-Loaded Dipole    "
1720    LOCATE 15,22:PRINT " (M)ain Menu                       "
1730    LOCATE 16,22:PRINT "                                   "
1740    LOCATE 17,22:PRINT "  Make Selection:                  ";
1750    DIPOLE$=INPUT$(1)
1760      IF DIPOLE$ = "R" THEN DIPOLE = 1
1770      IF DIPOLE$ = "r" THEN DIPOLE = 1
1780      IF DIPOLE$ = "I" THEN DIPOLE = 2
1790      IF DIPOLE$ = "i" THEN DIPOLE = 2
1800      IF DIPOLE$ = "T" THEN DIPOLE = 3
1810      IF DIPOLE$ = "t" THEN DIPOLE = 3
1820      IF DIPOLE$ = "S" THEN DIPOLE = 4
1830      IF DIPOLE$ = "s" THEN DIPOLE = 4
1840      IF DIPOLE$ = "M" THEN DIPOLE = 5
1850      IF DIPOLE$ = "m" THEN DIPOLE = 5
1860        IF DIPOLE < 1 THEN 1630 ELSE 1870
```

```
1870        IF DIPOLE > 5 THEN 1630 ELSE 1880
1880        IF DIPOLE = 0 THEN 1630 ELSE 1890
1890        IF INT(DIPOLE) = DIPOLE THEN 1900 ELSE 1630
1900      IF DIPOLE = 1 THEN KD = 468:'Dipole numerator constant
1910      IF DIPOLE = 2 THEN KD = 496:'Dipole numerator constant
1920      IF DIPOLE = 3 THEN KD = 702:'Dipole numerator constant
1930      IF DIPOLE = 4 THEN KD = 0:'Dipole selected special loaded type
1940      IF DIPOLE = 4 THEN GOSUB 2090
1950   IF DIPOLE = 4 THEN GOTO2080
1960      IF DIPOLE = 5 THEN GOTO 4470:'Exit from subroutine
1970        F = FREQ:LENGTH = KD/F:'Calculate length of dipole
1980 'Dipole results print-out
1990      CLS:LINE (460,198)-(145,150),3,BF
2000      LOCATE 12,20:PRINT " Design Center Frequency: ";F*1000;" KHz "
2010      LOCATE 13,20:PRINT "          Overall Length: ";
2020      PRINT USING "###.##";LENGTH;:PRINT " Feet "
2030 IF DIPOLE = 3 THEN 2060 ELSE 2040
2040      LOCATE 14,20:PRINT "     Length each Element: ";
2050      PRINT USING "###.##";LENGTH/2;:PRINT " Feet "
2060      LOCATE 22,20
2070 GOSUB 13910
2080 CLS:RETURN:'End of Dipole Subroutine
2090 'Shortened Loaded Dipole Sub-Subroutine
2100    K=234:F = FREQ
2110 ' EXECUTION SUBROUTINE
2120     GOSUB 3730:' Get dipole graphic screen for 5 seconds
2130     GOSUB 2310:' Get opening announcment
2140     GOSUB 2410:' Get overall length of antenna in feet (returns A)
2150     GOSUB 3050:' Go test A for correct value referenced to F
2160       IF LL = 1 THEN CLS
2170       IF LL = 1 THEN 4460
2180     GOSUB 2540:' Get position of loading coil (returns B)
2190     GOSUB 3160:' Get antenna element conductor diameter (Return D)
2200     GOSUB 4170:' Go do calculations
2210   GOSUB 4340:' Go printout results
2220 GOTO 4460
2230 ' END OF PROGRAM SUBROUTINE
2240 CLS
2250 LINE (320,180)-(220,140),3,BF
2260 LINE (330,190)-(210,130),2,B
2270 LOCATE 12,30:PRINT " GOODBYE "
2280 TIMELOOP=TIMER:WHILE TIMER<TIMELOOP+3:WEND
2290 CLS:SCREEN 0
2300 END
2310 ' OPENING ANNOUNCEMENT
2320 CLS:COLOR 14
2330    LINE (555,240)-(134,125),3,BF:' Make colored text box
2340    LOCATE 11,20:PRINT "                                                   "
2350    LOCATE 12,20:PRINT " This program calculates the inductive reactance "
2360    LOCATE 13,20:PRINT " and inductance required for loading coils in a  "
2370    LOCATE 14,20:PRINT " shortened dipole antenna.                         "
2380    LOCATE 15,20:PRINT "                                                   "
2390 LOCATE 16,30:GOSUB 4130
2400 RETURN:' End of subroutine
2410 ' OVERALL ANTENNA LENGTH SUBROUTINE
2420      CLS
2430      LINE (580,195)-(125,140),3,BF
2440      LOCATE 13,20:PRINT " and then press ENTER                 "
2450      LOCATE 12,20:PRINT " Input overall antenna length in feet ";
2460      INPUT A$:' Get overall length in feet
2470        ' Check for good input
2480            IF A$="" THEN BEEP
2490            IF A$="" THEN 2410
2500            A = VAL(A$)
```

```
2510            IF A = 0 THEN BEEP
2520            IF A = 0 THEN 2410
2530 RETURN:' End of subroutine
2540 ' SUBROUTINE TO DETERMINE COIL LOCATION
2550     CLS
2560     LINE (550,280)-(120,130),3,BF
2570     LOCATE 11,20:PRINT "                                                 "
2580     LOCATE 12,20:PRINT " Please select location of coil                  "
2590     LOCATE 13,20:PRINT "                                                 "
2600     LOCATE 14,20:PRINT " (C)enter of each element (50-percent)           "
2610     LOCATE 15,20:PRINT " (O)ne-third way on each element (33-percent) "
2620     LOCATE 16,20:PRINT " (F)eedpoint of antenna (0-percent)              "
2630     LOCATE 17,20:PRINT " (S)elect different location                     "
2640     LOCATE 18,20:PRINT "                                                 "
2650     LOCATE 19,20:PRINT " Make selection please...                        ";
2660     B$ = INPUT$(1)
2670       ' Check for good input
2680         IF B$ = "" THEN BEEP
2690         IF B$ = "" THEN 2540
2700         BCHEK=VAL(B$)
2710         IF BCHEK > 0 THEN BEEP
2720         IF BCHEK > 0 THEN 2540
2730         IF B$="0" THEN BEEP
2740         IF B$="0" THEN 2540
2750       ' Convert B$ to B numeric
2760          IF B$ = "C" THEN B = .5*(A/2)
2770          IF B$ = "c" THEN B = .5*(A/2)
2780          IF B$ = "O" THEN B = .333*(A/2)
2790          IF B$ = "o" THEN B = .333*(A/2)
2800          IF B$ = "F" THEN B = .0001*(A/2)
2810          IF B$ = "f" THEN B = .0001*(A/2)
2820          IF B$ = "S" THEN B = 1
2830          IF B$ = "s" THEN B = 1
2840      ' Test value of B numeric
2850         IF B = 0 THEN BEEP
2860         IF B = 0 THEN 2540
2870      ' Decide what to do based on value of B
2880         IF B = 1 THEN 2900 ELSE 3040
2890      ' Select own percentage for loading coil
2900        CLS:LINE (550,240)-(120,130),3,BF
2910        LOCATE 11,20:PRINT "                                              "
2920        LOCATE 12,20:PRINT " Enter location of loading coil in feet       "
2930        LOCATE 13,20:PRINT " from center feed point of antenna. Must      "
2940        LOCATE 14,20:PRINT " be less than overall length entered before. "
2950        LOCATE 15,20:PRINT "                                              "
2960        LOCATE 16,20:PRINT " Input value and press ENTER                  ";
2970         INPUT B$
2980         B = VAL(B$)
2990       ' Check for good input
3000         IF B = 0 THEN BEEP
3010         IF B = 0 THEN 2900
3020         IF B > A THEN BEEP
3030         IF B > A THEN 2900
3040 RETURN:' End of subroutine
3050 ' SUBROUTINE TO TEST FOR VALUE OF "A" RELATIVE TO "F"
3060     L = 468/F:' Calculate regular length of fullsize dipole
3070     IF A > L THEN 3080 ELSE 3150:'Compare to full size dipole
3080     BEEP:CLS:LINE (550,220)-(130,130),3,BF:' Message for L>A error
3090     LOCATE 11,20:PRINT "                                                 "
3100     LOCATE 12,20:PRINT " Shortened dipole not needed because selected "
3110     LOCATE 13,20:PRINT " length is longer than half-wavelength at the "
3120     LOCATE 14,20:PRINT " selected frequency.                             "
3130     LOCATE 15,20:PRINT "                                                 "
3140 LL=1:TIMELOOP=TIMER:WHILE TIMER<TIMELOOP+3:WEND
```

```
3150 RETURN:' End of subroutine
3160 ' SUBROUTINE TO DETERMINE ANTENNA CONDUCTOR SIZE
3170    CLS:' Draw screen
3180    LINE (520,340)-(125,80),3,BF
3190    LOCATE  8,20:PRINT "                                          "
3200    LOCATE 10,20:PRINT "                                          "
3210    LOCATE  9,20:PRINT " Select antenna element conductor size "
3220    LOCATE 11,20:PRINT " 1.  #10 wire                             "
3230    LOCATE 12,20:PRINT " 2.  #12 wire                             "
3240    LOCATE 13,20:PRINT " 3.  #14 wire                             "
3250    LOCATE 14,20:PRINT " 4.  #16 wire                             "
3260    LOCATE 15,20:PRINT " 5.  #18 wire (not recommended)           "
3270    LOCATE 16,20:PRINT " 6.  #20 wire (not recommended)           "
3280    LOCATE 17,20:PRINT " 7.  #22 wire (not recommended)           "
3290    LOCATE 18,20:PRINT " 8.  Aluminum or copper tubing            "
3300    LOCATE 19,20:PRINT "                                          "
3310    LOCATE 21,20:PRINT " Make selection...                        ";
3320    D$ = INPUT$(1)
3330     'Check for good input
3340        D = VAL(D$)
3350        IF D < 1 THEN BEEP
3360        IF D < 1 THEN 3160
3370        IF D > 8 THEN BEEP
3380        IF D > 8 THEN 3160
3390        IF D = 1 THEN DD$ = " #10 wire "
3400        IF D = 2 THEN DD$ = " #12 wire "
3410        IF D = 3 THEN DD$ = " #14 wire "
3420        IF D = 4 THEN DD$ = " #16 wire "
3430        IF D = 5 THEN DD$ = " #18 wire "
3440        IF D = 6 THEN DD$ = " #20 wire "
3450        IF D = 7 THEN DD$ = " #22 wire "
3460     ' Select aluminum/copper tubing size
3470        IF D = 8 THEN 3600 ELSE 3490
3480        IF D = 8 THEN DD$ = "Alum/Copper Tubing "
3490        LOCATE 22,20:PRINT DD$
3500        TIMELOOP=TIMER:WHILE TIMER<TIMELOOP+.5:WEND:CLS
3510        IF D = 1 THEN D = .1019
3520        IF D = 2 THEN D = .0808
3530        IF D = 3 THEN D = .0641
3540        IF D = 4 THEN D = .0508
3550        IF D = 5 THEN D = .0403
3560        IF D = 6 THEN D = .032
3570        IF D = 7 THEN D = .0253
3580        IF D = 8 THEN D = D
3590 GOTO 3710:' Go to end of routine
3600     ' Subroutine to select tubing diameter
3610        CLS:LINE (500,200)-(120,120),3,BF:' Draw screen
3620        LOCATE 12,20:PRINT "                                          "
3630        LOCATE 13,20:PRINT " Select tubing outside diameter (o.d.) "
3640        LOCATE 14,20:PRINT " 0.5 inch to 2 inch                       ";
3650        INPUT D:'Enter tubing size
3660        IF D < .5 THEN BEEP
3670        IF D < .5 THEN 3600
3680        IF D > 2 THEN BEEP
3690        IF D > 2 THEN 3600
3700        GOTO 3480
3710 RETURN:' End of subroutine
3720 LINE (52,105)-(58,110)
3730 'SUBROUTINE FOR GRAPHIC OPENING
3740 CLS
3750 LINE (600,150)-(50,150)
3760 LINE (600,149)-(50,149)
3770 LINE (335,150)-(315,150),0,BF
3780 LINE (335,149)-(315,149),0,BF
```

```
3790 LINE (335,225)-(335,150)
3800 LINE (315,225)-(315,150)
3810 LINE (198,155)-(178,145),3,BF
3820 LINE (473,155)-(453,145),3,BF
3830 LINE (600,142)-(600,90)
3840 LINE (50,142)-(50,90)
3850 LINE (52,105)-(598,105)
3860 LINE (340,105)-(310,105),0,BF
3870 LOCATE 8,41:PRINT "A"
3880 LINE (52,105)-(58,100)
3890 LINE (52,105)-(58,110)
3900 LINE (598,105)-(592,100)
3910 LINE (598,105)-(592,110)
3920 LINE (449,130)-(202,130)
3930 LINE (449,145)-(449,125)
3940 LINE (201,145)-(201,125)
3950 LINE (337,135)-(313,125),0,BF
3960 LOCATE 10,32:PRINT " B "
3970 LOCATE 10,48:PRINT " B "
3980 LINE (335,143)-(335,125)
3990 LINE (315,143)-(315,125)
4000 LINE (315,130)-(310,125)
4010 LINE (315,130)-(310,135)
4020 LINE (202,130)-(207,125)
4030 LINE (202,130)-(207,135)
4040 LINE (449,130)-(444,125)
4050 LINE (449,130)-(444,135)
4060 LINE (335,130)-(340,125)
4070 LINE (335,130)-(340,135)
4080 LOCATE 13,24:PRINT "L1"
4090 LOCATE 13,58:PRINT "L2"
4100 LOCATE 18,18:PRINT " Form of the inductor loaded shortened dipole "
4110 TIMELOOP=TIMER:WHILE TIMER<TIMELOOP+5:WEND
4120 CLS:RETURN:' End of subroutine
4130 ' SUBROUTINE: Press Any Key
4140     PRINT " Press any key to continue "
4150     AA$=INKEY$:IF AA$="" THEN 4150
4160 RETURN:' End of subroutine
4170 'CALCULATIONS SUBROUTINE
4180     CLS:LINE (500,150)-(130,120),3,BF:' Draw screen
4190      LOCATE 10,30:PRINT " Doing Arithmetic "
4200     LOCATE 12,30:PRINT "                    "
4210        'Arithmetic
4220           MA# = (10^6)/(34*PI#*F)
4230           MB# = (LOG(((24*(K/F))-B)/(D)) - 1)
4240           MC# = (K/F) - B
4250           MD# = (((1 - ((F*B)/(K)) )^2) - 1)
4260           ME# = (MB#*MD#)/MC#
4270           MF# = (LOG((1/D)*24*((A/2)-B))) - 1
4280           MG# = ((((F*A)/2)-(F*B))/K)^2 - 1
4290           MH# = ((A/2) - B)
4300           MI# = (MF#*MG#)/MH#
4310           XL# = MA#*(ME# - MI#)
4320           LUH# = XL#/(2*PI#*F)
4330 RETURN:' End of subroutine
4340 ' RESULTS PRINTOUT SUBROUTINE
4350 IF B<1 THEN B = 0
4360 CLS:LINE (550,255)-(120,130),3,BF
4370 LOCATE 12,20:PRINT " Operating frequency: ";F;" MHz          "
4380 LOCATE 13,20:PRINT " Overall length of antenna: ";A;" Feet        "
4390 LOCATE 14,20:PRINT " Distance from center to each coil: ";B;" Feet "
4400 LOCATE 15,20:PRINT " Inductive reactance of coil : ";
4410 PRINT USING "#####.#";XL#;:PRINT " Ohms  "
4420 LOCATE 16,20:PRINT " Inductance of coil: ";
```

```
4430 PRINT USING "####.##";LUH#;:PRINT " uH          "
4440 LOCATE 18,20:GOSUB 4130
4450 CLS:RETURN:' End of subroutine
4460 CLS:RETURN
4470 CLS:RETURN:'End of Dipole Subroutine
4480 'Gateway subroutine
4490      CLS
4500      LOCATE 7,20:PRINT "OOOOps! The Gateway is not yet enabled!"
4510      LOCATE 8,20:PRINT "It is a worm hole to a parallel universe"
4520      LOCATE 9,20:PRINT "called SOCKENIA, which is populated only"
4530      LOCATE 10,20:PRINT "by socks and stockings. The gateway is a"
4540      LOCATE 11,20:PRINT "worm hole at the bottom of a cosmic black"
4550      LOCATE 12,20:PRINT "hole. Across the opening of the worm hole"
4560      LOCATE 13,20:PRINT "is covered with a structure like a semi-"
4570      LOCATE 14,20:PRINT "permeable membrane that passes only socks."
4580      LOCATE 15,20:PRINT "The terminus of the Sockenian worm hole in"
4590      LOCATE 16,20:PRINT "our universe is apparently somewhere inside"
4600      LOCATE 17,20:PRINT "my clothes closet. No dirty sock that enters"
4610      LOCATE 18,20:PRINT "ever returns to this universe. Mysterious!"
4620      LOCATE 21,20:GOSUB 13910
4630 RETURN
4640   LINE (470,246)-(145, 12),3,BF
4650   LOCATE 2,20:PRINT "                                          "
4660   LOCATE 3,20:PRINT " Based on two books by Joseph J. Carr: "
4670   LOCATE 4,20:PRINT "                                          "
4680   LOCATE 5,20:PRINT "        RECEIVER ANTENNA HANDBOOK         "
4690   LOCATE 6,20:PRINT "          HighText Publications           "
4700   LOCATE 7,20:PRINT "          7128 Miramar Road, #15          "
4710   LOCATE 8,20:PRINT "          San Diego, CA, 92121            "
4720   LOCATE 9,20:PRINT "             (619) 693-5900               "
4730   LOCATE 10,20:PRINT "                                          "
4740   LOCATE 11,20:PRINT "                  -and-                   "
4750   LOCATE 12,20:PRINT "                                          "
4760   LOCATE 13,20:PRINT "       PRACTICAL ANTENNA HANDBOOK         "
4770   LOCATE 14,20:PRINT "             TAB/McGraw-Hill              "
4780   LOCATE 15,20:PRINT "    Blue Ridge Summit, PA, 17294          "
4790   LOCATE 16,20:PRINT "             (717) 794-2191               "
4800   LOCATE 17,20:PRINT "                                          "
4810   LOCATE 18,27:GOSUB 13910
4820   CLS:RETURN
4830 RETURN
4840 'FREQUENCY SELECTION SUBROUTINE
4850      CLS
4860      LINE (417,215)-(158,106),3,BF
4870      LOCATE  9,22:PRINT "       SELECT FREQUENCY      "
4880      LOCATE 10,22:PRINT "                              "
4890      LOCATE 11,22:PRINT " (E)nter Specific Frequency  "
4900      LOCATE 12,22:PRINT " (S)elect Standard Band       "
4910      LOCATE 13,22:PRINT " (M)ain Menu                  "
4920      LOCATE 14,22:PRINT "                              "
4930      LOCATE 15,22:PRINT " Make Selection:              ";
4940      FRQ$ = INPUT$(1)
4950        IF FRQ$ = "" THEN 4840 ELSE
4960:'Test input 4960       IF FRQ$ = "E" THEN FRQ = 1
4970        IF FRQ$ = "e" THEN FRQ = 1
4980        IF FRQ$ = "S" THEN FRQ = 2
4990        IF FRQ$ = "s" THEN FRQ = 2
5000        IF FRQ$ = "M" THEN FRQ = 3
5010        IF FRQ$ = "m" THEN FRQ = 3
5020        IF FRQ < 1 THEN 4840 ELSE 5030
5030        IF FRQ > 3 THEN 4840 ELSE 5040
5040        IF FRQ = 0 THEN 4840 ELSE 5050
5050        IF FRQ = 1 THEN GOTO 5080:'Branch to correct subroutine
5060        IF FRQ = 2 THEN GOTO 5330
```

```
5070       IF FRQ = 3 THEN GOTO 6520
5080 'Sub-subroutine for entering frequency by keyboard
5090       CLS
5100       LINE (384,215)-(143,133),3,BF
5110       LOCATE 11,20:PRINT "                              "
5120       LOCATE 12,20:PRINT " Enter Desired Frequency in "
5130       LOCATE 13,20:PRINT " Kilohertz (KHz) And Then   "
5140       LOCATE 14,20:PRINT " Press ENTER...             "
5150       LOCATE 15,20:PRINT "                              "
5160       LOCATE 19,20
5170       INPUT  FREQ$:'Input the desired frequency
5180       FREQ = VAL(FREQ$)
5190         IF FREQ < 10 THEN BEEP:'Test frequency limits
5200         IF FREQ < 10 THEN GOTO 5240
5210         IF FREQ > 50000! THEN BEEP
5220         IF FREQ > 50000! THEN GOTO 5280
5230       GOTO 6460
5240 'ERROR MESSAGE (Frequency too low)
5250     CLS:LOCATE 12,25:PRINT " FREQUENCY TOO LOW - TRY AGAIN "
5260     TIMELOOP = TIMER:WHILE TIMER < TIMELOOP + 2:WEND
5270     GOTO 5080:'Return to start of subroutine
5280 'ERROR MESSAGE (Frequency too high)
5290     CLS:LOCATE 12,25:PRINT " FREQUENCY TCO HIGH - TRY AGAIN "
5300     TIMELOOP = TIMER:WHILE TIMER < TIMELOOP + 2:WEND
5310     GOTO 5080:'Return to start of subroutine
5320 GOTO 5950
5330 'Sub-subroutine for standard band selection
5340     CLS:LINE (412,255)-(187,135),3,BF
5350     LOCATE 11,25:PRINT "                            "
5360     LOCATE 12,25:PRINT "  Select A Standard Band    "
5370     LOCATE 13,25:PRINT "                            "
5380     LOCATE 14,25:PRINT " (H)am Radio Bands          "
5390     LOCATE 15,25:PRINT " (I)nternational Broadcast "
5400     LOCATE 16,25:PRINT " (O)ther Bands              "
5410     LOCATE 17,25:PRINT " (M)ain Menu                "
5420     LOCATE 18,25:PRINT "                            ";
5430     LOCATE 22,25 5440     WW$ = INPUT$(1)
5450       IF WW$ = "H" THEN WW = 1:'Test input value
5460       IF WW$ = "h" THEN WW = 1
5470       IF WW$ = "I" THEN WW = 2
5480       IF WW$ = "i" THEN WW = 2
5490       IF WW$ = "O" THEN WW = 3
5500       IF WW$ = "o" THEN WW = 3
5510       IF WW$ = "M" THEN WW = 4
5520       IF WW$ = "m" THEN WW = 4
5530       IF WW < 1 THEN 5330 ELSE 5540
5540       IF WW > 4 THEN 5950:'Go to exit
5550     ON WW GOTO 5560,5860,6160,6460
5560 'Ham band Menu
5570       CLS:LINE (333,298)-(186,108),3,BF
5580       LOCATE 9,25:PRINT "                   "
5590       LOCATE 10,25:PRINT " 1. 160-Meters   "
5600       LOCATE 11,25:PRINT " 2. 75/80-Meters "
5610       LOCATE 12,25:PRINT " 3. 40-Meters    "
5620       LOCATE 13,25:PRINT " 4. 30-Meters    "
5630       LOCATE 14,25:PRINT " 5. 20-Meters    "
5640       LOCATE 15,25:PRINT " 6. 17-Meters    "
5650       LOCATE 16,25:PRINT " 7. 15-Meters    "
5660       LOCATE 17,25:PRINT " 8. 12-Meters    "
5670       LOCATE 18,25:PRINT " 9. 10-Meters    "
5680       LOCATE 19,25:PRINT "                  "
5690       LOCATE 20,25:PRINT " Select One      "
5700       LOCATE 21,25:PRINT "                  ";
5710       WY$ = INPUT$(1)
```

```
5720        WY =  VAL(WY$)
5730        IF WY < 1 THEN 5560
5740        IF WY > 9 THEN 5560
5750        IF WY = 0 THEN 5560
5760          IF WY = 1 THEN FREQ = 1900
5770          IF WY = 2 THEN FREQ = 3750
5780          IF WY = 3 THEN FREQ = 7200
5790          IF WY = 4 THEN FREQ = 10125
5800          IF WY = 5 THEN FREQ = 14175
5810          IF WY = 6 THEN FREQ = 18118
5820          IF WY = 7 THEN FREQ = 21225
5830          IF WY = 8 THEN FREQ = 24940
5840          IF WY = 9 THEN FREQ = 29000
5850  GOTO 6460
5860 'International Broadcast Menu
5870        CLS:LINE (333,298)-(186,108),3,BF
5880        LOCATE 9,25:PRINT "                  "
5890        LOCATE 10,25:PRINT " 1. 60-Meters     "
5900        LOCATE 11,25:PRINT " 2. 49-Meters     "
5910        LOCATE 12,25:PRINT " 3. 41-Meters     "
5920        LOCATE 13,25:PRINT " 4. 31-Meters     "
5930        LOCATE 14,25:PRINT " 5. 25-Meters     "
5940        LOCATE 15,25:PRINT " 6. 22-Meters     "
5950        LOCATE 16,25:PRINT " 7. 19-Meters     "
5960        LOCATE 17,25:PRINT " 8. 16-Meters     "
5970        LOCATE 18,25:PRINT " 9. 13-Meters     "
5980        LOCATE 19,25:PRINT "                  "
5990        LOCATE 20,25:PRINT " Select One       "
6000        LOCATE 21,25:PRINT "                  ";
6010        WY$ = INPUT$(1)
6020        WY =  VAL(WY$)
6030        IF WY < 1 THEN 5860
6040        IF WY > 9 THEN 5860
6050        IF WY = 0 THEN 5860
6060          IF WY = 1 THEN FREQ = 4875
6070          IF WY = 2 THEN FREQ = 6075
6080          IF WY = 3 THEN FREQ = 7200
6090          IF WY = 4 THEN FREQ = 9700
6100          IF WY = 5 THEN FREQ = 11813
6110          IF WY = 6 THEN FREQ = 13700
6120          IF WY = 7 THEN FREQ = 15350
6130          IF WY = 8 THEN FREQ = 17725
6140          IF WY = 9 THEN FREQ = 21650
6150        GOTO 6460
6160 'Other Bands Menu
6170        CLS:LINE (397,297)-(186,108),3,BF
6180        LOCATE 9,25:PRINT "                          "
6190        LOCATE 10,25:PRINT " 1. Low AM Broadcast      "
6200        LOCATE 11,25:PRINT " 2. High AM Broadcast     "
6210        LOCATE 12,25:PRINT " 3. 2182 KHz Marine       "
6220        LOCATE 13,25:PRINT " 4. 70-Meter Marine/Air   "
6230        LOCATE 14,25:PRINT " 5. 35-Meter Marine/Air   "
6240        LOCATE 15,25:PRINT " 6. 10-MHz WWV/WWVH       "
6250        LOCATE 16,25:PRINT " 7. 15-MHz WWV/WWVH       "
6260        LOCATE 17,25:PRINT " 8. 16-Meter Air          "
6270        LOCATE 18,25:PRINT " 9. 11-Meter C.B.         "
6280        LOCATE 19,25:PRINT "                          "
6290        LOCATE 20,25:PRINT " Select One               "
6300        LOCATE 21,25:PRINT "                          ";
6310        WY$ = INPUT$(1)
6320        WY =  VAL(WY$)
6330        IF WY < 1 THEN 6160
6340        IF WY > 9 THEN 6160
6350        IF WY = 0 THEN 6160
```

```
6360        IF WY = 1 THEN FREQ = 750
6370        IF WY = 2 THEN FREQ = 1350
6380        IF WY = 3 THEN FREQ = 2182
6390        IF WY = 4 THEN FREQ = 4285
6400        IF WY = 5 THEN FREQ = 8575
6410        IF WY = 6 THEN FREQ = 10000
6420        IF WY = 7 THEN FREQ = 15000
6430        IF WY = 8 THEN FREQ = 18000
6440        IF WY = 9 THEN FREQ = 27110
6450  GOTO 6460
6460    CLS:LINE (357,185)-(162,150),3,BF
6470    LOCATE 12,22:PRINT " Frequency Selected Is "
6480    LOCATE 13,22:PRINT FREQ;" KHz           "
6490    FREQ = FREQ/1000
6500    FREQHZ = FREQ*10^6
6510    TIMELOOP = TIMER:WHILE TIMER < TIMELOOP + 2:WEND
6520      IF FRQ = 3 THEN 230
6530 CLS:RETURN:'End of subroutine
6540 'SELECT TYPE OF VERTICAL
6550     CLS
6560     LINE (440,243)-(162,105),3,BF
6570     LOCATE 9,22:PRINT "          VERTICALS MENU         "
6580     LOCATE 10,22:PRINT "                                  "
6590     LOCATE 11,22:PRINT " (Q)uarter Wavelength Vertical    "
6600     LOCATE 12,22:PRINT " (F)ive-Eighths Wavelength Vert. "
6610     LOCATE 13,22:PRINT " (T)hree-Quarter Wavelength Vert."
6620     LOCATE 14,22:PRINT " (S)hortened (Coil Loaded) Vert. "
6630     LOCATE 15,22:PRINT " (M)ain Menu                      "
6640     LOCATE 16,22:PRINT "                                  "
6650     LOCATE 17,22:PRINT "  Make Selection:                 ";
6660       VERT$ = INPUT$(1)
6670        IF VERT$ = "Q" THEN VERT = 1:'Set value of variable VERT
6680        IF VERT$ = "q" THEN VERT = 1
6690        IF VERT$ = "F" THEN VERT = 2
6700        IF VERT$ = "f" THEN VERT = 2
6710        IF VERT$ = "T" THEN VERT = 3
6720        IF VERT$ = "t" THEN VERT = 3
6730        IF VERT$ = "S" THEN VERT = 4
6740        IF VERT$ = "s" THEN VERT = 4
6750        IF VERT$ = "M" THEN VERT = 5
6760        IF VERT$ = "m" THEN VERT = 5
6770      IF VERT < 1 THEN 6540 ELSE 6780:'Test value of variable VERT
6780      IF VERT > 5 THEN 6540 ELSE 6790
6790      IF VERT = 0 THEN 6540 ELSE 6800
6800      IF INT(VERT) = VERT THEN 6810 ELSE 1630
6810       IF VERT = 1 THEN KD = 234:'Quarter wavelength constant
6820       IF VERT = 2 THEN KD = 585:'5/8-wavelength constant
6830       IF VERT = 3 THEN KD = 702:'3/4-wavelength constant
6840       IF VERT = 4 THEN KD = 1:'Loaded dipole selected (spec.routine)
6850       IF VERT = 5 THEN GOTO 7080:'End subroutine to main menu
6860       IF VERT = 4 THEN GOSUB 7100:'Get loaded vertical subroutine
6870       IF VERT = 4 GOTO 6960:'Printout routine for short vertical
6880       F = FREQ:LENGTH = KD/F:'Calculate vertical height
6890 'Vertical Results Print-Out
6900      CLS:LINE(460,188)-(145,150),3,BF
6910      LOCATE 12,20:PRINT " Design Center Frequency: ";F*1000;" KHz "
6920      LOCATE 13,20:PRINT "       Overall Length: ";
6930      PRINT USING "###.##";LENGTH;:PRINT " Feet   "
6940      LOCATE 15,20:GOSUB 13910:'Press any ley legend
6950      GOTO 7090:'Go to end of subroutine
6960      CLS:LINE(460,258)-(145,150),3,BF:'short vert. print out
6970      LOCATE 12,20:PRINT " Design Center Frequency: ";F*1000;" KHz "
6980      LOCATE 13,20:PRINT " Overall Length:          ";
6990      PRINT USING "####.##";LENGTH;:PRINT " Feet   "
```

```
7000      LOCATE 14,20:PRINT " Antenna Capacitance: ";
7010      PRINT USING "####.##";CA;:PRINT " pF "
7020      LOCATE 15,20:PRINT " Required Inductance: ";
7030      PRINT USING "####.##";LUH;:PRINT " uH "
7040      LOCATE 16,20:PRINT " Reactance: ";
7050      PRINT USING "####.##";XL;:PRINT " Ohms "
7060      LOCATE 17,20:PRINT "                          "
7070      LOCATE 18,20:GOSUB 13910
7080 '
7090 CLS:RETURN:'End of subroutine
7100 'Sub-subroutine for loaded verticals
7110     CLS:F = FREQ
7120     LINE(440,180)-(135,100),3,BF
7130     LOCATE 9,20:PRINT " Coil loaded short verticals are  "
7140     LOCATE 10,20:PRINT " shorter than quarter wavelength, "
7150     LOCATE 11,20:PRINT " and use an inductor to make the  "
7160     LOCATE 12,20:PRINT " antenna resonant.                "
7170     TIMELOOP=TIMER:WHILE TIMER < TIMELOOP + 3:WEND:CLS
7180     LINE (410,150)-(135,100),3,BF
7190     LOCATE 9,20:PRINT " Input Antenna LENGTH in Feet "
7200     LOCATE 10,20:PRINT " And then Press ENTER        ";
7210     INPUT LENGTH
7220     TEST = 234/F
7230     IF LENGTH > .9*TEST THEN BEEP
7240     IF LENGTH > .9*TEST GOTO 7670:'failed short test
7250 IF LENGTH < 1 THEN 7180 ELSE 7260
7260     CLS:'Select radiator conductor size
7270     LINE(460,290)-(155,100),3,BF
7280     LOCATE 9,22:PRINT " Select Antenna Radiator Conductor "
7290     LOCATE 10,22:PRINT " Size From Table Below.            "
7300     LOCATE 11,22:PRINT "                                   "
7310     LOCATE 12,22:PRINT " 1. #10 Wire                       "
7320     LOCATE 13,22:PRINT " 2. #12 Wire                       "
7330     LOCATE 14,22:PRINT " 3. #14 Wire                       "
7340     LOCATE 15,22:PRINT " 4. #16 Wire                       "
7350     LOCATE 16,22:PRINT " 5. 1/2 Inch Aluminum Tubing       "
7360     LOCATE 17,22:PRINT " 6. 3/4 Inch Aluminum Tubing       "
7370     LOCATE 18,22:PRINT " 7. 7/8 Inch Aluminum Tubing       "
7380     LOCATE 19,22:PRINT " 8. 1 Inch Aluminum Tubing         "
7390     LOCATE 20,22:PRINT " 9. 1-1/2 Inch Aluminum Tubing     ";
7400     DIAM$ = INPUT$(1):'Get conductor diameter
7410     DIAM = VAL(DIAM$):'Strip off diameter selection as number
7420       IF DIAM < 1 THEN 7260 ELSE 7430:'Test input value
7430       IF DIAM > 9 THEN 7260 ELSE 7440
7440         IF DIAM = 1 THEN D = .1019
7450         IF DIAM = 2 THEN D = .0808
7460         IF DIAM = 3 THEN D = .0641
7470         IF DIAM = 4 THEN D = .0508
7480         IF DIAM = 5 THEN D = .5
7490         IF DIAM = 6 THEN D = .75
7500         IF DIAM = 7 THEN D = .875
7510         IF DIAM = 8 THEN D = 1!
7520         IF DIAM = 9 THEN D = 1.5
7530    CLS:'Clear mess off screen
7540 NUMER = 17*LENGTH
7550 DENOMA = (24*LENGTH)/D
7560 DENOMA = LOG(DENOMA)
7570 DENOMA = DENOMA - 1
7580 DENOMB = (F*LENGTH)/234
7590 DENOMB = (DENOMB)^2
7600 DENOMB = 1 - DENOMB
7610 DENOM = DENOMA*DENOMB
7620 CA = NUMER/DENOM
7630 XC = 1/(2*3.1415*FREQ*10^6*(CA/10^12) )
```

```
7640 XL = XC
7650 LUH = XL/(2*3.1415*F):'Inductance in microhenrys
7660     GOTO 7750:'Jump to return statement
7670     CLS
7680     LINE(440,160)-(135,100),3,BF
7690     LOCATE 9,20:PRINT " ERROR! Antenna Too Long For Coil "
7700     LOCATE 10,20:PRINT " Loading. Use Regular Quarter     "
7710     LOCATE 11,20:PRINT " Wavelength Vertical.             "
7720     TIMELOOP=TIMER:WHILE TIMER < TIMELOOP + 2.5:WEND
7730     LENGTH = 234/F
7740     GOTO 7750
7750     CLS:RETURN:'Return to vertical subroutine
7760 'SUBROUTINE FOR RDF LOOP ANTENNAS
7770    CLS:LINE (380,198)-(148,137),3,BF
7780    LOCATE 11,20:PRINT "                            "
7790    LOCATE 12,20:PRINT " Radio Direction Finding    "
7800    LOCATE 13,20:PRINT " Small Loop Option Selected "
7810    LOCATE 14,20:PRINT "                            "
7820    TIMELOOP=TIMER:WHILE TIMER<TIMELOOP+2.5:WEND
7830 CLS:'Draw Loop Antenna Picture
7840    LINE (316,300)-(165,300)
7850    LINE (315,300)-(315,330)
7860    LINE (165,300)-(165,110)
7870    LINE (475,110)-(165,110)
7880    LINE (475,290)-(475,110)
7890    LINE (475,290)-(180,290)
7900    LINE (180,290)-(180,120)
7910    LINE (460,120)-(180,120)
7920    LINE (460,280)-(460,120)
7930    LINE (460,280)-(195,280)
7940    LINE (195,280)-(195,130)
7950    LINE (445,130)-(197,130)
7960    LINE (445,270)-(445,130)
7970    LINE (445,270)-(350,270)
7980    LINE (350,270)-(350,330)
7990    LINE (165,245)-(120,245)
8000    LINE (165,245)-(160,240)
8010    LINE (165,245)-(160,250)
8020    LINE (195,245)-(235,245)
8030    LINE (195,245)-(200,240)
8040    LINE (195,245)-(200,250)
8050    LINE (474,90)-(166,90)
8060    LINE (474,90)-(469,85)
8070    LINE (474,90)-(469,95)
8080    LINE (165,90)-(170,95)
8090    LINE (165,90)-(170,85)
8100    LINE (165,107)-(165,70)
8110    LINE (475,107)-(475,70)
8120      LOCATE 12,30:PRINT " FORM OF THE WIRE LOOP "
8130      LOCATE 14,30:PRINT " Shape: Square "
8140      LOCATE 16,30:PRINT " A/B > 5 "
8150      LOCATE 24,38:PRINT "To Receiver
8160      LOCATE 17,14:PRINT "B"
8170      LOCATE 6,40:PRINT " A "
8180    TIMELOOP=TIMER:WHILE TIMER<TIMELOOP+7:WEND
8190      CLS:'Clear screen
8200 'LOOP INDUCTANCE PROGRAM
8210 'Initialization Routine
8220    CLS:'Clear screen if not already clear. Hiya Harry Helms
8230 'Execution Routine
8240      GOSUB 8450
8250      ON TYPE GOSUB 8730,8760,8790,8820:'Get constants
8260      GOSUB 8850:'Determine loop dimensions
8270        IF TYPE = 1 THEN GOSUB 9140:'Calculation for other loops
```

```
8280        IF TYPE = 2 THEN GOSUB 9140
8290        IF TYPE = 3 THEN GOSUB 9140
8300        IF TYPE = 4 THEN GOSUB 9140
8310      F = FREQ*1000
8320      GOSUB 9210:'Calculate capacitance needed for resonance
8330      CLS:LINE (493,250)-(100,145),3,BF
8340        LOCATE 12,16:PRINT "                                  "
8350        LOCATE 13,16:PRINT " Loop Type: ";A$                "
8360        LOCATE 14,16:PRINT "                                "
8370        LOCATE 15,16:PRINT " Inductance: ";:PRINT USING "####.#";L;
8380        PRINT " uH       "
8390        LOCATE 16,16:PRINT "                                "
8400        LOCATE 17,16:PRINT " Capacitance to Resonate ";F;" KHz: ";
8410        PRINT USING "####.#";C;:PRINT " pF"
8420      LOCATE 20,16:GOSUB 13910:'Press any key...
8430      GOSUB 9240:'Determine next step
8440 GOTO 9410:'Skip to end of program
8450 'Opening Screen
8460     LINE (450,250)-(100,100),3,BF
8470     LOCATE 9,25:PRINT "                       "
8480     LOCATE 10,25:PRINT " SELECT TYPE OF LOOP "
8490     LOCATE 11,25:PRINT "                     "
8500     LOCATE 12,25:PRINT "      (T)riangle     "
8510     LOCATE 13,25:PRINT "      (S)quare       "
8520     LOCATE 14,25:PRINT "      (H)exagon      "
8530     LOCATE 15,25:PRINT "      (O)ctagon      "
8540     LOCATE 16,25:PRINT "                     "
8550     LOCATE 17,25:PRINT "   Make Selection:   ";
8560     TYPE$ = INPUT$(1)
8570     'Test TYPE$ variable
8580        IF TYPE$ = "T" THEN TYPE = 1
8590        IF TYPE$ = "t" THEN TYPE = 1
8600        IF TYPE$ = "s" THEN TYPE = 2
8610        IF TYPE$ = "H" THEN TYPE = 3
8620        IF TYPE$ = "h" THEN TYPE = 3
8630        IF TYPE$ = "O" THEN TYPE = 4
8640        IF TYPE$ = "o" THEN TYPE = 4
8650          IF TYPE < 1 THEN 8450 ELSE 8660
8660          IF TYPE > 4 THEN 8450 ELSE 8670
8670          IF INT(TYPE) = TYPE THEN 8680 ELSE 8450
8680        IF TYPE = 1 THEN A$ = "Triangle "
8690        IF TYPE = 2 THEN A$ = "Square "
8700        IF TYPE = 3 THEN A$ = "Hexagon "
8710        IF TYPE = 4 THEN A$ = "Octagon "
8720 CLS:RETURN:'End of subroutine
8730 'Subroutine for Triangle Antennas
8740     K1 = .006:K2 = 1.1547:K3 = .65533:K4 = .1348
8750 RETURN
8760 'Subroutine for Square Antennas
8770     K1 = 8.000001E-03:K2 = 1.4142:K3 = .37942:K4 = .3333
8780 RETURN
8790 'Subroutine for Hexagon Antennas
8800     K1 = .012:K2 = 2:K3 = .65533:K4 = .1348
8810 RETURN
8820 'Subroutine for Octagon Antennas
8830     K1 = .016:K2 = 2.613:K3 = .75143:K4 = .07153
8840 RETURN
8850 'Subroutine to determine loop size
8860     CLS:LINE (458,211)-(149,152),3,BF
8870     LOCATE 12,20:PRINT "                                      "
8880     LOCATE 13,20:PRINT " What is the length of each loop SIDE "
8890     LOCATE 14,20:PRINT " in centimeters (cm)?                 "
8900     LOCATE 15,20:PRINT "                                      ";
8910        INPUT A
```

```
8920     CLS:'Now go get depth
8930     CLS:LINE (403,211)-(149,152),3,BF
8940     LOCATE 12,20:PRINT "                              "
8950     LOCATE 13,20:PRINT " What is the WIDTH of the loop "
8960     LOCATE 14,20:PRINT " in centimeters (cm)?          "
8970     LOCATE 15,20:PRINT "                              ";
8980        INPUT B
8990     CLS
9000     CLS:LINE (394,211)-(149,152),3,BF
9010     LOCATE 12,20:PRINT "                              "
9020     LOCATE 13,20:PRINT " Enter the number of turns in "
9030     LOCATE 14,20:PRINT " the loop                     "
9040     LOCATE 15,20:PRINT "                              ";
9050        INPUT N
9060     CLS
9070 RETURN:'End of subroutine
9080 'Calculation for flat wound square loops
9090     A = A/100:B = B/100
9100     XA = 2*A*4*PI*10^-7*N^2/PI
9110     XB = LOG((16*A)/B)
9120     L = XA*XB:L = L*10^6
9130 RETURN:' End of subroutine
9140 'Subroutine for other types of antenna
9150     NN = N+1:XA = K1*N^2*A
9160     XB = ((K2*A*N)/(NN*B))
9170     XB = LOG(XB)
9180     XB = XB + K3 + (K4*NN*B)/(A*N)
9190     L = XA*XB
9200 RETURN:'End of subroutine
9210 'Calculate resonance
9220     C = (1)/(4*PI^2*F^2*L):C = C*10^12
9230 RETURN:'End of subroutine
9240 'Subroutine for determing what's next
9250     CLS:LINE (450,250)-(100,100),3,BF
9260     LOCATE 10,20:PRINT "                     "
9270     LOCATE 11,20:PRINT " (D)o Another?       "
9280     LOCATE 12,20:PRINT " (F)inished?         "
9290     LOCATE 13,20:PRINT "                     "
9300     LOCATE 14,20:PRINT "  Make Selection:    ";
9310     BB$ = INPUT$(1)
9320        IF BB$ = "D" THEN BB = 1
9330        IF BB$ = "d" THEN BB = 1
9340        IF BB$ = "F" THEN BB = 2
9350        IF BB$ = "f" THEN BB = 2
9360          IF BB < 1 THEN 9240 ELSE 9370
9370          IF BB > 2 THEN 9240 ELSE 9380
9380          IF INT(BB) = BB THEN 9390 ELSE 9240
9390     CLS
9400     ON BB GOTO 8230,8440
9410 RETURN:'END OF LOOP SUBROUTINE (Return to main program)
9420 'Quad and Yagi Beam Subroutine
9430    CLS:LINE (378,211)-(165,123),3,BF
9440    LOCATE 10,22:PRINT "                          "
9450    LOCATE 11,22:PRINT " Select Beam Antenna Type "
9460    LOCATE 12,22:PRINT "                          "
9470    LOCATE 13,22:PRINT "         (Y)agi           "
9480    LOCATE 14,22:PRINT "         (Q)uad           "
9490    LOCATE 15,22:PRINT "                          ";
9500       BEAM$ = INPUT$(1)
9510         IF BEAM$ = "Y" THEN BEAM = 1
9520         IF BEAM$ = "y" THEN BEAM = 1
9530         IF BEAM$ = "Q" THEN BEAM = 2
9540         IF BEAM$ = "q" THEN BEAM = 2
9550             IF BEAM < 1 THEN 9420 ELSE 9560
```

```
9560              IF BEAM > 2 THEN 9420 ELSE 9570
9570              IF INT(BEAM) = BEAM THEN 9580 ELSE 9420
9580          ON BEAM GOSUB 9600,9820
9590      GOTO 10010:START CALCULATING LENGTHS
9600 'Subroutine to draw Yagi picture
9610      CLS:LINE (200,275)-(200, 80):'Draw director
9620      LINE (280,172)-(280,75):'Draw driven element A
9630      LINE (280,280)-(280,182):'Draw driven element B
9640      LINE (360,285)-(360,70):'Draw reflector
9650      LINE (280,172)-(260,172)
9660      LINE (280,182)-(260,182)
9670       LOCATE 21,22:PRINT "Director"
9680       LOCATE 21,42:PRINT "Reflector"
9690       LOCATE 5,32:PRINT "Element"
9700       LOCATE 4,32:PRINT "Driven"
9710      LINE (150,180)-(75,180)
9720      LINE (150,180)-(145,175)
9730      LINE (150,180)-(145,185)
9740       LOCATE 15,10:PRINT "Direction of"
9750       LOCATE 16,10:PRINT "Maximum"
9760       LOCATE 17,10:PRINT "Response"
9770       LOCATE 12,28:PRINT "To"
9780       LOCATE 13,28:PRINT "Rcvr"
9790       LOCATE 23,25:PRINT "FORM OF SELECTED ANTENNA"
9800      TIMELOOP=TIMER:WHILE TIMER<TIMELOOP+4:WEND
9810       CLS:RETURN
9820 'Subroutine to draw quad picture
9830       CLS:LINE (450,270)-(100,70),3,B
9840           LINE (285,270)-(265,270),0
9850           LINE (285,300)-(285,270),3
9860           LINE (265,300)-(265,270),3
9870           LINE (480,220)-(130,20),3,B
9880           LINE (420,330)-(70,130),3,B
9890       LOCATE 2,25:PRINT "Reflector"
9900       LOCATE 5,25:PRINT "Driven ELement"
9910       LOCATE 9,25:PRINT "Director"
9920       LOCATE 22,32:PRINT "To Rcvr"
9930           LINE (460,330)-(560,200),3
9940           LINE (460,330)-(460,323),3
9950           LINE (460,330)-(470,323),3
9960       LOCATE 22,62:PRINT "Direction of"
9970       LOCATE 23,62:PRINT "Maximum Response"
9980       LOCATE 1,25:PRINT "MANY PEOPLE USE ONLY TWO ELEMENTS"
9990     TIMELOOP=TIMER:WHILE TIMER<TIMELOOP+4:WEND
10000 CLS:RETURN:'Return to BEAM subroutine
10010 'Start calculations
10020     F = FREQ
10030     IF BEAM = 1 THEN KD = 465
10040     IF BEAM = 1 THEN KDE = 473
10050     IF BEAM = 1 THEN KR = 501
10060     IF BEAM = 1 THEN KS = 195
10070     IF BEAM = 2 THEN KD = 975
10080     IF BEAM = 2 THEN KDE = 1005
10090     IF BEAM = 2 THEN KR = 1030
10100     IF BEAM = 2 THEN KS = 180
10110     IF BEAM = 1 THEN W$ = "Yagi    "
10120     IF BEAM = 2 THEN W$ = "Quad    "
10130     LD = KD/F:LKDE=KDE/F:LKR=KR/F:LS=KS/F
10140 'Print to screen routine
10150     CLS:LINE (387,241)-(148,122),3,BF
10160     LOCATE 10,20:PRINT "                                    "
10170     LOCATE 11,20:PRINT " Type of Antenna: ";W$;"            "
10180     LOCATE 12,20:PRINT "                                    "
```

```
10190     LOCATE 13,20:PRINT " Driven Element: ";:PRINT USING "###.##";LKDE;
10200     PRINT " Feet "
10210     LOCATE 14,20:PRINT "      Reflector: ";:PRINT USING "###.##";LKR;
10220     PRINT " Feet "
10230     LOCATE 15,20:PRINT "       Director: ";:PRINT USING "###.##";LD;
10240     PRINT " Feet "
10250     LOCATE 16,20:PRINT "        Spacing: ";:PRINT USING "###.##";LS;
10260     PRINT " Feet "
10270     LOCATE 17,20:PRINT "                                "
10280     IF BEAM = 2 THEN 10310 ELSE 10290
10290     LOCATE 21,20:GOSUB 13910
10300 CLS:RETURN:'Return to main menu
10310     LOCATE 19,13:PRINT " For quads, make each side 1/4 above lengths "
10320     GOTO 10290
10330 'OTHER WIRE ANTENNA SUBROUTINE
10340     CLS:'Clear Screen
10350     LINE (402,268)-(180, 82),3,BF:'Draw box for text
10360       LOCATE 7,24:PRINT " OTHER WIRE ANTENNAS MENUS "
10370       LOCATE  8,24:PRINT "                           "
10380       LOCATE 9,24:PRINT " (I)nverted-L              "
10390       LOCATE 10,24:PRINT " (L)ong-Wire               "
10400       LOCATE 11,24:PRINT " (C)ollinear Array         "
10410       LOCATE 12,24:PRINT " (D)ouble Extended Zepp    "
10420       LOCATE 13,24:PRINT " (M)arconi                 "
10430       LOCATE 14,24:PRINT " (F)olded Tee              "
10440       LOCATE 15,24:PRINT " (Q)uarter Wave Sky Loop   "
10450       LOCATE 16,24:PRINT " (B)i-Square Loop          "
10460       LOCATE 17,24:PRINT " (T)horne Array/Bobtail    "
10470       LOCATE 18,24:PRINT "                           "
10480       LOCATE 19,24:PRINT " Make Selection:           "
10490         OTHER$=INPUT$(1):'Input selection from menu above
10500         IF OTHER$ = "I" THEN OTHER = 1
10510         IF OTHER$ = "i" THEN OTHER = 1
10520         IF OTHER$ = "L" THEN OTHER = 2
10530         IF OTHER$ = "l" THEN OTHER = 2
10540         IF OTHER$ = "C" THEN OTHER = 3
10550         IF OTHER$ = "c" THEN OTHER = 3
10560         IF OTHER$ = "D" THEN OTHER = 4
10570         IF OTHER$ = "d" THEN OTHER = 4
10580         IF OTHER$ = "M" THEN OTHER = 5
10590         IF OTHER$ = "m" THEN OTHER = 5
10600         IF OTHER$ = "F" THEN OTHER = 6
10610         IF OTHER$ = "f" THEN OTHER = 6
10620         IF OTHER$ = "Q" THEN OTHER = 7
10630         IF OTHER$ = "q" THEN OTHER = 7
10640         IF OTHER$ = "B" THEN OTHER = 8
10650         IF OTHER$ = "b" THEN OTHER = 8
10660         IF OTHER$ = "T" THEN OTHER = 9
10670         IF OTHER$ = "t" THEN OTHER = 9
10680           IF OTHER < 0 THEN 10330:'Test selection for sanity
10690           IF OTHER = 0 THEN 10330
10700           IF OTHER > 9 THEN 10330
10710           IF INT(OTHER) = OTHER THEN 10720 ELSE 10330
10720           IF OTHER = 1 THEN KA = 246
10730           IF OTHER = 1 THEN KB = 0
10740           IF OTHER = 2 THEN KA = 492
10750           IF OTHER = 2 THEN KB = 0
10760           IF OTHER = 3 THEN KA = 468
10770           IF OTHER = 3 THEN KB = 234
10780           IF OTHER = 4 THEN KA = 600
10790           IF OTHER = 4 THEN KB = 103
10800           IF OTHER = 5 THEN KA = 246
10810           IF OTHER = 5 THEN KB = 0
10820           IF OTHER = 6 THEN KA = 164
```

```
10830          IF OTHER = 6 THEN KB = 246
10840          IF OTHER = 7 THEN KA = 1003
10850          IF OTHER = 7 THEN KB = 0
10860          IF OTHER = 8 THEN KA = 2006
10870          IF OTHER = 8 THEN KB = 0
10880          IF OTHER = 9 THEN KA = 473
10890          IF OTHER = 9 THEN KB = 225
10900       ON OTHER GOT10910,11140,11560,11890,12120,12420,12900,13160,13380
10910  'Subroutine for Inverted-L Wire Antenna
10920   CLS:'Clear screen
10930   'Draw Inverted-L antenna on screen
10940       LINE (550, 25)-(250, 25)
10950       LINE (250,240)-(250,25)
10960       LINE (250,240)-(220,240)
10970       LINE (250,250)-(220,250)
10980       LINE (250,250)-(250,260)
10990       LINE (259,260)-(241,260)
11000       LINE (256,264)-(244,264)
11010       LINE (253,268)-(247,268)
11020         LOCATE 18,20:PRINT " XMTR "
11030         LOCATE 12,35:PRINT " Each Leg is Quarter Wavelength "
11040         LOCATE 10,35:PRINT " Inverted-L Antenna "
11050  TIMELOOP = TIMER:WHILE TIMER < TIMELOOP + 6:WEND
11060       CLS:F = FREQ:LENGTH = KA/F
11070  'Printout results for Inverted-L
11080         LINE (458,185)-(148,151),3,BF
11090           LOCATE 12,20:PRINT " Design Center Frequency: ";FREQ;" KHz  "
11100           LOCATE 13,20:PRINT "          Length Each Leg: ";
11110           PRINT USING "###.##";LENGTH;:PRINT " Feet "
11120         LOCATE 15,20:GOSUB 13910
11130  CLS:GOTO 13710:'Go to select course of action
11140  'Subroutine for Long-Wire Antenna
11150   CLS:'Clear Screen then draw antenna
11160     LINE (600,100)-(200,100)
11170     LINE (200,150)-(200,100)
11180     LINE (190,150)-(190,100)
11190     LINE (190,100)-( 50,100)
11200 LOCATE 1,20:PRINT "Long-Wire with Radial Counterpoise Ground"
11210       LOCATE 12,24:PRINT "XMTR"
11220       LOCATE 4,15:PRINT "Radial (Quarter Wavelength)"
11230       LOCATE 12,50:PRINT "Radiator  > Two Wavelengths"
11240     LINE (70,100)-(70, 50)
11250     LINE (100,50)-(70,50)
11260     LINE (70,100)-(65,95)
11270     LINE (75, 95)-(70,100)
11280     LINE (345,160)-(345,100)
11290     LINE (387,160)-(345,160)
11300     LINE (345,100)-(340,110)
11310     LINE (350,110)-(345,100)
11320  TIMELOOP = TIMER:WHILE TIMER < TIMELOOP + 7:WEND
11330  CLS:'End of display of long-wire form
11340       LINE (411,185)-(187,122),3,BF
11350         LOCATE 10,25:PRINT " Enter the number of        "
11360         LOCATE 11,25:PRINT " HALF Wavelengths proposed "
11370         LOCATE 12,25:PRINT " for the long-wire antenna "
11380         LOCATE 13,25:PRINT "                           ";
11390            INPUT  NNN$
11400            NNN = VAL(NNN$)
11410               IF NNN < 2 THEN BEEP
11420               IF NNN < 2 THEN GOTO 11340
11430               IF INT(NNN) = NNN THEN 11440 ELSE BEEP
11440               IF INT(NNN) = NNN THEN 11450 ELSE 11330
11450        CLS:F=FREQ:LENGTH = (492*(NNN - .025))/F
11460        RADIAL = 246/F
```

```
11470 ' Print-out Long-Wire Results
11480        LINE (475,198)-(148,150),3,BF
11490        LOCATE 12,20:PRINT " Design Center Frequency: ";F*1000;" KHz  "
11500        LOCATE 13,20:PRINT "     Length of Radiator:  ";
11510        PRINT USING "###.##";LENGTH;:PRINT " Feet  "
11520        LOCATE 14,20:PRINT "       Length of Radial:  ";
11530        PRINT USING "###.##";RADIAL;:PRINT " Feet  "
11540        LOCATE 16,20:GOSUB 13910
11550  CLS:GOTO 13710
11560 'Subroutine for Collinear Array
11570    CLS:'Clear the screen and draw antenna
11580        LINE (600,100)-(450,100)
11590        LINE (440,100)-(365,100)
11600        LINE (355,100)-(280,100)
11610        LINE (270,100)-(120,100)
11620        LINE (450,150)-(450,100)
11630        LINE (440,150)-(440,100)
11640        LINE (450,150)-(440,150)
11650        LINE (270,150)-(270,100)
11660        LINE (280,150)-(280,100)
11670        LINE (280,150)-(270,150)
11680        LINE (355,250)-(355,100)
11690        LINE (365,250)-(365,100)
11700          LOCATE 19,44:PRINT "XMTR"
11710 LOCATE 3,20:PRINT "Collinear Wire Array Antenna"
11720 LOCATE  7,20:PRINT "A                  B          B"
11730 LOCATE  7,65:PRINT "A"
11740 LOCATE 10,33:PRINT "B"
11750 LOCATE 10,54:PRINT "B"
11760 LOCATE 15,10:PRINT "A = Half Wavelength"
11770 LOCATE 16,10:PRINT "B = Quarter Wavelength"
11780 TIMELOOP = TIMER:WHILE TIMER < TIMELOOP + 7:WEND
11790 CLS:F = FREQ:LENGTHA = 468/F:LENGTHB = 246/F
11800 'Print out results for Collinear Antenna
11810   LINE (508,198)-(108,151),3,BF
11820   LOCATE 12,15:PRINT "           Design Center Frequency: ";FREQ;" KHz "
11830   LOCATE 13,15:PRINT "    Length of Half Wavelength Parts: ";
11840   PRINT USING "###.##";LENGTHA;:PRINT " Feet "
11850   LOCATE 14,15:PRINT " Length of Quarter Wavelength Parts: ";
11860   PRINT USING "###.##";LENGTHB;:PRINT " Feet "
11870   LOCATE 16,15:GOSUB 13910
11880   CLS:GOTO 13710;'Go to end of subroutine selection
11890 'Subroutine for Double Extended Zepp Antenna
11900    CLS:'Clear screen and draw antenna
11910    LINE (550,100)-(325,100)
11920    LINE (315,100)-(90 ,100)
11930    LINE (325,150)-(325,100)
11940    LINE (315,150)-(315,100)
11950 LOCATE 4,20:PRINT "Double-Extended Zepp Antenna"
11960    LOCATE 7,25:PRINT "A                          A"
11970    LOCATE 10,38:PRINT "B"
11980    LOCATE 14,20:PRINT "A is radiator element"
11990    LOCATE 15,20:PRINT "B is a matching stub"
12000    LOCATE 17,20:PRINT "Feed at bottom of stub"
12010    LOCATE 18,20:PRINT "Feedpoint impedance is 150 ohms"
12020    TIMELOOP=TIMER:WHILE TIMER < TIMELOOP + 7:WEND
12030    CLS:F=FREQ:LENGTHA = KA/F:LENGTHB = KB/F
12040    LINE (459,198)-(149,150),3,BF
12050    LOCATE 12,20:PRINT " Center Design Frequency: ";FREQ;" KHz "
12060    LOCATE 13,20:PRINT "  Length of radiator (A): ";
12070    PRINT USING "###.##";LENGTHA;:PRINT " Feet "
12080    LOCATE 14,20:PRINT " Length of matching stub: ";
12090    PRINT USING "###.##";LENGTHB;:PRINT " Feet "
12100    LOCATE 16,20:GOSUB 13910
```

```
12110    CLS:GOTO 13710
12120 'Subroutine for Marconi Antenna
12130    CLS:' Clear screen and draw picture
12140      LINE (640,100)-(250,300)
12150      LINE (250,300)-(230,300)
12160      LINE (250,310)-(230,310)
12170      LINE (250,320)-(250,310)
12180      LINE (260,320)-(240,320)
12190      LINE (255,325)-(245,325)
12200      LINE (252,330)-(247,330)
12210      LINE (230,320)-(230,290)
12220      LINE (230,290)-(160,290)
12230      LINE (230,320)-(160,320)
12240      LINE (160,320)-(160,290)
12250      LINE (160,305)-(100,305)
12260      LINE (160,304)-(100,304)
12270       LOCATE 22,9:PRINT "XMTR"
12280       LOCATE 22,24:PRINT "ATU"
12290       LOCATE 12,10:PRINT "Marconi Antenna"
12300       LOCATE 13,10:PRINT "Typically quarter wavelength or less"
12310       LOCATE 14,10:PRINT "and fed with an L-section coupler ATU"
12320 TIMELOOP=TIMER:WHILE TIMER < TIMELOOP + 7:WEND:'Hold screen 7 sec.
12330 CLS:F=FREQ:LENGTH = KA/F
12340 'Print out results of Marconi antenna calculations
12350       LINE (460,184)-(148,152),3,BF
12360       LOCATE 12,20:PRINT " Design Center Frequency: ";FREQ;" KHz "
12370       LOCATE 13,20:PRINT "       Length of antenna: ";
12380       PRINT USING "###.##";LENGTH;:PRINT " Feet "
12390       LOCATE 15,20:GOSUB 13910
12400 CLS
12410 GOTO 13710
12420 'Folded-Tee Antenna
12430       CLS:'Clear screen and draw antenna picture
12440       LINE (400, 80)-(150, 80)
12450       LINE (400,100)-(150,100)
12460       LINE (150,100)-(150,80)
12470       LINE (400,100)-(400,120)
12480       LINE (400,120)-(150,120)
12490       LINE (278,102)-(272, 98),14,BF
12500       LINE (275,320)-(275,100)
12510       LINE (275,320)-(255,320)
12520       LINE (275,330)-(255,330)
12530       LINE (284,340)-(266,340)
12540       LINE (280,343)-(270,343)
12550       LINE (278,346)-(272,346)
12560       LINE (275,340)-(275,330)
12570       LINE (175,340)-(175,310)
12580       LINE (255,340)-(175,340)
12590       LINE (255,310)-(175,310)
12600       LINE (255,340)-(255,310)
12610       LINE (175,324)-(100,324)
12620       LINE (175,326)-(100,326)
12630       LOCATE 18,10:PRINT "To XMTR"
12640       LINE (150,324)-(150,245)
12650       LINE (150,245)-(135,245)
12660       LINE (150,324)-(145,319)
12670       LINE (150,324)-(155,319)
12680       LOCATE 19,25:PRINT "ATU"
12690       LINE (245,309)-(245,260)
12700       LINE (245,260)-(225,260)
12710       LINE (245,309)-(240,304)
12720       LINE (245,309)-(250,304)
12730 LOCATE 17,40:PRINT "Folded-Tee Antenna"
12740       LOCATE 14,33:PRINT "B"
```

```
12750       LOCATE 7,33:PRINT "A"
12760       LOCATE  2,14:PRINT "A = 3-element folded horizontal radiator"
12770       LOCATE  3,14:PRINT "    (spacing 8-10 inches)"
12780       LOCATE  5,14:PRINT "B = Quarter wavelength vertical section"
12790       TIMELOOP=TIMER:WHILE TIMER < TIMELOOP + 7:WEND
12800       CLS:F=FREQ:LENGTHA = KA/F:LENGTHB = KB/F
12810  'Print out results for folded tee antenna
12820       LINE (484,198)-(148,151),3,BF
12830       LOCATE 12,20:PRINT " Design Center Frequency: ";FREQ;" KHz "
12840       LOCATE 13,20:PRINT "   Length of Tee section: ";
12850       PRINT USING "###.##";LENGTHA;:PRINT " Feet "
12860       LOCATE 14,20:PRINT " Length of vertical section: ";
12870       PRINT USING "###.##";LENGTHB;:PRINT " Feet "
12880       LOCATE 16,20:GOSUB 13910
12890       CLS:GOTO 13710
12900 'Subroutine for quarter wavelength skyloop
12910     CLS:'Clear screen and draw antenna picture
12920     LINE (500,100)-(220,100)
12930     LINE (500,300)-(500,100)
12940     LINE (500,300)-(220,300)
12950     LINE (220,300)-(220,210)
12960     LINE (220,200)-(220,100)
12970     LINE (220,210)-(180,210)
12980     LINE (220,200)-(180,200)
12990     LINE (180,220)-(160,190),,B
13000     LINE (160,204)-(50,204)
13010     LINE (160,205)-(50,205)
13020     LOCATE 3,30:PRINT "SKYLOOP ANTENNA"
13030     LOCATE 4,30:PRINT "Each side is quarter wavelength"
13040     LOCATE 15,1:PRINT "XMTR"
13050     LOCATE 13,17:PRINT "4:1 BALUN"
13060     LOCATE 5,30:PRINT "May be installed vertically or horizontally"
13070     TIMELOOP=TIMER:WHILE TIMER < TIMELOOP + 7:WEND
13080     CLS:F=FREQ:LENGTH=KA/F
13090     'Display results of skyloop antenna
13100     LINE (460,184)-(148,151),3,BF
13110     LOCATE 12,20:PRINT " Design Center Frequency: ";FREQ;" KHz "
13120     LOCATE 13,20:PRINT "          Length overall: ";
13130     PRINT USING "###.##";LENGTH;:PRINT " Feet "
13140     LOCATE 16,20:GOSUB 13910
13150     CLS:GOTO 13710
13160 'Subroutine for bisquare loop
13170     CLS:'Clear screen and draw antenna picture
13180     LINE (500,100)-(220,100)
13190     LINE (500,300)-(500,100)
13200     LINE (500,300)-(220,300)
13210     LINE (220,300)-(220,210)
13220     LINE (220,200)-(220,100)
13230     LINE (220,210)-(180,210)
13240     LINE (220,200)-(180,200)
13250     LOCATE 3,30:PRINT "BI-SQUARE LOOP ANTENNA"
13260     LOCATE 4,30:PRINT "Each side is HALF wavelength"
13270     LOCATE 15,19:PRINT "XMTR"
13280     LOCATE 5,30:PRINT "Should be installed horizontally"
13290     TIMELOOP=TIMER:WHILE TIMER < TIMELOOP + 7:WEND
13300     CLS:F=FREQ:LENGTH=KA/F
13310     'Display results of bisquare antenna
13320     LINE (460,184)-(148,151),3,BF
13330     LOCATE 12,20:PRINT " Design Center Frequency: ";FREQ;" KHz "
13340     LOCATE 13,20:PRINT "          Length overall: ";
13350     PRINT USING "###.##";LENGTH;:PRINT " Feet "
13360     LOCATE 15,20:GOSUB 13910
13370     CLS:GOTO 13710
13380 'Subroutine for Thorne array or Bobtail Curtain
```

```
13390    CLS:'Clear screen and draw antenna picture
13400    LINE (550,280)-(550,150)
13410    LINE (100,280)-(100,150)
13420    LINE (325,280)-(325,150)
13430    LINE (550,150)-(100,150)
13440    LINE (325,280)-(305,280)
13450    LINE (325,290)-(305,290)
13460    LINE (325,300)-(325,290)
13470    LINE (334,300)-(317,300)
13480    LINE (330,304)-(319,304)
13490    LINE (328,308)-(321,308)
13500    LOCATE 21,34:PRINT "XMTR"
13510    LOCATE 15,12:PRINT "B"
13520    LOCATE 15,40:PRINT "B"
13530    LOCATE 15,68:PRINT "B"
13540    LOCATE 10,28:PRINT "A"
13550    LOCATE 10,54:PRINT "A"
13560    LOCATE 2,10:PRINT "THORNE ARRAY OR BOBTAIL CURTAIN"
13570    LOCATE 3,10:PRINT "Bobtail curtain is shown. Thorne Array"
13580    LOCATE 4,10:PRINT "is upside-down Bobtail Curtain. Place"
13590    LOCATE 5,10:PRINT "the A-element along bottom. Feed with ATU"
13600    TIMELOOP=TIMER:WHILE TIMER < TIMELOOP + 7:WEND
13610    CLS:F=FREQ:LENGTHA = KA/F:LENGTHB = KB/F
13620    'Print-out results of Thorne/Bobtail antenna
13630    LINE (508,200)-(147,150),3,BF
13640    LOCATE 12,20:PRINT "       Design Center Frequency: ";FREQ;" KHz "
13650    LOCATE 13,20:PRINT " Length of horizontal elements: ";
13660    PRINT USING "###.##";LENGTHA;:PRINT " Feet "
13670    LOCATE 14,20:PRINT "   Length of vertical elements: ";
13680    PRINT USING "###.##";LENGTHB;:PRINT " Feet "
13690    LOCATE 17,20:GOSUB 13910
13700    CLS:GOTO 13710
13710 'Select end routine course of action
13720         LINE (348,199)-(147,95),3,BF
13730         LOCATE 8,20:PRINT  "                       "
13740         LOCATE  9,20:PRINT " Select One From Below  "
13750         LOCATE 10,20:PRINT "                       "
13760         LOCATE 11,20:PRINT " (A)nother Wire Antenna "
13770         LOCATE 12,20:PRINT " (R)eturn to Main Menu  "
13780         LOCATE 13,20:PRINT "                       "
13790         LOCATE 14,20:PRINT " Selection:            "
13800         ACTION$ = INPUT$(1)
13810         IF ACTION$ = "A" THEN ACTION = 1
13820         IF ACTION$ = "a" THEN ACTION = 1
13830         IF ACTION$ = "R" THEN ACTION = 2
13840         IF ACTION$ = "r" THEN ACTION = 2
13850           IF ACTION < 1 THEN 13710
13860           IF ACTION > 2 THEN 13710
13870           IF INT(ACTION) = ACTION THEN 13880 ELSE 13710
13880             IF ACTION = 1 THEN 10330 ELSE 13890
13890 CLS:'Returning to main menu
13900 RETURN
13910 'PRESS ANY KEY SUBROUTINE
13920     PRINT " Press Any Key To Continue "
13930     A$=INKEY$:IF A$="" THEN 13930
13940 RETURN:'End of subroutine
```

ZMATCH program listing (BASIC language)

(See beginning of this appendix for a description of ZMATCH)

```
10 CLS:KEY OFF:SCREEN 9
1000 'Execution subroutine
1010 CLS:LINE (483,297)-(147,123),3,BF
1015 LOCATE 10,20:PRINT "                                          "
1017 LOCATE 12,20:PRINT "                                          "
1020 LOCATE 11,20:PRINT "               ZMATCH                     "
1030 LOCATE 13,20:PRINT " This program is a test version of a new "
1040 LOCATE 14,20:PRINT " program to calculate the values for L-C "
1050 LOCATE 15,20:PRINT " impedance matching networks. It is used "
1060 LOCATE 16,20:PRINT " with the program called ANTLERS, but is "
1070 LOCATE 17,20:PRINT " a standalone program.                   "
1080 LOCATE 18,20:PRINT "                                          "
1090 LOCATE 20,20:PRINT "                                          "
1100 LOCATE 21,20:PRINT "                                          "
1120 LOCATE 4,20:PRINT "Copyright 1992 Joseph J. Carr"
1130 LOCATE 6,20:PRINT "Joseph J. Carr"
1140 LOCATE 7,20:PRINT "P.O. Box 1099"
1150 LOCATE 8,20:PRINT "Falls Church, VA 22041"
1160 LOCATE 22,26:GOSUB 30000
1200 CLS:LINE (462,185)-(147,150),3,BF
1210     LOCATE 12,20:PRINT " Enter Desired Operating Frequency in "
1220     LOCATE 13,20:PRINT " KILOHERTZ (KHz):                     ";
1230     INPUT FREQ$
1240     FREQ = VAL(FREQ$)
1245        IF FREQ < 50 THEN 1200 ELSE 1250
1250        IF FREQ = 0 THEN 1200 ELSE 1260
1260        IF FREQ > 50000! THEN 1200 ELSE 1270
1270 CLS:LINE (435,170)-(147,150),3,BF
1280 LOCATE 12,20:PRINT " Selected Frequency is ";FREQ;" KHz "
1290 TIMELOOP = TIMER:WHILE TIMER < TIMELOOP + 3:WEND:CLS
2000 GOSUB 34000
10000 END
30000 PRINT " Press Any Key To Continue "
30010 A$ = INKEY$:IF A$ = "" THEN 30010 ELSE 30020
30020 RETURN
34000 'IMPEDANCE MATCHING SUBROUTINE
34010 '  Provides for calculation of either BALUN parameters or
34020 '  L-C network component values.
34030    CLS:LINE (468,283)-(147,150),3,BF
34040    LOCATE 12,20:PRINT "                                    "
34050    LOCATE 13,20:PRINT " Select Type of Impedance Matching   "
34060    LOCATE 14,20:PRINT " Circuit Desired:                    "
34070    LOCATE 15,20:PRINT "                                    "
34080    LOCATE 16,20:PRINT " (N)etwork Antenna Tuner (L-C)       "
34090    LOCATE 17,20:PRINT "                                    "
34100    LOCATE 18,20:PRINT "                                    "
34110    LOCATE 19,20:PRINT " Select From Above List...           ";
34115    LOCATE 20,20:PRINT "                                    "
34120    FRED$ = INPUT$(1):'Input selection from menu
34130       IF FRED$ = "N" THEN FRED = 1
34140       IF FRED$ = "n" THEN FRED = 1
34150       IF FRED$ = "B" THEN FRED = 2
34160       IF FRED$ = "b" THEN FRED = 2
34170       IF FRED < 1 THEN 34000 ELSE 34180
34180       IF FRED > 2 THEN 34000 ELSE 34190
34190       IF INT(FRED) = FRED THEN 34200 ELSE 34000
34200    CLS:ON FRED GOSUB 35000,49000:'Branch to subprogram selected
34990 PRINT " END OF PROGRAM - THANKS FOR YOUR PATIENCE "
34998 TIMELOOP = TIMER:WHILE TIMER < TIMELOOP + 3:WEND
34999 CLS:RETURN:'End of subprogram. Return to main menu
35000 'Subroutine for impedance matching networks
35020     CLS:SETFLAG = 0:F = FREQ*1000:PI = 3.141592654#
35040     LINE (388,185)-(147,150),3,BF
35060     LOCATE 12,20:PRINT " Impedance Matching Networks "
```

```
35080     LOCATE 13,20:PRINT "       Option Selected       "
35100     TIMELOOP=TIMER:WHILE TIMER < TIMELOOP + 2:WEND
35120     CLS
35140     LINE (450,200)-(444,160),,B
35160     LINE (150,200)-(144,160),,B
35180     LINE (447,160)-(447,140)
35200     LINE (447,220)-(447,200)
35220     LINE (147,160)-(147,140)
35240     LINE (147,220)-(147,200)
35260     LINE (447,140)-(400,140)
35280     LINE (447,220)-(400,220)
35300     LINE (197,220)-(147,220)
35320     LINE (197,140)-(147,140)
35340     LINE (400,240)-(197,120),,B
35360      LOCATE 14,8:PRINT "Source End"
35380      LOCATE 13,12:PRINT "R1"
35400      LOCATE 13,61:PRINT "R2"
35420      LOCATE 14,58:PRINT "Load End"
35440      LOCATE 13,30:PRINT "Impedance Matching"
35460      LOCATE 14,35:PRINT "Network"
35480      LOCATE 12,37:PRINT "L-C"
35500      LOCATE 20,24:PRINT "Source end is for XMTR or RCVR"
35520      LOCATE 21,24:PRINT "Load end is for ANTENNA"
35540 TIMELOOP = TIMER:WHILE TIMER < TIMELOOP + 7:WEND
35560      CLS:LINE (395,142)-(148,123),3,BF:'Input impedance
35580      LOCATE 10,20:PRINT " Enter Source Impedance (R1): ";
35600      INPUT R1$
35620        R1 = VAL(R1$):CLS
35640        IF R1 < 1 THEN 35560 ELSE 35660
35660        IF R1 > 5000 THEN 35560 ELSE 35680
35680      CLS:LINE (395,142)-(148,123),3,BF:'Input impedance
35700      LOCATE 10,20:PRINT " Enter Load Impedance (R2): ";
35720      INPUT R2$
35740        R2 = VAL(R2$):CLS
35760        IF R2 < 1 THEN 35680 ELSE 35780
35780        IF R2 > 5000 THEN 35680 ELSE 35800
35800      CLS
35820 IF R2/R1 > .9 AND R2/R1 < 1.1 THEN GOSUB 40800
35840        IF SETFLAG = 1 THEN 35920 ELSE 35860:'Test R1=R2 flag
35860        SETFLAG = 0:'Reset R1=R2 flag
35880 IF R2 > R1 GOTO 35940
35900 IF R2 < R1 GOTO 39220
35920 RETURN 35940  'Sub-subroutine for R2 > R1
35960 CLS:LINE (460,297)-(147,123),3,BF
35980     LOCATE 10,20:PRINT "                                         "
36000     LOCATE 11,20:PRINT " Network Options Available if R2 > R1 "
36020     LOCATE 12,20:PRINT "                                         "
36040     LOCATE 14,20:PRINT "                                         "
36060     LOCATE 15,20:PRINT " (1) L-Section                           "
36080     LOCATE 16,20:PRINT " (2) Reverse L-Section                   "
36100     LOCATE 17,20:PRINT " (3) Three-Element L-Section             "
36120     LOCATE 18,20:PRINT " (4) Split-Capacitor Network             "
36140     LOCATE 19,20:PRINT "                                         "
36160     LOCATE 20,20:PRINT " Select one from above...                "
36180     LOCATE 21,20:PRINT "                                         ";
36200        NETWORK$ = INPUT$(1)
36220          NETWORK = VAL(NETWORK$)
36240          IF NETWORK < 1 THEN 35940 ELSE 36260
36260          IF NETWORK > 4 THEN 35940 ELSE 36280
36280          IF INT(NETWORK) = NETWORK THEN 36300 ELSE 35940
36300     IF NETWORK = 1 THEN RON$ = " L-Section Coupler "
36320     IF NETWOKR = 2 THEN RON$ = " Reverse L-Section Coupler "
36340     IF NETWORK = 3 THEN RON$ = " Three-Element L-Section Coupler "
36360     IF NETWORK = 4 THEN RON$ = " Split-Capacitor Network "
```

```
36380 ON NETWORK GOSUB 36760, 37260, 37740, 38420
36400 'End of subroutine section of program
36420       CLS:LINE (397,240)-(147,151),3,BF
36440       LOCATE 12,20:PRINT "                              "
36460       LOCATE 13,20:PRINT " What Is Your Pleasure?       "
36480       LOCATE 14,20:PRINT "                              "
36500       LOCATE 15,20:PRINT " (D)o Another Z-Match Problem "
36520       LOCATE 16,20:PRINT " (R)eturn to Main Menu        "
36540       LOCATE 17,20:PRINT "                              ";
36560       BILL$ = INPUT$(1)
36580          IF BILL$ = "D" THEN BILL = 1
36600          IF BILL$ = "d" THEN BILL = 1
36620          IF BILL$ = "R" THEN BILL = 2
36640          IF BILL$ = "r" THEN BILL = 2
36660            IF BILL < 1 THEN 36400 ELSE 36680
36680            IF BILL > 2 THEN 36400 ELSE 36700
36700            IF INT(BILL) = BILL THEN 36720 ELSE 36400
36720       ON BILL GOTO 35560, 36740
36740 CLS:RETURN:'End of Z-match subroutine
36760 'Subroutine to calculate values of L-section components
36770 GOSUB 41800
36780      Q = SQR((R2/R1)-1)
36800         IF Q < 1 THEN GOSUB 40940:'Display message
36820         IF Q < 1 THEN GOTO 39200:'Go to end of subroutine
36840         IF Q > 10 THEN GOSUB 41160:'Display message
36860         IF Q > 10 THEN GOTO 39200:'Go to end of subroutine
36880  XL = Q*R1:XC = R2/Q
36900  L = XL/(2*PI*F):L = L*10^6
36920  C = 1/(2*PI*F*XC): C = C*10^12
36940         CLS:LINE (478,260)-(146,120),3,BF:'Display results
36960         LOCATE 10,20:PRINT "                                       "
36980         LOCATE 11,20:PRINT " Component values for an L-Section ATU: "
37000         LOCATE 12,20:PRINT "                                       "
37020         LOCATE 13,20:PRINT " Frequency of Operation: ";FREQ;" KHz   "
37040         LOCATE 14,20:PRINT " Inductive reactance: ";
37060         PRINT USING "####.#";XL;:PRINT " Ohms      "
37080         LOCATE 15,20:PRINT " Capacitive reactance: ";
37100         PRINT USING "####.#";XC;:PRINT " Ohms      "
37120         LOCATE 16,20:PRINT " Inductance: ";
37140         PRINT USING "####.#";L;:PRINT " uH        "
37160         LOCATE 17,20:PRINT " Capacitance: ";
37180         PRINT USING "####.#";C;:PRINT " pF        "
37200         LOCATE 18,20:PRINT "                                       "
37220         LOCATE 20,20:GOSUB 30000:'Go get press any key routine
37240 CLS:RETURN:'End of L-section subroutine
37260 'Subroutine for Reverse L-Section Coupler Network
37270 GOSUB 42000
37280  CLS:TEST = R2-R1
37300  IF TEST < .75 THEN GOSUB 40800
37320  IF TEST < .75 THEN GOTO 39200
37340      XL = R2*(SQR(R1/(R2-R1)))
37360      XC = (R1*R2)/XL
37380       L = XL/(2*PI*F):L = L*10^6
37400       C = 1/(2*PI*F*XC):C = C*10^12
37420      LINE (478,282)-(147,123),3,BF
37440       LOCATE 10,20:PRINT "                                        "
37460       LOCATE 11,20:PRINT " Component values for Reverse L-Section "
37480       LOCATE 12,20:PRINT " Antenna Coupler:                       "
37500       LOCATE 13,20:PRINT "                                        "
37520       LOCATE 14,20:PRINT " Inductive Reactance: ";
37540       PRINT USING "####.#";XL;:PRINT " Ohms      "
37560       LOCATE 15,20:PRINT " Capacitive Reactance: ";
37580       PRINT USING "####.#";XC;:PRINT " Ohms      "
37600       LOCATE 16,20:PRINT "                                        "
```

```
37620       LOCATE 17,20:PRINT " Inductance:          ";
37640       PRINT USING "####.#";L;:PRINT " uH      "
37660       LOCATE 18,20:PRINT " Capacitance:         ";
37680       PRINT USING "####.#";C;:PRINT " pF      "
37700       LOCATE 20,20:GOSUB 30000
37720 CLS:RETURN:'End of reverse L-section subroutine
37740 'Subroutine for Three-Element L-Section Network
37750 GOSUB 42200
37760       CLS:LINE (379,212)-(148,123),3,BF
37780       LOCATE 10,20:PRINT "                              "
37800       LOCATE 11,20:PRINT " Input Value of Q (1 to 9)  "
37820       LOCATE 12,20:PRINT " and press <ENTER>            "
37840       LOCATE 13,20:PRINT " Or...Press <ENTER> for the "
37860       LOCATE 14,20:PRINT " default value of Q = 5       "
37880       LOCATE 15,20:PRINT "                              "
37900       Q$ = INPUT$(1):IF Q$ = "" THEN 37940 ELSE 37920
37920       Q = VAL(Q$)
37940          IF Q = 0 THEN Q = 5:'Test Q for default value
37960       XC1 = Q*R1:XC2 = R2*(SQR(R1/(R2-R1)))
37980       XL = ((R1*R2)/XC2) + XC1
38000       L = XL/(2*PI*F):C1 = 1/(2*PI*F*XC1):C2 = 1/(2*PI*F*XC2)
38020       L = L*10^6:C1 = C1*10^12:C2 = C2*10^12
38040       CLS:LINE (483,296)-(147,122),3,BF
38060       LOCATE 10,20:PRINT "                                          "
38080       LOCATE 11,20:PRINT " Component values for Three-Element     "
38100       LOCATE 12,20:PRINT " L-Section Network                      "
38120       LOCATE 13,20:PRINT "                                          "
38140       LOCATE 14,20:PRINT " Inductive Reactance:      ";
38160       PRINT USING "####.#";L;:PRINT " Ohms  "
38180       LOCATE 15,20:PRINT " Capacitive Reactance XC1:  ";
38200       PRINT USING "####.#";XC1;:PRINT " Ohms  "
38220       LOCATE 16,20:PRINT " Capacitive Reactance XC2:  ";
38240       PRINT USING "####.#";XC2;:PRINT " Ohms  "
38260       LOCATE 17,20:PRINT " Inductance: ";
38280       PRINT USING "####.#";L;:PRINT " uH    "
38300       LOCATE 18,20:PRINT " Capacitance C1: ";
38320       PRINT USING "####.#";C1;:PRINT " pF     "
38340       LOCATE 19,20:PRINT " Capacitance C2: ";
38360       PRINT USING "####.#";C2;:PRINT " pF     "
38380       LOCATE 21,20:GOSUB 30000
38400 CLS:RETURN:'End of three-element L-section subroutine
38420 'Subroutine for Split-Capacitor Network
38430 GOSUB 42500
38440   CLS:LINE (428,213)-(147,123),3,BF
38460   LOCATE 10,20:PRINT "                                 "
38480   LOCATE 11,20:PRINT " Select a value of Q (1 to 9) and "
38500   LOCATE 12,20:PRINT " then press <ENTER>. Or...if you  "
38520   LOCATE 13,20:PRINT " wish to accept the default value "
38540   LOCATE 14,20:PRINT " of Q = 5 just press <ENTER>      "
38560   LOCATE 15,20:PRINT "                                 "
38580   Q$ = INPUT$(1)
38600   Q = VAL(Q$)
38620   IF Q = 0 THEN Q = 5
38640   TESTQ = SQR ((R2/R1) - 1)
38660       IF Q < TESTQ THEN GOSUB 41380
38680       IF Q < TESTQ THEN 38440 ELSE 38700
38700   XL = R2/Q
38720   XC2 = R1/( SQR((R1*(Q^2 + 1))/R2))
38740   XC1 = ((R2*Q)/(Q^2 + 1))*(1 - (R1/(Q*XC2)))
38760   L = XL/(2*PI*F):L = L*10^6
38780   C1 = 1/(2*PI*F*XC1):C1 = C1*10^12
38800   C2 = 1/(2*PI*F*XC2):C2 = C2*10^12
38820   CLS:LINE (468,297)-(148,123),3,BF
38840    LOCATE 10,20:PRINT "                                          "
```

```
38860    LOCATE 11,20:PRINT " Component values for Split-Capacitor "
38880    LOCATE 12,20:PRINT " Network:                            "
38900    LOCATE 13,20:PRINT "                                     "
38920    LOCATE 14,20:PRINT " Inductive Reactance: ";
38940    PRINT USING "####.#";XL;:PRINT " Ohms "
38960    LOCATE 15,20:PRINT " Capacitive Reactance XC1: ";
38980    PRINT USING "####.#";XC1;:PRINT " Ohms "
39000    LOCATE 16,20:PRINT " Capacitive Reactance XC2: ";
39020    PRINT USING "####.#";XC2;:PRINT " Ohms "
39040    LOCATE 17,20:PRINT " Inductance: ";
39060    PRINT USING "####.#";L;:PRINT " uH      "
39080    LOCATE 18,20:PRINT " Capacitance C1: ";
39100    PRINT USING "####.#";C1;:PRINT " pF      "
39120    LOCATE 19,20:PRINT " Capacitance C2: ";
39140    PRINT USING "####.#";C2;:PRINT " pF      "
39160    LOCATE 21,20:GOSUB 30000
39180 CLS:RETURN
39200 END
39220  'Sub-subroutine for R2 < R1
39240  CLS:LINE (460,268)-(147,123),3,BF
39260     LOCATE 10,20:PRINT "                                     "
39280     LOCATE 11,20:PRINT " Network Options Available if R2 < R1 "
39300     LOCATE 12,20:PRINT "                                     "
39320     LOCATE 14,20:PRINT "                                     "
39340     LOCATE 15,20:PRINT " (1) Pi-Network                      "
39360     LOCATE 16,20:PRINT " (2) Inverted L-Section              "
39380     LOCATE 17,20:PRINT "                                     "
39400     LOCATE 18,20:PRINT " Select one from above...            "
39420     LOCATE 19,20:PRINT "                                     "
39440     NETWORK$ = INPUT$(1)
39460      NETWORK = VAL(NETWORK$)
39480      IF NETWORK < 1 THEN 39220 ELSE 39500
39500      IF NETWORK > 2 THEN 39220 ELSE 39520
39520      IF INT(NETWORK) = NETWORK THEN 39540 ELSE 39220
39540     ON NETWORK GOSUB 39600, 40380:'39600 is pi-net, 40380 is Inv. L 39541
'End of subroutine section of program
39542      CLS:LINE (397,240)-(147,151),3,BF
39543      LOCATE 12,20:PRINT "                              "
39544      LOCATE 13,20:PRINT " What Is Your Pleasure?       "
39545      LOCATE 14,20:PRINT "                              "
39546      LOCATE 15,20:PRINT " (D)o Another Z-Match Problem "
39547      LOCATE 16,20:PRINT " (R)eturn to Main Menu        "
39548      LOCATE 17,20:PRINT "                              ";
39549      BILL$ = INPUT$(1)
39550         IF BILL$ = "D" THEN BILL = 1
39551         IF BILL$ = "d" THEN BILL = 1
39552         IF BILL$ = "R" THEN BILL = 2
39553         IF BILL$ = "r" THEN BILL = 2
39554           IF BILL > 2 THEN 36400 ELSE 36700
39555           IF INT(BILL) = BILL THEN 36720 ELSE 36400
39556      ON BILL GOTO 35560, 36740
39557 CLS:RETURN:'End of Z-match subroutine
39560    '
39580 CLS:RETURN
39600 'Subroutine for Pi-Network calculations
39610 GOSUB 42800
39620   CLS:LINE (412,212)-(148,123),3,BF
39640   LOCATE 10,20:PRINT "                              "
39660   LOCATE 11,20:PRINT " Input a value of Q (3 - 20)    "
39680   LOCATE 12,20:PRINT " and press <ENTER>. Or...to     "
39700   LOCATE 13,20:PRINT " accept default value of Q = 12 "
39720   LOCATE 14,20:PRINT " just press <ENTER>             "
39740   LOCATE 15,20:PRINT "                              ";
39760   INPUT Q$:Q = VAL(Q$)
```

```
39780   IF Q = 0 THEN Q = 12
39800   IF Q < 3 THEN 39600 ELSE 39820
39820   IF Q > 20 THEN 39600 ELSE 39840
39840   IF INT(Q) = Q THEN 39860 ELSE 39600
39860   TESTQ = SQR ((R1/R2) - 1)
39880   IF Q < TESTQ THEN GOSUB 41620 ELSE 39900
39900   IF Q < TESTQ THEN 39600 ELSE 39920
39920   XC1 = R1/Q:C1 = 1/(2*PI*F*XC1):C1 = C1*10^12
39940   XC2 = R2/(SQR(((R2/R1)*(1+Q^2)-1))):C2 = 1/(2*PI*F*XC2):C2 = C2*10^12
39960   XL = R1*((Q + (R2/XC2))/(Q^2 + 1)):L = XL/(2*PI*F):L = L*10^6
39980   'Display results of Pi-network calculations
40000     CLS:LINE (468,297)-(148,122),3,BF
40020     LOCATE 10,20:PRINT "                                   "
40040     LOCATE 11,20:PRINT " Component values for Pi-Network "
40060     LOCATE 12,20:PRINT " Antenna Coupler:                 "
40080     LOCATE 13,20:PRINT "                                   "
40100     LOCATE 14,20:PRINT " Inductive Reactance: ";
40120     PRINT USING "####.#";XL;:PRINT " Ohms "
40140     LOCATE 15,20:PRINT " Capacitive Reactance XC1: ";
40160     PRINT USING "####.#";XC1;:PRINT " Ohms "
40180     LOCATE 16,20:PRINT " Capacitive Reactance XC2: ";
40200     PRINT USING "####.#";XC2;:PRINT " Ohms "
40220     LOCATE 17,20:PRINT " Inductance: ";
40240     PRINT USING "####.#";L;:PRINT " uH "
40260     LOCATE 18,20:PRINT " Capacitance C1: ";
40280     PRINT USING "####.#";C1;:PRINT " pF "
40300     LOCATE 19,20:PRINT " Capacitance C2: ";
40320     PRINT USING "####.#";C2;:PRINT " pF "
40340     LOCATE 21,20:GOSUB 30000
40360 CLS:RETURN
40380 'Subroutine for Inverted L-Section
40390 GOSUB 43500
40400    XL = SQR((R1*R2) - R2^2):XC = (R1*R2)/XL
40420    L = XL/(2*PI*F):L = L*10^6
40440    C = 1/(2*PI*F*XC):C = C*10^12
40460 'Display results of Inverted L-section
40480    CLS:LINE (485,268)-(148,122),3,BF
40500    LOCATE 10,20:PRINT "                                        "
40520    LOCATE 11,20:PRINT " Component values for Inverted L-Section "
40540    LOCATE 12,20:PRINT " Antenna Coupler:                        "
40560    LOCATE 13,20:PRINT "                                        "
40580    LOCATE 14,20:PRINT " Inductive Reactance: ";
40600    PRINT USING "####.#";XL;:PRINT " Ohms "
40620    LOCATE 15,20:PRINT " Capacitive Reactance: ";
40640    PRINT USING "####.#";XC;:PRINT " Ohms "
40660    LOCATE 16,20:PRINT " Inductance: ";
40680    PRINT USING "####.#";L;:PRINT " uH "
40700    LOCATE 17,20:PRINT " Capacitance: ";
40720    PRINT USING "####.#";C;:PRINT " pF "
40740    LOCATE 19,20:GOSUB 30000
40760 CLS:RETURN
40780 END
40800 'Impedances Equal Message
40820  SETFLAG = 1:CLS:BEEP:LINE (500,199)-(108,150),3,BF
40840  LOCATE 12,15:PRINT " Source and Load Impedances are equal (R1 = R2) "
40860  LOCATE 13,15:PRINT " or nearly equal, so no impedance matching is   "
40880  LOCATE 14,15:PRINT " necessary.                                      "
40900  LOCATE 16,15:GOSUB 30000
40920  CLS:RETURN
40940  'Message for L-section coupler when R1 and R2 are close
40960      BEEP:CLS:LINE (460,282)-(147,150),3,BF
40980      LOCATE 12,20:PRINT "                                          "
41000      LOCATE 13,20:PRINT " Q is TOO LOW, indicating that R1 and    "
41020      LOCATE 14,20:PRINT " R2 are too close to each other. It      "
```

```
41040      LOCATE 15,20:PRINT " may be wise to consider not using an   "
41060      LOCATE 16,20:PRINT " antenna coupler in this situation. The "
41080      LOCATE 17,20:PRINT " R2/R1 ratio should be 2:1 or more.     "
41100      LOCATE 18,20:PRINT "                                        "
41120      LOCATE 20,20:GOSUB 30000
41140   CLS:RETURN:'End of subroutine
41160  'Message when Q is not correct value
41180      BEEP:CLS:LINE (387,282)-(147,150),3,BF
41200      LOCATE 12,20:PRINT "                              "
41220      LOCATE 13,20:PRINT " Q is out of range. Either    "
41240      LOCATE 14,20:PRINT " select a different type of   "
41260      LOCATE 15,20:PRINT " antenna coupler, or make     "
41280      LOCATE 16,20:PRINT " sure that the impedance are  "
41300      LOCATE 17,20:PRINT " correctly entered.           "
41320      LOCATE 18,20:PRINT "                              "
41340      LOCATE 20,20:GOSUB 30000
41360   CLS:RETURN:'End of subroutine
41380 'Message for Split-Capacitor network when Q out of range
41400      BEEP:CLS:LINE (403,282)-(147,150),3,BF
41420      LOCATE 12,20:PRINT "                                "
41440      LOCATE 13,20:PRINT " Q is out of range for the      "
41460      LOCATE 14,20:PRINT " values of R1 and R2 selected.  "
41480      LOCATE 15,20:PRINT " Try another value of Q. The    "
41500      LOCATE 16,20:PRINT " correct value is greater than  "
41520      LOCATE 17,20:PRINT "     ";
41540      PRINT USING "##.#";TESTQ;:PRINT "                    "
41560      LOCATE 18,20:PRINT "                                "
41580      LOCATE 20,20:GOSUB 30000
41600      CLS:RETURN:'End of message for Split-capacitor network
41620 'Message for Pi-Network When Q is too low
41640   BEEP:CLS:LINE (500,255)-(147,151),3,BF
41660   LOCATE 12,20:PRINT "                                              "
41680   LOCATE 13,20:PRINT " Selected value of Q is too low. Either      "
41700   LOCATE 14,20:PRINT " select another value of Q or make sure      "
41720   LOCATE 15,20:PRINT " that the impedances R1 and R2 are correct   "
41740   LOCATE 16,20:PRINT "                                              "
41760   LOCATE 18,20:GOSUB 30000
41780 CLS:RETURN
41800 'Graphic for L-section coupler
41810  CLS:LINE (290,205)-(220,195),3,B
41820      LINE (440,200)-(290,200)
41830      LINE (220,200)-(170,200)
41840      LINE (440,280)-(170,280)
41850      LINE (175,260)-(165,220),,B
41860      LINE (445,260)-(435,220),,B
41870      LINE (170,220)-(170,200)
41880      LINE (170,280)-(170,260)
41890      LINE (440,220)-(440,200)
41900      LINE (440,280)-(440,260)
41910      LINE (365,235)-(365,200)
41920      LINE (365,280)-(365,245)
41930      LINE (375,235)-(355,235)
41935      LINE (375,245)-(355,245)
41940      LOCATE 18,19:PRINT "R1"
41945      LOCATE 10,23:PRINT "L-Section Coupler"
41950      LOCATE 18,43:PRINT "C"
41960      LOCATE 18,58:PRINT "R2"
41970      LOCATE 14,32:PRINT "L"
41980      LOCATE 22,23:GOSUB 30000
41995 CLS:RETURN
42000 'Graphic for Reverse L-Section coupler
42010  CLS:LINE (440,200)-(250,200)
42020      LINE (240,200)-(170,200)
42030      LINE (440,280)-(170,280)
```

```
42040      LINE (240,210)-(240,190)
42050      LINE (250,210)-(250,190)
42060      LINE (175,260)-(165,220),,B
42070      LINE (445,260)-(435,220),,B
42080      LINE (170,280)-(170,260)
42090      LINE (170,220)-(170,200)
42100      LINE (440,280)-(440,260)
42110      LINE (440,220)-(440,200)
42120      LINE (350,265)-(340,215),,B
42130      LINE (345,280)-(345,265)
42140      LINE (345,215)-(345,200)
42150      LOCATE 10,23:PRINT "Reverse L-Section Filter"
42160      LOCATE 18,18:PRINT "R1"
42170      LOCATE 18,41:PRINT "L"
42180      LOCATE 18,58:PRINT "R2"
42190      LOCATE 14,33:PRINT "C"
42194      LOCATE 22,24:GOSUB 30000
42197 CLS:RETURN
42200 'Graphic for Three-Element L-Section Network
42210 CLS:LINE (500,200)-(365,200)
42220     LINE (365,205)-(295,195),,B
42230     LINE (295,200)-(240,200)
42240     LINE (230,200)-(170,200)
42250     LINE (500,280)-(170,280)
42260     LINE (230,210)-(230,190)
42270     LINE (240,210)-(240,190)
42280     LINE (402,280)-(402,245)
42290     LINE (402,235)-(402,200)
42300     LINE (415,235)-(389,235)
42310     LINE (415,245)-(389,245)
42320     LINE (500,280)-(500,260)
42330     LINE (500,220)-(500,200)
42340     LINE (170,280)-(170,260)
42350     LINE (170,220)-(170,200)
42360     LINE (505,260)-(495,220),,B
42370     LINE (175,260)-(165,220),,B
42380     LOCATE 18,19:PRINT "R1"
42390     LOCATE 18,47:PRINT "C2"
42400     LOCATE 18,60:PRINT "R2"
42410     LOCATE 13,30:PRINT "C1"
42420     LOCATE 13,42:PRINT "L"
42425     LOCATE 10,24:PRINT "Three-Element L-Section Network"
42430     LOCATE 22,24:GOSUB 30000
42440  CLS:RETURN
42500 'Graphics for Split-Capacitor Network
42510  CLS:LINE (250,280)-(250,238)
42520      LINE (250,230)-(250,190)
42530      LINE (250,182)-(250,140)
42540      LINE (262,238)-(238,238):'Draw C2
42550      LINE (262,230)-(238,230):' "    "
42560      LINE (262,182)-(238,182):'Draw C1
42570      LINE (262,190)-(238,190):' "    "
42580      LINE (440,140)-(250,140)
42590      LINE (350,245)-(340,175),,B
42600      LINE (440,280)-(250,280)
42610      LINE (345,280)-(345,245)
42620      LINE (345,175)-(345,140)
42640      LINE (440,280)-(440,230)
42650      LINE (440,190)-(440,140)
42660      LINE (250,210)-(150,210)
42670      LINE (250,280)-(150,280)
42680      LINE (155,265)-(145,225),,B
42690      LINE (150,225)-(150,210)
42700      LINE (150,280)-(150,265)
```

```
42702      LINE (445,230)-(435,190),,B
42710      LOCATE 18,16:PRINT "R1"
42720      LOCATE 17,28:PRINT "C2"
42730      LOCATE 14,28:PRINT "C1"
42740      LOCATE 16,42:PRINT "L"
42750      LOCATE 16,58:PRINT "R2"
42760      LOCATE  8,24:PRINT "Split-Capacitor Network"
42770      LOCATE 22,24:GOSUB 30000
42780 CLS:RETURN
42800 'Graphics for Pi-Network
42810 CLS:LINE (500,280)-(140,280)
42820     LINE (355,205)-(285,195),,B
42830     LINE (500,200)-(355,200)
42840     LINE (285,200)-(140,200)
42850     LINE (145,260)-(135,220),,B
42860     LINE (505,260)-(495,220),,B
42870     LINE (500,280)-(500,260)
42880     LINE (500,220)-(500,200)
42890     LINE (140,280)-(140,260)
42900     LINE (140,220)-(140,200)
42910     LINE (213,280)-(213,244)
42920     LINE (213,236)-(213,200)
42930     LINE (427,280)-(427,244)
42940     LINE (427,236)-(427,200)
42950     LINE (440,236)-(414,236)
42960     LINE (440,244)-(414,244)
42970     LINE (226,244)-(200,244)
42980     LINE (226,236)-(200,236)
42990     LOCATE 18,15:PRINT "R1"
43000     LOCATE 18,23:PRINT "C1"
43010     LOCATE 18,60:PRINT "C2"
43020     LOCATE 18,50:PRINT "C2"
43030     LOCATE 13,41:PRINT "L"
43040     LOCATE 11,26:PRINT "Pi-Network"
43050     LOCATE 22,25:GOSUB 30000
43060  CLS:RETURN
43500 ' Graphics for Inverted L-Section Coupler
43510  CLS:LINE (500,200)-(350,200)
43520      LINE (350,205)-(265,195),,B
43530      LINE (265,200)-(110,200)
43540      LINE (500,280)-(110,280)
43550      LINE (505,260)-(495,220),,B
43560      LINE (115,260)-(105,220),,B
43570      LINE (110,280)-(110,260)
43580      LINE (110,220)-(110,200)
43590      LINE (500,280)-(500,260)
43600      LINE (500,220)-(500,200)
43630      LINE (187,280)-(187,244)
43640      LINE (187,236)-(187,200)
43650      LINE (200,236)-(174,236)
43660      LINE (200,244)-(174,244)
43690      LOCATE 18,11:PRINT "R1"
43700      LOCATE 18,21:PRINT "C1"
43720      LOCATE 18,60:PRINT "R2"
43730      LOCATE 13,39:PRINT "L"
43740      LOCATE 10,25:PRINT "Inverted L-Section Coupler"
43750      LOCATE 22,25:GOSUB 30000
43760 CLS:RETURN
```

Index

E

F

G

H

I

J

K

L

M

N

O

S

T

U

V

W

Y

Z

About the Author

An experienced electrical engineer, technician, and ham radio operator, Joseph J. Carr holds a Certified Electronics Technician (CET) certificate in both consumer electronics and communications. He is a columnist for *Popular Electronics, Popular Communications*, and *73 Amateur Radio Today*, and author of *Mastering Radio Frequency Circuits through Projects & Experiments.*